Encyclopaedia of GENETIC RESOURCES, CHROMOSOME ENGINEERING, AND CROP IMPROVEMENT

Encyclopaedia of GENETIC RESOURCES, CHROMOSOME ENGINEERING, AND CROP IMPROVEMENT

Volume I: Chromosome Engineering in Plants

Contributors

Teresa Cuellar,Chia-Lin Chung,and Luciano Lourenço Nass et al.

AURIS
Reference

www.aurisreference.com

Encyclopaedia of Genetic Resources, Chromosome Engineering, and Crop Improvement

Volume I: Chromosome Engineering in Plants

Contributors: Teresa Cuellar,Chia-Lin Chung,and Luciano Lourenço Nass et al.

Published by Auris Reference Limited

www.aurisreference.com

United Kingdom

Printed in 2017 for Sale in the Indian Subcontinent

Chromosome Engineering in Plants

ISBN: 978-1-78154-957-5 (Set of 3 Volumes)

British Library Cataloguing in Publication Data
A CIP record for this book is available from the British Library

Printed in the United Kingdom

Exclusively distributed by CBS Publishers & Distributors Pvt. Ltd.

Sales & Distribution Rights only for India, Pakistan, Bangladesh, Sri Lanka, Nepal and Bhutan.This book is not to be sold outside these territories.

Contents

List of Abbreviations

AEs	Axial Elements
ALS	Acetolactate Synthase
BFB	Breakage-Fusion-Bridge
BIBAC	Binary Bacterial Artificial Chromosome
CCD	Chilled Charge-Coupled Device
ChIP	Chromatin Immunoprecipitation
CPC	Chromosome Passenger Complex
CS	Chinese Spring
CSC	Cellulose Synthase Complex
CSILs	Chromosome Segment Introgression Lines
DEGs	Differentially Expressed Genes
DSB	Double-Stranded Break
ESTs	Expressed Sequence Tags
FISH	Fluorescence In Situ Hybridization
HFDs	Histone Fold Domains
LEs	Lateral Elements
LINE	Long Interspersed Nuclear Element
LSC	Large Single Copy
MET	Multi-Environment Trials
MTOC	Microtubule-Organizing Centers
NCBI	National Center for Biotechnology Information
NORs	Nucleolar Organizer Regions
PEG	Polyethylene Glycol
RB	Right Border
RE	Relative Enrichment
SCMT	Silicon Carbide Mediated Transformation
SSR	Simple Sequence Repeats
ZFNs	Zinc-Finger Endonucleases

List of Contributors

Mansoor Zoveidavianpoor
Universiti Teknologi, Faculty of Petroleum & Renewable Energy Engineering, Malaysia

Ariffin Samsuri
Universiti Teknologi, Faculty of Petroleum & Renewable Energy Engineering, Malaysia

Seyed Reza Shadizadeh
Petroleum University of Technology, Abadan Faculty of Petroleum Engineering, Iran

Lixin Chen
Key Laboratory Soil and Water Conservation and Desertification Combating, Ministry of Education, College of Soil and Water Conservation, Beijing Forestry University, Beijing, People's Republic of China

Zhiqiang Zhang
Key Laboratory Soil and Water Conservation and Desertification Combating, Ministry of Education, College of Soil and Water Conservation, Beijing Forestry University, Beijing, People's Republic of China

Brent E. Ewers
Program in Ecology, Department of Botany, University of Wyoming, Laramie, Wyoming, United States of America

Stefan Wirtz
Department of Physical Geography, Trier University, Trier, Germany

Manuel Seeger
Department of Physical Geography, Trier University, Trier, Germany
Department of Land Degradation and Development, Wageningen University, Wageningen, The Netherlands

Andreas Zell
Department 7.3- Technical Physics, Saarland University, Saarbrücken, Germany

Christian Wagner
Department 7.3- Technical Physics, Saarland University, Saarbru¨ cken, Germany

Jean-Frank Wagner
Department of Geology, Trier University, Trier, Germany

Johannes B. Ries
Department of Physical Geography, Trier University, Trier, Germany

Stanislaus J. Schymanski
Department of Environmental Systems Sciences, ETH Zurich, Zurich, Switzerland

Dani Or
Department of Environmental Systems Sciences, ETH Zurich, Zurich, Switzerland

Maciej Zwieniecki
Department of Plant Sciences, University of California Davis, Davis, California, United States of America

Rebecca T. Barnesa
Department of Earth Science, Rice University, Houston, Texas, United States of America

Morgan E. Gallagher
Department of Earth Science, Rice University, Houston, Texas, United States of America

Caroline A. Masiello
Department of Earth Science, Rice University, Houston, Texas, United States of America

Zuolin Liu
Department of Earth Science, Rice University, Houston, Texas, United States of America

Brandon Dugan
Department of Earth Science, Rice University, Houston, Texas, United States of America

Yifeng Chen
State Key Laboratory of Water Resources and Hydropower Engineering Science, Key Laboratory of Rock Mechanics in Hydraulic Structural Engineering, Wuhan University, P. R. China

Chuangbing Zhou
State Key Laboratory of Water Resources and Hydropower Engineering Science, Key Laboratory of Rock Mechanics in Hydraulic Structural Engineering, Wuhan University, P. R. China

Wenzel Kröber
Institute of Biology, Geobotany and Botanical Garden, Martin-Luther-University Halle-Wittenberg, Halle (Saale), Germany

Shouren Zhang
State Key Laboratory of Vegetation and Environmental Change, Institute of Botany, the Chinese Academy of Sciences, Beijing, China

Merten Ehmig
Institute of Biology, Geobotany and Botanical Garden, Martin-Luther-University Halle-Wittenberg, Halle (Saale), Germany

Helge Bruelheide
Institute of Biology, Geobotany and Botanical Garden, Martin-Luther-University Halle-Wittenberg, Halle (Saale), Germany
German Centre for Integrative Biodiversity Research (iDiv) Halle-Jena-Leipzig, Leipzig, Germany

Julia L. Barringer
U.S. Geological Survey, USA

Pamela A. Reilly
U.S. Geological Survey, USA

Arshad Ashraf
Water Resources Research Institute, National Agricultural Research Center, Islamabad, Pakistan

Fereshte Haghighi
Soil Conservation and Watershed Management Institute, Tehran,

Mirmasoud Kheirkhah
Soil Conservation and Watershed Management Research Institute, Tehran, Iran

Bahram Saghafian
Soil Conservation and Watershed Management Research Institute, Tehran, Iran

Preface

Encyclopaedia of Genetic Resources, Chromosome Engineering, and Crop Improvement summarizes landmark research and describes medicinal plants as nature's pharmacy. Genetic resources (GRs) refer to genetic material of actual or potential value. Genetic material is any material of plant, animal, microbial or other origin containing functional units of heredity. With an increasing human population and a decreasing amount of arable land, creative improvements in agriculture will be a necessity in the coming decades to maintain or improve the standard of living. The text Chromosome Engineering in Plants presents techniques for the modification of crops and other plant species in order to achieve the goal of developing the much needed novel approaches to the production of food, feed, fuel, fiber, and pharmaceuticals. First chapter focuses on chromosomal differentiation in Helianthus annuus var.macrocarpus. Recent advances in plant transformation have been presented in second chapter. In third chapter, we isolate a cDNA encoding CENH3 from sunflowers and raise a peptide antibody against CENH3. In fourth chapter, we review the current knowledge on the processes taking place during chromosome segregation in plant meiosis, focusing on the characterization of the molecular factors involved. Principles and applications of immunolocalization technique of chromosome-associated proteins in plants have been discussed in fifth chapter. Sixth chapter discusses about chromosomal location of HWA1 and HWA2, complementary hybrid weakness genes in rice. Structure and stability of telocentric chromosomes in wheat have been described in seventh chapter. Major locus for chloride accumulation on chromosome 5A in bread wheat has been outlined in eighth chapter. In last chapter, we identify three CSILs with stronger fiber or high fiber strength that carried different G. barbadense chromosome segment(s) in the recurrent parent TM-1.

Chapter 1

CHROMOSOMAL DIFFERENTIATION IN HELIANTHUS ANNUUS VAR.MACROCARPUS: HETEROCHROMATIN CHARACTERIZATION AND RDNA LOCATION

Teresa Cuellar[1], Eric Belhassen[2], Begoña Fernández-Calvín[3], Juan Orellana[3], and Jose L Bella[1]

[1]Departemento de Biología, Unidad de Genética, Facultad de Ciencias, Universidad Autónomo de Madrid, E-28049 Madrid, Spain

[2]Génétique et Amélioration des Plantes, INRA, Place Pierre Viala 2, F-34060 Montpellier, France

[3]Departamento de Biotecnología, ETSI Agrónomos, Universidad Politécnica de Madrid, E-28040 Madrid, Spain

ABSTRACT

The 2n = 34 chromosomes of the inbred line HA89, and the Flamme and Mirasol hybrids of *Helianthus annuus* var. *macrocarpus* possess centromeric heterochromatin as established by Giemsa C-banding. This heterochromatin can not be differentiated by fluorochromes such as DAPI or Chromomycin A3, with selective affinity for specific DNA base pairs. This situation probably results from either a balanced AT/GC composition of the involved repeat or the existence of alternating repetitive sequences of opposite base pair composition in these heterochromatic areas. However, there is also heterochromatin associated with the secondary constrictions on three pairs of chromosomes. This heterochromatin appears to be GC-rich according to its response to the fluorochrome treatments, thus indicating heterochromatin heterogeneity in *H. annuus*. Silver staining reveals the existence of active NORs associated with these secondary constrictions. *In situ* hybridization with an rDNA probe confirms these results and makes the existence of other inactive rDNA sites unlikely. These results are relevant to evolutionary and breeding studies on sunflowers.

INTRODUCTION

Although the genus Helianthus is of clear economic and evolutionary interest, the karyotypic status of the 50 or so species that make up the genus (Schilling & Heiser, 1981) has not been analysed to any great extent (Chandler, 1991). Sunflower chromo- somes have not been individually identified and numbered, so there are no cytological markers and no work on chromosome banding has been reported. Although some genetic linkage groups have been identified, they have yet to be localized to specific chromosomes (Rieseberg et al., 1993; Gentzbittel et at., 1995). Most of the karyological information about this taxon has come from the analysis of meiotic pairing in interspecific hybrids in the study of hybrid fertility, introgression and genome relationships (Chandler et al., 1986; Chandler, 1991). Our limited knowledge is a consequence of the *Correspondence. intrinsic difficulties of studying an organism which has a relatively large number of comparatively small chromosomes. Furthermore, their occurrence in an oily cytoplasm makes good quality preparations difficult to obtain.

On the other hand, the use of banding techniques is a general procedure by which markers for individual chromosomes may be obtained, thereby permitting discrimination of species and varieties at the karyological level. In the sunflower the lack of such markers has made it difficult to analyse the introgression of economically interesting traits and the evolutionary relationships of the genus are not well understood. In fact, it has even been proposed that the recognized species of Helianthus are ecotypes of a single ecospecies (Kulshreshtha & Gupta, 1979). As we show in this study, the use of pectinase treatment improves the quality of sunflower chromo- somal preparations. This has allowed us to characterize the chromosomes of one of the commonest inbred lines and two varieties of the cultivated sunflower species Helianthus annuus L. We describe the distribution of the constitutive heterochromatin, its response to specific DNA base pair ligands such as 4'-6-diamidino-2-phenylindole (DAPI) or Chromomycin A3, and the number and position of the Nucleolar Organizer Regions (NORs), as demonstrated by silver staining and fluorescence in situ hybridization (FISH) with an rDNA probe.

MATERIALS AND METHODS

Seeds from the inbred line HA89 of H. annuus var. macrocarpus and the commercial hybrids Mirasol and Flamme were grown in the dark at 25°C. To obtain mitotic chromosomes arrested at metaphase, roots were collected after 3 days and placed in a 0.04 per cent solution of colchicine for 90 mi fixed in 3:1 absolute ethanol: glacial acetic acid and stored at 4°C. Roots were pretreated with a 0.5 per cent pectinase solution in standard pH 4.2 citrate buffer for 1 h at 37°C. The meristems were squashed in 45 per cent glacial acetic acid on microscope slides using a coverslip, the coverslip was removed with a razor blade after immersion in liquid nitrogen and the slides were air dried. Additionally, some root tips were stained employing the standard Feulgen technique and squashed on slides, according to Darlington & La Cour (1976, pp. 125—127). A minimum of 15 roots per line or variety was studied for each technique employed

C-banding and silver staining treatments were carried out according to Schwarzacher et al. (1980) and Lacadena et a!. (1984), respectively, with minor modifications. Air-dried slides were stained after a minimum of 3 days with 4'-6-diamidino-2-phenylindole (DAPI) or Chromomycin A3 (CMA3) and counterstained with Distamycin-A (DA) or Actinomycin-D (AD), as described by Schweizer (1976, 1980).

The 9 kbp EcoRI rDNA fragment of wheat containing the 18S, 5.8S and 26S rRNA coding regions (plus the spacers), obtained from the pTA71 clone (Gerlach & Bedbrook, 1979) was nick translated with digoxigenin-11-dUTP and in situ hybridized using the method of Pendás et a!. (1993). Hybridization sites were detected with antidigoxigenm antibody conjugated to fluorescein isothiocyanate (FISH). Some slides were counterstained with propidium iodide (2.5 jig mL1) and mounted in Vecta Shield antifade solution (Vector Laboratories).

Preparations were examined at 100 x magnification under a microscope equipped with a 100 W epifluorescence system and the appropriate filters for the observation of DAPI- and CMA3-stained chromosomes, and a double filter for the simultaneous observation of fluorescein and propidium iodide in the in situ hybridized chromosomes. Light field photographs were taken with Kodak Imagelink HO film and Kodak Plus-X film was used for fluorescence and in situ hybridization photographs.

RESULTS

The techniques used revealed no chromosomal differences among the line HA89 and the two hybrids of H. annuus studied.

Feulgen Staining

Helianthus annuus has 2n = 34 chromosomes of similar size which, following Al-Allaf & Godward (1979), can be grouped as four pairs of metacentric (M1—M4), eight pairs of submetacentric (SM5—SM12) and five pairs of subtelocentric (ST13—ST17) chromosomes (Fig. la). The similarity in size and morphology of the chromosomes within each group makes further differentiation difficult. However, in all the cases considered, two pairs of submetacentric (SM7 and SM1O) and one pair of subtelocentric (ST13) chromosomes show secondary constrictions (Figs la,b) which allow them to be distinguished.

C-banding

All the chromosomes possess small amounts of centromeric constitutive heterochromatin, and the three pairs of chromosomes bearing secondary constrictions (satellite chromosomes) also have associated heterochromatin (Fig. ib). Distal or interstitial C-heterochromatin has not been revealed in any member of the complement. Variation in size between different individuals of these heterochromatic regions has not been observed.

Fluorochrome Staining

DA-DAPI staining shows no positive response, but instead the chromosomes stain uniformly (Fig. 2a). Similarly, AD-DAPI does not reveal any clear differentiation, although some centromeric regions yield a diffuse, slightly positive response, although this cannot be reproduced photographically. However, in both cases the heterochromatin associated with the secondary constrictions appears negative, in contrast with the positive response it shows with DA-CMA3.

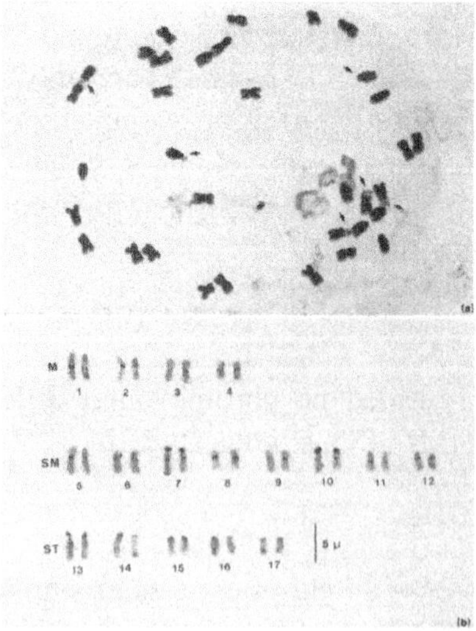

Figure 1. (a) Feulgen-stained mitotic metaphase chromosomes of the hybrid Flamme of Helianthus annuus. Note the existence of three pairs of chromosomes bearing sec-

ondary constrictions (arrows). (b) C-banded karyotype of the inbred line HA89. The heterochromatin is restricted to the centromeric areas and the secondary constrictions of the sate!- lite chromosomes (pairs SM7, SM1O and ST13).

No other differentiation is revealed with this latter treatment (Fig. 2b). None of these fluorescence techniques has differentiated non-C-heterochromatic regions.

Silver Staining

Silver precipitates attached to the secondary constrictions are shown by the three pairs of satellite chromosomes: SM7, SM1O and ST13 (Fig. 3a). Interphase nuclei show between one and six spots of silver. In a sample of 537 silver-stained interphase meristem cells, 39.8 per cent had one nucleolus, 25.6 per cent had two, 28.3 percent had three, 4.4 per cent four, 1.1 per cent five and 0.5 per cent six nucleoli.

FISH

The chromosomal in situ hybridization of the rDNA probe confirms the existence of ribosomal clusters associated with the secondary constrictions of the three pairs of chromosomes (Fig. 3b). Hybridization does not occur in any other location and interphase nuclei consistently show six hybridization sites (close to 99 per cent nuclei in a sample of 214 cells).

DISCUSSION

Despite the economic and evolutionary importance of the genus Helianthus its cytogenetics has hitherto not been investigated in any great detail. This is partially because of the intrinsic difficulty of obtain ing mitotic chromosome preparations of sufficient quality. The enzymatic pretreatment of the meristematic root tissues with pectinase helps to solve this problem, as is commonly observed in other plants. The chromosome number and karyomorphology of the line and varieties of H. annuus var. macrocarpus studied here agree with those described by other authors for other varieties (Al-Allaf & Godward, 1979; Chandler, 1991). However, there have been no previous studies of heterochromatin distribution, chromosomal response to specific DNA sequence fluorochromes, or rDNA activity and location.

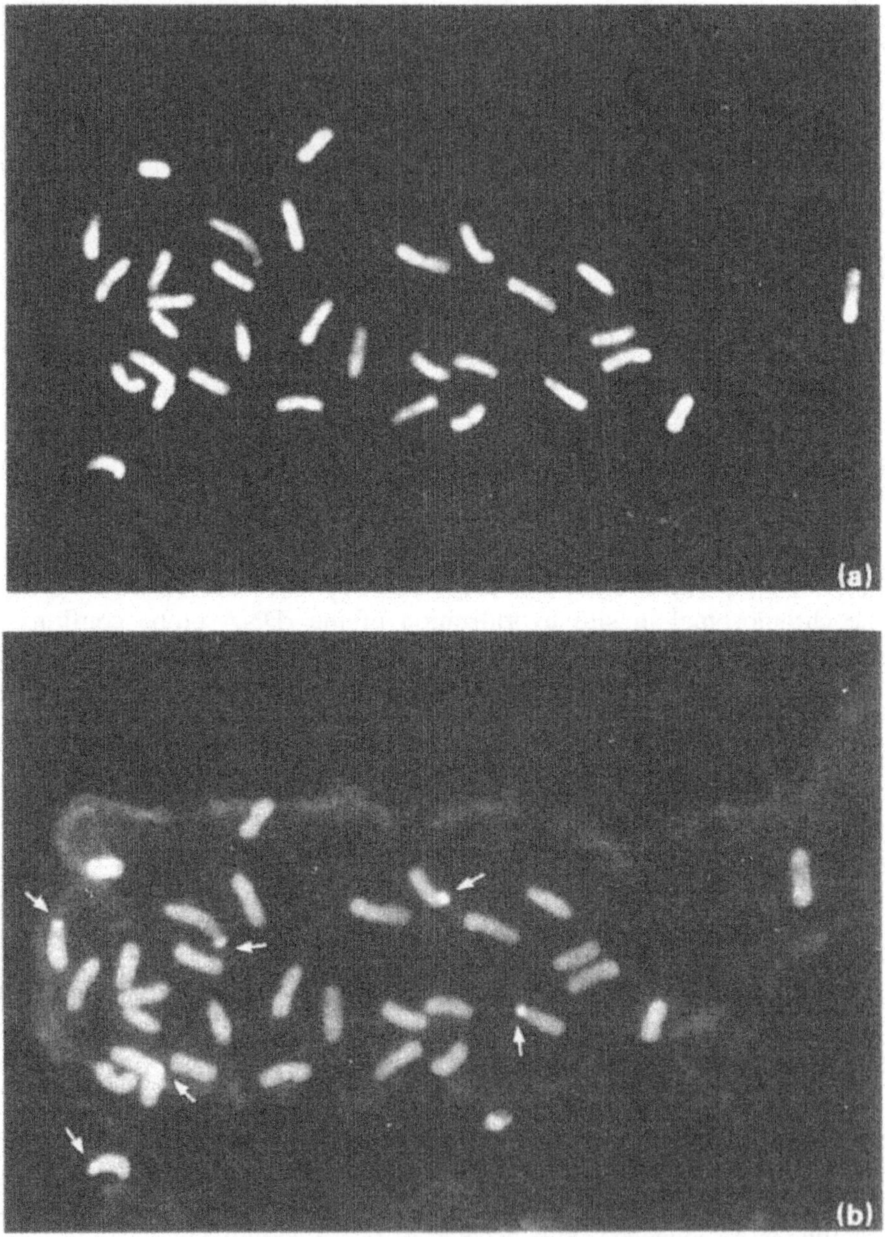

Figure 2. Fluorescent tn-staining of the mitotic metaphase chromosomes of the commercial hybrid Flamme of Helianthus annuus. The DA-DAPI treatment (a) does not show clear differentiation whereas there is positive response to the DA-CMA3 staining of the heterochromatin associated with the secondary constrictions (b).

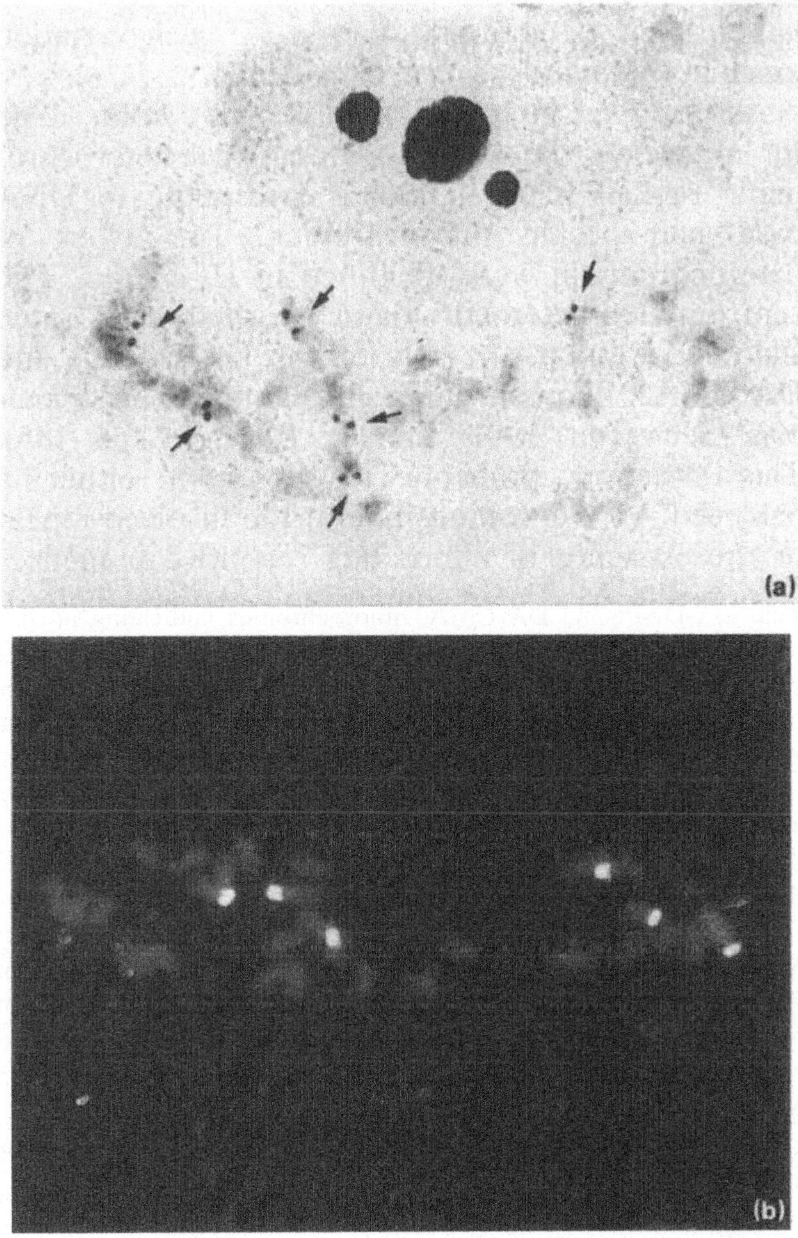

Figure 3. In the inbred line HA89 of Helianthus annuus the silver staining (a) shows six silver spots (NOR5) attached to the secondary constrictions of pairs SM7, SM1O and ST13 (arrows) in the mitotic metaphase chromosomes. These results are confirmed by FISH with an rDNA probe (b). Note also the three nucleoli in the interphase nucleus in (a).

The absence of variation in the C-heterochromatin distribution within and between the inbred line and varieties studied here suggests that this is very probably the standard heterochromatin pattern of the macro carpus variety, which is the most commonly cultivated sunflower. However, this does not imply that this pattern is invariable within H. annuus. It is reasonable to expect that in natural populations of H. annuus there exist polymorphisms for the distribution, size or even composition of the heterochromatin. The lack of karyological information concerning this genus precludes comparison of the C-band distribution reported here with the banding pattern of other Helianthus species and varieties. C-band variation can provide valuable cytotaxonomic information as has been shown in work on the related Anthemidea (Schweizer & Ehrendorfer, 1983). It would therefore be useful to survey a larger range of germplasm to see if such variation exists in H. annuus.

As observed in other organisms (see John, 1988 for a review), the H. annuus equilocal heterochromatic regions have an identical response to DNA base pair-specific fluorochromes. Two types of heterochromatin were identified in H. annuus. The centromeric heterochromatin showed the same staining response with the DA-DAPI and DA-CMA3 fluorochromes, indicating no base pair bias (Schweizer, 1981; Sumner, 1990, pp. 155—185). This situation probably results from either a balanced ATIGC composition of the involved repeat or the existence of alternating repetitive sequences of opposite base pair composition in these heterochromatic areas. However, the heterochromatin in the secondary constrictions of the satellited chromosomes was DA-CMA3-positive, showing that it was GC-rich. In any case, the two types of heterochromatin revealed by these treatments confirm, at the cytological level, the coexistence of distinct families of DNA repetitive sequences in the genome of H annuus that have also been distinguished by molecular analysis (K. Sossey, personal communication). DNA content within the genus Helianthus exhibits a four-fold range of variation largely because of polyploidy; however, there is also considerable variation in DNA content and chromosome size among sunflower species with the same chromosome number. For example, H annuus (HA89) has the fourth lowest DNA content of the 19 species with 2n = 34 chromosomes (Sims & Price, 1985). Intra- specific variation in DNA content amongst cultivated varieties and inbred lines of H. annuus has also been described (Cavallini et al., 1986; Chandler, 1991; Michaelson et a!., 1991). To what extent different amounts of heterochromatin may be involved in this DNA content variation remains to be elucidated, but from our results we know that HA89 has little constitutive heterochromatin, which is confined to small bands in the centromeric areas and secondary constrictions. Further molecular and cytological studies will investigate this relationship.

Silver staining reveals active ribosomal genes (Hubbell, 1985). In root meristem cells from HA89 and the two hybrids of H. annuus, the three pairs of satellite chromosomes show silver deposits attached to the secondary constrictions, revealing the position of active rDNA clusters. This coincidence with secondary constrictions as well as the GC richness of the associated heterochromatin are common features of NORs (Schweizer, 1980). The in situ hybridization with an rDNA probe confirms the number and location of the rDNA clusters and makes the existence of other inactive ribosomal sites unlikely. The existence of one to three silver spots in most of the interphase nuclei of root meristem cells (93.7 per cent) is probably the result of nucleolar fusion (Giménez-MartIn et a!., 1977) although FISH with the rDNA probe shows six clearly separated fluorescent dots in the majority of the interphase cells. The six chromosomes with rDNA clusters show silver deposits attached to them at metaphase, indicating that all of them were active during the preceding interphase (Miller et at., 1976).

The search for chromosomal markers in other species of the genus Helianthus is helpful in a taxon where (i) there are diploid, tetraploid and hexaploid species, all of which have a basic chromosome number of x = 17 and are probably of polyploid origin (Jackson & Murray, 1983; Chandler et a!., 1986); (ii) different races of the same Helianthus species may be chromosomally distinct, although successful crosses may be produced between perennial and annual species, and between diploid and polyploid species (Kulshreshtha & Gupta, 1979); (iii) the cytogenetical information comes almost exclusively from the analysis of interspecific hybrid meiotic configurations (Chandler, 1991). These chromosomal markers could be useful either from the taxonomic point of view to identify chromosomes from different genomes or to follow the introgression of alien chromosomes in cultivated sunflowers, as happens in other plant species (Gustafson & Dille, 1992; Werner et al., 1992). In any case, the existence of linkage maps in the sunflower (Rieseberg et al., 1993; Gentzbittel et a!., 1995) where ribosomal gene polymorphisms have been described (Choumane & Heizmann, 1988) might permit the chromosomes with NORs to be assigned to linkage groups. Current work on sunflower RFLP maps may also yield in situ hybridization probes of use in assigning linkage groups to chromosomes.

The study of the constitutive heterochromatin distribution and fluorochrome response, as well as the number and location of the rDNA sequences in other species and other varieties of Helianthus will provide complementary information on their evolutionary status and, hopefully, suitable markers for breeding studies. Combined research on breeding, cytogenetics and molecular techniques are required in the sunflower to understand natural and artificial interspecific gene transfer (introgression) and the selective forces involved (Belhassen et at., 1994), as well as to provide new tools to evaluate variability in this taxon.

ACKNOWLEDGEMENTS

We are grateful to Drs J. Gosálvez, P. L. Mason and M. Navarrete for their very valuable help. This work has been partially supported by the Spanish-French programme of 'Acciones Integradas', and grants AGF93—0869 and SAF93—0093 from the Spanish CICYT.

REFERENCES

1. Al-Allaf, S, and Godward, M B E. 1979. Karyotype analysis of four varieties of*Helianthus annuus* L. *Cytologia*, **44**, 319–323.
2. Belhassen, E, Auge, G, Ji, J, Billot, C, Frenández-Martínez, J, Ruso, J, and Vares, D. 1994. Dynamic management of genetic resources: first generation analysis of sunflower artificial populations. *Génét Sél Évol*, **26**, 241–253.
3. Cavallini, A, Zolfino, C, Cionini, G, Cremonini, R, Natali, L, Sassoli, O, and Cionini, P G. 1986. Nuclear DNA changes within *Helianthus annuus* L.: cytophotometric, karyological and biochemical analyses. *Theor Appl Genet*,**73**, 20–26.
4. Chandler, J M. 1991. Chromosome evolution in sunflower. In: Tsuchiya, P. and Gupta, P. K (eds) *Chromosome Engineering in Plants: Genetics, Breeding, Evolution*, part B, pp. 229–249. Elsevier, Amsterdam.
5. Chandler, J M, Jan, C, and Beard, B H. 1986. Chromosomal differentiation among the annual *Helianthus* species. *Syst Bot*, **11**, 354–371.
6. Choumane, W, and Heizmann, P. 1988. Structure and variability of nuclear ribosomal genes in the genus *Helianthus*. *Theor Appl Genet*, **76**, 481–489.
7. Darlington, C D, and La Cour, L F. 1976. *The Handling of Chromosomes*, 6th edn. George Allen & Unwin Ltd, London.
8. Gentzbittel, L, Vear, F, Zhang, Y X, Berville, A, and Nicolas, P. 1995. Development of a consensus linkage map of cultivated sunflower (*Helianthus annuus* L.). *Theor Appl Genet*, **90**, 1079–1086.
9. Gerlach, W L, and Bedbrook, J R. 1979. Cloning and characterization of ribosomal RNA genes from wheat and barley. *Nucl Acids Res*, **7**, 1869–1885.

10. Giménez-Martín, G, De La Torre, C, López-Sáez, J F, and Esponda, P. 1977. Plant nucleolus: structure and physiology. *Cytobiologie*, **14**, 421–462.

11. Gustafson, J P, and Dille, J E. 1992. Chromosome location of *Oryza sativa*recombination linkage groups. *Proc Natl Acad Sci USA*, **89**, 8646–8650.

12. Hubbell, H R. 1985. Silver staining as an indicator of active ribosomal genes.*Stain Technol*, **60**, 285–294.

13. Jackson, R C, and Murray, B G. 1983. Colchicine induced quadrivalent formation in *Helianthus*: evidence for ancient polyploidy. *Theor Appl Genet*,**64**, 219–222.

14. John, B. 1988. The biology of heterochromatin. In: Verma, R. S. (ed) *Molecular and Structural Aspects of Heterochromatin*, pp. 1–147. Cambridge University Press, Cambridge.

15. Kulshreshtha, V B, and Gupta, P K. 1979. Cytogenetic studies in the genus*Helianthus* L. *Cytologia*, **44**, 325–334.

16. Lacadena, J R, Cermeño, M C, Orellana, J, and Santos, J L. 1984. Evidence for wheat-rye nucleolar competition (amphiplasty) in triticale by silver staining procedure. *Theor Appl Genet*, **67**, 207–213.

17. Michaelson, M J, Price, H J, Spencer-Johnston, J, and Ellison, J R. 1991. Variation of nuclear DNA content in *Helianthus annuus* (Asteraceae). *Am J Bot*, **78**, 1238–1243.

18. Miller, D A, Dev, V G, Tantravahi, R, and Miller, O J. 1976. Suppression of human nucleolus organizer activity in mouse-human somatic hybrid cells. *Exp Cell Res*, **101**, 135–243.

19. Pendás, A M, Morán, P, and García-Vázquez, E. 1993. Ribosomal genes are interspersed throughout a heterochromatic arm in Atlantic salmon. *Cytogenet Cell Genet*, **63**, 128–130.

20. Rieseberg, L H, Choi, H, Chan, R, and Spore, C. 1993. Genomic map of a diploid hybrid species. *Heredity*, **70**, 285–293.

21. Schilling, E E, and Heiser, C B, Jr, 1981. Infrageneric classification of*Helianthus* compositae. *Taxon*, **30**, 393–403.

22. Schwarzacher, T, Ambros, P, and Schweizer, D. 1980. Application of Giemsa banding to orchid karyotype analysis. *Plant Syst Evol*, **134**, 293–297.

23. Schweizer, D. 1976. Reverse fluorescent chromosome banding with Chromomycin and DAPI. *Chromosoma*, **58**, 307–324.

24. Schweizer, D. 1980. Fluorescent chromosome bands in plants: applications, mechanisms and implications for chromosome structure. In: Davies, D. R. and Hopwood, R. A (eds) *The Plant Genome: Proceedings of the 4th John Innes Symposium*, pp. 61–72. John Innes Charity, Norwich.
25. Schweizer, D. 1981. Counterstaining-enhanced chromosome banding. *Hum Genet*, **57**, 1–14.
26. Schweizer, D, and Ehrendorfer, F. 1983. Evolution of C-band patterns in Asteraceae-Anthemideae. *Biol Zbl*, **102**, 637–655.
27. Sims, L E, and Price, H J. 1985. Nuclear DNA content variation in *Helianthus*(Asteraceae). *Am J Bot*, **72**, 1213–1219.
28. Sumner, A T. 1990. *Chromosome Banding*. Unwin Hyman Ltd., London.
29. Werner, J E, Endo, T R, and Gill, B S. 1992. Towards a cytogenetically based physical map of the wheat genome. *Proc Natl Acad Sci USA*, **89**, 11307–11311.

Chapter 2

RECENT ADVANCES IN PLANT TRANSFORMATION

INTRODUCTION

The world's agriculture and farming are heavily dependent on crops that provide food and fibers for human use, either directly or through livestock. For the past two centuries, modern technology has improved agricultural practices, thereby augmenting conventional plant breeding methods to achieve improved yield and quality of crops. However, multiple factors such as population growth, environmental stress, ecological considerations, and demand for renewable energy have led to the demand for further improvements in the quality and quantity of crops. Plant genetic engineering offers new avenues in this regard and has become one of the most important molecular tools in the modern molecular breeding of crops.

Advancements in plant genetic engineering have made it possible to transfer genes into crop plants from unrelated plants and even from nonplant organisms; as a result, many crop species are being genetically modified for better agronomical traits, including disease resistance, insect tolerance, better nutritional values, and other desirable qualities.. Presently, foreign genes from various origins and production of products in transgenic plants represent a new aspect of the molecular agriculture revolution. In addition, transgenic plants have great impact on nonagricultural applications and represent an alternative for the production of medicinally useful and recombinant proteins and vaccines. However, in some crop plants, the lack of efficient transformation methods to introduce foreign DNA remains an obstacle to the application of plant genetic engineering.

Over the last decade, some significant achievements have been made in the development of new and efficient transformation methods in plants. Methods for delivering exogenous DNA to plant cells and gene transformation in general can be divided into two major categories: indirect and direct DNA deliveries. In the former approach, genes of interest are introduced into the target cell via bacteria, e.g., Agrobacterium tumefaciens or Agrobacterium rhizogenes. In contrast, the latter approach does not employ bacterial cells as mediators to transfer DNA to plant cells. Although various delivery methods have been reported, including the use of other bacterial strains (i.e., TransBacter™ Technology), Agrobacterium-mediated transformation remains the method of choice for plant transformation. The enduring success of the Agrobacterium-mediated approach is primarily attributable to the continuous improvements in plant tissue culture and T-DNA transfer processes. As a result, reproducible and efficient Agrobacterium-mediated protocols have been developed for many dicot and some monocot crops. Agrobacterium-mediated transformation possesses intrinsic advantages over direct DNA delivery systems. These advantages include the ability to transfer large intact segments of DNA, simple transgene insertions with defined ends and low copy number, stable integration and inheritance, and consistent gene expression over the generations.

The lower rates of success achieved with Agrobacterium in monocots and recalcitrant plant species have led to the development of specific direct DNA transfer methods, one of which is microparticle bombardment. This technology, first developed by Sanford and coworkers, is often termed biolistics or gene gun. Particle bombardment has a high success rate in monocot species in which agroinfection is limited. However, biolistic technology possesses several intrinsic disadvantages, including a low transformation efficiency as compared with Agrobacterium (when agroinfection works), a high frequency integration of the vector backbone and a loss of transgene cassette integrity, and transgene silencing due to multicopy insertions. Alternative delivery systems have also been used for gene transfer in plants, including electroporation, microinjection, silicon carbide, and chloroplast transformation. Of these, silicon carbide-mediated transformation represents one of the least complicated methods of plant transformation. Yet, each of these methods has their own limitations in the successful production of transgenic plants.

To date, Agrobacterium-mediated and particle bombardment transformation are the most commonly used methods for plant transformation. Nevertheless, the resulting transgenic plants are of course subject to biosafety issues related to the presence of vector backbone sequences and/or selectable marker genes, irrespective of the delivery method used. Transgenic plants produced by Agrobacterium-mediated transformation are likely subject to integration of vector backbone sequences. This leads to multiple transgene copies in the transgenic plants, complicating the regulatory process of genetically engineered plants. In recent years, there has been an important advancement in generation of vector-backboneand selectable-marker-free transgenic plants while still enabling the use of marker genes to select and identify transgenic plants.

Several strategies have been proposed and are in use for the production of marker-free transgenic plants. The methods that have been developed include simultaneous transformation of two marker genes (cotransformation), the movement of a transgene segment within the genome (transposition), and recombination between two specific sequences that are not necessarily homologous (site-specific recombination). Currently, there is a great demand for simultaneous expression of multiple genes for expression of complex traits in plants. With recent advancement in molecular biology and vector construction technology, it is possible to achieve stable expression of multiple transgenes in a single genome. The recent development of minichromosome technology might represent a strategy for gene stacking in plants. The minichromosome technique can be used to incorporate desirable traits such as insect, bacterial, or fungal resistance, herbicide tolerance, and increased crop quality. Hence, new techniques are in demand to boost yield or to improve crop traits. Finally, the genetic engineering of plants has already begun to play a crucial role in the production of biofuels. This chapter discusses various methods and recent advances in plant transformation technology.

PLANT TRANSFORMATION METHODS

Recent advances in genetic transformation have made it possible to transfer genes of both academic and agronomic importance into various crop species. A prerequisite for successful transformation system is an efficient regeneration protocol when tissue culture-based transformation process is employed. The very basis of plant regeneration relies on the realization that plant somatic cells are totipotent and can be stimulated to regenerate into whole plants. However, this insight is limited because, in reality, only a limited number of plant species and certain types of explant tissues have been found to be capable of regenerating whole plants under appropriate culture conditions. Therefore, much effort has been aimed at establishing and improving plant regeneration systems. Yet, efficient regeneration alone does not necessarily lead to efficient transformation.

There is a need to develop advanced transformation methods that would not only incorporate the required characteristics (stable and desirable transgene integration and expression) into plants but also enable generation of transgenic events in a high-throughput manner. These requirements are particularly relevant now in the crop post-genome era in which ever-increasing amounts of genome sequence information, BAC clones, ESTs, and full-length cDNAs are available. This situation presents both new challenges and opportunities for plant transformation research. At present, Agrobacterium and microprojectile are the commonly used methods for this purpose; other methods, such as electroporation and microinjection, are still used only rarely. The following sections discuss in detail the recent advances in each of the plant transformation methods.

Agrobacterium Mediated Transformation

In this method, A. tumefaciens or A. rhizogenes is employed to introduce foreign genes into plant cells. A. tumefaciens is a soilborne gram-negative bacterium that causes crown-gall, a plant tumor. The tumor-inducing capability of this bacterium is due to the presence of a large Ti (tumor-inducing) plasmid in its virulent strains. Similarly, Ri (root-inducing) megaplasmids are found in virulent strains of A. rhizogenes, the causative agent of "hairy root" disease. Both Ti- and Ri-plamids contain a form of "T-DNA" (transferred DNA). The T-DNA contains two types of genes: oncogenic genes, encoding enzymes involved in the synthesis of auxins and cytokinins (causing tumor formation), and genes involved in opine production. The T-DNA element is flanked by two 25-bp direct repeats called the left border (LB) and right border (RB), respectively, which act as a cis element signal for the T-DNA transfer. Both oncogenic and opine catabolism genes are located inside the T-DNA of the Ti plasmid whereas the virulence (vir) genes are situated outside the T-DNA on the Ti plasmid and bacterial chromosome. These vir genes are organized into several operons (virA, virB, virC, virD, virE, virF, virG, and virH) on the Ti-plasmid and other operons (chvA, chvB, and chvF) that are chromosomal and are essential for T-DNA transfer.

The mechanism of gene transfer from A. tumefaciens to plant cells involves several steps, which include bacterial colonization, induction of the bacterial virulence system, generation of the T-DNA transfer complex, T-DNA transfer, and integration of the T-DNA into the plant genome. The process of T-DNA transfer is initiated upon receipt of specific signals (e.g., phenolic compounds) received from host cells. Previous observations suggested that wounding or vigorous cell division also promotes T-DNA transfer, presumably due to induction by phenolic compounds produced during cell repair or during the formation of new cells. In response, a signal received by virA activates a cascade of other vir protein machinery genes. However, very little is known about the nature and function of the factors that Agrobacterium utilizes, for instance, specific receptors on the host cell surface and/or cell wall. Subsequently, virD1 and virD2 proteins nick both the left and right borders on the bottom strand of the T-DNA. The resulting single-stranded T-DNA molecule (T-strand), together with several vir proteins, is then exported into the host cell cytoplasm through a channel formed by the Agrobacterium VirD4 and VirB protein complex. Before its entry into the host cell cytoplasm, the VirD2–T-strand conjugate is most likely coated by VirE2, forming the T-complex. VirE2 is a single-stranded DNA-binding Agrobacterium protein that is transported into the plant cell, where it presumably functions to protect the T-DNA from degradation.

The Agrobacterium T-complex is likely transported through the host cell cytoplasm by a cellular-motor-assisted mechanism. In a recent report, a dynein-like Arabidopsis protein (DLC3), coupled with another protein (VIP1), has been proposed to function in the intracellular transport of the Agrobacterium T-complex. Recently, an additional Arabidopsis protein, VIP2 (VirE2 interacting protein2), has been demonstrated to play a major role in T-DNA integration into the plant genome. The T-complex then enters the cell nucleus by an active mechanism mediated by the nuclear import machinery of the host cell. This facilitates integration of the T-strand into the host genome at random positions by a process of nonhomologous, or more precisely, illegitimate recombination

Many recent reviews have addressed mechanisms related to T-DNA transfer.Characterization of the mechanisms governing the T-DNA transfer process is very important for plant transformation studies and should facilitate the identification of conditions to maximize T-DNA transfer. The best example of this is the use of a phenolic compound (e.g., acetosyringone) as well as a low-pH media and temperature to induce T-DNA transfer during the Agrobacterium infection stage.

Advancement in molecular biology techniques have enabled the development of binary Ti vectors that are compatible with utilization of both Agrobacterium strains and Escherichia coli. Development of the binary vector and bacterial strain systems for plant transformation is achieved by placing virulence genes on a separate plasmid (the large Ti-plasmid) and the gene to be transferred on separate vector (the small binary vector). Since most gene manipulations are carried out in E. coli, the binary plasmids are designed to replicate in both E. coli and Agrobacterium. Recent advancements in vector-cloning techniques have led to the development of binary bacterial artificial chromosome (BIBAC) vectors that enhance the frequency of T-DNA transfer of large-sized DNA fragments. The key features that make BIBAC vectors useful include its extremely low copy number and high stability when they are replicated in either Agrobacterium or E. coli cells.

The development of superbinary vectors made it possible to transform monocot plants for the first time. A superbinary vector represents an improved version of a binary vector; the vector carries a 14.8 kb KpnI fragment containing the virB, virG, vir C genes derived from pTiBo542. These genes are responsible for the supervirulence phenotype of A. tumefaciens strain A281.

The superbinary vector has been highly efficient in transforming various plants, particularly recalcitrant species, such as important cereal crops.

The integration and enhancement of gene expression in the plant genome greatly depends on the promoter that is fused at the 5¢ end of the gene of interest. The most widely used foreign regulatory elements include the 35S promoter of the cauliflower mosaic virus and the transcriptional terminator of the Agrobacterium nopaline synthase gene (nos), which together promote high-level gene expression in transgenic plants. The 35S promoter is a constitutive promoter that is used in vector constructs to drive target gene expression in many plant species. Recently, a new, stronger promoter has been developed. This "super promoter" is a hybrid construct combining a triple repeat of the octopine synthase (ocs) activator sequence plus the mannose synthase (mas) activator elements fused to the mas promoter. An initial study performed with this construct in maize (Zea mays) and tobacco (Nicotiana tabacum) confirmed the stable expression of superpromoter – GUS fusion gene in both the plant species. In tobacco, activity of the superpromoter is higher in mature leaves than young leaves, whereas in maize, the activity differed little among the tested aerial portions of the plant.

In order to achieve efficient Agrobacterium-mediated T-DNA transfer, several factors must be taken into consideration, including the plant genotypes, sources of explants, Agrobacterium strains, medium salt strength and pH, duration and temperature of Agrobacterium–explant interactions (inoculation and cocultivation), and use of T-DNA-inducing compounds. Agrobacterium-mediated transformation of higher plants is now well established for dicotyledonous species. In recent years, the frequency of gene transfer to monocotyledonous species has also been greatly improved (Table 1). A variety of explants can be used as target material for Agrobacterium-mediated transformation, including embryonic cultures, immature or mature zygotic embryos, mature seed-derived calli, meristems, shoot apices, primary leaf nodes, excised leaf blades, roots, cotyledons (including or excluding nodal areas), stem segments, and callus suspension cultures. These explant tissues have been able to regenerate through either a somatic embryogenesis or organogenesis regime. As an alternative to the organogenesis regeneration regime, somatic embryogenesis offers the advantage of single cell regeneration. However, the types and physiological conditions of explants used are critical to successful regeneration and subsequent recovery of whole transgenic plants. For example, in sorghum, a higher transformation efficiency was achieved in immature embryos taken from field grown plants than in the immature embryos from greenhouse-grown plants (34).

Table 1: Major gene transfer methods used for transformation of monocot and dicotyledonous plant species

Plant species	Types of explants	Transfer method	Gene transferred	Transformation efficiency (%)	Reference
Monocots					
Oryza sativa	C	*Ag*	*cry1Ac*	2	(152)
	C	MpB	*hGH*	79.5	(153)
	SA	EP	*npt* II, *ppt*	13.8	(154)
Hordeum vulgare	ImE	*Ag*	*hpt* II, *luc*	25	(155)
	EC	MpB	*AtNDPK2*	0.15	(156)
Saccharum sp.	AxB	*Ag*	*npt* II, *bar*, *gusA*	50	(157)
	C	EP	*gusA*	80	(158)
Sorghum bicolor	C	*Ag*	*Man*, *gfp*	8.3	(159)
Triticum sp.	ImE	*Ag*	*bar*, *gusA*	9.7	(160)
Zea mays	ImE	*Ag*	*gusA*, *bar*	12.2	(41)
	ImE	MPB	*gusA*, *hpt* II	31	(73)
Dicots					
Arabidopsis thaliana	S	*Ag*	T-DNA	26	(161)
	L	MPB	*gusA*	–	(162)
Arachis hypogea	CN	*Ag*	*gusA*	38	(163)
	SE	MPB	*VP2*, *gusA*	12.3	(164)
	EL	EP	*gusA*	3	(165)
Brassica oleracea	ML	Cl-MPB	*cry1Ab*	11.1	(166)
Cajanus cajan	CN	*Ag*	*npt* II, *H*	51	(167)
Eucalyptus sp.	ApS	*Ag*	*gusA*	9	(168)
Glycine max	CN	*Ag*	*bar*, *gusA*	5.5	(169)
	SE	MPB	*Os-mALS*	60	(170)
	Fl	PTP	*phyA*	13	(171)
Gossypium hirsutum	EC	*Ag*	*cry1Ia5*	83	(7)
	EC	SCW	*AVP1*, *npt* II	64	(172)
Malus domestica	IVS	*Ag*	*Lc*	50	(173)
Pinus sp.	EC	*Ag*	*npt* II, *bar*, *gusA*	65–98	(174)

Ag Agrobacterium-mediated transformation, AtNDPK2 Arabidopsis nucleoside diphosphate kinase gene, ApS apical shoot, AVP1 Arabidopsis vacuolar pyrophosphatase, AxB axilllary bud, bar bialaphos-resistance gene, C callus, Cl-MPB chloroplast-mediated MPB, CN cotyledonary node, cry1Ia5 Bacillus thuringiensis toxins, EC embryogenic callus, EL embryonic leaflets, EP electroporation, Fl flower, gusA b-glucuronidase, H hemagglutinin protein, hGH human growth hormone, hpt II hygromycin phosphotransferase II, ImE immature embryo, IVS in vitro shoot, L leaf, Lc maize leaf color regulatory gene, ML mature leaf, MPB microprojectile bombardment-mediated transformation, npt II neomycin phosphotransferase II, Os-mALS acetolactate synthase derived from rice, PTP pollen tube pathway transformation, phyA phytase A, S seed, SA shoot apex, SCW silicon-carbide-whiskers-mediated transformation, SE somatic embryo, T-DNA transfer DNA

Competence for transformation can be enhanced in recalcitrant explants by phytohormone treatments. The maximal percent of calli showed higher transient GUS activity when picloram was used in Typha latifolia. Similarly, in Hibiscus cannabinus, preculturing of explants for 2 days in benzyladenine (BA) containing medium enhanced transient GUS expression . Phytohormone treatment activates cell division and dedifferentiation in many tissues. The stimulation of cell division by phytohormones suggests that efficient Agrobacterium transformation may occur at a particular stage of the plant cell cycle.

The procedures that promote Agrobacterium cells to come into close contact with the plant cells around wounded tissue sites have been found to enhance T-DNA transfer. For example, "dip-wounding," which is prewounding of the explants prior to cocultivation with blade dipped in Agrobacterium suspension, increases transformation frequency as high as 10-fold (Table 2, Xinlu Chen, Xiujuan Su, and Zhanyuan J. Zhang, unpublished data). When "dip-wounding" is combined with the use of phenolic compounds in inoculation and cocultivation media, the attraction of Agrobacterium is presumably enhanced at wounded sites, which facilitates increased access of bacteria to plant cells.

Table 2: Impact of dip-wounding on soybean regeneration and transformation

Inoculation procedure	Number of explants			Percentage of explants		Number of events	Percentage of events
	To start	Good	Green	Good	Green		
Standard	350	172	129	49.0	37.0a	1	0.3a
Dip-wounding	350	198	236	57.0	67.0b	12	3.4b

In a number of plant species, explants are hypersensitive to Agrobacterium infection, forming necrotic barriers; this can be overcome by the use of antioxidants to reduce the oxidative burst. Tissue browning/necrosis associated with Agrobacteriummediated transformation has been reported in various types of explants of both dicotyledonous and monocotyledonous species. Antioxidants such as polyvinylpyrrolidone (PVPP), dithiothreitol (DTT), cysteine, glutathione, lipoic acid, ascorbic acid, and citric acid are now commonly used to reduce tissue browning/necrosis of explants during plant transformation. In sugarcane, pretreatment of explants on media containing ascorbic acid and cysteine prior to transformation results in higher transformation efficiency. In grapes and rice, the addition of antioxidants, such as polyvinylpyrrolidone and dithiothreitol also increases the transformation efficiency. In soybean, which is difficult to transform, a higher transformation rate was achieved by including l-cysteine, DTT, and sodium thiosulphate in cocultivation media. These antioxidants also enhanced the transformation in maize Hi-II. Such increased T-DNA transfer enabled the use of standard binary vectors to routinely achieve efficient transformation without the use of a superbinary vector. In contrast, lipoic acid was found to enhance GUS transient expression and transformation efficiency in tomato.

In addition to the strategies discussed above, other treatment conditions have been devised more recently to promote Agrobacterium-mediated transformation. Desiccation of explants prior to Agrobacterium infection enhances transformation efficiency in sugarcane, whereas addition of surfactants such as Silwet-L77 or pluronic acid F68 enhances transformation in wheat. Such desiccation helps to reduce cell damage due to the reduction of cell turgidity, whereas the use of surfactants may induce wounds and thinner cell walls in the explant tissues, thereby promoting Agrobacterium attachment to explants and ultimately T-DNA transfer.Another method is treatment of plant tissues and Agrobacterium to brief sonication, which allows Agrobacterium and T-DNA entry into the tissues. In loblolly pine, sonication was found to enhance not only transient transformation but also the recovery of hygromycin-resistant lines.

Plant transformation frequency is also associated with cell division or dedifferentiation of the host explants. Recent studies have revealed the phase of the plant cell cycle at the time of transformation to be a major determinant of transformation and regeneration efficiency. To achieve a stable transgenic event, the differentiated cell for regeneration should adopt a "stem-cell-like" state for pluripotentiality to renter the S phase of the cell cycle. The transformation competence of the cells is high in S and G2 phase/M phase, and lower in G0 and G1 phases. A cell cycle study identified the RepA, HP1, E2Fa, CycD3, and CycD1 genes to be involved in the pluripotency of cells in G1-S phase and further progression through S and G2 phases . Gordon-Kamm et al. demonstrated substantial improvement in maize transformation by overexpressing the RepA gene. Arabidopsis histone H2A, which is expressed at higher levels in tissues that are more susceptible to Agrobacterium infection, is essential for T-DNA integration in somatic cells. Co-overexpressing E2Fa together with its dimerization partner, DPa, resulted in increased cell proliferation in cotyledons, leading to approximately twice the number of cells as wild type.

Transcriptome analysis of E2Fa–DPa-overexpressing plants showed upregulation of 14 genes that are involved in DNA replication and S phase onset (54). VIP1 (VirE2 interacting protein 1) in Arabidopsis; those genes were also found to be upregulated during the pluripotent stage. Overexpression of VIP1 increased the rate of transient and stable plant transformation and predicted its role in interacting with histone proteins. A recent study in Phaseolous coccineus has identified the proteins PIN and CUC to be involved in shoot apical meristem formation. The use of transgenic marker genotypes, such as WUSCHEL (WUS)- reporter or STM-reporter (SHOOTMERISTEMLESS) construct should be useful in identifying meristemoids in early stages in development.

Agrobacterium rhizogenes strains contain a T-DNA region located on the Ri plasmid that carries genes involved in root initiation, which are essential for production of hairy roots. Studies on the function of Ri T-DNA-encoded genes performed using the agropine-type Ri plasmids led to the identification of 18 ORF, including rolA, rolB, rolC, and rolD genes. It was evident that these genes also participate in the production of hairy roots. Several of the experiments in this study were carried out to inactivate or overexpress various rol genes, generating stable transgenic lines with various alterations in plant phenotypes and root morphology.

In general, however, A. rhizogenes-mediated root transformation has received considerably less attention than A. tumefaciens transformation. The main reason for this is the difficulty in regenerating plants from hairy roots transformed by A. rhizogenes. Therefore, this delivery system has been predominantly used to generate transgenic roots for transient assays. One of the most advanced systems in such type of assay is the production of "composite plants". The important characteristic feature of hairy roots is their ability to grow in plant hormone-free media. These growth characteristics have made hairy roots a useful tool for secondary metabolite production, use in metabolic engineering, and studies of root biology in general. Recently, it has been demonstrated that hairy-root cultures can be adapted for T-DNAactivation tagging studies. Hairy roots are also used to genetically investigate root–bacterial interactions in soybean. For example, various studies have revealed that hairy roots derived from various soybean cultivars maintain their cognate nematode resistance or susceptible phenotypes. Recently, hairy roots have been used to produce recombinant proteins. This system is also suitable for high-throughput analysis of root-related transgene expression in Medicago or soybean root tissues and is now expected to be applied to root transformation for highthroughput functional analysis of certain gene expressions in other plant species.

In summary, Agrobacterium-mediated transformation has been very successfully employed recently in transformation of both dicots and particularly monocots, the latter of which had long been thought to be unable to host Agrobacterium. These successes are attributable to the development of the superbinary vector, the use of antioxidants, and optimization of the composition of inoculation and cocultivation media. This trend of success in transformation of various plant species will continue not just because of new ideas and approaches in improving Agrobacterium transformation but also because of the obvious advantages of such a natural gene delivery system.

Microprojectile (Particle) Bombardment Transformation

Microprojectile bombardment is one of the direct gene transfer methods for development of transgenics. This method was developed in 1980s to genetically engineer plants that were recalcitrant to transformation with Agrobacterium. Subsequently, the technique has been widely used to produce transgenic plants in a wide range of plant species. The first particle delivery method was developed by Sanford and coworkers. The Sanford device was extensively modified to produce the PDS-1000/He machine, which was licensed to DuPont. The technique involves coating microcarriers (gold or tungsten particles approx 0.6–1.0 mm in diameter) with the DNA of interest and then accelerating them at high velocities, to penetrate into the cell of essentially any organism.

Briefly, the microcarriers are spread evenly on circular plastic film (macrocarrier). The entire unit is then placed below the rupture disk in the main vacuum chamber of the biolistic device. A variety of rupture disks are available that burst at pressures ranging from 450 to 2,200 psi. Below the macrocarrier is a stopping screen, in which a wire-mesh is designed to retain the macrocarrier, while allowing the microcarriers to pass through. The target tissue is placed below the launch assembly unit. Under a partial vacuum, the microprojectile is fired, and helium is then allowed to fill the gas-acceleration tube. The helium pressure builds up behind a rupture disk, which bursts at a specific pressure, thus releasing a shock wave of helium that forces the macrocarriers down onto the stopping screen. The microcarriers leave the circular plastic film and continue flying down the chamber to hit and penetrate the target tissue, thus delivering the DNA.

Several factors must be considered for successful gene transfer using particle bombardment technology. These factors include the design of a suitable vector with a small size and high copy number, as well as the quantity and quality of the delivered DNA. The entire process must be performed under sterile conditions to prevent contaminations of target tissue during subsequent tissue culture. The types and sizes of microcarriers are important choices because they affect the depth of penetration of the accelerated microcarrier as well as the degree of damage to the target cells. Gold particles ranging from 0.6 to 3.0 μm in diameter are commercially available. The degree of penetration required will depend on the thickness of the cell wall, the type of tissue being transformed, and the depth of the target cell layers. Variation in the helium pressures, the level of vacuum generated, the size of the particles, and the position of target tissues will dictate the momentum and penetrating power at which the microprojectiles strike the tissue. All of these parameters are under the experimenter's control and must be optimized for a given target tissue.

Treatment of the target tissues prior to and after particle bombardment has a significant effect on the frequency of recoverable transgenic cell lines and plants. An attractive feature of particle bombardment is its ability to transfer foreign DNA into any cell or tissue type whose cell wall and plasma membrane can be penetrated. Embryogenic and meristematic tissues are the most commonly employed target tissues for the production of genetically transformed plants. Particle bombardment of embryogenic tissues has been successfully exploited to produce transgenic plants in a wide range of agronomically important plants, including legumes, tuber crops, starchy staples, trees, commodity crops, and all of the major cereals. In the case of bombarding apical meristems, the treatment, physiological status, and age of the mother plants prior to excision of the explants must be taken into consideration. Use of an osmotic pretreatment or partial drying of the target cells prior to bombardment is a commonly used strategy to increase the frequency of successful transformation.

One of the advantages of particle bombardment is the possible expression of multiple transgenes in the target tissue, which can be achieved by fusion of genes within the same plasmid that is then bombarded into the target tissues. In recent years, multiple independent gene expression cassettes have been successfully transferred using particle bombardment in wheat, rice, and soybean. The use of microprojectile bombardment has made it easy to transfer large DNA fragments into the plant genome, although the integrity of the DNA is a concern. Integration of yeast artificial chromosomes (YACs) into the plant genome by particle bombardment has been successful with inserts of up to 150 kb. Although requiring further development, integration of large DNA fragments promises to be an important tool in future plant research and crop biotechnology.

In summary, over the last two decades, many published studies have utilized microprojectile technology in both monocot and dicot plant species (Table 1). In the process, a diverse range of agronomic traits have been transferred and imparted to crop plants via particle bombardment, which remains the most promising technique to genetically engineer plastids. However, this technology is limited due to several drawbacks, such as the integration of multiple copies of the desired transgene, in addition to superfluous DNA sequences that are associated with the plasmid vector. Multicopy integrations and superfluous DNA can lead to silencing of the gene of interest in the transformed plant. This problem was overcome by transferring the desired coding region only with its control elements into the target cells of plant genome.

Electroporation Mediated Transformation

Electroporation-mediated transformation requires the application of strong electric field pulses to cells and tissues and is known to cause some type of structural rearrangement of the cell membrane. In vitro introduction of DNA into cells is now the most common application of electroporation. The technique was originally developed for protoplast transformation but has subsequently been shown to work with intact plant cells as well. A voltage of 25 mV and an amperage of 0.5 mA for 15 min are the most often used parameters. However, factors such as surface concentration of DNA and tolerance of cells to membrane permeation may affect electroporation efficiency. Using the electroporation method, successful transformation has been achieved with protoplasts of both monocot and dicot plants. The first report of fertile transgenic rice utilized electroporation of DNA into embryogenic protoplasts. However, using protoplasts as explants for regeneration of transformants limits the use of electroporation for stable transformation because the protoplastto-plant regeneration system has not been developed in most plant species. The electroporation of plant cells and tissues is very similar in its principles to the electroporation of protoplasts. This approach enabled the recovery of the first transgenic plants in barley. In sugarcane, a gene was transferred into intact meristem tissue using electroporation-mediated transformation (Table 1). While electroporation was proposed as an alternative to biolistics, it is not nearly as efficient. Compared to biolistics, it is inexpensive and simple, but the technique has only been successful in a few plant species. The thick cell walls of intact tissues represent key physical barriers to electroporation.

PEG/Liposome Mediated Transformation

Polyethylene glycol (PEG)-mediated transformation is a method used to deliver DNA using protoplasts as explants. The method is similar to electroporation in that the DNA to be introduced is simply mixed with the protoplast, and uptake of DNA is then stimulated by the addition of PEG, rather than an electrical pulse. PEG-mediated transformation has several advantages because it is easy to handle and no specialized equipment is required. However, the technique is rarely used due to the low frequency of transformation and because many species cannot be regenerated into whole plants from protoplasts. In addition, fertility may be a concern because of the somaclonal variation of the transgenic plants derived from protoplast cultures. Nonetheless, using this method, transgenic maize and barley have been produced. Thus, protoplast transformation is feasible in cereals, even though fertility problems in the regenerants are often encountered. In cotton, transformation was achieved using combined polybrene–spermidine-based callus treatment.

Related to PEG-mediated transformation is the liposomemediated transformation technique. In this method, DNA enters protoplasts via endocytosis of liposomes. Generally, this process involves three steps: adhesion of liposomes to the protoplast surface, fusion of liposomes at the site of adhesion, and release of the plasmid inside the cell. Liposomes are microscopic spherical vesicles that form when phospholipids are hydrated. Liposomes being positively charged tend to attract negatively charged DNA and cell membrane. In this process, the engulfed DNA is free to integrate into the host genome. However, there have been very few successful reports on the application of this technique in plant species because the technique is very laborious and is associated with low efficiency. In tobacco, intact YACs were transformed via lipofection-PEG technique.

Silicon Carbide Mediated Transformation (SCMT)

Kaeppier et al. first reported the use of silicon-mediated transformation, which is one of the least complicated methods. In this method, small needle-type silicon carbide whiskers are mixed with plant cells and the gene of interest, and the mixture is then vortexed. In the process, the whiskers pierce the cells, permitting DNA entry into the cells. The fibers most often used in this procedure have an elongated shape, a length of 10–80 mm and a diameter of 0.6 mm and show high resistance to expandability. The method is simple, inexpensive, and effective on a variety of cell types. The efficiency of SCMT depends on fiber size, vortexing, the shape of vortexing vessel, as well as the plant material used for transformation. The SCMT technique has been used in a variety of plants, including maize, rice, wheat, tobacco, etc. Furthermore, silicon carbide fibers have been found to improve the efficiency of Agrobacterium-mediated transformation. The main disadvantages of SCMT include low transformation efficiency and damage to cells, thereby negatively affecting their regeneration capacity. Furthermore, this method imposes health hazards due to fiber inhalation if not performed properly.

More recently, two related technologies have been developed: silicon fibers have been reported to increase callus transformation by 30–50% in rice, and mesoporous silica nanoparticles have been used to deliver DNA and chemicals into both plant cells and intact leaves. Mesoporous silica nanoparticles are synthesized from a reaction between tetraethyl orthosilicate and a template made of micellar rods.

Microinjection

Microinjection is the direct mechanical introduction of DNA into the nucleus or cytoplasm using a glass microcapillary injection pipette. Using a microscope, cells (protoplasts) are immobilized in low-melting-point agar with a holding pipette and gentle suction; DNA is then injected into the cytoplasm or nucleus. The microinjection technique requires relatively expensive technical equipment for the micromanipulation of single cells under a microscope and involves precise injection of small amounts of DNA solution; the procedure is also very time-consuming. The injected cells or clumps of cells are subsequently cultured in tissue culture systems and regenerated into plants. Successful regeneration of microinjected rapeseed, tobacco, and barley has produced genome-integrated, stable transformants. However, microinjection has achieved only limited success in plant transformation due to the thick cell walls of plants and, more challengingly, to a lack of availability of a single-cell-to-plant regeneration system in most plant species.

Chloroplast Mediated Transformation

In genetically modified plants, the gene of interest usually integrates into the nucleus; however, it is also possible to transfer the gene into the plastid. The chloroplast genome is highly conserved among plant species and typically consists of double-stranded DNA of 120–220 kb, arranged in monomeric circles or in linear molecules. In most higher plant species, the chloroplast genome has two similar inverted repeat (IR_A and IR_B) regions of 20–30 kb, that separate a large single copy (LSC) region and small single copy (SSC) region. Both the microprojectile or protoplastmediated transformation methods are capable of delivering DNA to plastids, but to achieve successful transformation, chloroplast-specific vectors are required. The basic plastid transformation vector is comprised of chloroplast-specific expression cassettes and target-specific flanking sequences. Integration of the transgene into the chloroplast occurs via homologous recombination of the flanking sequences used in the chloroplast vectors. The first successful chloroplast transformation (of the aada gene, which confers spectinomycin resistance) was reported in Chlamydomonas. In higher plants, plastid transformation has been accomplished in tobacco with various foreign genes. In recent years, several crop chloroplast genomes have been transformed through organogenesis, and maternal inheritance has been observed. In economically important crops such as cotton, efficient plastid transformation has been achieved through somatic embryogenesis by bombarding embryogenic cell cultures.

Several transgenes engineered through chloroplast transformation have conferred valuable agronomic traits in plants, including insect and pathogen resistance, and both drought and salt resistance. In soybean expressing the Cry1Ab gene, insecticidal activity against velvet bean caterpillar was conferred to the transformed plant (Table 1). Advancement in chloroplast engineering has made it possible to use chloroplasts as bioreactors for the production of recombinant proteins and biopharmaceuticals. Because plastid genes are maternally inherited, transgenes inserted into these plastids are not disseminated by pollen. Additional advantages of this transformation system include the ability to express several genes as a polycistronic unit, thereby potentially eliminating position effects and gene silencing in chloroplast genetic engineering.

Native Gene Transfer

Transformation of native genes (including regulatory elements) into plants without a selectable marker is highly desirable to overcome consumers' concerns about GM crops. Historically, the development of transgenics with important agronomic traits depended on the use of genes derived from other organisms. Over the past decade, however, rapid advances in plant molecular biology have resulted in a major shift from bacteria and viruses to plants as important gene sources. A broad variety of plant genes associated with agronomically important traits have now been identified. For example, plants containing modified acetolactate synthase (ALS) genes displayed the same high levels of sulfonylurea tolerance as transgenic plants that expressed bacterial ALS tolerance genes. Likewise, the occurrence of glyphosate tolerance in a goosegrass (Eleusine indica) biotype has been associated with a mutated 5-enolpyruvylshikimate-3-phosphate synthase (EPSPS) gene. With respect to native gene transfer research, the application of transposon tagging and map-based cloning methods have resulted in the identification of more than 50 functionally active resistance (R) genes, several of which are currently being used as viable alternatives to foreign antimicrobial genes in crop improvement programs. One of the most agronomically important R-genes isolated is the Solanum bulbocastanum RB gene, which provides resistance to the potato late blight fungus Phytophthora infestans.

Rapid progress has also been made in the development of plant-based gene alternatives and recovery of various insecticidal proteins that are involved in insect resistance. It has been suggested that a 30-kDa maize cysteine protease can be used to enhance maize tolerance to caterpillars and armyworms. Alternatively, silencing or overexpressing key biosynthetic genes can enhance a plant's ability to produce insecticidal secondary metabolites. In tobacco, suppression of a P450 hydroxylase gene resulted in a 19-fold increase in cembratrieneol levels in trichomes, dramatically enhancing aphid resistance.

About 40 diverse plant genes have been used to enhance tolerance to abiotic stresses. In Arabidopsis, overexpression of C-repeat-binding transcription factor (CBF) increased survival resistance to freezing, drought, and salt.

Various methods such as promoter trapping and RNA fingerprinting have facilitated the identification of hundreds of plant promoters (e.g., ubiquitin and actin gene promoters), many of which contain regulatory elements that support high-level gene expression in most tissues of transgenic plants. Use of the same genetic material available in the plant benefits genetic engineering approaches into existing plant breeding programs.

MARKER GENES AND METHODS TO REMOVE SELECTABLE MARKERS FROM TRANSGENICS

After explants are transformed with the requisite vector having the gene of interest, the transformed cells/tissues need to be selected, which requires a selectable marker gene in the vector. In current transformation systems, a selectable marker gene is codelivered with the gene of interest to identify and resolve rare transgenic cells from nontransgenic cells. However, during transformations, only a few plant cells accept the integration of foreign DNA; most cells remain nontransgenic. Several selectable marker genes are currently in use, and most utilize antibiotic or herbicide selection (Table 3). Of these, the npt II gene (encoding neomycin phosphotransferase II), which imparts kanamycin resistance, is commonly used in the development of transgenics. Visual observation of gene expression is achieved using reporter genes that can represent important components for transient and stable expression studies. The most commonly used reporter genes in plant transformation are the ß-glucuronidase (GUS) gene and the green fluorescent protein gfp. Selectable marker genes are required to recover stably transformed plants. However, due to environmental concerns, as well as human health risks, the use of nonselectable markers needs to be promoted. In recent years, the following techniques have been used for the production of selectable-marker-free plants

Cotransformation

Cotransformation is a strategy that utilizes two plasmid vectors: one containing the gene of interest and the other containing a selectable marker gene. With this approach, integration of genes may be either within a single locus or at unlinked loci. The method achieves a high cotransformation efficiency and is usually carried out using biolistic or Agrobacterium-mediated transformation. The approach may permit the simultaneous introduction of many genes, independent of gene sequence, with a limited number of selectable marker genes. For example, in one report, nine genes were transferred into the rice genome by biolistics. Because the cotransformed genes are integrated at a single locus, they cosegregate. However, Agrobacterium-mediated cotransformation has the advantage over biolistic transformation that cotransformed genes often integrate into different loci, which results in the segregation of unlinked selectable marker genes from the gene of interest, thereby permitting production of marker-free transgenic plants. Cotransformation via Agrobacterium uses either two mixed Agrobacterium populations (each carrying a different binary vector) or a single Agrobacterium population (carrying two different binary vectors) (Fig. 1).

Table 3: Various selectable markers and reporter genes commonly used in transgenic plants

Gene	Enzymes encoded	Substrate	Gene source	Reference
Selectable markers				
bar	Phosphinothricin acetyl transferase	Phosphinothricin	*Streptomyces hygroscopicus*	(175)
BADH	Betaine aldehyde dehydrogenase	Betaine aldehyde	*Spinacia oleracea*	(176)
bxn	Bromoxynil nitrilase	Oxynils	*Klebsiella pneumonia*	(177)
cat	Chloramphenicol acetyl transferase	Chloramphenicol	*Escherichia coli* Tn5	(178)
dhfr	Dihydrofolate reductase	Methotrexate	*Candida albicans*	(179)
EPSPS	5-Enolpyruvyl shikimate-3-phosphate synthase	Glyphosate	*Petunia × hybrida*	(180)
gox	Glyphosate oxidoreductase	Glyphosate	*Ochrobactrum anthropi*	(181)
hpt II	Hygromycin phosphotransferase II	Hygromycin B	*E. coli*	(182)
ManA	Phosphomannose isomerase	D-Mannose	*E. coli*	(183)
npt II	Neomycin phosphotransferase II	Kanamycin	*E. coli* Tn5	(184)
xylA	Xylose isomerase	D-Xylose	*Streptomyces rubignosus*	(185)
Reporter genes				
uidA/GUS	β-Glucuronidase	X-gluc	*E. coli*	(186)
gfp	Geen fluorescent protein		*Aequorea victoria*	(187)
lacZ	Galactosidase	X-gal	*E. coli*	(188)
luc	Luciferase	Luciferin	*Photinus pyralis*	(189)
	Oxalate oxidase	Oxalic acid	*Triticum aestivum*	(190)

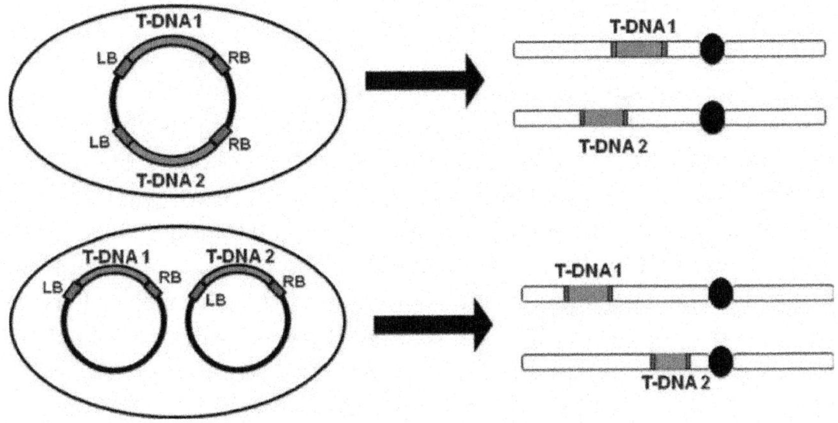

Figure 1. Schematic representation of two T-DNA on single or two separate vectors in Agrobacterium. LB and RB: T-DNA left and right borders, respectively. After transformation, the cotransformed T-DNA integrates, along with its flanking sequences (the LB and RB), into the plant genome at different loci.

Transposon Mediated Transformation

Recent developments in genetic engineering have made it possible to use transposon-based transformation to eliminate selectable marker genes in transgenics; for example, the maize Ac/Ds transposable element system has been used to excise selectable marker genes from plants. In this approach, the marker gene is flanked by the inverted repeat sequences of the Ds element. Subsequent to the transformation and T-DNA integration, expression of the Ac transposase from within the T-DNA results in excision of the gene of interest from the T-DNA insert containing the selectable marker gene. As a result, the gene of interest is transferred from the T-DNA site to another chromosomal location. Successful application of the system therefore requires the activity of the Ac transposase for the development of markerfree transgenic plants.

Repositioning within the genome can also enhance the expression profile of the gene of interest. In tomato, transposition of the GUS reporter gene and subsequent generation of npt II-free plants has been accomplished with both single and multiple T-DNA insertions. For example, hpt II-free rice plants were created that expressed Bt endotoxin encoded by the cry1B gene excision; reinsertion of the transgene occurred at very high frequencies (25–37%), preserving high levels of resistance to striped stem borer. However, because this technology relies on crossing plants to segregate the gene of interest from the marker gene and transposase, it is of limited use in plants that are vegetatively propagated or which have a long reproductive cycle.

Multiautosystem

To improve the selectable-marker-free plant technology, an IPT-type MAT (multiautotransformation) vector system was developed as an alternative to the Ac/Ds transposable element system. A MAT vector system uses the IPT (isopentenyltransferase) gene and Ac element under the control of the CaMV35S promoter. However, IPT expression in the transformed plant generates an abnormal phenotype called the "extreme shooty phenotype." Thus, subsequent to transformation, the IPT gene is removed using the Ac transposable element from the T-DNA, leaving only the gene of interest in the inserted copy of the T-DNA. This results in marker-free transgenic plants with a normal phenotype. However, there are several drawbacks of using a MAT system for marker gene removal. Specifically, there is the possibility of variable rates of transposition efficiency andalso of reinsertion of the transposable element.

Because of potential biosafety concerns, attempts have been made to generate transgenic plants devoid of selectable marker genes and vector backbone sequences from the T0 generation. Thus, methods have been devised to reduce the frequency of vector backbone sequence integration during plant transformation. In maize, the ovary-drip method was used to increase transformation frequency using linear green fluorescent protein (GFP) cassettes (Ubi-GFP-nos) flanked by 25-bp T-DNA borders as the transfer gene. In another approach, multiple tandem LB repeats were used to suppress the transfer of vector backbone in rice transformation. And more recently, soybean transformation experiments have been conducted to reduce vector backbone sequences by using nonlethal genes that interfere with plant development (113), resulting in a high frequency of vector-backbone-free transgenics. Multiple border sequences can also be used to generate vector-backbonefree transgenics. And finally, in maize, backbone-free transgenics were developed using vectors having a selectable marker gene in the vector backbone and the gene of interest in the T-DNA.

Site-Specific Recombination

The most widely used site-specific recombination system is the Cre–lox system from bacteriophage P1, which is very effective in the generation of marker-free plants. In this system, the plant is transformed with a T-DNA vector carrying the gene of interest with lox sites (34 bp repeats in direct orientation) flanking the selectable marker. In the second round of transformation, Cre recombinase is introduced to achieve precise excision of the marker gene. Specifically, Cre catalyzes the recombination between the lox repeat sequences, thereby eliminating the marker gene in the progeny. This system has been used in various plant species to generate marker-free transgenics. Using the Cre– lox system, it is also possible to resolve multiple transgene copies into only a single copy per recipient genome. Recent advances in this technology have led to the use of different inducible promoter systems; for example, a strategy has been developed using the B estradiol-inducible promoter system in which an artificial transcription factor, XVE, was constructed for use in plants with its cognate promoter. Using the Cre/lox system under the control of an inducible promoter, successful marker-free transgenic plants were developed. Recently, successful excision of loxP-flanked selectable marker has been achieved using a flowerspecific promoter in rice. However, to achieve a high transformation efficiency, more refinement is needed to improve this technology.

Two T-DNA Mediated Transformation

Using two T-DNAs, one bearing a selectable marker and the other containing the gene of interest, in a single vector, markerfree transgenic plants have been produced. This approach yields higher frequencies of cotransformation than a strategy using a mixture of A. tumefaciens strains carrying separate vectors. In one study, GUS and hpt II genes cotransformed into tobacco showed segregation of both the genes at unlinked loci, resulting in hpt marker-free plants. In another report, a 100% cotransformation frequency was achieved in tobacco when the selected T-DNA was twice as large as the nonselected T-DNA. In maize, cotransformation with an octopine strain carrying a binary vector with two T-DNAs yielded cotransformation frequencies of 93% for the bar and GUS genes in the T0 generation. In barley, a similar approach using smaller vectors yielded a cotransformation frequency of 66%, but only 24% of these segregated as marker-free plants. These studies clearly demonstrate that marker-free plants can be generated with variable efficiencies using Agrobacterium-mediated cotransformation followed by segregation of the genes in the subsequent sexual generations.

GATEWAY PLANT TRANSFORMATION VECTORS

Gateway cloning technology offers a fast and reliable highthroughput, restriction-enzyme-free cloning strategy for plasmid construction. The Gateway technology is based on the sitespecific recombination reaction mediated by bacteriophage l DNA fragments flanked by recombination sites (att). These sites can be transferred into vectors containing compatible recombination sites (attL attR or attB, attP) in a reaction mixture mediated by the Gateway clonase mix. The backbone of all described Gateway-compatible plant transformation vectors is the plasmid pPZP200. Two recombination reactions, catalyzed by LR and BP recombinases (clonases), respectively, are used in Gateway cloning. The first step, catalyzed by LR, inserts the gene of interest into the

Gateway vector at the attL and attR sites. The resulting construct is called the entry clone. All entry clones have attLs flanking the gene of interest. These are necessary in the Gateway system because these attL sites are cut to form sticky ends by the Gateway clonase. These sticky ends match with the sticky ends of the destination vector, which contains attR restriction sites. This process is called a LR reaction and is mediated by LR clonase mix, which contains the recombination proteins necessary for excision and incision. The product formed in the LR reaction is called the expression clone, which represents a subclone of the starting DNA sequence, correctly positioned in a new vector backbone. The second Gateway step is the BP reaction, which is the reverse of the LR reaction. In the BP reaction, the DNA insert flanked by 25 bp attB sites is transferred from the expression clone into a vector donated by a plasmid containing the attP sites. The final product is termed the destination clone and contains the transferred DNA sequence. Alternatively, these two sequential steps can be reversed to meet specific cloning needs (Fig. 2). The BP reaction thus allows rapid, efficient, directional PCR Cloning.

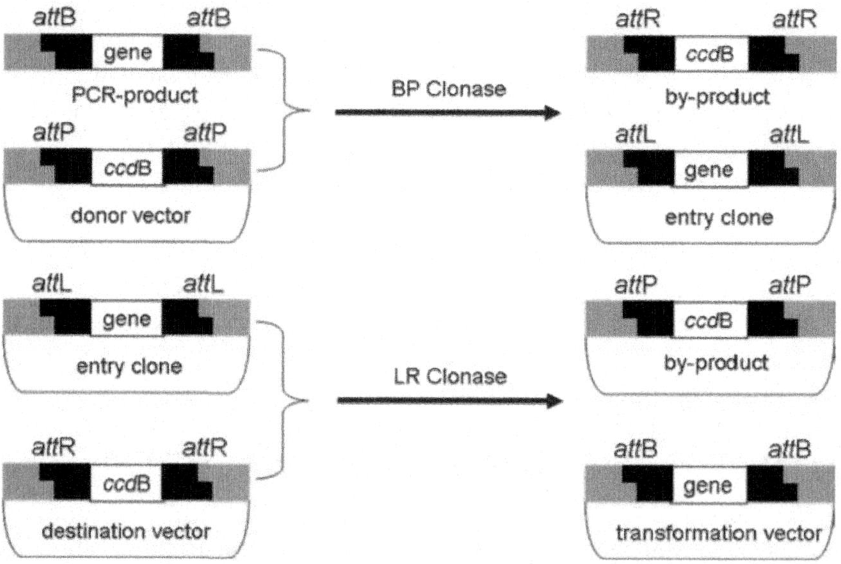

Figure 2. Diagram of gateway cloning technology to clone a PCR-product containing the gene of interest into plant transformation vector, without the use of restriction or ligation enzymes. ccdB, a bacterial suicide gene whose replacement by the incoming transgene-containing PCR product, will allow the transformed bacterial cells to survive, ensuring the presence of the transgene in either the entry or destination clone.

With advances in Gateway technology, it is possible to create many vectors with fused or directly linked multiple transgenes. Recently, for RNA interference, a high throughput cloning system has been designed using pHELLSGATE vectors. In these vectors, the recombination sites do not affect gene silencing performance, in contrast to the conventional restriction enzyme cloning vectors. In Arabidopsis, overexpression of large DNA fragments was found to be effective using Gateway technology. Molecular analysis of T0 and T1 generations confirmed that Gateway-compatible constructs were active and also that the att recombination sites did not inhibit transgene activity or interfere with enhancer activity in vivo.

However, the use of Gateway-RNAi transformation vectors can be a challenge due to the requirement for generating doublestranded RNA via inverted repeats. Thus, because the att recombination sites need to be duplicated, there is an increased chance of altering the orientation of the intron spacer whose splicing efficiency may be adversely impacted, causing reduced RNAimediated silencing efficacy. One solution to this problem is to use a double-intron spacer in the opposite orientation. However, the effectiveness of such a design needs to be validated in more experiments with different plant species.

Recently, the "pEarleyGate" vectors have been designed for Agrobacterium-mediated plant transformations; these vectors translationally fuse FLAG, HA, cMyc, AcV5, or tandem affinity purification epitope tags onto target proteins, with or without an adjacent fluorescent protein. A high-throughput protocol has recently been developed using the Gateway binary vector R4pGWBs for transformation of Arabidopsis thaliana. This vector is designed for the one-step construction of chimeric genes between any promoter and any cDNA. Also, autofluorescent protein tags are useful due to their ability to visualize cellular processes in vivo. For example, the pSAT vectors are useful for both autofluorescent protein tagging and multiple gene transfer. In many plant laboratories, Gateway binary vector technology has enabled in planta expression of recombinant proteins fused to fluorescent tags. Efficient expression of fluorescent protein has been achieved in Nicotiana benthamiana. For easy manipulation and efficient cloning of DNA fragments for gene expression studies, a new Gateway expression vector has been developed by combining the Gateway system and a recombineering system. The recombineering system uses bacteriophage-based homologous recombination in which genomic DNA in a bacterial artificial chromosome (BAC) is modified or subcloned without restriction enzymes or DNA ligase, thereby permitting the direct cloning of gDNA fragments from BACs to plant transformation vectors. The construct is converted into a novel Gateway Expression vector that incorporates cognate 5' and 3' regulatory regions, using recombineering, to replace the intervening coding region with the Gateway Destination cassette. Using this approach, efficient transformation has been achieved in Arabidopsis.

GENE TARGETING (ZINC-FINGER)

Zinc-finger endonucleases (ZFNs), or "molecular scissors," have recently been developed to target genes in plant systems. ZFNs are engineered proteins that have highly specific ZF domains fused to a sequence-independent nuclease domain. The current generation of ZFNs combines the nonspecific cleavage endonuclease domain of the Fok I restriction enzyme with several (usually three) zinc fingers domains that provide cleavage specificity. Subsequent to transformation, ZFNs introduce targeted double-stranded breaks in genomic DNA, thereby inducing recombination and repair processes at specific sites. In this process, homologous recombination will occur using the homologous sequence of the transferred gene to repair the double-strand DNA break (Fig. 3).

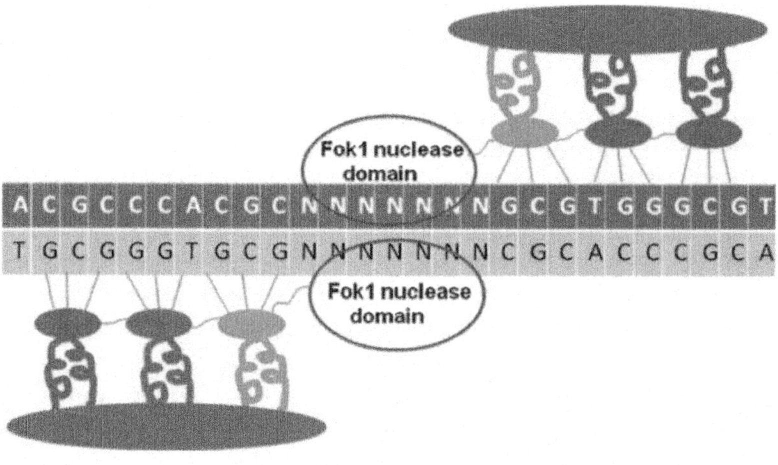

Figure 3. Schematic representation of a zinc finger nuclease (ZFN) binding to its target site. The three different zinc finger domain indicate different sequences. The zinc-finger domains are fused to the cleavage domain of endonuclease FokI to create the ZFN. The three-finger ZFN requires two copies of the 9 bp recognition site in an inverted orientation to produce a double-stranded break (DSB).

ZFN-mediated gene targeting promises to be a powerful tool in the development of novel crop species possessing beneficial agronomic traits. ZFNs have been utilized successfully in both tobacco and Arabidopsis. In Arabidopsis, the ZFN have been shown to cleave and stimulate mutation at specific genomic sites, resulting in a mutation frequency of 0.2 per target. Sitespecific cleavage and transgene integration has also been demonstrated in an engineered tobacco cell culture system. Such studies demonstrate the utility of zinc finger technology in plants and hold much promise for the development of novel transgenic plants.One major limitation in ZFN technology is the potential for cellular toxicity due to off-target DNA cleavage; thus, ZFNs are being modified to reduce this effect.

CHROMOSOME TRUNCATION

The recent development of plant artificial chromosome technology provides an opportunity for the stable expression and maintenance of multiple transgenes in a single genome. Plant artificial chromosomes are produced through telomere-mediated truncation of endogenous chromosomes. Telomeres are comprised of arrays of tandemly repeat sequences present at each end of a plant chromosome. To facilitate in vivo telomere-mediated truncation, an array of telomere sequences is cloned into a vector and then delivered into the genome via Agrobacterium-mediated T-DNA transformation or particle bombardment. Upon integration of the construct into the genome, the telomeric sequences can at some frequency be recognized as a bona fide chromosome terminus in the recipient chromosome, thereby truncating the chromosome at the insertion site. Recently, a 2.6-kb telomere repeat array isolated from Arabidopsis has been successfully transformed in maize to generate telomere-truncated chromosomes whose structure was verified by marker gene expression and FISH (fluorescent in situ hybridization)-based karyotyping. Plant minichromosome technology can also be combined with site-specific recombination systems to facilitate the stacking of multiple transgenes. This strategy for construction of engineered chromosomes should easily be extended to other plant species because it does not rely on species-specific cloned centromere sequences.

FUTURE PROSPECTS

Over the last two decades numerous transformation techniques have been developed for plants. Both Agrobacterium-mediated and microprojectile bombardment-based transformations are now standard laboratory techniques in plant labs. Recent advancements in genetic engineering techniques have led to the identification of genes and proteins involved in the DNA transfer mechanism accomplished by Agrobacterium transformation. Yet, despite this progress, many economically important crops and tree species remain highly recalcitrant to Agrobacterium infection. Thus, much effort is being made to develop regeneration protocols that can efficiently integrate exogenous genes to develop stable transgenics using both Agrobacterium- and biolisticmediated transformation. Novel techniques are also under development to develop genetically transformed plants with desired characteristics.

The generation of transgenic plants requires the use of various selectable marker genes that are introduced together with the gene of interest. Yet, it is abundantly clear that marker-free transgenic plants will be required in the future, thereby requiring more progress in genomics, cloning technology, and vector design, so as to eliminate the need for residual bacterial selectable marker genes in the future. Genetically engineered plants also play an important role in the ongoing study of gene function and metabolic pathways. Promising research has led to the identification of genes that control organogenesis or somatic embryogenesis which may function as selectable marker genes. The reintroduction of native genes and regulatory elements into plants therefore represents a viable means to achieve exogenous marker-free plants. A wide variety of plant genes associated with agronomically important traits have now been identified.

Chloroplast transformation technology is gaining importance due to its unique advantage of gene stacking without concern about gene silencing and of creating an opportunity to produce multivalent vaccines in a single transformation step. The recently developed zinc finger nuclease technology holds much promise in both basic and applied agricultural biotechnology. ZFN-assisted gene targeting and chromatin remodeling studies should also aid in characterizing gene function in plants. Resources are now available for engineering ZFNs in numerous plant species. Advancements in genetic engineering have also led to the development of minichrosome technology, which may provide a solution to gene stacking. Large DNA sequences, including multiple genes, could be introduced into the genome using this technology. Therefore, advances in transgenic technology would provide a solution for production of improved crop species to meet the world's demands for food, feed, fiber, and fuel. Hence, it is hoped that in the near future, GM plants with minimal genomic modifications can be developed.

REFERENCES

1. Job, D. (2002) Plant biotechnology in agriculture. Biochimie 84, 1105–1110.

2. Vain, P. (2007) Thirty years of plant transformation technology development. Plant Biotechnol. J. 5, 221–229.

3. Fischer, R., Stoger, E., Schillberg, S., Christou, P., and Twyman, R.M. (2004) Plant based production of biopharmaceuticals. Curr. Opin. Plant Biol. 7, 152–158.

4. Tzfira, T., and Citovsky, V. (2006) Agrobacterium-mediated genetic transformation of plants: biology and biotechnology. Curr. Opin. Biotechnol. 17, 147–154.

5. Broothaerts, W., Mitchell, H.J., Weir, B., Kaines, S., Smith, L.M., Yang, W., Mayer, J.E., Roa-Rodríguez, C., and Jefferson, R.A. (2005) Gene transfer to plants by diverse species of bacteria. Nature 433, 629–633.
6. Hiei, Y., Ohta, S., Komari, T., and Kumashiro, T. (1994) Efficient transformation of rice (Oryza sativa) mediated by Agrobacterium and sequence analysis of the boundaries of the T-DNA. Plant J. 6, 271–282.
7. Leelavathi, S., Sunnichan, V.G., Kumria, R., Vijaykanth, G.P., Bhatnagar, R.K., and Reddy, V.S. (2004) A simple and rapid Agrobacterium-mediated transformation protocol for cotton (Gossypium hirsutum L.): embryogenic calli as a source to generate large numbers of transgenic plants. Plant Cell Rep. 22, 465–470.
8. Zhang-Hua, H., Jin-Qing, C., Guan-Ting, W., Wei, J., Chun-Xiu, L., Rui-Zhi, H., Fu-Lin, W., Zhi-Hong, L., and Xiao-Yun, C. (2005) Highly efficient transformation and plant regeneration of tall fescue mediated by Agrobacterium tumefaciens. J. Plant. Physiol. Mol. Biol. 31, 149–159.
9. Sanford, J.C. (1990) Biolistic plant transformation. Physiol. Plant. 79, 206–209.
10. Taylor, N.J., and Fauquet, C.M. (2002) Microparticle bombardment as a tool in plant science and agricultural biotechnology. DNA Cell Biol. 21, 963–977.
11. Rakoczy-Trojanowska, M. (2002) Alternative methods of plant transformation – a short review. Cell Mol. Biol. Lett. 7, 849–858.
12. Scutta, C.P., Zubko, E., and Meyer, P. (2002) Techniques for the removal of marker genes from transgenic plants. Biochimie 84, 1119–1126.
13. Darbani, B., Eimanifar, A., Stewart, C.N., Jr., and Camargo, W.N. (2007) Methods to produce marker-free transgenic plants. Biotechnol. J. 2, 83–90.
14. Yu, W., Lamb, J.C., Han, F., and Birchler, J.A. (2006) Telomere-mediated chromosomal truncation in maize. Proc. Natl. Acad. Sci. USA 103, 17331–17336.
15. Zupan, J., Muth, T.R., Draper, O., and Zambryski, P. (2000) The transfer of DNA from Agrobacterium tumefaciens into plants: a feast of fundamental insights. Plant J. 23, 11–28.
16. Christie, P.J. (1997) Agrobacterium tumefaciens T-complex transport apparatus: a paradigm for a new family of multifunctional transporters in Eubacteria. J. Bacteriol. 179, 3085–3094.
17. Tzfira, T., Vaidya, M., and Citovsky, V. (2002) Increasing plant susceptibility to Agrobacterium infection by overexpression of the

Arabidopsis nuclear protein VIP1 Proc. Natl. Acad. Sci. USA 99, 10435–10440.

18. Anand, A., Krichevsky, A., Schornack, S., Lahaye, T., Tzfira, T., Tang, Y., Citovsky, V., and Kirankumar, S.M. (2007) Arabidopsis VIRE2 INTERACTING PROTEIN2 is required for Agrobacterium T-DNA Integration in plants. Plant Cell 19, 1695–1708.
19. Tzfira, T., and Citovsky, V. (2002) Partnersin-infection: host proteins involved in the transformation of plant cells by Agrobacterium. Trends Cell Biol. 12, 121–129.
20. Bako, L., Umeda, M., Tiburcio, A.F., Schell, J., and Koncz, C. (2003) The VirD2 pilot protein of Agrobacterium-transferred DNA interacts with the TATA box-binding protein and a nuclear protein kinase in plants. Proc. Natl. Acad. Sci. USA 100, 10108–10113.
21. Gelvin, S.B. (2000) Agrobacterium and plant genes involved in T-DNA transfer and integration. Annu. Rev. Plant. Physiol. 51, 223–256.
22. Veena, J.H., Doerge, R.W., and Gelvin, S.B. (2003) Transfer of T-DNA and vir proteins to plant cells by Agrobacterium tumefaciens induces expression of host genes involved in mediating transformation and suppresses host defense gene expression. Plant J. 35, 219–236.
23. Lacroix, B., Kozlovosky, S.V., and Citovsky, V. (2008) Recent patents on Agrobacteriummediated gene and protein transfer for research and biotechnology. Recent Pat. DNA Gene Seq. 2, 69–81.
24. Hoekema, A., Hirsch, P.R., Hooykaas, P.J.J., and Schilperoort, R.A. (1983) A binary plant vector strategy based on separation of vir and T-region of the Agrobacterium tumefaciens Ti plasmid. Nature 303, 179–180.
25. Hamilton, C.M. (1997) A binary-BAC system for plant transformation with high-molecular-weight DNA. Gene 200, 107–116.
26. Rui-Feng, H., Yuan-Yuan, W., Bo, D., Ming, T., Ai-Qing, Y., Li-Li, Z., and Guang-Cun, H. (2006) Development of transformation system of rice based on binary bacterial artificial chromosome (BIBAC) vector. Acta Gen. Sin. 33, 269–276.
27. Komari, T. (1990) Transformation of cultured cells of Chenopodium quinoa by binary vectors that carry a fragment of DNA from the virulence region of pTiBo542. Plant Cell Rep. 9, 303–306.
28. Lee, L., and Gelvin, S.B. (2008) T-DNA binary vectors and systems. Plant Physiol. 146, 325–332.

29. Ishida, Y., Saito, H., Ohta, S., Hiei, Y., Komari, T., and Kumashiro, T. (1996) High efficiency transformation of maize (Zea mays) mediated by Agrobacterium tumefaciens. Nat. Biotechnol. 14, 745–750.

30. Khanna, H.K., and Daggard, G.E. (2003) Agrobacterium tumefaciens-mediated transformation of wheat using a super binary vector and a polyamine supplemented regeneration medium. Plant Cell Rep. 21, 429–436.

31. Lee, L.-Y., Kononov, M.E., Bassuner, B., Frame, B.R., Wang, K., and Gelvin, S.B. (2007) Novel pant transformation vectors containing the superpromoter. Plant Physiol. 145, 1294–1300.

32. Cheng, M., Lowe, B.A., Spencer, T.M., Ye, X., and Armstrong, C.L. (2004) Factors influencing Agrobacterium-mediated transformation of monocotyledonous species. In Vitro Cell Dev. Biol. Plant 40, 31–45.

33. Opabode, J.T. (2006) Agrobacterium mediated transformation of plants: emerging factors that influence efficiency. Biotech. Mol. Biol. Rev. 1, 12–20.

34. Zhao, Z., Cai, T., Tagliani, L., Miller, M., Wang, N., Pang, H., Rudert, M., Schroeder, S., Hondred, D., Seltzer, J., and Pierce, D. (2000) Agrobacteriummediated sorghum transformation. Plant Mol. Biol. 44, 789–798.

35. Nandakumar, R., Chen, L., and Rogers, S.M.D. (2004) Factors affecting the Agrobacterium-mediated transient transformation of the wetland monocot, Typha latifolia. Plant Cell Tiss. Organ Cult. 79, 31–38.

36. Herath, S.P., Suzuki, T., and Hattori, K. (2005) Factors influencing Agrobacteriummediated genetic transformation of kenaf. Plant Cell Tiss. Organ Cult. 82, 201–206.

37. Chateau, S., Sangwa, R.S., and SangwanNorreel, B.S. (2000) Competence of Arabidopsis thaliana genotypes and mutants for Agrobacterium tumefaciens-mediated gene transfer: role of phytohormones. J. Exp. Bot. 51, 1961–1968.

38. Dan, Y. (2008) Biological functions of antioxidants in plant transformation. In Vitro Cell Dev. Biol. Plant 44, 149–161.

39. Enriquez-Obregon, G.A., Vazquez-Padron, R.I., Prieto-Samsonov, D.L., de la Riva, G.A., and Selman-Housein, G. (1998) Herbicide-resistant sugarcane (Saccharum officinarum L.) plants by Agrobacteriummediated transformation. Planta 205, 20–27.

40. Olhoft, P.M., and Somers, D.A. (2001) l-Cysteine increases Agrobacterium mediated T-DNA delivery into soybean cotyledonary-node cells. Plant Cell Rep. 20, 706–711.

41. Vega, J., Yu, W., Kennon, A.M., Chen, X., and Zhang, Z.J. (2008) Improvement of Agrobacterium-mediated transformation in Hi-II maize (Zea mays L.) using standard binary vectors. Plant Cell Rep. 27, 297–305.

42. Frame, B.R., Shou, H., Chikwamba, R.K., Zhang, Z., Xiang, C.I., Fonger, T.M., Pegg, S.E.K., Li, B., Nettleton, D.S., Pei, D., and Wang, K. (2002) Agrobacterium tumefaciens-mediated transformation of maize embryos using a standard binary vector system. Plant Physiol. 129, 13–22.

43. Dan, Y.A. (2004) A novel plant transformation technology-Lipoic acid. In vitro Cell Dev. Biol. Plant 42, 18.

44. Dan, Y., Armstrong, C.L., Dong, J., Feng, X., Fry, J.E., Keithly, G.E., Martinell, B.J., Roberts, G.A., Smith, L.A., Tan, L.J., and Duncan, D.R. (2009) Lipoic acid – a unique plant transformation enhancer. In Vitro Cell Dev. Biol. Plant (DOI: 10.1007/s11627- 009-9227-5).

45. Cheng, M., Fry, J.E., Pang, S., Zhou, I., Hironaka, C., Duncan, D.R.I., Conner, T.W.L., and Wang, Y. (1997) Genetic transformation of wheat mediated by Agrobacterium tumefacien. Plant Physiol. 115, 971–980.

46. Tang, W. (2003) Additional virulence genes and sonication enhance Agrobacterium tumefaciens-mediated loblolly pine transformation. Plant Cell Rep. 21, 555–562.

47. Pena, L., Perez, R.M., Cervera, M., Juarez, J.A., and Navarro, L. (2004) Early events in Agrobacterium-mediated genetic transformation of citrus explants. Ann. Bot. (Lond.) 94, 67–74.

48. Arias, R.S., Filichkin, S.A., and Strauss, S.H. (2006) Arias divide and conquer: development and cell cycle genes in plant transformation. Trends Biotechnol. 24, 267–273.

49. Grafi, G. (2004) How cells dedifferentiate: a lesson from plants. Dev. Biol. 268, 1–6.

50. Riou-Khamlichi, C., Huntley, R., Jacqmard, A., and Murray, J.A. (1999) Cytokinin activation of Arabidopsis cell division through a D-type cyclin. Science 283, 1541–1544.

51. De Veylder, L., Beeckman, T., Beemster, G.T.S., Engler, J.D.A., Ormenese, S., Maes, S., Naudts, M., Der Schueren, E.V., Jacqmard, A., Engler, G., and Inze, D. (2002) Control of proliferation, endoreduplication and differentiation by the Arabidopsis E2Fa–DPa transcription factor. EMBO J. 21, 1360–1368.

52. Gordon-Kamm, W., Dilkes, B.P., Lowe, K., Hoerster, G., Sun, X., Ross, M., Church, L., Bunde, C., Farrell, J., Hill, P., Maddock, S., Snyder, J., Sykes, L., Li, Z., Woo, Y.-M., Bidney, D., and Larkins, B.A. (2002)

Stimulation of the cell cycle and maize transformation by disruption of the plant retinoblastoma pathway. Proc. Natl. Acad. Sci. USA 99, 11975–11980.

53. Mysore, K.S., Nam, J., and Gelvin, S. (2000) An Arabidopsis histone H2A mutant is deficient in Agrobacterium T-DNA integration. Proc. Natl. Acad. Sci. USA 97, 948–953.

54. Vlieghe, K., Vuylsteke, M., Florquin, K., Rombauts, S., Maes, S., Ormenese, S., Hummelen, P.V., de Peer, Y.V., Inze, D., and De Veylder, L. (2003) Microarray analysis of E2Fa–DPa-overexpressing plants uncovers a cross-talking genetic network between DNA replication and nitrogen assimilation. J. Cell Sci. 116, 4249–4259.

55. Zambre, M., Geerts, P., Maquet, A., Montagu, M.V., Dillen, W., and Angenon, G. (2001) Regeneration of fertile plants from callus in Phaseolus polyanthus Greenman (year bean). Ann. Bot. (Lond.) 88, 371–377.

56. Taoka, K.-I., Yanagimoto, Y., Daimon, Y., Hibara, K.-I., Aida, M., and Tasaka, M. (2004) The NAC domain mediates functional specificity of CUP-SHAPED COTYLEDON proteins. Plant J. 40, 462–473.

57. Slightom, J.L., Durand-Tardif, M., Jouanin, L., and Tepfer, D. (1986) Nucleotide sequence analysis of TL-DNA of Agrobacterium rhizogenes agropine type plasmid. Identification of open reading frames. J. Biol. Chem. 261, 108–121.

58. Collier, R., Fuchs, B., Walter, N., Kevin, L.W., and Taylor, C.G. (2005) Ex vitro composite plants: an inexpensive, rapid method for root biology Plant J. 43, 449–457.

59. Giri, A., and Narasu, M.L. (2000) Transgenic hairy roots: recent trends and applications. Biotechnol. Adv. 18, 1–22.

60. Georgiev, M.I., Pavlov, A.I., and Bley, T. (2007) Hairy root type plant in vitro systems as sources of bioactive substances. Appl. Microbiol. Biotechnol. 74, 1175–1185.

61. Seki, H., Nishizawa, T., Tanaka, N., Niwa, Y., Yoshida, S., and Muranaka, T. (2005) Hairy root-activation tagging: a highthroughput system for activation tagging in transformed hairy roots. Plant Mol. Biol. 59, 793–807.

62. Narayanan, R.A., Atz, R., Denny, R., Young, N.D., and Somers, D.A. (1999) Expression of soybean cyst nematode resistance in transgenic hairy roots of soybean. Crop Sci. 39, 1680–1686.

63. Skarjinskaia, M., Karl, J., Araujo, A., Ruby, K., Rabindran, S., Streatfield, S.J., and Yusibov, V. (2008) Production of recombinant proteins in clonal

root cultures using episomal expression vectors. Biotechnol. Bioeng. 100, 814–819.

64. Kuster, H., Vieweg, M.F., Manthey, K., Baier, M.C., Hohnjec, N., and Perlick, A.M. (2007) Identification and expression regulation of symbiotically activated legume genes. Phytochemistry 68, 8–18.

65. Breitler, J.C., Labeyrie, A., Meynard, D., Legavre, T., and Guiderdoni, E. (2002) Efficient microprojectile bombardmentmediated transformation of rice using gene cassettes. Theor. Appl. Genet. 104, 709–719.

66. Christou, P. (1995) Strategies for varietyindependent genetic transformation of important cereals, legumes and woody species utilising particle bombardment. Euphyica 85, 13–27.

67. Campbell, B.T., Baeziger, P.S., Mitra, A., Sato, S., and Clemente, T. (2000) Inheritance of multiple genes in wheat. Crop Sci. 40, 1133–1141.

68. Schmidt, M.A., Lafayette, P.R., Artelt, B.A., and Parrott, W.A. (2008) A comparison of strategies for transformation with multiple genes via microprojectile-mediated bombardment. In Vitro Cell Dev. Biol. Plant 44, 162–168.

69. Agrawal, P.K., Kohli, A., Twyman, R.M., and Christou, P. (2005) Transformation of plants with multiple cassettes generates simple transgene integration patterns and high expression levels. Mol. Breed. 16, 247–260.

70. Schmidt, M.A., Tucker, D.M., Cahoon, E.B., and Parrott, W.A. (2005) Towards normalization of soybean somatic embryo maturation. Plant Cell Rep. 24, 383–391.

71. Pawlowski, W.P., and Somers, D.A. (1996) Transgene inheritance in plants genetically engineered by microprojectile bombardment. Mol. Biotechnol. 6, 17–30.

72. Kohli, A., Gahakwa, D., Vain, P., Laurie, D.A., and Christou, P. (1999) Transgene expression in rice engineered through particle bombardment: molecular factors controlling stable expression and transgene silencing. Planta 208, 88–97.

73. Lowe, B.A., Prakash, N.S., Melissa, W., Mann, M.T., Spencer, T.M., and Boddupalli, R.S. (2009) Enhanced single copy integration events in corn via particle bombardment using low quantities of DNA. Transgenic Res. 18, 831–840 (DOI: 10.1007/s11248-009-9265-0).

74. Shimamoto, K., Teralda, R., Izawa, T., and Fujimoto, H. (1989) Fertile transgenic rice plants regenerated from transformed protoplasts. Nature 338, 274–276.

75. Salmenkallio-Marttila, M., Aspegren, K., Kerman, S., Kurt, U., Mannonen, L., Ritala, A., Teeriz, T.H., and Kauppinen, V. (1995) Transgenic barley (Hordeum vulgare L.) by electroporation of protoplasts Plant Cell Rep. 15, 301–304.

76. Daveya, M.R., Anthonya, P., Powera, J.B., and Loweb, K.C. (2005) Plant protoplasts: status and biotechnological perspectives. Biotechnol. Adv. 23, 131–171.

77. Sawahel, W.A. (2001) Stable genetic transformation of cotton plants using polybrene spermidine treatment. Plant Mol. Biol. Rep. 19, 377a–377f.

78. Gad, A.E., Rosenberg, N., and Altman, A. (1990) Liposome-mediated gene delivery into plant cells. Physiol. Plant. 79, 177–183.

79. Wordragen, M.V., Roshani, S., Ruud, V., Regis, P., Abvan, K., and Pim, Z. (1997) Liposome-mediated transfer of YAC DNA to tobacco cells. Plant Mol. Biol. Rep. 15, 170–178.

80. Kaeppler, H., Somers, D.A., Rines, H.W., and Cockburn, A.F. (1992) Silicon carbide fiber-mediated stable transformation of plant cells. Theor. Appl. Genet. 84, 560–566.

81. Frame, B.R., Drayton, P.R., Bagnall, S.V., Lewnau, C.J., Bullock, W.P., Wilson, H.M., Dunwell, J.M., Thompson, J.A., and Wang, K. (1994) Production of fertile transgenic maize plants by silicon carbide whisker-mediated transformation. Plant J. 6, 941–948.

82. Singh, N., and Chawla, H.S. (1999) Use of silicon carbide fibers for Agrobacteriummediated transformation in wheat. Curr. Sci. 76, 1483–1485.

83. Nagatani, N., Honda, H., Shimada, T., and Kobayashi, T. (1997) DNA delivery into rice cells and transformation using silicon carbide whiskers. Biotechnol. Tech. 11, 781–786.

84. Torney, F., Trewyn, B.G., Lin, V.S.Y., and Wang, K. (2007) Mesoporous silica nanoparticles deliver DNA and chemicals into plants. Nature Nanotech. 2, 295–300.

85. Nandiyanto, A.B.D., Kim, S.G., Iskandar, F., and Okuyama, K. (2009) Synthesis of silica nanoparticles with nanometer-size controllable mesopores and outer diameters. Microporous Mesoporous Mater. 120, 447–453.

86. Crossway, A., Oakes, J.W., Irvine, J.M., Ward, B., Knauf, V.C., and Shewmaker, C.K. (1986) Integration of foreign DNA following microinjection of tobacco mesophyll protoplasts. Mol. Gen. Genet. 202, 179–185.

87. Jones-Villeneuve, E., Huang, B., Prudhome, I., Bird, S., Kemble, R., Hattori, J., and Miki, B. (1995) Assessment of microinjection for introducing DNA into uninuclear microspores of rapeseed. Plant Cell Tiss. Organ Cult. 40, 97–100.

88. Holm, P.B., Olsen, O., Schnorf, M., BrinchPederse, H., and Knudsen, S. (2000) Transformation of barley by microinjection into isolated zygote protoplasts. Transgenic Res. 9, 21–32.

89. Lilly, J.W., Havey, M.J., Jackson, S.A., and Jiang, J. (2001) Cytogenomic analysis reveal the structural plasticity of the chloroplast genome in higher plants. Plant Cell 13, 245–254.

90. Maliga, P. (2004) Plastid transformation in higher plants. Annu. Rev. Plant Biol. 55, 289–313.

91. Verma, D., and Daniell, H. (2007) Chloroplast vector systems for biotechnology applications. Plant Physiol. 145, 1129–1143.

92. Boynton, J.E., Gillham, N.W., Harris, E.H., Hosler, J.P., Johnson, A.M., and Jones, A.R. (1988) Chloroplast transformation in Chlamydomonas with high velocity microprojectiles. Science 240, 1534–1538.

93. Lee, S.M., Kang, K., Chung, H., Yoo, S.H., Xu, X.M., Lee, S.B., Cheong, J.J., Daniell, H., and Kim, M. (2006) Plastid transformation in the monocotyledonous cereal crop, rice (Oryza sativa) and transmission of transgenes to their progeny. Mol. Cells 21, 401–410.

94. Kumar, S., Dhingra, A., and Daniell, H. (2004) Stable transformation of the cotton plastid genome and maternal inheritance of transgenes. Plant Mol. Biol. 56, 203–216.

95. Daniell, H., Chebolu, S., Kumar, S., Singleton, M., and Falconer, R. (2005) Chloroplast derived vaccine antigens and other therapeutic proteins. Vaccine 23, 1779–1783.

96. Rommens, C.M. (2004) All-native DNA transformation: a new approach to plant genetic engineering. Trends Plant Sci. 9, 457–464.

97. Bernasconi, P. et al. (1995) A naturally occurring point mutation confers broad range tolerance to herbicides that target acetolactate synthase. J. Biol. Chem. 270, 17381–17385.

98. Baerson, S.R., Rodriguez, D.J., Tran, M., Feng, M., Biest, N.A., and Dill, G.M. (2002) Glyphosate-resistant goosegrass identification of a mutation in the target enzyme 5-enolpyruvylshikimate-3-phosphate synthase. Plant Physiol. 129, 1265–1275.

99. Song, J., Bradeen, J.M., Naess, S.K., Raasch, J.A.,Wielgus, S.M., Haberlach, G.T., Liu, J., Kuang, H., Austin-Phillips, S., Buell, C.R.,

Helgeson, J.P., and Jiang, J. (2003) Gene RB cloned from Solanum bulbocastanum confers broad spectrum resistance to potato late blight. Proc. Natl. Acad. Sci. USA 100, 9128–9133.

100. Pechan, T., Cohen, A., Williams, W.P., and Luthe, D.S. (2002) Insect feeding mobilizes a unique plant defense protease that disrupts the peritrophic matrix of caterpillars. Proc. Natl. Acad. Sci. USA 99, 13319–13323.

101. Wang, E., Wang, R., DeParasis, J., Loughrin, J.H., Gan, S., and Wagner, G.J. (2001) Suppression of a P450 hydroxylase gene in plant trichome glands enhances naturalproduct-based aphid resistance. Nat. Biotechnol. 19, 371–374.

102. Kasuga, M., Liu, Q., Miura, S., YamaguchiShinozaki, K., and Shinozaki, K. (1999) Improving plant drought, salt, and freezing tolerance by gene transfer of a single stressinducible transcription factor. Nat. Biotechnol. 17, 287–291.

103. Miki, B., and Mc Hugh, S. (2004) Selectable marker genes in transgenic plants: applications, alternatives and biosafety. J. Biotechnol. 107, 193–232.

104. Daley, M., Knauf, V.C., Summerfelt, K.R., and Turner, J.C. (1998) Co-transformation with one Agrobacterium tumefaciens strain containing two binary plasmids as a method for producing marker-free transgenic plants. Plant Cell Rep. 17, 489–496.

105. Komari, T., Hiei, Y., Saito, Y., Murai, N., and Kumashiro, T. (1996) Vectors carrying two separate T-DNAs for co-transformation of higher plants mediated by Agrobacterium tumefaciens and segregation of transformants free from selection markers. Plant J. 10, 165–174.

106. Hua, Y., and Rommens, C.M. (2007) Transposition-based plant transformation. Plant Physiol. 143, 570–578.

107. Cotsaftis, O., Sallaud, C., Breitler, J.C., Meynard, D., Greco, R., Pereira, A., and Guiderdoni, E. (2002) Transposon-mediated generation of T-DNA- and marker-free rice plants expressing a Bt endotoxin gene. Mol. Breed. 10, 165–180.

108. Saelim, L., Phansiri, S., Suksangpanomrung, M., Netrphan, S., and Narangajavana, J. (2009) Evaluation of a morphological marker selection and excision system to generate marker-free transgenic cassava plants. Plant Cell Rep. 28, 445–455.

109. Ebinuma, H., and Komamine, A. (2001) MAT (Multi-Auto-Transformation) vector system. The oncogenes of Agrobacterium as

positive markers for regeneration and selection of marker-free transgenic plants. In Vitro Cell Dev. Biol. Plant 37, 103–113.

110. Kuraya, Y., Ohta, S., Fukuda, M., Hiei, Y., Murai, N., Hamada, K., Ueki, T., Imaseki, H., and Komari, T. (2004) Suppression of transfer of non-T-DNA "vector backbone" sequences by multiple left border repeats in vectors for transformation of higher plants mediated by Agrobacterium tumefaciens. Mol. Breed. 14, 309–320.

111. Yang, A., Su, Q., and An, L. (2009) Ovarydrip transformation: a simple method for directly generating vector- and marker-free transgenic maize (Zea mays L.) with a linear GFP cassette transformation. Planta 229, 793–801.

112. Yang, A., Su, Q., An, L., Liu, J., Wu, W., and Qiu, Z. (2009) Detection of vector- and selectable marker-free transgenic maize with a linear GFP cassette transformation via the pollen-tube pathway. J. Biotechnol. 139, 1–5.

113. Ye, X., Williams, E.J., Shen, J., Esser, J.A., Nichols, A.M., Petersen, M.W., and Gilbertson, L.A. (2008) Plant development inhibitory genes in binary vector backbone improve quality event efficiency in soybean transformation. Transgenic Res. 17, 827–838.

114. Lu, H.J., Zhou, X.R., Gong, Z.X., and Upadhyaya, N.M. (2001) Generation of selectable marker-free transgenic rice using double right-border (DRB) binary vectors. Aust. J. Plant Physiol. 28, 241–248.

115. Hajdukiewicz, P., Svab, Z., and Maliga, P. (1994) The small, versatile pPZP family of Agrobacterium binary vectors for plant transformation. Plant Mol. Biol. 25, 989–994.

116. Fu, X., Duc, L.T., Fontana, S., Bong, B.B., Tinjuangjun, P., Sudhakar, D., Twyman, R.M., Christou, P., and Kohli, A. (2000) Linear transgene constructs lacking vector backbone sequences generate low-copynumber transgenic plants with simple integration patterns. Transgenic Res. 9, 11–19.

117. Dale, E., and Ow, D. (1991) Gene transfer with subsequent removal of the selection gene from the host genome. Proc. Natl. Acad. Sci. USA 88, 10558–10562.

118. Wang, Y., Chen, B., Hu, Y., Li, J., and Lin, Z. (2005) Inducible excision of selectable marker gene from transgenic plants by the Cre/lox site-specific recombination system. Transgenic Res. 14, 605–614.

119. Srivastava, V., and Ow, D.W. (2001) Singlecopy primary transformants of maize obtained through the co-introduction of a recombinase-expressing construct. Plant Mol. Biol. 46, 561–566.

120. Zuo, J., Niu, Q.-W., Moller, S.G., and Chua, N.-H. (2001) Chemical-regulated, site-specific DNA excision in transgenic plants. Nat. Biotechnol. 19, 157–161.

121. Bai, X., Qiuyun, W., and Chu, C. (2008) Excision of a selective marker in transgenic rice using a novel Cre/loxP system controlled by a floral specific promoter. Transgenic Res. 17, 1035–1043.

122. McCormac, A.C., Fowler, M.R., Chen, D.F., and Elliott, M.C. (2001) Efficient co-transformation of Nicotiana tabacum by two independent T-DNAs, the effect of T-DNA size and implications for genetic separation. Transgenic Res. 10, 143–155.

123. Miller, M., Tagliani, L., Wang, N., Berka, B., Bidney, D., and Zhao, Z.Y. (2002) Highefficiency transgene segregation in co-transformed maize plants using an Agrobacterium tumefaciens 2T-DNA binary system. Transgenic Res. 11, 381–396.

124. Matthews, P.R., Waterhouse, P.M., Thornton, S., Fieg, S.J., Gubler, F., and Jacobsen, J.V. (2001) Marker gene elimination from transgenic barley, using co-transformation with adjacent "twin T-DNAs" on a standard Agrobacterium transformation vector. Mol. Breed. 7, 195–202.

125. Karimi, M., Inze, D., and Depicker, A. (2002) GATEWAY vectors for Agrobacteriummediated plant transformation. Trends Plant Sci. 7, 193–195.

126. Karimi, M., Bjorn, D.M., and Hilson, P. (2005) Modular cloning in plant cells. Trends Plant Sci. 10, 103–105.

127. Karimi, M., Depicker, A., and Hilson, P. (2007) Recombinational cloning with plant gateway vectors. Plant Physiol. 145, 1144–1154.

128. Chen, Q.-J., Zhou, H.-M., Chen, J., and Wang, X.C. (2006) A gateway-based platform for multigene plant transformation. Plant Mol. Biol. 62, 927–936.

129. Helliwell, C., and Waterhouse, P. (2003) Constructs and methods for high-throughput gene silencing in plants. Methods 30, 289–295.

130. Daxinger, L., Hunter, B., Sheikh, M., Jauvion, V., Gasciolli, V., Vaucheret, H., Matzke, M., and Furner, I. (2008) Unexpected silencing effects from T-DNA tags in Arabidopsis. Trends Plant Sci. 13, 4–6.

131. Heilersig, B.H.J.B., Loonen, A.E.H.M., Wolters, A.-M.A., and Visser, R.G.F. (2006) Presence of an intron in inverted repeat constructs does not necessarily have an effect on efficiency of post-transcriptional gene silencing. Mol. Breed. 17, 307–316.

132. Earley, K.W., Haag, J.R., Pontes, O., Opper, K., Juehne, T., Song, K., and Pikaard, C.S. (2006) Gateway-compatible vectors for plant functional genomics and proteomics. Plant J. 45, 616–629.

133. Nakagawa, T., Kurose, T., Hino, T., Tanaka, K., Kawamukai, M., Niwa, Y., Toyooka, K., Matsuoka, K., Jinbo, T., and Kimura, T. (2007) Development of series of gateway binary vectors, pGWBs, for realizing efficient construction of fusion genes for plant transformation. J. Biosci. Bioeng. 104, 34–41.

134. Nakagawa, T., Nakamura, S., Tanaka, K., Kawamukai, M., Suzuki, T., Nakamura, K., Kimura, T., and Ishiguro, S. (2008) Development of R4 gateway binary vectors (R4pGWB) enabling high-throughput promoter swapping for plant research. Biosci. Biotechnol. Biochem. 72, 624–629.

135. Curtis, M.D., and Grossniklaus, U. (2003) A gateway cloning vector set for high-throughput functional analysis of genes in planta. Plant Physiol. 133, 462–469.

136. Tzfira, T., Tian, G.-W., Lacroix, B., Vyas, S., Li, J., Leitner-Dagan, Y., Krichevsky, A., Taylor, T., Vainstein, A., and Citovsky, V. (2005) pSAT vectors: a modular series of plasmids for autofluorescent protein tagging and expression of multiple genes in plants. Plant Mol. Biol. 57, 503–516.

137. Zhong, S., Lin, Z., Fray, R.G., and Grierson, D. (2008) Improved plant transformation vectors for fluorescent protein tagging. Transgenic Res. 17, 985–989.

138. Martin, K., Kopperud, K., Chakrabarty, R., Banerjee, R., Brooks, R., and Goodin, M.M. (2009) Transient expression in Nicotiana benthamiana fluorescent marker lines provides enhanced definition of protein localization, movement and interactions in planta. Plant J. 59, 150–162.

139. Copeland, N.G., Jenkins, N.A., and Court, D.L. (2001) Recombineering: a powerful new tool for mouse functional genomics. Nat. Rev. Genet. 2, 769–779.

140. Rozwadowski, K., Yang, W., and Kagale, S. (2008) Homologous recombination-mediated cloning and manipulation of genomic DNA regions using gateway and recombineering systems. BMC Biotechnol. 8, 88.

141. Cai, C.Q., Doyon, Y.W., Ainley, M., Miller, J.C., et al. (2009) Targeted transgene integration in plant cells using designed zinc finger nucleases. Plant Mol. Biol. 69, 699–709.

142. Durai, S., Mani, M., Kandavelou, K., Wu, J., Porteus, M.H., and Chandrasegaran, S. (2005) Zinc finger nucleases: customdesigned

molecular scissors for genome engineering of plant and mammalian cells. Nucleic Acids Res. 33, 5978–5990.

143. Lloyd, A., Plaisier, C.L., Carroll, D., and Drews, G.N. (2005) Targeted mutagenesis using zinc-finger nucleases in Arabidopsis. Proc. Natl. Acad. Sci. USA 102, 2232–2237.

144. Ordiz, M.I., Barbas, C.F., and Beachy, R.N. (2002) Regulation of transgene expression in plants with polydactyl zinc finger transcription factors Proc. Natl. Acad. Sci. USA 99, 13290–13295.

145. Szxzepek, M., Brondani, V., Buchd, I., Serrano, L., Segal, D., and Cathomen, T. (2007) Structure-based redesign of the dimerization interface reduces the toxicity of zinc-finger nucleases. Nat. Biotechnol. 25, 786–793.

146. Weichang, Y., Fangpu, H., Gao, Z., Vega, J.M., and Birchler, J.A. (2007) Construction and behavior of engineered minichromosomes in maize. Proc. Natl. Acad. Sci. USA 104, 8924–8929.

147. Yu, W., Han, F., and Birchler, J.A. (2007) Engineered minichromosomes in plants. Curr. Opin. Biotechnol. 18, 425–431.

148. Kato, A., Zheng, Y.Z., Auger, D.L., PhelpsDurr, T., Bauer, M.J., Lamb, J.C., and Birchler, J.A. (2005) Minichromosomes derived from the B chromosome of maize. Cytogenet. Genome. Res. 109, 156–165.

149. Veena (2008) Engineering plants for future: tools and options Physiol. Mol. Biol. Plants 14, 131–135.

150. Vain, P. (2005) Plant transgenic science knowledge. Nat. Biotechnol. 23, 1348–1349.

151. Chapotin, S.M., and Wolt, J.D. (2007) Genetically modified crops for the bioeconomy: meeting public and regulatory expectations. Transgenic Res. 16, 675–688.

152. Kim, E.H., Suh, S.C., Park, B.S., Shin, K.S., Kweon, S.J., Han, E.J., Park, S.-H., Kim, Y.-S., and Kim, J.-K. (2009) Chloroplasttargeted expression of synthetic cry1Ac in transgenic rice as an alternative strategy for increased pest protection. Planta 230, 397–405.

153. Kim, T.-G., Baek, M.-Y., Lee, E.-K., Kwon, T.-H., and Yang, M.-S. (2008) Expression of human growth hormone in transgenic rice cell suspension culture. Plant Cell Rep. 27, 885–891.

154. De Padua, V.L.M., Ferreira, R.P., Meneses, L., Uchoa, L., Marcia, M.-P., and Mansur, E. (2001) Transformation of Brazilian elite Indica-Type Rice (Oryza Sativa L.) by electroporation of shoot apex explants. Plant Mol. Biol. Rep. 19, 55–64.

155. Cho, M.-J., Choi, H.W., Okamoto, D., Zhang, S., and Lemaux, P.G. (2003) Expression of green fluorescent protein and its inheritance in transgenic oat plants generated from shoot meristematic cultures. Plant Cell Rep. 21, 467–474.

Chapter 3

SUNFLOWER CENTROMERES CONSIST OF A CENTROMERE-SPECIFIC LINE AND A CHROMOSOME-SPECIFIC TANDEM REPEAT

Kiyotaka Nagaki[1], Keisuke Tanaka[2], Naoki Yamaji[1], Hisato Kobayashi[2], and Minoru Murata[1]

[1]Applied Genomics Unit, Institute of Plant Science and Resources, Okayama University, Kurashiki, Japan

[2]NODAI Genome Research Center, Tokyo University of Agriculture, Setagaya, Japan

ABSTRACT

The kinetochore is a protein complex including kinetochore-specific proteins that plays a role in chromatid segregation during mitosis and meiosis. The complex associates with centromeric DNA sequences that are usually species-specific. In plant species, tandem repeats including satellite DNA sequences and retrotransposons have been reported as centromeric DNA sequences. In this study on sunflowers, a cDNA-encoding centromere-specific histone H3 (CENH3) was isolated from a cDNA pool from a seedling, and an antibody was raised against a peptide synthesized from the deduced cDNA. The antibody specifically recognized the sunflower CENH3 (HaCENH3) and showed centromeric signals by immunostaining and immunohistochemical staining analysis. The antibody was also applied in chromatin immunoprecipitation (ChIP)-Seq to isolate centromeric DNA sequences and two different types of repetitive DNA sequences were identified. One was a long interspersed nuclear element (LINE)-like sequence, which showed centromere-specific signals on almost all chromosomes in sunflowers. This is the first report of a centromeric LINE sequence, suggesting possible centromere targeting ability. Another type of identified repetitive DNA was a tandem repeat sequence with a 187-bp unit that was found only on a pair of chromosomes. The HaCENH3 content of the tandem repeats was estimated to be much higher than that of the LINE, which implies centromere evolution from LINE-based centromeres to more stable tandem-repeat-based centromeres. In addition, the epigenetic status of the sunflower centromeres was investigated by immunohistochemical staining and ChIP, and it was found that centromeres were heterochromatic.

INTRODUCTION

Asterales is the most diverged order of dicots and includes 11 families and 27,000 species. Among the 11 families, Asteraceae is the largest and includes the sunflower and daisy. Sunflowers (*Helianthus annuus* L., $2n = 2x = 34$, genome size = 2.43 Gb/haploid) are one of the most important crops in Asterales because their seeds can be used for oil production (Bennett et al., 1982). Sunflowers have been genetically and cytogenetically investigated (Feng et al., 2013), and a genome sequencing project is now in progress (http://sunflowergenome.org/). For karyotypic analyses, repetitive DNA sequences including rDNA, sunflower-specific tandem repeats and bacterial artificial chromosome (BAC) clones have been used as probes, but none of these probes showed centromeric localization (Ceccarelli et al., 2007; Talia et al., 2010; Feng et al., 2013). In other words, at present, there is no genetic and cytogenetic marker for sunflower centromeres.

The kinetochore is a special protein complex formed on centromeric regions that ensures equal and accurate distribution of chromatids to daughter cells during mitosis and meiosis (Amor et al., 2004). Among the constitutive proteins of kinetochores, centromere-specific histone H3 (CENH3) acts as a base for assembling other kinetochore proteins, and its presence epigenetically determines the kinetochore position (Perpelescu and Fukagawa, 2011). The first identified CENH3 was CENP-A in humans (Earnshaw and Rothfield, 1985), and its orthologs have been isolated from 11 of 63 APG III orders, including Poales, Asparagales, Rosales, Fabales, Malpighiales, Malvales, Brassicales, Myrtales, Solanales, Asterales, and Apiales, in this decade (Talbert et al., 2002; Zhong et al., 2002; Nagaki et al., 2004,2005, 2012b; Nagaki and Murata, 2005; Sanei et al., 2011; Wang et al., 2011; Neumann et al., 2012, 2015; Dunemann et al., 2014; Masonbrink et al., 2014; He et al., 2015; Maheshawari, 2015). However, no CENH3 has been isolated from Asterales species.

The CENH3s possess relatively conserved histone fold domains (HFDs) and highly variable N-terminal tails, and the HFD has been shown to be important for centromere localization (Vermaak et al., 2002; Black et al., 2004; Lermontova et al., 2006). The Loop 1 region on the HFD is longer than that of canonical histone H3, which is common among CENH3s. This region allows more compact packing of nucleosomes with CENH3 compared to those with canonical histone H3 (Black et al., 2004; Tek et al., 2010, 2011; Nagaki et al., 2012a; Maheshawari, 2015).

CENH3 is a component of core histones in the centromeric regions, and it directly binds to DNA. Additionally, CENH3 localizes to functional centromeres (Warburton et al., 1997; Nasuda et al., 2005; Han et al., 2006). Therefore, centromeric DNA sequences have been identified in plant species by chromatin immunoprecipitation (ChIP) using anti-CENH3 antibodies (Zhong et al., 2002; Nagaki et al., 2003, 2004, 2009, 2011, 2012a,b; Nagaki and Murata, 2005;Houben et al., 2007; Tek et al., 2010, 2011; Wang et al., 2011; Gong et al., 2012; Neumann et al., 2012; He et al., 2015). In most cases, species-specific microsatellites, minisatellites, macrosatellites, and retrotransposons have been identified as the centromeric sequences, and these sequences are located on all centromeric regions in the species (Zhong et al., 2002; Nagaki et al., 2003, 2009, 2011; Nagaki and Murata, 2005; Houben et al., 2007; Tek et al., 2010,2011; Wang et al., 2011; Neumann et al., 2012; He et al., 2015).

Histone modifications are the key components in epigenetics, and they have been much investigated in this decade (Desvoyes et al., 2014; Sharma et al., 2015). In plant species, two different types of heterochromatin distribution are reported to be related to the genome size of the species (Houben et al., 2003). In plant species with a genome size smaller than 510 Mb, chromocenters appear on interphase nuclei, and heterochromatic modifications occur specifically on the chromocenters. *A. thaliana* is a good example for small-genome species, as it shows heterochromatic modifications on centromeric and pericentromeric regions not only at interphase but also at metaphase. Euchromatic modifications of the species were observed in the regions with no heterochromatic modifications (Houben et al., 2003;Jasencakova et al., 2003). In contrast, no chromocenters appear on interphase nuclei in plant species with genomes larger than 530 Mb. However, in these species, heterochromatic and euchromatic modifications are observed to disperse on interphase nuclei (Houben et al., 2003), and neither modification appears on centromeric and pericentromeric regions during metaphase.

The epigenetic modifications around centromeric regions have also been analyzed by ChIP using centromeric DNA sequences. In human cells, the centromere-specific histone H3 variants (CENP-A) coexist with heterochromatic modified histones at almost all stages during the cell cycle, but the histones are instantaneously modified with a euchromatic modification from anaphase to early G1 (Ohzeki et al., 2012). In rice, a CENH3-binding region in a centromere, Cen8, was revealed by ChIP to be heterochromatic (Nagaki et al., 2004). However, euchromatic markers were also detected in genic regions and 167-bp CentO variants in rice centromeres (Yan et al., 2006; Zhang et al., 2013).

In this study, we isolated a cDNA encoding CENH3 from sunflowers and raised a peptide antibody against CENH3. The antibody showed centromere-specific signals in immunostaining and immunohistochemical staining experiments. A ChIP-Seq experiment was also conducted to isolate DNA sequences that coexist with CENH3. Additionally, the epigenetic status of the centromeres was successfully revealed by immunohistochemical staining using anti-CENH3 and anti-modified histone antibodies as well as by ChIP-qPCR using these antibodies and isolated DNA sequences from ChIP-Seq.

MATERIALS AND METHODS

Plant Material

Sunflower seeds (*H. annuus* L., $2n = 2x = 34$, number 1802-065684) were obtained from a commercial source (LIC, Okayama, Japan).

Identification of a Sunflower Expressed Sequence Tag (EST) Encoding CENH3

An EST sequence encoding sunflower CENH3 (HaCENH3) was identified from the gene indices using the tblastn program (http://compbio.dfci.harvard.edu/tgi/) and the amino acid sequence of NtCENH3-1 (GenBank accession number: BAH03514, Nagaki et al., 2009) as a query.

RNA Isolation and PCR

Total RNA was isolated from a 3-day-old sunflower seedling using the RNeasy Plant Mini kit (Qiagen, Hilden, Germany). To determine the full-length cDNA sequence of a sunflower CENH3 gene, rapid amplification of cDNA ends (RACE) was conducted. For 3′RACE, the primer HaCENH3-3RACE was designed from an EST encoding a putative sunflower CENH3 sequence found during the BLAST search; this primer was used with the SMARTer RACE cDNA Amplification Kit (Clontech, CA, USA). Another primer, HaCENH3-5RACE, was designed from the sequences determined by 3′RACE and was used to determine the 5′ end.

Sequencing and Sequence Analyses

The 3′- and 5′-RACE products were cloned into a pGEM-T easy vector (Promega, WI, USA) and sequenced from both ends using a BigDye Terminator v1.1 cycle sequencing kit and an ABI PRISM 3130xl genetic analyzer (Applied Biosystems, CA, USA). A putative amino acid sequence was deduced from the DNA sequences and used as a query sequence for a protein

BLAST search on the NCBI website (http://blast.ncbi.nlm.nih.gov/Blast.cgi?CMD=Web&PAGE_TYPE=BlastHome). The deduced HaCENH3 amino acid sequence was aligned with orthologs identified by the protein BLAST search and canonical histone H3 of rice using the Clustal X software program (Thompson et al., 1997). Phylogenetic relationships among the CENH3s were analyzed by the neighbor-joining method (Saitou and Nei, 1987).

Immunostaining

Based on the deduced HaCENH3 amino acid sequence, a peptide corresponding to the N-terminus of HaCENH3 (H_2N-ARTKHPAKRSSGIPADGRSS-COOH) was synthesized and injected into two rabbits. The raised antisera were purified using an affinity Sepharose column consisting of the aforementioned peptide.

Immunostaining was conducted as previously described (Nagaki et al., 2012b). In brief, root tips of 3-day-old sunflowers were fixed in microtubule stabilizing buffer (50 mM PIPES, pH 6.9, 5 mM MgSO4, and 5 mM EGTA) containing 3% (w/v) paraformaldehyde and 0.2% (v/v) Triton X-100. The fixed tips were washed and digested with a mixture of 1% (w/v) cellulase Onozuka RS (Yakult Pharmaceutical Industry, Tokyo, Japan) and 0.5% (w/v) pectolyase Y-23 (Seishin Pharmaceuticals, Tokyo, Japan) and then compressed onto slides coated with poly-L-lysine (Matsunami, Osaka, Japan). A 1:100 dilution of the purified anti-HaCENH3 antibody and monoclonal anti-modified-histone antibodies produced in mice (anti-histone H3 dimethyl K4 (H3K4me2): MBL (Nagoya, Japan) MABI0303 and anti-histone H3 dimethyl K9 (H3K9me2): MBL MABI0317) were applied to the slides. To detect acetylations of histone H4, an anti-histone H4 acetyl (H4Ac) antibody raised in rabbits (Millipore, MA, USA: #06-598) was used. The antibodies were detected using 1:1000 dilutions of Alexa Fluor 555-labeled anti-rabbit antibodies (Molecular Probes, OR, USA) and Alexa Fluor 488-labeled anti-mouse antibodies (Molecular Probes), respectively. Chromosomes were counterstained with 0.1 μg/ml 4,6-diamino-2-phenylindole (DAPI). Immunosignals and stained chromosomes were captured using a chilled charge-coupled device (CCD) camera, AxioCam HR (Carl Zeiss, Oberkochen, Germany), and images were pseudo-colored and processed using AxioVision software (Carl Zeiss).

Immunohistochemical Staining

Immunohistochemical staining was conducted as described with minor modifications (Yamaji and Ma, 2007; Nagaki et al., 2012b). Three-day-old sunflower roots were fixed as described above for immunostaining, and the fixed roots were sectioned at 100-μm thickness using a microslicer (LinearSlicer PRO10; Dosaka EM). These sections were transferred onto slides and then macerated. For three-color detection, 1:100 dilutions of the purified anti-HaCENH3 rabbit antibody and anti-α-tubulin mouse antibody (Sigma, MO, USA: T6199) were applied to the slides. For four-color detection, 1:100 dilutions of the purified anti-HaCENH3 rabbit antibody, anti-α-tubulin rat antibody (Abcam, Cambridge, UK: ab64332) and monoclonal anti-H3K9me2 mouse antibody (MBL MABI0317) were applied to the slides. After washing in PBS, the primary antibodies were detected using 1:1000 diluted secondary antibodies, the Alexa Fluor 555-labeled anti-rabbit antibodies and the Alexa Fluor 488-labeled anti-mouse antibodies for the three-color detection and Alexa Fluor 647-labeled anti-rabbit antibodies (Molecular Probes), Alexa Fluor 488-labeled anti-rat antibodies (Molecular Probes), and Alexa Fluor 546-labeled anti-mouse antibodies (Molecular Probes) for the four-color detection. Then, nuclei and chromosomes were counterstained with DAPI. Immunosignals and stained chromosomes were observed with a laser-scanning confocal microscope (LSM700; Carl Zeiss). The obtained data were analyzed using AxioVision software.

Chromatin Immunoprecipitation (ChIP)

ChIP was performed as previously described (Nagaki et al., 2012b) with minor modifications using the anti-HaCENH3 antibody and the anti-modified histone antibody (anti-H3K9me2: MBL #MABI0317, anti-H3K4me2: MBL #MABI0303 and anti-H4Ac: Millipore #06-598). Nuclei were isolated from the leaves of 1-month-old sunflowers and then digested with micrococcal nuclease (Sigma) to produce chromatin. Following overnight incubation of the chromatin with the antibodies at 4°C, the antibodies were captured using Dynabeads Protein G (Invitrogen, CA, USA). For mock experiments, a normal rabbit serum was used instead of the antibodies. DNA was purified from the chromatin with the captured antibodies by phenol/chloroform extraction followed by ethanol precipitation.

ChIP-Seq and Repeatexplorer Analysis

ChIP-Seq was conducted using precipitated DNA from the input and HaCENH3 fractions in the ChIP. Libraries were constructed using the NEBNext ChIP-Seq Library Prep Reagent Set for Illumina (New England Biolabs, MA, USA), and the libraries were read by MiSeq (Illumina, CA, USA) with the paired-end 2 × 300 bp protocol. Conversion of raw base-call data to sequence data in the fastq format, identification of reads derived from each sample by index sequences, and adapter trimming were performed using MiSeq reporter 2.3.32. The sequence data were analyzed by a similarity-based clustering program,

RepeatExplorer (http://www.repeatexplorer.org) (Novák et al., 2010) with default parameters.

qPCR

qPCR was conducted using SYBR Premix Ex Taq II (Tli RNaseH Plus) (Takara, Shiga, Japan) and primers for qPCR with a StepOne instrument (Applied Biosystems). The primers were designed based on the sequences in the clusters of the RepeatExplorer analysis. The precipitated DNA in the ChIP experiment was used as a template, and the mock was used as a negative control. Relative enrichment (RE) was calculated by the following formula: RE = amount of the sequence in the antibody fraction/amount of the sequence in the mock. The qPCR results were assessed by Student's *t*-test.

Fluorescence *in Situ* Hybridization (FISH)

Probes were amplified using primer sets designed based on the sequences in clusters of the RepeatExplorer analysis using sunflower genomic DNA as a template. The amplified DNA was cloned into pGEM T-easy, and sequences were confirmed using a BigDye Terminator v1.1 cycle sequencing kit and an ABI PRISM 3130xl genetic analyzer. To detect nucleolar organizing regions, an 18S-5.8S-28SrDNA clone from wheat (pTa71) was used (Gerlach and Bedbrook, 1979). To characterize sunflower chromosomes, a reported 386-bp tandem repetitive sequence, HAG004N15 (Ceccarelli et al., 2007), was amplified using specific primers and cloned. The sequence was confirmed as described above.

FISH analysis of mitotic chromosomes was performed as previously described (Nagaki et al., 2011). Chromosomes were prepared from the root tips of 3-day-old sunflowers. The plasmid DNA was labeled by nick translation using a DIG-Nick Translation Mix (Roche, Basel, Switzerland) or a Biotin-Nick Translation Mix (Roche). The digoxigenin- and biotin-labeled probes were visualized using rhodamine-conjugated anti-digoxigenin antibody (Roche) and Alexa Fluor 488-conjugated streptavidin (Molecular Probes), respectively.

RESULTS

Isolation and Sequence Analyses of CENH3 in Sunflowers

To identify the HaCENH3 gene, a BLAST search was conducted in a sunflower EST database of the gene index project using the amino acid sequence of NtCENH3-1 as a query. One EST group (TC48348) containing a 675-bp sequence showed 68% identity to the query sequence. However, because of low quality in the first 150 bp of the EST, no start codon was found in the sequence. To obtain a full-length cDNA of the gene, 5′- and 3′-RACE experiments were performed using sunflower seedling cDNA as a template. As a result, a putative full-length cDNA containing a 429-bp ORF that encodes 143 amino acids (GenBank accession number LC075743) was obtained.

The amino acid sequence deduced from the ORF showed similarity to the sequences of some other plants, with CENH3 from *Daucus muricatus* showing the highest similarity (70%). The amino acid sequence of HaCENH3 was aligned with those of CENH3s from other plant species and rice canonical histone H3. The alignment indicated that the N-terminal amino acid at position 14–44 of HaCENH3 did not show any similarity to those of the other CENH3s or the canonical histone H3, and it also indicated that HaCENH3 possessed a longer loop 1 domain than canonical histone H3. The longer loop 1 is a feature of CENH3s.

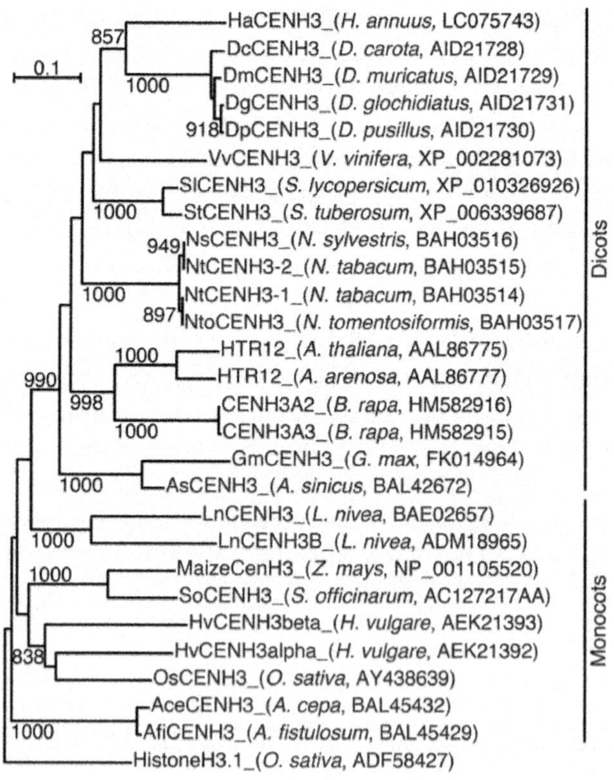

Figure 1. Phylogenetic tree based on the amino acid sequences of plant CENH3. The species name and GenBank accession number are indicated in parentheses. Rice ca-

nonical histone H3 was used as an outgroup. Bootstrap values greater than 800 in 1000 tests are indicated on the branches.

In a phylogenetic analysis using the alignment, HaCENH3 was classified into a dicot clade (Figure 1). As expected, HaCENH3 was found to be most closely related to the CENH3s of *Daucus* species, but the sequence was placed outside of a clade containing the *Daucus* CENH3s.

Centromere Localization of HaCENH3

An anti-HaCENH3 antibody was raised against a synthetic peptide comprising N-terminal amino acid residues 2–21 of the deduced HaCENH3 amino acid sequence. To confirm its specificity to the centromeres, immunostaining using the antibody was conducted, and centromere-specific immunosignals appeared on all sunflower chromosomes (Figure 2). Additionally, in immunohistochemical staining, microtubule signals on all chromosomes were associated with all of the anti-HaCENH3 immunosignals at metaphase.

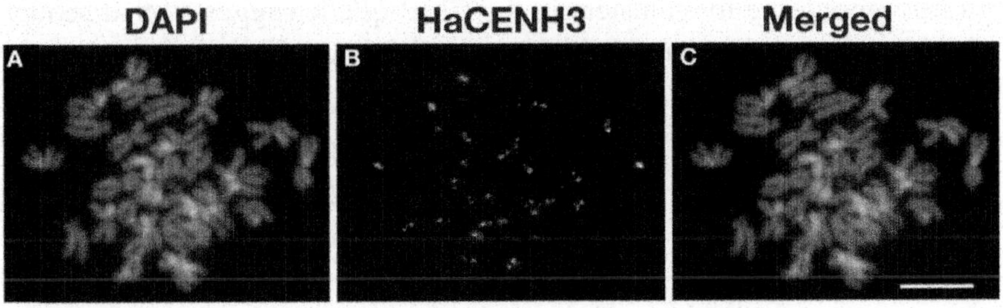

Figure 2. Immunostaining of sunflower metaphase chromosomes using an anti-Ha-CENH3 antibody.(A) DAPI-stained chromosomes. (B) Immunosignals of an anti-Ha-CENH3 antibody. (C) Merged image of (A,B). Scale bar, 10 μm.

DNA Sequences that Interact with Centromeric Nucleosomes

To investigate DNA sequences that coexist with HaCENH3, a ChIP-Seq experiment was conducted using the anti-HaCENH3 antibody and chromatin extracted from sunflower leaves. DNA fragments from the input and the HaCENH3 fraction in the ChIP were sequenced using MiSeq with the paired-end 2 × 300 bp protocol and deposited in DDBJ (Accession number: DRA003719). After the initial quality checks, 1,795,002 paired reads from the input and 2,012,000 paired reads from the HaCENH3 fraction were analyzed using the RepeatExplorer program. By this analysis, a total of 372 clusters containing at least 0.01% of the used sequences were generated. Then, enrichment ratios (ERs) were calculated by the following formula: ER = HaCENH3 ChIP reads/

input reads in the each cluster, and the clusters were sorted by ER. Out of the 372 clusters formed, 41 clusters showed an ER higher than 2.0.

To confirm the ChIP-Seq results, ChIP-qPCR was conducted using the DNA sequences in five clusters (HaCENH3CL1, 20, 22, 124, and 289) selected from the 41 clusters having an ER higher than 2.0 (Figure 3). In the ChIP-qPCR, a sunflower ubiquitin gene (GenBank accession number: X14333) was used as a negative (non-centromeric) control. The DNA sequences from all five clusters were significantly increased (27-fold for HaCENH3CL1, 154-fold for HaCL289, 289-fold for HaCENH3CL22, 344-fold for HaCENH3CL22, and 370-fold for HaCENH3CL124) compared with the negative control in the CENH3 fractions ($P < 0.01$ via Student's t-test, $n = 4$), whereas a non-centromeric control sequence, HAG004N15, did not increase ($P = 0.14$ via Student's t-test, $n = 4$). Since the ER in the ChIP-Seq and the RE in the ChIP-qPCR showed a high correlation coefficient ($r = 0.78$) according to the Pearson product-moment correlation, the ChIP-qPCR data support the ChIP-Seq results. The significant increase in HaCENH3CL1 showing the minimum ER (2.0) among the five clusters in the ChIP-qPCR suggested that sequences in the 41 clusters having an ER higher than 2.0 in the ChIP-Seq coexisted with HaCENH3 in the sunflower genome.

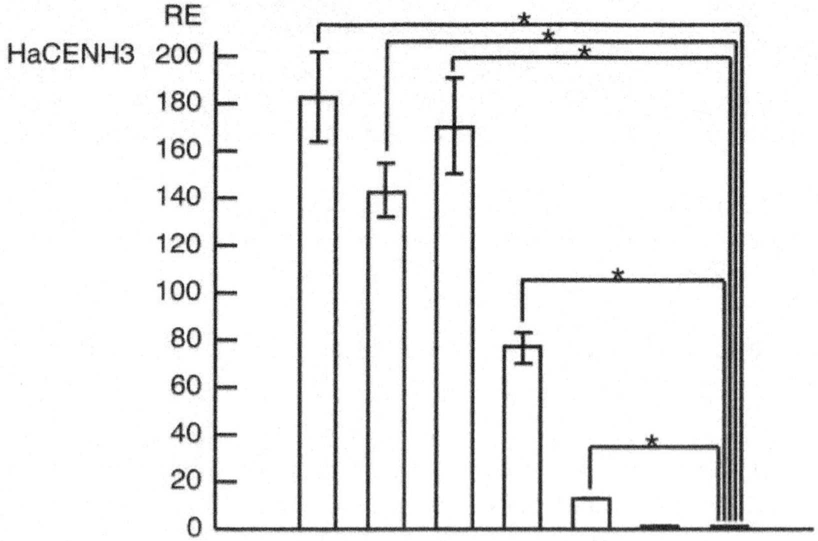

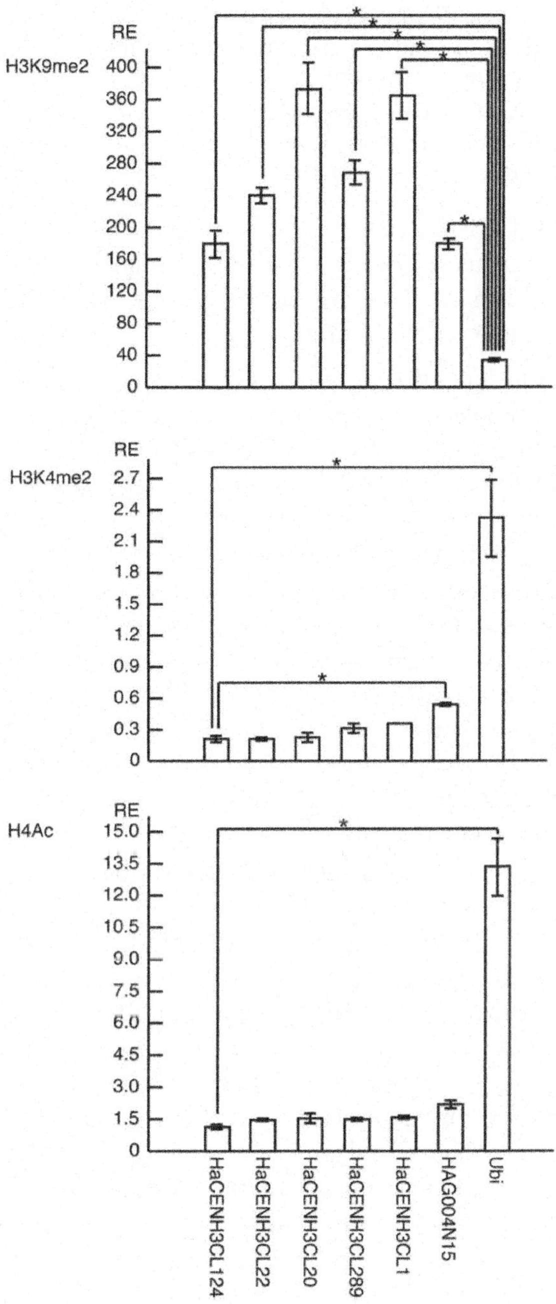

Figure 3. ChIP-qPCR analysis of centromeric DNA sequences of sunflowers. The columns and error bars represent the average relative enrichments (REs) and the standard errors from four independent ChIP reactions, respectively. The coding region of the

sunflower ubiquitin gene (X14333) was used as a non-centromeric (negative) control in the anti-HaCENH3 ChIP and as a negative control in the anti-H3K9me2 ChIP. Since HaCENH3CL124 showed the lowest REs among the sequences in the anti-H3K4me2 and H4Ac ChIP, HaCENH3CL124 was used as a negative control. The statistical significance of differences between the negative controls and other sequences was determined using Student's *t*-test ($^*P < 0.01$).

The HaCENH3CL124 had the highest ER (52.0) and contained 187-bp tandem repeat sequences. The consensus DNA sequence of the tandem repeat was relatively AT-rich (58.3%), and no sequence similarity to DNA sequences in the GenBank database was found.

Of the 41 clusters, the program suggested that 25 clusters contained long interspersed nuclear element (LINE)-like sequences, and the sequences involve an ORF encoding an endonuclease (ENDO) and reverse transcriptase (RT) complex of LINE. These sequences show approximately 60% similarity to each other. Almost all RepeatExplorer cluster graphs of the LINE showed typical patterns of retrotransposons (line shape) rather than typical patterns of tandem repeats (star or ring shape). As exceptional cases, two clusters, HaCENH3CL78 and HaCENH3CL115, involved 12 and four ENDO/RT-related reads, and showed star shape. Since RepeatExplorer splits some satellite repeats with long monomers, e.g., rDNA, into multiple clusters, connections of the LINE-like sequences clusters were investigated. In the investigation, rDNA was used as a positive control of a satellite repeat with long monomers, and 17 clusters related to rDNA were found in the investigation. Ends of the rDNA clusters showed similarity to other clusters. On the other hand, almost all of the LINE clusters did not show frequent similarity hits observed among the rDNA clusters. As exceptional cases, two clusters, HaCENH3CL78 and HaCENH3CL115, showed frequent similarity hits to HaCENH3CL8, suggesting these are not LINE-like elements. Twenty-two of the remaining 23 LINE-like clusters showed ERs higher than 10.0. Additionally, junction of the elements and insertion sites were surveyed. If clusters include ends of mobile elements, junctions should be visible as sites with heterologous sequences at ends of contigs in the cluster. For example, a clear junction was observed in an end of HaCENH3CL189 showing a typical cluster graph of LTR-retrotransposon, and similar junctions were also observed in some of the LINE-like clusters. However, sunflower DNA sequences that were similar to the LINE-like sequences were not found in the GenBank DNA database. These results suggested that most centromeric regions of sunflowers are not involved in the current sunflower genome sequencing project because the project has not involved repetitive DNA sequence-rich regions (http://sunflowergenome.org/). An additional 12 of 41 clusters contained transposable element-like sequences, whereas DNA sequences in the remaining four clusters showed no homology to those registered in the GenBank DNA database.

To confirm the centromeric localization of the DNA sequences immunoprecipitated with anti-HaCENH3 in the ChIP-Seq and ChIP-qPCR, these sequences were used as probes in FISH analysis (Figure 4). A probe containing the HaCENH3CL124 sequence (pHaCENH3CL124-1, GenBank accession number: LC075744) showed centromeric signals on a pair of chromosomes with a secondary constriction (Figures 4A–D). To identify the chromosomes showing the HaCENH3CL124 signal, another FISH experiment in which the repeat and a reported FISH marker, HAG004N15, were used as probes was conducted; it revealed that the chromosome in question was chromosome 8 in Ceccarelli's report (Ceccarelli et al., 2007) (Figures 4E–M). A cloned probe containing the LINE-like sequence from HaCENH3CL20 (pHaCENH3CL20-1, GenBank accession number:

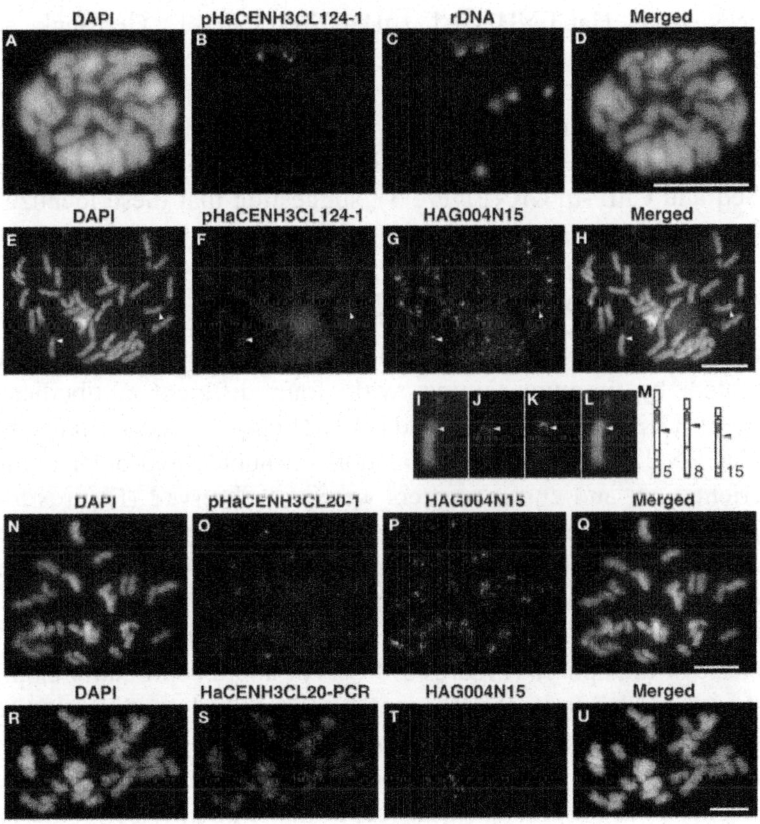

Figure 4. FISH using the enriched sequence from ChIP-Seq. (A,E,I,N,R) DAPI-stained sunflower chromosomes. (B,F,J) FISH signals of pHaCENH3CL124-1. (C)

FISH signals of rDNA. (G,K,P,T) FISH signals of HAG004N15. (O) FISH signals of pHaCENH3CL20-1. (S) FISH signals of the PCR products of HaCENH3CL20. (D) A merged image of (A–C). (H) A merged image of (E–G). (L) A merged image of (I–K). (Q) A merged image of (N–P). (U) A merged image of (R–T). (I–L) Enlarged images of (E–H). (M) Karyograms from Ceccarelli's report (2007). White arrowheads in (E–H) and black arrowheads in (M) indicate centromeres on the HaCENH3CL124-positive chromosomes and the reported chromosomes, respectively. Scale bar, 10 μm.

LC075745) showed centromeric signals on almost all of the chromosomes (Figures 4N–Q). Since the LINE-like sequence showed centromeric signals, we named it HaCEN-LINE. In FISH using HaCENH3CL20-PCR products as probes, much stronger signals on all the centromeric and pericentromeric regions were found, but faint signals were also observed on the arm regions (Figures 4R–U). These results imply that the PCR products contain not only some HaCEN-LINE variants but also some non-centromeric LINE variants; centromeric variants are on all of the centromeres. Two Ty3-related DNA sequences from HaCENH3CL1 (pHaCENH3CL1-1, GenBank accession number: LC075747) and HaCENH3CL189 (pHaCENH3CL189-1, GenBank accession number: LC075746), when used as FISH probes, showed dispersed patterns with some strong spots on centromeres. These partial centromere localizations coincided with the partial enrichment of the sequences in the ChIP-Seq and ChIP-qPCR (Figure 3), suggesting that these localize on both centromeric and non-centromeric arm regions.

Epigenetic Status of Sunflower Centromeres

Post translational histone modifications of sunflower centromeres were investigated by immunostaining with four different antibodies against H3K4me2, H3K9me2, H3K9Ac, and H4Ac (Figures 5 and 6). In the interphase cells, HaCENH3 signals scattered as dots on nuclei, and polar organization, Rabl orientation, and chromocenters were not observed (Figures 5A–E and 6A–E). H3K4me2, one of the representative euchromatic modifications, was detected on almost all regions of nuclei with many small dot signals, but the dot signals did not overlap with the HaCENH3 signals (Figures 5A–E). Similarly, the H3K4me2 signals did not co-localize with the HaCENH3 signals at prophase or metaphase (Figures 5F–O). Another euchromatic modification, H3K9Ac, showed a similar tendency at interphase and prophase, but almost all H3K9Ac signals disappeared at metaphase. Additionally, another euchromatic modification, H4Ac, showed stronger signals on the arms of metaphase chromosomes, but the signals on the centromeric and pericentromeric regions were weaker than those on the arms.

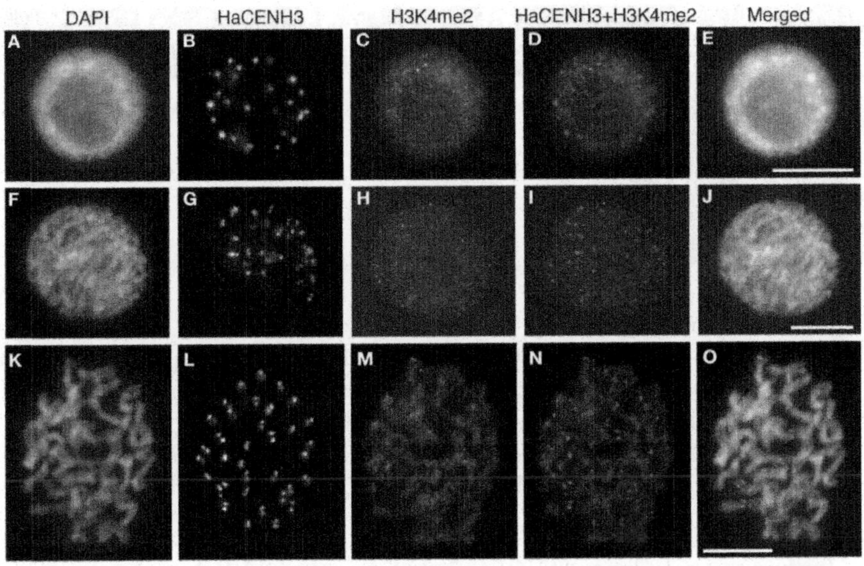

Figure 5. Immunostaining using the anti-HaCENH3 and anti-H3K4me2 antibodies. (A–E) An interphase nucleus. (F–J) Prophase chromosomes. (K–O) Metaphase chromosomes. Scale bar, 10 μm.

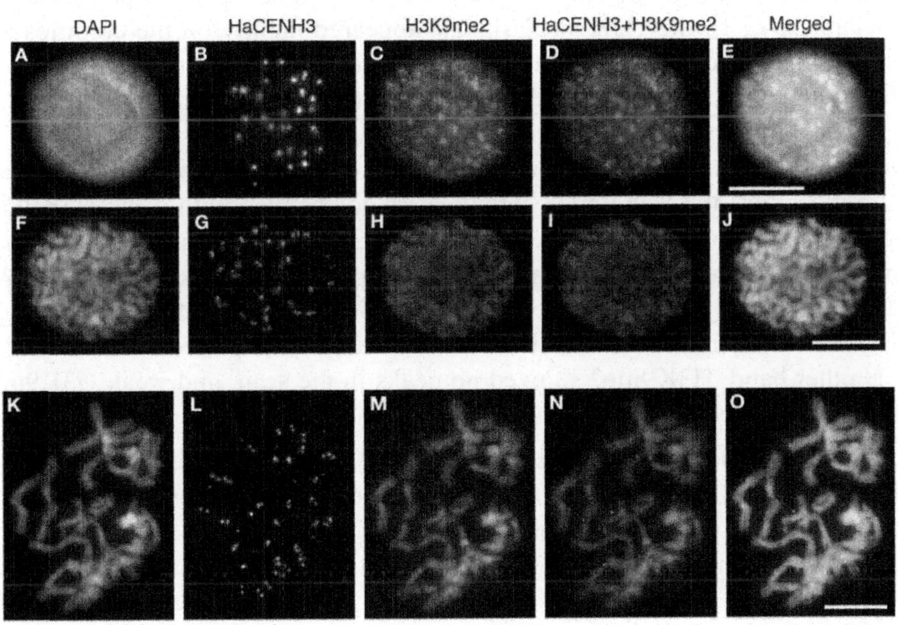

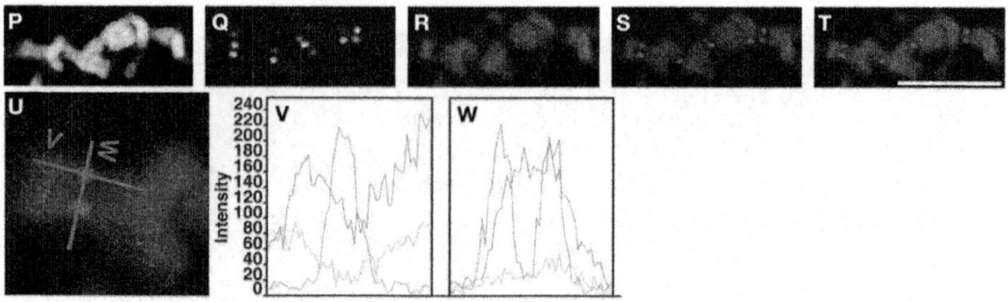

Figure 6. Immunostaining and immunohistochemical staining using the anti-HaCENH3 and anti-H3K9me2 antibodies. (A–O) Immunostaining images. (A–E) An interphase nucleus. (F–J) Prophase chromosomes. (K–O) Metaphase chromosomes. (P–T) Immunohistochemical staining images of metaphase chromosomes. Scale bar, 10 μm. (U) Scanning of metaphase chromosomes. Scanned positions are indicated as red lines. (V) A vertical scan of the chromosome. (W) A horizontal scan of the chromosome.

Immunosignals from H3K9me2 that represented heterochromatin status were distributed on an entire region of nuclei with dots in the interphase cells, but the dot sizes were larger than these of H3K4me2 (Figures 6A–E). Although the larger dot signals did not overlap with the HaCENH3 signals, the HaCENH3 signals were colocalized with faint H3K9me2 signals. In the prophase and metaphase cells, the H3K9me2 signals appeared mainly on the chromosome arms, whereas centromeres and pericentric regions of chromosomes showed faint signals for histone modification (Figures 6F–O). Co-localization of HaCENH3 and H3K9me2 was confirmed by immunohistochemical staining using a laser-scanning confocal microscope (Figures 6P–W). On metaphase chromosomes, the HaCENH3 signals were colocalized with weak H3K9me2 signals, and both arms showed strong H3K9me2 (Figures 6U,V). In a horizontal-scanning of centromeres on a pair of chromatids (Figures 6U,W), a DAPI-staining peak appeared at the middle of the paired chromatids, and HaCENH3 signals were detected on both sides of the paired chromatids. On the other hand, H3K9me2 showed no peaks in the scan, and weak H3K9me2 signals appeared constantly on the centromeric region.

Histone modifications on centromeric DNA sequences were also investigated by ChIP-qPCR (Figure 3). In the analyses, the ubiquitin gene was used as a negative control for a heterochromatin marker, H3K9me2, and as a positive control for two euchromatin markers, H3K4me2 and H4Ac. In ChIP-qPCR with the anti-H3K9me2 antibody, all five HaCENH3-positive sequences and a non-centromeric repetitive sequence (HAG004N15) significantly accumulated compared with the negative control ($P < 0.01$ via Student's t-test, $n = 4$, Figure 3). For modification of H3K4me2, HaCENH3CL124 showed the lowest RE

(0.21), and the REs of the positive control (ubiquitin) and HAG004N15 were 2.32 and 0.54, respectively (Figure 3). These two sequences significantly accumulated compared with HaCENH3CL124 ($P < 0.01$ via Student's *t*-test, $n = 4$). Similarly, HaCENH3CL124 showed the lowest RE (1.14) for H4Ac, and only the positive control (ubiquitin, RE = 13.36) increased significantly ($P = 0.003$ via Student's *t*-test, $n = 4$) compared with HaCENH3CL124 (Figure 3). Since leaf cells usually do not divide, these data suggest that all HaCENH3-positive sequences were involved with heterochromatin at interphase.

DISCUSSION

In this study, a CENH3-encoding cDNA was identified in sunflower. Based on the amino acid sequence deduced from the cDNA sequence, a peptide corresponding to the N-terminal 20 amino acids was synthesized. An antibody against the synthesized peptide recognized centromeres on all sunflower chromosomes. Using this antibody, ChIP-Seq analysis was applied and succeeded in isolating centromeric DNA sequences from sunflowers. Additionally, the epigenetic status at the centromeric DNA sequences was investigated by ChIP-qPCR with antibodies against modified histones.

Usually, centromeric DNA sequences in plant species consist of species-specific tandem repeats and retrotransposons, and these sequences are located on all centromeric regions (Zhong et al., 2002; Nagaki et al., 2003, 2009, 2011; Nagaki and Murata, 2005; Houben et al., 2007; Tek et al., 2010, 2011; Wang et al., 2011; Neumann et al., 2012; He et al., 2015). In this study of sunflower, we identified two types of centromeric DNA sequences (Figure 4). One was a LINE-like sequence, HaCEN-LINE, which showed centromeric signals on all of the chromosomes (Figures 4N–U). Although some tandem repeats evolved from retrotransposons were reported in potato (Gong et al., 2012), no centromere-specific LINE-like elements have been reported. The cluster graphs of HaCEN-LINEs in RepeatExplorer showed line shape, implying their retroelement form rather than modified tandem repeat form. Therefore, this is the first report describing a centromere-specific LINE. Although the centromere-targeting mechanisms of transposable elements remain unknown, the present study indicated that the LINE-like sequence can also target centromeric regions in the same manner as retrotransposons. Another centromeric DNA sequence in sunflowers was a 187-bp tandem repeat, which was located on a single pair of chromosomes (Figure 4). Such chromosome-specific centromeric DNA sequences have been reported in chickens (Shang et al., 2010), tobacco (Nagaki et al., 2012a), and potatoes (Gong et al., 2012); however, they are not very common. The enrichment of the tandem repeats in HaCENH3 ChIP was much higher than that of the LINE (Figure 3), suggesting that the tandem repeat may be more useful for building more stable centromeres than the LINE. As discussed previously for other species (Nagaki et al., 2004, 2012a; Shang et al., 2010; Gong et al., 2012), sunflower centromeres may also be undergoing centromeric DNA evolution to equalize centromeric DNA sequences among chromosomes; the tandem repeats may form stabilized centromeres on the all sunflower chromosomes after this evolutionary event.

With the exception of phosphorylation, other histone modifications on the centromeric regions are not well characterized in plants (Desvoyes et al., 2014; Sharma et al., 2015). According to Houben et al. (2003), plant species with a genome smaller than 510 Mb formed chromocenters at interphase, and H3K9me2 preferentially occurred on the chromocenters. In *Arabidopsis*, H3K9me2 signals appeared on all centromeric and pericentromeric regions of metaphase chromosomes. Euchromatin-specific histone modifications, such as H3K4me2 and H4Ac, were observed in an inverse pattern compared to that of the H3K9me2 (Houben et al., 2003; Jasencakova et al., 2003). In plant species with genomes larger than 530 Mb, such as barley, no chromocenters were observed, and dispersed H3K9me2 signals were observed at interphase; their low level of modification appears on centromeric and pericentromeric regions of metaphase chromosomes (Fuchs et al., 2006). Occasionally, H4Ac were detected at centromeric and pericentromeric regions of metaphase chromosomes in barley and field beans (Jasencakova et al., 2000, 2001). In this study, sunflowers showed dispersed modification patterns for H3K4me2, H3K9Ac and H3K9me2 at interphase and a lower level of H3K9me2 modification on metaphase centromeres (Figures 5, 6). However, the irregular modification patterns observed at metaphase in barley and field bean were not detected at metaphase in sunflower. The genome size of sunflower was estimated to be 2.43 Gb (Bennett et al., 1982), and the observed distribution patters of the modified histones in this study (Figures 5 and 6) coincided with those in plants with large genomes (Houben et al., 2003).

To determine histone modification events on each different centromeric DNA sequence, ChIP analysis has higher resolution and quantitative capacity than immunostaining. In rice, ChIP revealed that centromeric DNA sequences excepting the genic regions on chromosome 8 were heterochromatic (Nagaki et al., 2004; Yan et al., 2006). Subsequently, different histone modifications were observed in CentO sequences (Zhang et al., 2013). CentO sequences with regular size units (155 bp) were enriched in OsCENH3 ChIP, whereas 167-bp CentO variants were enriched in a euchromatic modification, H3K4me2, ChIP rather than OsCENH3. In the present study with sunflowers, two centromeric DNA sequences showed heterochromatic status (Figure 3), suggesting that the status of plant centromeres at interphase was heterochromatic. Additionally, immunohistochemical staining revealed the existence of H3K9me2 on the centromeres at metaphase (Figure 6). Furthermore, immunostaining showed that the level of H3K9Ac in the interphase nuclei was higher than that on metaphase chromosomes. Although we could not quantify the H3K9Ac level on centromeres, the increased level of H3K9Ac in the interphase nuclei implies that it increased on centromeres as well. In the case of human cells, the combination of rigid cell cycle control and ChIP made it possible to detect H3K9Ac within a short range of time during the cell cycle (Ohzeki et al., 2012). To investigate H3K9Ac and other modifications on plant centromeres, a more accurate quantification system, such as that utilized in human cells, is required in the future.

CONFLICT OF INTEREST STATEMENT

The authors declare that the research was conducted in the absence of any commercial or financial relationships that could be construed as a potential conflict of interest.

ACKNOWLEDGMENTS

This research was supported by the Cooperative Research Grant of the Genome Research for BioResource, NODAI Genome Research Center, Tokyo University of Agriculture.

REFERENCES

1. Amor, D. J., Kalitsis, P., Sumer, H., and Choo, K. H. (2004). Building the centromere: from foundation proteins to 3D organization. *Trends Cell Biol.* 14, 359–368. doi: 10.1016/j.tcb.2004.05.009
2. Bennett, M. D., Smith, J. B., and Heslopharrison, J. S. (1982). Nuclear-DNA amounts in angiosperms. *Proc. R. Soc. Lond. B Biol. Sci.* 216, 179–199. doi: 10.1098/rspb.1982.0069
3. Black, B. E., Foltz, D. R., Chakravarthy, S., Luger, K., Woods, V. L. Jr., and Cleveland, D. W. (2004). Structural determinants for generating centromeric chromatin. *Nature* 430, 578–582. doi: 10.1038/nature02766
4. Ceccarelli, M., Sarri, V., Natali, L., Giordani, T., Cavallini, A., Zuccolo, A., et al. (2007). Characterization of the chromosome complement of *Helianthus annuus* by *in situ* hybridization of a tandemly repeated DNA sequence. *Genome* 50, 429–434. doi: 10.1139/G07-019
5. Desvoyes, B., Fernández-Marcos, M., Sequeira-Mendes, J., Otero, S., Vergara, Z., and Gutierrez, C. (2014). Looking at plant cell cycle from the chromatin window. *Front. Plant Sci.* 5:369. doi: 10.3389/fpls.2014.00369

6. Dunemann, F., Schrader, O., Budahn, H., and Houben, A. (2014). Characterization of centromeric histone H3 (CENH3) variants in cultivated and wild carrots (*Daucus* sp.). *PLoS ONE* 9:e98504. doi: 10.1371/journal.pone.0098504

7. Earnshaw, W. C., and Rothfield, N. (1985). Identification of a family of human centromere proteins using autoimmune sera from patients with scleroderma. *Chromosoma* 91, 313–321. doi: 10.1007/BF00328227

8. Feng, J., Liu, Z., Cai, X., and Jan, C. C. (2013). Toward a molecular cytogenetic map for cultivated sunflower (*Helianthus annuus* L.) by landed BAC/BIBAC clones. *G3 (Bethesda)* 3, 31–40. doi: 10.1534/g3.112.004846

9. Fuchs, J., Demidov, D., Houben, A., and Schubert, I. (2006). Chromosomal histone modification patterns–from conservation to diversity. *Trends Plant Sci.* 11, 199–208. doi: 10.1016/j.tplants.2006.02.008

10. Gerlach, W. L., and Bedbrook, J. R. (1979). Cloning and characterization of ribosomal RNA genes from wheat and barley. *Nucleic Acids Res.* 7, 1869–1885. doi: 10.1093/nar/7.7.1869

11. Gong, Z., Wu, Y., Koblízková, A., Torres, G. A., Wang, K., Iovene, M., et al. (2012). Repeatless and repeat-based centromeres in potato: implications for centromere evolution. *Plant Cell* 24, 3559–3574. doi: 10.1105/tpc.112.100511

12. Han, F., Lamb, J. C., and Birchler, J. A. (2006). High frequency of centromere inactivation resulting in stable dicentric chromosomes of maize. *Proc. Natl. Acad. Sci. U.S.A.* 103, 3238–3243. doi: 10.1073/pnas.0509650103

13. He, Q., Cai, Z., Hu, T., Liu, H., Bao, C., Mao, W., et al. (2015). Repetitive sequence analysis and karyotyping reveals centromere-associated DNA sequences in radish (*Raphanus sativus* L.). *BMC Plant Biol.* 15:105. doi: 10.1186/s12870-015-0480-y

14. Houben, A., Demidov, D., Gernand, D., Meister, A., Leach, C. R., and Schubert, I. (2003). Methylation of histone H3 in euchromatin of plant chromosomes depends on basic nuclear DNA content. *Plant J.* 33, 967–973. doi: 10.1046/j.1365-313X.2003.01681.x

15. Houben, A., Schroeder-Reiter, E., Nagaki, K., Nasuda, S., Wanner, G., Murata, M., et al. (2007). CENH3 interacts with the centromeric retrotransposon cereba and GC-rich satellites and locates to centromeric substructures in barley. *Chromosoma* 116, 275–283. doi: 10.1007/s00412-007-0102-z

16. Jasencakova, Z., Meister, A., and Schubert, I. (2001). Chromatin organization and its relation to replication and histone acetylation during the cell cycle in barley. *Chromosoma* 110, 83–92. doi: 10.1007/s004120100132

17. Jasencakova, Z., Meister, A., Walter, J., Turner, B. M., and Schubert, I. (2000). Histone H4 acetylation of euchromatin and heterochromatin is cell cycle dependent and correlated with replication rather than with transcription. *Plant Cell* 12, 2087–2100. doi: 10.1105/tpc.12.11.2087

18. Jasencakova, Z., Soppe, W. J. J., Meister, A., Gernand, D., Turner, B. M., and Schubert, I. (2003). Histone modifications in Arabidopsis - high methylation of H3 lysine 9 is dispensable for constitutive heterochromatin. *Plant J.* 33, 471–480. doi: 10.1046/j.1365-313X.2003.01638.x

19. Lermontova, I., Schubert, V., Fuchs, J., Klatte, S., Macas, J., and Schubert, I. (2006). Loading of *Arabidopsis* centromeric histone CENH3 occurs mainly during G2 and requires the presence of the histone fold domain. *Plant Cell* 18, 2443–2451. doi: 10.1105/tpc.106.043174

20. Maheshawari, S. (2015). Naturally occurring differences in CENH3 affect chromosome segregation in Zygotic Mitosis of hybrid. *PLoS Genet.* 11:e1004970. doi: 10.1371/journal.pgen.1004970

21. Masonbrink, R. E., Gallagher, J. P., Jareczek, J. J., Renny-Byfield, S., Grover, C. E., Gong, L., et al. (2014). CenH3 evolution in diploids and polyploids of three angiosperm genera. *BMC Plant Biol.* 14:383. doi: 10.1186/s12870-014-0383-3

22. Nagaki, K., Cheng, Z., Ouyang, S., Talbert, P. B., Kim, M., Jones, K. M., et al. (2004). Sequencing of a rice centromere uncovers active genes. *Nat. Genet.* 36, 138–145. doi: 10.1038/ng1289

23. Nagaki, K., Kashihara, K., and Murata, M. (2005). Visualization of diffuse centromeres with centromere-specific histone H3 in the holocentric plant*Luzula nivea*. *Plant Cell* 17, 1886–1893. doi: 10.1105/tpc.105.032961

24. Nagaki, K., Kashihara, K., and Murata, M. (2009). A centromeric DNA sequence colocalized with a centromere-specific histone H3 in tobacco. *Chromosoma* 118, 249–257. doi: 10.1007/s00412-008-0193-1

25. Nagaki, K., and Murata, M. (2005). Characterization of CENH3 and centromere-associated DNA sequences in sugarcane. *Chromosome Res.* 13, 195–203. doi: 10.1007/s10577-005-0847-2

26. Nagaki, K., Shibata, F., Kanatani, A., Kashihara, K., and Murata, M. (2012a). Isolation of centromeric-tandem repetitive DNA sequences by chromatin affinity purification using a HaloTag7-fused centromere-

specific histone H3 in tobacco. *Plant Cell Rep.* 31, 771–779. doi: 10.1007/s00299-011-1198-4

27. Nagaki, K., Shibata, F., Suzuki, G., Kanatani, A., Ozaki, S., Hironaka, A., et al. (2011). Coexistence of NtCENH3 and two retrotransposons in tobacco centromeres. *Chromosome Res.* 19, 591–605. doi: 10.1007/s10577-011-9219-2
28. Nagaki, K., Talbert, P. B., Zhong, C. X., Dawe, R. K., Henikoff, S., and Jiang, J. M. (2003). Chromatin immunoprecipitation reveals that the 180-bp satellite repeat is the key functional DNA element of *Arabidopsis thaliana* centromeres. *Genetics* 163, 1221–1225.
29. Nagaki, K., Yamamoto, M., Yamaji, N., Mukai, Y., and Murata, M. (2012b). Chromosome dynamics visualized with an anti-centromeric histone H3 antibody in Allium. *PLoS ONE* 7:e51315. doi: 10.1371/journal.pone.0051315
30. Nasuda, S., Hudakova, S., Schubert, I., Houben, A., and Endo, T. R. (2005). Stable barley chromosomes without centromeric repeats. *Proc. Natl. Acad. Sci. U.S.A.* 102, 9842–9847. doi: 10.1073/pnas.0504235102
31. Neumann, P., Navrátilová, A., Schroeder-Reiter, E., Koblížková, A., Steinbauerová, V., Chocholová, E., et al. (2012). Stretching the rules: monocentric chromosomes with multiple centromere domains. *PLoS Genet.* 8:e1002777. doi: 10.1371/journal.pgen.1002777
32. Neumann, P., Pavlíková, Z., Koblížková, A., Fuková, I., Jedličková, V., Novák, P., et al. (2015). Centromeres off the hook: massive changes in centromere size and structure following duplication of CenH3 gene in Fabeae species. *Mol. Biol. Evol.* 32, 1862–1879. doi: 10.1093/molbev/msv070
33. Novák, P., Neumann, P., and Macas, J. (2010). Graph-based clustering and characterization of repetitive sequences in next-generation sequencing data.*BMC Bioinformatics* 11:378. doi: 10.1186/1471-2105-11-378
34. Ohzeki, J., Bergmann, J. H., Kouprina, N., Noskov, V. N., Nakano, M., Kimura, H., et al. (2012). Breaking the HAC Barrier: histone H3K9 acetyl/methyl balance regulates CENP-A assembly. *EMBO J.* 31, 2391–2402. doi: 10.1038/emboj.2012.82
35. Perpelescu, M., and Fukagawa, T. (2011). The ABCs of CENPs. *Chromosoma* 120, 425–446. doi: 10.1007/s00412-011-0330-0
36. Saitou, N., and Nei, M. (1987). The neighbor-joining method: a new method for reconstructing phylogenic trees. *Mol. Biol. Evol.* 4, 406–425.
37. Sanei, M., Pickering, R., Kumke, K., Nasuda, S., and Houben, A. (2011). Loss of centromeric histone H3 (CENH3) from centromeres precedes

uniparental chromosome elimination in interspecific barley hybrids. *Proc. Natl. Acad. Sci. U.S.A*. 108, E498–E505. doi: 10.1073/pnas.1103190108

38. Shang, W. H., Hori, T., Toyoda, A., Kato, J., Popendorf, K., Sakakibara, Y., et al. (2010). Chickens possess centromeres with both extended tandem repeats and short non-tandem-repetitive sequences. *Genome Res*. 20, 1219–1228. doi: 10.1101/gr.106245.110
39. Sharma, S. K., Yamamoto, M., and Mukai, Y. (2015). Immuno-cytogenetic manifestation of epigenetic chromatin modification marks in plants. *Planta* 241, 291–301. doi: 10.1007/s00425-014-2233-9
40. Talbert, P. B., Masuelli, R., Tyagi, A. P., Comai, L., and Henikoff, S. (2002). Centromeric localization and adaptive evolution of an *Arabidopsis* histone H3 variant. *Plant Cell* 14, 1053–1066. doi: 10.1105/tpc.010425
41. Talia, P., Greizerstein, E., Quijano, C. D., Peluffo, L., Fernandez, L., Fernández, P., et al. (2010). Cytological characterization of sunflower by *in situ*hybridization using homologous rDNA sequences and a BAC clone containing highly represented repetitive retrotransposon-like sequences. *Genome* 53, 172–179. doi: 10.1139/G09-097
42. Tek, A. L., Kashihara, K., Murata, M., and Nagaki, K. (2010). Functional centromeres in soybean include two distinct tandem repeats and a retrotransposon. *Chromosome Res*. 18, 337–347. doi: 10.1007/s10577-010-9119-x
43. Tek, A. L., Kashihara, K., Murata, M., and Nagaki, K. (2011). Functional centromeres in Astragalus sinicus include a compact centromere-specific histone H3 and a 20-bp tandem repeat. *Chromosome Res*. 19, 969–978. doi: 10.1007/s10577-011-9247-y
44. Thompson, J. D., Gibson, T. J., Plewniak, F., Jeanmougin, F., and Higgins, D. G. (1997). The CLUSTAL_X windows interface: flexible strategies for multiple sequence alignment aided by quality analysis tools. *Nucleic Acids Res*. 25, 4876–4882. doi: 10.1093/nar/25.24.4876
45. Vermaak, D., Hayden, H. S., and Henikoff, S. (2002). Centromere targeting element within the histone fold domain of Cid. *Mol. Cell. Biol.* 22, 7553–7561. doi: 10.1128/MCB.22.21.7553-7561.2002
46. Wang, G., He, Q., Cheng, Z., Talbert, P. B., and Jin, W. (2011). Characterization of CENH3 proteins and centromere-associated DNA sequences in diploid and allotetraploid Brassica species. *Chromosoma* 120, 353–365. doi: 10.1007/s00412-011-0315-z
47. Warburton, P. E., Cooke, C. A., Bourassa, S., Vafa, O., Sullivan, B. A., Stetten, G., et al. (1997). Immunolocalization of CENP-A suggests a distinct nucleosome structure at the inner kinetochore plate of active

centromeres. *Curr. Biol.* 7, 901–904. doi: 10.1016/S0960-9822(06)00382-4

48. Yamaji, N., and Ma, J. F. (2007). Spatial distribution and temporal variation of the rice silicon transporter Lsi1. *Plant Physiol.* 143, 1306–1313. doi: 10.1104/pp.106.093005
49. Yan, H., Ito, H., Nobuta, K., Ouyang, S., Jin, W., Tian, S., et al. (2006). Genomic and genetic characterization of rice Cen3 reveals extensive transcription and evolutionary implications of a complex centromere. *Plant Cell* 18, 2123–2133. doi: 10.1105/tpc.106.043794
50. Zhang, T., Talbert, P. B., Zhang, W., Wu, Y., Yang, Z., Henikoff, J. G., et al. (2013). The CentO satellite confers translational and rotational phasing on cenH3 nucleosomes in rice centromeres. *Proc. Natl. Acad. Sci. U.S.A.* 110, E4875–E4883. doi: 10.1073/pnas.1319548110
51. Zhong, C. X., Marshall, J. B., Topp, C., Mroczek, R., Kato, A., Nagaki, K., et al. (2002). Centromeric retroelements and satellites interact with maize kinetochore protein CENH3. *Plant Cell* 14, 2825–2836. doi: 10.1105/tpc.006106

Chapter 4

CHROMOSOME SEGREGATION IN PLANT MEIOSIS

Linda Zamariola[1], Choon Lin Tiang[2], Nico De Storme[1], Wojtek Pawlowski[2], and Danny Geelen[1]

[1]Department of Plant Production, Faculty of Bioscience Engineering, University of Ghent, Ghent, Belgium

[2]Department of Plant Breeding and Genetics, Cornell University, Ithaca, NY, USA

Faithful chromosome segregation in meiosis is essential for ploidy stability over sexual life cycles. In plants, defective chromosome segregation caused by gene mutations or other factors leads to the formation of unbalanced or unreduced gametes creating aneuploid or polyploid progeny, respectively. Accurate segregation requires the coordinated execution of conserved processes occurring throughout the two meiotic cell divisions. Synapsis and recombination ensure the establishment of chiasmata that hold homologous chromosomes together allowing their correct segregation in the first meiotic division, which is also tightly regulated by cell-cycle dependent release of cohesin and monopolar attachment of sister kinetochores to microtubules. In meiosis II, bi-orientation of sister kinetochores and proper spindle orientation correctly segregate chromosomes in four haploid cells. Checkpoint mechanisms acting at kinetochores control the accuracy of kinetochore-microtubule attachment, thus ensuring the completion of segregation. Here we review the current knowledge on the processes taking place during chromosome segregation in plant meiosis, focusing on the characterization of the molecular factors involved.

INTRODUCTION

Meiosis is a specialized cell division that generates four haploid daughter cells from a diploid parent cell after a single round of DNA replication and two consecutive rounds of nuclear division. In the first nuclear division, homologous chromosomes segregate (reductional cell division), and in the second one, sister chromatids segregate (equational cell division). As such, each daughter cell carries half the amount of the parental genetic material. The accurate segregation of chromosomes during meiosis is essential for the formation of haploid gametes. Failure in the proper execution of chromosome segregation inevitably leads to the formation of imbalanced gametes and aneuploid or polyploid progeny. In plants, aneuploidy is more tolerated than in animals and viable aneuploid plants have been observed, especially among the progeny of triploid individuals (Henry et al., 2005, 2010). Despite being affected in growth and reproduction (Birchler et al., 2001), aneuploids may have an evolutionary role, serving as a bridge to euploid polyploid plant formation through repeated generations of selfing (Ramsey and Schemske, 1998; Henry et al., 2005). Polyploid plants generated through aneuploids or by the polyploidization events of somatic doubling and unreduced gametes, are considered as a prominent driving force in plant genome evolution (Ramsey and Schemske, 2002; Adams and Wendel, 2005; Comai, 2005; Otto, 2007).

To ensure the correct completion of the meiotic cell division program, a sequence of coordinated steps must take place during the two phases of meiosis. In meiosis I, homologous chromosomes must pair and synapse and physically exchange genetic material through recombination. The resulting points of crossing-over, also termed chiasmata, form links between the two homologs in the bivalent configuration and ensure proper positioning of the bivalent relative to the division spindle and balanced segregation of homologs in anaphase I. Additionally, to achieve this, sister kinetochores from each homolog must attach to microtubules emanating from the same spindle pole, a process called monopolar kinetochore attachment, and cohesion must be lost in a stepwise manner. More specifically, at anaphase I, cohesion is released at chromosome arms but not at sister centromeres, allowing homologs to segregate without affecting the physical connection between both sister chromatids. In meiosis II, chromosome segregation in the two resulting haploid interphase nuclei occurs in an equational manner and hence strongly resembles the dynamics of a mitotic cell division. Cohesion at centromeres is retained until anaphase II to ensure bipolar attachment of sister kinetochores to microtubules and equational segregation of chromatids into four haploid daughter cells. Progression through the meiotic cell division is regulated at determined checkpoints by the activity of CDK (Cyclin-Dependent Kinase) - cyclin complexes and the Anaphase Promoting Complex/Cyclosome (APC/C) (Harper et al., 2002; Cooper and Strich, 2011). In particular, the Spindle Assembly Checkpoint (SAC) acts during the transition between metaphase and anaphase of the two meiotic cell divisions to ensure correct kinetochore-microtubule attachments and faithful chromosome segregation (Malmanche et al., 2006; Yamamoto et al., 2008). In plants, checkpoints appear to be less stringent compared to yeast and animals, since completion of meiosis is achieved in several meiotic mutants creating imbalanced gametes (Wijnker and Schnittger, 2013).

Most of the knowledge on the molecular biology of mitotic and meiotic chromosome segregation comes from studies in yeast (reviewed in Marston, 2014). However, the mechanisms of chromosome segregation are conserved in eukaryotes, including plants (Dawe, 1998; Bhatt et al., 2001). In the last decade, the increasing availability of genomic tools and the development of *Arabidopsis thaliana*, but also maize (*Zea mays*) and rice (*Oryza sativa*), as model systems, have led to the identification of a large number of conserved meiotic genes (Mercier and Grelon, 2008). Phenotypic and cytogenetic analyses of the corresponding mutants, have unraveled the function of several molecular factors required for proper chromosome segregation in plants (Bhatt et al., 2001; Ma, 2006). Therefore, the focus of this review will be on the major cellular processes that take place to ensure accurate chromosome segregation in plant meiosis and the related genes that have been yet identified in *Arabidopsis*, maize and rice. Other factors having an effect on chromosome segregation in plant meiosis, such as environmental stresses and changes in ploidy level have been described in recent reviews (Comai, 2005; Madlung and Wendel, 2013; De Storme and Geelen, 2014). After mentioning the importance of homologous chromosome pairing and recombination, two subjects extensively discussed in other reviews (Hamant et al., 2006; Edlinger and Schlögelhofer, 2011; Osman et al., 2011; Tiang et al., 2012; Da Ines et al., 2014), we describe the relevance of cohesion, focusing on the roles of the cohesin complex and on the cohesion dynamics (e.g., loading, release and protection) during meiotic cell division. Next, we discuss the role of centromeric and kinetochore proteins in establishing proper spindle attachment during meiosis I and II, and additionally describe what is currently known on the checkpoint control mechanisms acting at kinetochores. Finally, we report the molecular mechanisms underlying microtubule organization and we focus on the relevance of spindle orientation in plant meiosis.

HOMOLOGOUS PAIRING AND RECOMBINATION AS A BASIS FOR REDUCTIONAL CELL DIVISION IN MEIOSIS I

Homologous Chromosome Pairing and Synapsis

To ensure accurate segregation, chromosomes must first recognize their homologous partners and pair with them during early meiotic prophase I. This process leads to the formation of bivalents, which ensures correct bipolar attachment of homologous centromeres to the division spindle at metaphase I in a way that each of the chromosomes in the bivalent moves to a different pole at anaphase I. Bivalent formation is also required for proper positioning of chromosomes at the metaphase plate. Consequently, mutants with chromosome pairing problems exhibit chromosome segregation defects (Bozza and Pawlowski, 2008).

It is assumed that chromosome homology recognition is based on their DNA sequence. Although mechanisms that bring homologous chromosomes together have yet to be fully elucidated, studies in a variety of species, including plants, have shown that chromosome pairing is strongly dependent on their dynamics in early meiotic prophase as well as the initiation and progression through the early stages of the recombination pathway. Chromosome dynamics in prophase I is largely controlled by the behavior of telomeres, blocks of highly conserved repetitive DNA sequence at the ends of chromosomes (Siderakis and Tarsounas, 2007). Telomeres attach to the nuclear envelope before the onset of chromosome pairing, and gather on a small region, forming a unique structure that resembles a flower bouquet, the so called telomere bouquet (Bass et al., 2000; Golubovskaya et al., 2002; Harper et al., 2004; Richards et al., 2012). The bouquet arrangement has been observed in most eukaryotes (Klutstein and Cooper, 2014). The exact role of the bouquet is still being debated. However, mutants defective in bouquet formation are frequently also defective in chromosome pairing, which implies a role of the bouquet in this process (Harper et al., 2004; Klutstein and Cooper, 2014). One example of such mutant is *plural abnormality of meiosis 1* (*pam1*) in maize, which exhibits significant reduction in homologous pairing (Golubovskaya et al., 2002). In this mutant, telomeres attach to the nuclear envelope but fail to cluster. The bouquet formation has been, therefore, suggested to promote homologous paring by bringing chromosome ends together (Harper et al., 2004).

Alternative chromosome interaction mechanisms have been described in several species, including *Caenorhabditis elegans* and *Arabidopsis* (Armstrong et al., 2001; Phillips and Dernburg, 2006). In *C. elegans*, telomeres do not form the bouquet but pairing centers, short chromosome segments recognized by specific zinc-finger proteins, that attach to the nuclear envelope during early prophase I, also bringing homologous chromosomes together (Phillips and Dernburg, 2006). In *Arabidopsis*, telomeres cluster in meiotic interphase on the nucleolus rather than the nuclear envelope (Armstrong et al., 2001). Subtelomeric regions of *Arabidopsis* chromosomes start to pair before telomeres dissociate from the nucleolus, suggesting that the clustering on the nucleolus may play a role similar to that of the canonical bouquet. *Arabidopsis* telomeres establish their connections with the nuclear envelope during leptotene and zygotene, although without an obvious bouquet formation (Armstrong et al., 2001).

Interestingly, the connections used to attach chromosomes to the nuclear envelope in *C. elegans* and *Arabidopsis* are homologs of the same transmembrane proteins that are used in other species to tether telomeres to the nuclear envelope during bouquet formation. SUN domain proteins, identified in yeast, mammals, *C. elegans*, maize, as well as*Arabidopsis*, cross the inner nuclear membrane (Chikashige et al., 2007; Schmitt et al., 2007; Penkner et al., 2009; Sato et al., 2009; Graumann et al., 2010; Murphy et al., 2010). They interact at their N-termini with telomere binding proteins while their C-termini bind transmembrane proteins containing a conserved KASH domain that cross the outer membrane and interact with the cytoskeleton (Miki et al., 2004; Zhou et al., 2012). The commonality of the structures attaching telomeres to the nuclear envelope reinforces the notion that the telomere-nuclear membrane attachments in*C. elegans* and *Arabidopsis* may be functionally similar to the presence of the canonical bouquet.

It has been shown in several species that the cytoskeleton acts through the telomere-nuclear membrane attachments to induce dynamic motility of chromosomes (Bhalla and Dernburg, 2008; Koszul et al., 2009; Sheehan and Pawlowski, 2009; Woglar and Jantsch, 2013). The chromosome movements are thought to help the chromosomes to engage in finding their pairing partners as well as resolving their entanglements.

Another process, which is required for proper chromosome segregation, and closely follows chromosome pairing, is synapsis. Synapsis is installation of a proteinaceous structure, the synaptonemal complex (SC), between the paired homologous chromosomes, which stabilizes the pairing interactions. The SC consists of two lateral elements (LEs) which reside at the base of the chromosome loops and are held together in parallel by transverse filament proteins. In most eukaryotes, the LEs are derived from the axial elements (AEs) loaded on the chromosomal axis before synapsis. Installation of the synaptonemal complex is also closely linked with the formation of crossovers (see the following section), and so synapsis also affects chromosome segregation through its role in crossover formation. *Arabidopsis*mutants defective in synaptonemal complex formation exhibit univalents at metaphase I and improper chromosome segregation at anaphase I (Ross et al., 1997; Higgins et al., 2005).

Meiotic Recombination

Meiotic recombination affects segregation of chromosomes in at least two ways. First, studies in many species, including plants, mammals, and fungi, have indicated that homologous chromosome pairing is closely connected to meiotic recombination (Pawlowski and Cande, 2005). Second, crossovers, reciprocal chromosome segment exchanges formed as a result of meiotic recombination, form physical connections, known as chiasmata, between homologous chromosomes in each bivalents. Chiasmata keep bivalents together to ensure proper orientation and segregation of chromosomes during the first meiotic division.

Recombination in meiosis is initiated by the formation of double strand breaks (DSBs) in chromosomal DNA, triggered by Spo11, a conserved topoisomerase type-II-like protein (Keeney et al., 1997). The MRN complex (MRE11/RAD50/NBS1) then resects the breaks creating single-stranded DNA overhangs (Borde, 2007), which then invade appropriate regions on the homologous chromosomes. This process is promoted by two recombination proteins, Rad51 and Dmc1 (Masson and West, 2001). Rad51 is solely responsible for the repair of DNA breaks using sister chromatids as templates. However, this process is restrained and replaced by repair via the homologous chromosome when Dmc1 is localized to meiotic DNA break sites together with Rad51 (Bishop et al., 1992; Niu et al., 2009). In*Arabidopsis*, mutating *Rad51* results in chromosome fragmentation (Li et al., 2004). However, fragmentation is not observed in the *dmc1* mutant (Couteau et al., 1999). These observations suggest that the function of Dmc1 is distinct from Rad51, as Dmc1 promotes interhomolog recombination rather than intersister recombination (Kurzbauer et al., 2012; Pradillo et al., 2012).

Meiotic recombination results in formation of crossovers and non-crossovers (which include gene conversions). The number and location of crossovers are tightly regulated. In most plant species, only one to four crossovers are formed per bivalent (Crismani and Mercier, 2012). At least one crossover must be formed per bivalent to ensure correct chromosome segregation at anaphase I. However, the number of crossovers per chromosome is limited by crossover interference, a mechanism that prevents formation of crossovers next to each other (Jones, 1984). A group of proteins called ZMM, which contains Zip1, Zip2, Zip3, Zip4, Msh4, Msh5, and Mer3, have been identified as essential for the formation of interference-dependent crossovers in yeast (Börner et al., 2004). Homologs of several of these proteins have been studied in *Arabidopsis* and found to play similar roles in crossover formation (Higgins et al., 2004, 2005,2008; Chen et al., 2005; Mercier et al., 2005; Chelysheva et al., 2007). Loss of MSH4 in *Arabidopsis*, results in a reduction in crossover frequency to 15% of the wild-type level (Higgins et al., 2004). Similar effect was shown in the*Arabidopsis mer3* mutant (Chen et al., 2005; Mercier et al., 2005). Interestingly, the ZMM group includes proteins that are primary components of the synaptonemal complex, such as ZIP1. This interdependence indicates a link between crossover formation and synapsis. Overall, about 85% of *Arabidopsis* crossovers arise from the interference-dependent pathway (Higgins et al., 2004). The remaining crossovers are interference-independent, and are generated by a distinct group of proteins including MUS81 and EME1/MMS4 (Berchowitz et al., 2007).

Recombination events, including crossovers are not distributed randomly along chromosomes. Instead they tend to appear at certain chromosomal locations known as recombination hotspots (Drouaud et al., 2006). In plant species with large genomes, such as maize, barley, or wheat, crossovers are predominantly present in chromosome regions close to the telomeres (Akhunov et al., 2003; Gore et al., 2009). Crossover distribution affects the positions of chiasmata and may have implications for bivalent stability and chromosome segregation. However, neither mechanisms that control crossover distribution nor implications of crossover distribution for chromosome behavior in meiosis are well understood.

Early Defects in Chromatin Structure have an Impact on Homologous Chromosome Segregation: ASK1

ASK1 (*Arabidopsis* SKP1-like1) encodes one of the 21 predicted *Arabidopsis* homologs of the yeast and human Skp1 proteins (Yang et al., 1999; Zhao et al., 2003a,b). Skp proteins are an essential component of the Skp1-Cullin-F-box (SCF) complex, that belongs to a class of E3 ubiquitin ligases that target a variety of proteins for ubiquitin-mediated degradation via the 26S proteasome pathway (Petroski and Deshaies, 2005). ASK1 is the Skp homolog that has been best characterized in *Arabidopsis*. *ask1-1* mutants display defects in plant growth, flower development and male fertility (Yang et al., 1999; Zhao et al., 2001, 2003b). Male sterility arises from meiotic defects in prophase I that lead to erroneous homologous chromosome segregation in meiosis I and sister chromatid segregation in meiosis II, and to the subsequent formation of unbalanced spores. During prophase I, chromosomes maintain a leptotene-like structure with long and thin threads that do not synapse, as demonstrated by the absence of the typical SC structure (Wang et al., 2004). FISH experiments using a centromeric probe showed the presence of more than 5 signals in *ask1-1* meiocytes during pachytene, confirming lack of homologous pairing and bivalents formation (Zhao et al., 2006). The localization of the α-kleisin subunit of the cohesin complex SYN1 (described in the next paragraph) was also found to be altered in *ask1*meiocytes from zygotene to anaphase I. These observations together with a premature sister chromatid detachment detected by FISH in anaphase I, suggest that *ask1* mutation alters cohesin distribution and function, which is necessary for proper pairing and synapsis (Zhao et al., 2006). The abnormalities detected in *ask1* seem to derive from early defects in meiotic chromatin structure and chromosome reorganization in leptotene that cause a prolonged attachment of chromosomes to the nuclear membrane and the nucleolus, alterations in rDNA structure, prolonged attachment of the telomeres to the nucleolus, and defects in histone 3 acetylation, overall leading to the absence of homologous chromosome pairing (Yang et al., 2006). Hence, ASK1 is most likely required for chromosome conformation and remodeling of meiotic chromosomes by controlling the release of chromatin from the nucleolus and nuclear membrane starting from leptotene (Yang et al., 2006). Several hypotheses have been currently proposed to explain the potential role of ASK1 in meiosis, consistent with the meiotic defects observed in the mutant and the homology of ASK1 to Skp proteins (Yang et al., 2006; Zhao et al., 2006). ASK1 may control the degradation of a protein which inhibits the leptotene to zygotene transition, so that the alterations observed in chromatin structure and organization would be a consequence of the block of this transition. Alternatively, ASK1 might regulate the interaction of chromosomes to the nuclear membrane by degrading one or more proteins that link chromatin to the nuclear matrix, thus allowing a nuclear reorganization during leptotene and zygotene. ASK1 may also control chromatin structure by regulating chromatin remodeling proteins, as suggested by the alterations detected in histone 3 acetylation. However, the specific function of ASK1 in male meiosis is not yet defined.

SISTER CHROMATID COHESION IS ESSENTIAL FOR FAITHFUL CHROMOSOME SEGREGATION

The Cohesin Complex

Sister chromatids must be held together from the moment of their synthesis in S-phase until their separation in anaphase to ensure correct attachment of chromosomes to the spindle and accurate chromosome segregation in dividing cells. Cohesin is the multi-subunit protein complex that mediates sister chromatid cohesion in meiosis and mitosis by physically trapping them in a tripartite ring structure (Haering et al., 2008). The complex is highly conserved in eukaryotes and is composed of a core of four evolutionary conserved proteins, extensively studied in yeast and animals. In mitosis, the cohesin complex is composed of two members of the SMC family (structural maintenance of chromosomes), SMC1 and SMC3, and two auxiliary SCC subunits (sister chromatid cohesion), the α-kleisin RAD21/SCC1 and SCC3. In meiosis, the structure of the cohesin complex is highly similar, except for the RAD21/SSC1 component, which is replaced by its counterpart Rec8 (Stoop-Myer and Amon, 1999; Watanabe and Nurse, 1999). SMC1 and SMC3 consist, in their folded configuration, of a globular head and a hinge domain, connected by a long anti-parallel coiled coil. The proposed model of action of cohesin, the embrace model, requires the connection of the SMC hinge domains to form a SMC1/SMC3 heterodimer with a V-shaped structure, that can bind across sister chromatids and close, forming a ring, through a physical connection of the α-kleisin subunit to the C-terminal domain of SMC1 and the N-terminal domain of SMC3 (Gruber et al., 2003). The complex is stabilized by recruitment of SCC3 by the α-kleisin subunit (Figure 1) (for reviews on cohesin complex: Nasmyth and Haering, 2005; Onn et al., 2008; Peters et al., 2008).

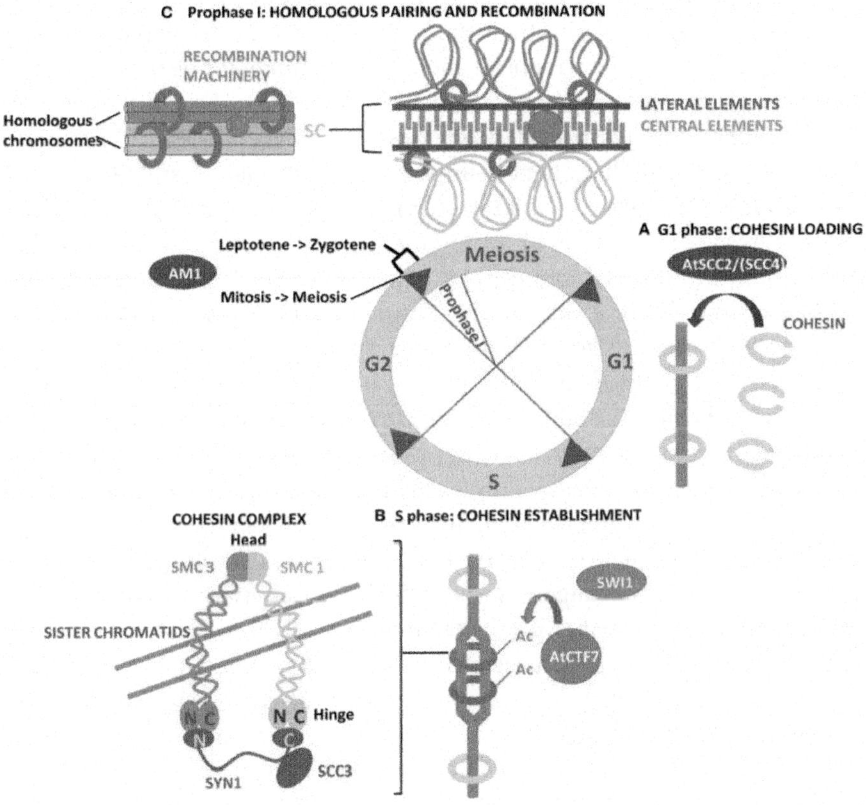

Figure 1. Overview of the events that allow the establishment of the cohesin complex on chromosomes, necessary for the successive steps of chromosome segregation, including homologous pairing and recombination in meiosis I. (A) Loading of cohesin on chromosomes requires the SCC2/SCC4 complex, only AtSCC2 has been characterized in *Arabidopsis*. (B) The establishment of chromosome cohesion takes place during DNA replication in S phase when Eco1/CTF7 acetylates SMC3 residues, ef-

fectively closing the cohesin ring. AtCTF7 has acetylatransferase activity *in vitro* and is required for the establishment of chromosome cohesion in *Arabidopsis*. SWI1 is an *Arabidopsis* protein with a role in cohesin establishment even if the mechanism of action is not yet known. Its maize homolog AM1 is required for the switch from mitosis to meiosis and for a putative checkpoint between leptotene and zygotene in prophase I. (C) Cohesion is required for SC elongation and polymerization and for meiotic recombination in prophase I.

SMC proteins

Similar as in yeast and animals, the sister chromatid connection in plants is also established through the cohesin complex. Homologs of the cohesin complex have been identified in some plant species and major progress on the understanding of their function has been achieved in the model plant Arabidopsis thaliana, in which all the components have been described. The Arabidopsis genome contains single copies of SMC1 and SMC3 cohesin subunits. Genetic studies revealed that loss of AtSMC1 or AtSMC3 functionality causes seedling lethality, hence impairing functional characterization (Liu et al., 2002). Localization studies using a specific antibody revealed that AtSMC3 is present in the cytoplasm and nucleus, on chromosomes and in the nuclear matrix of meiotic and mitotic cells, indicating a function in both types of cell divisions (Lam et al., 2005). At meiotic prophase, AtSMC3 localizes along sister chromatids to axial elements and lateral elements, similar to the Arabidopsis α-kleisin subunit SYN1. This observation confirms the conserved role of the cohesin complex in sister chromatid cohesion but also supports an additional function in SC formation, as proposed in yeast and mammals (Klein et al., 1999; Eijpe et al., 2000). By metaphase I, AtSMC3 localizes only to chromosome centromeres and, in addition, co-localizes to the spindle at metaphase I and anaphase I and II. The spindle localization is independent of SYN1 functionality and suggests that AtSMC3 might play an additional role as spindle associated protein, distinct from its conserved role in sister chromatid cohesion. This novel localization pattern is also conserved in mitosis and could be related to a role of AtSMC3 in spindle assembly and/or in the chromosome association with the spindle (Lam et al., 2005). A similar novel function has been suggested in human mitosis for the entire cohesin complex (Gregson et al., 2001).

Immunolocalization studies in tomato meiocytes (Solanum lycopersicum) revealed that SMC1 and SMC3 show a similar localization pattern as AtSMC3. In prophase I, SMC1 and SMC3 antibodies display a signal along AEs of the SC from leptotene to diplotene and a weak and diffuse signal on chromosomes at metaphase I and telophase II (Lhuissier et al., 2007). However, no localization to the spindle was documented, suggesting that the novel spindle function might be specific for Arabidopsis AtSMC3 and not conserved in other plant species.

Rec8 and SCC3

More intensive studies have been undertaken on the role of the meiotic α-kleisin subunit Rec8 in Arabidopsis (named SYN1 but also DIF1 and AtRec8), maize (AFD1), and rice (OsRad21-4/OsRec8) meiosis. In Arabidopsis, the homolog of Rec8, SYN1, is required for sister chromatid cohesion in meiosis (Cai et al., 2003; Chelysheva et al., 2005). SYN1 fully co-localizes with AtSCC3 at pachytene and is necessary for its proper loading on sister chromatids, confirming that they are indeed part of a complex (Chelysheva et al., 2005). FISH (fluorescence *in situ* hybridization) analysis on syn1 meiocytes using chromosome arm and centromeric probes show defective sister chromatid arm and centromere cohesion in meiosis I, confirming that SYN1 functions in cohesion (Cai et al., 2003).

In addition, SYN1 is required for synapsis of homologous chromosomes, being necessary for SC polymerization and elongation (Chelysheva et al., 2005). In syn1 meiocytes, synapsis is blocked and chromosome condensation and pairing are almost completely absent, leading to the presence of univalents at metaphase I (Bai et al., 1999; Bhatt et al., 1999). Localization of ASY1, a protein required for chromosome synapsis, recombination and SC assembly and widely used as a marker for chromosome axes in meiosis (Sanchez-Moran et al., 2008), is impaired in syn1 mutants, confirming the requirement of SYN1 for AE polymerization and elongation but not for their formation (Chelysheva et al., 2005). Synapsis is known to be closely related to meiotic recombination. Therefore, it is not surprising that SYN1 plays also a role in recombination, specifically in DSBs repair. Indeed, chromatin bridges and chromosome fragmentation are observed in syn1 meiosis I. They are suppressed by introducing into the syn1 mutant background the Atspo11 mutation, which abolishes DSBs formation and prevents recombination, confirming that SYN1 is required for DSBs repair (Chelysheva et al., 2005). Involvement of the cohesin complex in homologous chromosome pairing, assembly of the SC, and in meiotic recombination has been shown previously in other organisms. In yeast, Rec8 and SMC3 are required for SC formation and for repairing DSBs (Klein et al., 1999). In mouse, loss of Rec8 affects homologous recombination but does not affect SC formation and assembly. However, synapsis occurs between sister chromatids instead of homologous chromosomes, suggesting that Rec8 might define the chromosome unit and limit the SC binding sites to one single chromosome surface of a sister-chromatid pair in mammals (Xu et al., 2005).

Support for an additional role of Rec8 in homologous pairing and recombination in plants comes from studies on the maize α-kleisin subunit AFD1 and the rice OsRad21-4/OsRec8. A study on different afd1 alleles has revealed that AFD1 is required for AE installation, affecting the deposition of the recombination machinery on chromosomes (Golubovskaya et al., 2006). The rice OsRec8 regulates AE formation and may have a role in DNA DSBs repair, since localization of PAIR2 (homolog of Arabidopsis ASY1), ZEP1 (ZYP1 homolog), and MER3, involved in the formation of crossovers, is affected in Osrec8 mutants. As a consequence, no proper homologous pairing occurs (Zhang et al., 2006; Shao et al., 2011). Moreover, defective telomere bouquet formation is observed in Osrec8 and afd1 mutants, also preventing proper pairing of homologous chromosomes. Hence, OsRec8 regulates AE formation, homologous recombination and synapsis by affecting downstream proteins PAIR2, ZEP1, and MER3 (Shao et al., 2011).

Rec8 has a crucial role in the determination of kinetochore geometry for monopolar orientation in fission yeast, since rec8 mutants display loss of monopolar orientation at meiosis I and chromosome segregation defects (Yokobayashi et al., 2003; Sakuno et al., 2009). Similarly, Arabidopsis SYN1 is necessary for the monopolar attachment of sister kinetochores in meiosis I, as indicated by the observation of bipolar sister kinetochore attachment in meiosis I in the double syn1 Atspo11 mutant, in which syn1 chromosome fragmentation is suppressed allowing a clearer observation of chromosome segregation. However, the same defect in kinetochore orientation is observed for the other SCC cohesin subunit, AtSCC3, indicating that SYN1 is not sufficient for monopolar kinetochore orientation or, most likely, is inactive when the other members of the complex are not present. These data suggests that Rec8-containing cohesin complex is responsible for defining kinetochore geometry in meiosis I in plants, as proposed in yeast, Drosophila and mammals (Chelysheva et al., 2005; Watanabe, 2012).

AtSCC3 is the sole SCC3 homolog investigated in plants so far. It is required for normal plant growth and fertility and has a conserved role in proper sister chromatid cohesion, confirmed by the combination of univalents and bivalents observed in Atscc3 mutants (Chelysheva et al., 2005). However, in contrast to SYN1, AtSCC3 is not required for AE formation, since ASY1 localization in Atscc3 is normal and synapsis does not show major defects in the mutant. Moreover, only a low level of fragmentation is observed in Atscc3 and recombination is not defective, suggesting that the two SCC subunits, although being part of the same complex, may fulfill different additional functions (Chelysheva et al., 2005).

While AtSCC3 has no paralogs in the Arabidopsis genome, three α-kleisin homologs, SYN2, SYN3, and SYN4, are present that share about 38 % sequence similarity at their N-termini and 20 % at their C-termini with SYN1, and could partially compensate for each other (Schubert et al., 2009a). Two observations raise the hypothesis that the α-kleisin paralogs may be involved in cohesion in meiosis. First, in the syn1 Atspo11 double mutant, sister chromatid cohesion is only lost at anaphase I, suggesting that other homologs of the SYN1 family might be responsible for cohesion before that stage (Chelysheva et al., 2005). Second, SYN1 localization is only observed along chromosome axes but not at the core centromeres at metaphase I and metaphase II (Chelysheva et al., 2005; Cromer et al., 2013; Zamariola et al., 2013). It is known that SYN1, SYN2 and SYN4 may partially compensate for each other whereas SYN3 is required for plant viability, it localizes to the nucleolus and might have evolved a role in rDNA transcription and/or processing. A specific function in DNA repair in somatic cells has been suggested for SYN2, while SYN4 is required for centromere cohesion in mitosis (Schubert et al., 2009b). However, the role of the different paralogs is, at this time, not clear, and the creation of double or triple mutants might help unravelling their specific functions (Schubert et al., 2009b).

Loading and Establishment of Chromosome Cohesion

The loading of the cohesin complex onto chromosomes starts at telophase in humans and at the end of G1 in yeast and requires the evolutionary conserved SCC2/SCC4 complex (for reviews see Uhlmann, 2009; Ocampo Hafalla and Uhlmann, 2011). Cohesin loading has been shown to be enriched at centromeric and pericentromeric regions promoting high fidelity chromosome segregation (Eckert et al., 2007). Recent studies in budding yeast have revealed that the observed enrichment is defined by the presence of the kinetochore subcomplex Ctf19, that promotes SCC2/SCC4 centromere association (Fernius et al., 2013). Also in Angiosperms, interphase nuclei show a preferred alignment of sister chromatids at centromeres, which might facilitate kinetochore bipolar orientation in mitosis, essential for correct chromosome segregation (Schubert et al., 2007). In plants, only the *Arabidopsis* homolog of the adherin SCC2 has been described. AtSCC2 is essential for plant viability and *Atscc2* plants show defects in embryogenesis and endosperm development (Schubert et al., 2009b; Sebastian et al., 2009). Using an inducible RNAi (RNA interference) system,Sebastian et al. (2009) demonstrated that AtSCC2 is required for sister chromatid cohesion and loading of the cohesin complex in meiosis, as indicated by defects in AtSCC3 localization. Furthermore, *Atscc2* mutants show an irregular localization of ASY1 and chromosome fragmentation, indicating that

AtSCC2 is required for axial development and most likely for repair of DNA DSBs, supporting the notion that sister chromatid cohesion is a prerequisite for axial development and DSBs resolution (Sebastian et al., 2009).

The loading of cohesin is the first step through the establishment of sister chromatid cohesion that takes place during DNA replication. After the loading, cohesin is unstable due to the activity of the Wapl-Pds5 complex that promotes cohesin dissociation (Rowland et al., 2009). In yeast, cohesion is established during S-phase by the Eco1/CTF7 protein, that acetylates the SMC3 residues, effectively closing the cohesin ring (Rowland et al., 2009). In *Arabidopsis*, AtCTF7 exhibits acetyltranferase activity *in vitro* like its yeast and human homologs (Jiang et al., 2010). *Atctf7* homozygous mutants display a dwarf phenotype and aberrant microsporogenesis due to defects in chromosome segregation in mitosis and PMCs (pollen mother cells). FISH performed with a centromeric and a chromosome 4 arm probes on male meiocytes of *Atctf7* and AtCTF7 RNAi plants, revealed that the protein is required for both centromere and arm cohesion in meiosis (Bolaños-Villegas et al., 2013; Singh et al., 2013). Furthermore, localization of the cohesin complex subunits AtSMC3, AtSYN1 and AtSCC3 is impaired in *Atctf7* male meiosis, indicating that AtCTF7 is necessary for association of cohesin on chromatin in meiosis (Bolaños-Villegas et al., 2013; Singh et al., 2013). In addition, the level of expression of genes required for DNA repair is significantly altered in *Atctf7* mitotic and meiotic tissues, and the mutant plants show a lower ability to repair DNA double strand breaks *in vivo* in mitotic cells (Bolaños-Villegas et al., 2013). Taken together, these observations suggest that AtCTF7 is also required for DNA repair in *Arabidopsis*, as shown for Eco1 in yeast mitosis (Lu et al., 2010).

SWITCH1/DYAD (SWI1/DYAD) is an *Arabidopsis* protein with an essential role in the establishment of sister chromatid cohesion during early meiosis (Mercier et al., 2001, 2003). Different allelic mutations have been investigated for the*SWI1/DYAD* gene, all of them showing an impact on fertility due to different mechanisms affecting megasporogenesis (*swi1-1* and *dyad*; Motamayor et al., 2000; Siddiqi et al., 2000; Mercier et al., 2001; Agashe et al., 2002) or both mega and microsporogenesis (*swi1-2* and *dsy10*; Mercier et al., 2003; Boateng et al., 2008). *Swi1-1* and *swi1-2* alleles have been shown to have an effect on the female mitosis-meiosis switch, so that meiosis is converted into a mitotic cell division (Motamayor et al., 2000; Mercier et al., 2001). However, analysis of the *dyad* allele by Agashe et al. (2002) andSiddiqi et al. (2000) with a meiotic marker, provided evidence that the female megaspore enters the meiotic programme but does not progress into further meiotic divisions. Detailed studies of male meiosis for *swi1-2* and *dsy10*alleles, have shown that the mutants loose cohesion in a stepwise manner already in meiosis I, leading to the presence of 20 chromatids at metaphase I which segregate randomly in meiosis II, forming polyads (Mercier et al., 2001). Furthermore, the mutant lacks AE formation, leading to incorrect pairing and synapsis, and does not initiate recombination. These defects probably all derive from defective establishment of cohesion before the initiation of meiosis, since the protein is expressed exclusively in meiotic G1 and S phase (Mercier et al., 2003). Specifically, the localization of SYN1 in *swi1-2* meiocytes, indicates that SWI1 performs its function after the loading of the cohesin complex (Mercier et al., 2003). However, its specific function in chromosome cohesion is not yet understood.

Maize AM1 and rice OsAM1 are proteins closely related to SWI1. Mutants in *AM1* and *OsAM1* genes show defective sister chromatid cohesion, absence of homologous pairing and synapsis, and lack of homologous recombination (Pawlowski et al., 2009; Che et al., 2011). However, while *Arabidopsis swi1* mutants affect meiotic processes downstream of meiotic initiation and do not affect entrance in meiosis, maize *am1* mutants show typical features of mitotic division in the early steps of meiosis, indicating that AM1 is required for the transition from the mitotic cell cycle into meiosis. Meiocytes of a specific *am1* allele arrest during early meiotic prophase at the transition between leptotene and zygotene, suggesting the presence of a novel checkpoint in maize required for progression through prezygotene (Pawlowski et al., 2009). Similarly, in rice, OsAM1 is also likely involved in a checkpoint mechanism that regulates the transition from leptotene to zygotene (Che et al., 2011).

A schematic overview of the processes of cohesin loading and establishment and homologous chromosome pairing and recombination, is shown in Figure 1.

Release of Chromosome Cohesion: Separase

Cleavage of the α-klesin subunit occurs in a stepwise manner during meiosis. In meiosis I, Rec8 is cleaved at chromosome arms, allowing the resolution of chiasmata and homologous chromosome segregation in meiosis I, whereas in meiosis II cohesin is released at centromeres, enabling sister chromatid separation (Nasmyth, 2001). Cleavage of Rec8 is performed by the cysteine protease Separase, which is conserved in various organisms, including yeast and vertebrates (Kitajima et al., 2003; Kudo et al., 2009). Separase function is inhibited by a protein called Securin, which is degraded at the onset of anaphase by ubiquitylation by the APC/C (Uhlmann, 2001). Homologs of separase are present in many plant species. However, the studies undertaken so far have only focused on the *Arabidopsis* separase AESP (Liu and Makaroff, 2006). AESP is an essential gene but RNA interference of AESP under the control of the meiotic DMC1 promoter, and the finding of the temperature permissive mutant *rsw4* (radially swollen 4), have allowed to investigate AESP function in meiosis (Liu and Makaroff, 2006; Wu et al., 2010; Yang et al., 2011). *Aesp* and *rsw4*mutants display defective chromosome segregation in meiosis I, in which entangled chromosomes and chromosome fragments are observed, and in meiosis II, where bivalents are still present, indicating persistence of cohesion (Liu and Makaroff, 2006; Yang et al., 2011). In support of this, SYN1 and SMC3 signals persist on *aesp* and *rsw4* chromosomes at later stages after metaphase I, demonstrating that AESP is responsible for removal of the cohesin complex from chromosomes. The creation of a double mutant between *aesp* and *ask1*, in which homologous chromosomes prematurely separate in meiosis I due to defects in homologous synapsis, showed that sister chromatids did not separate in meiosis II. This observation confirms that AESP is responsible for sister chromatid separation also in anaphase II (Yang et al., 2009). In *Arabidopsis*, a large amount of cohesin is released from chromosome arms in prophase I and the residual arm cohesin is released at anaphase I (Cai et al., 2003). While AESP is required for the release of cohesin at anaphase I and in meiosis II, it does not participate in the first step of release in prophase I, suggesting that a separase-independent mechanism might exist at early stages in *Arabidopsis*, similar to budding yeast, in which the condensin complex SMC2/SMC4 and a Polo kinase are responsible for cohesin removal at chromosome arms before metaphase I (Sumara et al., 2002; Yu and Koshland, 2005; Liu and Makaroff, 2006).

Separase is a multifunctional protein that in various organisms possesses additional roles to sister chromatid separation mechanistically less understood, such as proteolytic cleavage of other target proteins in yeast and spindle assembly in humans (Moschou and Bozhkov, 2012). Also in *Arabidopsis* additional functions of separase have been reported (Yang et al., 2009, 2011). *Aesp* mutants show alterations in non-homologous centromere associations at zygotene, suggesting that AESP might play a role in the control/release of the transient centromere associations that occur during zygotene in *Arabidopsis* (Armstrong et al., 2001). Furthermore, in *aesp* male meiocytes the radial microtubule array (RMA) is disturbed at telophase II and phragmoplast-like structures are observed, suggesting that AESP might have a function in microtubule organization or cell polarity (Yang et al., 2009). Absence of AESP also causes the formation of multinucleate microspores as a consequence of defective RMA (Yang et al., 2009). In contrast to yeast, where separase is required for normal meiotic spindle formation (Jensen et al., 2001; Baskerville et al., 2008), in*Arabidopsis* only RMA formation is defective while AESP might be required for the proper interaction of microtubules with the nuclear envelope at the tetrad stage (Yang et al., 2009).

Protection of Centromere Cohesion: Shugoshin and Patronus

In meiosis, sister chromatid cohesion is controlled in a time- and space-dependent manner, with chromosome arm cohesion release at the start of anaphase I, and maintenance of centromeric cohesion up till anaphase II. Meiosis-specific protection of Rec8 at pericentromeric regions from anaphase I to anaphase II is performed by Shugoshin (Sgo), a protein first described in *Drosophila* (MEI-S332; Kerrebrock et al., 1995), and successively identified in yeast, mammals and plants (Yao and Dai, 2012). Studies from yeast and vertebrates have elucidated the mechanism of action of Sgo, which is recruited at pericentromeric heterochromatin regions where it associates with the phosphatase PP2A to dephosphorylate Rec8 and prevent its cleavage in meiosis I (Lee et al., 2008a; Xu et al., 2009). In yeast, Sgo1 localizes at centromeres until the end of anaphase I (Kitajima et al., 2004), whereas in vertebrates SGOL2 persists on the chromosomes also in meiosis II (Lee et al., 2008a). Currently, two hypotheses are postulated to explain the dynamic association of Shugoshin with centromeres. On the one hand, Sgo function may be controlled by microtubule attachment and deactivated by a spatial change of its localization in the peri-centromeric domain in response to a change in microtubule tension (Lee et al., 2008a). Alternatively, a PP2A inhibitor may block dephosphorylation thereby conferring loss of protection of centromeric cohesion in meiosis II (Chambon et al., 2013). Flies and budding yeast possess a single copy of Sgo, while fission yeast, mammal and plant genomes have two Sgo paralogs, Sgo1 and Sgo2. In*Drosophila*, yeasts and plants, Sgo1 is responsible for the protection of centromere-specific sister chromatid cohesion in meiosis I, while in mammals SGOL2 performs the function of protector (Gutiérrez-Caballero et al., 2012). Though they are homologs, Sgo genes share limited sequence similarity and display in the different organisms somewhat different functions which have been acquired during evolution (for a recent review on the Shugoshin protein family and the additional roles of Shugoshin see Clift and Marston, 2011; Gutiérrez-Caballero et al., 2012). The *Sgo1* paralog *Sgo2*possesses different properties depending on the species examined. In fission yeast, Sgo2 plays a role in chromosome segregation in mitosis (Kitajima et al., 2004), in particular it has been shown to control the localization of the CPC, a protein complex that senses lack of tension between kinetochores and microtubules (Kawashima et al., 2007;Vanoosthuyse et al., 2007; Tsukahara et al., 2010). In addition, fission yeast Sgo2 also plays a role in meiosis, as *Sgo2*deletion leads to a modest increase in non-disjunction of homologs at meiosis I (Kitajima et al., 2004). In humans, hSGOL1 protects centromeric cohesion in mitosis (Salic et al., 2004; McGuinness et al., 2005), whereas hSGOL2 is dispensable for sister chromatid cohesion in mitotic cell division but is essential for correcting erroneous kinetochore attachments by recruiting the microtubule depolymerase MCAK to the centromeres (Huang et al., 2007), a role that is consistent with the one shown for fission yeast Sgo2 (Kawashima et al., 2007).

In plants, the role of Sgo as protector of centromere cohesion in meiosis has been described for the maize ZmSGO1, the rice OsSGO1 as well AtSGO1 and AtSGO2 of *Arabidopsis* (Hamant et al., 2005; Wang et al., 2011; Cromer et al., 2013;Zamariola et al., 2013, 2014). FISH analysis performed on *sgo1* meiocytes with a centromeric probe revealed a premature detachment of sister chromatid centromeres in anaphase I, resulting in random chromosome segregation in meiosis II. However, monopolar orientation of sister kinetochores in meiosis I is not affected in the mutants and chromosomes normally segregate in the reductional division, indicating that SGO proteins are required for protection of cohesion at anaphase I but not for monopolar orientation of sister kinetochores. In fission yeast and mammals, which possess two Sgo homologs, one copy is generally required for protection of sister chromatid cohesion in meiosis, while the other has evolved additional roles, as previously mentioned. So far, no function in somatic cells has been described for any of the plant Sgo proteins. *Arabidopsis* is the only species in which the role of both Sgo paralogs has been investigated. Single mutants show no vegetative phenotype and a meiotic phenotype is detected exclusively for *Atsgo1*. However, *Atsgo1 Atsgo2* double mutants reveal a partially redundant role for the two SGOs, opposite to yeast and vertebrate (Cromer et al., 2013; Zamariola et al., 2014). Immunolocalization of ZmSGO1 and OsSGO1 has revealed that SGO1 is loaded on chromosomes at leptotene, earlier than in other organisms such as yeast or mammals in which loading occurs during late prophase I or at diplotene, respectively (Kitajima et al., 2004; Gómez et al., 2007). Thus, plant SGO proteins might have a function in prophase I. In support of this hypothesis, ZEP1 localization is defective in*Ossgo1* mutants in about 21% of meiocytes, indicating that OsSGO1 may be required for the timely assembly of the SC, even if not for its initial assembly (Wang et al., 2011). In contrast, *Arabidopsis* ZYP1 localizes normally in *Atsgo1*mutants (Zamariola et al., 2013).

Recently, a novel protein involved in the protection of sister chromatid cohesion during meiosis II has been identified in*Arabidopsis*, named PATRONUS (PANS1) (Cromer et al., 2013; Zamariola et al., 2014). PANS1 is a plant specific protein that shares homology with genes belonging to the Eudicots family. *Pans1* meiocytes show a premature release of sister chromatid cohesion at metaphase II but not at meiosis I, indicating that the protein is required for protection of cohesion during interkinesis, at a later stage than SGOs. Moreover, similar to SGOs, PANS1 is not required for monopolar attachment of sister kinetochores in meiosis I. TAP-TAG and Y2H experiments have revealed that PANS1 may be a regulator of the APC/C complex because of the interaction with some of the APC/C subunits. In addition, the presence of two destruction boxes in the PANS1 sequence may indicate that PANS1 is at the same time also targeted by the APC/C complex. Currently, three hypotheses have been suggested to explain how PANS1 maintains sister chromatid cohesion at interkinesis: (1) by protecting SGOs from destruction by the APC/C; (2) by protecting sister chromatid cohesion from Separase independently of SGOs, in the case SGOs are no longer present after anaphase I, and (3) by inhibiting via APC/C regulation the Wapl-dependent process of cohesin release, which is usually activated at the end of mitosis/G1 phase to allow dynamic cohesin renewal and that could be present also at the end of meiotic telophase I (Cromer et al., 2013). At the moment, AtSGOs and PANS1 localization, that could help unraveling the function of PANS1 in meiosis and the relation among the protectors, is lacking. Besides its role as protector of cohesion, PANS1 has also been shown to be required for spindle organization in meiosis since *pans1* meiocytes display defective spindles starting from telophase I. Defective spindles is probably the cause of the formation of an aberrant internuclear organelle band at interkinesis, detected in 7% of *pans1* meiocytes. Taken together, these phenotypes and the premature separation of sister chromatids observed in meiosis II, suggest a function of PANS1 in ensuring the coordinate organization of the cell organelles in accordance with the meiotic cell cycle phase and chromosome cohesion (Zamariola et al., 2014), which is in agreement with the interaction of PANS1 with the APC/C.

CENTROMERES AND KINETOCHORES

Role of Centromeres and Kinetochores in Chromosome Segregation

Centromeres are DNA-protein structures necessary to direct chromosome movement in cell division. Centromere DNA sequences are fast evolving and highly variable among species. However, centromeric regions in most plant species encompass mainly two domains. One is the core centromere, which contains satellite tandem repeats, usually 150–180 bp long, and specialized nucleosomes in which histone H3 is replaced by a centromere-specific H3 histone variant, CENH3. This region is required for the assembly of the kinetochore, a protein structure that binds to spindle microtubules allowing faithful chromosome segregation. The core centromere is flanked by pericentromeric heterochromatin domains containing retroelements and other transposons. In yeast the pericentromeric domains have been shown to have mainly a role in the recruitment of Shugoshin (Pidoux and Allshire, 2005; Yamagishi et al., 2008). In addition, epigenetic mechanisms may be involved in the specification of centromeric chromatin and propagation of centromeres (Houben and Schubert, 2003; Ekwall, 2007; Torras-Llort et al., 2009; Wang et al., 2009).

The specific centromeric variant Histone 3, CENH3, was first identified in human as CENP-A and subsequently in all eukaryotic model systems (De Rop et al., 2012), including *Arabidopsis* (also called HTR12; Talbert, 2002). Despite its essential and conserved role in ensuring proper chromosome segregation, CENH3 proteins are highly variable in their sequences and fast evolving, especially their N-terminal tail domain and a loop 1 region at the C-terminal domain, which are necessary for CENH3 localization to centromeres in *Arabidopsis* (Ravi et al., 2010; Moraes et al., 2011). The C-terminal part of the protein is sufficient for the centromeric localization of CENH3 in mitotic cells even when the N-terminal part is absent (Lermontova et al., 2006). In meiosis, a different loading mechanism for CENH3 is present, in which the N-terminal tail plays a critical role. *Arabidopsis* plants transformed with a N-terminally truncated YFP-CENH3(C) protein show meiotic defects and partial sterility and the YFP signal cannot be detected in meiotic nuclei (Lermontova et al., 2011). Similarly, the replacement of the N-terminal tail with a GFP tagged variant, GFP-tailswap, causes sterility due to defects during sporogenesis (Ravi et al., 2011). In GFP-tailswap plants, meiosis is disturbed starting from metaphase I, in which bivalents align on the division plate but are not subjected to tension from the spindle, which is confirmed by decreased interkinetochore distance and by defective spindles (Ravi et al., 2011). CENH3 protein signal is reduced or not detected in GFP-tailswap meiocytes and is again detected after meiosis on mitotic chromosomes at the microspore stage, indicating the existence of distinct mechanisms for CENH3 loading in meiosis and mitosis (Ravi et al., 2011). The work of Lermontova et al. (2011) also suggests a different loading mechanism in meiosis and mitosis, since the YFP-CENH3(C) variant is deposited to the centromeres in mitosis but not in meiotic nuclei.

Recently, the *Arabidopsis* homolog of KNL2 has been identified. It represents one of the components of the Mis18 complex, responsible for the initiation of CENH3 deposition at the centromeres in humans (Hayashi et al., 2004), *C. elegans* (De Rop et al., 2012) and fission yeast (Hayashi et al., 2004). In *Arabidopsis*, KNL2 is associated with centromeres at all stages of the cell cycle except from metaphase to mid-anaphase. *Arabidopsis* KNL2 knockout mutants show defects in mitosis and meiosis and reduced CENH3 loading at the centromeres (Lermontova et al., 2013). Furthermore, CENH3 gene expression is decreased in *knl2* mutants but KNL2 expression is stable in CENH3 RNAi transformants, indicating that KNL2 acts upstream of CENH3 and has a function in the assembly of CENH3 at the centromeres (Lermontova et al., 2013). Moreover, KNL2 is co-expressed with H3K9 histone methyltransferases genes, whose expression is reduced in *knl2* mutants. Also DNA methylation levels are lower in *knl2* mutant plants. The requirement of KNL2 for CENH3 expression and for DNA methylation, suggests that KNL2 may interact with methyltransferases to allow the maintenance of DNA methylation, in order to control the epigenetic status of centromeric chromatin and to control CENH3 loading (Lermontova et al., 2013).

Sister kinetochores must behave differently in meiosis I and II: in meiosis I are oriented toward the same pole (mono-orientation) to allow homologous chromosomes segregation, while in meiosis II they face opposite poles (bi-orientation) (Brar and Amon, 2009). The tension exerted at the kinetochores by microtubules during division, and the kinetochore geometry, defined in meiosis and mitosis by cohesion, are fundamental for stabilizing the monopolar attachment in MI and the bipolar in MII (for review see Watanabe, 2012). In contrast to the high variability of centromeric sequences, more than 20 active kinetochore proteins are conserved between humans and yeasts (Lampert and Westermann, 2011), and are specific either for the inner kinetochore, where they directly recognize and bind DNA, or for the outer kinetochore, being responsible for the interaction with microtubules (Santaguida and Musacchio, 2009; Wang et al., 2009). However, to date, only 7 kinetochore proteins have been reported to be conserved in *A. thaliana*, the majority of which has not been yet functionally characterized (Murata, 2013). The inability to identify homologs of many human and yeast kinetochore proteins in plants, may suggest the existence of different kinetochore structure in plants (Murata, 2013).

Kinetochore functionality depends on the presence of a functional centromere in meiosis. Indeed MIS-12, a kinetochore protein which co-localizes with CENH3 at the centromere regions (Sato et al., 2005), does not do it in *Arabidopsis* GFP-tailswap meiocytes. In contrast, CENP-C, another kinetochore protein which localizes at the centromeres in mitotic cells (Ogura et al., 2004), is not affected in CENH3 RNAi transformants, suggesting that its localization does not depend on the presence of a functional CENH3 (Lermontova et al., 2011).

In maize, kinetochore proteins have been more thoroughly investigated. CENPC is part of the inner kinetochore and interacts at one side with the DNA repeats located at the centromeric regions, and, on the other side, with the members of the outer kinetochore (Dawe et al., 1999; Zhong et al., 2002). At the outer kinetochore NCD80 and MIS12 are present. Homologs of these two proteins are known to be parts of the KMN (KNL-1/Mis12/Ndc80) complex that constitutes the core microtubule-binding site of the kinetochore in *C.*

elegans (Cheeseman et al., 2006). NDC80 is a constitutive kinetochore protein which localizes at kinetochores in all meiotic and mitotic stages (Du and Dawe, 2007). It does not bind DNA directly and interacts with MIS12, which is also present at kinetochores during all stages of the cell cycle. NCD80 and MIS12 form at metaphase I a bridge structure that links sister kinetochores, while CENH3 and CENPC appear at the inner side of sister kinetochores as two distinct signals (Li and Dawe, 2009). MIS12 has an important role in sister chromatid connection at meiosis I and is required for the initiation of reductional division (Li and Dawe, 2009). Knock-down of MIS12 by RNAi leads to a weakening of the MIS12-NCD80 bridge and aberrant chromosome segregation in meiosis I, where in 30% of the cells sister kinetochores separate and segregate in an equational division instead of reductional (Li and Dawe, 2009). In MIS12 RNAi cells, the signal of the centromere protector ZmSGO1 does not weaken (Li and Dawe, 2009). The protein lies between sister kinetochores but cannot restore kinetochore co-orientation, confirming that Shugoshin is not required for the monopolar orientation of kinetochores (Hamant et al., 2005; Li and Dawe, 2009). A model, in which axial elements and cohesin hold sister chromatids together during prophase I and create the base for fused sister kinetochore formation promoted by the MIS12-NCD80 bridge has been proposed (Li and Dawe, 2009). This structure would cooperate with Shugoshin to induce reductional segregation by co-orienting sister kinetochores (Figure 2). MIS12 and NCD80 are thought to be similar to the monopolin complex, which promotes sister kinetochore co-orientation in budding yeast (Corbett and Harrison, 2012).

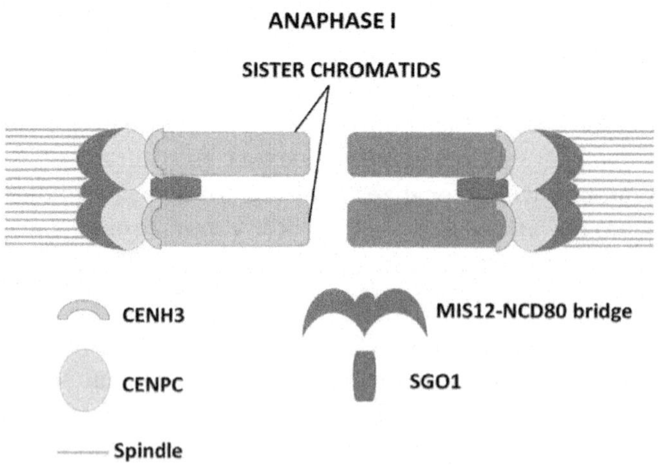

Figure 2. Model proposed in maize by Li and Dawe (2009) for reductional segregation in meiosis I. Sister kinetochores are fused in meiosis I by formation of the MIS12-

NCD80 bridge that, together with SGO1, allows monopolar attachment of sister chromatids to the spindle pole. The inner kinetochore proteins CENPC and CENH3 are visualized as two distinct signals.

Checkpoint Mechanisms Acting at Kinetochores

In eukaryotes, checkpoint mechanisms are present in meiosis and mitosis to prevent chromosome mis-segregation that would result in aneuploidy or apoptosis (Murray, 1994). The SAC is a conserved protein complex that controls proper attachment of microtubules to kinetochores in the metaphase to anaphase transition. In case of lacking or improper kinetochore-microtubule attachment, SAC creates a "wait anaphase" signal that stops anaphase progression. This response is promoted by APC/C together with its co-activator Cdc20 protein (for reviews see Peters, 2006; Vader et al., 2008; Musacchio, 2011). When all kinetochores are properly attached to microtubules, APC/C targets the destruction of Securin, the inhibitor of Separase as well as other cyclins, promoting chromosome segregation and exit from meiosis or mitosis. Evolutionary conserved proteins of SAC are MAD1, MAD2 (mitotic arrest deficient), Bub1, Bub3 (budding unhibited by benomyl), BubR1 kinase (bub-related1, Mad3 in yeast), and Mps1 (Monopolar kinase1) (May and Hardwick, 2006). The SAC proteins BubR1, Bub3, and MAD2 are also members of the Mitotic Checkpoint Complex (MCC), which is the effector of SAC that physically inhibits APC/C by binding to its co-activator Cdc20 until the moment all chromosomes are properly attached to kinetochores (Sudakin et al., 2001). SAC function has been investigated in depth in mitosis. However, a similar control mechanism is active also during meiosis (Malmanche et al., 2006; Sun and Kim, 2012).

Homologs of SAC proteins have been described in plants. MAD2 was first identified in maize where it localizes to the outer kinetochore in prometaphase I and II of meiosis, next to the inner kinetochore protein CENPC (Yu et al., 1999). During meiosis I and II, microtubule attachment is not sufficient for MAD2 dissociation from kinetochores, and the dissociation might occur in response to tension applied to the kinetochores (Yu et al., 1999). This hypothesis is supported by the concomitant staining of the MAD2 and 3F3/2 antibodies in maize meiosis. 3F3/2 recognizes a kinetochore phosphoepitope that is known to disappear in animal cells when tension is applied to the kinetochore (Nicklas et al., 1995). MAD2 homologs have been identified also in wheat and *Arabidopsis*, and their roles have been mainly investigated in mitotic checkpoint control. In wheat, intense MAD2 signal was observed at all centromeres in colchicine treated cells but not in untreated cells, confirming the function of MAD2 in the spindle checkpoint (Kimbara et al., 2004). In

Arabidopsis, MAD2 localization was studied together with BubR1 and Bub3.1, the others SAC proteins identified in the model plant (Caillaud et al., 2009). Interactions between the three proteins were observed in the nuclei of tobacco cells using bimolecular fluorescence complementation (Caillaud et al., 2009). During normal mitosis, localization of the SAC proteins to the kinetochores was not detected. However, by application of microtubule destabilizing drugs or of the proteasome inhibitor MG132, MAD2, BubR1, and Bub3.1 localized at the kinetochores, suggesting that SAC proteins are only recruited at kinetochores in case of defective spindle assembly in *Arabidopsis*(Caillaud et al., 2009). In contrast, a study by Ding et al. (2012) on the *Arabidopsis* MAD2 protein, showed that MAD2-GFP localizes at kinetochores also during normal mitotic progression from prophase to metaphase, as shown in maize. Moreover, AtMAD2 binds to AtMAD1, which interacts with the nucleoporin NUA, showing that SAC components interact with the nuclear pore. This interaction has been found in several other organisms, and it seems that the presence of SAC proteins at the nuclear pore mediates mitotic spindle checkpoint (Lee et al., 2008b).

Mps1 (Monopolar kinase 1) is also required for SAC function in the mitotic checkpoint in several eukaryotes and has been shown to be responsible for the recruitment of Mad1 and Mad2 at kinetochores in humans (Hewitt et al., 2010). Mps1 *Arabidopsis* homolog has conserved motifs which could mediate its interaction with MAD2 but also with cyclins, the APC/C and MAPK (mitogen-activated protein kinases), however, proof of its biological role in the checkpoint mechanism is still required (De Oliveira et al., 2012).

In most organisms, SAC is controlled by the chromosome passenger complex (CPC). In general, the CPC consists of the core enzyme Aurora B kinase, and three non-enzymatic subunits that control the targeting, enzymatic activity and stability of Aurora B: inner centromeric protein (INCENP), borealin and survivin (for review see Ruchaud et al., 2007). The major role of CPC is sensing incorrect kinetochore-microtubule attachments and generating, in response, unattached kinetochores, which allows new rounds of attachment until the correct configuration is obtained. The presence of unattached kinetochores activates the SAC that blocks the progression of cell divisions until all chromosomes are under tension. In plants, little is known about the role of the CPC in meiosis and few components of the complex have been identified. Like animals, *Arabidopsis* possesses three Aurora kinase homologs, which share a similar structure to the ones of other species (Kawabe et al., 2005). AtAurora1 and AtAurora2 display similar localization dynamics to Aurora B kinase in *Arabidopsis* mitosis, suggesting that they could function as chromosomal passenger proteins (Demidov et al., 2005). AtAurora1 interacts with SAC proteins BubR1 and MAD2 *in vivo* and phosphorylate them *in vitro*, which suggests that it functions in checkpoint mechanisms (Demidov D., personal communication). Furthermore, deregulation of AtAurora kinases activity, either by mutagenesis or by chemical treatment, results in defects in microsporogenesis and generation of polyploid and aneuploid progeny, suggesting that AtAurora may regulate correct chromosome segregation in *Arabidopsis* meiosis (Demidov D., personal communication).

A putative ortholog of the CPC subunit INCENP, WYR, has been identified in *Arabidopsis* (Kirioukhova et al., 2011). WYR shares with the INCENP homolog proteins a characteristic C-terminal domain, a coiled coil domain and a IN-box at the C-terminus, required for the binding of Aurora kinase. WYR is an essential gene with a role in cell cycle control and, independently, in cell fate and differentiation in *Arabidopsis*, since is required for both female and male gametogenesis. Similar functions have been reported also for the orthologs of INCENP in animals (Ruchaud et al., 2007). However, further genetic and biochemical analyses on WYR and Aurora kinases are required to establish the role of CPC proteins in plants.

MICROTUBULE ORGANIZATION AND SPINDLE DYNAMICS

In all eukaryotic cells, faithful chromosome segregation is accomplished by microtubule-based movement and requires a bipolar structure, the spindle, which consists of an antiparallel array of microtubules. The microtubules have their minus-end anchored at the spindle pole and their plus-end toward the chromosomes (Wittmann et al., 2001). They are highly dynamic polar polymers of noncovalently bound α and β tubulin heterodimers and represent the major components of the cytoskeleton in eukaryotic cells (Nogales, 2000). They rapidly polymerize and depolymerize while being continually translocated toward the poles. In animal and yeast cells, microtubules nucleate from microtubule-organizing centers (MTOC), such as the centrosome and the spindle pole body, which are responsible for the organization of the cortical astral arrays in interphase and mitotic spindles during cell division (Pereira and Schiebel, 1997; Jaspersen and Winey, 2004). γ-tubulin is enriched at the nucleation centers where it is recruited as a ring-shaped complex together with associated proteins, enhancing the nucleation of microtubules (O›Toole et al., 2012). In contrast to animals and yeast, plant microtubules lack conspicuous organizing centers. However, they are organized into ordered arrays that are associated with a growth pattern of the plant cell and relocate in a cell-cycle specific manner (Azimzadeh et al., 2001). During cell division, a succession of microtubule arrays is identified: radial arrays from the nuclear surface and cortical arrays of interphase, preprophase bands, spindles, and phragmoplasts (Wasteneys, 2002; De Storme and Geelen, 2013a). Like in animal and yeast, γ-tubulin is also required for microtubule nucleation in plants, being essential for the organization of the microtubule structures in interphase and cell division (Canaday et al., 2000; Shimamura et al., 2004; Pastuglia et al., 2006).

Microtubule motor proteins have an essential role in spindle assembly in both centrosomal and acentrosomal systems (Walczak et al., 1998). The best studied class of microtubule motor proteins are kinesins, proteins that participate in a variety of biological processes, including transport of vesicles, chromosomes or organelles, and organization of spindle microtubules, and chromosome segregation (Woehlke and Schliwa, 2000). They move unidirectionally along microtubules toward their plus or minus-ends. They use energy derived from ATP hydrolysis, in a processive or non-processive way, depending on their capacity of moving cargo long or only short distances before detaching from the microtubules. Several kinesins are known to be required for the structure, assembly and positioning of the mitotic and meiotic spindles in animals and fungi (Endow, 1999; Sharp et al., 2000). The *Arabidopsis* genome contains 61 predicted kinesins, one-third of them belonging to the kinesin-14 family that includes minus end-directed motor proteins (Reddy and Day, 2001). ATK1 is a member of this family and has been shown to support microtubule movement in an ATP-dependent manner and to be a non-processive, minus-end motor protein (Marcus et al., 2002). ATK1 has a specific role in male meiosis, in which *atk1-1* meiocytes display defective chromosome alignment and segregation in meiosis I and II due to aberrant formation of metaphase and anaphase spindles, leading to spore and pollen abortion and decreased plant fertility (Chen et al., 2002). ATK1 is involved in the assembly of the meiotic spindle and is needed for organizing microtubules at the two poles at metaphase and anaphase I and II, but not for the organization of microtubules for other structures, such as the interzonal microtubule array formed at telophase I (Chen et al., 2002). Studies in yeast and*Drosophila* have suggested that minus and plus-ended motor proteins could produce counteracting forces within the spindle to maintain its structure (Sharp et al., 1999, 2000). Thus, ATK1 might have a similar function in plant male meiosis, by producing inward-acting forces necessary for the assembly and maintenance of a bipolar spindle (Chen et al., 2002). The creation of a double heterozygote mutant between ATK1 and its homolog ATK5 (also named AtKIN14a and AtKIN14b, respectively), has shown that both proteins are required for proper chromosome segregation in female and male meiosis and for normal spindle morphogenesis in male meiosis (Quan et al., 2008). In addition to its male meiotic function, ATK1 localizes to the midzone of the mitotic spindle from metaphase through anaphase, suggesting a function also in the mitotic spindle apparatus (Liu et al., 1996).

AtPRD2/MPS1 (Multi-polar spindle1) is a putative *Arabidopsis* coiled-coil protein with homologs only among Embryophytes. Although having been identified as AtPRD2, an essential protein for DSBs formation, due to the presence of univalents in *Atprd2* mutant meiosis (De Muyt et al., 2009), the protein has also been found to be required for spindle organization and determination of spindle polarity in male meiosis (MPS1; Jiang et al., 2009). *Mps1*meiocytes display multiple focused spindles at metaphase I, indicating that spindle assembly is not defective, in contrast to *atk1* and *atk1/atk5* mutants, but spindle bipolarity is compromised in meiosis I and II, and chromosome segregation results more affected than in the kinesin mutants. This observations suggest that MPS1, ATK1, and ATK5 play a role in different mechanisms in plant meiosis. It has been proposed that MPS1 might guide microtubule minus-end migration in meiosis, maybe through binding to an unknown MAP (microtubule associated proteins) or, alternatively, could be a component of the spindle pole transmitting the signal to attract the minus-end of the spindle microtubules before spindle assembly (Jiang et al., 2009). However, whether the spindle defects observed in *mps1*meiocytes correspond to a primary function of the protein in spindle organization and polarity, or to a secondary effect caused by univalents formation in meiosis I, is not clear since conflicting observations on the relationship between unpaired chromosomes and spindle aberrations have been reported (Chan and Cande, 1998; Dawe, 1998).

In rice, a Kinesin-1-like protein, Pollen Semisterility 1 (PSS1), has been shown to have microtubule-stimulated ATPase activity and to be required for proper chromosome alignment and segregation in meiosis. However, spindle morphology is only slightly affected in *pss1* mutants, indicating that PSS1 might have a minor and not essential role in the formation of the meiotic spindle or alternatively might be involved in the regulation of chromosome movements along the spindles, as suggested by the delayed chromosomes observed in meiosis in *pss1* (Zhou et al., 2011).

Recently, the identification of a MATH-BTB domain protein, MAB1 (MATH-BTB1) in maize has been reported. This protein is required for organizing microtubule spindles and nuclei positioning in meiosis II and in the first mitotic division in both male and female germlines. Since no direct interaction between MAB1 and the spindles has been observed, it has been proposed that MAB1 may act through the control of a spindle apparatus regulator(s) (Juranič et al., 2012). Six MATH-BTB proteins have been currently identified in the *Arabidopsis* genome, however, no similar function has been reported (Weber and Hellmann, 2009).

The correct orientation of spindles in the second meiotic division is an essential requirement for faithful chromosome segregation. Alterations in the orthogonal configuration of the division planes in meiosis II lead to co-orientation of the spindles producing unreduced gametes, that represent the major route to polyploidization in plants (Brownfield and Köhler, 2011). Co-orientation can lead to the formation of three types of MII spindle defects which usually occur together in cells: parallel, tripolar or fused (Conicella et al., 2003; De Storme and Geelen, 2013b). This phenomenon only takes place in PMCs (pollen mother cells) of plants with simultaneous cytokinesis. In this type of cytokinesis, as opposed to the successive type, no cell plate is formed at the end of meiosis I and the two sets of chromosomes stay in the same cytoplasm and need to be perpendicularly oriented to create the tetrahedral configuration observed at the end of meiosis II (De Storme and Geelen, 2013a). They have been documented in many plant species, however, the molecular mechanisms behind their occurrence are still largely unknown. Two proteins involved in spindle orientation specifically in male meiosis II have been identified in *Arabidopsis*: AtPS1 and JASON (D'Erfurth et al., 2008; Erilova et al., 2009; De Storme and Geelen, 2011). Mutations in these genes produce at the end of meiosis II a high number of unreduced gametes (i.e., dyads and triads) instead of normal haploid gametes, leading to diploid pollen formation and triploid offspring. The biological mechanism causing 2n gamete formation in the mutants has been elucidated by tubulin immunostainings, which have shown the formation of parallel, tripolar and fused spindles in meiosis II. The defective spindles lead to 2n spores that retain parental heterozygosity at the centromeres, indicative of a FDR-type (first division restitution) of meiotic restitution (D'Erfurth et al., 2008; De Storme and Geelen, 2011). The introduction of *Atsp1* or *jason* mutations into the *Atspo11* mutant background has confirmed the model of 2n gametes formation through co-oriented spindles, since the unbalanced segregation caused by *Atspo11* at meiosis I is nullified by parallel spindles in meiosis II, leading to the formation of mainly balanced dyads as result of meiosis in the double mutants.*Atps1* and *jason* meiocytes lack the characteristic interzonal microtubule array (IMA) observed in simultaneous PMCs at telophase I, which physically separates the two new formed nuclei. They mostly show fused nuclei at metaphase II. In potato, the absence of IMA has also been proposed to cause alterations in cell polarity and the formation of fused spindles (Conicella et al., 2003), suggesting that also in the *Arabidopsis* mutants depending on the total, partial, or unipolar loss of IMA fused, parallel or tripolar spindles are formed (De Storme and Geelen, 2013b).

AtPS1 is a protein conserved in the plant kingdom (Cigliano et al., 2011), which contains two conserved domains in its structure: an N-terminal

Forkhead-associated (FHA) domain required for phosphoprotein interaction in many signaling pathways (Li et al., 2000) and a PINc domain that has RNA-binding properties associated with RNAse activity, and which is generally found in proteins involved in RNAi and in nonsense-mediated mRNA decay (NMRD) (Clissold and Ponting, 2000). JASON encodes a protein of unknown function and no known domains that is conserved in plants (Erilova et al., 2009). Expression analysis have demonstrated that *JASON* controls the *AtPS1* transcript level specifically in meiotic flower buds, suggesting the existence of a regulatory mini-network for the control of spindle orientation in meiosis II (De Storme and Geelen, 2011).

Defects in spindle orientation in the second meiotic division have been also reported in mutants in one of the*Arabidopsis* formins, AFH14 (Li et al., 2010). Formins are a class of proteins known to regulate the microfilament cytoskeleton (Blanchoin and Staiger, 2010), but have been recently shown to have also a prominent role in microtubule regulation and in the crosstalk between actin filaments and microtubules in higher eukaryotes (Bartolini and Gundersen, 2010). Indeed, microtubules and microfilaments have been shown to co-distribute and interact in the meiotic spindle and in the phragmoplast in maize (Staiger and Cande, 1991). AFH14 co-localizes with MTs and MFs arrays during cell division in *Arabidopsis* suspension cells and with MTs in meiotic cells, affecting their arrangement during microsporogenesis. *Afh14* mutants display abnormal MTs structures including defective RMS at telophase I, parallel spindles at metaphase II and the absence of phragmoplast structures at late cytokinesis.

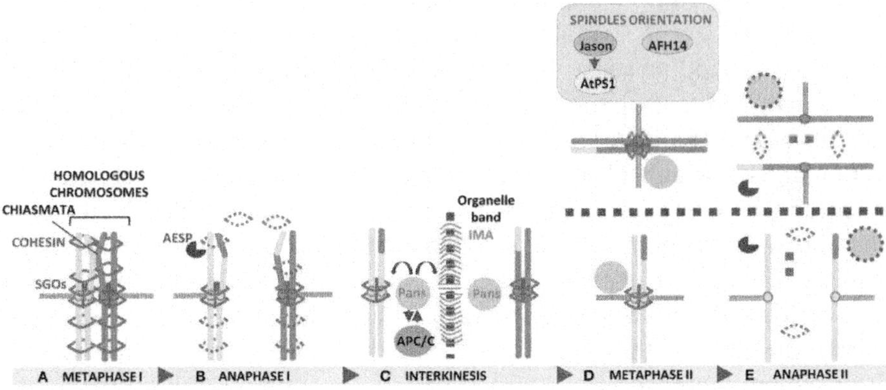

Figure 3. Overview of chromosome segregation in *Arabidopsis* meiosis. (A) At metaphase I, homologous chromosomes are connected by chiasmata and SHUGOSHINs (AtSGOs) are present at the centromeres. (B) At anaphase I, Separase AESP is activated and cleaves the cohesin at chromosome arms but not at centromeres, allowing

resolution of chiasmata and homologous chromosomes segregation by monopolar attachment to the spindles. (C) At interkinesis, an internuclear microtubule array (IMA) is formed at the site of the organelle band, to physically separate homologous chromosomes. PANS1 is active and protects centromere cohesin, probably in conjunction with SGOs. PANS1 also interacts with the APC/C, and it is probably also an APC/C target. In addition, PANS1 plays a role in spindle organization from telophase I to telophase II. (D) At metaphase II, the chromosomes orient perpendicularly to the metaphase plate through the perpendicular orientation of spindles regulated by Jason and AtPS1. Also the formin AFH14 influences spindle orientation by linking MTs and MFs. (E) Releasing or degradation of SGOs and PANS allows cleavage of centromeric cohesin by separase and sister chromatids segregation.

AFH14 has been shown to preferentially bind MTs and to link MTs and MFs *in vitro*, thus playing a key role in cytoskeletal dynamics and organization required for cell division, including MII spindle orientation (Li et al., 2010; De Storme and Geelen, 2013b).

An overview of the process of chromosome segregation between metaphase I and anaphase II, and of the molecular factors playing an essential role in *Arabidopsis* chromosome segregation, is displayed in Figure 3.

CONCLUSIONS AND PERSPECTIVES

In the past 15 years the identification and characterization of plant meiotic genes has seen a remarkable acceleration due to the forward and reverse genetics strategies used in the model plants *Arabidopsis*, maize and rice. In addition, investigation of the molecular mechanisms regulating meiosis in other kingdoms has enormously contributed to the development of plant research in this field. The coordinate events leading to accurate chromosome segregation have been elucidated in budding yeast and studies in plants have confirmed the conserved role of many proteins in the steps of meiotic chromosome segregation, such as cohesin and the dynamics of cohesin removal and protection, the machinery of homologous pairing and recombination, and the function of kinetochores and microtubules. However, even if the main genes have been identified and their function in chromosome segregation confirmed, not much is known about their regulation in accordance with the cell cycle. Further research should focus on investigating the molecular mechanisms regulating protein functions and the interaction between the proteins to define their role in the broader context of chromosome segregation.

CONFLICT OF INTEREST STATEMENT

The authors declare that the research was conducted in the absence of any commercial or financial relationships that could be construed as a potential conflict of interest.

REFERENCES

1. Adams, K. L., and Wendel, J. F. (2005). Polyploidy and genome evolution in plants. *Curr. Opin. Plant Biol.* 8, 135–141. doi: 10.1016/j.pbi.2005.01.001
2. Agashe, B., Prasad, C. K., and Siddiqi, I. (2002). Identification and analysis of DYAD: a gene required for meiotic chromosome organisation and female meiotic progression in Arabidopsis. *Development* 129, 3935–3943.
3. Akhunov, E. D., Goodyear, A. W., Geng, S., Qi, L.-L., Echalier, B., Gill, B. S., et al. (2003). The organization and rate of evolution of wheat genomes are correlated with recombination rates along chromosome arms. *Genome Res.* 13, 753–763. doi: 10.1101/gr.808603
4. Armstrong, S. J., Franklin, F. C. H., and Jones, G. H. (2001). Nucleolus-associated telomere clustering and pairing precede meiotic chromosome synapsis in Arabidopsis thaliana. *J. Cell Sci.* 114, 4207–4217.
5. Azimzadeh, J., Traas, J., and Pastuglia, M. (2001). Molecular aspects of microtubule dynamics in plants. *Curr. Opin. Plant Biol.* 4, 513–519. doi: 10.1016/S1369-5266(00)00209-0
6. Bai, X., Peirson, B. N., Dong, F., Xue, C., and Makaroff, C. A (1999). Isolation and characterization of SYN1, a RAD21-like gene essential for meiosis in Arabidopsis. *Plant Cell* 11, 417–430. doi: 10.1105/tpc.11.3.417
7. Bartolini, F., and Gundersen, G. G. (2010). Formins and microtubules. *Biochim. Biophys. Acta* 1803, 164–173. doi: 10.1016/j.bbamcr.2009.07.006
8. Baskerville, C., Segal, M., and Reed, S. I. (2008). The protease activity of yeast separase (esp1) is required for anaphase spindle elongation independently of its role in cleavage of cohesin. *Genetics* 178, 2361–2372. doi: 10.1534/genetics.107.085308
9. Bass, H. W., Riera-Lizarazu, O., Ananiev, E. V, Bordoli, S. J., Rines, H. W., Phillips, R. L., et al. (2000). Evidence for the coincident initiation of homolog pairing and synapsis during the telomere-clustering (bouquet) stage of meiotic prophase. *J. Cell Sci.* 113, 1033–1042.

10. Berchowitz, L. E., Francis, K. E., Bey, A. L., and Copenhaver, G. P. (2007). The role of AtMUS81 in interference-insensitive crossovers in *A.* thaliana. *PLoS Genet.* 3:e132. doi: 10.1371/journal.pgen.0030132

11. Bhalla, N., and Dernburg, A. F. (2008). Prelude to a division. *Annu. Rev. Cell Dev. Biol.* 24, 397–424. doi: 10.1146/annurev.cellbio.23.090506.123245

12. Bhatt, A. M., Canales, C., and Dickinson, H. G. (2001). Plant meiosis: the means to 1N. *Trends Plant Sci.* 6, 114–121. doi: 10.1016/S1360-1385(00)01861-6

13. Bhatt, A. M., Lister, C., Page, T., Fransz, P., Findlay, K., Jones, G. H., et al. (1999). The DIF1 gene of Arabidopsis is required for meiotic chromosome segregation and belongs to the REC8/RAD21 cohesin gene family. *Plant J.* 19, 463–472. doi: 10.1046/j.1365-313X.1999.00548.x

14. Birchler, J. A., Bhadra, U., Bhadra, M. P., and Auger, D. L. (2001). Dosage-dependent gene regulation in multicellular eukaryotes: implications for dosage compensation, aneuploid syndromes, and quantitative traits. *Dev. Biol.* 234, 275–288. doi: 10.1006/dbio.2001.0262

15. Bishop, D. K., Park, D., Xu, L., and Kleckner, N. (1992). DMC1: a meiosis-specific yeast homolog of E. coli recA required for recombination, synaptonemal complex formation, and cell cycle progression. *Cell* 69, 439–456. doi: 10.1016/0092-8674(92)90446-J

16. Blanchoin, L., and Staiger, C. J. (2010). Plant formins: diverse isoforms and unique molecular mechanism. *Biochim. Biophys. Acta* 1803, 201–206. doi: 10.1016/j.bbamcr.2008.09.015

17. Boateng, K. A., Yang, X., Dong, F., Owen, H. A., and Makaroff, C. A. (2008). SWI1 is required for meiotic chromosome remodeling events. *Mol. Plant* 1, 620–633. doi: 10.1093/mp/ssn030

18. Bolaños-Villegas, P., Yang, X., Wang, H.-J., Juan, C.-T., Chuang, M.-H., Makaroff, C. A., et al. (2013). Arabidopsis CHROMOSOME TRANSMISSION FIDELITY 7 (AtCTF7/ECO1) is required for DNA repair, mitosis and meiosis. *Plant J.* 75, 927–940. doi: 10.1111/tpj.12261

19. Borde, V. (2007). The multiple roles of the Mre11 complex for meiotic recombination. *Chromosome Res.* 15, 551–563. doi: 10.1007/s10577-007-1147-9

20. Börner, G. V., Kleckner, N., and Hunter, N. (2004). Crossover/noncrossover differentiation, synaptonemal complex formation, and regulatory surveillance at the leptotene/zygotene transition of meiosis. *Cell* 117, 29–45. doi: 10.1016/S0092-8674(04)00292-2

21. Bozza, C. G., and Pawlowski, W. P. (2008). The cytogenetics of homologous chromosome pairing in meiosis in plants. *Cytogenet. Genome Res*. 120, 313–319. doi: 10.1159/000121080

22. Brar, G. A., and Amon, A. (2009). Emerging roles for centromeres in meiosis I chromosome segregation. *Nat. Rev. Genet*. 9, 899–910. doi: 10.1038/nrg2454

23. Brownfield, L., and Köhler, C. (2011). Unreduced gamete formation in plants: mechanisms and prospects. *J. Exp. Bot*. 62, 1659–1668. doi: 10.1093/jxb/erq371

24. Cai, X., Dong, F., Edelmann, R. E., and Makaroff, C. A. (2003). The Arabidopsis SYN1 cohesin protein is required for sister chromatid arm cohesion and homologous chromosome pairing. *J. Cell Sci*. 116, 2999–3007. doi: 10.1242/jcs.00601

25. Caillaud, M.-C., Paganelli, L., Lecomte, P., Deslandes, L., Quentin, M., Pecrix, Y., et al. (2009). Spindle assembly checkpoint protein dynamics reveal conserved and unsuspected roles in plant cell division. *PLoS ONE* 4:e6757. doi: 10.1371/journal.pone.0006757

26. Canaday, J., Stoppin-Mellet, V., Mutterer, J., Lambert, A. M., and Schmit, A. C. (2000). Higher plant cells: gamma-tubulin and microtubule nucleation in the absence of centrosomes. *Microsc. Res. Tech*. 49, 487–495. doi: 10.1002/(SICI)1097-0029(20000601)49:5<487::AID-JEMT11>3.0.CO;2-I

27. Chambon, J.-P., Touati, S. A., Berneau, S., Cladière, D., Hebras, C., Groeme, R., et al. (2013). The PP2A inhibitor I2PP2A is essential for sister chromatid segregation in oocyte meiosis II. *Curr. Biol*. 23, 485–490. doi: 10.1016/j.cub.2013.02.004

28. Chan, A., and Cande, W. Z. (1998). Maize meiotic spindles assemble around chromatin and do not require paired chromosomes. *J. Cell Sci*. 111, 3507–3515.

29. Che, L., Tang, D., Wang, K., Wang, M., Zhu, K., Yu, H., et al. (2011). OsAM1 is required for leptotene-zygotene transition in rice. *Cell Res*. 21, 654–665. doi: 10.1038/cr.2011.7

30. Cheeseman, I. M., Chappie, J. S., Wilson-Kubalek, E. M., and Desai, A. (2006). The conserved KMN network constitutes the core microtubule-binding site of the kinetochore. *Cell* 127, 983–997. doi: 10.1016/j.cell.2006.09.039

31. Chelysheva, L., Diallo, S., Vezon, D., Gendrot, G., Vrielynck, N., Belcram, K., et al. (2005). AtREC8 and AtSCC3 are essential to the

monopolar orientation of the kinetochores during meiosis. *J. Cell Sci.* 118, 4621–4632. doi: 10.1242/jcs.02583

32. Chelysheva, L., Gendrot, G., Vezon, D., Doutriaux, M.-P., Mercier, R., and Grelon, M. (2007). Zip4/Spo22 is required for class I CO formation but not for synapsis completion in Arabidopsis thaliana. *PLoS Genet.* 3:e83. doi: 10.1371/journal.pgen.0030083
33. Chen, C., Marcus, A., Li, W., Hu, Y., Calzada, J.-P. V., Grossniklaus, U., et al. (2002). The Arabidopsis ATK1 gene is required for spindle morphogenesis in male meiosis. *Development* 129, 2401–2409.
34. Chen, C., Zhang, W., Timofejeva, L., Gerardin, Y., and Ma, H. (2005). The Arabidopsis ROCK-N-ROLLERS gene encodes a homolog of the yeast ATP-dependent DNA helicase MER3 and is required for normal meiotic crossover formation. *Plant J.* 43, 321–334. doi: 10.1111/j.1365-313X.2005.02461.x
35. Chikashige, Y., Haraguchi, T., and Hiraoka, Y. (2007). Another way to move chromosomes. *Chromosoma* 116, 497–505. doi: 10.1007/s00412-007-0114-8
36. Cigliano, R. A., Sanseverino, W., Cremona, G., Consiglio, F. M., and Conicella, C. (2011). Evolution of parallel spindles like genes in plants and highlight of unique domain architecture. *BMC Evol. Biol.* 11:78. doi: 10.1186/1471-2148-11-78
37. Clift, D., and Marston, A. L. (2011). The role of shugoshin in meiotic chromosome segregation. *Cytogenet. Genome Res.* 133, 234–242. doi: 10.1159/000323793
38. Clissold, P. M., and Ponting, C. P. (2000). PIN domains in nonsense-mediated mRNA decay and RNAi. *Curr. Biol.* 10, 888–890. doi: 10.1016/S0960-9822(00)00858-7
39. Comai, L. (2005). The advantages and disadvantages of being polyploid. *Nat. Rev. Genet.* 6, 836–846. doi: 10.1038/nrg1711
40. Conicella, C., Capo, A., Cammareri, M., Errico, A., Shamina, N., and Monti, L. M. (2003). Elucidation of meiotic nuclear restitution mechanisms in potato through analysis of microtubular cytoskeleton. *Euphytica*, 107–115. doi: 10.1023/A:1025636321757
41. Cooper, K. F., and Strich, R. (2011). Meiotic control of the APC/C: similarities & differences from mitosis. *Cell Div.* 6:16. doi: 10.1186/1747-1028-6-16
42. Corbett, K. D., and Harrison, S. C. (2012). Molecular Architecture of the Yeast Monopolin Complex. *Cell. Rep.* 1, 583–589. doi: 10.1016/j.celrep.2012.05.012

43. Couteau, F., Belzile, F., Horlow, C., Grandjean, O., Vezon, D., and Doutriaux, M. P. (1999). Random chromosome segregation without meiotic arrest in both male and female meiocytes of a dmc1 mutant of Arabidopsis. *Plant Cell* 11, 1623–1634. doi: 10.1105/tpc.11.9.1623

44. Crismani, W., and Mercier, R. (2012). What limits meiotic crossovers? *Cell Cycle* 11, 3527–3528. doi: 10.4161/cc.21963

45. Cromer, L., Jolivet, S., Horlow, C., Chelysheva, L., Heyman, J., De Jaeger, G., et al. (2013). Centromeric cohesion is protected twice at meiosis, by SHUGOSHINs at anaphase I and by PATRONUS at interkinesis. *Curr. Biol.* 23, 2090–2099. doi: 10.1016/j.cub.2013.08.036

46. Da Ines, O., Gallego, M. E., and White, C. I. (2014). Recombination-independent mechanisms and pairing of homologous chromosomes during meiosis in plants. *Mol. Plant* 7, 492–501. doi: 10.1093/mp/sst172

47. Dawe, R. K. (1998). Meiotic chromosome organization and segregation in plants. *Annu. Rev. Plant Physiol. Plant Mol. Biol.* 49, 371–395. doi: 10.1146/annurev.arplant.49.1.371

48. Dawe, R. K., Reed, L. M., Yu, H.-G., Muszynski, G., and Hiatt, E. N. (1999). A maize homolog of mammalian CENPC is a constitutive component of the inner kinetochore. *Plant Cell* 11, 1227–1238. doi: 10.1105/tpc.11.7.1227

49. Demidov, D., Van Damme, D., Geelen, D., Blattner, F. R., and Houben, A. (2005). Identification and dynamics of two classes of aurora-like kinases in arabidopsis and other plants. *Plant* 17, 836–848. doi: 10.1105/tpc.104.029710.1

50. De Muyt, A., Pereira, L., Vezon, D., Chelysheva, L., Gendrot, G., Chambon, A., et al. (2009). A high throughput genetic screen identifies new early meiotic recombination functions in Arabidopsis thaliana. *PLoS Genet.* 5:e1000654. doi: 10.1371/journal.pgen.1000654

51. De Oliveira, E. A. G., Romeiro, N. C., Ribeiro, E. D. S., Santa-Catarina, C., Oliveira, A. E. A., Silveira, V., et al. (2012). Structural and functional characterization of the protein kinase Mps1 in Arabidopsis thaliana. *PLoS ONE* 7:e45707. doi: 10.1371/journal.pone.0045707

52. D'Erfurth, I., Jolivet, S., Froger, N., Catrice, O., Novatchkova, M., Simon, M., et al. (2008). Mutations in AtPS1 (Arabidopsis thaliana parallel spindle 1) lead to the production of diploid pollen grains. *PLoS Genet.* 4:e1000274. doi: 10.1371/journal.pgen.1000274

53. De Rop, V., Padeganeh, A., and Maddox, P. S. (2012). CENP-A: the key player behind centromere identity, propagation, and kinetochore assembly.*Chromosoma* 121, 527–538. doi: 10.1007/s00412-012-0386-5

54. De Storme, N., and Geelen, D. (2011). The Arabidopsis mutant jason produces unreduced first division restitution male gametes through a parallel/fused spindle mechanism in meiosis II. *Plant Physiol.* 155, 1403–1415. doi: 10.1104/pp.110.170415

55. De Storme, N., and Geelen, D. (2013a). Cytokinesis in plant male meiosis. *Plant Signal. Behav.* 8:e23394. doi: 10.4161/psb.23394

56. De Storme, N., and Geelen, D. (2013b). Sexual polyploidization in plants—cytological mechanisms and molecular regulation. *New Phytol.* 198, 670–684. doi: 10.1111/nph.12184

57. De Storme, N., and Geelen, D. (2014). The impact of environmental stress on male reproductive development in plants: biological processes and molecular mechanisms. *Plant. Cell Environ.* 37, 1–18. doi: 10.1111/pce.12142

58. Ding, D., Muthuswamy, S., and Meier, I. (2012). Functional interaction between the Arabidopsis orthologs of spindle assembly checkpoint proteins MAD1 and MAD2 and the nucleoporin NUA. *Plant Mol. Biol.* 79, 203–216. doi: 10.1007/s11103-012-9903-4

59. Drouaud, J., Camilleri, C., Bourguignon, P.-Y., Canaguier, A., Bérard, A., Vezon, D., et al. (2006). Variation in crossing-over rates across chromosome 4 of Arabidopsis thaliana reveals the presence of meiotic recombination "hot spots." *Genome Res.* 16, 106–114. doi: 10.1101/gr.4319006

60. Du, Y., and Dawe, R. K. (2007). Maize NDC80 is a constitutive feature of the central kinetochore. *Chromosome Res.* 15, 767–775. doi: 10.1007/s10577-007-1160-z

61. Eckert, C. A., Gravdahl, D. J., and Megee, P. C. (2007). The enhancement of pericentromeric cohesin association by conserved kinetochore components promotes high-fidelity chromosome segregation and is sensitive to microtubule-based tension. *Genes Dev.* 21, 278–291. doi: 10.1101/gad.1498707.somes

62. Edlinger, B., and Schlögelhofer, P. (2011). Have a break: determinants of meiotic DNA double strand break (DSB) formation and processing in plants. *J. Exp. Bot.* 62, 1545–1563. doi: 10.1093/jxb/erq421

63. Eijpe, M., Heyting, C., Gross, B., and Jessberger, R. (2000). Association of mammalian SMC1 and SMC3 proteins with meiotic chromosomes and synaptonemal complexes. *J. Cell Sci.* 113(Pt 4), 673–682.

64. Ekwall, K. (2007). Epigenetic control of centromere behavior. *Annu. Rev. Genet.* 41, 63–81. doi: 10.1146/annurev.genet.41.110306.130127

65. Endow, S. A. (1999). Microtubule motors in spindle and chromosome motility. *Eur. J. Biochem.* 262, 12–18. doi: 10.1046/j.1432-1327.1999.00339.x

66. Erilova, A., Brownfield, L., Exner, V., Rosa, M., Twell, D., Mittelsten Scheid, O., et al. (2009). Imprinting of the polycomb group gene MEDEA serves as a ploidy sensor in Arabidopsis. *PLoS Genet.* 5:e1000663. doi: 10.1371/journal.pgen.1000663

67. Fernius, J., Nerusheva, O. O., Galander, S., Alves, F. D. L., Rappsilber, J., and Marston, A. L. (2013). Cohesin-dependent association of scc2/4 with the centromere initiates pericentromeric cohesion establishment. *Curr. Biol.* 23, 599–606. doi: 10.1016/j.cub.2013.02.022

68. Golubovskaya, I. N., Hamant, O., Timofejeva, L., Wang, C.-J. R., Braun, D., Meeley, R., et al. (2006). Alleles of afd1 dissect REC8 functions during meiotic prophase I. *J. Cell Sci.* 119, 3306–3315. doi: 10.1242/jcs.03054

69. Golubovskaya, I. N., Harper, L. C., Pawlowski, W. P., Schichnes, D., and Cande, W. Z. (2002). The pam1 gene is required for meiotic bouquet formation and efficient. *Genetics* 1993, 1979–1993.

70. Gómez, R., Valdeolmillos, A., Parra, M. T., Viera, A., Carreiro, C., Roncal, F., et al. (2007). Mammalian SGO2 appears at the inner centromere domain and redistributes depending on tension across centromeres during meiosis II and mitosis. *EMBO Rep.* 8, 173–180. doi: 10.1038/sj.embor.7400877

71. Gore, M. A., Chia, J. M., Elshire, R. J., Sun, Q., Ersoz, E. S., Hurwitz, B. L., et al. (2009). A first-generation haplotype map of maize. *Science* 326, 1115–1117. doi: 10.1126/science.1177837

72. Graumann, K., Runions, J., and Evans, D. E. (2010). Characterization of SUN-domain proteins at the higher plant nuclear envelope. *Plant J.* 61, 134–144. doi: 10.1111/j.1365-313X.2009.04038.x

73. Gregson, H. C., Schmiesing, J. A., Kim, J. S., Kobayashi, T., Zhou, S., and Yokomori, K. (2001). A potential role for human cohesin in mitotic spindle aster assembly. *J. Biol. Chem.* 276, 47575–47582. doi: 10.1074/jbc.M103364200

74. Gruber, S., Haering, C. H., and Nasmyth, K. (2003). Chromosomal cohesin forms a ring. *Cell* 112, 765–777. doi: 10.1016/S0092-8674(03)00162-4

75. Gutiérrez-Caballero, C., Cebollero, L. R., and Pendás, A. M. (2012). Shugoshins: from protectors of cohesion to versatile adaptors at the centromere.*Trends Genet.* 28, 351–360. doi: 10.1016/j.tig.2012.03.003

76. Haering, C. H., Farcas, A.-M., Arumugam, P., Metson, J., and Nasmyth, K. (2008). The cohesin ring concatenates sister DNA molecules. *Nature* 454, 297–301. doi: 10.1038/nature07098

77. Hamant, O., Golubovskaya, I., Meeley, R., Fiume, E., Timofejeva, L., Schleiffer, A., et al. (2005). A REC8-dependent plant Shugoshin is required for maintenance of centromeric cohesion during meiosis and has no mitotic functions. *Curr. Biol.* 15, 948–954. doi: 10.1016/j.cub.2005.04.049

78. Hamant, O., Ma, H., and Cande, W. Z. (2006). Genetics of meiotic prophase I in plants. *Annu. Rev. Plant Biol.* 57, 267–302. doi: 10.1146/annurev.arplant.57.032905.105255

79. Harper, J. W., Burton, J. L., and Solomon, M. J. (2002). The anaphase-promoting complex: it's not just for mitosis any more. *Genes Dev.* 16, 2179–2206. doi: 10.1101/gad.1013102

80. Harper, L., Golubovskaya, I., and Cande, W. Z. (2004). A bouquet of chromosomes. *J. Cell Sci.* 117, 4025–4032. doi: 10.1242/jcs.01363

81. Hayashi, T., Fujita, Y., Iwasaki, O., Adachi, Y., Takahashi, K., and Yanagida, M. (2004). Mis16 and Mis18 are required for CENP-A loading and histone deacetylation at centromeres. *Cell* 118, 715–729. doi: 10.1016/j.cell.2004.09.002

82. Henry, I. M., Dilkes, B. P., Miller, E. S., Burkart-Waco, D., and Comai, L. (2010). Phenotypic consequences of aneuploidy in Arabidopsis thaliana. *Genetics* 186, 1231–1245. doi: 10.1534/genetics.110.121079

83. Henry, I. M., Dilkes, B. P., Young, K., Watson, B., Wu, H., and Comai, L. (2005). Aneuploidy and genetic variation in the Arabidopsis thaliana triploid response. *Genetics* 170, 1979–1988. doi: 10.1534/genetics.104.037788

84. Hewitt, L., Tighe, A., Santaguida, S., White, A. M., Jones, C. D., Musacchio, A., et al. (2010). Sustained Mps1 activity is required in mitosis to recruit O-Mad2 to the Mad1-C-Mad2 core complex. *J. Cell Biol.* 190, 25–34. doi: 10.1083/jcb.201002133

85. Higgins, J. D., Armstrong, S. J., Franklin, F. C. H., and Jones, G. H. (2004). The Arabidopsis MutS homolog AtMSH4 functions at an early step in recombination: evidence for two classes of recombination in Arabidopsis. *Genes Dev.* 18, 2557–2570. doi: 10.1101/gad.317504

86. Higgins, J. D., Sanchez-Moran, E., Armstrong, S. J., Jones, G. H., and Franklin, F. C. H. (2005). The Arabidopsis synaptonemal complex protein ZYP1 is required for chromosome synapsis and normal fidelity of crossing over. *Genes Dev.* 19, 2488–2500. doi: 10.1101/gad.354705

87. Higgins, J. D., Vignard, J., Mercier, R., Pugh, A. G., Franklin, F. C. H., and Jones, G. H. (2008). AtMSH5 partners AtMSH4 in the class I meiotic crossover pathway in Arabidopsis thaliana, but is not required for synapsis. *Plant J.* 55, 28–39. doi: 10.1111/j.1365-313X.2008.03470.x

88. Houben, A., and Schubert, I. (2003). DNA and proteins of plant centromeres. *Curr. Opin. Plant Biol.* 6, 554–560. doi: 10.1016/j.pbi.2003.09.007

89. Huang, H., Feng, J., Famulski, J., Rattner, J. B., Liu, S. T., Kao, G. D., et al. (2007). Tripin/hSgo2 recruits MCAK to the inner centromere to correct defective kinetochore attachments. *J. Cell Biol.* 177, 413–424. doi: 10.1083/jcb.200701122

90. Jaspersen, S. L., and Winey, M. (2004). The budding yeast spindle pole body: structure, duplication, and function. *Annu. Rev. Cell Dev. Biol.* 20, 1–28. doi: 10.1146/annurev.cellbio.20.022003.114106

91. Jensen, S., Segal, M., Clarke, D. J., and Reed, S. I. (2001). A novel role of the budding yeast separin Esp1 in anaphase spindle elongation: evidence that proper spindle association of Esp1 is regulated by Pds1. *J. Cell Biol.* 152, 27–40. doi: 10.1083/jcb.152.1.27

92. Jiang, H., Wang, F.-F., Wu, Y.-T., Zhou, X., Huang, X.-Y., Zhu, J., et al. (2009). MULTIPOLAR SPINDLE 1 (MPS1), a novel coiled-coil protein of Arabidopsis thaliana, is required for meiotic spindle organization. *Plant J.* 59, 1001–1010. doi: 10.1111/j.1365-313X.2009.03929.x

93. Jiang, L., Yuan, L., Xia, M., and Makaroff, C. A. (2010). Proper levels of the Arabidopsis cohesion establishment factor CTF7 are essential for embryo and megagametophyte, but not endosperm, development. *Plant Physiol.* 154, 820–832. doi: 10.1104/pp.110.157560

94. Jones, G. H. (1984). The control of chiasma distribution. *Symp. Soc. Exp. Biol.* 38, 293–320.

95. Juranič, M., Srilunchang, K., Krohn, N. G., Leljak-Levanic, D., Sprunck, S., and Dresselhaus, T. (2012). Germline-specific MATH-BTB substrate adaptor MAB1 regulates spindle length and nuclei identity in maize. *Plant Cell* 24, 4974–4991. doi: 10.1105/tpc.112.107169

96. Kawabe, A., Matsunaga, S., Nakagawa, K., Kurihara, D., Yoneda, A., Hasezawa, S., et al. (2005). Characterization of plant Aurora kinases during mitosis.*Plant Mol. Biol.* 58, 1–13. doi: 10.1007/s11103-005-3454-x

97. Kawashima, S. A., Tsukahara, T., Langegger, M., Hauf, S., Kitajima, T. S., and Watanabe, Y. (2007). Shugoshin enables tension-generating

attachment of kinetochores by loading Aurora to centromeres. *Genes Dev.* 21, 420–435. doi: 10.1101/gad.1497307

98. Keeney, S., Giroux, C. N., and Kleckner, N. (1997). Meiosis-specific DNA double-strand breaks are catalyzed by Spo11, a member of a widely conserved protein family. *Cell* 88, 375–384. doi: 10.1016/S0092-8674(00)81876-0
99. Kerrebrock, A. W., Moore, D. P., Wu, J. S., and Orr-Weaver, T. L. (1995). Mei-S332, a Drosophila protein required for sister-chromatid cohesion, can localize to meiotic centromere regions. *Cell* 83, 247–256. doi: 10.1016/0092-8674(95)90166-3
100. Kimbara, J., Endo, T. R., and Nasuda, S. (2004). Characterization of the genes encoding for MAD2 homologues in wheat. *Chromosome Res.* 12, 703–714. doi: 10.1023/B:CHRO.0000045760.42880.8c
101. Kirioukhova, O., Johnston, A. J., Kleen, D., Kägi, C., Baskar, R., Moore, J. M., et al. (2011). Female gametophytic cell specification and seed development require the function of the putative Arabidopsis INCENP ortholog WYRD. *Development* 138, 3409–3420. doi: 10.1242/dev.060384
102. Kitajima, T. S., Kawashima, S. A., and Watanabe, Y. (2004). The conserved kinetochore protein shugoshin protects centromeric cohesion during meiosis.*Nature* 427, 510–517. doi: 10.1038/nature02312
103. Kitajima, T. S., Miyazaki, Y., Yamamoto, M., and Watanabe, Y. (2003). Rec8 cleavage by separase is required for meiotic nuclear divisions in fission yeast.*EMBO J.* 22, 5643–5653. doi: 10.1093/emboj/cdg527
104. Klein, F., Mahr, P., Galova, M., Buonomo, S. B., Michaelis, C., Nairz, K., et al. (1999). A central role for cohesins in sister chromatid cohesion, formation of axial elements, and recombination during yeast meiosis. *Cell* 98, 91–103. doi: 10.1016/S0092-8674(00)80609-1
105. Klutstein, M., and Cooper, J. P. (2014). The chromosomal courtship dance-homolog pairing in early meiosis. *Curr. Opin. Cell Biol.* 26, 123–131. doi: 10.1016/j.ceb.2013.12.004
106. Koszul, R., Kim, K. P., Prentiss, M., Kleckner, N., and Kameoka, S. (2009). Actin-mediated motion of meiotic chromosomes. *NIH* 133, 1188–1201. doi: 10.1016/j.cell.2008.04.050
107. Kudo, N. R., Anger, M., Peters, A. H., Stemmann, O., Theussl, H.-C., Helmhart, W., et al. (2009). Role of cleavage by separase of the Rec8 kleisin subunit of cohesin during mammalian meiosis I. *J. Cell Sci.* 122, 2686–2698. doi: 10.1242/jcs.035287

108. Kurzbauer, M.-T., Uanschou, C., Chen, D., and Schlögelhofer, P. (2012). The recombinases DMC1 and RAD51 are functionally and spatially separated during meiosis in Arabidopsis. *Plant Cell* 24, 2058–2070. doi: 10.1105/tpc.112.098459

109. Lam, W. S., Yang, X., and Makaroff, C. A. (2005). Characterization of Arabidopsis thaliana SMC1 and SMC3: evidence that AtSMC3 may function beyond chromosome cohesion. *J. Cell Sci.* 118, 3037–3048. doi: 10.1242/jcs.02443

110. Lampert, F., and Westermann, S. (2011). A blueprint for kinetochores—new insights into the molecular mechanics of cell division. *Nat. Rev. Mol. Cell. Biol.* 12, 407–412. doi: 10.1038/nrm3133

111. Lee, J., Kitajima, T. S., Tanno, Y., Yoshida, K., Morita, T., Miyano, T., et al. (2008a). Unified mode of centromeric protection by shugoshin in mammalian oocytes and somatic cells. *Nat. Cell Biol.* 10, 42–52. doi: 10.1038/ncb1667

112. Lee, S. H., Sterling, H., Burlingame, A., and McCormick, F. (2008b). Tpr directly binds to Mad1 and Mad2 and is important for the Mad1-Mad2-mediated mitotic spindle checkpoint. *Genes Dev.* 22, 2926–2931. doi: 10.1101/gad.1677208

113. Lermontova, I., Koroleva, O., Rutten, T., Fuchs, J., Schubert, V., Moraes, I., et al. (2011). Knockdown of CENH3 in Arabidopsis reduces mitotic divisions and causes sterility by disturbed meiotic chromosome segregation. *Plant J.* 68, 40–50. doi: 10.1111/j.1365-313X.2011.04664.x

114. Lermontova, I., Kuhlmann, M., Friedel, S., Rutten, T., Heckmann, S., Sandmann, M., et al. (2013). Arabidopsis KINETOCHORE NULL2 is an upstream component for centromeric histone H3 variant cenH3 deposition at centromeres. *Plant Cell* 25, 3389–3404. doi: 10.1105/tpc.113.114736

115. Lermontova, I., Schubert, V., Fuchs, J., Klatte, S., Macas, J., and Schubert, I. (2006). Loading of Arabidopsis centromeric histone CENH3 occurs mainly during G2 and requires the presence of the histone fold domain. *Plant Cell* 18, 2443–2451. doi: 10.1105/tpc.106.043174

116. Lhuissier, F. G. P., Offenberg, H. H., Wittich, P. E., Vischer, N. O. E., and Heyting, C. (2007). The mismatch repair protein MLH1 marks a subset of strongly interfering crossovers in tomato. *Plant Cell* 19, 862–876. doi: 10.1105/tpc.106.049106

117. Li, J., Lee, G. I., Van Doren, S. R., and Walker, J. C. (2000). The FHA domain mediates phosphoprotein interactions. *J. Cell Sci.* 113(Pt 23), 4143–4149.

118. Li, W., Chen, C., Markmann-Mulisch, U., Timofejeva, L., Schmelzer, E., Ma, H., et al. (2004). The Arabidopsis AtRAD51 gene is dispensable for vegetative development but required for meiosis. *Proc. Natl. Acad. Sci. U.S.A.* 101, 10596–10601. doi: 10.1073/pnas.0404110101

119. Li, X., and Dawe, R. K. (2009). Fused sister kinetochores initiate the reductional division in meiosis I. *Nat. Cell Biol.* 11, 1103–1108. doi: 10.1038/ncb1923

120. Li, Y., Shen, Y., Cai, C., Zhong, C., Zhu, L., Yuan, M., et al. (2010). The type II Arabidopsis formin14 interacts with microtubules and microfilaments to regulate cell division. *Plant Cell* 22, 2710–2726. doi: 10.1105/tpc.110.075507

121. Liu, B., Cyr, R. J., and Palevitz, B. A. (1996). A kinesin-like protein, KatAp, in the cells of arabidopsis and other plants. *Plant Cell* 8, 119–132. doi: 10.1105/tpc.8.1.119

122. Liu, C. M., McElver, J., Tzafrir, I., Joosen, R., Wittich, P., Patton, D., et al. (2002). Condensin and cohesin knockouts in Arabidopsis exhibit a titan seed phenotype. *Plant J.* 29, 405–415. doi: 10.1046/j.1365-313x.2002.01224.x

123. Liu, Z., and Makaroff, C. A. (2006). Arabidopsis separase AESP is essential for embryo development and the release of cohesin during meiosis. *Plant Cell*18, 1213–1225. doi: 10.1105/tpc.105.036913.2001

124. Ma, H. (2006). A molecular portrait of Arabidopsis meiosis. *Arabidopsis Book* 4:e0095. doi: 10.1199/tab.0095

125. Madlung, A., and Wendel, J. F. (2013). Genetic and epigenetic aspects of polyploid evolution in plants. *Cytogenet. Genome Res.* 140, 270–285. doi: 10.1159/000351430

126. Malmanche, N., Maia, A., and Sunkel, C. E. (2006). The spindle assembly checkpoint: preventing chromosome mis-segregation during mitosis and meiosis. *FEBS Lett.* 580, 2888–2895. doi: 10.1016/j.febslet.2006.03.081

127. Marcus, A. I., Ambrose, J. C., Blickley, L., Hancock, W. O., and Cyr, R. J. (2002). Arabidopsis thaliana protein, ATK1, is a minus-end directed kinesin that exhibits non-processive movement. *Cell Motil. Cytoskeleton* 52, 144–150. doi: 10.1002/cm.10045

128. Marston, A. L. (2014). Chromosome segregation in budding yeast: sister chromatid cohesion and related Mechanisms. *Genetics* 196, 31–63. doi: 10.1534/genetics.112.145144

129. Masson, J. Y., and West, S. C. (2001). The Rad51 and Dmc1 recombinases: a non-identical twin relationship. *Trends Biochem. Sci.* 26, 131–136. doi: 10.1016/S0968-0004(00)01742-4

130. May, K. M., and Hardwick, K. G. (2006). The spindle checkpoint. *J. Cell Sci.* 119, 4139–4142. doi: 10.1242/jcs.03165

131. McGuinness, B. E., Hirota, T., Kudo, N. R., Peters, J.-M., and Nasmyth, K. (2005). Shugoshin prevents dissociation of cohesin from centromeres during mitosis in vertebrate cells. *PLoS Biol.* 3:e86. doi: 10.1371/journal.pbio.0030086

132. Mercier, R., Armstrong, S. J., Horlow, C., Kackson, N. P., Makaroff, C. A., Vezon, D., et al. (2003). The meiotic protein SWI1 is required for axial element formation and recombination initiation in Arabidopsis. *Development* 130, 3309–3318. doi: 10.1242/dev.00550

133. Mercier, R., and Grelon, M. (2008). Meiosis in plants: ten years of gene discovery. *Cytogenet. Genome Res.* 120, 281–290. doi: 10.1159/000121077

134. Mercier, R., Jolivet, S., Vezon, D., Huppe, E., Chelysheva, L., Giovanni, M., et al. (2005). Two meiotic crossover classes cohabit in Arabidopsis: one is dependent on MER3,whereas the other one is not. *Curr. Biol.* 15, 692–701. doi: 10.1016/j.cub.2005.02.056

135. Mercier, R., Vezon, D., Bullier, E., Motamayor, J. C., Sellier, A., Lefèvre, F., et al. (2001). SWITCH1 (SWI1): a novel protein required for the establishment of sister chromatid cohesion and for bivalent formation at meiosis. *Genes Dev.* 15, 1859–1871. doi: 10.1101/gad.203201

136. Miki, F., Kurabayashi, A., Tange, Y., Okazaki, K., Shimanuki, M., and Niwa, O. (2004). Two-hybrid search for proteins that interact with Sad1 and Kms1, two membrane-bound components of the spindle pole body in fission yeast. *Mol. Genet. Genomics* 270, 449–471. doi: 10.1007/s00438-003-0938-8

137. Moraes, I. C. R., Lermontova, I., and Schubert, I. (2011). Recognition of *A.* thaliana centromeres by heterologous CENH3 requires high similarity to the endogenous protein. *Plant Mol. Biol.* 75, 253–261. doi: 10.1007/s11103-010-9723-3

138. Moschou, P. N., and Bozhkov, P. V (2012). Separases: biochemistry and function. *Physiol. Plant.* 145, 67–76. doi: 10.1111/j.1399-3054.2011.01550.x

139. Motamayor, J. C., Vezon, D., Bajon, C., Sauvanet, A., Grandjean, O., Marchand, M., et al. (2000). Switch (swi1), an Arabidopsis thaliana mutant affected in the female meiotic switch. *Sex. Plant Reprod.* 12, 209–218. doi: 10.1007/s004970050002

140. Murata, M. (2013). "Arabidopsis centromeres," in *Plant Centromere Biology*, eds J. Jiang and J. A. Birchler (Oxford, UK: Wiley-Blackwell). doi: 10.1002/9781118525715.ch1

141. Murphy, S. P., Simmons, C. R., and Bass, H. W. (2010). Structure and expression of the maize (Zea mays L.) SUN-domain protein gene family: evidence for the existence of two divergent classes of SUN proteins in plants. *BMC Plant Biol.* 10:269. doi: 10.1186/1471-2229-10-269

142. Murray, A. (1994). Cell cycle checkpoints. *Curr. Opin. Cell Biol.* 6, 872–876. doi: 10.1016/0955-0674(94)90059-0

143. Musacchio, A. (2011). Spindle assembly checkpoint: the third decade. *Philos. Trans. R. Soc. Lond. B. Biol. Sci.* 366, 3595–604. doi: 10.1098/rstb.2011.0072

144. Nasmyth, K. (2001). Disseminating the genome: joining, resolving, and separating sister chromatids during mitosis and meiosis. *Annu. Rev. Genet.* 35, 673–745. doi: 10.1146/annurev.genet.35.102401.091334

145. Nasmyth, K., and Haering, C. H. (2005). The structure and function of SMC and kleisin complexes. *Annu. Rev. Biochem.* 74, 595–648. doi: 10.1146/annurev.biochem.74.082803.133219

146. Nicklas, R. B., Ward, S. C., and Gorbsky, G. J. (1995). Kinetochore chemistry is sensitive to tension and may link mitotic forces to a cell cycle checkpoint.*J. Cell Biol.* 130, 929–939. doi: 10.1083/jcb.130.4.929

147. Niu, H., Wan, L., Busygina, V., Kwon, Y., Allen, J. A., Li, X., et al. (2009). Regulation of meiotic recombination via Mek1-mediated Rad54 phosphorylation. *Mol. Cell* 36, 393–404. doi: 10.1016/j.molcel.2009.09.029

148. Nogales, E. (2000). Structural insights into microtubule function. *Annu. Rev. Biochem.* 69, 277–302. doi: 10.1146/annurev.biochem.69.1.277

149. Ocampo-Hafalla, M. T., and Uhlmann, F. (2011). Cohesin loading and sliding. *J. Cell Sci.* 124, 685–691. doi: 10.1242/jcs.073866

150. Ogura, Y., Shibata, F., Sato, H., and Murata, M. (2004). Characterization of a CENP-C homolog in Arabidopsis thaliana. *Genes Genet. Syst.* 79, 139–144. doi: 10.1266/ggs.79.139

151. Onn, I., Heidinger-Pauli, J. M., Guacci, V., Unal, E., and Koshland, D. E. (2008). Sister chromatid cohesion: a simple concept with a complex reality.*Annu. Rev. Cell Dev. Biol.* 24, 105–129. doi: 10.1146/annurev.cellbio.24.110707.175350

152. Osman, K., Higgins, J. D., Sanchez-Moran, E., Armstrong, S. J., and Franklin, F. C. H. (2011). Pathways to meiotic recombination in

Arabidopsis thaliana.*New Phytol.* 190, 523–544. doi: 10.1111/j.1469-8137.2011.03665.x

153. O'Toole, E., Greenan, G., Lange, K. I., Srayko, M., and Müller-Reichert, T. (2012). The role of γ-tubulin in centrosomal microtubule organization. *PLoS ONE* 7:e29795. doi: 10.1371/journal.pone.0029795

154. Otto, S. P. (2007). The evolutionary consequences of polyploidy. *Cell* 131, 452–462. doi: 10.1016/j.cell.2007.10.022

155. Pastuglia, M., Azimzadeh, J., Magali, G., Christine, C., Belcram, K., Evrard, J., et al. (2006). g-Tubulin is essential for microtubule organization and development in Arabidopsis. *Plant Cell* 18, 1412–1425. doi: 10.1105/tpc.105.039644.2

156. Pawlowski, W. P., and Cande, W. Z. (2005). Coordinating the events of the meiotic prophase. *Trends Cell Biol.* 15, 674–681. doi: 10.1016/j.tcb.2005.10.005

157. Pawlowski, W. P., Wang, C.-J. R., Golubovskaya, I. N., Szymaniak, J. M., Shi, L., Hamant, O., et al. (2009). Maize AMEIOTIC1 is essential for multiple early meiotic processes and likely required for the initiation of meiosis. *Proc. Natl. Acad. Sci. U.S.A.* 106, 3603–3608. doi: 10.1073/pnas.0810115106

158. Penkner, A. M., Fridkin, A., Gloggnitzer, J., Baudrimont, A., Machacek, T., Woglar, A., et al. (2009). Meiotic chromosome homology search involves modifications of the nuclear envelope protein Matefin/SUN-1. *Cell* 139, 920–933. doi: 10.1016/j.cell.2009.10.045

159. Pereira, G., and Schiebel, E. (1997). Centrosome-microtubule nucleation. *J. Cell Sci.* 110(Pt 3), 295–300.

160. Peters, J.-M. (2006). The anaphase promoting complex/cyclosome: a machine designed to destroy. *Nat. Rev. Mol. Cell Biol.* 7, 644–656. doi: 10.1038/nrm1988

161. Peters, J.-M., Tedeschi, A., and Schmitz, J. (2008). The cohesin complex and its roles in chromosome biology. *Genes Dev.* 22, 3089–3114. doi: 10.1101/gad.1724308

162. Petroski, M. D., and Deshaies, R. J. (2005). Function and regulation of cullin-RING ubiquitin ligases. *Nat. Rev. Mol. Cell Biol.* 6, 9–20. doi: 10.1038/nrm1547

163. Phillips, C. M., and Dernburg, A. F. (2006). A family of zinc-finger proteins is required for chromosome-specific pairing and synapsis during meiosis in *C. elegans. Dev. Cell* 11, 817–829. doi: 10.1016/j.devcel.2006.09.020

164. Pidoux, A. L., and Allshire, R. C. (2005). The role of heterochromatin in centromere function. *Philos. Trans. R. Soc. Lond. B. Biol. Sci.* 360, 569–579. doi: 10.1098/rstb.2004.1611

165. Pradillo, M., López, E., Linacero, R., Romero, C., Cuñado, N., Sánchez-Morán, E., et al. (2012). Together yes, but not coupled: new insights into the roles of RAD51 and DMC1 in plant meiotic recombination. *Plant J.* 69, 921–933. doi: 10.1111/j.1365-313X.2011.04845.x

166. Quan, L., Xiao, R., Li, W., Oh, S.-A., Kong, H., Ambrose, J. C., et al. (2008). Functional divergence of the duplicated AtKIN14a and AtKIN14b genes: critical roles in Arabidopsis meiosis and gametophyte development. *Plant J.* 53, 1013–1026. doi: 10.1111/j.1365-313X.2007.03391.x

167. Ramsey, J., and Schemske, D. W. (1998). Pathways, mechanisms, and rates of polyploid formation in flowering plants. *Annu. Rev. Ecol. Syst.* 29, 467–501.

168. Ramsey, J., and Schemske, D. W. (2002). Neopolyploidy in flowering plants. *Annu. Rev. Ecol. Syst.* 33, 589–639. doi: 10.1146/annurev.ecolsys.33.010802.150437

169. Ravi, M., Kwong, P. N., Menorca, R. M. G., Valencia, J. T., Ramahi, J. S., Stewart, J. L., et al. (2010). The rapidly evolving centromere-specific histone has stringent functional requirements in Arabidopsis thaliana. *Genetics* 186, 461–471. doi: 10.1534/genetics.110.120337

170. Ravi, M., Shibata, F., Ramahi, J. S., Nagaki, K., Chen, C., Murata, M., et al. (2011). Meiosis-specific loading of the centromere-specific histone CENH3 in Arabidopsis thaliana. *PLoS Genet.* 7:e1002121. doi: 10.1371/journal.pgen.1002121

171. Reddy, A. S. N., and Day, I. S. (2001). Kinesins in the Arabidopsis genome?: a comparative analysis among eukaryotes. *BMC Genomics* 2:2. doi: 10.1186/1471-2164-2-2

172. Richards, D. M., Greer, E., Martin, A. C., Moore, G., Shaw, P. J., and Howard, M. (2012). Quantitative dynamics of telomere bouquet formation. *PLoS Comput. Biol.* 8:e1002812. doi: 10.1371/journal.pcbi.1002812

173. Ross, K. J., Fransz, P., Armstrong, S. J., Vizir, I., Mulligan, B., Franklin, F. C., et al. (1997). Cytological characterization of four meiotic mutants of Arabidopsis isolated from T-DNA-transformed lines. *Chromosome Res.* 5, 551–559. doi: 10.1023/A:1018497804129

174. Rowland, B. D., Roig, M. B., Nishino, T., Kurze, A., Uluocak, P., Mishra, A., et al. (2009). Building sister chromatid cohesion: smc3 acetylation

counteracts an antiestablishment activity. *Mol. Cell* 33, 763–774. doi: 10.1016/j.molcel.2009.02.028

175. division. *Nat. Rev. Mol. Cell Biol.* 8, 798–812. doi: 10.1038/nrm2257
176. Sakuno, T., Tada, K., and Watanabe, Y. (2009). Kinetochore geometry defined by cohesion within the centromere. *Nature* 458, 852–858. doi: 10.1038/nature07876
177. Salic, A., Waters, J. C., and Mitchison, T. J. (2004). Vertebrate shugoshin links sister centromere cohesion and kinetochore microtubule stability in mitosis. *Cell* 118, 567–578. doi: 10.1016/j.cell.2004.08.016
178. Sanchez-Moran, E., Osman, K., Higgins, J. D., Pradillo, M., Cuñado, N., Jones, G. H., et al. (2008). ASY1 coordinates early events in the plant meiotic recombination pathway. *Cytogenet. Genome Res.* 120, 302–312. doi: 10.1159/000121079
179. Santaguida, S., and Musacchio, A. (2009). The life and miracles of kinetochores. *EMBO J.* 28, 2511–2531. doi: 10.1038/emboj.2009.173
180. Sato, A., Isaac, B., Phillips, C. M., Rillo, R., Carlton, P. M., Wynne, D. J., et al. (2009). Cytoskeletal forces span the nuclear envelope to coordinate meiotic chromosome pairing and synapsis. *Cell* 139, 907–919. doi: 10.1016/j.cell.2009.10.039
181. Sato, H., Shibata, F., and Murata, M. (2005). Characterization of a Mis12 homologue in Arabidopsis thaliana. *Chromosome Res.* 13, 827–834. doi: 10.1007/s10577-005-1016-3
182. Schmitt, J., Benavente, R., Hodzic, D., Hoog, C., Stewart, C. L., and Alsheimer, M. (2007). Transmembrane protein Sun2 is involved in tethering mammalian meiotic telomeres to the nuclear envelope. *Proc. Natl. Acad. Sci. U.S.A.* 104, 7426–7431. doi: 10.1073/pnas.0609198104
183. Schubert, V., Kim, Y.-M., Berr, A., Fuchs, J., Meister, A., Marschner, S., et al. (2007). Random homologous pairing and incomplete sister chromatid alignment are common in angiosperm interphase nuclei. *Mol. Genet. Genomics* 278, 167–176. doi: 10.1007/s00438-007-0242-0
184. Schubert, V., Weissleder, A., Ali, H., Fuchs, J., Lermontova, I., Meister, A., et al. (2009a). Cohesin gene defects may impair sister chromatid alignment and genome stability in Arabidopsis thaliana. *Chromosoma* 118, 591–605. doi: 10.1007/s00412-009-0220-x
185. Schubert, V., Weissleder, A., Ali, H., Fuchs, J., Lermontova, I., Meister, A., et al. (2009b). Cohesin gene defects may impair sister chromatid alignment and genome stability in Arabidopsis thaliana. *Chromosoma* 118, 591–605. doi: 10.1007/s00412-009-0220-x

186. Sebastian, J., Ravi, M., Andreuzza, S., Panoli, A. P., Marimuthu, M. P. A., and Siddiqi, I. (2009). The plant adherin AtSCC2 is required for embryogenesis and sister-chromatid cohesion during meiosis in Arabidopsis. *Plant J.* 59, 1–13. doi: 10.1111/j.1365-313X.2009.03845.x

187. Shao, T., Tang, D., Wang, K., Wang, M., Che, L., Qin, B., et al. (2011). OsREC8 is essential for chromatid cohesion and metaphase I monopolar orientation in rice meiosis. *Plant Physiol.* 156, 1386–1396. doi: 10.1104/pp.111.177428

188. Sharp, D. J., Rogers, G. C., and Scholey, J. M. (2000). Microtubule motors in mitosis. *Nature* 407, 41–47. doi: 10.1038/35024000

189. Sharp, D. J., Yu, K. R., Sisson, J. C., Sullivan, W., and Scholey, J. M. (1999). Antagonistic microtubule-sliding motors position mitotic centrosomes in*Drosophila* early embryos. *Nat. Cell Biol.* 1, 51–54. doi: 10.1038/9025

190. Sheehan, M. J., and Pawlowski, W. P. (2009). Live imaging of rapid chromosome movements in meiotic prophase I in maize. *Proc. Natl. Acad. Sci. U.S.A.* 106, 20989–20994. doi: 10.1073/pnas.0906498106

191. Shimamura, M., Brown, R. C., Lemmon, B. E., Akashi, T., Mizuno, K., Nishihara, N., et al. (2004). γ-Tubulin in basal land plants?: characterization, localization, and implication in the evolution of acentriolar microtubule. *Plant Cell* 16, 45–59. doi: 10.1105/tpc.016501.bules

192. Siddiqi, I., Ganesh, G., Grossniklaus, U., and Subbiah, V. (2000). The dyad gene is required for progression through female meiosis in Arabidopsis.*Development* 127, 197–207.

193. Siderakis, M., and Tarsounas, M. (2007). Telomere regulation and function during meiosis. *Chromosome Res.* 15, 667–679. doi: 10.1007/s10577-007-1149-7

194. Singh, D. K., Andreuzza, S., Panoli, A. P., and Siddiqi, I. (2013). AtCTF7 is required for establishment of sister chromatid cohesion and association of cohesin with chromatin during meiosis in Arabidopsis. *BMC Plant Biol.* 13:117. doi: 10.1186/1471-2229-13-117

195. Staiger, C. J., and Cande, W. Z. (1991). Microfilament distribution in maize meiotic mutants correlates with microtubule organization. *Plant Cell* 3, 637–644. doi: 10.1105/tpc.3.6.637

196. Stoop-Myer, C., and Amon, A. (1999). Meiosis: rec8 is the reason for cohesion. *Nat. Cell Biol.* 1, E125–E127. doi: 10.1038/12956

197. Sudakin, V., Chan, G. K., and Yen, T. J. (2001). Checkpoint inhibition of the APC/C in HeLa cells is mediated by a complex of BUBR1,

BUB3, CDC20, and MAD2. *J. Cell Biol.* 154, 925–936. doi: 10.1083/jcb.200102093

198. Sumara, I., Vorlaufer, E., Stukenberg, P. T., Kelm, O., Redemann, N., Nigg, E. A., et al. (2002). The dissociation of cohesin from chromosomes in prophase is regulated by Polo-like kinase. *Mol. Cell* 9, 515–525. doi: 10.1016/S1097-2765(02)00473-2
199. Sun, S.-C., and Kim, N.-H. (2012). Spindle assembly checkpoint and its regulators in meiosis. *Hum. Reprod. Update* 18, 60–72. doi: 10.1093/humupd/dmr044
200. Talbert, P. B. (2002). Centromeric localization and adaptive evolution of an arabidopsis histone H3 variant. *Plant Cell Online* 14, 1053–1066. doi: 10.1105/tpc.010425
201. Tiang, C.-L., He, Y., and Pawlowski, W. P. (2012). Chromosome organization and dynamics during interphase, mitosis, and meiosis in plants. *Plant Physiol.* 158, 26–34. doi: 10.1104/pp.111.187161
202. Torras-Llort, M., Moreno-Moreno, O., and Azorín, F. (2009). Focus on the centre: the role of chromatin on the regulation of centromere identity and function. *EMBO J.* 28, 2337–2348. doi: 10.1038/emboj.2009.174
203. Tsukahara, T., Tanno, Y., and Watanabe, Y. (2010). Phosphorylation of the CPC by Cdk1 promotes chromosome bi-orientation. *Nature* 467, 719–723. doi: 10.1038/nature09390
204. Uhlmann, F. (2001). Chromosome cohesion and segregation in mitosis and meiosis. *Curr. Opin. Cell Biol.* 13, 754–761. doi: 10.1016/S0955-0674(00)00279-9
205. Uhlmann, F. (2009). A matter of choice: the establishment of sister chromatid cohesion. *EMBO Rep.* 10, 1095–1102. doi: 10.1038/embor.2009.207
206. Vader, G., Maia, A. F., and Lens, S. M. (2008). The chromosomal passenger complex and the spindle assembly checkpoint: kinetochore-microtubule error correction and beyond. *Cell Div.* 3:10. doi: 10.1186/1747-1028-3-10
207. Vanoosthuyse, V., Prykhozhij, S., and Hardwick, K. G. (2007). Shugoshin 2 regulates localization of the chromosomal passenger proteins in fission yeast mitosis. *Mol. Biol. Cell* 18, 1657–1669. doi: 10.1091/mbc.E06
208. Walczak, C. E., Vernos, I., Mitchison, T. J., Karsenti, E., and Heald, R. (1998). A model for the proposed roles of different microtubule-based motor proteins in establishing spindle bipolarity. *Curr. Biol.* 8, 903–913. doi: 10.1016/S0960-9822(07)00370-3

209. Wang, G., Zhang, X., and Jin, W. (2009). An overview of plant centromeres. *J. Genet. Genomics* 36, 529–537. doi: 10.1016/S1673-8527(08)60144-7

210. Wang, M., Tang, D., Wang, K., Shen, Y., Qin, B., Miao, C., et al. (2011). OsSGO1 maintains synaptonemal complex stabilization in addition to protecting centromeric cohesion during rice meiosis. *Plant J.* 67, 583–594. doi: 10.1111/j.1365-313X.2011.04615.x

211. Wang, Y., Wu, H., Liang, G., and Yang, M. (2004). Defects in nucleolar migration and synapsis in male prophase I in the ask1-1 mutant of Arabidopsis.*Sex. Plant Reprod.* 16, 273–282. doi: 10.1007/s00497-004-0206-z

212. Wasteneys, G. O. (2002). Microtubule organization in the green kingdom: chaos or self-order? *J. Cell Sci.* 115, 1345–1354.

213. Watanabe, Y. (2012). Geometry and force behind kinetochore orientation: lessons from meiosis. *Nat. Rev. Mol. Cell Biol.* 13, 370–382. doi: 10.1038/nrm3349

214. Watanabe, Y., and Nurse, P. (1999). Cohesin Rec8 is required for reductional chromosome segregation at meiosis. *Nature* 400, 461–464. doi: 10.1038/22774

215. Weber, H., and Hellmann, H. (2009). Arabidopsis thaliana BTB/ POZ-MATH proteins interact with members of the ERF/AP2 transcription factor family.*FEBS J.* 276, 6624–6635. doi: 10.1111/j.1742-4658.2009.07373.x

216. Wijnker, E., and Schnittger, A. (2013). Control of the meiotic cell division program in plants. *Plant Reprod.* 26, 143–158. doi: 10.1007/s00497-013-0223-x

217. Wittmann, T., Hyman, A., and Desai, A. (2001). The spindle: a dynamic assembly of microtubules and motors. *Nat. Cell Biol.* 3, E28–E34. doi: 10.1038/35050669

218. Woehlke, G., and Schliwa, M. (2000). Walking on two heads: the many talents of kinesin. *Nat. Rev. Mol. Cell Biol.* 1, 50–58. doi: 10.1038/35036069

219. Woglar, A., and Jantsch, V. (2013). Chromosome movement in meiosis I prophase of Caenorhabditis elegans. *Chromosoma* 123, 15–24. doi: 10.1007/s00412-013-0436-7

220. Wu, S., Scheible, W.-R., Schindelasch, D., Van Den Daele, H., De Veylder, L., and Baskin, T. I. (2010). A conditional mutation in Arabidopsis thaliana separase induces chromosome non-disjunction, aberrant morphogenesis and cyclin B1;1 stability. *Development* 137, 953–961. doi: 10.1242/dev.041939

221. Xu, H., Beasley, M. D., Warren, W. D., van der Horst, G. T. J., and McKay, M. J. (2005). Absence of mouse REC8 cohesin promotes synapsis of sister chromatids in meiosis. *Dev. Cell* 8, 949–961. doi: 10.1016/j.devcel.2005.03.018

222. Xu, Z., Cetin, B., Anger, M., Cho, U. S., Helmhart, W., Nasmyth, K., et al. (2009). Structure and function of the PP2A-shugoshin interaction. *Mol. Cell* 35, 426–441. doi: 10.1016/j.molcel.2009.06.031

223. Yamagishi, Y., Sakuno, T., Shimura, M., and Watanabe, Y. (2008). Heterochromatin links to centromeric protection by recruiting shugoshin. *Nature* 455, 251–255. doi: 10.1038/nature07217

224. Yamamoto, A., Kitamura, K., Hihara, D., Hirose, Y., Katsuyama, S., and Hiraoka, Y. (2008). Spindle checkpoint activation at meiosis I advances anaphase II onset via meiosis-specific APC/C regulation. *J. Cell Biol.* 182, 277–288. doi: 10.1083/jcb.200802053

225. Yang, M., Hu, Y., Lodhi, M., McCombie, W. R., and Ma, H. (1999). The Arabidopsis SKP1-LIKE1 gene is essential for male meiosis and may control homologue separation. *Proc. Natl. Acad. Sci. U.S.A.* 96, 11416–11421. doi: 10.1073/pnas.96.20.11416

226. Yang, X., Boateng, K. A., Strittmatter, L., Burgess, R., and Makaroff, C. A. (2009). Arabidopsis separase functions beyond the removal of sister chromatid cohesion during meiosis. *Plant Physiol.* 151, 323–333. doi: 10.1104/pp.109.140699

227. Yang, X., Boateng, K. A., Yuan, L., Wu, S., Baskin, T. I., and Makaroff, C. A. (2011). The radially swollen 4 separase mutation of Arabidopsis thaliana blocks chromosome disjunction and disrupts the radial microtubule system in meiocytes. *PLoS ONE* 6:e19459. doi: 10.1371/journal.pone.0019459

228. Yang, X., Timofejeva, L., Ma, H., and Makaroff, C. A. (2006). The Arabidopsis SKP1 homolog ASK1 controls meiotic chromosome remodeling and release of chromatin from the nuclear membrane and nucleolus. *J. Cell Sci.* 119, 3754–3763. doi: 10.1242/jcs.03155

229. Yao, Y., and Dai, W. (2012). Shugoshins function as a guardian for chromosomal stability in nuclear division. *Cell Cycle* 11, 2631–2642. doi: 10.4161/cc.20633

230. Yokobayashi, S., Yamamoto, M., and Watanabe, Y. (2003). Cohesins determine the attachment manner of kinetochores to spindle microtubules at meiosis I in fission yeast. *Mol. Cell Biol.* 23, 3965–3973. doi: 10.1128/MCB.23.11.3965-3973.2003

231. Yu, H.-G., and Koshland, D. (2005). Chromosome morphogenesis: condensin-dependent cohesin removal during meiosis. *Cell* 123, 397–407. doi: 10.1016/j.cell.2005.09.014

232. Yu, H., Muszynski, M. G., and Dawe, R. K. (1999). The maize homologue of the cell cycle checkpoint protein MAD2 localization Patterns. *J. Cell Biol.* 145, 425–435.

233. Zamariola, L., De Storme, N., Tiang, C. L., Armstrong, S. J., Franklin, F. C. H., and Geelen, D. (2013). SGO1 but not SGO2 is required for maintenance of centromere cohesion in Arabidopsis thaliana meiosis. *Plant Reprod.* 26, 197–208. doi: 10.1007/s00497-013-0231-x

234. Zamariola, L., De Storme, N., Vannerum, K., Vandepoele, K., Armstrong, S. J., Franklin, F. C., et al. (2014). SHUGOSHINs and PATRONUS protect meiotic centromere cohesion in Arabidopsis thaliana. *Plant J.* 77, 782–794. doi: 10.1111/tpj.12432

235. Zhang, L., Tao, J., Wang, S., Chong, K., and Wang, T. (2006). The rice OsRad21-4, an orthologue of yeast Rec8 protein, is required for efficient meiosis.*Plant Mol. Biol.* 60, 533–554. doi: 10.1007/s11103-005-4922-z

236. Zhao, D., Han, T., Risseeuw, E., Crosby, W. L., and Ma, H. (2003a). Conservation and divergence of ASK1 and ASK2 gene functions during male meiosis in Arabidopsis thaliana. *Plant Mol. Biol.* 53, 163–173. doi: 10.1023/B:PLAN.0000009273.81702.b5

237. Zhao, D., Ni, W., Feng, B., Han, T., Petrasek, M. G., and Ma, H. (2003b). Members of the arabidopsis-SKP1-like gene family exhibit a variety of expression patterns and may play diverse roles in arabidopsis 1. 133, 203–217. doi: 10.1104/pp.103.024703.to

238. Zhao, D., Yang, X., Quan, L., Timofejeva, L., Rigel, N. W., Ma, H., et al. (2006). ASK1, a SKP1 homolog, is required for nuclear reorganization, presynaptic homolog juxtaposition and the proper distribution of cohesin during meiosis in Arabidopsis. *Plant Mol. Biol.* 62, 99–110. doi: 10.1007/s11103-006-9006-1

239. Zhao, D., Yu, Q., Chen, M., and Ma, H. (2001). The ASK1 gene regulates B function gene expression in cooperation with UFO and LEAFY in Arabidopsis.*Development* 128, 2735–2746.

240. Zhong, C. X., Marshall, J. B., Topp, C., Mroczek, R., Kato, A., Nagaki, K., et al. (2002). Centromeric retroelements and satellites interact with maize kinetochore Protein CENH3. *Plant Cell* 14, 2825–2836. doi: 10.1105/tpc.006106

241. Zhou, S., Wang, Y., Li, W., Zhao, Z., Ren, Y., Wang, Y., et al. (2011). Pollen semi-sterility1 encodes a kinesin-1-like protein important for male

meiosis, anther dehiscence, and fertility in rice. *Plant Cell* 23, 111–129. doi: 10.1105/tpc.109.073692

242. Zhou, X., Graumann, K., Evans, D. E., and Meier, I. (2012). Novel plant SUN-KASH bridges are involved in RanGAP anchoring and nuclear shape determination. *J. Cell Biol.* 196, 203–211. doi: 10.1083/jcb.201108098

Chapter 5

IMMUNOLOCALIZATION OF CHROMOSOME-ASSOCIATED PROTEINS IN PLANTS – PRINCIPLES AND APPLICATIONS

Cristina Maria Pinto de Paula and Vânia Helena Techio

Department of Biology, Federal University of Lavras, Lavras, Minas Gerais, Brazil

ABSTRACT

The use of the immunolocalization technique combined with cytogenetic and epigenetic studies is an indispensable tool and has contributed significantly to the analysis of the structure and function of chromosomes, since it can provide information about the spatial or temporal distribution of a given protein in the nucleus and chromosomes. Several chromosome-associated proteins in plant cells have already been identified by immunolocalization, such as histone and non-histone proteins and cell division-related protein (mitosis and meiosis). The principle of the immunolocalization technique in plants basically involves fixation and permeabilization of cells, the use of monoclonal or polyclonal antibodies attached to a signaling molecule, usually a fluorochrome and detection of the target molecule by using an epifluorescence microscope.

INTRODUCTION

Considerable advances in plant biology have been accompanied by the use of microscopy of higher resolution and improvement of the detection limit of analyses through use of specific antibodies and fluorochromes, that provide an interaction between knowledge of cell biology, classical cytogenetics and molecular cytogenetics. From the development of molecular cytogenetic techniques, understanding the organization of genomes had a large impact on taxonomic and evolutionary studies (Fedak and Kim 2008; Figueroa and Bass 2010), in breeding and genetic engineering (Seijo et al. 2010).

Recent research on the architecture of the interphase nucleus and chromatin remodeling, mediated primarily by changes in histones, has prospected the era of epigenetics and epigenomics. In this new scientific background, the use of the technique of immunolocalization in chromosome analysis has become an indispensable tool for understanding the organization and expression of the genome of species. As a result of this interaction, significant improvements were obtained for the analysis of chromosome structure and function, as it can provide information about the spatial and temporal distribution of a given protein in the nucleus and chromosomes.

These analyses have been widely applied in recent decades to study and locate proteins in cells or cell populations and comprise the use of specific antibodies to the protein of interest (Hoshi et al. 2008). In the context of chromosomal studies, advances are more recent, but have already made possible to identify several chromosome-associated proteins in plant cells, including histones (Fuchs et al. 1998), non-histones (Stepinski 2009), proteins associated with mitosis (Caperta et al. 2006) and with meiosis (Franklin et al. 1999; Armstrong et al. 2002; Qiao et al. 2011).

Given the importance of the topic and potential use, this review aims to present the principles of the immunolocalization technique and its application in the detection of proteins associated with chromosomes in plants.

REVIEW

Principles of Immunolocalization

The application of the immunolocalization technique was facilitated through the development of fluorochrome-labeled antibodies, in the pioneering work of Albert Coons (Coons et al. 1941), thus allowing visualization of specific molecules in a cell. Therefore, it has been possible to observe the spatial or temporal distribution of a given protein in the whole cell, in the nucleus, in specific chromosome or organelles.

The procedures started to be used in plants from the experiences observed in animal cells. One of the first descriptions of the use of immunolocalization in plant cytology occurred in 1970, with the location of antigens of the cell wall in pollen (Knox 1970) and proteins of cotyledons (Graham and Gunning 1970). Currently, the immunolocalization has allowed, among other studies, the identification of chromosomal proteins, that is, the study of how the activity of these proteins is regulated during cell divisions (Seijo et al. 2010; Houben et al.2013), and also the assessment of post-translational changes in histones, which are reflected in the chromatin structure, known as epigenetic regulation (Chen et al. 2010).

The immunolocalization technique is based on the principles of antigen-antibody reactions to find highly specific molecules in a cell or tissue. The common reagent for immunolocalization are polyclonal or monoclonal antibodies, both produced in animals such as rabbits, goats, rats, mice, among others, by means of specific procedures (Boenisch 2009).

Polyclonal antibodies are a heterogeneous mixture of antibodies that bind to different epitopes of the same protein. Monoclonal antibodies are homogeneous populations, more specific for binding to only one epitope of the protein (Boenisch 2009).

There are many commercially available antibodies, some of universal use for chromosomal proteins, such as anti-tubulin (α, β and γ) or for modified DNA regions, such as the anti-5-methylcytosine (Guerra 2012; Birchler and Han 2013). Other antibodies are specific for a species or group of species, e.g., phosphorylated H2AThr120, which is used for cytological detection of centromeric chromatin of plants (Dong and Han2012; Demidov et al. 2014).

In addition to factors related to the type of antibody, storage conditions and dilution of antibodies, which are usually indicated by the manufacturers, the immunolocalization reactions must take into account factors like affinity, avidity and cross-reactivity.

Affinity refers to the interaction strength between the antigen epitope and the paratope of an antibody. Similarly, avidity is the term used to describe the overall strength of the interactions between the antigen and the various antibodies that recognize it (Steward and Steensgaard1983; Boenisch 2009). High affinity antigen-antibody means greater tendency to hold them together (Boenisch 2009). However, factors such as high salt concentration, high temperature and low pH during the washing cycles applied in the technique, may result in the dissociation of the antigen-antibody complex.

Cross-reactivity occurs when an antibody reacts with two or more antigens, or when an antigen reacts with several different antibodies. In this case there is a sharing of at least one common epitope among the various antigens (Boenisch 2009).

For detection of proteins by immunolocalization, the cells must be fixed and permeabilized to facilitate the access of antibodies to the cytoplasm and nucleus. Fixation of the nucleus or chromosomes must preserve the arrangement of molecules as they occur in vivo, and treatment should not affect the cytological integrity of the material to be analyzed. In general, aldehyde fixatives are used, which make a link between proteins, nucleic acids and phospholipids, thus forming a network which retains molecules in their original position. Formaldehyde, paraformaldehyde and glutaraldehyde are the most frequently used in immunolocalization (Marttila and Santén 2007).

There are some cases in which acid fixatives can be effective to the meiosis, for example, Carnoy (ethyl alcohol: acetic acid 3:1) (Chelysheva et al. 2010, 2013). In this last case, the fixation is associated with microwave procedures. This method combines a strong fixation to preserve the chromosomal structures, acetic acid spreading to remove the cytoplasm and treatment in microwave (850 W) to increase the access of the antibody to the protein.

Another possibility for meiosis is use of the methanol/acetone fixative, which extracts the lipids and rapidly dehydrates the cell, thus disrupting hydrophobic interactions and resulting in aggregation and precipitation of proteins. This fixative preserves cell architecture, but can result in reduced antigenicity of some proteins. Nevertheless, it can also allow better access of the antibody to certain antigens without requiring cell permeabilization (Yang et al. 2013).

Permeabilization is usually carried out by incubation with detergents or organic solvents, which solubilize or remove lipids from the plasma membrane and thus allow these antibodies to have access to the structure of interest. The most commonly used permeabilizing agents include Triton X-100, Tween 20 and also Lipsol. These are hydrophilic nonionic detergents, considered mild surfactants, in other words they break lipid-protein and lipid-lipid interactions, but not protein-protein bonds, maintaining their active form (Johnson 2013).

In the case of plant cells, it is necessary to overcome the barrier of the cell wall, so that permeabilization usually requires a prior cell wall digestion using enzyme cocktails containing pectinase, cellulase, pectoliase and others, whose concentration and exposure time are adjusted according to the species analyzed.

The usual procedure for plants is indirect immunolocalization. This technique includes the use of at least two antibodies: the primary which recognizes and binds to epitopes of the antigen (protein) to be located, and the secondary, often conjugated to a fluorochrome, which recognizes the corresponding primary antibody, the reason why the procedure is also referred to immunofluorescence. Alternatively, gold conjugated secondary antibodies can be used, in this case, the visualization is performed in a transmission electron microscope and the procedure is called immunogold (Holgate et al. 1983; Marttila and Santén 2007). The direct method employs only one antibody linked to a fluorochrome. The advantage of indirect labeling is the amplification of the signal obtained, since several secondary antibodies may bind to a single primary antibody (Brown and Lemmon 1995; Marttila and Santén 2007).

Fluorochromes are chemicals that, upon excitation at a specific wave length, fluoresce at another wavelength, and the signal is observed with an epifluorescence microscope (Brown and Lemmon 1995). Fluorescein (FITC) is the fluorochrome most used in immunolocalization, it can be excited by UV light (wavelength below 400 nm) or blue light (490 nm) and emits green light (500–550 nm), as well as Rhodamine (TRITC) and Texas red, fluorochromes which are excited between 520 and 590 nm and emit red light (550–620 nm) (McNamara 2006). There are other types of fluorochromes with higher fluorescence intensity and photostability, compared with those aforementioned and may also be used in immunolocalization such as Dylight, Alexa Fluor (Jensen 2012) and quantum dots, the latter a class of inorganic fluorochromes made up of nanoscale crystals of a semiconductor material (Ioannou et al. 2009).

An alternative methodology proposed by (Wang 2013) analyzes meiotic chromosomes by three-dimensional structured illumination microscopy (3D-SIM) using high resolution images and can be applied for the localization of proteins or fine structures by immunofluorescence. The protocol (Acrylamide sandwich) was developed aimed at better preservation of 3D chromosome structure and spatial organization of the nucleus; in turn the purpose of 3D-SIM is to illuminate a sample with a light distribution pattern produced by a diffraction grating. With this technique, several optical sections are captured and a 3D image is reconstructed by computer analysis.

Major Chromosome-Associated Proteins Localized in Plants

Study of Non-Histone Protein

A large number of studies have already identified several proteins associated with chromosomes in plant cells so far. The study of the behavior of these proteins provide information about the function of chromosomes, since part of the control of the genetic information encoded by DNA depends on interactions with proteins. The immunolocalization of chromosomal proteins in plants has been widely used for different purposes in cytogenetic and epigenetic studies such as identification of proteins that are directly or indirectly involved with mitosis and meiosis and associated with chromatin modeling.

The proper chromosome condensation during mitosis and meiosis is essential for the correct segregation of the genomic information contained in the DNA. Besides histones, other non-histone proteins are also involved in the process of condensation and cohesion, such as a topoisomerase II, condensins and cohesin (Moser and Swedlow 2011).

Cohesin complex proteins are essential for sister chromatid cohesion and proper chromosome segregation during mitosis and meiosis. Cohesin proteins are also components of axial elements (AEs) and lateral elements (LEs) of the synaptonemal complex during meiosis. The cytological behavior of four cohesin proteins (SMC1, SMC3, SCC and REC8/SYN1), assessed by immunofluorescence during prophase I in tomato microsporocyte revealed that the four cohesins are distributed unevenly and are not co-located along the AE/LEs in diplotene (Qiao et al. 2011). Nevertheless, based on current models of the cohesin complex, these proteins must be present at the same time and place in equivalent amounts. These results indicate that cohesin proteins studied can form different complexes and/or perform additional functions during meiosis in plants.

The success of meiosis depends on the regulation of several cellular processes that ensure proper chromosome segregation. By employing the immunolocalization technique it is possible to visualize a large number of proteins related to different aspects of meiosis, including sister chromatid cohesion, synapsis, recombination, and chromosome segregation (Yang et al. 2013).

A number of proteins involved in the pairing and recombination are expressed during meiosis, including RAD51 and MLH1. Studies with maize have shown that RAD51 is involved in the homology search during chromosomal pairing and recombination. Through immunolocalization, it was observed, in zygotene, 500 RAD51 signals per nucleus, clearly associated with the partially paired and paired chromosomes, which is consistent with its role in the homology search. In pachytene, the number of RAD51 signals decreased, ranging from 7 to 22 per nucleus, with the largest number corresponding approximately to the number of chiasmata found in maize, thus coinciding with the expected number of recombination during meiosis (Franklin et al. 1999).

Likewise, RAD51 protein in *Arabidopsis* is related to homology search and meiotic recombination (Kurzbauer et al. 2012). *Atrad51-2* mutants express low amounts of the RAD51 protein, meiosis with partial synapsis and 51% of bivalents between non-homologous chromosomes, thus demonstrating that RAD51 is required for effective chromosome pairing (Pradillo et al. 2012).

Similar studies were developed with *Lily*, being detected the presence of RAD51 and LIM15 proteins as discrete signals in leptotene and zygotene. This localization suggests that meiotic recombination is initiated in leptotene with the cooperation of these two proteins and remains in zygotene. The location of the signals on chromosomes or in their vicinity suggests that these proteins must also act in the pairing of homologous chromosomes in this species (Terasawa et al. 1995; Anderson et al. 1997).

The ASY1 gene is essential for synapsis in homologous chromosomes (Caryl et al. 2000). Studies using specific antibodies against the ASY1 protein were applied to investigate the temporal expression and localization of this protein in *Arabidopsis thaliana* (L.) Heynh. Detection of the immuno signal associated with electron microscopy data showed that ASY1 is not a component of the synaptonemal complex, but is associated with the axial/lateral elements of meiotic chromosomes. The authors suggest that this protein possibly defines the regions of chromatin associated with the development of the synaptonemal complex structure (Armstrong et al. 2002).

Cytoskeleton microtubules play a crucial role during the cell cycle and meiosis. The use of antimitotic substances, such as colchicine, promotes depolymerization of microtubules and prevents the formation of spindle fibers, this methodology is used to block cell cycle progression and thus induce polyploidy (Pereira et al. 2012). Through immunolocalization of alpha-tubulin, (Caperta et al. 2006) analyzed the effect of different colchicine concentrations on spindle fibers in *Secale cereale* L. At a low concentration of colchicine (0.5 mM) most of the cells showed arrays of discontinuous microtubule or no microtubule were detected at C-metaphase. Cells exposed to high concentrations (5.0 mM) showed a different effect, the C- metaphase presented fibrous and branched cortical filaments that allowed reconstitution of 4C nuclei and cell cycle progression. These results demonstrate the contrasting and opposite effects at different concentrations of colchicine.

Study of Histone Modifications

Histones form another group of proteins associated with eukaryotic DNA, comprising the nucleosome, which consists of an octamer of the four major histones (H3, H4, H2A, H2B) (Luger et al. 1997; Kouzarides 2007).

Each histone that comprises the nucleosome has a terminal amino acid chain, called N-terminal tail, which is subjected to various post-translational modifications (Kouzarides 2007; Jin et al. 2008). These epigenetic marks affect the structure and function of chromatin, thereby regulating gene expression and can be identified by immunolocalization (Bannister and Kouzarides 2011).

There are at least eight different types of alterations in histones, standing out acetylation, methylation and phosphorylation, which are the most studied (Kouzarides 2007; Chen et al. 2010).

Plant chromosomes usually exhibit a lower amount of acetylated histones in heterochromatin and often form significant markings in the nucleolar organizing region (NOR), which allows a correlation with active gene transcription. In *Vicia faba* L., antibodies against acetylated H4 at lysine 5, 8 and 12 labeled the entire chromosome complement, with the exception of large blocks of heterochromatin (Fuchs et al. 1998), since the NOR was the most strongly acetylated of all lysine residues investigated (Houben et al. 1996).

One of the most studied post-translational modification is the phosphorylation of histone H3, which seems to be decisive for cell cycle in relation to the chromosome condensation and segregation, activation of apoptosis, transcription and repair of DNA damage (Houben et al.2007). In plants, phosphorylation of histone H3 at serine 10 (H3S10ph) and 28 (H3S28ph) is restricted to the pericentromeric region in mitosis (Germand et al. 2003). In meiosis, the chromosomes are phosphorylated along the entire arm during the first division, however, the phosphorylation is restricted to the pericentromeric region during the second division (Manzanero et al. 2000).

In studies with forage grasses Paula et al. (2013) evaluated the phosphorylation pattern of H3S10 in diploid and tetraploid genotypes of*Brachiaria* (Trin.) Griseb. species in order to investigate the dynamics of this post-translational modification and correlate it with its regulatory function during mitosis and meiosis. In meiosis, the chromosomes are phosphorylated at H3S10 along its entire length during the first division; in the second division, phosphorylation is restricted to the pericentromeric region, as well as in mitosis. Alltogether, these observations indicated that H3S10ph is necessary for the maintenance of sister chromatid cohesion in mitosis and meiosis.

Phosphorylation of threonine in histone H3 in *Secale cereale* L. and threonine 11 in H3 in *Vicia faba* L. was observed throughout the chromosome during mitosis and meiosis. Phosphorylation began in prophase and ended in telophase, thus correlating with chromosome condensation (Houben et al. 2005, 2007).

By means of immunolocalization, it was possible to detect that methylation of some isoforms of histones in plants may be associated with heterochromatin (H3K9, H3K27 and H4K20) and also euchromatin (H3K4, H3K36 and H3K79) (Fuchs et al. 2006).

Different types of chromatin associated with histone modifications were analyzed by immunolocalization in *Costus spiralis*. Antibodies against components characteristic of histones of euchromatin (acetylated H4 histone at lysine 5 - H4K5ac and dimethylated H3 histone at lysine 4 - H3K4me2) and heterochromatin (dimethylated H3 histone at lysine 9 - H3K9me2) were used to characterize the centromeric chromatin during meiosis. The centromeric region was highly enriched only with H4k5ac and only in pachytene. Still in pachytene, the terminal decondensed euchromatin of chromosome arms were labeled with H4K5ac and H3K4me2, whereas the condensed proximal region was labeled with H3K9me2 (Feitoza and Guerra 2011a).

In plant species *Costus spiralis* and *Eleutherine bulbosa* during mitosis, labeling with antibodies against components characteristic of histones of euchromatin were observed in the regions of late condensation while the components of heterochromatin was more intense in the early condensation chromatin. These data indicate that there are different chromatin domains in these species, the late condensing prophase chromatin localized on the distal region and strongly enriched in H4K5ac and H3K4me2 and early condensing chromatin, localized on the proximal chromosome region in *C. spiralis* and mainly in the larger pair I of *E. bulbosa* (Feitoza and Guerra 2011b).

Intraspecific hybrids and parental acessions (Col-0 and C24) of *Arabidopsis thaliana* were used as a model to investigate the relationship between changes in DNA methylation, chromatin structure, endopolyploidy and gene expression in heterotic genotypes. Through immunolocalization, the distribution of methylation of histone H3 (H3K27me3, H3K4me2 and H3K9me2) was compared between parental and hybrid plants. For all modifications of the chromatin analyzed, no clear difference was found in the distribution and intensity of signals between Col-0, C24 and their reciprocal hybrids, revealing that the parental pattern profile was kept in the hybrid progeny (Moghaddam et al. 2010).

Besides that, the correlation between the distribution of histone H3 methylated at lysine 4 (typical of transcriptionally active euchromatin) and lysine 9 (inactive heterochromatin) was analyzed in relation to genome size for 24 species. In species with small genome, strong H3K9me was restricted to constitutive heterochromatin, whereas species with large genomes exhibited uniform distribution of this modification. Unlike and independent of the genome size, H3k4me was enriched in euchromatin of all species. The authors concluded that large genomes with large amounts of dispersed repetitive sequences (mainly retroelements) silence these sequences through epigenetic modifications such as H3K9me (Houben et al. 2003).

Centromeres are responsible for several important events in meiosis and mitosis. In contrast to the telomeres, centromeres do not have highly conserved DNA sequence and are determined by epigenetic modifications (Houben and Schubert 2003). In eukaryotes, the centromeres have a variant of histone H3, the CENH3 (CENPA in humans) which determines the chromosomal position of kinetochore assembly, forming a bond between the centromeric DNA and the kinetochore (Blower et al. 2002). Thus, the anti-CENH3 antibody has been used in studies related to the centromeric regions in plants like corn (Jin et al. 2005; Zhang et al. 2005), wheat (Zhang et al. 2010) and rye (Houben et al.2011).

Recently, Demidov et al. (2014) evidenced that the antibody against phosphorylated histone H2A at threonine 120 can be used as a universal maker for the cytological detection of centromeres in species of monocentric and holocentric plants. These results were obtained from studies on 20 different species of mono- and eudicots.

Although histones and their modifications are highly conserved, their distribution in chromosomes, observed by immunolocalization, showed a variation over the differentiation and cell cycle processes, as well as within and between groups of eukaryotes (Hans and Dimitrov 2001).

CONCLUSION

Immunolocalization is an important and efficient tool in studies involving the understanding of genome organization and analysis of the structure and function of chromosomes. Furthermore, immunolocalization studies allowed the understanding of epigenetic mechanisms involved in chromatin modeling and modifications related to different states of condensation, such as transcriptional activity and silencing.

AUTHORS' CONTRIBUTIONS

All authors were involved in drafting the manuscript, providing guidelines for the review, modification and preparation the final version of the manuscript. Both authors read and approved the final manuscript.

REFERENCES

1. Anderson LK, Offenberg HH, Verkuijlen WMHC, Heyting C (1997) RecA-like proteins are components of early meiotic nodules in Lily. Proc Natl Acad Sci U S A 94:6868–6873
2. Armstrong SJ, Caryl AP, Jones GH, Franklin FCH (2002) Asy1, a protein required for meiotic chromosome synapsis, localizes to axis-associated chromatin in *Arabidopsis* and *Brassica*. J Cell Sci 18:3645–3655
3. Bannister AJ, Kouzarides T (2011) Regulation of chromatin by histone modifications. Cell Res 21(3):381–395
4. Birchler JA, Han F (2013) Centromere epigenetics in plants. J Genet Genomics 40:201–204
5. Blower MD, Sullivan BA, Karpen GH (2002) Conserved organization of centromeric chromatin in flies and humans. Dev Cell 2:319–330

6. Boenisch T (2009) Antibodies. In: Kummar LG, Rudbeck L (eds) Immunohistochemical (IHC) Staining Methods. Dako North America, California, pp 1–9

7. Brown RC, Lemmon BE (1995) Methods in plant immunolight microscopy. Methods Cell Biol 49:85–107

8. Caperta AD, Delgado M, Ressurreição F, Meister A, Jones RN, Viegas W et al (2006) Colchicine-induced polyploidization depends on tubulin polymerization in c-metaphase cells. Protoplasma 227:147–153

9. Caryl AP, Armstrong SJ, Jones GH, Franklin FCH (2000) A homologue of the yeast *HOP1* gene is inactivated in the *Arabidopsis* meiotic mutant *asy1*. Chromosoma 159:62–71

10. Chelysheva L, Grandont L, Vrielynck N, Le Guin S, Mercier R, Grelon M (2010) An Easy Protocol for studying chromatin and recombination protein dynamics during *Arabidopsis thaliana* meiosis: Immunodetection of cohesins, histones and MLH1. Cytogenet Genome Res 129:143–153

11. Chelysheva LA, Grandont L, Grelon M (2013) Immunolocalization of meiotic proteins in Brassicaceae: Method 1. In: Pawlowski WP, Grelon M, Armstrong S (eds) Plant Meiosis: Methods and Protocols. Springer Science+Business Media, New York, pp 103–108

12. Chen M, Shaolei LV, Meng Y (2010) Epigenetic performers in plants. Dev Growt Differ 52:555–566

13. Coons AH, Creech HJ, Jones RN (1941) Immunological properties of an antibody containing a fluorescent group. Proc Soc Exp Biol Med 47:200–202

14. Demidov D, Schubert V, Kumbe K, Weiss O, Karimi-Ashtiyani R, Butlar J et al (2014) Anti-phosphorylated histone H2AThr120 - a universal marker for centromeric chromatin of mono- and holocentric plant species. Cytogenet Genome Res. doi:10.1159/000360018

15. Dong Q, Han F (2012) Phosphorylation of histone H2A is associated with centromere function and maintenance in meiosis. Plant J 71:800–809

16. Fedak G, Kim NS (2008) Tools and methodologies for cytogenetic studies of plant chromosomes. Cytology Genet 42:189–203

17. Feitoza L, Guerra M (2011a) The centromeric heterochromatin of *Costus spiralis*: poorly methylated and transiently acetylated during meiosis. Cytogenet Genome Res 135:160–166

18. Feitoza L, Guerra M (2011b) Different types of plant chromatin associated with modified histones H3 and H4 and methylated DNA. Genetica 139(3):305–314

19. Figueroa DM, Bass HW (2010) A historical and modern perspective on plant cytogenetics. Brief Funct Genomics 9:95–102

20. Franklin AE, Mcelver J, Sunjevaric I, Rothstein R, Bowen B, Cande WZ (1999) Three-dimensional microscopy of the Rad51 recombination protein during meiotic prophase. Plant Cell 11:809–824

21. Fuchs J, Strehl S, Brandes A, Schweizer D, Schubert I (1998) Molecular-cytogenetic characterization of *Vicia faba* genome-heterochromatin differentiation, replication patterns and sequence localization. Chromosome Res 6:219–230

22. Fuchs J, Demidov D, Houben A, Schubert I (2006) Chromosomal histone modification patterns: from conservation to diversity. Trends Plant Sci 11:199–208

23. Germand D, Demidov D, Houben A (2003) The temporal and spatial pattern of histone H3 phosphorylation at serine 28 and serine 10 is similar in plants but differs between mono- and polycentric chromosomes. Cytogenet Genome Res 101:172–176

24. Graham TA, Gunning BES (1970) Localization of legumin and vicilin in bean cotyledon cells using fluorescent antibodies. Nature 228:81–82

25. Guerra M (2012) Citogenética molecular: Protocolos comentados. Sociedade Brasileira de Genética, Ribeirão Preto

26. Hans F, Dimitrov S (2001) Histone H3 phosphorylation and cell division. Oncogene 20:3021–3027

27. Holgate CS, Jackson P, Cowen PN, Bird CC (1983) Immunogold-silver staining: New method of immunostaining with enhanced sensitivity. J Histochem Cytochem 31:938–944

28. Hoshi O, Hirota T, Kimura E, Komatsubara N, Ushiki T (2008) Immunocytochemistry for analyzing chromosomes. In: Fukui K, Ushiki T (eds) Chromosome nanoscience and technology. CRC press, New York, pp 81–89

29. Houben A, Schubert I (2003) DNA and proteins of plant centromeres. Curr Opin Plant Biol 6:554–560

30. Houben A, Belyaev ND, Turner BM, Schubert I (1996) Differential immunostaining of plant chromosomes by antibodies recognizing acetylated histone H4 variants. Chromosome Res 4:191–194

31. Houben A, Demidov D, Gernand D, Meister A, Leach CR, Schubert I (2003) Methylation of histone H3 in euchromatin of plant chromosomes depends on basic nuclear DNA content. Plant J 33:967–973

32. Houben A, Demidov D, Rutten T, Scheidtmann KH (2005) Novel phosphorylation of histone H3 at threonine 11 that temporally correlates with condensation of mitotic and meiotic chromosomes in plant cells. Cytogenet Genome Res 109:148–155
33. Houben A, Demidov D, Caperta AD, Karimi R, Aqueci F, Vlasenko L (2007) Phosphorylation of histone H3 in plants: a dynamic affair. Biochim Biophys Acta 1769:308–315
34. Houben A, Kumke K, Nagaki K, Hause G (2011) CENH3 distribution and differential chromatin modifications during pollen development in rye (*Secale cereale* L.). Chromosome Res 19:471–480
35. Houben A, Demidov D, Karimi-Ashtiyani R (2013) Epigenetic control of cell division. In: Grafi G, Ohad N (eds) Epigenetic Memory and Control in Plants. Springer-Verlag, Berlin, pp 155–175
36. Ioannou D, Tempest HG, Skinner BM, Thornhill AR, Ellis M, Griffin DK (2009) Quantum dots as new-generation fluorochromes for FISH: an appraisal. Chromosome Res 17:519–530
37. Jensen EC (2012) Types of imaging, Part 2: An overview of Fluorescence Microscopy. Anat Rec 295:1621–1627
38. Jin WW, Lamb JC, Vega JM, Dawe RK, Birchler JA, Jiang J (2005) Molecular and functional dissection of the maize B centromere. Plant Cell 17:1412–1423
39. Jin W, Lamb JC, Zhang W, Kolano B, Birchler JA, Jiang J (2008) Histone modifications associated with both A and B chromosomes of maize. Chromosome Res 16:1203–1214
40. Johnson M (2013) Detergents: Triton X-100, Tween-20, and more. Mater Methods 3:163
41. Knox RB (1970) Freeze-sectioning of plant tissues. Stain Technol 45:265–272
42. Kouzarides T (2007) Chromatin modifications and their function. Cell 128:693–705
43. Kurzbauer M-T, Uanschou C, Chen D, Schlögelhofer P (2012) The Recombinases DMC1 and RAD51 Are Functionally and Spatially Separated during Meiosis in *Arabidopsis*. Plant Cell 24:2058–2070
44. Luger K, Mader AW, Richmond RK, Sargent DF, Richmond TJ (1997) Crystal structure of the nucleosome core particle at 2.8A° resolution. Nature 389:251–260

45. Manzanero S, Arana P, Puertas MJ, Houben A (2000) The chromosomal distribution of phosphorylated histone H3 differs between plants and animals at meiosis. Chromosoma 109:308–317
46. Marttila S, Santén K (2007) Practical aspects of Immunomicroscopy on plant material. In: Méndez-Vilas A, Díaz J (eds) Modern Research and Educational Topics in Microscopy. Formatex, Badajoz, pp 1015–1021
47. McNamara G (2006) Introduction to immunofluorescence microscopy. In: Spector D, Goldman RD (eds) Basic methods in microscopy. Cold Spring Harbor, New York, pp 145–154
48. Moghaddam AMB, Fucs J, Czauderna T, Houben A, Mette MF (2010) Intraspecific hybrids of *Arabidopsis thaliana* revealed no gross alterations in endopolyploidy, DNA methylation, histone modifications and transcript levels. Theor Appl Genet 120:215–226
49. Moser SC, Swedlow JR (2011) How to be a mitotic chromosome. Chromosome Res 19:307–319
50. Paula CMP, Techio VH, Souza Sobrinho F, Freitas AS (2013) Distribution pattern of histone H3 phosphorylation at serine 10 during mitosis and meiosis in *Brachiaria* species. J Genet 92:259–266
51. Pereira RC, Davide LC, Techio VH, Timbó ALO (2012) Duplicação cromossômica de gramíneas forrageiras: uma alternativa para programas de melhoramento genético. Cienc Rural 42:1278–1285
52. Pradillo M, López E, Linacero R, Romero C, Cuñado N, Sánchez-Morán E, Santos JL (2012) Together yes, but not coupled: new insights into the roles of RAD51 and DMC1 in plant meiotic recombination. Plant J 69:921–933
53. Qiao H, Lohmiller LD, Anderson LK (2011) Cohesin proteins load sequentially during prophase I in tomato primary microsporocytes. Chromosome Res 19:193–207
54. Seijo G, Lavia GI, Robledo G, Fernández A, Neffa VGS (2010) La citogenética molecular e inmunocitogenética em el estúdio de los genomas vegetales. In: Levitus G, Echenique V, Rubinstein C, Hopp E, Mroginski L (eds) Biotecnología y mejoramiento vegetal II. Instituto Nacional de tecnologia Agropecuária, Argentina, pp 34–46
55. Stepinski D (2009) Immunodetection of nucleolar proteins and ultrastructure of nucleoli of soybean root meristematic cells treated with chilling stress and after recovery. Protoplasma 235:77–89
56. Steward MW, Steensgaard J (1983) Antibody affinity: Termodynamic aspects and biological significance 1ed. CRC Press, Boca Raton

57. Terasawa M, Shinohara A, Hotta Y, Ogawa H, Ogawa T (1995) Localization of RecA-like recombination proteins on chromosomes of the lily at various meiotic stages. Genes Dev 9:925–934
58. Wang C-JR (2013) Analyzing maize meiotic chromosomes with super-resolution structured illumination microscopy. In: Pawlowski WP, Grelon M, Armstrong S (eds) Plant Meiosis: Methods and Protocols. Springer Science+Business Media, New York, pp 109–118
59. Yang X, Yuan L, Makaroff CA (2013) Immunolocalization protocols for visualizing meiotic proteins in *Arabidopsis thaliana*: Method 3. In: Pawlowski WP, Grelon M, Armstrong S (eds) Plant Meiosis: Methods and Protocols. Springer Science+Business Media, New York, pp 109–118
60. Zhang X, Li X, Marshall JB, Zhong CX, Dawe RK (2005) Phosphoserines on maize Centromeric histone H3 and Histone H3 demarcate the centromere and pericentromere during chromosome segregation. Plant Cell 17:572–583
61. Zhang WL, Friebe B, Gill BS, Jiang JM (2010) Centromere inactivation and epigenetic modifications of a plant chromosome with three functional centromeres. Chromosoma 119:553–563

Chapter 6

CHROMOSOMAL LOCATION OF HWA1 AND HWA2, COMPLEMENTARY HYBRID WEAKNESS GENES IN RICE

Katsuyuki Ichitani[1], Satoru Taura[2], Takahiro Tezuka[3], Yuuya Okiyama[4], and Tsutomu Kuboyama[4]

[1]Faculty of Agriculture, Kagoshima University, 1-21-24 Korimoto, Kagoshima, Kagoshima 890–0065, Japan

[2]Institute of Gene Research, Kagoshima University, 1-21-24 Korimoto, Kagoshima, Kagoshima 890–0065, Japan

[3]Graduate School of Life and Environmental Sciences, Osaka Prefecture University, 1–1 Gakuen-cho, Nakaku, Sakai, Osaka 599–8531, Japan

[4]College of Agriculture, Ibaraki University, 3-21-1 Chuo, Ami, Ibaraki 300–0393, Japan

ABSTRACT

Hybrid weakness phenomena in rice reportedly have two causes: those of *HWC1* and *HWC2* genes and those of *HWA1* and *HWA2* genes. No detailed study of the latter has been reported. For this study, we first produced crosses among cultivars carrying the weakness-causing allele on the *HWA1* and *HWA2* loci to confirm the phenotype of the hybrid weakness and the genotypes of the cultivars on the two loci, as reported earlier. We then confirmed that these cultivars belong to Indica. Subsequent linkage analysis of *HWA1* and *HWA2* genes conducted using DNA markers revealed that both genes are located in the 1,637-kb region, surrounded by the same DNA markers on the long arm of chromosome 11. The possibility of allelic interaction inducing hybrid weakness is discussed.

INTRODUCTION

Many postzygotic reproductive barrier forms have been reported in rice (*Oryza sativa*), such as hybrid weakness (e.g., Oka 1957), hybrid pollen sterility (e.g., Long et al. 2008), and hybrid sterility causing female gamete abortion (e.g.,

Chen et al. 2008). Among them, hybrid weakness is definable as weak growth occurring in F_1 hybrids derived from crosses between two normal strains. According to its degree or symptom, it is also called hybrid lethality, hybrid abnormality, or hybrid necrosis. Hybrid weakness is also apparent in many other plant species including *Arabidopsis thaliana* (Bomblies et al. 2007), *Phaseolus vulgaris* (Shii et al. 1980), interspecific crosses among *Gossypium*(Lee 1981), and interspecific crosses among *Nicotiana* (Tezuka et al. 2007).

In rice, two hybrid weakness phenomena from different intraspecific cross combinations have been reported: one by Oka (1957) and the other by Amemiya and Akemine (1963). The hybrid weakness reported by Oka (1957) resulted from use of a set of complementary dominant genes: L_1 and L_2. They were renamed *L-1-a* and *L-1-b* by Kinoshita (1984); then *Hwa-1* and *Hwa-2* by Sato et al. (1987). According to the new gene nomenclature system for rice (McCouch and Committee on Gene Symbolization, Nomenclature and Linkage, Rice Genetics Cooperative (CGSNL) 2008), the new gene symbols *HWA1* and *HWA2* are used respectively for reference to *Hwa-1* and *Hwa-2*, as presented in Table 1. The latter phenomenon results from the use of a set of complementary dominant genes: *Hybrid weakness C1* (*HWC1*) and *Hybrid weakness C2* (*HWC2*). We have detected the chromosomal locations of these genes and have performed fine mapping (Ichitani et al. 2001; Ichitani et al. 2007; Kuboyama et al. 2009). However, the chromosomal locations of *HWA1* and *HWA2* have remained unknown.

Table 1. Gene symbols frequently used in this study according to the new gene nomenclature system for rice (McCouch and CGSNL 2008)

	Gene symbol					
	Oka (1957)	**Kinoshita (1984)**	**Sato et al. (1987)**	**This study**	**Gene full name**	**Carrier of weakness-causing gene**
Locus/gene				*HWA1*	*HYBRID WEAKNESS A1*	
Dominant allele	L_1	*L-1-a*	*Hwa-1*	*Hwa1-1*	*Hybrid weakness a1*-1	P.T.B.10, A.D.T.4, A.D.T.14, M.T.U.9, P.T.B.8
Recessive allele	+	+	*Hwa-1*$^+$	*hwa1-2*	*hybrid weakness a1-2*	
Locus/gene				*HWA2*	*HYBRID WEAKNESS A2*	

Dominant allele	L_2	*L-1-b*	*Hwa-2*	*Hwa2-1*	*Hybrid weakness a2*-1	P.T.B.7, accession 418
Recessive allele	+	+	*Hwa-2* $^+$	*hwa2-2*	*hybrid weakness a2-2*	

According to Oka (1957), hybrids carrying both *Hwa1-1* and *Hwa2-1* showed the following symptoms: germination and growth in the seedling stage were quite normal until the seedlings had developed three to four leaves, at which time growth stopped and the leaves yellowed. The phenotype differs considerably from that of hybrids carrying both *Hwc1-1* and *Hwc2-1*, of which the symptoms are characterized by root growth inhibition appearing just 5 days after germination (Ichitani et al. 2001; Saito et al. 2007).

The distribution of *Hwa1-1* and *Hwa2-1* genes was limited to some Indian cultivars (Oka 1957). Based on results of three-way cross combinations, Oka (1957) reported that P.T.B.10 (accession 414) and P.T.B.7 (accession 419) carry *Hwa1-1* and *Hwa2-1*, respectively, whereas Pei-ku (accession 108), Padi ase banda (accession 647), and Kissin (accession 521) carry neither *Hwa1-1* nor *Hwa2-1*. Oka (1957) examined many different cultivars for the distribution of these genes. The four Indian cultivars—A.D.T.4 (accession 415), A.D.T.14 (accession 417), M.T.U.9 (accession 420), and P.T.B.8 (accession 421)—showed weakness in hybrids with P.T.B.7, indicating that these four cultivars also carry *Hwa1-1*. Actually, accession 418, an Indian cultivar whose name is unknown, showed weakness in the hybrid with P.T.B.10, indicating that accession 418 carries *Hwa2-1*. All cultivars relating to this hybrid weakness belong to Indica. The other cultivars that were examined carry neither *Hwa1-1* nor *Hwa2-1*. Information about the gene carriers of *Hwa1-1* and *Hwa2-1* is presented in Table 1.

The molecular mechanism of the hybrid weakness has remained unknown. Elucidating the mechanism necessitates clarification of the causal genes and their gene products. We selected a map-based cloning strategy to identify the two causal genes. As a starting point of map-based cloning, this report describes the chromosomal locations of *HWA1* and *HWA2* genes. It is particularly interesting that both genes are located in the 1,637-kb region, surrounded by the same DNA markers on the long arm of chromosome 11. The possibility of allelic interaction inducing hybrid weakness is discussed in this report.

RESULTS

Confirmation of the Experiment by Oka (1957)

Indian cultivars carrying *Hwa1-1* or *Hwa2-1* gene, P.T.B.10, P.T.B.7, and A.D.T.14, were provided by the Genebank of the National Institute of Agrobiological Sciences, Japan. We produced crosses among these three cultivars and confirmed that the F_1 hybrids from the cross between P.T.B.7 and P.T.B.10, and those from the cross between P.T.B.7 and A.D.T.14 showed the weakness symptom, as reported by Oka (1957) (Fig. 1). The F_1 hybrids from the cross between P.T.B.10 and A.D.T.14 showed normal growth. These results indicate that these Indian cultivars carry one of the hybrid-weakness-causing genes reported by Oka (1957).

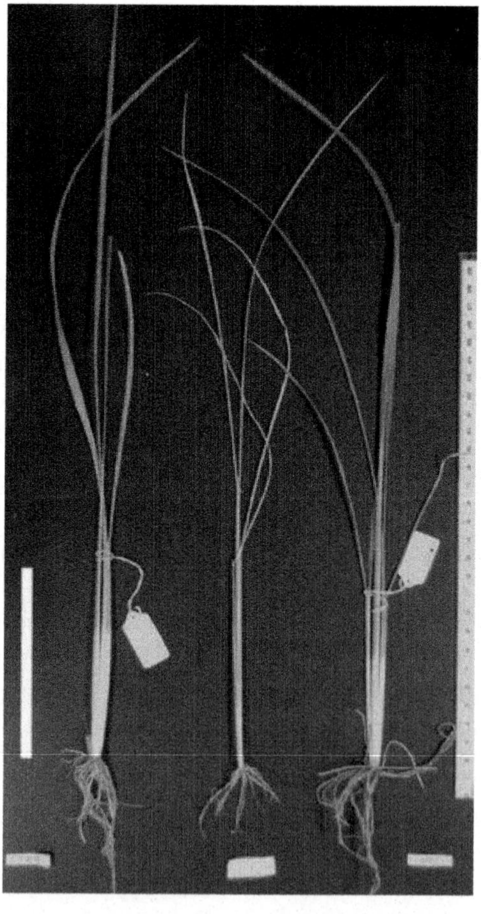

Figure 1. Hybrid weakness caused by two complementary genes: *Hwa1-1* and *Hwa2-1*. Seedlings are shown 40 days after sowing date: *from left to right*, P.T.B.10, F_1 de-

rived from the cross between P.T.B.10 and P.T.B.7, and P.T.B.7. The F_1's developed leaves have turned yellow from the tip. *Bar* shows 10 cm.

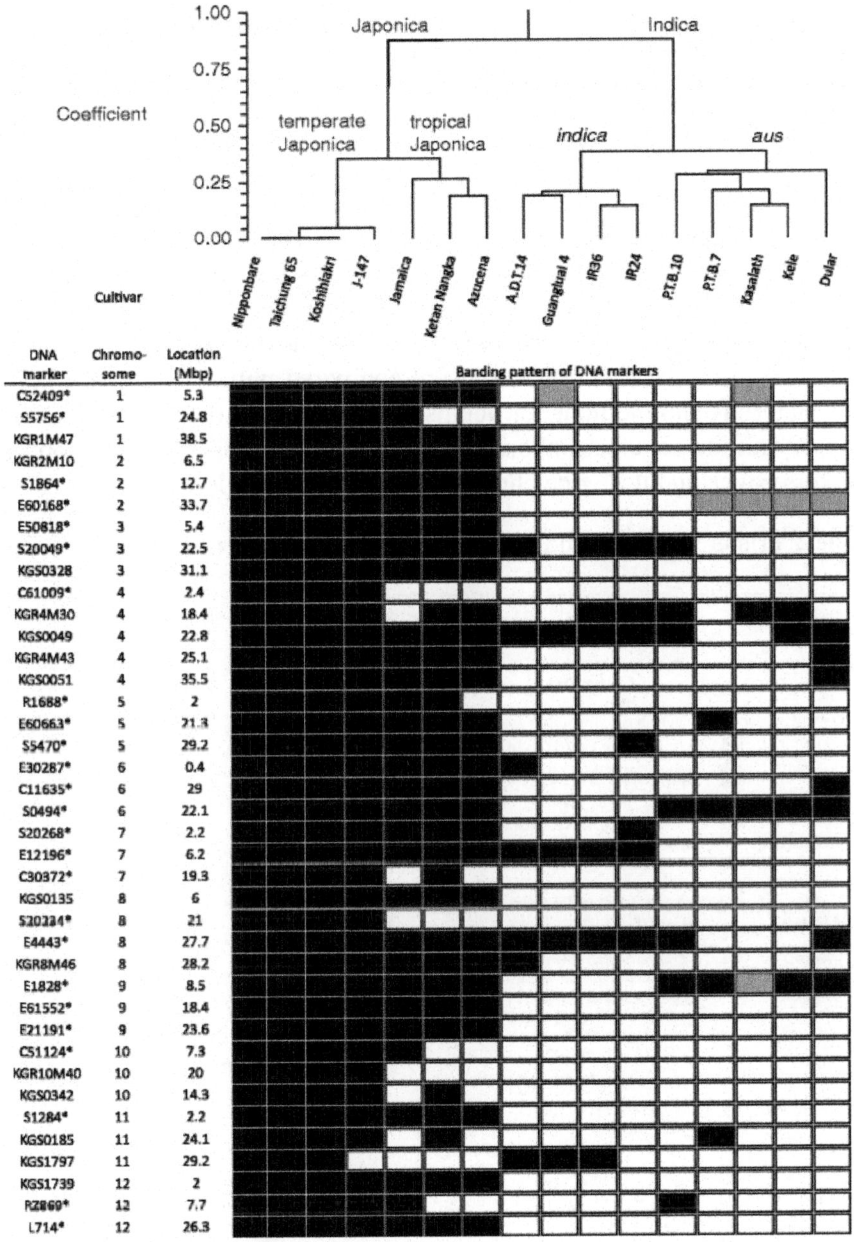

Figure 2. Classification of the Indian cultivars carrying *Hwa1-1* or *Hwa2-1* gene and 13 reference cultivars into four varietal groups based on banding patterns of 39 DNA

markers covering all 12 rice chromosomes. DNA markers with an *asterisk* (*) were developed by RGP (http: //rgp.dna.affrc.go.jp/E/publicdata/caps/index.html); those *without an asterisk* were developed in this study (Table 2). The same banding patterns as Nipponbare are indicated as *solid rectangles*. Other patterns are shown as *open* and *shaded rectangles*. Shown*above* are genetic relations among the 16 cultivars obtained using UPGMA cluster analysis. Each fragment size of the 39 markers was treated as a unique characteristic and scored as present (1) or absent (0). The data matrix was used to calculate genetic similarities using the Jaccard (1908) coefficient.

Oka (1957) reported that all cultivars related to this hybrid weakness belong to Indica. If that is true, then the mapping population of *HWA1* and*HWA2* should be constructed from the cross between these cultivars and Japonica cultivars because DNA polymorphism between Indica cultivars and Japonica cultivars is frequently observed. To confirm the classification by Oka (1957), the three Indian cultivars were compared with 13 reference cultivars (see Materials and methods for details) with respect to the banding patterns of polyacrylamide gel electrophoresis of 39 PCR-based insertion/deletion (indel) markers covering all 12 rice chromosomes (Fig. 2, Table 2).

Table 2. Primer sequences and locations of the indel markers designed for this study

Marker name		Primer sequences (5′–3′)	Location on IRGSP pseudomolecules Build05			Source of DNA polymorphism information
			Chromosome	Position		
				From	To	
KGR1M47	F	CACAAATAGAATTACTGATGAAACCTT	1	38529588	38529747	Shen et al. 2004
	R	CGTTACCGCTTATGTAGAGTCATC				
KGR2M10	F	CATACATGGATGTCTAGTCGAAGA	2	6465517	6465683	Shen et al. 2004
	R	CAGTTCCAGTAAGTACATGGGTTT				
KGS0328	F	ATCTAGCAAAATTATTCGAGCAGAA	3	31067890	31068058	Monna et al. 2006
	R	ACTTTACAGTAACAAGGGGTGCAA				
KGR4M30	F	CAAAATAGGGAGGCAGATAGACA	4	18395635	18395794	Shen et al. 2004
	R	CTCCTGGTTGTATGCTCGTAAAT				
KGS0049	F	AAATTATACTTCCAATCAAGCATCAAG	4	22809557	22809823	Monna et al. 2006
	R	AATTGTATTGGATTGGAGCTGGT				
KGR4M43	F	GAGGTTATCCTCCCTAACACCAG	4	25116441	25116556	Shen et al. 2004
	R	TGCCAAATACAATATGACCACAA				
KGS0051	F	CGAAAACAGTTAGGTGTTTGTTAGG	4	35483948	35484104	Monna et al. 2006
	R	CTTACATGTAGTACAAACAACCCACA				
KGS0494	F	GATGGCTAGCTTGACTCCTGAATA	6	22083785	22084001	Monna et al. 2006
	R	CTCTAGGATACAATCATGGCAAACT				
KGS0135	F	ACCTCATTTTATTTCAACATTGCAG	8	6028192	6028300	Monna et al. 2006
	R	CTTCCTGACCAAGTTAAACCCCTA				
KGR8M46	F	CGAATAATTTGTAGCCGAGAAAA	8	28228891	28229086	Shen et al. 2004
	R	GCAGAGTCCAGAGAAGATCCAT				
KGR10M40	F	CGTTTAATTTACGTGCGAATAGG	10	19966648	19966807	Shen et al. 2004
	R	AACCTGAGGCACTAGTTCGTTATC				
KGS0342	F	ATACACACAGCAGACATAAGGTGAT	10	14305320	14305584	Monna et al. 2006
	R	AATAACCGTTTGATTGGACTAAAAA				
KGS0185	F	TGACTACAGAAATAGTGCAGCTTCT	11	24091407	24091582	Monna et al. 2006
	R	CCCCCTCCTTGACTTTGG				
KGC11M1	F	GGCAGGAGAGGAGAAACTGA	11	25263891	25263963	This study
	R	GAAGAAAGTGACCATGGATGAA				
KGS1797	F	AGTGGTGAGCTGCTAACAAATCTCT	11	29167452	29167572	Monna et al. 2006
	R	GCCGAGCGAGCTGAGTATC				
KGS1739	F	AGAGACGCAGGAGCTGCTTA	12	1999528	1999818	Monna et al. 2006
	R	CATGACCCTTCTATGGCAATTAT				

The relations among the 16 cultivars were revealed by UPGMA cluster analysis (Fig. 2). The 16 cultivars were classified clearly into two groups corresponding to Japonica and Indica. All cultivars related to this hybrid weakness belong to Indica, which is consistent with the report by Oka (1957). The Japonica group was divided into temperate Japonica and tropical Japonica. The four cultivars in temperate Japonica showed the same banding patterns with 39 DNA markers, except that J-147 showed a different pattern with KGS1797. The three cultivars in tropical Japonica showed the same banding patterns as temperate Japonica does with 28 DNA markers. The Indica group was divided into two subgroups, with one subgroup comprising IR36, IR24, Guangluai 4, and A.D.T.14, and the other comprising Kasalath, Kele, Dular, P.T.B.10, and P.T.B.7. Garris et al. (2005) classified 234 rice cultivars into five groups: *indica*, *aus*, *aromatic*, *temperate japonica*, and *tropical japonica*. Then IR36 was categorized into *indica*, and Kasalath into *aus*. In the present study, according to Garris et al. (2005), the subgroup comprising Kasalath and the other subgroup comprising IR36 will be designated respectively hereinafter as *aus* and *indica* (Fig. 2). Regarding the three cultivars related to this hybrid weakness, A.D.T.14 was found to belong to *indica*, whereas P.T.B.7 and P.T.B.10 belong to *aus*.

Mapping of *HWA1* and *HWA2*

The three cultivars proved to belong to Indica, as reported by Oka (1957). They showed polymorphism with Japonica cultivars with high frequency (Fig. 2). We selected a Taiwanese temperate Japonica cultivar Taichung 65 (T65) and produced hybrids among the three Indian cultivars and T65 because much polymorphism has been noted among the three Indian cultivars and T65. Furthermore, because it is weakly photoperiod sensitive, hybridization between T65 and other cultivars can be accomplished more easily. All hybrids from crosses between T65 and the three cultivars showed normal growth, indicating that T65 carries neither *Hwa1-1* nor *Hwa2-1*.

Then, linkage analysis of *HWA1* and *HWA2* was conducted using a three-way cross population. For the location of *HWA1*, the F_1 plants derived from the cross between T65 and P.T.B.10 were crossed with P.T.B.7. The three-way cross population [P.T.B.7 × (T65 × P.T.B.10)] ($n=176$) was divided into 105 weak plants carrying both *Hwa1-1* and *Hwa2-1*, and 71 normal plants carrying only the *Hwa2-1* gene. The observed ratio did not fit the expected ratio 1: 1 ($^2=6.57$, $P=0.01038$). We inferred that the deviation from the expected ratio was probably caused by the linkage between *HWA1* gene and gametophytic reproductive barrier gene(s), which is often found in crosses between Indica and Japonica (see Discussion). Six typical weak plants and six typical normal plants were selected from the three-way cross population. They were subjected to preliminary linkage analysis with 83 DNA markers distributed on the whole rice genome (for details, see Materials and methods). A clear linkage was detected between *HWA1* and DNA markers on the long arm of chromosome 11, RM5349 (McCouch et al. 2002), and RM224 (Chen et al. 1997). Linkage of *HWA1* with RM5349 and RM224 was confirmed by adding the 82 three-way cross plants (Fig. 3).

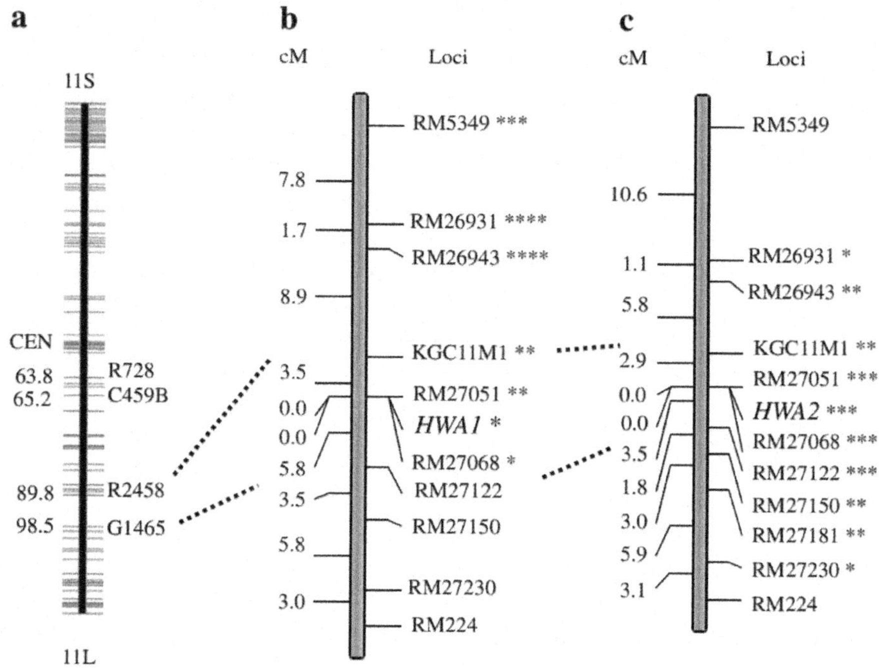

Figure 3. Linkage map showing the location of *HWA1* and *HWA2* on rice chromosome 11. **a** RFLP framework map of chromosome 11 quoted from Harushima et al. (1998). **b** Linkage map of *HWA1* constructed from the three-way cross population [P.T.B.7 × (T65 × P.T.B.10)] (n=176). **c** Linkage map of *HWA2* constructed from the three-way cross population [A.D.T.14 × (T65 × P.T.B.7)] (n=175). Genotypes of RM5349 and RM224 were analyzed for 94 plants in *HWA1* mapping and for 96 plants in *HWA2* mapping to know if these genes are located between the two markers. *, **, ***, and **** mean that the gene segregation was significantly deviated from the expected ratio (1: 1) at 0.05, 0.01, 0.001, and 0.0001 levels, respectively.

Along with the mapping of *HWA1*, linkage analysis of *HWA2* was performed using the same strategy. The F_1 plants derived from the cross between T65 and P.T.B.7 were crossed with A.D.T.14. The three-way cross population [A.D.T.14 × (T65 × P.T.B.7)] (n=175) was divided into 110 weak plants carrying both *Hwa1-1* and *Hwa2-1*, and 65 normal plants carrying only the *Hwa1-1* gene. The observed ratio did not fit the expected ratio 1: 1 (^{2}=11.57, P=0.00067), probably because of the reproductive barrier, as described above. Six typical weak plants and six typical normal plants that had been selected from the three-way cross population were subjected to preliminary linkage analysis with 84 DNA markers distributed throughout the rice genome (see Materials and methods), indicating that *HWA2* was also linked with RM5349 and RM224. Linkage of *HWA2* with RM5349 and RM224 was confirmed by adding the 84 three-way cross plants (Fig. 3).

We selected seven and eight polymorphic simple sequence repeats (SSR) markers reported by the International Rice Genome Sequencing Project (IRGSP) (2005), respectively, for mapping of *HWA1* and *HWA2*. Linkage analysis using these markers revealed a large gap separating RM26943 and RM27051 (Fig. 3). Although we also examined several published SSR markers in IRGSP (2005), they were not applicable to the linkage analysis. To fill the gap, we detected a four-base-pair indel between Nipponbare and Kasalath by comparing the genome sequence of Nipponbare and the BAC-end sequence of Kasalath in the chromosomal region. The PCR products from the primer pair encompassing the indel showed clear banding patterns. Polymorphism between Nipponbare and Kasalath was conserved between T65 and the three Indian cultivars. The DNA marker using this indel was named KGC11M1 (Table 2), which we added to the linkage analysis. The linkage analysis of *HWA1* and eight DNA markers using 176 three-way cross plants showed that *HWA1* was located between KGC11M1 and RM27122, and that it cosegregated with RM27051 and RM27068 (Fig. 3). Linkage analysis of *HWA2* and nine DNA markers using 175 three-way cross plants showed that *HWA2* was also located between KGC11M1 and RM27122, and that it cosegregated with RM27051 and RM27068 (Fig. 3). To sum up, both *HWA1* and *HWA2* genes cosegregated with RM27051 and RM27068, located on the 1,637-kb region surrounded by KGC11M1 and RM27122 on the long arm of chromosome 11.

DISCUSSION

In this study, both *HWA1* and *HWA2* were located between the two DNA markers on chromosome 11: KGC11M1 and RM27122. Harushima et al. (1998) constructed an often-cited restriction fragment length polymorphism (RFLP) marker-based high-density linkage map for rice, in which some RFLP markers have been sequenced. Based on Nipponbare genome sequence, KGC11M1 and RM27122 were close to the two RFLP markers R2458 and G1465, which were reported respectively at positions 89.8 and 98.4 in the map by Harushima et al.

(1998). That location indicates that *HWA1* and *HWA2* are in the chromosomal region surrounded by these markers (Fig. 3).

In the three-way cross-mapping populations, gene segregation was not fitted to the expected ratio, but it was skewed toward weakness-causing alleles for both *HWA1* and *HWA2* loci. Sawamura and Sano (1997) reported that P.T.B.10 carries a gamete eliminator of S_{11}(t), which is linked to *la* on chromosome 11. The P.T.B.10 allele S_{11}(t) induces abortion of gametes carrying the T65 allele S_{11}^{a}(t) only in heterozygotes. The recombination value between S_{11} and *la* (lazy habit) was estimated to be 0.06 ± 0.03 to 0.23 ± 0.23. Miura et al. (2003) conducted linkage analysis of *la* with RFLP markers. The *la* gene was located between R728 and C459B, of which the positions were 63.8 and 65.2, respectively, in the map by Harushima et al. (1998). These experimentally obtained results suggest that distorted segregation at the*HWA1* locus in the three-way cross [P.T.B.7 × (T65 × P.T.B.10)] is attributable to S_{11}(t) linked to *Hwa1-1*. Distorted segregation at the *HWA2*locus in the three-way cross [A.D.T.14 × (T65 × P.T.B.7)] reports that P.T.B.7 might carry the S_{11}(t) allele at S_{11}(t) locus. The heterozygote (S_{11}(t)S_{11}^{a}(t)) induces female and male infertility (Sawamura and Sano 1997). The fact that F_1 plants from the cross (T65 × P.T.B.10) and those from the cross (T65 × P.T.B.7) both showed semi-sterile panicles (Ichitani et al. unpublished results) also suggests that a gamete eliminator such as S_{11}(t) is attributable to the segregating distortion in both cross combinations. However, the peaks of distortion differed between the two populations, implying that a gene(s) other than S_{11}(t) was responsible for the distortion in the three-way cross [A.D.T.14 × (T65 × P.T.B.7)].

Two genetic models can explain hybrid weakness. According to one model, hybrid weakness results from interaction among non-allelic genes: the causal genes of another hybrid weakness in rice, *HWC1* and *HWC2*, are located on the different chromosomes (Ichitani et al.2007; Kuboyama et al. 2009). This case is true for the *A. thaliana* case (Bomblies et al. 2007). The candidate genes in the mapped regions of*HWC1* and *HWC2* do not mutually overlap; they belong to distinct gene families. The other model suggests allelic interaction: a hybrid necrosis symptom observed in interspecific crosses among *Gossypium* was explained by interlocus and intralocus interaction, shown respectively by the dominant Le_2^{dav} gene from *Gossypium davidsonii* at Le_2 locus and Le_1 and/or Le_2, dominant alleles from *Gossypium hirsutum* at the Le_1 and Le_2 loci (Lee 1981). These two loci are thought to be on mutually homoeologous chromosomal segments on the *G. hirsutum* allotetraploid genome (Samora et al. 1994). Increased gene dosage of Le_1 and Le_2 with Le_2^{dav} reportedly hastens necrosis (Rooney and Stelly 1990). In the *Gossypium* case, no candidate gene information was reported. Regarding the reproductive barrier in rice, causal genes were cloned in an allelic interaction case: heterozygotes of Indica allele *S5-i* and Japonica allele *S5-j* at the *S5* locus induce embryo-sac semi-sterility caused by partial abortion of female gametes carrying the Japonica alleles. Chen et al. (2008) cloned the *S5* gene, demonstrating that *S5* encodes an aspartic protease conditioning embryo-sac fertility, the allelic difference being very small at the sequence level. Chen et al. (2008) did not explain the molecular mechanism caused by the heterozygous form at the *S5* locus. In another case, causal genes of reproductive barrier in rice, which had been thought to be allelic before, proved to be a complex of two very tightly linked genes. At the hybrid male sterility locus *Sa*, heterozygotes of Indica allele Sa^i and Japonica allele Sa^j exhibited male semi-sterility. Long et al. (2008) cloned the *Sa* gene, finding that it comprises two adjacently located genes, *SaM* and *SaF*, which respectively encode a small ubiquitin-like modifier E3 ligase-like protein and an F-box protein. Most Indica cultivars contain a haplotype SaM^+SaF^+, whereas all Japonica cultivars have SaM^-SaF^-. Interaction of SaM^- and SaF^+ in the presence of SaM^+ is necessary for this male sterility.

In this study, the physical distance between the KGC11M1 and RM27122 markers was found to be 1,637 kb. According to the Rice Genome Annotation Project (Ouyang et al. 2007, http: //rice.plantbiology.msu.edu/), more than 100 genes are located in the region. We were unable to determine whether *Hwa1-1* and *Hwa2-1* are allelic or not, but the possibility exists that heterozygotes induce the hybrid weakness symptom. To elucidate their relations, whether by tight linkage or allelism, larger mapping populations must be analyzed.

A causal gene controlling hybrid necrosis in *Arabidopsis* was cloned. Reportedly, it encodes an NB-LRR protein (Bomblies et al. 2007). We fine-mapped *HWC2* gene in rice, and narrowed down the area of interest to 19 kb, identifying five candidate genes, one of which encodes an NB-LRR protein (Kuboyama et al. 2009). Alcázar et al. (2009) fine mapped a causal gene of incompatibility or hybrid breakdown to a cluster of TIR-NB-LRR genes in *Arabidopsis*. Yamamoto et al. (2010) reported that *hbd3*, a causal gene controlling hybrid breakdown in rice, is located on the cluster of NB-LRR genes. These reports suggest that plant immune systems conditioned by NB-LRR protein(s) induce hybrid weakness, as reported by Bomblies and Weigel (2007). According to the Rice Genome Annotation Project (Ouyang et al. 2007, http: //rice.plantbiology.msu.edu/), two NB-LRR disease-resistance gene homologs, LOC_Os11g39160 and LOC_Os11g39260, and ten disease resistance-related genes containing NB-ARC domain are located on the target region of *HWA1* and *HWA2*, implying that these genes are causal genes. However, many other genes are also located in the target region. No NB-LRR disease resistance gene homolog is located in the *HWC1* target region (Ichitani et al. 2007). To clarify the fundamental mechanism causing hybrid weakness, many cross combinations causing hybrid weakness in various species should be studied at the molecular level. The cloning of *HWA1* and *HWA2* might provide new information related to hybrid weakness.

Results of cluster analysis show that A.D.T.14 carrying *Hwa1-1* belongs to *indica*, whereas P.T.B.10 carrying *Hwa1-1* and P.T.B.7 carrying*Hwa2-1* belong to *aus*, which suggests a lack of relation between varietal differentiation and the genotypes at the *HWA1* and *HWA2*. However, the three cultivars are insufficient to support discussion of the distribution of these genes. Having learned recently that all the plant materials used by Oka (1957) have been maintained in the National Institute of Genetics, Japan, we plan to use these cultivars to find *Hwa1-1-* or*Hwa2-1*-specific conserved haplotypes of DNA markers. Regarding *HWC2* genes, we found that the DNA marker banding patterns of 14 markers surrounding the *HWC2* locus were conserved completely among the 13 *Hwc2-1* carriers (Kuboyama et al. 2009). We will find potential carriers with the aid of DNA markers from our rice germplasm collection and make test crosses to determine whether they really carry the genes if such a haplotype is found in the *Hwa1-1* or *Hwa2-1* carriers. We plan to add these gene carriers in the cluster analysis to confirm the relation between varietal differentiation and these genes. The increase in the number of gene carriers might disclose conserved haplotypes in the *HWA1* or *HWA2* locus, which help to identify the causal genes.

To identify the causal genes and to confirm whether they are allelic or not, we are undertaking high-resolution mapping and linkage disequilibrium analysis of both *HWA1* and *HWA2* genes. We have also started physiological and close morphological analyses of the hybrid weakness. In three-way cross populations, there existed a few intermediate type plants between normal and weak ones, partly because of diverse genetic backgrounds in the populations (see Materials and methods). We are introducing the *Hwa1-1* gene from P.T.B.10 and A.D.T.14, and the *Hwa2-1* gene from P.T.B.7 in T65 genetic background by backcrossing to develop near-isogenic lines of these genes. Using these near-isogenic lines, we will evaluate the effects of these genes more exactly.

MATERIALS AND METHODS

Plant Materials

The three cultivars related to this hybrid weakness, P.T.B.7, P.T.B.10, and A.D.T.14, were compared with the 13 reference cultivars with regard to the banding patterns of polyacrylamide gel electrophoresis of 39 PCR-based indel markers (Fig. 2). Among the cultivars, Nipponbare and Koshihikari are improved Japanese cultivars, and T65 is an improved Taiwanese cultivar. These three cultivars are generally classified as temperate Japonica. J-147 is a native cultivar in Japan (Sato and Morishima 1988). Ketan Nangka, Azucena, and Jamaica are native cultivars respectively originating in Malaysia, the Philippines, and Peru. They are generally classified as tropical Japonica. Three cultivars generally classified as Indica are Guangluai 4, an improved Chinese cultivar, and IR36 and IR24 developed by the International Rice Research Institute. Kele, Dular, and Kasalath cultivars are native to India. Kele and Dular are categorized as *aus* in Wan and Ikehashi (1997). Kasalath has been categorized as Indica in many studies; DNA marker polymorphism has been observed with high frequency between Kasalath and Japonica cultivars. According to classification by Garris et al. (2005) based on SSR markers or Zhao et al. (2010) based on single nucleotide polymorphism—which divided rice cultivars into *temperate japonica*, *tropical japonica*, *indica*, *aus*, and one more group—Nipponbare and Koshihikari were classified as *temperate japonica*, Azucena as *tropical japonica*, IR36 as *indica*, Kasalath as *aus*, and Dular as admixed.

For hybridization, flowering times of plant materials were mutually synchronized by the use of many sowing dates. Seeds were sown in nurseries with plant spacing of 3 × 3 cm. One month after sowing, seedlings were transplanted in a paddy field in the experimental farm of Kagoshima University.

The three-way cross populations in which *HWA1* or *HWA2* gene segregated were grown in nurseries along with parental cultivars and F_1 plants inducing the weakness. Plant spacing was 3 × 6 cm.

DNA Marker Analysis

The DNA of the cultivars used in the cluster analysis was extracted according to Dellaporta et al. (1983) with some modifications. The DNA of the mapping populations was extracted according to the experimental protocols of Rice Genome Research Program (RGP) (http: //rgp.dna.affrc.go.jp/E/rgp/protocols/index.html, written in Japanese) with some modifications. Briefly, each young leaf tip, 2 cm long from a single plant, was put on a well in a 96-deep well plate. Then 100 μl of extraction buffer (100 mM Tris–HCl (pH 8.0), 1 M KCl, and 10 mM EDTA) were added with a 5-mm-diameter stainless steel ball in a well. After covering with a hard lid, the plate was shaken hard (ShakeMaster ver. 1.2; BioMedical Science Inc.) for a minute to grind the leaves. After centrifuging, a plate was incubated at 70°C for an hour. Then 9 μl of supernatant was recovered and 7 μl of 2-propanol was added. After centrifuging, the supernatant was discarded and the DNA pellet was rinsed with 50 μl of 70% ethanol. The DNA pellet was dried and dissolved in 50 μl of sterilized distilled water.

The PCR conditions for indel and SSR markers used for this study were 95°C for 10 min, 40 cycles of 94°C for 30 s, 55°C for 30 s, and 72°C for 30 s with subsequent final extension of 72°C for 1 min. The PCR mixture (5 μl) contained 1 ml of template DNA, 200 mM of each dNTP, 0.2 μM of primers, 0.25 units of Taq polymerase (AmpliTaq Gold; Applied BioSystems), and 1× buffer containing $MgCl_2$. The PCR products were analyzed using electrophoresis in 10% (29: 1) polyacrylamide gel with subsequent ethidium bromide staining and were viewed under ultraviolet light irradiation. Most PCR–based DNA markers used for this study have already been published: Markers with an asterisk in Fig. 2were designed by RGP (http: //rgp. dna.affrc. go.jp/E/publicdata/caps/index.html). Information related to SSR markers was obtained from Panaud et al. (1996), Chen et al. (1997), Temnykh et al. (2001), McCouch et al. (2002), and IRGSP (2005). Some primer pairs did not perform well. Therefore, we redesigned them or developed new DNA markers.

Most of the new markers were based on the indel between Nipponbare and an indica cultivar 93–11 (Shen et al. 2004) or those between Nipponbare and Kasalath or an indica cultivar Guangluai 4 (Monna et al. 2006). Markers were named by adding KG to the original indel information ID, R*** (Shen et al. 2004) or S**** (Monna et al. 2006). In redesigning or developing DNA markers, we made alignments of Nipponbare genome sequence (IRGSP 2005), 93–11 genome sequence (Yu et al. 2002) and/or BAC-end sequence of Kasalath (http: //rgp.dna.affrc.go.jp/blast/runblast.html, Katagiri et al. 2004) using DNAsis Pro (Hitachi Software Engineering Co.). Primer design was performed using Primer 3 (Rozen and Skaletsky 2000). The sequences of primers designed in this study are presented in Table 2.

Cluster Analysis

Genetic relations among the 16 cultivars (Fig. 2) were evaluated using the 39 DNA markers. Each fragment size was treated as a unique characteristic, and scored as present (1) or absent (0). The data matrix was used to calculate genetic similarities using the Jaccard coefficient (Jaccard 1908). A phenogram was constructed using the unweighted pair-group method with the arithmetic mean (UPGMA) with software (NTSYS-pc ver. 2.2; Exeter Software).

Mapping Strategy

About 1 month after the sowing date, the seedlings in the three-way cross populations, [P.T.B.7 × (T65 × P.T.B.10)] (n=176) and [A.D.T.14 × (T65 × P.T.B.7)] (n=175), developed three to four leaves. Plants of two types appeared. One type was characterized by the yellowing of leaf tips and weak growth as F_1 plants (Fig. 1). It was categorized as a weak plant carrying both *Hwa1-1* and *Hwa2-1* genes. Another type showed normal growth as parental cultivars. It was categorized as a normal plant carrying only *Hwa1-1* or *Hwa2-1*. The tips of lower leaves of normal plants sometimes yellowed also, but the tips of upper leaves remained green. In contrast, in weak plants, all leaf tips except those of the youngest (still expanding) ones yellowed. Therefore, the difference between weak plants and normal plants was clear in most cases. A few intermediate type plants could not be categorized into the above two types. They were excluded from analyses.

For *HWA1* mapping, a total of 83 DNA markers distributed on the whole rice genome were used: 24 polymorphic indel markers in Fig. 2, an indel marker named S21074 designed by RGP, another indel marker named Knindel1 (Matsubara et al. 2007), a sequence characterized amplified region (SCAR) marker named d (Ueda et al. 2005), and 56 published SSR markers. For *HWA2* mapping, a total of 84 DNA markers distributed on the whole rice genome were used: the 23 polymorphic indel markers presented in Fig. 2, 59 SSR markers, and two indel markers designed by RGP—S21074 and S5865.

In each mapping population, six typical weak plants and six typical normal plants were selected and subjected to preliminary linkage analysis with DNA markers, with the result that the linkage of *HWA1* and *HWA2* with RM5349 (McCouch et al. 2002) and RM224 (Chen et al. 1997) was visible on the long arm of chromosome 11. Then linkage analysis using all the plants in the mapping populations and polymorphic DNA markers surrounded by RM5349 and RM224 was conducted using a computer program (MapDisto ver. 1.7; Lorieux 2007). Map distances were estimated using the Kosambi function (Kosambi 1944).

ACKNOWLEDGMENTS

We gratefully acknowledge Dr. Atsushi Yoshimura of Kyushu University and Dr. Lisa Monna of the Plant Genome Center for valuable information related to DNA polymorphism in rice. We are also grateful to Dr. Yo-Ichiro Sato of the Research Institute for Humanity and Nature, the Genebank of the National Institute of Agrobiological Sciences for the kind provision of rice cultivars. This work was funded by the Ministry of Education, Culture, Sports, Science and Technology.

REFERENCES

1. Alcázar R, García AV, Parker JE, Reymond M. Incremental steps toward incompatibility revealed by Arabidopsis epistatic interactions modulating salicylic acid pathway activation. Proc Natl Acad Sci USA. 2009; 106: 334–9.
2. Amemiya A, Akemine H. Biochemical genetic studies on the root growth inhibiting complementary lethals in rice plant (Studies on the embryo culture in rice plant. 3). Bull Nat Inst Agric Sci D. 1963; 10: 139–226. in Japanese with English summary.
3. Bomblies K, Lempe J, Epple P, Warthmann N, Lanz C, Dangl JL, et al. Autoimmune response as a mechanism for a Dobzhansky-Muller-type incompatibility syndrome in plants. PLoS Biol. 2007; 9: 1962–72.
4. Bomblies K, Weigel D. Hybrid necrosis: autoimmunity as a potential gene-flow barrier in plant species. Nat Rev Genet. 2007; 8: 382–93.
5. Chen J, Ding J, Ouyang Y, Du H, Yang J, Cheng K, et al. A triallelic system of *S5* is a major regulator of the reproductive barrier and compatibility of indica–japonica hybrids in rice. Proc Natl Acad Sci USA. 2008; 105: 11436–41.
6. Chen X, Temnykh S, Xu Y, Cho YG, McCouch SR. Development of a microsatellite framework map providing genome-wide coverage in rice (*Oryza sativa* L.). Theor Appl Genet. 1997; 95: 553–67.
7. Dellaporta SL, Wood J, Hicks JB. A plant DNA minipreparation: version II. Plant Mol Biol Rep. 1983; 1: 19–21.
8. Garris AJ, Tai TH, Coburn J, Kresovich S, McCouch S. Genetic structure and diversity in *Oryza sativa* L. Genetics. 2005; 169: 1631–8.
9. Harushima Y, Yano M, Shomura A, Sato M, Shimano T, Kuboki Y, et al. A high-density rice genetic linkage map with 2275 markers using a single F_2 population. Genetics. 1998; 148: 479–94.
10. Ichitani K, Fukuta Y, Taura S, Sato M. Chromosomal location of *Hwc2*, one of the complementary hybrid weakness genes, in rice. Plant Breeding. 2001; 120: 523–5.
11. Ichitani K, Namigoshi K, Sato M, Taura S, Aoki M, Matsumoto Y, et al. Fine mapping and allelic dosage effect of *Hwc1*, a complementary hybrid weakness gene in rice. Theor Appl Genet. 2007; 114: 1407–15.
12. International Rice Genome Sequencing Project (IRGSP). The map-based sequence of the rice genome. Nature. 2005; 436: 793–800.
13. Jaccard P. Nouvelles recherches sur la distribution florale. Bull Soc Vaudoise Sci Nat. 1908; 44: 223–70.

14. Katagiri S, Wu J, Ito Y, Karasawa W, Shibata M, Kanamori H, et al. End sequencing and chromosomal *in silico* mapping of BAC clones derived from an *indica* rice cultivar, Kasalath. Breed Sci. 2004; 54: 273–9.

15. Kinoshita T. Proposal for rules of gene symbolization. Rice Genet Newsl. 1984; 1: 4–15.

16. Kosambi D. The estimation of map distance from recombination values. Ann Eugen. 1944; 12: 172–5.

17. Kuboyama T, Saito T, Matsumoto T, Wu J, Kanamori H, Taura S, et al. Fine Mapping of *HWC2*, a complementary hybrid weakness gene, and haplotype analysis around the locus in rice. Rice. 2009; 2: 93–103.

18. Lee JA. Genetics of D_3 complementary lethality in *Gossypium hirsutum* and *G. barbadense*. J Hered. 1981; 72: 299–300.

19. Long Y, Zhao L, Niu B, Su J, Wu H, Chen Y, et al. Hybrid male sterility in rice controlled by interaction between divergent alleles of two adjacent genes. Proc Natl Acad Sci USA. 2008; 105: 18871–6.

20. Lorieux M. 2007. MapDisto, a free user-friendly program for computing genetic maps. Computer demonstration (p958) given at the plant and animal genome XV conference, Jan 13–17 2007, San Diego, CA. URL: http: //mapdisto.free.fr/.

21. Matsubara K, Ito S, Nonoue Y, Ando T, Yano M. A novel gene responsible for hybrid breakdown found in a cross between *japonica* and*indica* cultivars in rice. Rice Genet Newsl. 2007; 23: 11–3.

22. McCouch SR, Teytelman L, Xu Y, Lobos KB, Clare K, Walton M, et al. Development and mapping of 2240 new SSR markers for rice (*Oryza sativa* L.). DNA Res. 2002; 9: 199–207.

23. McCouch SR, Committee on Gene Symbolization, Nomenclature and Linkage, Rice Genetics Cooperative (CGSNL). Gene nomenclature system for rice. Rice. 2008; 1: 72–84.

24. Miura K, Ashikari M, Matsuoka M, Hsing YIC. Mapping of the lazy gene, *la*. Rice Genet Newsl. 2003; 20: 29–30.

25. Monna L, Ohta R, Masuda H, Koike A, Minobe Y. Genome-wide searching of single-nucleotide polymorphisms among eight distantly and closely related rice cultivars (*Oryza sativa* L.) and a wild accession (*Oryza rufipogon* Griff.). DNA Res. 2006; 13: 43–51.

26. Oka HI. Phylogenetic differentiation of cultivated rice. XV. Complementary lethal genes in rice. Jpn J Genet. 1957; 32: 83–7.

27. Ouyang S, Zhu W, Hamilton J, Lin H, Campbell M, Childs K, et al. The TIGR rice genome annotation resource: improvements and new features. Nucleic Acid Res. 2007; 35: D883–7.

28. Panaud O, Chen X, McCouch SR. Development of microsatellite markers and characterization of simple sequence length polymorphism (SSLP) in rice (*Oryza sativa* L.). Mol Gen Genet. 1996; 252: 597–607.

29. Rooney WL, Stelly DM. Genetic effects on the timing of Le_2^{dav} induced necrosis of cotton. Crop Sci. 1990; 30: 70–4.

30. Rozen S, Skaletsky HJ. Primer3 on the WWW for general users and for biologist programmers. In: Krawetz S, Misener S, editors. Bioinformatics methods and protocols: methods in molecular biology. Totowa: Humana Press; 2000. p. 365–86.

31. Saito T, Ichitani K, Suzuki T, Marubashi W, Kuboyama T. Developmental observation and high temperature rescue from hybrid weakness in a cross between Japanese rice cultivars and Peruvian rice cultivar 'Jamaica'. Breed Sci. 2007; 57: 281–8.

32. Samora PJ, Stelly DM, Kohel RJ. Localization and mapping of the Le_1 and Gl_2 loci of cotton (*Gossypium hirsutum* L.). J Hered. 1994; 85: 152–7.

33. Sato YI, Sano Y, Nakagahra M. Gene symbols for gametic effect, sterility and weakness. Rice Genet Newsl. 1987; 4: 46–51.

34. Sato YI, Morishima H. Distribution of the genes causing F_2 chlorosis in rice cultivars of the Indica and Japonica types. Theor Appl Genet. 1988; 75: 723–727.

35. Sawamura N, Sano Y. Chromosomal location of gamete eliminator, S_{11}(t), found in an Indica-Japonica hybrid. Rice Genet Newsl. 1996; 13: 70–2.

36. Shen YJ, Jiang H, Jin JP, Zhang ZB, Xi B, He YY, et al. Development of genome-wide DNA polymorphism database for map-based cloning of rice genes. Plant Physiol. 2004; 135: 1198–205.

37. Shii CT, Mok MC, Temple SR, Mok DWS. Expression of developmental abnormalities of *Phaseolus vulgaris* L. Interaction between temperature and allelic dosage. J Hered. 1980; 71: 219–22.

38. Temnykh S, DeClerck G, Lukashova A, Lipovich L, Cartinhour S, McCouch S. Computational and experimental analysis of microsatellites in rice (*Oryza sativa* L.): frequency, length variation, transposon associations, and genetic marker potential. Genome Res. 2001; 11: 1441–52.

39. Tezuka T, Kuboyama T, Matsuda T, Marubashi W. Possible involvement of genes on the Q chromosome of *Nicotiana tabacum* in expression of hybrid lethality and programmed cell death during interspecific hybridization to *Nicotiana debneyi*. Planta. 2007; 226: 753–64.

40. Ueda T, Sato T, Hidema J, Hirouchi T, Yamamoto K, Kumagai T, et al. *qUVR-10*, a major quantitative locus for ultraviolet-B resistance in rice, encodes cyclobutane pyrimidine dimer photolyase. Genetics. 2005; 171: 1941–50.

41. Wan J, Ikehashi H. Identification of two types of differentiation in cultivated rice (*Oryza sativa* L.) detected by polymorphism of isozymes and hybrid sterility. Euphytica. 1997; 94: 151–61.

42. Yamamoto E, Takashi T, Morinaka Y, Lin S, Wu J, Matsumoto T, et al. Gain of deleterious function causes an autoimmune response and Bateson–Dobzhansky–Muller incompatibility in rice. Mol Genet Genomics. 2010; 283: 305–15.

43. Yu J, Hu S, Wang J, Wong GK, Li S, Liu B, et al. A draft sequence of the rice genome (*Oryza sativa* L. spp. *indica*). Science. 2002; 296: 79–92.

44. Zhao K, Wright M, Kimball J, Eizenga G, McClung A, Kovach M, et al. Genomic diversity and introgression in *O. sativa* reveal the impact of domestication and breeding on the rice genome. PLoS One. 2010; 5: e10780. doi: 10.1371/journal.pone.0010780.

Chapter 7

STRUCTURE AND STABILITY OF TELOCENTRIC CHROMOSOMES IN WHEAT

Dal-Hoe Koo[1], Sunish K. Sehgal[2], Bernd Friebe[1], and Bikram S. Gill[1]

[1]Department of Plant Pathology, Wheat Genetics Resource Center, Throckmorton Plant Sciences Center, Kansas State University, Manhattan, KS, 66506–5502, United States of America

[2]Department of Plant Science, South Dakota State University, Brookings, SD, 57007, United States of America

ABSTRACT

In most eukaryotes, centromeres assemble at a single location per chromosome. Naturally occurring telocentric chromosomes (telosomes) with a terminal centromere are rare but do exist. Telosomes arise through misdivision of centromeres in normal chromosomes, and their cytological stability depends on the structure of their kinetochores. The instability of telosomes may be attributed to the relative centromere size and the degree of completeness of their kinetochore. Here we test this hypothesis by analyzing the cytogenetic structure of wheat telosomes. We used a population of 80 telosomes arising from the misdivision of the 21 chromosomes of wheat that have shown stable inheritance over many generations. We analyzed centromere size by probing with the centromere-specific histone H3 variant, CENH3. Comparing the signal intensity for CENH3 between the intact chromosome and derived telosomes showed that the telosomes had approximately half the signal intensity compared to that of normal chromosomes. Immunofluorescence of CENH3 in a wheat stock with 28 telosomes revealed that none of the telosomes received a complete CENH3 domain. Some of the telosomes lacked centromere specific retrotransposons of wheat in the CENH3 domain, indicating that the stability of telosomes depends on the presence of CENH3 chromatin and not on the presence of CRW repeats. In addition to providing evidence for centromere shift, we also observed chromosomal aberrations including inversions and deletions in the short arm telosomes of double ditelosomic 1D and 6D stocks. The role of centromere-flanking, pericentromeric heterochromatin in mitosis is discussed with respect to genome/chromosome integrity.

INTRODUCTION

The centromeres, an essential part of all chromosomes, are responsible for chromosome segregation at mitosis and meiosis. Centromeres usually contain highly repetitive DNA, e.g. satellite DNA, which is associated with proteins in higher eukaryotes [1,2]. Although centromeres are not conserved at the DNA sequence level, many core centromeres analyzed to date contain nucleosomes with a histone H3 variant, CENPA in humans [3] and CENH3 in plants [4]. Thus, these specialized nucleosomes serve as the primary marker of centromere identity. In most eukaryotes, centromere proteins assemble at a single location in each chromosome [5] and are visible as a primary constriction in mitotic metaphase chromosomes. The centromere within a chromosome can move to a new location and form a neocentromere, which is often associated with chromosomal rearrangements in humans [6]. Such neocentromeres also have been reported in plants [7–10].

Each chromosome can be identified based on the centromere position; metacentric, submetacentric, acrocentric, or telocentric [11]. Naturally occurring telocentric chromosomes (hereafter referred as telosomes) are rare in plants. Darlington [12] suggested that the absence of telosomes in plants was caused by their instability. Evidence for the existence of stable telosomes has been provided by Marks [13], Strid [14] and Schubert and Rieger [15] in plants and by Southern [16] and Takagi and Sasaki [17] in animals. According to White [18], experimentally produced telosomes are unstable; and thus, all telosomes are unstable. Experimentally produced telosome stocks in plants, such as wheat (*Triticum aestivum* L.), barley (*Hordeum vulgare* L.), rye (*Secale cereal* L.), and rice (*Oryza sativa* L.), are reported [19,20]. Barley telotrisomics are fairly stable, except for triplo 1L, which shows chimaerism [21]. The question arises, why are some telosomes stable whereas others are not?

Centromeres divide quite regularly at mitosis and meiosis. However, when a chromosome is univalent, centric misdivision may occur during meiosis giving rise to telosomes [12]. Such a chromosome aberration is lethal in diploid organisms but can be tolerated in polyploids. The polyploidy nature of hexaploid wheat, *T. aestivum* (2n = 6x = 42), tolerates aneuploidy with either the addition or deletion of chromosomes. Sears and Sears [19] developed several aueuploid stocks in Chinese Spring (CS) wheat including ditelosomic stocks (one chromosome pair is substituted by a pair of either short or long arm telosomes, 2n = 40+2t; such stocks are nullisomic for one of the arms) and double ditelocentric stocks (one chromosome pair is substituted by a pair of short and long arm telosomes, 2n = 40+4t; such stocks are euploid). These stocks were used intensively for centromere mapping and allocating genes and markers to specific chromosome arms [22,23]. Moreover, the telosomes in the ditelosomic (hereafter Dt) and double ditelosomic (hereafter dDt) stocks are about half the size of a metacentric chromosome, making them amenable for flow sorting [24–26].

Flow-sorted telosomes are the foundation material for chromosome-arm-based BAC libraries and the wheat physical maps developed under the auspices of the International Wheat Genome Sequencing Consortium project [27]. Recently, individual flow-sorted chromosome arms were used to generate a draft sequence of the 17-Gb wheat genome [28]. Even with the extensive use of telosomic stocks for genetic and genomic studies in wheat, their detailed cytogenetic nature is poorly understood. The cytological stability of a telosome depends on the structure of its kinetochore. Steinitz-Sears [29] reported that the relative instability of a telosome may be attributed to the degree of completeness of its kinetochore. Because wheat telosomic stocks were developed by centric misdivision and stably transmitted to progeny, they are supposed to have either complete or nearly complete kinetochores.

In this study, we first developed a wheat CENH3 antibody (see experimental procedure) and then used it to identify and characterize the functional centromeric region of the telosomes. Second, centromeric-specific retrotransposons of wheat (CRWs) [30,31] were used to study the structure of intact and telosomic chromosomes. D-genome-specific, repetitive DNA (pAs1) [32] and single-copy DNA probes [33] were used to identify chromosomes and characterize chromosomal rearrangements. In addition, chromosome arm-specific molecular markers, derived from the wheat deletion bin map (http://probes.pw.usda.gov:8080/snpworld/Map) [34], were used to detect chromosomal aberrations. The data provide new insights into the structure and stability of telocentric chromosomes and their centromeres; the implications of these results to genetic and genomic studies of wheat are discussed.

MATERIALS AND METHODS

Plant Material and Chromosome Preparation

The cytogenetic stocks used in this study are listed (Table 1). Wheat cultivars, Chinese Spring (CS), Jagger, and TAM111, also were used in molecular and cytogenetical studies. For chromosome preparations, seeds were germinated in petri dishes on moist filter paper. Root tips (1–2 cm long) were treated overnight in ice water. The root tips were fixed overnight in a 3:1 ethanol:glacial acetic acid and then squashed in a drop of 45% acetic acid. For the immunofluorescence of CENH3, ice-cold-treated root tips were fixed immediately using 4% paraformaldehyde in PHEM (60 mM Pipes, 25 mM

Hepes, 10 mM EGTA, 2 mM $MgCl_2$, and 0.3 mM sorbitol, pH 6.8) for 40 min. After washing with 1x PBS (10 mM sodium phosphate, pH 7.0, and 140 mM NaCl), the root tips were treated with 2% cellulase, 1% pectinase (Sigma, St. Louis, MO) and 1% pectolyase in PHEM for 1 h and then squashed on poly-l-lysine coated slides (Sigma). All preparations were stored at -70°C until use.

Table 1. Wheat cytogenetic stocks used in this study.

Cytogenetic stock	Source/Plant ID
Dt1DS	WGRC, TA3087
Dt1DL	WGRC, TA3131
dDt1D	WGRC, TA3158
Dt2DS	WGRC, TA3123
Dt2DL	WGRC, TA3124
dDt2D	WGRC, TA3146
Dt3DS	WGRC, TA3193
Dt3DL	WGRC, TA3192
dDt3D	WGRC, TA3147
Dt4DS	WGRC, TA3125
Dt4DL	WGRC, TA3126
dDt4D	WGRC, TA3148
Dt5DS·Mt5DL	U.C Riverside
Dt5DL	WGRC, TA3127
dDt5D	WGRC, TA3149
Dt6DS	WGRC, TA3128
Dt6DL	WGRC, TA3129
dDt6D	WGRC, TA3150
Dt7DS	WGRC, TA3130
Dt7DL	WGRC, TA3071
dDt7D	WGRC, TA3151
dDt1A	WGRC, TA3132
dDt2A	WGRC, TA3133
dDt3A	WGRC, TA3134
dDt5A	WGRC, TA3136
dDt6A	WGRC, TA3137
dDt7A	WGRC, TA3138
CS dDt1B-dDt2B-dDt3B-dDt4A-dDt5B-dDt6B-dDt7B	WGRC, TA3356

Dt: ditelosome, one chromosome is represented by a pair of either short or long arm telosomes (2n = 40 +2t). dDt: double ditelosome, one chromosome is represented by one pair each of short (S) and long (L) arm telosomes (2n = 40+4t).

doi:10.1371/journal.pone.0137747.t001

Immuno-Detection of CENH3 and Fluorescence *in situ* Hybridization (FISH)

Wheat *CENH3* genes were described previously [31]. A peptide antigen, 'RTKHPAVRKTKALPKK', was synthesized and used to immunize rabbits at Thermo Fisher Scientific (www.thermofisher.com). The raised antisera were purified using an affinity sepharose column comprising the aforementioned peptide. The specificity of the antibody was checked by immunostaining of root tip and pollen mother cells of wheat (data not shown). Immunodetection of CENH3 and FISH procedures followed previously published protocols [35–37]. The rabbit antibodies to CENH3 were diluted to 1:1000 in TNB buffer (0.1 M Tris-HCl, pH 7.5, 0.15 M NaCl, and 0.5% blocking reagent). Approximately 100 μL of the diluted antibodies was added to each slide, and the slides were incubated in a humid chamber at 37°C for 2–3 h. After three washes in 1x PBS, 100 μL of rhodamine-conjugated goat anti-rabbit antibody (Jackson ImmunoResearch, West Crove, PA) (1:100 in TNB buffer) was added to the slides. Incubation and washes were the same as for the primary antibody. DNA probes of the CRWs, pAs1, pSc119, and the other single-gene probes were labeled with digoxigenin-11-dUTP, biotin-16-dUTP, and/or DNP-11-dUTP, depending on whether two or three probes were used in the FISH experiment. The cDNA clones used in this study were supplied by the National BioResource Project-Wheat, Japan. After post-hybridization washes, the probes were detected with Alexafluor 488 streptavidin for biotin-labeled probes, and rhodamine-conjugated anti-digoxigenin for dig-labeled probe. The DNP-labeled probe was detected with rabbit anti-DNP, followed by amplification with a chicken anti-rabbit Alexafluor 647 antibody.

Multicolor immuno-FISH detection was described previously [36]. Chromosomes were counterstained with 4′,6-diamidino-2-phenylindole (DAPI) in Vectashield antifade solution (Vector Laboratories, Burlingame, CA). The images were captured with a Zeiss Axioplan 2 microscope (Carl Zeiss Microscopy LLC, Thornwood, NY) using a cooled CCD camera CoolSNAP HQ2 (Photometrics, Tucson, AZ) and AxioVision 4.8 software. The final contrast of the images was processed using Adobe Photoshop CS5 software.

Sequential Detection of CENH3, CRWs, pSc119 and pAs1

For sequential detection the slides were first incubated with anti-CENH3 overnight at 4°C in a wet chamber. After washes in 1x PBS, the slides were incubated with the appropriate secondary antibody at 37°C for 50 min. Then slides were re-fixed with 4% paraformaldehyde at RT for 30 min. The slides were then denatured in 70% formamide in 2x SSC, 80°C for 2 min, washed in ice-cold 1x PBS for 5 min, and then DNA probe, CRWs, was applied to the slides. Post-hybridization wash and signal detection were the same as FISH procedure. After recording the both CENH3 and CRWs signals, the slides were washed in 4T (4x SSC/0.05% Tween 20) buffer for 1 h at 37°C and re-fixed with 4% paraformaldehyde and dehydrated in an ethanol series. The slides were re-probed with pAs1 and pSc119 to detect additional sequences on the same chromosome.

Genome-Specific Markers and PCR

The genome-specific primers used are listed in Table 2. PCR was performed with 15 μL of the reaction mixture containing 1x PCR buffer (Bioline USA Inc., Taunton, MA), 2 mM MgC12, 0.25 mM dNTPs, 5 pmol forward primer and reverse primer, respectively, 0.02 U/μL Taq DNA polymerase (Bioline USA Inc., Taunton, MA), and 20 ng genomic DNA. PCR amplification was according to Liu et al. [38]. PCR products were resolved on 2.5% agarose gels and visualized by ethidium bromide staining under UV light.

Table 2. Wheat genome specific EST markers used in this study.

EST marker	Forward primer	Reverse primer	Deletion bin
BE405518	GTCTCAGGTATTGATTGATCCC	GCTGATGCTCCTTGATCTCC	1DS0.70–1.00
BE637971	TGCCTGATGTTTGATGCTCC	CAAAGCGAAGTGACTGTCCA	1DS0.70–1.00
BE444846	TCTTCGCCACAGGAGTACCTA	GGCTCGTAGCGGGTATACAA	1DS0.00–0.48
BE591601	GTTAGTGGCACTCCTACCTG	GATGTCCAACCATAATGCCC	1DS0.00–0.48
BE637864	TCCTCATTTTGTAATCCTTCTCTC	TTTTGTTCCCACCATCAGGT	1DS0.00–0.48
BF202643	GAATAGCAACAGTGCTCATGAAT	GAAGAACAGCAGGGCGTTAC	1DS0.00–0.48
BF474569	CGTACCAACTCAACCCCTC	TGAAGGGTGAGAGAACTCCG	1DS0.00–0.48
BF478737	CTCTTCACAGTTACAACATCAGC	TGAGGCTCAATGATGACCAG	1DS0.00–0.48
BE424523	CAGTAAGGAAATATGGCCGAT	TTGATGCAGAAAAAGTTGGAT	6DS0.79–0.99
BE490604	AAGCGGTTCCATCTCTCC	CTGCCATTGCTTGTCGTAGA	6DS0.79–0.99
BE500768	ACCTCGACCACTCACTCCA	TCAGCGGTCTCAGTTTGTTG	6DS0.79–0.99
BE517858	CCGGTGATGACCGAACTGAT	CCGGATGATCTCGCTGCTCTC	6DS0.79–0.99
BE444631	CTCCAGTTTCAGGGAGCAAG	GTTCTCTGGCAAGTACTTCAAATCC	6DS0.45–0.79
BE445201	AATGAATTGCTACCATTATTCTCA	ACAGCCGTGAACGTTAGTAAGT	6DS0.45–0.79
BF478958	TCACCTGTACAACAACATGATTTCAA	TGTGCTCATATGTTTTAACT	6DS0.45–0.79
BF483025	ATTCTGTAAGCATGACGGC	AAGGAACTAAGGCCAAGCAATT	6DS0.45–0.79
BE405809	AACGATGCAAGGCTAAAATCTGTGT	GAAGCTGCTGGTTTCTTTGG	6DS0.00–0.45
BE426591	CAGACAATCTTCTTGCCGCT	GTTAGAAATACCGTAAAGCTTTTACCATTAC	6DS0.00–0.45

doi:10.1371/journal.pone.0137747.t002

RESULTS

Immunofluorescence of CENH3, coupled with centromeric DNAs (CRWs), was used to identify the centromeric regions of intact chromosomes and telosomes of wheat. The localization pattern of CENH3 and the CRWs on mitotic metaphase chromosome of CS and their derived telosomes is shown in Fig 1. Consistent with previous reports [30,31], the two probes co-localized in most chromosomes of CS wheat. However, in one chromosome pair of CS, identified as 4D, the two probes were clearly separate from each other (Fig 1a), suggesting that not all CRWs are located within the functional centromere.

Interestingly, in the two winter wheat cultivars Jagger and TAM111, the two signals co-localized on chromosome 4D, indicating that the 4D centromeres in CS underwent repositioning (S1 Fig).

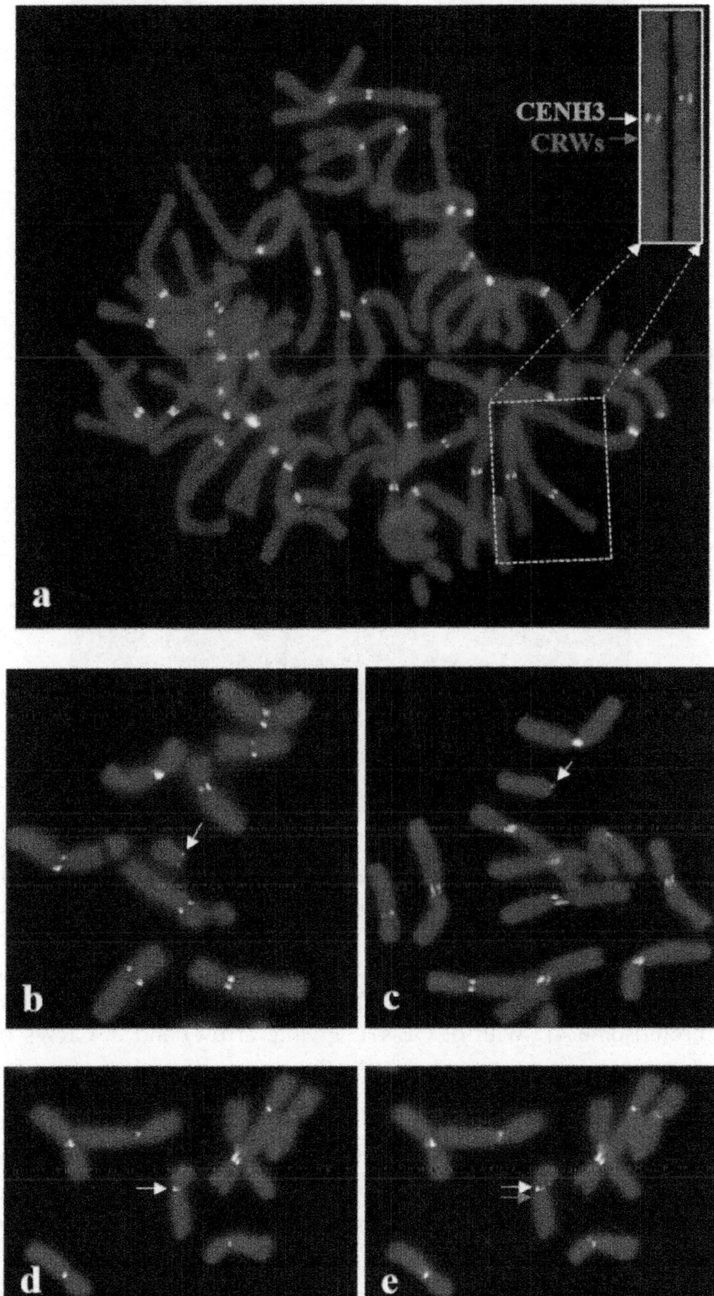

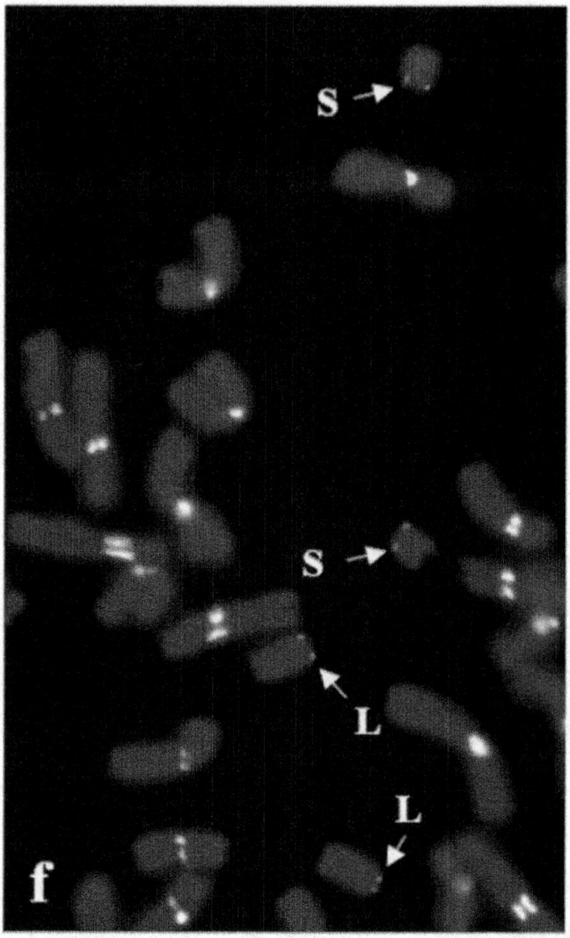

Figure 1. Detection of CENH3 and CRWs on mitotic chromosome of Chinese Spring (CS) wheat and the derived 4D telosomic stocks; a: the insert shows chromosome 4D probed with CENH3 (green) and CRWs (red), straightening was performed using the 'straighten-curved-objects' command of the Image J software, the signals are clearly separated from each other; Relative CENH3 signal intensity of b: t4DS; c: t4DL; f. dDt4D; Chromosome 4D with d: CENH3 (white arrow) and e: CRWs (red arrow) signals.

Next, we applied the wheat CENH3 antibodies to study the structure of t4DS (telosome for 4DS arm) and t4DL, which arose independently from centric misdivison of chromosome 4D. The results showed that CENH3 was detected at the extreme ends of t4DS and t4DL, indicating that they are true telosomes (Fig 1b and 1c). The immunofluorescence of CENH3 on the telosomes was weaker compared to that of the other chromosomes. In order to compare the signal intensity between the intact chromosome 4D and t4DS and to minimize measurement error, root tips from euploid CS and the t4DS stock were squashed on the same slide, and chromosome images probed with CENH3 were captured from the same preparation. The measurement data showed that t4DS had a signal intensity of 43±4.8% (n = 4), compared with that of an intact 4D chromosome. To compare the signal intensity between t4DS and t4DL, we used the dDt4D stock, which contains a pair each of t4DS and t4DL (Fig 1f). The result showed that both telosomes had a similar amount of signal intensity, approximately a 1:1 ratio (79.6±3.8% in t4DS: 80.2±1.7% in t4DL, n = 4) (S2 Fig). Although the dDt4D stock was developed by intercrossing the t4DS and t4DL telosomes, which arose independently from centric misdivison, they showed approximately half of the signal intensity compared with that of the intact 4D chromosome.

We also studied the telosomic derivatives of the seven, D-genome chromosomes of wheat (Figs 2 and 3). Data on the detection of CENH3, the CRWs, pAs1, and single-copy DNA probes on Dt1DS and dDt1DS are shown (Fig 2 and S3 Fig). Signals for CENH3 (arrow in Fig 2a) and the CRWs (arrow in Fig 2b) on Dt1DS were located at the end of the chromosome, indicating that it is a true telosome. In the t1DS of the reconstituted dDt1D stock, however, CENH3 was localized interstitially, forming a small acrocentric chromosome (arrows in Fig 2f). To further discern the chromosomal rearrangement, we performed FISH using pAs1, 1S-1, and 1S-3 as probes. Hybridization signals for pAs1 (arrows in Fig 2c and 2d) and 1S-3 (red dots in Fig 2d) were detected on the terminal region of Dt1DS. However, the pAs1 FISH pattern in dDt1DS showed multiple localizations (arrows in Fig 2g) to both telomeric regions and the interstitial region of the chromosome arm. Probe 1S-3 (arrows in Fig 2i) was detected in the middle of the arm, instead of in the telomeric region, indicating the presence of a paracentric inversion (Fig 2l). Moreover, the FISH signal for 1S-1 (arrow in Fig 2h) was not detected in the pericentromeric region of dDt1DS, implying that dDt1DS has lost the original centromere and now has a *de novo* centromere in a new position (Fig 2l). Supporting evidence came from PCR analysis using six genome-specific markers derived from the 1D proximal bin, which failed to amplify in the dDt1D stock but did amplify in CS and Dt1DS (S4 Fig). Thus, dDt1DS contains multiple chromosomal rearrangements, including a centromere shift, a paracentric inversion, and a deletion. The labeling patterns of CENH3, the CRWs, and pAs1 on dDt1DL were similar with those of Dt1DL (Fig 3).

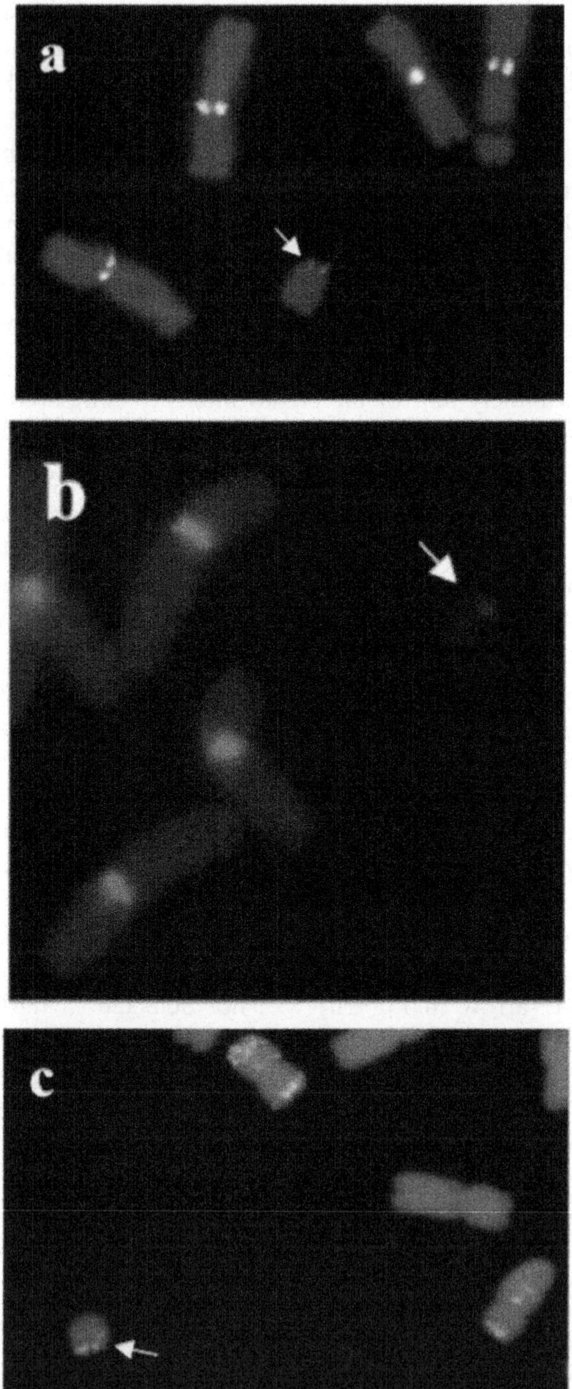
a
b
c

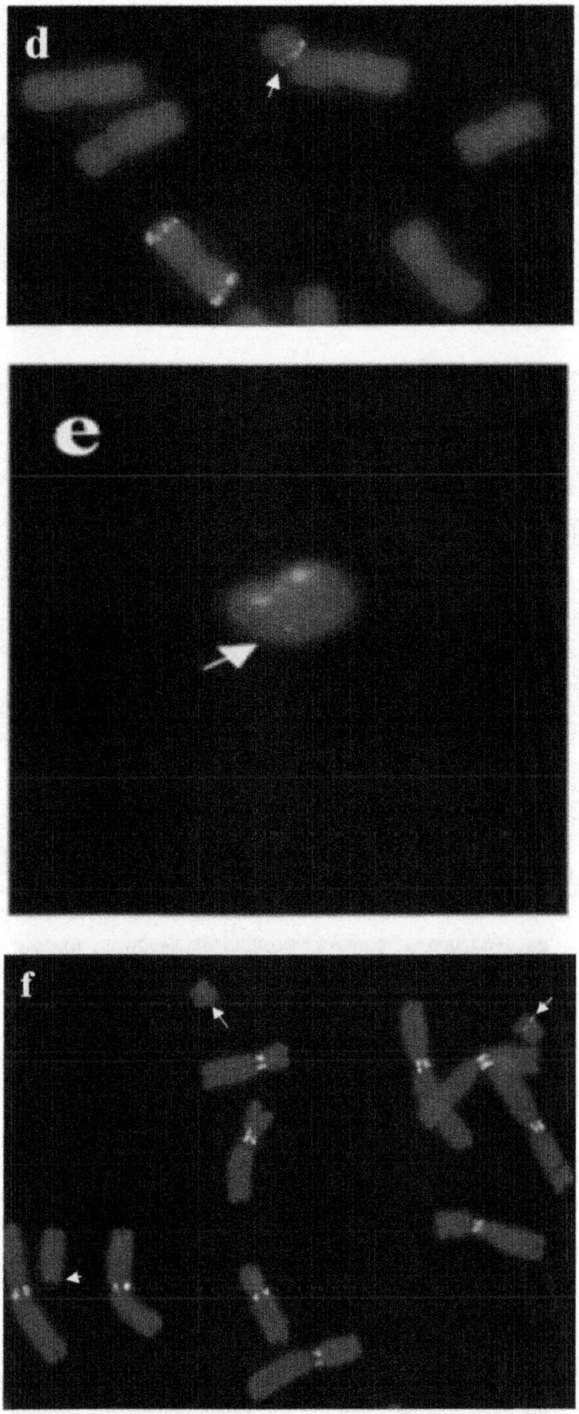
d
e
f

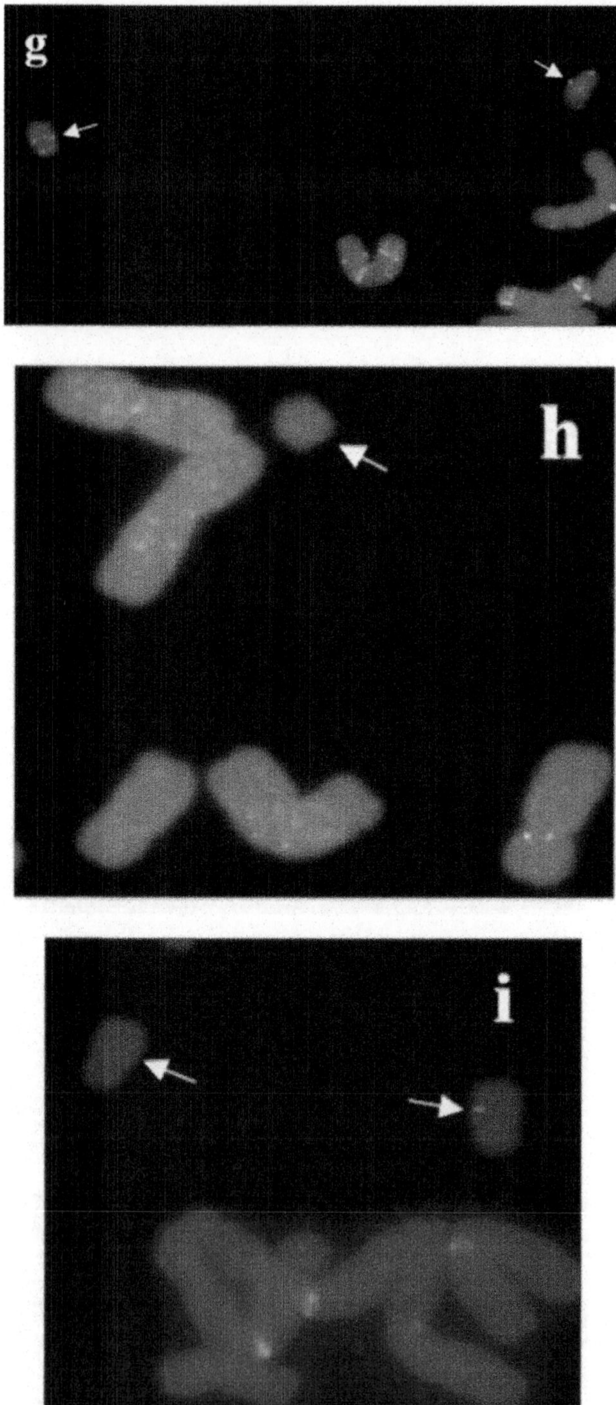
g
h
i

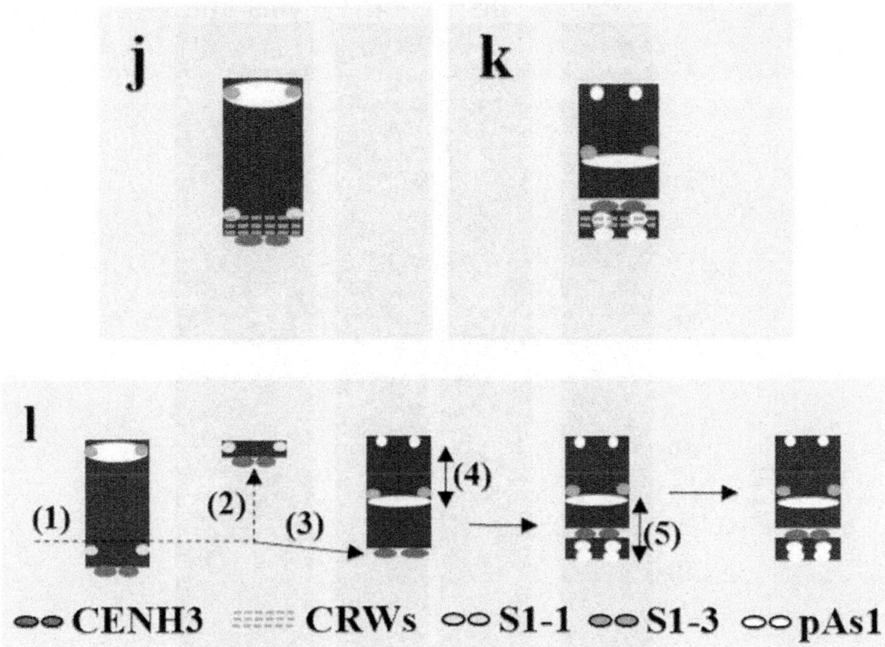

Figure 2. Probing of t1DS telosomes present in the Dt1DS stock with a: CENH3, b: CRWs, c: pAs1, d: 1S-3 (red dots) and pAs1 (green), e: 1S-1 (red dots) and pAs1 (green); Probing of 1DS telosomes present in the dDt1DS stock with f: CENH3, g: pAs1, h: 1S-1 (not detected), and i: 1S-3 (red dots) and CRWs (faint green signals). Simultaneous detection of CENH3, CRWs and pAs1 on dDt1DS also provided in S3 Fig. Ideograms depicting localization of each probe on telosomes, Dt1DS and dDt1DS (j-k, and S3 Fig). Possible scenario of the origin of the chromosomal rearrangements observed in the dDt1DS: (1) chromatin breakage, (2) loss of original centromere, (3) *de novo* formation of a centromere, (4) paracentric inversion, (5) pericentric inversion (s).

The same approach was used to analyze all the D-genome telosomes, and the results are presented (Fig 3). The FISH pattern of pAs1 on chromosome 2D showed multiple localizations, with four FISH sites in the long arm and a single hybridization site on the telomere of the short arm. CENH3 and the CRWs co-localized on the primary constriction (Fig 3). Applying these probes to the Dt and dDt stocks revealed largely identical hybridization patterns with those of an intact 2D chromosome, indicating that there are no rearrangements in these telosomes. Similar patterns were observed for all the remaining D-genome telosome stocks, except for chromosome 6D.

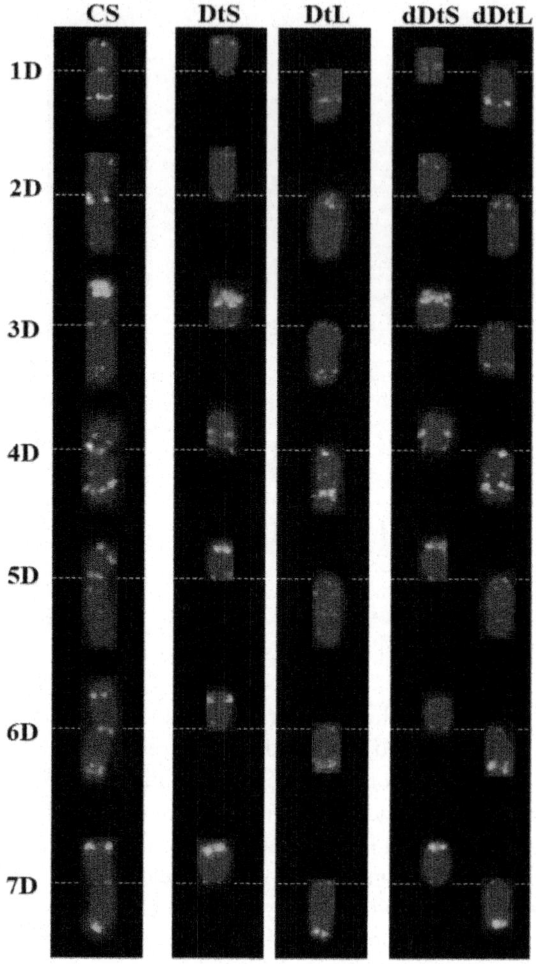

Figure 3. Immuno-FISH based karyotype of D-genome chromosomes of wheat and their derived telosomes using CENH3 (white), CRWs (red) and pAs1 (green) as probes. CRWs (red signals) co-localized with CENH3 (white signals) in most of the chromosome except dDt1DS, 4D, Dt4DS, dDt4DS, Dt5DL and dDt5DL. The centromeric regions of chromosome or chromosome arm were seen as pinkish red colors because the CRWs (red signals) are abundant in centromeric region and much brighter than CENH3 signals except in the above mentioned telosomes. The dDt1DS stock contained multiple chromosome rearrangement including inversion, deletion and centromere shift. Note that the CRWs were not detected in Dt4DS and dDt4DS, instead the pAs1 signal was overlapped with the CENH3 signal in these telosomes. A very faint pAs1 FISH site was detected in the terminal region of dDt6DS, indicating a terminal deletion. Short arm and long arm telosomes present in the ditelosomic stocks are represented as (DtS) and (DtL), respectively and short arm and long arm telosomes present in the double ditelosomic stocks are represented as (dDtS) and (dDtL), respectively.

CENH3 and the CRWs were detected at the end of chromosome arms in Dt6DS (arrows inFig 4a and 4b) and dDt6DS (arrows in Fig 4d and 4e). Whereas a prominent pAs1 signal was located on the subtelomeric region of Dt6DS (arrows in Fig 4c and 4g), a faint hybridization signal was detected on dDt6DS (Fig 4f and 4h) indicating that the chromosome arm of dDt6DS has suffered from a terminal deletion. The 6S-2 FISH signal on Dt6DS was detected at 61.0±3.1% (n = 3) from the telomere (Fig 4g). However, in dDt6DS, the 6S-2 FISH signal was detected 39.9±3.2% (n = 3) from the telomere (Fig 4h), thus about 20% of the distal region was deleted. In order to verify the deletion at the molecular level, we used genome-specific PCR primers, which were derived from the terminal deletion bin of 6DS (S5 Fig). Four markers derived from terminal bin had no amplification; the other six markers derived from interstitial and proximal bins had amplification (S5 Fig), confirming that about the distal 20% of the telosome was deleted. We made a blastn search of ten EST sequences against the sequence assembly from flow-sorted chromosome arm 6DS [28] and found no hit for markers *XBE424523*, *XBE490604*, *XBE500768*, and *XBE517858* derived from terminal deletion bin. We further analyzed the genome zipper maps of wheat group 6 [28] and observed that this region is deleted in t6DS, whereas the corresponding region is present in t6AS and t6BS. This region corresponds to rice locus *Os02g0116800-Os02g0128800* [39], which is about 595 kb in rice and 702 kb in *Brachypodium Bradi3g01540.1-Bradi3g02817.1* [40]. This region is syntenic with the terminal tip of rice chromosome 2. Nearly 100 genes are annotated in this missing region in the rice genome, 25 of which are syntenic to wheat.

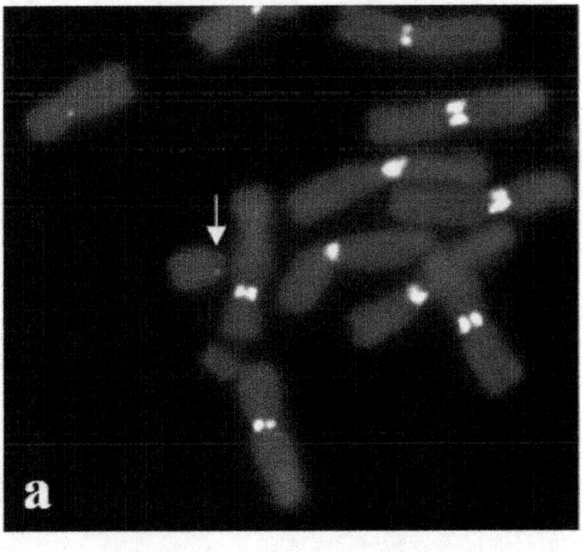

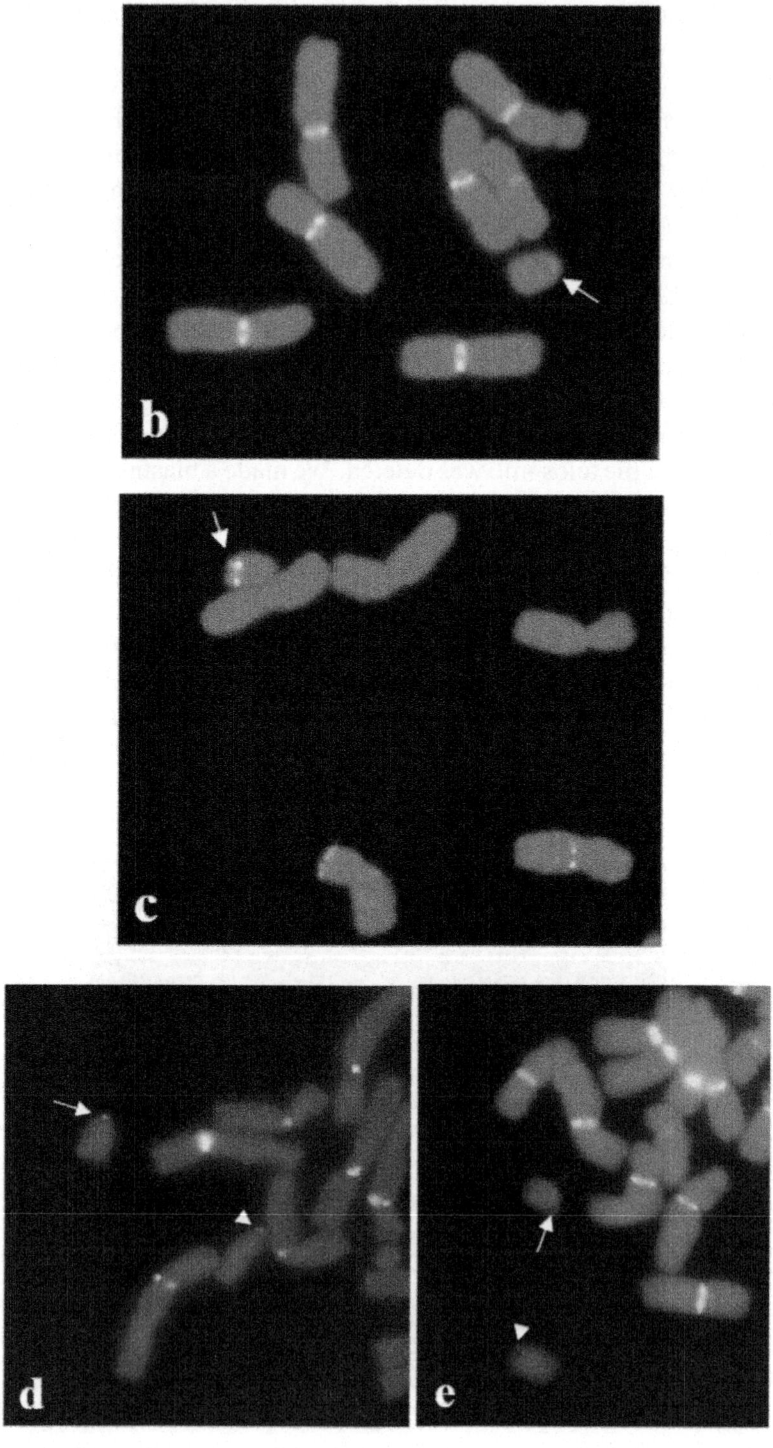
b
c
d
e

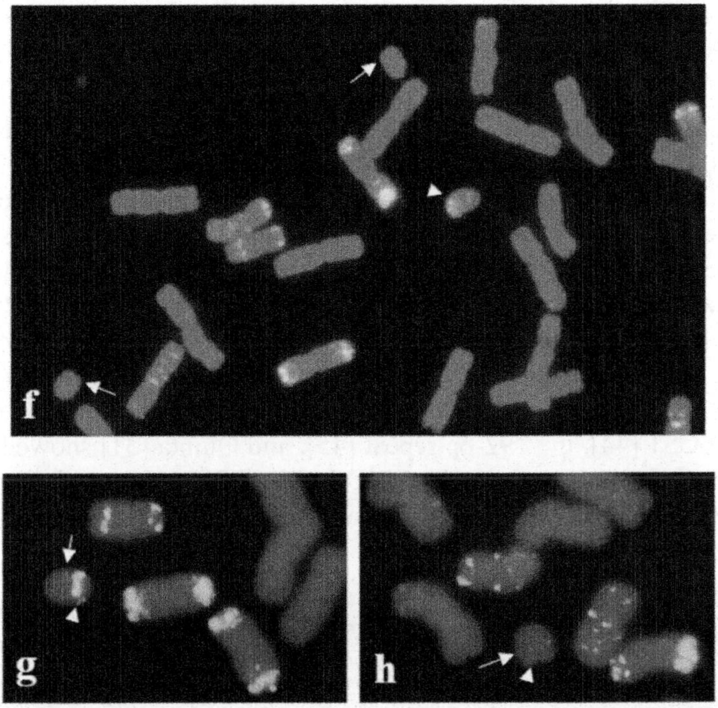

Figure 4. Localization of CENH3 (arrow in a), CRWs (arrow in b) and pAs1 (arrow in c) on Dt6DS; localization of CENH3 (d), CRWs (e) and pAs1 (f) on dDt6D, arrows and arrowhead indicate the 6DS and 6DL telosomes, respectively. Two color detection of single gene probe, 6S-2 and pAs1 on Dt6DS (g) and dDt6DS (h); the hybridization signal for single copy probe 6S-2 (red dots) is indicated by arrows and the pAs1 was labeled with green colors by arrowheads.

Thus, sequence analysis further confirms the loss of a segment from dDt6DS. Because the available dDt1D and dDt6D stocks are rearranged, we re-isolated both stocks, which are now intact and similar to the Dt1DS and Dt6DS stocks. The labeling patterns of CENH3 in the new stocks were similar to those of the DtS lines (data not shown). Cytogenetic analysis is in progress to investigate the mitotic behavior on the newly developed dDt stocks.

To further understand the CENH3 deposition on other telosomes, we used dDt lines that were derived from all the A- and B-genome chromosomes. Our results on immunofluorescence of CENH3 in dDt lines derived from A-genome chromosome showed that the position of CENH3 signals on telosomes is terminal with a signal intensity weaker than that of the other regular chromosomes (S6 Fig). Consistent with dDt4D, signal intensity for CENH3 between the DtS and DtL arms in dDt lines showed similar intensity except in dDt4AS (arrowheads in S6 Fig). The telosome, dDt4AS is known to be acrocentric and, thus, contains a complete centromere [41].

To study the B-genome telosomes, we used a line that has all the B-genome chromosome arms except 4B as telosomes (2n = 28+28t) [42]. This line was obtained by intercrossing the appropriate dDt stocks. Chromosomes t4BS and t4BL in this stock were later identified as t4AS and t4AL telosomes [43]. Likewise, we observed that the CENH3 signal on all B-genome telosomes was smaller compared with that in the complete chromosomes (S6 Fig). In addition, the telosomes, including Dt1BS and Dt6BS in this stock, lack CRWs in the CENH3 region (Fig 5), indicating that the stability of telocentric chromosomes depends on the presence of CENH3 chromatin but not centromeric DNA repeats. Further analysis using the previously known wheat centromeric repeats CCS1 [44], the 192-bp repeat [45], and Quinta [31] showed no signal on these telosomes together with Dt4DS, dDt4DS, Dt5DL, and dDt5DL (data not shown), further indicating that centromeric DNAs are not essential for normal centromere function.

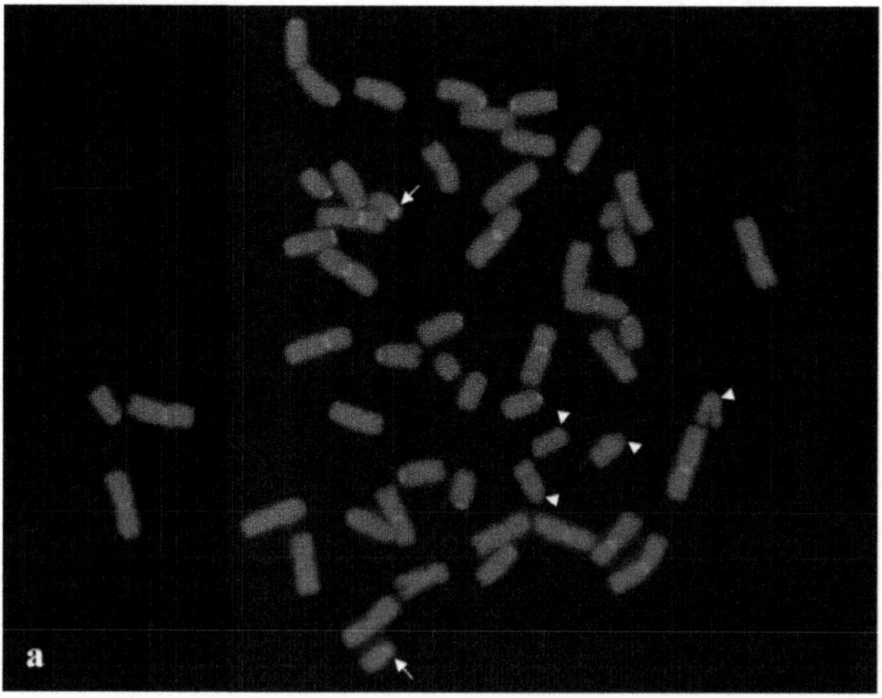

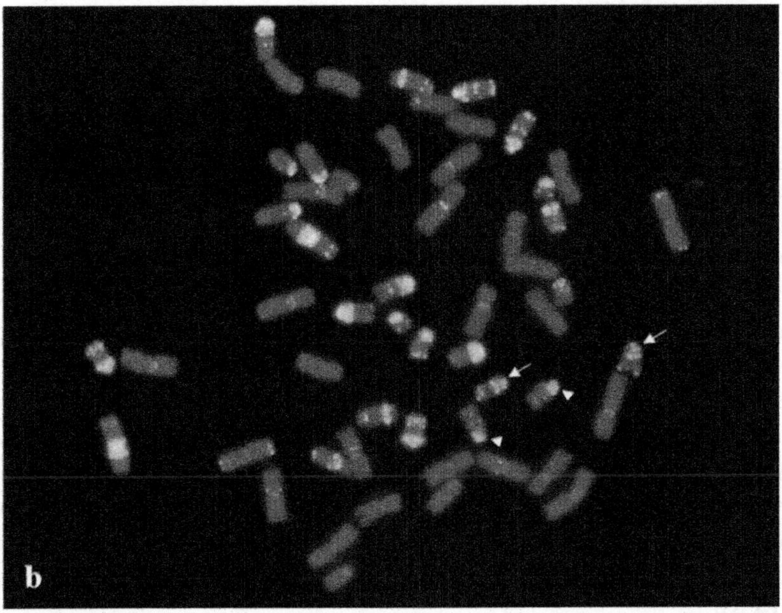

Figure 5. Two color FISH mapping of CRWs (red) and GAAn (green) repeats on mitotic metaphase chromosome of CS containing 28 (24 telosomes from B-genome telosomes and 4 telosomes from t4AS and t4AL). The signals for CRWs were not detected on two pairs of t1BS and t6BS (arrowheads) (a). Instead GAAn repeats were presented in their centromeric region. Arrows indicate the t4AS (a). Arrows and arrowheads indicate the t1BS and t6BS, respectively (b).

DISCUSSION

In many plant species, centromeres consist of complex DNA, including satellite DNA and reterotransposons [2], that are species- and chromosome-specific [46,47]. Some of the centromeric repeats are located within the functional centromere, but centromere-associated repeats also are observed in subtelomeric and interstitial chromosome regions [46,48].

Our study, using immuno-FISH probed with CENH3 and CRWs, shows that CENH3 and CRW elements co-localized on most of the A-, B- and D-genome chromosomes, including their telosomic derivatives. These results indicate that wheat centromeres contain CRW elements that interact with wheat CENH3 [30,31]. However, our results further show that wheat CRWs do not always co-localize with CENH3, as was the case with the complete chromosome 4D and some of the telosomes, including Dt4DS, dDt4DS, Dt5DL, dDt5DL, Dt1BS, and Dt6BS, because these chromosome or chromosome arms lack CRWs in their CENH3.

The CRWs located on chromosome 4D in CS did not overlap with the CENH3 signals but in other wheat cultivars, such as Jagger and TAM111, they co-localized in the same region, providing evidence for centromere repositioning in chromosome 4D of CS (S1 Fig). Comparison of the pAs1 FISH pattern in the pericentromeric region of chromosomes 4D between CS, Jagger, and TAM111 revealed that their pericentromeric localizations were different from each other, implying that the pericentromeric region of 4D in CS underwent a structural rearrangement. The pAs1 FISH site overlapped with CENH3 in CS but not in Jagger or TAM111 (S1 Fig). Lo et al. [49] and Lomiento et al. [50] have reported that neocentromeres form in gene-desert regions containing many repetitive DNAs in many species. These results also indicate that neocentromere function is independent from the presence of original centromeric DNAs [51]. Our immuno-FISH results on telosomes also support their stabilization without centromeric DNAs, because we did not observe CRW signals in Dt1BS and Dt6BS arms. In these telosomes, the GAA repeats co-localized with CENH3 (Fig 5). Thus, wheat centromeres consist of complex DNAs, which may contain CRWs or satellite DNAs, such as pAs1 and GAAn.

Univalent chromosomes at metaphase-I have the tendency to misdivide at the centromere in a transverse manner, which gives rise to telocentrics or potential isochromosomes. Steinitz-Sears [29], suggested that the transverse misdivision split can occur in different regions of the centromere, resulting in telosomes that differ in the completeness of their centromeric regions, and that chromosome arms with incomplete or partial centromeres behave like acentric fragments and are lost during cell division. Because the set of wheat telosomes was produced by centric misdivision and they are stably transmitted to the offspring, they must have received a complete or nearly complete kinetochore. Similarly, barley and rye telocentric chromosomes are cytologically stable. Giemsa N-banding on barley telotrisomics revealed that they contain half of a diamond-shaped kinetochore, whereas complete chromosomes contain an intact, diamond-shaped kinetochore [52]. Rice telocentric chromosomes also contain half of CentO, compared with its normal centromeres [53].

Our immunostaining analysis using the CENH3 antibody suggests that the signals in most Dt and dDt telosomes had approximately half or even less the CENH3 signal intensity compared with that of a complete chromosome. None of the derived telosomes received a complete CENH3 region except dDt4AS, which is an acrocentric chromosome (S6 Fig). These results indicate that telosomes, which only receive half or less than half of the complete CENH3 region, are cytologically stable and transmitted to the offspring. Because most of the wheat telosomes received a partial CENH3 region, it is possible that the transverse misdivision split is not random and may preferentially occur in the middle of the CENH3 chromatin. However, we cannot rule out that breakage also occurs in the entire functional centromere region and that telocentric chromosomes with insufficient CENH3 are mitotically unstable and lost. This is also supported by analyzing the centromere structure of wheat-rye Robertsonian translocations derived from repeated centric breakage-fusion events, which revealed that breakage can occur in different regions of the centromere resulting in wheat-rye hybrid centromeres with different sizes of wheat and rye centromeric repeats [54].

We observed chromosomal aberrations in dDt1DS and dDt6DS. Double labeling of CENH3 and CRWs on the Dt1DS telosomes showed co-localization at the terminal region. In dDt1DS, however, a very faint CRW hybridization signal was observed at the telomere, and CENH3 was localized proximal to it. Single-copy, 1S-1 FISH was not detected in the proximal region of dDt1DS, indicating the presence of a proximal deletion and supporting a previous finding that ESTs mapped in the wheat 1DS deletion bin were absent in dDt1DS [55]. The deletion placed the centromere in a new position. Thus, dDt1DS contains a *de novo* formed centromere (Fig 2l).

The formation of a *de novo* centromere in dDt1DS supports earlier reports in *Drosophila* [56] and chicken (*Gallus gallus*) [57], where neocentromeres formed in regions close to the original centromeres. For instance, when the Z centromere was deleted, neocentromeres most frequently formed near the original Z centromere [57]. CENP-A/CENH3 enrichment in the flanking regions is low but still more enriched compared with that in the rest of the genome [58]. In our study, the formation of *de novo* centromeres or centromere shift was observed in chromosome 4D and in the 1DS telosomes present in the dDt1DS stock. The preference of the formation of the *de novo* centromere near the original centromere is likely caused by the presence of CENH3 in the flanking pericentromeric regions. Chromosome rearrangements after chromatid breaks are a common cause of neocentromere formation in humans [6]. Likewise, neocentromeres reported in plants, maize chromosomes in an oat background and barley chromosomes, also were associated with loss of endogenous centromeres by chromatid breakages [7,8]. We also found a paracentric inversion in telosome dDt1DS using single-copy FISH mapping. Interestingly, the pAs1 signal was observed near the *de novo* centromere region and in an interstitial region of dDt1DS but was absent in Dt1DS, indicating the possibility of another chromosome rearrangement (Fig 2l).

We also identified a deletion in dDt6DS, which comprised about 20% of the terminal deletion bin of chromosome 6D. This deletion was not reported in a recent whole-genome, sequencing analysis [28]. However, our results show that several EST sequences from the terminal deletion bin were missing from the 6DS assembly but present in 6AS and 6BS shotgun sequence assemblies [28]. Thus, when employing these lines for genetic studies, it is important to be aware of the potential presence of chromosomal aberrations in the telocentric chromosome lines to avoid misinterpretation of experimental results. Further relocation of centromeres in chromosomes 4D in CS (the cultivar being used to sequence the wheat genome) compared to other wheat cultivars, Jagger and TAM111, suggests that*de novo* sequencing of more wheat genotypes might provide further insight in the structural organization in the wheat genome.

The wheat dDt stocks were developed by intercrossing the appropriate Dt stocks followed by selection. Therefore, chromosomal rearrangements, including centromere shifts, deletions, and inversions observed in dDt1DS and dDt6DS, might have formed after hybridization of the Dt lines. Rhoades [59] found that telocentric chromosomes in maize undergo structural changes during somatic cell divisions leading to loss or diminution in size. Steinitz-Sears [29] also found that a telocentric chromosome is often unstable and may be lost during plant development in wheat. During mitosis, the structural integrity of the centromeric and flanking pericentric heterochromatic regions is essential for proper assembly of the kinetochore and genome stability [60]. In fission yeast, pericentromeric heterochromatin is an absolute requirement for the establishment of the centromere [61]. In addition to fission yeast, pericentromeric heterochromatin seems to be required for the accurate segregation of chromosome during mitosis in many eukaryotes, including mammals [62]. The implication is that mono-arm oriented pericentromeric heterochromatin in telosomes might be relatively insufficient for maintaining chromosome stability compared to chromosomes with bi-arms oriented pericentromeric heterochromatin. This conclusion is supported by recent findings of Wanner et al. [63] that in monocentrics microtubules attach via CENH3 to both pericentromeres to stabilize the chromosomes during anaphase against the pulling forces.

SUPPORTING INFORMATION

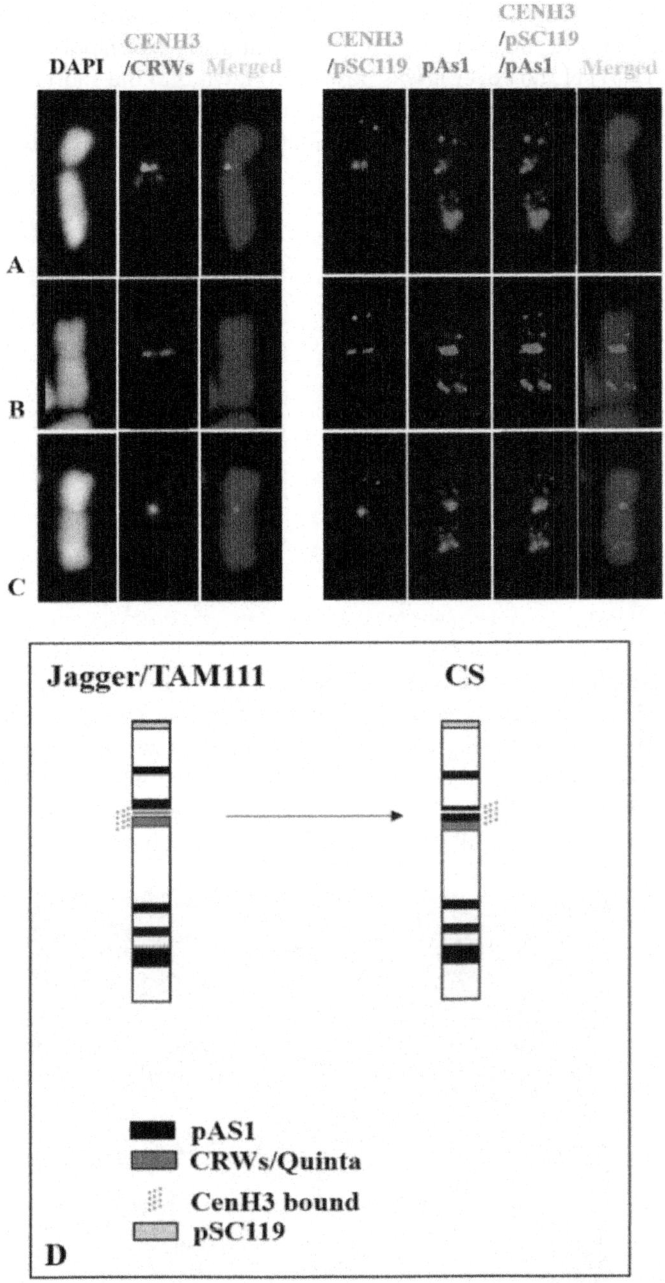

S1 Figure. Sequential detection of CENH3, CRWs, pSc119 and pAs1 on the chromosomes of 4D in CS (A), Jagger (B) and TAM111 (C). The hybridization signals for CENH3 and CRWs were clearly separated from each other in 4D of CS but these were co-localized on the chromosomes of 4D in Jagger and TAM11, indicating the centromere repositioning in 4D of CS. In 4D of CS, pAs1 localization pattern tend to be positioned toward log arm and which is completely overlapped with CENH3. In Jagger and TAM111, however it was positioned toward to the short arm. The pSc119 was used for additional FISH marker to identify the chromosomes 4D in CS, Jagger and TAM111. D, Ideogram depicting distribution of each probe on the chromosomes of 4D in three wheat cultivars.

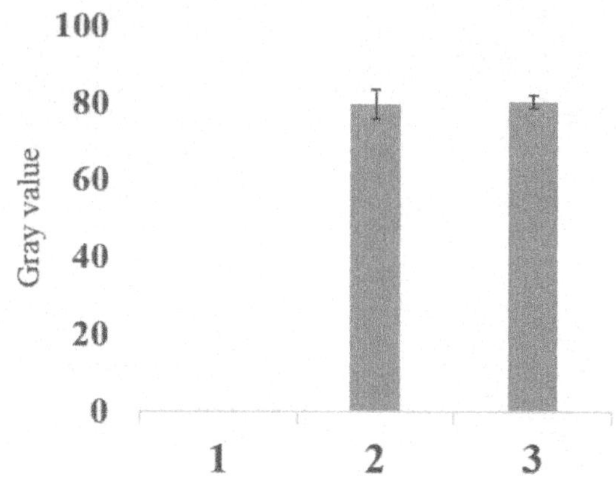

S2 Figure. Graph showing the measurements of the immunofluorescence signal intensity of CENH3. Numbers at y axis represent the gray value (relative signal intensity of antibody to background, background was normalized as zero). 1: background signal, 2: CENH3 signal intensity in dDt4DS, 3: CENH3 signal intensity in dDt4DL. Measurements were done by Image J software. The gray value of CENH3 was 79.6±3.8 (n = 4) and 80.2±1.7 (n = 4) in dDt4DS and dDt4DL, respectively.

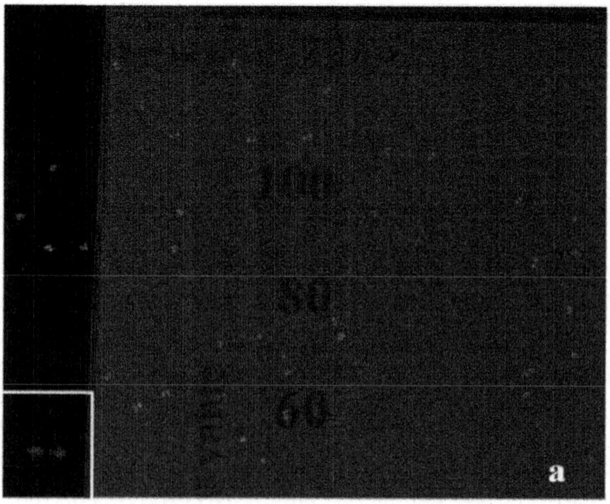

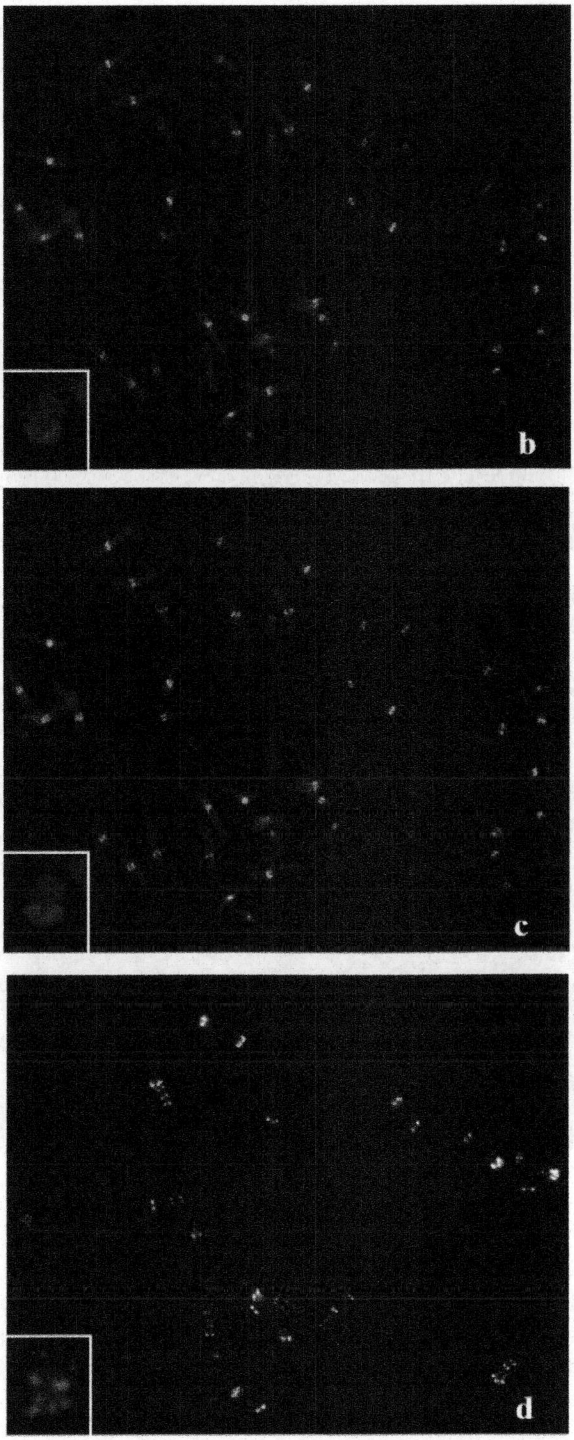

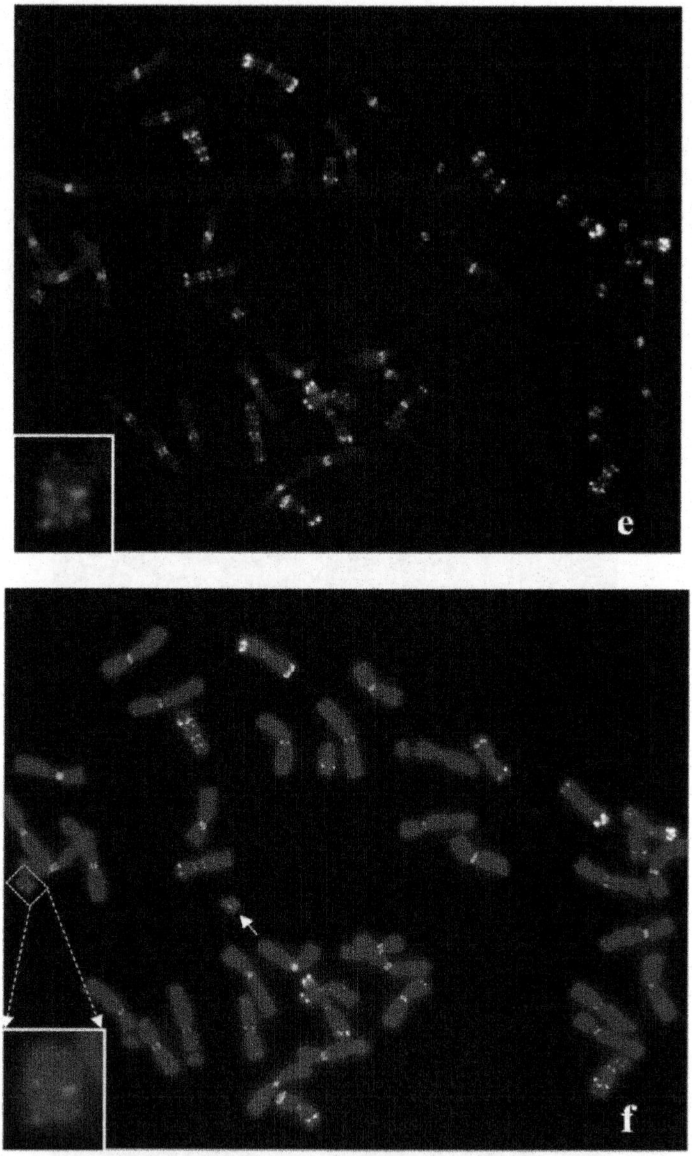

S3 Figure. Multicolor immuno-FISH detection of CENH3 (a), CRWs (b) and pAs1 (d) on telosome, dDt1D. Merged images, CENH3 and CRWs (c), and CENH3, CRWs and pAs1 (e) with DAPI stained metaphase chromosome (f). The inserts show telosome, dDt1DS probed with CENH3 (red), CRWs (green) and pAs1 (white). CENH3 was detected by rhodamine-conjugated anti-rabbit antibodies (red), and the signals were fixed with 4% paraformaldehyde. The same metaphase cell was probed with CRWs (green) and pAs1 (far red, the signals were pseudocolored in white).

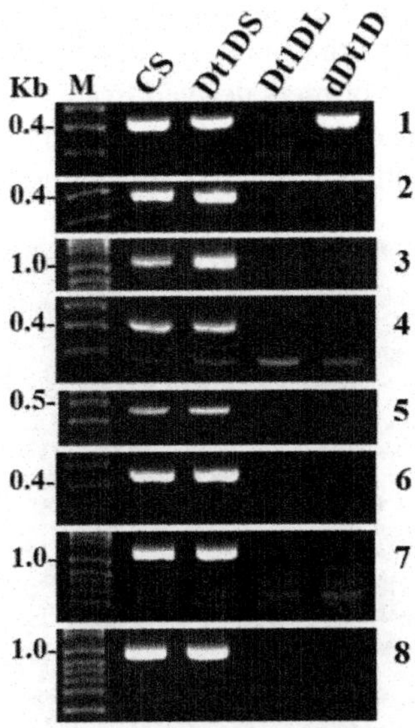

S4 Figure. PCR patterns of CS, Dt1DS, Dt1DL and dDt1D by using genome specific primers: two primers, BE405518 and BE637971, derived from the terminal deletion bin, BE405518 was not amplified (2); six primers, BE444846, BE591601, BE637864, BF202643, BF474569 and BF478737, derived from proximal bin had no amplification (3–8) indicating proximal deletion in dDt1DS.

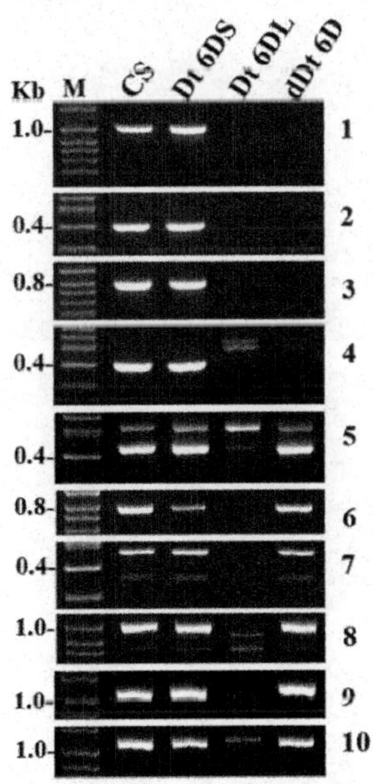

S5 Figure. PCR patterns of CS, Dt6DS, Dt6DL and dDt6D by using genome specific primers: four primers, BE424523, BE490604, BE500768 and BE517858 derived from the terminal deletion bin; four primers, BE444631, BE445201, BF478958 and BF483025 derived from interstitial bin; two primers, BE405809 and BE426591 derived from proximal bin. Four primers derived from terminal deletion bin had no amplification (1–4) while six primers derived from interstitial (5–8) and proximal deletion bin (9–10) had amplification in dDt6DS indicating terminal deletion in dDt6DS.

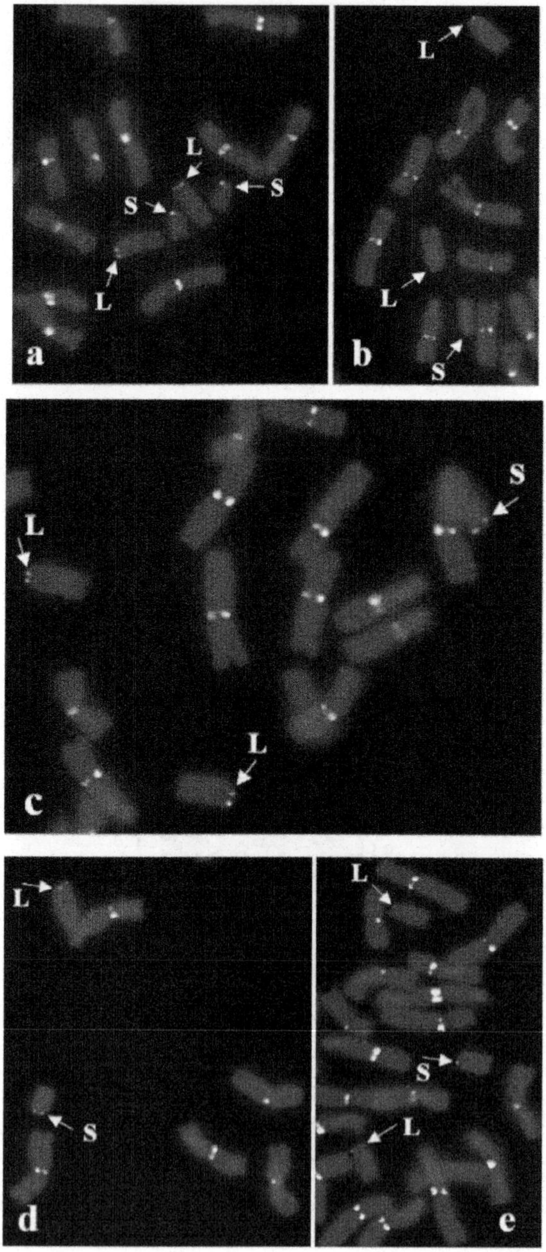

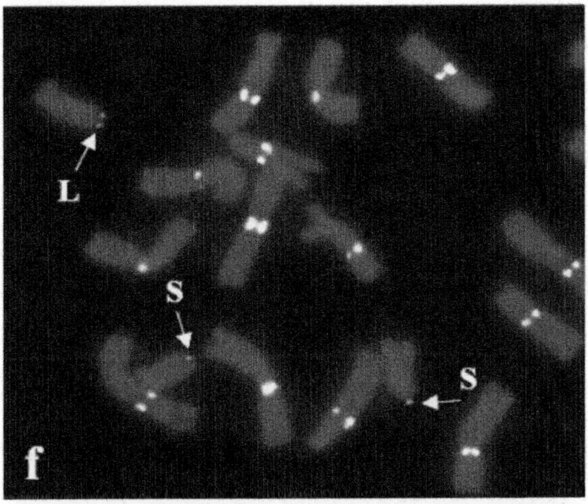

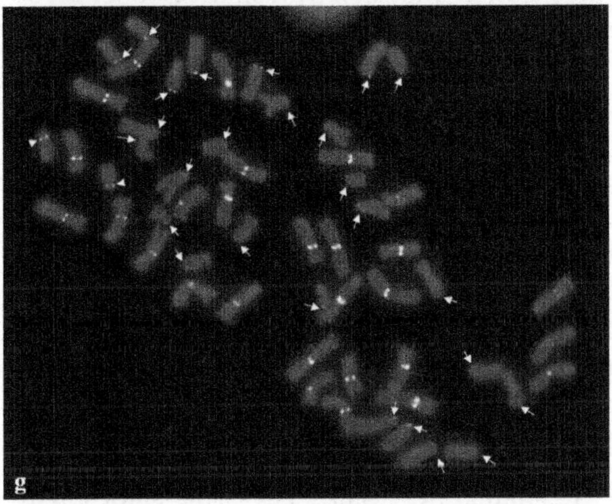

S6 Figure. Partial metaphase cells probed with CENH3 in dDt lines derived from A-genome chromosomes: a, dDt1A; b, dDt2A; c, dDt3A; d, dDt5A; e, dDt6A; f, dDt7A. Telosomes are indicated by arrows. Immunofluorescence of CENH3 on CS containing 28 telosomes (g). The CENH3 fluorescent signals on 24 telosomes (derived from B-genome chromosomes are indicated by arrows) + one pair of t4AL were smaller than those of other intact chromosomes except 4AS (arrowhead) which is an acrocentric chromosome.

ACKNOWLEDGMENTS

We thank W. John Raupp for critical editorial review of the manuscript and Duane Wilson for technical assistance. We are grateful to Dr. Jiming Jiang for his valuable comments on this manuscript. We also thank Adam Lukaszewski

(University of California Riverside) for providing the Dt5DS/Mt5DL line. This is contribution number 15-427-J from the Kansas Agricultural Experiment Station, Kansas State University, Manhattan, KS 66506–5502, U.S.A.

AUTHOR CONTRIBUTIONS

Conceived and designed the experiments: DHK BF BSG. Performed the experiments: DHK SKS. Analyzed the data: DHK BF BSG. Contributed reagents/materials/analysis tools: DHK SKS BF BSG. Wrote the paper: DHK BF BSG.

COMPETING INTERESTS

The authors have the following interests: This research was financially supported by the Kansas Crop Association and by members of the Wheat Genetics Resource Center Industry/University Cooperative Research Center. The following companies and associations make up the membership: Bayer Cropscience, Dow AgroSciences, General Mills, Limagrain, Syngenta, Heartland Plant Innovations, DuPont Pioneer, Kansas Wheat Commission, Kansas Wheat Alliance, and the Kansas Department of Agriculture. The National Science Foundation also provided financial support. The public/private consortium is aimed at pre-competitive wheat genetics research. Dal-Hoe Koo, Bernd Friebe and Bikram S. Gill are affiliated to the Wheat Genetics Resource Center, Kansas State University. There are no patents, products in development or marketed products to declare. This does not alter the authors' adherence to all the PLoS ONE policies on sharing data and materials.

REFERENCES

1. Henikoff S, Ahmad K, Malik HS. The centromere paradox: stable inheritance with rapidly evolving DNA. Science. 2001; 293: 1098–1102. pmid:11498581 doi: 10.1126/science.1062939
2. Jiang JM, Birchler JA, Parrott WA, Dawe RK. A molecular view of plant centromeres. Trends Plant Sci. 2003; 8: 570–575. pmid:14659705 doi: 10.1016/j.tplants.2003.10.011
3. Earnshaw WC, Rothfield N. Identification of a family of human centromere proteins using autoimmune sera from patients with scleroderma. Chromosoma. 1985; 91: 313–321. pmid:2579778 doi: 10.1007/bf00328227
4. Talbert PB, Masuelli R, Tyagi AP, Comai L, Henikoff S. Centromeric localization and adaptive evolution of an *Arabidopsis* histone H3

variant. Plant Cell. 2002; 14: 1053–1066. pmid:12034896 doi: 10.1105/tpc.010425

5. Yu H-G, Hiatt EN, Dawe RK. The plant kinetochore. Trends Plant Sci. 2000; 5: 1360–1382. doi: 10.1016/s1360-1385(00)01789-1

6. Marshall OJ, Chueh AC, Wong LH, Choo KHA. Neocentromeres: new insights into centromere structure, disease development, and karyotype evolution. Am J Hum Genet. 2008; 82: 261–282. doi: 10.1016/j.ajhg.2007.11.009. pmid:18252209

7. Nasuda S, Hudakova S, Schubert I, Houben A, Endo TR. Stable barley chromosomes without centromeric repeats. Proc Natl Acad Sci. 2005; 102: 9842–9847. pmid:15998740 doi: 10.1073/pnas.0504235102

8. Topp CN, Okagaki RJ, Melo JR, Kynast RG, Phillips RL, Dawe RK. Identification of a maize neocentromere in an oat-maize addition line. Cytogenet Genome Res. 2009; 124: 228–238. doi: 10.1159/000218128. pmid:19556776

9. Fu SL, Lv ZL, Gao Z, Wu HJ, Pang JL, Zhang B, et al. De novo centromere formation on a chromosome fragment in maize. Proc Natl Acad Sci. 2013; 110: 6033–6036. doi: 10.1073/pnas.1303944110. pmid:23530217

10. Wang K, Wu YF, Zhang WL, Dawe RK, Jiang JM. Maize centromeres expand and adopt a uniform size in the genetic background of oat. Genome Res. 2014; 24: 107–116. doi: 10.1101/gr.160887.113. pmid:24100079

11. Levan A, Fredgra K, Sandberg AA. Nomenclature for centromeric position on chromosomes. Hereditas. 1964; 52: 201–220. doi: 10.1111/j.1601-5223.1964.tb01953.x

12. Darlington CD. Misdivision and the genetics of the centromere. J Genet. 1939; 37: 343–365. doi: 10.1007/bf02982733

13. Marks GE. Telocentric chromosomes. Am Nat. 1957; 91: 223–232. doi: 10.1086/281981

14. Strid A. Stable telocentric chromosomes formed by spontaneous mid-divison in*Nigella doerfleri* (Ranunculaceae). Bot Notiser. 1968; 121: 153–164.

15. Schubert I, Rieger R. Alteration by centric fission of the diploid chromosome number in *Vicia faba* L. Genetica. 1990; 81: 67–69. doi: 10.1007/bf00055238

16. Southern DI. Stable telocentric chromosomes produced following centric misdivision in *Myrmeleotettix maculatus* (Thunb.). Chromosoma. 1969; 26: 140–147. doi: 10.1007/bf00326451

17. Takagi N, Sasaki M. A phylogenetic study of bird karyotypes. Chromosoma. 1974; 46: 91–120. pmid:4134896 doi: 10.1007/bf00332341
18. White MJD. Animal cytology and evolution. 3rd Edition. London: Cambridge University Press. 1973
19. Sears ER, Sears L. The telocentric chromosomes of common wheat. In Proceedings of the 5th International Wheat Genet Symposium (Ramanujam S, ed). New Delhi: Indian Society of Genetics and Plant Breeding. 1978; pp: 389–407.
20. Cheng ZK, Yan H, Yu H, Tang S, Jiang JM, et al. Development and applications of a complete set of rice telotrisomics. Genetics. 2001; 157: 361–368. pmid:11139516
21. Tsuchiya T. Cytogenetics of the telocentric chromosome of the long arm of chromosome 1 in barley. Seiken Ziho. 1972; 23: 47–62.
22. Erayman M, Sandhu D, Sidhu D, Dilbirigi M, Baenziger PS, Gill KS. Demarcating the gene-rich regions of the wheat genome. Nucleic Acids Res. 2004; 12: 3546–3565. doi: 10.1093/nar/gkh639
23. Mutti JS, Sandhu D, Sidhu D, Gill KS. Dynamic nature of a wheat centromere with a functional gene. Mol Breeding. 2010; 26: 177–187. doi: 10.1007/s11032-009-9389-1
24. Gill KS, Arumuganathan K, Lee JH. Isolating individual wheat (*Triticum aestivum*) chromosome arm by flow cytometric analysis of ditelosomic lines. Theor Appl Genet. 1999; 98: 1248–1252. doi: 10.1007/s001220051190
25. Dolezel J, Kubalakova M, Bartos J, Macas J. Flow cytogenetics and plant genome mapping. Chromosome Res. 2004; 12: 77–91. pmid:14984104 doi: 10.1023/b:chro.0000009293.15189.e5
26. Dolezel J, Kubalakova M, Paux E, Bartos J, Feuillet C. Chromosome-based genomics in the cereals. Chromosome Res. 2007; 15: 51–66. pmid:17295126 doi: 10.1007/s10577-006-1106-x
27. Gill BS, Appels R, Botha-Oberholster AM. A workshop report on wheat genome sequencing: international genome research on wheat consortium. Genetics. 2004: 168, 1087–1096. pmid:15514080 doi: 10.1534/genetics.104.034769
28. The international wheat genome sequencing consortium (IWGSC) A chromosome-based draft sequence of the hexaploid bread wheat (Triticum aestivum) genome. Science. 2014; 345: 286. doi: 10.1126/science.1251788

29. Steinitz-Sears LM. Somatic instability of telocentric chromosomes in wheat and the nature of the centromere. Genetics. 1966; 54: 241–248. pmid:17248316

30. Liu Z, Yue W, Li D, Wang RR, Kong X, Lu K, et al. Structure and dynamics of retrotransposons at wheat centromeres and pericentromeres. Chromosoma. 2008; 117: 445–456. doi: 10.1007/s00412-008-0161-9. pmid:18496705

31. Li B, Choulet F, Heng Y, Hao W, Paux E, Liu Z, et al. Wheat centromeric retrotransposons: the new ones take a major role in centromeric structure. Plant J. 2013; 73: 952–965. doi: 10.1111/tpj.12086. pmid:23253213

32. Rayburn AL, Gill BS. Isolation of a D-genome specific repeated DNA sequence from*Aegilops tauschii*. Plant Mol Biol Rep. 1987; 4: 102–109. doi: 10.1007/bf02732107

33. Danilova TV, Friebe F, Gill BS. Development of a wheat single gene FISH map for analyzing homoeologous relationship and chromosomal rearrangements within the Triticeae. Theor Appl Genet. 2014; 127: 715–730. doi: 10.1007/s00122-013-2253-z. pmid:24408375

34. Qi LL, Echalier B, Chao S, Lazo GR, Butler GE, Anderson OD, et al. A chromosome bin map of 10,000 expressed sequence tag loci and distribution of genes among the three genomes of polyploid wheat. Genetics. 2004; 168: 701–712. pmid:15514046 doi: 10.1534/genetics.104.034868

35. Jin WW, Melo JR, Nagaki K, Talbert PB, Henikoff S, Dawe RK, et al. Maize centromeres: organization and functional adaptation in the genetic background of oat. Plant Cell. 2004; 16: 571–581. pmid:14973167 doi: 10.1105/tpc.018937

36. Koo D-H, Jiang JM. Super-stretched pachytene chromosomes for fluorescence *in situ* hybridization mapping and immunodetection of DNA methylation. Plant J. 2009; 59: 509–516. doi: 10.1111/j.1365-313X.2009.03881.x. pmid:19392688

37. Koo D-H, Han F, Birchler JA, Jiang JM. Distinct DNA methylation patterns associated with active and inactive centromeres of the maize B chromosome. Genome Res. 2011; 21: 908–914. doi: 10.1101/gr.116202.110. pmid:21518739

38. Liu W, Rouse M, Friebe B, Jin Y, Gill BS, Pumphrey MO. Discovery and molecular mapping of a new gene conferring resistance to stem rust, *Sr53*, derived from*Aegilops geniculata* and characterization of spontaneous translocation stocks with reduced alien chromatin. Chromosome Res. 2011; 19: 669–682. doi: 10.1007/s10577-011-9226-3. pmid:21728140

39. International Rice Genome Sequencing Project. The map-based sequence of the rice genome. Nature. 2005; 436: 793–800. pmid:16100779 doi: 10.1038/nature03895

40. International Brachypodium Initiative Genome sequencing and analysis of the model grass *Brachypodium distachyon*. Nature. 2010; 463: 763–768. doi: 10.1038/nature08747. pmid:20148030

41. Gill BS, Friebe B, Endo TR. Standard karyotype and nomenclature system for description of chromosome bands and structural aberrations in wheat (*Triticum aestivum*). Genome. 1991; 34: 830–839. doi: 10.1139/g91-128

42. Sears ER, Muramatsu M. A line with all B-genome chromosomes as doubleditelocentric. In Proceedings of the 7th International Wheat Genet Symposium (Miller T.E. and Koebner R.M.D., eds). Institute of Plant Science Research, Cambridge, UK. 1988; pp: 427–431.

43. Naranjo T, Roca A, Goicoechea PG, Giraldez R. Chromosome structure of common wheat: genome reassignment of chromosome 4A and 4B. In Proceedings of the 7thInternational Wheat Genet Symposium (Miller T.E. and Koebner R.M.D., eds). Institute of Plant Science Research, Cambridge, UK. 1988; pp: 115–120.

44. Aragon-Alcaide L, Miller T, Schwarzacher T, Reader S, Graham M. A cereal centromeric sequence. Chromosoma. 1996; 105: 261–268. pmid:8939818 doi: 10.1007/bf02524643

45. Ito H, Nasuda S, Endo TR. A direct repeat sequence associated with the centromeric retrotransposons in wheat. Genome. 2004; 47: 747–756. pmid:15284880 doi: 10.1139/g04-034

46. Lee HR, Zhang W, Langdon T, Jin WW, Yan H, et al. Chromatin immunoprecipitation cloning reveals rapid evolutionary patterns of centromeric DNA in *Oryza* species. Proc Natl Acad Sci. 2005; 102: 11793–11798. pmid:16040802 doi: 10.1073/pnas.0503863102

47. Gong ZY, Wu YF, Koblížková A, Torres GA, Wang K, Iovene M, et al. Repeatless and repeat-based centromeres in potato: Implications for centromere evolution. Plant Cell. 2012; 24: 3559–3574. doi: 10.1105/tpc.112.100511. pmid:22968715

48. Neumann P, Naratilova A, Schroeder-Reiter E, et al. Stretching the rules: monocentric chromosomes with multiple centromere domains. PLoS Genet. 2012; 8: e1002777. doi: 10.1371/journal.pgen.1002777. pmid:22737088

49. Lo AW, Craig JM, Saffery R, Kalitsis P, Irvine DV, et al. A 330 kb CENP-A binding domain and altered replication timing at a human

neocentromere. EMBO J. 2001; 20: 2087–2096. pmid:11296241 doi: 10.1093/emboj/20.8.2087

50. Lomiento M, Jiang ZS, D'Addabbo P, Eichler EE, Rocchi M. Evolutionary-new centromeres preferentially emerge within gene deserts. Genome Biol. 2008; 9: R173. doi: 10.1186/gb-2008-9-12-r173. pmid:19087244

51. Ketel C, Wang HSW, McClellan M, Bouchonville K, Selmecki A, et al. Neocentromeres form efficiently at multiple possible loci in *Candida albicans*. PLoS Genet. 2009; 5: e1000400. doi: 10.1371/journal.pgen.1000400. pmid:19266018

52. Singh RJ. Plant cytogenetics, Second Edition. CRC Press, Boca Raton, FL. 2003; pp: 216–217.

53. **53.**Cheng ZK, Dong FG, Langdon T, Ouyang S, Buell CR, Gu MH, et al. Functional rice centromeres are marked by a satellite repeat and a centromere-specific retrotransposon. Plant Cell. 2002; 14: 1691–1704. pmid:12172016 doi: 10.1105/tpc.003079

54. Zhang P, Friebe B, Lukaszewski AJ, Gill BS. The centromere structure in Robertsonian wheat-rye translocation chromosomes indicates that centric breakage-fusion can occur at different positions within the primary constriction. Chromosoma. 2001; 110: 335–344. pmid:11685533 doi: 10.1007/s004120100159

55. Wicker T, Mayer KFX, Gundlach H, Martis M, Steuernagel B, Scholz U, et al. Frequent gene movement and pseudogene evolution is common to the large and complex genomes of wheat, barley, and their relatives. Plant Cell. 2011; 23: 1706–1718. doi: 10.1105/tpc.111.086629. pmid:21622801

56. Maggert KA, Karpen GH. The activation of a neocentromere in *Drosophila* requires proximity to an endogenous centromere. Genetics. 2001; 158: 1615–1628. pmid:11514450

57. Shang W-H, Hori T, Martins NMC, Toyoda A, Misu S, Monma N, et al. Chromosome engineering allows the efficient isolation of vertebrate neocentromeres. Dev Cell. 2013; 24: 635–648. doi: 10.1016/j.devcel.2013.02.009. pmid:23499358

58. Scott KC, Sullivan BA. Neocentromeres: a place for everything and everything in its place. Trends Genet. 2014; 30: 66–74. doi: 10.1016/j.tig.2013.11.003. pmid:24342629

59. Rhoades MM. Studies of a telocentric chromosome in maize with reference to the stability of its centromere. Genetics. 1940; 25: 483–520. pmid:17246983

60. Sharp JA, Kaufman PD. Chromatin proteins are determinants of centromere function. Curr Top Mirobiol Immunol. 2003; 274: 23–52. doi: 10.1007/978-3-642-55747-7_2

61. Folco HD, Pidoux AL, Urano T, Allshire RC. Heterochromatin and RNAi are required to establish CENP-A chromatin at centromeres. Science. 2008; 319: 94–97. doi: 10.1126/science.1150944. pmid:18174443

62. Peters AH, O'Carroll D, Scherthan H, Mechtler K, Sauer S, Schofer C, et al. Loss of the Suv39h histone methyltransferases impairs mammalian heterochromatin and genome stability. Cell. 2001; 107: 323–337 pmid:11701123 doi: 10.1016/s0092-8674(01)00542-6

63. Wanner G, Schroeder-Reiter E, MA W, Houben A, Schubert V. The ultrastructure of mono- and holocentric plant centromeres: an immunological investigation by structured illumination microscopy and scanning electron microscopy. Chromosoma. 2015; doi: 10.1007/s00412-015-0521-1.

Chapter 8

A MAJOR LOCUS FOR CHLORIDE ACCUMULATION ON CHROMOSOME 5A IN BREAD WHEAT

Yusuf Genc[1,2], Julian Taylor[1], Jay Rongala[2], and Klaus Oldach[1,2]

[1]School of Agriculture, Food and Wine, University of Adelaide, Waite Campus, Glen Osmond, South Australia, Australia

[2]South Australian Research and Development Institute, Plant Genomics Centre, Waite Campus, Glen Osmond, South Australia, Australia

ABSTRACT

Chloride (Cl^-) is an essential micronutrient for plant growth, but can be toxic at high concentrations resulting in reduced growth and yield. Although saline soils are generally dominated by both sodium (Na^+) and Cl^- ions, compared to Na^+ toxicity, very little is known about physiological and genetic control mechanisms of tolerance to Cl^- toxicity. In hydroponics and field studies, a bread wheat mapping population was tested to examine the relationships between physiological traits [Na^+, potassium (K^+) and Cl^- concentration] involved in salinity tolerance (ST) and seedling growth or grain yield, and to elucidate the genetic control mechanism of plant Cl^- accumulation using a quantitative trait loci (QTL) analysis approach. Plant Na^+ or Cl^- concentration were moderately correlated (genetically) with seedling biomass in hydroponics, but showed no correlations with grain yield in the field, indicating little value in selecting for ion concentration to improve ST. In accordance with phenotypic responses, QTL controlling Cl^- accumulation differed entirely between hydroponics and field locations, and few were detected in two or more environments, demonstrating substantial QTL-by-environment interactions. The presence of several QTL for Cl^- concentration indicated that uptake and accumulation was a polygenic trait. A major Cl^- concentration QTL (5A; *barc56/gwm186*) was identified in three field environments, and accounted for 27–32% of the total genetic variance. Alignment between the 5A QTL interval and its corresponding physical genome regions in wheat and other grasses has enabled the search for candidate genes involved in Cl^- transport, which is discussed.

INTRODUCTION

Worldwide, salinity poses a serious threat to agricultural production, with globally salt-affected soils (including saline and sodic soils) totalling around 830 million hectares [1]. Saline soils are generally dominated by sodium and chloride ions, and the ability of crop plants to exclude these ions is often equated to salinity tolerance-ST (i.e. improved growth or yield under salinity stress). Despite intensive research and numerous scientific reports over several decades, there is a lack of consensus on whether Na^+ exclusion is always a useful trait to select for to improve ST, at least in the case of bread wheat. Few studies have shown good phenotypic correlation between Na^+ exclusion and ST [2]–[5], while others have reported weak [6], [7] or no correlation [8]–[10]. In contrast to Na^+ exclusion, there has been very little research on the role of Cl^- exclusion and its contribution to ST [3], [11]–[13]. Similar to Na^+ exclusion, these few studies on Cl^- reported either significant phenotypic correlation or no correlation with ST. It is clear that future research needs to place equal emphasis on the impact of Cl^- in ST as not only Na^+ but also Cl^- is present at toxic concentrations in saline-affected growth media used in the assessment of ST [13], [14].

In screening studies for ST, sodium chloride is the most commonly used salt. Despite both Na^+ and Cl^- ions being present at toxic concentrations in growth media, there has been a lack of interest in Cl^-, which may be attributed to the earlier reports that Na^+ was more toxic than Cl^- [15]. Although these authors cautioned about assumptions made from extrapolations of two lines of bread wheat, their findings were not verified with a large set of genotypes differing in ST. This was later questioned by Martin and Koebner [16] on the basis that while attempting to separate the toxic effects of Na^+ and Cl^-, the authors used phytotoxic concentrations of nitrate in their experiments. Martin and Koebner [16] concluded that Cl^- was more toxic than Na^+, but the full toxic effect was apparent only when Na^+ and Cl^- were present simultaneously. However, more recent studies in barley found that Na^+ and Cl^- had a similar effect upon plant growth [17]. While the issue of whether Na^+ or Cl^- is more detrimental to plant growth remains controversial, it is appropriate to measure the concentration of both ions in the plant and to determine their relevance to ST for reasons mentioned earlier. Munns and Tester [18] argued that toxicity of Na^+ versus Cl^- can be best studied through genetic approaches as the alternative method of using different salts has to date produced equivocal results. The potential of the genetic approach has not been employed extensively.

Chloride is an essential micronutrient and has several functions in plant metabolism; enzyme activation, photosynthesis, a counter ion for cation transport, osmoregulation, and movement of stomata. Like most anions, it is weakly bound to soil particles, mostly in a soluble form in soil solution, and at high concentrations can be toxic to plant growth. Critical toxicity concentration in plants was estimated at 110–200 mmol kg^{-1} DW (4,000–7,000 mg kg^{-1} DW) and 420–1400 mmol kg^{-1} DW (15,000–50,000 mg kg^{-1} DW) for sensitive and tolerant species, respectively [19]. At present there is very little information on genetic control mechanisms of Cl^- exclusion in plant species. A better understanding of genetic control of Cl^- exclusion and identification of molecular markers has the potential to speed up breeding for a complex trait such as ST. To date, rice and barley appear to be the only cereal species in which genetic control of Cl^- exclusion has been investigated. In two studies involving rice F_2 and RIL populations derived from a cross between salt tolerant and salt-sensitive varieties CSR27 and MI48 respectively, few Cl^- concentration QTL were co-located with Na^+ concentration QTL, while others mapped to different regions [20], [21]. Similar findings were also reported in barley [22], [23]. If QTL for Na^+ and Cl^- concentration map to different locations, this would suggest separate genetic control mechanisms for regulation of these ions. At present, we are not aware of any reports on QTL mapping of Cl^-concentration in wheat.

Screening for ST is usually conducted in one of three main environments: hydroponics, soil-based pot assays or field trials. Due to inherent difficulties with field screening such as non-uniform distribution of salinity throughout the experimental area, fluctuations in rainfall and potential nutritional deficiencies, controlled environments are often preferred. However, as reported in recent studies [24], the results of controlled environments can be quite different from those of field environments and, therefore, require verification. The only two genetic studies on Cl^- accumulation to date have been conducted in hydroponics, and how these results correlate to field environments is unknown. It is important to note that the ability of controlled environmental studies to predict yield responses in the field is rarely addressed in the scientific literature, and the validation of controlled environmental studies in the field is important for plant breeding.

In previous studies in hydroponics and field trials [24], [25], we reported QTL for grain yield, grain number m^{-2}, 1000-grain weight, maturity, plant height, seedling biomass, tiller number, chlorophyll content, leaf symptoms, Na^+ and K^+ concentrations of leaves and shoots for a bread wheat mapping population (Berkut/Krichauff). With the recent renewed interest in Cl^- [13], we revisited this population to (i) elucidate the genetic control mechanisms of Cl homeostasis via a QTL approach, and (ii) investigate the relationships among seedling biomass, grain yield and plant Na^+, K^+ and Cl^- concentrations. Here we report for the first time identification of a major QTL for Cl^- concentration in bread wheat and discuss its importance for marker-assisted selection and fine mapping/discovery of genes involved in Cl^- transport in bread wheat. We also demonstrate that Cl^- accumulation is a polygenic trait, but does not appear to be a reliable predictor of ST based on grain yield alone.

MATERIALS AND METHODS

Plant Material

A doubled-haploid (DH) population (152 lines) from a cross between bread wheat (*Triticum aestivum* L.) genotypes Berkut [Irene/Babax//Pastor] and Krichauff [Wariquam//Kloka/Pitic62/3/Warimek/Halberd/4/3Ag3/Aroona] was used in this study. The rationale for screening this population for plant Cl^- concentration and subsequent QTL detection was that a previous study revealed significantly lower shoot Cl^- accumulation in the Krichauff parent than in the Berkut parent [26].

Phenotyping and Trait Analysis

Growth room and field studies were described previously [24], [25]. The data for grain yield, grain number m^{-2}, 1000-grain weight, maturity, plant height, seedling biomass, tiller number, chlorophyll content, leaf symptoms and Na^+ and K^+ concentrations of penultimate leaves and shoots were reported earlier [24], [25]. In the present study shoot and leaf Cl^-concentrations of DH population grown in hydroponics and field trials (Roseworthy, Balaklava and Georgetown in South Australia) characterized by low, moderate and high salinity [24] were determined. Either single (hydroponics, two replicates) or 15–20 plants per entry (field trials, two replicates) were sampled for elemental analysis. Shoot (hydroponics) and leaf samples (field trials) were dried at 65°C for 48 h and dry weights recorded. The dried plant samples were then ground and analysed for Cl^-, calcium (Ca^{2+}) and magnesium (Mg^{2+}) concentration using either Inductively Coupled Plasma Optical Emission Spectrometry (ICP-OES) (ARL 3580 B, Appl. Res Lab. SA, Ecublens, Switzerland)[27], [28] or a chloride meter (Model 926, Sherwood, Cambridge, UK). For analysis of Cl^- using the ICP-OES method, 0.1 g of ground shoot sample was extracted with hot (95°C) 4% HNO_3 acid in 50 mL capped polypropylene tubes for 90 minutes, whereas for measurements of Cl^- using the chloride meter, 0.5 g of ground sample was digested in 40 mL of 1% HNO_3 at 85°C for 5 hours in a 54well HotBlock (Environmental Express, Mt. Pleasant, South Carolina, USA). ICP-OES was used initially for analysis of the Cl^-concentration of hydroponically-grown plants, but due to prohibitive cost, leaf samples of field-grown plants were analyzed using a chloride meter. As two different analytical methods were used for determination of Cl^- concentration in plant tissues, a number of samples were analyzed using both methods, and the high correlation of the measurements between the two methods (hydroponically-grown samples, $r^2=0.98$, n=45; field-grown samples, $r^2=0.97$, n=15) indicated that the methods were comparable. As single measurements were taken from each extraction, duplicate analysis (one sample per batch of 24) was carried out to determine the homogeneity of the samples, and relative standard deviations between the two measurements were below 5% in all cases. Chloride concentration was expressed on a dry mass basis (mmol kg^{-1} DW).

Linkage Map and Interval Construction

Genotyping and construction of a genetic linkage map for the Berkut/Krichauff population was described earlier [24], [25]. The constructed linkage map initially comprised 557 markers across 21 chromosomes. After omission of co-locating markers this was reduced to 403 markers with an average interval distance of 9.16 cM. For computational purposes the alleles of the Berkut (A)/ Krichauff (B) population were then converted into 1 and -1 respectively and missing marker scores were imputed using the flanking marker method of Martinez and Curnow [29]. A total of 384 inferred interval markers were then constructed using the mid-point interval method of Verbyla et al. [30].

Multi-Environment Analysis

To understand the genetic relationships of the DH lines across the field trials and growth room a multi-environment trials (MET) analysis was conducted for each of the traits Na^+, $K^+$and Cl^-. For grain yield the MET analysis was restricted to field sites only. The analysis approach follows Smith et al. [31] which involves a linear mixed model including the parsimonious modelling of genetic effects of the DH lines through an appropriate genotype by environment interaction model and also captures non-genetic sources of variation through the use of separate spatial models for the plot errors at each site. The method initially involves the assessment of the spatial or environmental variation occurring at each site through the investigation of the assumption of variance homogeneity, detection of outliers, and the identification of global trends that may exist across the rows or columns of the experiment. For trends existing due to adjacency of plots in a trial, the model included a separable row by column autoregressive correlation structure. Stronger linear trends in either direction are fitted as fixed effects in the model. Design parameters such as genotypic replication or blocking structures were fitted as separate random effects.

The genotype by environment interaction model involved the use of an unstructured heterogeneous correlation matrix that appropriately captures the genetic relationship of the DH lines between trials. If a strong positive genetic correlation exists between any two trials then the relative performance or rankings of the varieties at each of the trials will be similar. As a consequence these trials will most likely share common or co-locating QTL with a common parent being favoured at each locus. Trials that exhibit little or no genetic correlation between them would exhibit different relative rankings for the varieties and most likely have unshared QTL. Each multi-environment analysis was performed using residual maximum likelihood (REML) and the estimated genetic correlations between sites were extracted for interpretation.

Multivariate Analysis

A multivariate linear mixed model analysis of four traits (Na^+, K^+, Cl^- and grain yield) was conducted for each field trial in order to estimate genetic correlations of the DH lines between traits. A multivariate analysis was also conducted for the traits in the growth room and included Na^+, K^+, Cl^-, and seedling biomass. For each of the trials, estimates of the genetic relationships between DH lines were modelled through the use of a trait by genotype interaction with an unstructured heterogeneous correlation matrix. The model also included a separable trait by row by column spatial model for the plot errors with an unstructured heterogeneous correlation matrix for the trait component of the model. This unstructured correlation matrix ensures that traits collected from the same trial are connected phenotypically. This separable structure for the spatial model also assumes that the traits have a common row by column separable autoregressive correlation structure. Similar to the MET models, strong trends for any given trait were captured using the appropriate fixed effects, and random effects were used to model genotypic replication as well as blocking structures existing within each trial. Each multivariate analysis was performed using REML and the estimated genetic correlations between traits were extracted for interpretation.

QTL Analysis

For each of the field sites and the growth room the detection and estimation of QTL for the measured traits was accomplished using the R [32] package wgaim [33]. The package is a computational implementation of Verbyla et al. [30], [34]. In this approach, an initial base linear mixed model is established that contains non-genetic effects that account for extraneous variation. These include an appropriate spatial model for the errors as well as fixed and random effects that are relevant to the trial being examined. The base model also includes a random effect term that captures the genetic variation between the DH lines. Following Verbyla et al. [34] the base model is then extended by including the complete set of inferred interval markers into the base linear mixed model as a contiguous block of random effects with a single variance parameter. The significance of this variance parameter is then checked using a simple residual likelihood ratio test. If found significant, an alternative outlier model is formulated and the inferred interval marker with the largest outlier statistic is chosen as a putative QTL. This inferred interval marker is then removed from the contiguous block of random effects and placed as a separate random covariate in the original base model as well as the extended model that includes the remaining set of inferred interval markers. The forward selection process is then repeated until the variance parameter associated with the remaining inferred interval markers is not significant. The complete set of putative QTL selected appears additively as random covariates and is summarised using the methods of Verbyla et al. [34]. The summary includes the left and right flanking markers of the individual QTL, their effect sizes, approximate LOD scores as well as individual contributions to the overall genetic variance.

For all traits analysed in the present study, the best linear unbiased predictions of the genotypes were extracted from the base linear model and were used to calculate a generalized (broad-sense) heritability using the formula developed by Cullis et al. [35].

Comparative Analysis and Candidate Genes Co-located with the 5A Cl^--Concentration QTL

To align the genetic position of the 5A Cl^- QTL to its physical position in the wheat genome, the sequence of RFLP markers with nearby location to the QTL-flanking SSR markers*gwm304* and *barc141* were identified using the database GrainGenes 2.0 (wheat.pw.usda.gov). The sequences of co-located RFLP markers *bcd21* and *psr128* were 451 bp and 430 bp, respectively, and were both derived from ESTs (GrainGenes 2.0). The sequences were used for homology searches using the BLAST tool athttps://urgi.versailles.inra.fr/blast/blast.php. The search was carried out against the sequences of all bread wheat chromosomes showing the best hits on 5AL, as expected. The gene sequence hits were used for synteny analysis using the Genome Zipper v5 across wheat, rice, *Brachypodium* and sorghum revealing the syntenic regions between 64 non-rendundant wheat ESTs, *Brachypodium* chromosome 4, rice chromosome 9 and sorghum chromosome 2.

To assign potential functions to the wheat genes underlying the 5A Cl^- QTL interval, all 64 non-redundant wheat ESTs were used as queries for homology searches at the National Center for Biotechnology Information (NCBI) using BLASTN against the non-redundant nucleotide database.

RESULTS

Responses to Salinity Stress, Distributions and Relationships between Traits

Data on Cl^- concentration provided an opportunity to re-examine the relationships amongst traits associated with ST such Na^+ and K^+ accumulation, seedling biomass and grain yield in low (Balaklava), moderate (Roseworthy) and high (Georgetown and growth room) saline environments [24]. Krichauff had 10–20% lower Na^+ concentration than Berkut in field trials, and these differences diminished in hydroponics, whereas Krichauff had 14–25% lower Cl^-concentration than Berkut in all environments ranging from 362–669 and 482–775 mmol kg^{-1} DW for Krichauff and Berkut, respectively (Table 1), similar to previous studies [26]. It was interesting to observe that Cl^- concentrations were much higher than Na^+ concentrations. Berkut had slightly higher K^+ concentration than Krichauff but only at low to moderately saline field trials at Roseworthy and Balaklava (812–934 and 729–882 mmol kg^{-1} DW for Berkut and Krichauff respectively; Table 1). As for seedling biomass and/or grain yield production under salinity stress, Berkut produced 18% higher seedling biomass than Krichauff in hydroponics, whilst Krichauff had 6–7% higher grain yield in field trials (Table 1). In all traits, there was evidence of transgressive segregation.

To determine whether selection for Na^+ and/or Cl^- exclusion or even K^+ accumulation would lead to improved ST, multivariate analysis of concentrations of Na^+, Cl^-, K^+, seedling biomass and grain yield was performed for each of the environments (Table 2). The analysis showed that there was generally a good estimated correlation between Cl^- and either Na^+ or K^+ with two exceptions: there were negligible correlations for Na^+ *vs* Cl^-, and K^+ *vs* Cl^- at Balaklava and in hydroponics respectively. A negative but moderate correlation was observed between seedling biomass and either Na^+ (r=0.479) or Cl^- (0.527).

Table 1. Parental means, population mean and range for Na^+, Cl^-, K^+, Ca^{2+} and Mg^{2+} concentrations (mmol kg^1 DW) in penultimate leaves (field trials) and whole shoots (hydroponics), seedling biomass [shoot DW (g plant^1)] and grain yield (t ha^1) in Berkut/Krichauff DH population tested for ST in hydroponics and field trials (Roseworthy, Balaklava and Georgetown).

		Parental Lines		DH population		
Test environment	**Trait**	**Berkut**	**Krichauff**	**mean**	**range**	**Heritability h^2**
Hydroponics	Na^+ conc.	298	295	270	194–361	0.58
(100 mM NaCl ~10 dS m^1)	Cl^- conc.	457	376	378	236–480	0.67
	K^+ conc.	771	779	749	661–869	0.82
	Shoot DW	1.523	1.296	1.537	1.210–1.835	0.57
	Ca^{2+} conc.	107.2	87.3	97.9	77.4–116.7	0.83
	Mg^{2+} conc.	94.7	77.5	84.1	73.2–95.1	0.77
Roseworthy	Na^+ conc.	15.2	12.6	12.4	8.8–15.9	0.57
(low salinity, ECe<4 dS m^1)	Cl^- conc.	482	362	393	272–510	0.82
	K^+ conc.	812	729	739	621–847	0.64
	Grain yield	2.237	2.392	2.116	1.451–2.489	0.76
Balaklava	Na^+ conc.	13.1	11.6	10.9	7.4–14.5	0.78
(Moderate salinity, ECe = 4–8 dS m^1)	Cl^- conc.	557	447	474	301–598	0.82
	K^+ conc.	934	882	865	749–1011	0.77
	Grain yield	2.714	2.888	2.671	1.758–3.035	0.81
	Ca^{2+} conc.	121.3	97.2	110.7	71.9–151.4	0.82
	Mg^{2+} conc.	77.2	62.2	73.5	52.0–95.1	0.79
Georgetown	Na^+ conc.	20.5	16.3	18.7	10.9–30.4	0.60
(High salinity, ECe>8 dS m^1)	Cl^- conc.	775	669	688	571–884	0.84
	K^+ conc.	1363	1332	1328	1064–1523	0.76
	Grain yield	0.645	0.686	0.573	0.277–0.792	0.72

The means represent predicted values from MET (Na^+, Cl^-, K^+) and single environment (Ca^{2+} and Mg^{2+}) analysis of each trait. Broad-sense heritability is also given for individual traits at each environment.

doi:10.1371/journal.pone.0098845.t001

Table 2. Estimated genetic correlations between shoot DW (hydroponics), grain yield (field) Na^+, K^+, and Cl^- (field and hydroponics) extracted from the fitted multi-trait model at each environment.

Environment	Na^+ *vs* Cl^-	K^+ *vs* Cl^-	Shoot DW or yield *vs* Na^+	Shoot DW or yield *vs* K^+	Shoot DW or yield *vs* Cl^-
Hydroponics	0.875	0.195	−0.486	−0.146	−0.531
Roseworthy	0.451	0.634	0.200	0.097	0.053
Balaklava	0.122	0.517	−0.200	0.060	0.097
Georgetown	0.319	0.763	0.066	0.148	0.082

doi:10.1371/journal.pone.0098845.t002

To analyse genotype by environment interaction for each of the measured traits, multi-environment analyses and genetic correlations were estimated between environments. Representing low (Balaklava) to moderate (Roseworthy) saline environments, correlations for Na^+, Ka^+, Cl^- concentrations and grain yield were consistently high between these sites whereas low to moderate correlations were observed for these traits between other higher saline environments (Table 3). Therefore, it is reasonable to assume that selection for similar trait values may not be consistent between higher saline sites, indicating environmental effects controlling ion accumulation and grain yield.

Table 3. Estimated genetic correlations extracted from the fitted multi-environment model for individual traits.

		Roseworthy	Balaklava	Georgetown
Na^+ concentration	Balaklava	0.861		
	Georgetown	0.016	0.167	
	Growth room	0.182	0.186	0.464
		Roseworthy	Balaklava	Georgetown
K^+ concentration	Balaklava	0.794		
	Georgetown	0.568	0.598	
	Growth room	0.361	0.322	0.225
		Roseworthy	Balaklava	Georgetown
Cl^- concentration	Balaklava	0.897		
	Georgetown	0.582	0.462	
	Growth room	0.276	0.427	0.339
		Roseworthy	Balaklava	
Grain yield	Balaklava	0.706		
	Georgetown	0.130	0.338	

doi:10.1371/journal.pone.0098845.t003

Generalized (Broad-sense) Heritability (h^2)

Estimates of h^2 differed with trait and environment, ranging from moderate to high (0.6–0.8) (Table 1). Amongst the mineral elements, Na^+ concentration had the lowest h^2, while Cl^-concentration had the highest h^2. Heritability of shoot DW (0.6) was lower than that of grain yield (0.7–0.8). Heritability of individual traits across environments showed relatively consistent (Ca^{2+}, Mg^{2+} and Cl^- concentration) to inconsistent patterns (Na^+ concentration). As reported earlier [24], consistently higher h^2 values indicate greater ability for selection for the traits, while lower and variable h^2 suggest the presence of substantial environmental effects, difficulty for direct selection, and the need for more replications.

QTL for Cl^- Concentration

In the initial analysis in which differences in phenology were not included, there were few Cl^- concentration QTL co-locating with QTL for maturity on 5A and 5D (data not shown). Similar to Bonneau et al. [36], after the known differences in the genetic component of the phenology were addressed by fixing the maturity genes in the analyses, 14 QTL were identified (Table 4). However, most QTL were specific to single environments, and only three QTL were detected in two or more environments (3A, 5A, 7D) indicating some genotype by environment interaction. Interestingly, there were no co-located QTL controlling trait variation from hydroponics and field trials. The most significant QTL in hydroponics on chromosome 2A explained 20% of the total genetic variance and the Krichauff allele was responsible for increased Cl^- concentration. QTL on chromosome 3A and 7D were detected from multi-location trials and explained 4–11% of the total genetic variance, while QTL on 5A accounted for 27-32% of the total genetic variance (Table 4). Either the Berkut (3A, 5A) or the Krichauff (7D) allele was associated with increased Cl^- concentration at these loci. As the two QTL on 5A appear in tandem, these loci were further investigated to determine whether there may just be one rather than two separate QTL. The plot of outlier statistics (Figure 1) shows that there is in fact just one QTL on 5A expressed at all field locations. The QTL detected at one location only accounted for a small proportion of the total genetic variance, varying from 3.6 to 9.9% with either the Berkut or Krichauff allele being associated with increased Cl^- concentration (Table 4).

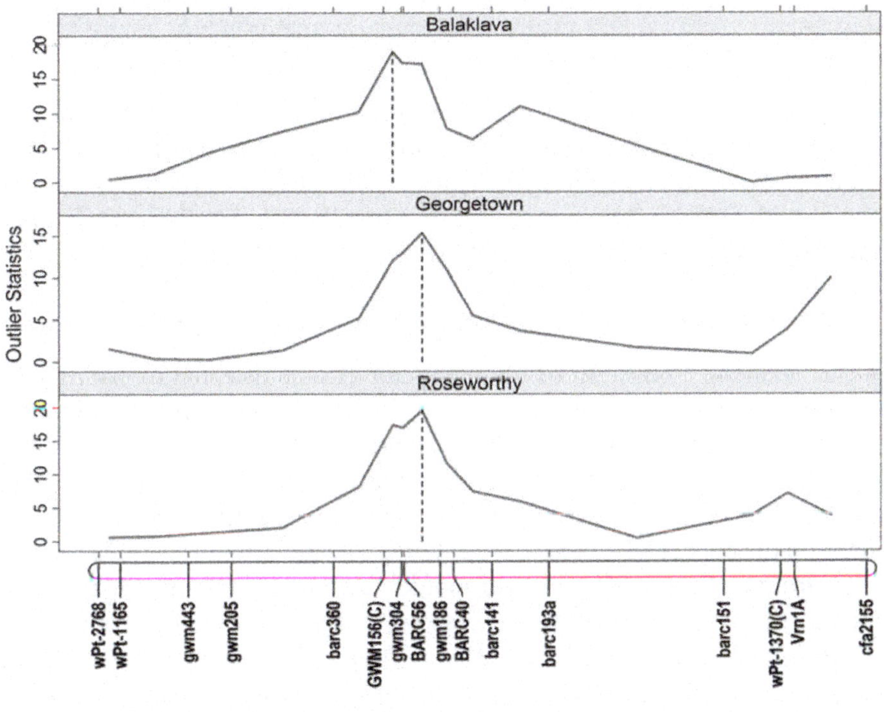

Figure 1. Location of Cl^- concentration QTL (*barc56*/*gwm186*) on chromosome 5A detected in field trials (Balaklava, Georgetown and Roseworthy) with varying salinity levels. The outlier statistics represent LOD scores (Table 4).

Table 4. QTL associated with Cl^- concentration in hydroponics and field trials (Roseworthy, Balaklava and Georgetown) under varying degrees of salinity stress.

Environment	Ch	Interval	Distance (cM)	Size*	Prob.	% Var.	LOD	**Co-locations with Na^+ or K^+ in field locations and hydroponics reported earlier [24,25]
Roseworthy	2B	*cfd011- gwm374(C)*	88.0–95.8	−12.5	0.001	4.1	2.4	
	3A	*wmc343- cfa2262*	75.7–79.7	20.6	0.000	11.1	6.4	
	3B	*wPt-9049(C)- wPt-0021*	124.4–154.0	13.4	0.002	3.6	2.1	
	5A	*barc56/gwm186*	90.0–100.5	36.8	0.000	32.4	18.3	Na^+-GT, K^+-RW, K^+-BA, K^+-GT
	6D	*cfd005(C)- gpw95010(C)*	172.8–174.1	−11.7	0.001	3.9	2.4	
	7D	*barc214- gwm437*	130.8–133.5	−11.3	0.002	3.6	2.0	Na^+-RW, Na^+- RH, Na^+-WT
	7D	*wmc014a- wPt-0695*	243.5–246.4	−11.3	0.002	3.6	2.1	
Balaklava	3A	*barc324- wmc343*	67.0–75.7	9.4	0.002	4.4	2.2	
	3B	*gwm247- wPt-4412*	195.1–196.1	13.9	0.000	9.9	4.8	
	5A	*gwm156(C)- gwm304*	84.1–89.3	25.4	0.000	30.8	14.8	
Georgetown	5A	*barc56- gwm186*	90.0–100.5	30.7	0.000	27	11.1	
	7B	*bcg1- wPt-7318*	0.0–22.7	−14.5	0.000	5.4	2.9	
	7D	*barc214- gwm437*	130.8–133.5	13.2	0.001	5.7	2.6	
Hydroponics	1D	*wmc216- wPt-1799*	80.7–131.2	21.3	0.002	5.8	2.0	
	2A	*wPt-3114- wmc170(C)*	92.5–109.9	−33.5	0.000	21.1	7.1	Na^+- HYDRO, Na^+-RW, Na^+-BA
	2B	*wmc272- barc349*	106.5–112.7	−19.4	0.000	8.5	2.7	Na^+-HYDRO
	2B	*wPt-7859- wPt-7161*	198.9–200.2	18.9	0.000	8.4	2.9	Na^+-HYDRO
	3A	*barc324- wmc343*	67.0–75.7	17.4	0.001	6.7	2.3	Na^+-RW
	7A	*gwm282- wPt-0961(C)*	170.2–183.3	−19.3	0.000	7.8	2.7	Na^+-HYDRO

Only those intervals with *P* values ≤0.01 and LOD>2.0 are presented. Field locations Roseworthy, Balaklava and Georgetown were classified as low, moderate and high salinity, respectively. Please see Genc et al. [24] for soil salinity classification.

*Positive and negative values indicate that Berkut and Krichauff alleles increased the phenotypic values, respectively. **The same co-locations appearing at multiple environments were presented only once. Abbreviations; Hydroponics = HYDRO, Balaklava = BA; Georgetown = GT; Roseworthy = RW. QTL names with letter C indicate several co-locating markers at those loci.

doi:10.1371/journal.pone.0098845.t004

Some QTL for Cl^- concentration were co-located with QTL for either Na^+ or K^+ concentration from hydroponics and field trials (Table 4). It was interesting to observe that in hydroponics QTL for Cl^- concentration were co-located with QTL for Na^+ concentration (2A, 2B, 2B, 3A, 7A), while in field trials they were co-located with both Na^+ (3A, 5A, 7D) and K^+ (5A).

QTL for Calcium (Ca^{2+}) and Magnesium (Mg^{2+}) Concentration

As Ca^{2+} and Mg^{2+} nutrition of the plant can also be affected by salinity [26], [37]–[40], their genetic control under salinity stress was also investigated, in a limited way, using a QTL mapping approach. A total of 13 QTL for Ca^{2+} concentration and 6 QTL for Mg^{2+}concentration were identified in hydroponics and the field (Table 5). At those loci, either the Berkut or Krichauff allele increased the concentration of these cations. The most significant QTL for Ca^{2+} concentration was the QTL in hydroponics accounting for 13% of the genetic variance (Table 5). This QTL was also co-located with the Cl^- concentration QTL detected in hydroponic conditions (Table 4). For Mg^{2+} concentration, the QTL on 3A in hydroponics was the most significant, explaining 15% of the genetic variance (Table 5). This QTL was also co-located with a QTL for Cl^- concentration detected in environments with low to moderately saline conditions (Roseworthy and Balaklava) and hydroponics e (Table 4). However, there were no common QTL that were detected under hydroponics and the field for either of these two cations.

Table 5. QTL associated with Ca^{2+} and Mg^{2+} concentrations in hydroponics and in a field trial (Balaklava) under varying degrees of salinity stress.

Environment	Trait	Ch	Interval	distance (cM)	Size*	Prob.	% Var.	LOD
Balaklava	Ca^{2+} conc.	3A	*cfa2262-wPt-3816*	79.7–100.95	5.7	0.000	10.6	3.6
		4A	*gwm165a-wmc420*	37.8–49.4	5.1	0.000	9.9	3.2
		5B	*gwm499-wPt-5851*	43.9–46.1	−5.2	0.000	11.4	2.9
		7D	*wmc436b-barc214*	96.8–130.8	4.6	0.001	6.2	2.2
	Mg^{2+} conc.	2A	*gwm294-gdm093(C)*	113.5–150.8	2.4	0.002	6.2	2.1
		4B	*gwm149(C)-wPt-1505(C)*	48.0–48.7	2.3	0.001	9.1	2.3
		5B	*wPt-3457-wPt-1250*	51.5–54.3	−2.4	0.001	10.1	2.2
Hydroponics	Ca^{2+} conc.	1D	*wmc216-wPt-1799*	80.7–131.2	4.9	0.000	13.3	10.1
		2B	*wPt-5707-wPt-3561*	58.4–61.0	−1.9	0.001	3.9	2.5
		3B	*gwm389-wPt-8093(C)*	0.0–0.7	−1.9	0.000	4.0	3.4
		3B	*wPt-4412-wPt-8352(C)*	196.1–197.1	1.6	0.002	2.8	2.1
		4A	*wmc106-gwm165a*	28.4–37.8	2.5	0.000	6.3	4.4
		4D	*gpw95001-gwm165b*	49.7–50.5	−2.9	0.000	9.0	6.5
		5B	*gwm371-gwm499*	37.6–43.9	−2.5	0.000	6.3	3.1
		6B	*cfd076b-wPt-4924*	119.8–124.4	3.4	0.000	11.8	6.6
		7B	*wPt-6372(C)-wPt-3833*	63.2–66.8	2.2	0.000	5.1	3.4
	Mg^{2+} conc.	2A	*gwm294-gdm093(C)*	113.5–150.8	1.6	0.000	6.6	2.8
		3A	*gwm732a-barc1121(C)*	2.2–8.8	−1.4	0.000	7.6	2.9
		3A	*barc324-wmc343*	67.0–75.7	2.1	0.000	15.4	5.5
		4B	*gwm513(C)-gwm495*	43.4–44.7	1.6	0.000	9.9	3.4

Only those intervals with P values ≤0.01 and LOD>2.0 are presented. Balaklava location was classified as moderate salinity. Please see Genc et al. [24] for soil salinity classification.
*Positive and negative values indicate that Berkut and Krichauff alleles increased the phenotypic values, respectively. QTL names with letter C indicate several co-locating markers at those loci.
doi:10.1371/journal.pone.0098845.t005

Known Chloride Transporters/Channel in Grass Genomes and the 5A Cl⁻QTL′

In previous studies, employing transcription analysis under salt stress, cellular localization and transgenesis for functional characterization, CLC (chloride channel) and CCC (cation chloride co-transporter) genes had been identified to play a role in Cl^- homoeostasis in plants; examples are CLC1 in tobacco and soybean [13], [41], [42], [43] and CCC in Arabidopsis [44].

To investigate the presence of candidate genes such as CLCs, CCCs and other ion transporters within the QTL interval on chromosome 5A, we physically positioned the 5A Cl^-concentration QTL in the wheat genome sequence. For this purpose, the gene-based sequences of RFLP markers, *bcd21* and *psr128* with close linkage to the QTL-flanking SSR markers *gwm304* and *barc141* (Figure 2) were used to find wheat genome sequences. As expected, both RFLP sequences had their best hits in bread wheat chromosome 5AL and allowed to retrieve matching contigs of 4.2 kb and 9.1 kb for *bcd21* and *psr128*, respectively. As both RFLPs had originally been derived from ESTs, the corresponding wheat genome contigs (4.2 and 9.1 kb) identified gene hits in rice chromosome 9 (*Os09g0321900* and *Os09g0412200*). These rice genes functioned as borders of the physical interval in the comparative analysis between rice, *Brachypodium* and wheat using the alignment in Genome Zipper v5 (Figure 2). In rice (MSU Release 7 at rice.plantbiology.msu.edu/cgi-bin/gbrowse/rice/#search), the syntenic interval contained 547 genes.

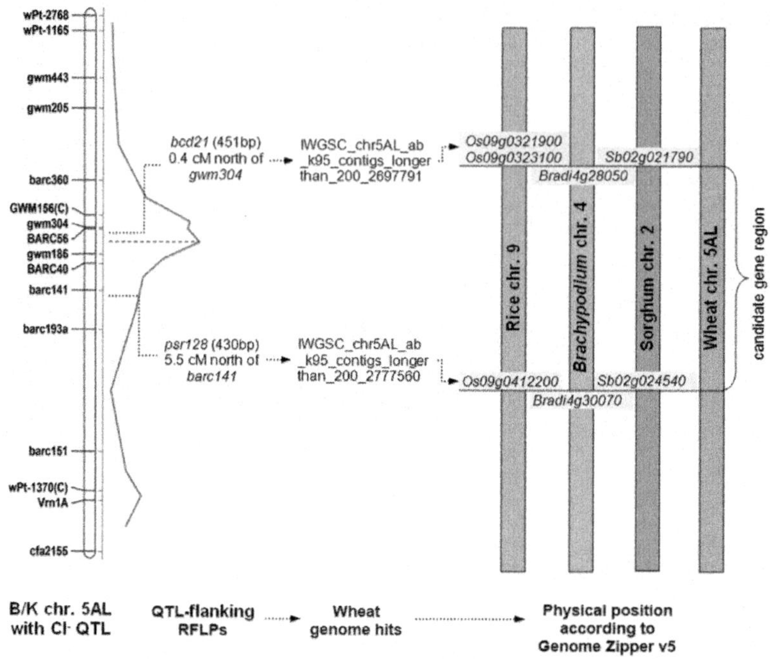

Figure 2. Inferred physical position of the Cl^- concentration QTL on 5AL in Berkut/Krichauff identified at Roseworthy field location onto 5AL in wheat.

Table 6. Candidate genes underlying the physical interval of the Cl^- QTL on chromosome 5A.

Brachypodium (v1.2)	Rice (Gene ID at MSU[a]/at IRGSP v2)	Sorghum (v1.4)	Wheat (v5[b])	Predicted protein function[a]
-	-	*Sb02g022750*	-	ABC transporter
Bradi4g28660	*LOC_Os09g19734/OS09G0361400*	*Sb02g022910*	*WHE0957_E03_J05ZT; Traes_5AL_99BEC1C3B*	Voltage dependent anion channel 1 (VDAC1)
Bradi4g29102	*LOC_Os09g20480/Os09g0371000*	*Sb02g023340*	*Traes_5AL_F8B48EC59*	ABC transporter
Bradi4g29110	*LOC_Os09g20490/Os09g0371100*	*Sb02g023370*	-	ABC transporter
Bradi4g29120	"	*Sb02g023380*	-	"
Bradi4g29110	*LOC_Os09g20500/Os09g0371200*	*Sb02g023370*	-	ABC transporter
Bradi4g29120	"	*Sb02g023380*	-	"
Bradi4g29140	*LOC_Os09g20510/Os09g0371300*	*Sb02g023360*	-	ABC transporter
Bradi4g29110	*LOC_Os09g20520/Os09g0371400*	*Sb02g023370*	-	ABC transporter
Bradi4g29120	"	*Sb02g023380*	-	"
Bradi4g29347	*LOC_Os09g21000/Os09g0376900*	*Sb02g023620*	*Traes_5AL_B64648FE6*	Potassium transporter family (HKT23-like)
Bradi4g29440	*LOC_Os09g21340/Os09g0381100*	*Sb02g023720*	*WHE1104_A05_B10ZS; WHE0807_A06_B11ZS; Traes_5AL_3E0C865DF*	Nucleobase-ascorbate transporter
Bradi4g29650	*LOC_Os09g23110/Os09g0394500*	*Sb02g024060*	*Traes_5AL_01A13992D; Traes_5AL_B8B668113*	ABC transporter
Bradi4g29720	*LOC_Os09g23300/Os09g0396900*	*Sb02g024130*	*WHE1787_E02_I03ZS; Traes_5AL_F80B422BA*	Vacuolar iron transporter 1 (VIT1)
Bradi4g29810	*LOC_Os09g23640/Os09g0401100*	-	*Traes_5AL_678EA44B2*	ABC transporter

[a]according to MSU Rice Genome Annotation Project release 7, Ensembl Plants release 22 or NCBI;
[b]Genome Zipper v5.
doi:10.1371/journal.pone.0098845.t006

Within the physical interval underlying the 5A Cl^- QTL, there were five different classes of genes encoding different transporters or channels: eleven ABC-transporter genes, and single genes encoding a nucleobase-ascorbate transporter, a vacuolar iron transporter (VIT1), a voltage-dependent anion channel (VDAC1) and a potassium transporter (HKT23-like) (Table 6).

DISCUSSION

Implications of Screening for Cl^- Concentration in Hydroponics and Field Environments

Phenotyping for ST is generally conducted under hydroponic conditions, and results are rarely validated in relevant field environments. Our recent studies in bread wheat demonstrated that phenotyping for Na^+ exclusion or ST in hydroponics had limited value in predicting field responses, and QTL differed vastly between hydroponics and field locations[24]. Given the renewed interest in Cl^- exclusion and ST [13], we analysed Cl^- concentration of hydroponically- and field-grown plants of Berkut/Krichauff DH population to determine the value of hydroponics for QTL analysis of Cl^- concentration, and elucidate genetic control mechanisms of Cl^- accumulation. The results demonstrated that plant Cl^- accumulation varied significantly between hydroponics and field trials and as a result, different QTL were identified between the two systems. As was the case with Na^+ concentration [24], [25], this was most probably due to the two systems being vastly different [24]. These results suggest that there may be very little value in hydroponics testing to predict field responses to Cl^- concentration in bread wheat, and future studies should consider field testing or soil-based-pot assays as an alternative.

ST and Na^+ or Cl^- Exclusion

In most studies to date Na^+ exclusion, and to a limited extent Cl^- exclusion, are traits contributing to ST. However, studies that also examined relationships between ST (absolute or relative growth) and Na^+ or Cl^- concentration reported inconsistent correlations [3]–[8],[10], [12], [45]–[48]. It is noteworthy that studies reporting high correlations ($r^2>0.5$) analysed either a small number of genotypes or were conducted under controlled environmental conditions [4], [7], [12], [46], [49]. In the present study genetic correlations were investigated using multivariate analysis. Moderate correlations in controlled environments (generally phenotypic) are not uncommon [13], [21]–[23], [25], while such correlations involving a large number of genotypes in the field are almost non-existent [24]. The lack of genetic correlation between Na^+ and/or Cl^- exclusion and grain yield in field studies suggests that other biochemical and physiological processes need to be taken into consideration for identifying mechanisms associated with ST. These results also indicate that a reductionist approach, such as selection for Na^+ and/or Cl^- exclusion only, may not substantially improve ST in bread wheat. A more reliable approach would be to select for grain yield, unless specific physiological traits are shown to have significant and consistent correlations with grain yield under saline environments.

Ion Channels and Transporters Involved in Na^+, K^+ and Cl^- Homeostasis

Here we aim to introduce a brief discussion on transport of these ions from the soil solution into the root cell and their movements within the plant with respect to ion channels and transporters. However, a greater focus will be placed on Cl^-, and for Na^+ and K^+, readers are referred to recent reviews [41], [50], [51]. It is well established that the initial entry of $Na^+$from the soil solution into the root cell is passive along the concentration gradient [52], and Na^+ uptake occurs primarily via non-selective cation channels and transporters [41], [51],[53]. However, Na^+ efflux (from cytosol into vacuole or removal from xylem) as a tolerance mechanism has to be active as it requires energy [41]. Potassium uptake from the soil solution into the root cell is an active process (i.e. moving across the membrane against its concentration gradient) and largely mediated by genes encoding channels and transporters [54]. Chloride uptake can be both active and passive depending on the external concentration. Under non-saline conditions transport of negatively charged Cl^-across negatively charged plasma membrane requires energy, therefore is an active process and mediated by transporters, while under saline conditions most Cl^- influx across the plasma membrane becomes passive [14]. However, its movement within or out of the plant must be active and aided by transporters [13]. Despite being the most abundant anion in the plant cells, compared to a number of well characterised Na^+ and K^+ transporters and encoding genes [41], very little is known about Cl^- transport mechanisms and the genes involved. To date two groups of gene families have repeatedly been discussed in relation to Cl^- homeostasis: CLCs and CCCs. From limited studies, it appears that CLCs, with their location in endomembranes, are involved in turgor regulation, stomatal movement and NO_3^- transport [55] but not root Cl^- uptake [41], while CCCs are involved in long distance transport of Na^+, K^+ and Cl^- and function as K^+:Cl^-, Na^+:Cl^- or Na^+: K^+:Cl^- co-transporters[44], [56]. However, there may be other gene families involved in Cl^- transport, as discussed by Teakle and Tyerman [13].

Genetic Control Mechanisms of Na^+, K^+ and Cl^-Homeostasis under Salinity Stress

A good understanding of inheritance of homeostasis of these ions is required for a successful breeding strategy. Our present knowledge of mechanisms of their inheritance is a mere reflection of the number of studies conducted on them. For instance, due to greater focus on Na^+, we know more about Na^+ than K^+ or Cl^-. The limited studies in rice and wheat suggest that Na^+ and K^+ homeostasis under salinity stress are under separate genetic control [25], [57]–[59], although studies in barley [22], [23] found that QTL for Na^+ co-located with QTL for K^+, suggesting one or more genes regulating Na^+ and/or K^+ transport such as vacuolar sodium-hydrogen antiporter (NHX) genes [22]. As for Cl^-, to date there have been only four quantitative genetic studies that reported QTL for Cl^- concentration; two in rice[20], [21] and two in barley [22], [23]. In those studies as well as in the present study, moderate correlations between Cl^- and either Na^+ or K^+, and QTL for different ions mapping to the same and/or different regions indicate the presence of common (i.e. CCCs) and specific transporters for the uptake of these ions [i.e. high-affinity potassium (HKT) and (CLCs)] which in turn suggests common and separate genetic control. For instance, in the present study, Cl^- QTL on 2A and 5A co-located with QTL for Na^+ and K^+ concentration, respectively [24]. A physiological explanation for these co-locations and correlations between cations (Na^+ and K^+) and anions (Cl^-) may be the charge balance between these two groups of ions since the net movement of ions must be balanced so that there is charge equivalence with small difference [13]. Whereas there were no CCC and CLC genes physically close to the Cl^- concentration QTL on chromosome 5A there are 15 transporter and channel genes in the physical interval as candidates for the observed Cl^-accumulation. Although we have used the latest release of the comparison between wheat ESTs and contigs with sequenced grass genomes (Genome Zipper v5), it is possible that other genes reside in this region in wheat and the microsynteny is less well preserved as it appears to be so far.

Nguyen et al. [22] recently reported a Cl^- concentration QTL under salt stress in barley but used an incorrect barley chromosome nomenclature so that the actual chromosome 5H was mislabelled as 7H (other chromosomes were also mislabelled). The physical position of the RFLP markers *ABC324* and *ABC302* that flank their barley Cl^- QTL on 5HL [22] suggest a position close to the physical chromosomal region corresponding to the 5AL Cl^- concentration QTL reported here. In fact, the physical position of the northern flanking marker *ABC324* (position chromosome 5H: 399,222,591) slightly overlaps with the position of the southern end of the5A QTL flanked by RFLP marker *psr128* (position chromosome 5H: 412,653,548). It is possible that both QTL are caused by orthologous genes in wheat and barley, although this is far from certain as the QTL intervals contain hundreds of genes and the barley QTL was identified under salt stress in a hydroponics system whereas the QTL in wheat was repeatedly observed under salt stressed field conditions but not in hydroponics. Only further work such as fine mapping and gene expression analysis of the candidate genes will prove unequivocally the identity of the underlying gene for the differential Cl^- accumulation between Berkut and Krichauff. The present study provides a compelling case that the 5A QTL contains a K^+:Cl^- co-transporter gene several other candidates capable of moving Cl^- ions across membranes. However, as this is the first report in wheat, there is clearly a need for testing other mapping populations and genetics resources to identify other Cl^- transporter gene(s) to gain a better understanding of genetic control mechanisms of Cl^- homoeostasis in crops.

Genetic Control Mechanisms of Ca^{2+} and Mg^{2+} Accumulation under Salinity Stress

The inheritance of plant Ca^{2+} and Mg^{2+} accumulation was also investigated, given the reports of salinity-induced nutritional deficiencies such as Ca^{2+} and Mg^{2+} [26], [37], [39], [40]and the importance of maintenance of adequate nutrition for these essential elements to plant growth and yield under salinity stress. To our knowledge, there have only been two studies in barley [22], [23] that investigated inheritance of Ca^{2+} and Mg^{2+} uptake or accumulation under salinity stress. However, only in one study [23] QTL for Ca^{2+} and Mg^{2+}were detected under salinity stress; one QTL for Mg^{2+} concentration on 6H, and three QTL for Ca^{2+} concentration on 1H, 6H and 7H. The QTL on 6H was common not only to Ca^{2+} and Mg^{2+} but also to ST. To our knowledge, this is the first time in the literature that a QTL for a nutrient other than Na^+ was co-located with ST, providing evidence for the role of Ca^{2+} and Mg^{2+} nutrition in growth and yield under salinity stress. These results also indicate that $Ca^{2+}$and Mg^{2+} uptake may occur through common as well as independent pathways. In contrast to the barley study, in the present study, none of the Ca^{2+} and Mg^{2+} concentration QTL co-located with each other or ST (measured as seedling biomass or grain yield), suggesting independent genetic control. However, further studies are required to enable better understanding of their genetic control mechanisms.

CONCLUSIONS

As was the case with Na^+ [24], plant Cl^- responses and related QTL differed widely between hydroponics and field tests, indicating substantial genotype and QTL interactions with environments. The results also indicated that hydroponics-based seedling assays may be very limited in their ability to predict field responses to salinity, and soil-based assays may be the second best option after field testing. As Cl^- concentration in the plant correlated only moderately with seedling biomass and showed no correlation with grain yield in the field, it does not appear, on its own, to be a reliable physiological parameter to select for in a breeding context, at least in bread wheat. Further research involving other mapping populations/genetic resources is warranted to be definitive. In the short term, selection for grain yield, which is integrative of all tolerance mechanisms, appears a more reliable strategy, while in the long term identification of donors for various physiological traits and subsequently combining them in a genotype (pyramiding) is likely to be the way forward[60]. This latter process can be fast-tracked via marker assisted selection. Finally the presence of several QTL for Cl^- concentration indicates that Cl^- uptake/accumulation is a polygenic trait. The discovery of a major QTL for Cl^- concentration on 5A that co-locates with several candidate genes that could be involved in Cl^- transport in bread wheat provides a starting point for further analysis through fine mapping and functional studies.

ACKNOWLEDGMENTS

We would like to thank Dr Hugh Wallwork (South Australian Research and Development Institute) for access to the Berkut/Krichauff DH population and the coordination of the field trials, the staff at Australian Grain Technologies, the LongReach Plant Breeders, the University of Adelaide Barley Breeding Program and Mr Jim Lewis (South Australian Research and Development Institute) for sowing, harvesting and general maintenance of the field trials, Mr Robin Hosking (the Australian Centre for Plant Functional Genomics) for his construction and creative modification of the supported hydroponic system and technical support throughout this project, Mrs Teresa Fowles and Mr Lyndon Palmer (the University of Adelaide) for their help with ICP-OES analysis, Prof. Mark Tester (Australian Centre for Plant Functional Genomics) for the use of a chloride meter, Drs Graham Lyons and John Harris (the University of Adelaide) for their critical review of the manuscript prior to submission, and the editor and the reviewers for their helpful comments.

AUTHOR CONTRIBUTIONS

Conceived and designed the experiments: YG JT JR KO. Performed the experiments: YG JR KO. Analyzed the data: YG JT KO. Contributed reagents/materials/analysis tools: YG JT KO. Wrote the paper: YG JT KO.

REFERENCES

1. Martinez-Beltran J, Manzur CL (2005) Overview of salinity problems in the world and FAO strategies to address the problem. In Proceedings of the international salinity forum; Riverside, California. pp. 311–313.
2. Schachtman DP, Munns R, Whitecross MI (1991) Variation in sodium exclusion and salt tolerance in *Triticum tauschii*. Crop Science 31: 992–997. doi: 10.2135/cropsci1991.0011183x003100040030x
3. Ashraf M, O'Leary JW (1996) Responses of newly developed salt-tolerant genotype of spring wheat to salt stress: yield components and ion distribution. Journal of Agronomy and Crop Science 176: 91–101. doi: 10.1111/j.1439-037x.1996.tb00451.x
4. Rashid A, Querishi RH, Hollington PA, Wyn Jones RG (1999) Comparative responses of wheat cultivars to salinity at the seedling stage. Journal of Agronomy and Crop Science 182: 199–207. doi: 10.1046/j.1439-037x.1999.00295.x
5. Poustini K, Siosemardeh A (2004) Ion distribution in wheat cultivars in response to salinity stress. Field Crops Research 85: 125–133. doi: 10.1016/s0378-4290(03)00157-6
6. Hollington PA (2000) Technological breakthroughs in screening/breeding wheat varieties for salt tolerance. In: Gupta SK, Sharma SK, Tyagi NK, editors. National Conference on Salinity Management in Agriculture. Central Soil Salinity Research Institute, Karnal, India. pp. 273–289.
7. Huang Y, Zhang G, Wu F, Chen J, Zhou M (2006) Differences in physiological traits among salt-stressed barley genotypes. Communications in Soil Science and Plant Analysis 37: 567–570. doi: 10.1080/00103620500449419
8. Ashraf M, McNeilly T (1988) Variability in salt tolerance of nine spring wheat cultivars. Journal of Agronomy and Crop Science 160: 14–21. doi: 10.1111/j.1439-037x.1988.tb01160.x
9. Bagci SA, Ekiz H, Yilmaz A (2007) Salt tolerance of sixteen wheat genotypes during seedling growth. Turkish Journal of Agriculture and Forestry 31: 363–372.
10. Genc Y, McDonald GK, Tester M (2007) Reassessment of tissue Na^+ concentration as a criterion for salinity tolerance in bread wheat. Plant, Cell and Environment 30: 1486–1498. doi: 10.1111/j.1365-3040.2007.01726.x
11. Royo A, Aragüés R (1999) Salinity-yield response functions on barley genotypes assessed with a triple line source sprinkler system. Plant and

Soil 209: 9–20.

12. El-Hendawy S, Hu Y, Schmidhalter U (2005) Growth, ion content, gas exchange and water relations of wheat genotypes differing in salt tolerances. Australian Journal of Agricultural Research 56: 123–134. doi: 10.1071/ar04019
13. Teakle NL, Tyerman SD (2010) Mechanisms of Cl^- transport contributing to salt tolerance. Plant, Cell and Environment 33: 566–589. doi: 10.1111/j.1365-3040.2009.02060.x
14. White PJ, Broadley MR (2001) Chloride in soils and its uptake and movement within the plant: A Review. Annals of Botany 88: 967–988. doi: 10.1006/anbo.2001.1540
15. Kingsbury R, Epstein E (1986) Salt sensitivity in wheat. A case for specific ion toxicity. Plant Physiology 80: 651–654. doi: 10.1104/pp.80.3.651
16. Martin P, Koebner R (1995) Sodium and chloride ions contribute synergistically to salt toxicity in wheat. Biologia Plantarum 37: 265–271. doi: 10.1007/bf02913224
17. Tavakkoli E, Rengasamy P, McDonald GK (2010) The response of barley to salinity stress differs between hydroponics and soil systems. Functional Plant Biology 37: 621–633. doi: 10.1071/fp09202
18. Munns R, Tester M (2008) Mechanisms of salinity tolerance. Annual Review of Plant Biology 59: 651–681. doi: 10.1146/annurev.arplant.59.032607.092911
19. Xu G, Magen H, Tarchitzky J, Kafkafi U (2000) Advances in chloride nutrition. Advances in Agronomy 68: 96–150. doi: 10.1016/s0065-2113(08)60844-5
20. Ammar MHM, Pandit A, Singh RK, Sameena S, Chaucan MS, et al. (2009) Na^+, $K^+$and Cl^- ion concentrations in salt tolerant *Indica* rice variety CSR27. Journal of Plant Biochemistry and Biotechnology 18: 139–150. doi: 10.1007/bf03263312
21. Pandit A, Rai V, Bal S, Sinha S, Kumar V, et al. (2010) Combining QTL mapping and transcriptome profiling of bulked RILs for identification of functional polymorphism for salt tolerance genes in rice (*Oryza sativa* L.). Molecular Genetics and Genomics 284: 121–136. doi: 10.1007/s00438-010-0551-6
22. Nguyen VL, Ribot SA, Dolstra O, Niks RE, Visser RGF, et al. (2013) Identification of quantitative trait loci for ion homeostasis and salt tolerance in barley (*Hordeum vulgare* L). Molecular Breeding 31: 137–152. doi: 10.1007/s11032-012-9777-9

23. Nguyen VL, Dolstra O, Malosetti M, Kilian B, Graner A, et al. (2013) Association mapping of salt tolerance in barley (*Hordeum vulgare* L.). Theoretical and Applied Genetics. DOI.10.1007/s00122-013-2139-0.

24. Genc Y, Oldach K, Gogel B, Wallwork H, McDonald GK, et al. (2013) Quantitative trait loci for agronomical and physiological traits for a bread wheat population grown in environments with a range of salinity levels. Molecular Breeding 32: 39–59 DOI: 10.1007/s110320139851y.

25. Genc Y, Oldach K, Verbyla A, Lott G, Hassan M, et al. (2010) Sodium exclusion QTL associated with improved seedling growth in bread wheat under salinity stress. Theoretical and Applied Genetics 121: 877–894. doi: 10.1007/s00122-010-1357-y

26. Genc Y, Tester M, McDonald GK (2010) Calcium requirement of wheat in saline and non-saline conditions. Plant and Soil 327: 331–345. doi: 10.1007/s11104-009-0057-3

27. Wheal MS, Palmer LT (2010) Chloride analysis of botanical samples by ICPOES. Journal of Analytical Atomic Spectrometry 25: 1946–1952. doi: 10.1039/c0ja00059k

28. Wheal MS, Fowles TO, Palmer LT (2011) A cost effective acid digestion method using closed polypropylene tubes for inductively coupled plasma optical emission spectrometry (ICP-OES) analysis of plant essential elements. Analytical Methods 3: 2854–2863. doi: 10.1039/c1ay05430a

29. Martinez O, Curnow RN (1992) Estimating the locations and sizes of the effects of quantitative trait loci using flanking markers. Theoretical and Applied Genetics 85: 480–488. doi: 10.1007/bf00222330

30. Verbyla AP, Cullis BR, Thompson R (2007) The analysis of QTL by simultaneous use of the of the full linkage map. Theoretical and Applied Genetics 116: 95–111. doi: 10.1007/s00122-007-0650-x

31. Smith A, Cullis BR, Thompson R (2001) Analysing variety by environment data using multiplicative mixed models. Biometrics 57: 1138–1147. doi: 10.1111/j.0006-341x.2001.01138.x

32. R Development Core Team (2013) R: A language and environment for statistical computing, R Foundation for Statistical Computing. Vienna, Austria.

33. Taylor JD, Diffey S, Verbyla AP, Cullis BC (2013) wgaim: Whole Genome Average Interval Mapping for QTL detection using mixed models, R package version 1.3–0.

34. Verbyla AP, Taylor JD, Verbyla KL (2012) RWGAIM: An efficient high dimensional random whole average (QTL) interval mapping approach. Genetics Research 94: 291–306. doi: 10.1017/s0016672312000493

35. Cullis BR, Smith AB, Coombes NE (2006) On the design of early generation variety trials with correlated data. Journal of Agricultural, Biological and Environmental Statistics 11: 381–393. doi: 10.1198/108571106x154443
36. Bonneau J, Taylor J, Parent B, Bennett D, Reynolds M, et al. (2013) Multi-environment analysis and improved mapping of a yield-related QTL on chromosome 3B of wheat. Theoretical and Applied Genetics 126: 747–761. doi: 10.1007/s00122-012-2015-3
37. Ehret DI, Redmann RE, Harvey BL, Cipywnyk A (1990) Salinity-induced calcium deficiencies in wheat and barley. Plant and Soil 128: 143–151. doi: 10.1007/bf00011103
38. Bergmann W (1992) Nutritional disorders of plants: Developments, visual and analytical diagnosis. Gustav Fisher, Jena, Stuttgart, New York.
39. Cramer GR (2002) Sodium-calcium interactions under salinity stress. In: Lauchli A, Luttge U, editors. Salinity: Environment- plants- molecules. Kluwer, Dordrecht. pp. 205–227.
40. Adcock KG, Gartrell JW, Brennan RF (2001) Calcium deficiency of wheat grown in acidic sandy soil from Southwestern Australia. Journal of Plant Nutrition 24: 1217–1227. doi: 10.1081/pln-100106977
41. Mian AA, Senadheera P, Maathuis FJ (2011) Improving crop salt tolerance: anion and cation transporters as genetic engineering targets. Plant Stress 5: 64–72.
42. Lurin C, Geelen D, Barbier-Brygoo H, Guern J, Maurel C (1996) Cloning and functional expression of a plant voltage-dependent chloride channel. Plant Cell 8: 701–711. doi: 10.2307/3870345
43. Wong TH, Li MW, Yao XQ, Lam HM (2013) The GmCLC1 protein from soybean functions as a chloride ion transporter. Journal of Plant Physiology 170: 101–104. doi: 10.1016/j.jplph.2012.08.003
44. Colemenero-Flores JM, Martinez G, Gamba G, Vazuez N, Iglesias DJ, et al. (2007) Identification and functional characterization of cation-chloride cotransporters in plants. Plant Journal 50: 278–292. doi: 10.1111/j.1365-313x.2007.03048.x
45. Isla R, Royo A, Aragüés R (1997) Field screening of barley cultivars to soil salinity using a sprinkler and a drip irrigation system. Plant and Soil 197: 105–117. doi: 10.1023/a:1004240622652
46. Khan MA, Shirazi MU, Khan MA, Mujtaba SM, Islam E, et al. (2009) Role of proline, K/Na ratio and chlorophyll content in salt tolerance of wheat (*Triticum aestivum* L.). Pakistan Journal of Botany 41: 633–638.

47. Rawson HM, Richards RA, Munns R (1988) An examination of selection criteria for salt tolerance in wheat, barley and triticale genotypes. Australian Journal of Agricultural Research 39: 759–772. doi: 10.1071/ar9880759

48. Salam A, Hollington PA, Gorham J, Wyn Jones RG, Gliddon C (1999) Physiological genetics of salt tolerance in wheat (*Triticum aestivum* L): Performance of wheat varieties, inbred lines and reciprocal F_1 hybrids under saline conditions. Journal of Agronomy and Crop Science 183: 145–156. doi: 10.1046/j.1439-037x.1999.00361.x

49. Munns R, James RA (2003) Screening methods for salinity tolerance: a case study with tetraploid wheat. Plant and Soil 253: 201–218. doi: 10.1023/a:1024553303144

50. Maathuis FJM, Amtmann A (1999) K^+ Nutrition and Na^+ Toxicity: The Basis of Cellular K^+/Na^+ Ratios. Annals of Botany 84: 123–133. doi: 10.1006/anbo.1999.0912

51. Tester M, Davenport R (2003) Na^+ tolerance and Na^+ transport in higher plants. Annals of Botany 91: 503–527. doi: 10.1093/aob/mcg058

52. Cheeseman JM (1982) Pump-leak sodium fluxes in low salt corn roots. Journal of Membrane Biology 70: 157–164. doi: 10.1007/bf01870225

53. Byrt CS, Platten JD, Spielmeyer W, James RA, Lagudah ES, et al. (2007) HKT1;5-like cation transporters linked to Na^+ exclusion loci in wheat, *Nax2* and *Kna1*. Plant Physiology 143: 1918–1928. doi: 10.1104/pp.106.093476

54. Horie T, Brodsky DE, Costa A, Kaneko T, Schiavo FL, et al. (2011) K^+ transport by the OsHKT2;4 transporter from rice (*Oryza sativa*) with atypical Na^+ transport properties and competition in permeation of K^+ over Mg^{2+} and Ca^{2+} ions. Plant Physiology 156: 1493–1507. doi: 10.1104/pp.110.168047

55. Hechenberger M, Schwappach B, Fischer WN, Frommer WB, Jentsch TJ, et al. (1996) A family of putative chloride channels from Arabidopsis and functional complementation of a yeast strain with a CLC gene disruption. Journal of Biological Chemistry 271: 33632–33638. doi: 10.1074/jbc.271.52.33632

56. Kong X-Q, Gao X-H, Sun W, An J, Zhao Y-X, et al. (2011) Cloning and functional characterization of a cation-chloride cotransporter gene OsCCC1. Plant Molecular Biology 75: 567–578. doi: 10.1007/s11103-011-9744-6

57. Garcia A, Rizzo CA, Ud-Din J, Bartos SL, Senadhira D, et al. (1997) Sodium and potassium transport to the xylem are inherited independently

in rice and the mechanisms of sodium: potassium selectivity differs between rice and wheat. Plant, Cell and Environment 20: 1167–1174. doi: 10.1046/j.1365-3040.1997.d01-146.x

58. Koyama ML, Levesley A, Koebner RMD, Flowers TJ, Yeo AR (2001) Quantitative trait loci for component physiological traits determining salt tolerance in rice. Plant Physiology 125: 406–422. doi: 10.1104/pp.125.1.406

59. Lin HX, Zhu MZ, Yano M, Gao JP, Liang ZW, et al. (2004) QTLs for Na^+ and K^+uptake of the shoots and roots controlling rice salt tolerance. Theoretical and Applied Genetics 108: 253–260. doi: 10.1007/s00122-003-1421-y

60. Peng S, Ismail AM (2004) Physiological basis of yield and environmental adaptation in rice. In: Nguyen HT, Blum A, editors. Physiology and biotechnology integration for plant breeding. CRC Press, New York, Basel. pp. 72–118.

Chapter 9

TRANSCRIPTOMIC ANALYSIS OF FIBER STRENGTH IN UPLAND COTTON CHROMOSOME INTROGRESSION LINES CARRYING DIFFERENT GOSSYPIUM BARBADENSE CHROMOSOMAL SEGMENTS

Lei Fang, Ruiping Tian, Jiedan Chen, Sen Wang, Xinghe Li, Peng Wang, and Tianzhen Zhang

National Key Laboratory of Crop Genetics and Germplasm Enhancement, Cotton Hybrid R & D Engineering Center (the Ministry of Education), Nanjing Agricultural University, Nanjing, China

ABSTRACT

Fiber strength is the key trait that determines fiber quality in cotton, and it is closely related to secondary cell wall synthesis. To understand the mechanism underlying fiber strength, we compared fiber transcriptomes from different *G. barbadense* chromosome introgression lines (CSILs) that had higher fiber strengths than their recipient, *G. hirsutum* acc. TM-1. A total of 18,288 differentially expressed genes (DEGs) were detected between CSIL-35431 and CSIL-31010, two CSILs with stronger fiber and TM-1 during secondary cell wall synthesis. Functional classification and enrichment analysis revealed that these DEGs were enriched for secondary cell wall biogenesis, glucuronoxylan biosynthesis, cellulose biosynthesis, sugar-mediated signaling pathways, and fatty acid biosynthesis. Pathway analysis showed that these DEGs participated in starch and sucrose metabolism (328 genes), glycolysis/gluconeogenesis (122 genes), phenylpropanoid biosynthesis (101 genes), and oxidative phosphorylation (87 genes), etc. Moreover, the expression of MYB- and NAC-type transcription factor genes were also dramatically different between the CSILs and TM-1. Being different to those of CSIL-31134, CSIL-35431 and CSIL-31010, there were many genes for fatty acid degradation and biosynthesis, and also for carbohydrate metabolism that were down-regulated in CSIL-35368. Metabolic pathway analysis in the CSILs showed that different pathways were changed, and some changes at the same developmental stage in some pathways. Our results extended our understanding that carbonhydrate metabolic pathway and secondary cell wall biosynthesis can affect the fiber strength and suggested more genes and/or pathways be related to complex fiber strength formation process.

INTRODUCTION

The cotton fiber is a terminally differentiated single cell derived from the epidermal cell of the developing ovule. After initiation, the fiber cell undergoes 1000- to 3000-fold elongation during its development. The development of cotton fibers involves four partially overlapping stages: initiation (−3 to +3 days post-anthesis; DPA), elongation and primary cell wall formation (3–23 DPA), secondary cell wall formation (16–40 DPA) and maturation (40–50 DPA) [1]–[6]. The most rapid period of fiber cell elongation begins around 10–16 DPA and continues to ~20 DPA. Primary and secondary cell wall synthesis overlaps during the period of 16–25 DPA. During the secondary cell wall formation stage, the speed of cell elongation slows down and even stops.

Fiber strength is an important indicator of cotton fiber quality, and depends on formation of the secondary cell wall. Cellulose synthesis plays a predominant role in fiber cells, and cellulose accounts for >95% of the dry weight of the mature cotton fiber [3], [7]. Genome and EST sequencing have revealed that there are at least ten different CesA genes for cellulose synthase in *Arabidopsis*; CesA-like genes have also been reported in rice and barley[8]–[10]. In cotton (*Gossypium raimondii*), at least 15 cellulose synthase (CESA) sequences are required for cellulose synthesis [11]. A recent investigation in *Arabidopsis thaliana* using microarrays led to the identification of genes that are highly co-expressed with cellulose synthase genes and two mutants, irx8 and irx13, that have irregular xylem phenotypes, were also identified [12]. Sucrose synthase (Susy) is the enzyme that catalyzes the hydrolysis of sucrose to UDP-glucose that is then used as a substrate for cellulose synthesis. In cotton, the expression of Susy is higher at 16–32 DPA, and this enzyme plays a major role in partitioning carbon toward cellulose synthesis in the fiber [13]. SusC is another new sucrose synthase gene with a high level of expression during secondary cell wall synthesis [14]. Peroxide, mainly as H_2O_2, promotes cellulose synthesis as a signal of secondary cell wall synthesis [15], [16].

At present, many ovule- and fiber-specific cDNA libraries have been constructed and sequenced, and more than 268,000 expressed sequence tags (ESTs) from *Gossypium* are deposited in the NCBI database (http://www.ncbi.nlm.nih.gov). For genetic characterization of rapid cell elongation in cotton fibers, approximately 14,000 unique genes were assembled from 46,603 expressed sequence tags (ESTs) from developmentally-staged fiber cDNAs of a cultivated diploid species (*G. arboreum* L.). Eighty-one genes that were significantly up-regulated during secondary cell wall synthesis were found to be involved in cell wall biogenesis and energy/carbohydrate metabolism, which is consistent with the stage of cellulose synthesis during secondary cell wall modification in developing fibers[17]. Transcriptome profiling of the cotton fiber early in development by high-throughput tag-sequencing (Tag-seq) analysis using the Solexa Genome Analyzer reveals significant differential expression of genes in a fuzzless/lintless mutant [18]. High-throughput, genome-wide transcriptomic analysis of cotton under drought stress revealed a significant down-regulation of genes and pathways involved in fiber elongation, and an up-regulation of defense response genes [19]. More research have been processed in fiber initiation and elongation stage [20]–[24]. Saturated very-long-chain fatty acids (VLCFAs; C20:0–C30:0) exogenously applied in ovule culture medium significantly promoted fiber cell elongation in cotton (*G. hirsutum L.*) by activating ethylene biosynthesis [25], [26]. Previous investigations into cotton fiber development mainly focused on the elongation stage, and the number of genes reported from the later stages is quite small. Most of the genes up-regulated during secondary cell wall synthesis were related to cellulose synthesis, cell wall biosynthesis, and carbohydrate metabolism [17], [22], [27].

Chromosome segment introgression lines (CSILs) consist of a battery of near-isogenic lines that have been developed to cover the entire genomes of some crops, including tomato, rice, wheat, and cotton [28]–[31]. With the exception of a single, homozygous chromosome segment transferred from a donor parent, the remaining genome of each CSIL is the same as the recipient parent [31]. We used *G. barbadense* CSILs in the background of the standard genetic line of *G. hirsutum*, cv. TM-1, in order to understand the molecular mechanism behind superior quality fiber formation. Multi-point tests showed that three CSILs produced stronger fibers when compared to the recipient parent TM-1, but one CSIL produced weaker fibers. Using Solexa Genome sequencing, we analyzed transcriptome profiles from the CSILs and TM-1. We found that many genes were either up- or down-regulated at the stage of secondary cell wall synthesis, and that many metabolic pathways were altered in the CSILs.

MATERIALS AND METHODS

Plant Materials

G. hirsutum cv. TM-1, the genetic standard for Upland cotton, was obtained from the Southern Plains Agricultural Research Center, USDA-ARS, College

Station/Texas, USA [32].*G. barbadense* cv. Hai7124, an extra-long staple cotton that is widely grown in China, is descended from a selected individual in a study of inheritance of resistance to *Verticillium dahlia* [33], [34]. In this study, we identified three CSILs with stronger fiber or high fiber strength that carried different *G. barbadense* chromosome segment(s) in the recurrent parent TM-1. The detailed method of developing CSILs has been described previously [31]. We selected three CSILs, CSIL-35431, CSIL-31134, and CSIL-31010, in which the average fiber strength were 35.1, 34.73 and 34.28 cN/tex, respectively, significantly higher than TM-1, and also CSIL-35368 which had poorer fiber strength than TM-1(28.71 cN/tex) (Table S1). The introgressed *G. barbadense* chromosomal segments were different in the four lines[35]. Fiber samples were collected at 15, 20, and 25 DPA, frozen in liquid nitrogen, and stored at −70°C.

RNA Isolation and Evaluation

Total RNA was extracted from frozen tissue using an improved CTAB extraction protocol [36]. RNAs were evaluated for quality using RNA Pico Chips on an Agilent 2100 Bioanalyzer (Agilent Technologies, Santa Clara, CA, USA). All RNA samples were quantified and qualified with an RNA Integrity Number (RIN) >8, and 28S/18S rRNA band intensity (2:1).

Library Construction and Sequencing

Digital gene expression libraries were constructed using the Illumina Gene Expression Sample Preparation Kit according to the manufacturer's instructions. We constructed and sequenced 14 libraries derived from immature fibers at 15, 20, and 25 DPA using the Solexa Genome Sequencing Analyzer system provided by BGI (Beijing Genomics Institute at Shenzhen, China), which gave 21 bp tags. The process was described in detail previously [18].

Data Processing, Statistical Evaluation, and Selection of Differentially Expressed Genes (DEGs)

Raw data reads were filtered by the Illumina pipeline to produce clean data. All low-quality data, such as short tags (<21 nt) and singletons, were removed. A database of 21-base-long sequences was produced beginning with CATG using 37,505 reference genes from the diploid species *G. raimondii* (http://www.phytozome.net). The remaining high quality sequences were then mapped to this database; only a single mismatch was allowed, and more than one match was excluded. Gene expression levels were the summation of tags aligned to the different positions of the same gene. Expression levels are expressed as TPM, transcripts per million. To identify DEGs during fiber elongation, we compared pairs of DEG profiles from different libraries. Three fiber development periods for the four CSILs were compared with the same period for TM-1, and 11 comparisons were obtained. P- and Q-values were also calculated for every comparison [37]. DEGs were defined as FDR≤0.001 with an absolute value of $|\log_2 \text{Ratio}| \geq 1$ to judge the significance of differences in transcript abundance.

Digital Tag Profiling Analysis

DEG clustering in CSILs at different developmental stages were performed with Cluster3.0 (http://bonsai.hgc.jp/~mdehoon/software/cluster/software.htm). We also performed clustering with the 'Self-organizing tree algorithm' (SOTA, Multiple Array Viewer software, MeV 4.9.0)[38].

GO enrichment and KEGG (Kyoto Encyclopedia of Genes and Genomes) pathway analysis was done using BLAST2GO (http://www.blast2go.com/b2ghome). Mapman was also used to analyze metabolic pathway base on KEGG database [39].

Quantitative RT-PCR

Quantitative RT-PCR assays were performed on a 7500 Real-Time PCR system (Applied Biosystems, San Francisco, CA, USA). Reactions were performed in a final volume of 15 μL and contained 2 μL of diluted cDNA, 7.5 μL of 2× SYBR mix (Roche, Basel, Switzerland), and 200 nM of the forward and reverse primers. Primer lengths were designed to be between 18 and 24 nt using Beacon Designer 7, and PCR amplicon lengths were designed to be between 100 bp and 150 bp (Table S2). The thermal cycling conditions were 40 cycles of 95°C for 15 s, 60°C for 30 s, and 72°C for 30 s. All reactions were run in triplicate, and the cotton *histone3* gene (ACC NO. AF024716) was used as an internal control for normalization of expression levels (F: 5′-GGTGGTGTGAAGAAGCCTCAT-3′, and R: 5′-AATTTCACGAACAAGCCTCTGGAA-3′). The relative gene expression levels were presented as $2^{-\Delta CT}$.

RESULTS

Statistical Analysis of Transcriptome Data

The total number of sequence tags per library ranged from 7.0 to 8.5 million, and the number of distinct sequence tags was between 1.8 and 2.2 million. Approximately 50% of the clean tags were mapped to reference genes, and

60% of the reference genes were mapped with unambiguous tag (Table 1 and Table S3).

Table 1. The distribution of total and distinct tags.

Summary	TM-1			CSIL-35431			CSIL-31010		
	15DPA	20DPA	25DPA	15DPA	20DPA	25DPA	15DPA	20DPA	25DPA
Raw Data	8374304	7231305	14267931	7461562	7389820	7298224	7471212	7007447	7203054
Distinct Raw Tag	389162	394430	494345	375999	371748	317147	283130	337525	386302
Clean Tag	8144920	7013147	14002534	7264426	7213791	7124575	7369012	6840101	7002766
Distinct Clean Tag	169700	177887	251173	182629	199257	148121	182184	172775	188639
Unique Clean Tag Mapping to Gene	3983215	3206328	6829327	3961228	3411249	3463314	3671014	3381955	3552566
Total % of clean tag	48.90%	45.72%	48.77%	54.53%	47.29%	48.61%	49.82%	49.44%	50.73%
Unique Distinct Clean Tag Mapping to Gene	45380	40475	56172	53187	48453	41065	46720	45762	48147
Total % of distinct tag	26.74%	22.75%	22.36%	29.12%	24.32%	27.72%	25.64%	26.49%	25.52%
Unambiguous Tag-mapped Genes	21498	20901	22594	21781	22325	20590	21900	22457	22811
Percentage of reference genes	57.32%	55.73%	60.24%	58.07%	59.53%	54.90%	58.39%	59.88%	60.82%

Clean tags: tags after filtering dirty tags (low quality tags) from raw data.
Distinct tags: different kinds of tags.
Unambiguous tags: the clean tags after removing tags mapped to reference sequences from multiple genes.
doi:10.1371/journal.pone.0094642.t001

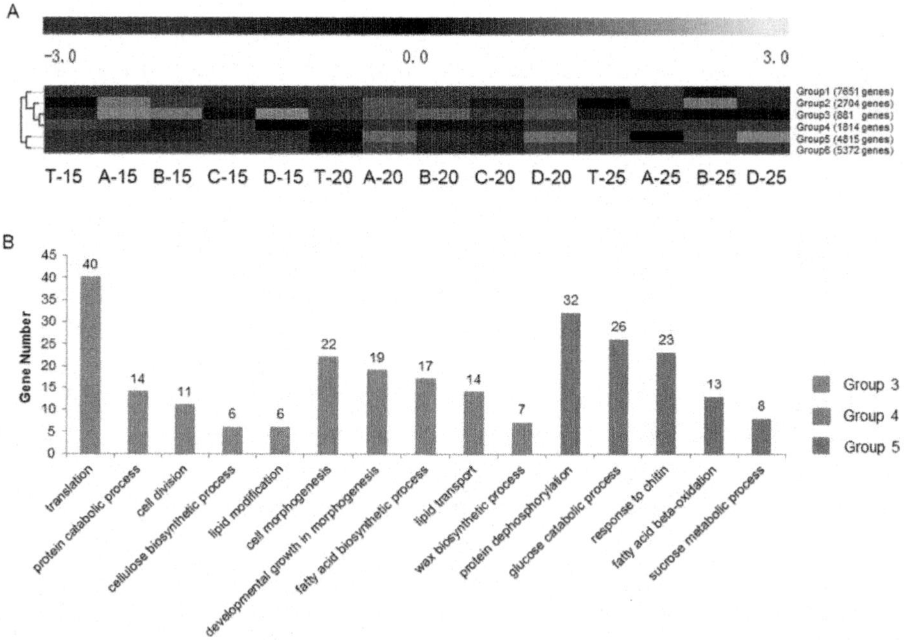

Figure 1. Statistical analysis of transcriptome data. (A) SOTA clustering of the different genes using Log2(TPM). T, TM-1; A, CSIL-35431; B, CSIL-31010; C, CSIL-31134; D, CSIL-35368. 15, 15 DPA; 20, 20 DPA; 25, 25 DPA. (B) Distribution of functions of genes in different clusters. Yellow square indicated group 3, green square indicated group 4 and blue square indicated group 5. X-axis indicated different enriched process and Y-axis indicated number of hit-found genes in these processes.

To see whether the fiber transcriptomes at different developmental stages were different, the 23,237 genes which were expressed in at least three libraries at one stage (15 DPA, 20 DPA, or 25 DPA) were classified into six groups using the Multiple Array Viewer using TPM value (Figure 1A). Genes in Group 3 had higher expression levels at 15 DPA and 20 DPA than at the later stage (25 DPA). Genes in Group 4 had higher expression levels at 15 DPA than at either 20 DPA or 25 DPA. Genes in Group 5 showed the opposite expression pattern, with higher expression levels at 20 DPA and 25 DPA compared to 15 DPA. The other groups also showed distinct expression patterns (Figure 1A).

Classification by gene function revealed that Group 3 is enriched in genes involved in protein catabolism, cell division, and cellulose biosynthesis, Group 4 is enriched in genes for cell morphogenesis, fatty acid biosynthesis, lipid transport, and wax biosynthesis, and Group 5 has more genes involved in glucose catabolism, response to chitin, and sucrose metabolism (Figure 1B). The unbalanced pattern of the expressed-gene functional distribution could possibly reflect some physiological events involved in secondary cell wall biosynthesis.

Cluster Analysis of Differentially Expressed Genes (DEGs) Between and/or Among CSILs

We specifically looked for DEGs in secondary cell wall fibers from 15 to 25 DPA, because previous studies have reported that the different sets of transcripts responsible for fiber secondary cell wall formation may be enriched at these stages of development [17], [22],[27]. Three fiber development periods for the four CSILs were compared with TM-1 at the same period. DEGs were defined as FDR≤0.001 with an absolute value of $|\log_2\text{Ratio}|\geq 1$. Analysis of the data indicated that many genes showed differential expression in the 11 comparison groups. The number of DEGs were about 6,000–8,000 in CSILs from 15 DPA to 25 DPA (Figure 2A). But the number of DEGs in CSIL-31010 at 20 DPA, CSIL-31010 at 25 DPA, and CSIL-31134 at 15 DPA, were 4,600, 10,106 and 2,060, respectively. We also found that the DEGs that were up-regulated or down-regulated were different in CSILs. There were ~1,500–3,500 DEGs in common from 15 DPA to 25 DPA between CSIL-35431, CSIL-31010, and CSIL-35368 (Figure 2B).

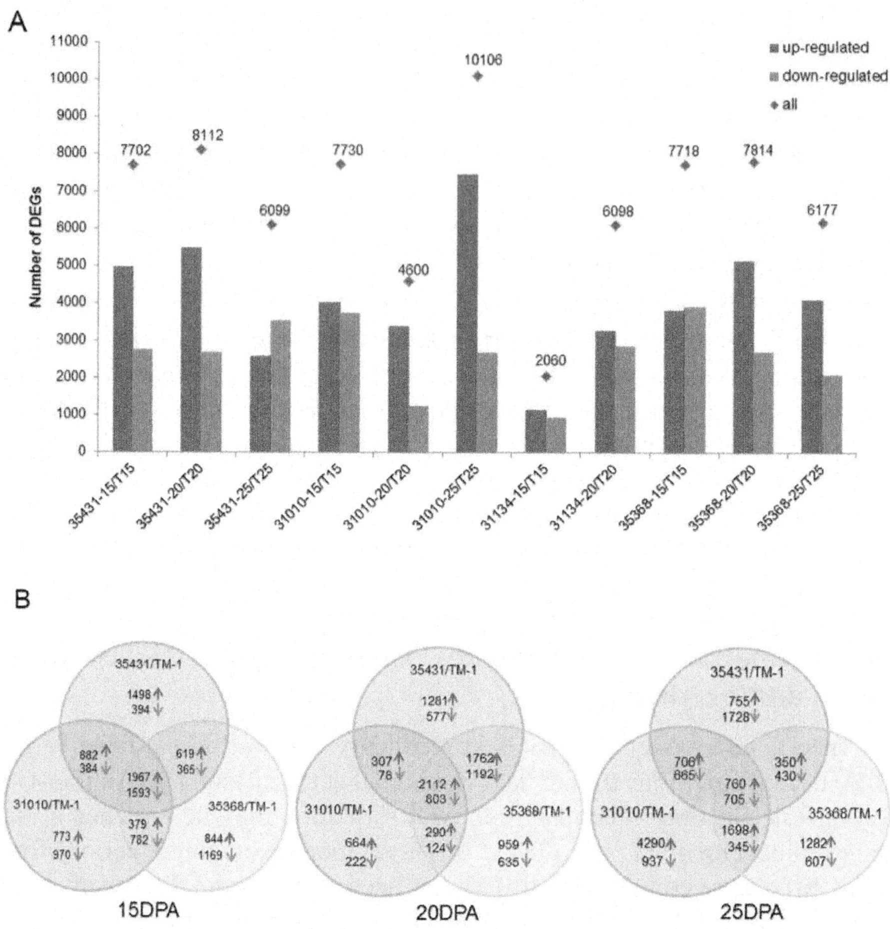

Figure 2. Statistical of DEGs between CSILs and TM-1 at 15, 20 and 25 DPA. (A) Up-regulated and down-regulated genes in different comparison. Red bar, up-regulated genes compared to TM-1; green bar, stand for down-regulated genes compared to TM-1, Blue square, total DEGs. CSILs included CSIL-35431, CSIL-3010, CSIL-31134, CSIL-35368 and TM-1. 15, 15DPA; 20, 20DPA; 25, 25DPA. (B) Common and special DEGs at 15 DPA, 20 DPA and 25 DPA.

To understand the mechanisms behind the changes in fiber strength observed in the CSILs, we also analyzed the common DEGs among CSIL-35431, CSIL-31010 and CSIL-31134. A total of 727 and 1796 common DEGs were selected at 15 and 20 DPA in three stronger fiber CSILs, respectively (Figure 3). More functional enrichment were shown at 15 DPA, including major CHO metabolism (carbohydrate), cell wall biosynthesis, amino acid metabolism and secondary metabolism (Figure 3E). Among these genes,

321 and 998 common upregulated DEGs between the same CSILs at 15 and 20 DPA were indentified, respectively (Figure 3). These common DEGs or processes maybe directly related to the fiber strength. However, these DEGs maybe function as downstream genes altered by the introgressed segments since these CSILs were inserted different *G. barbadense* segments in recipient TM-1.

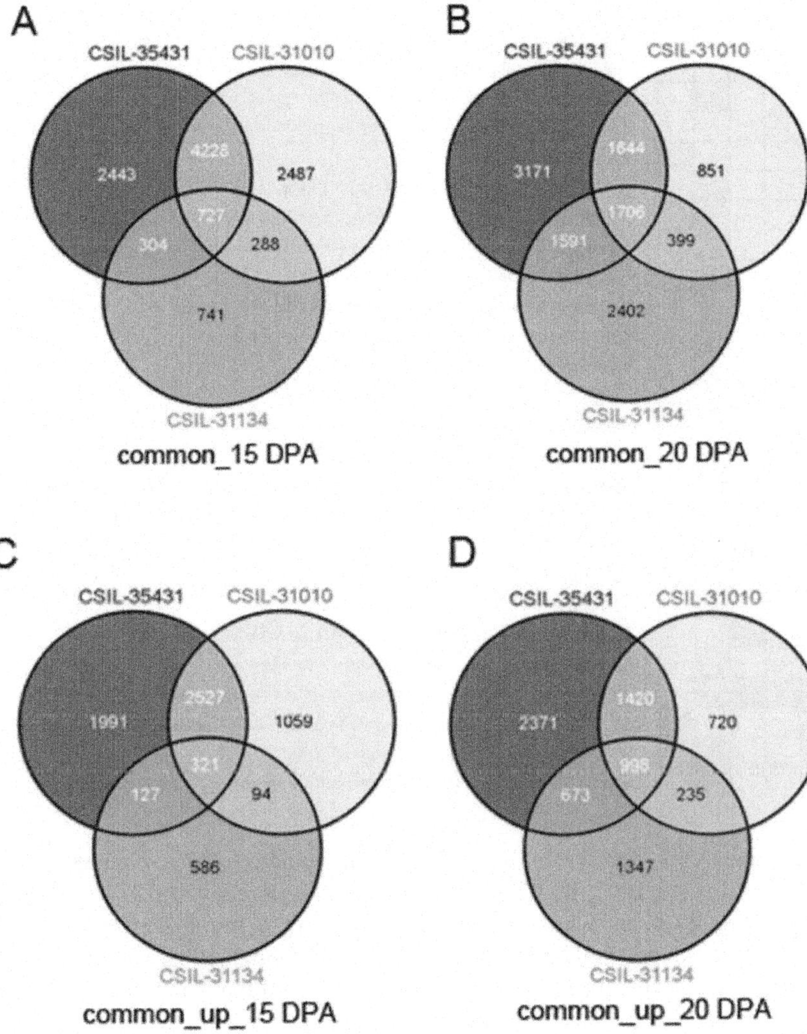

E

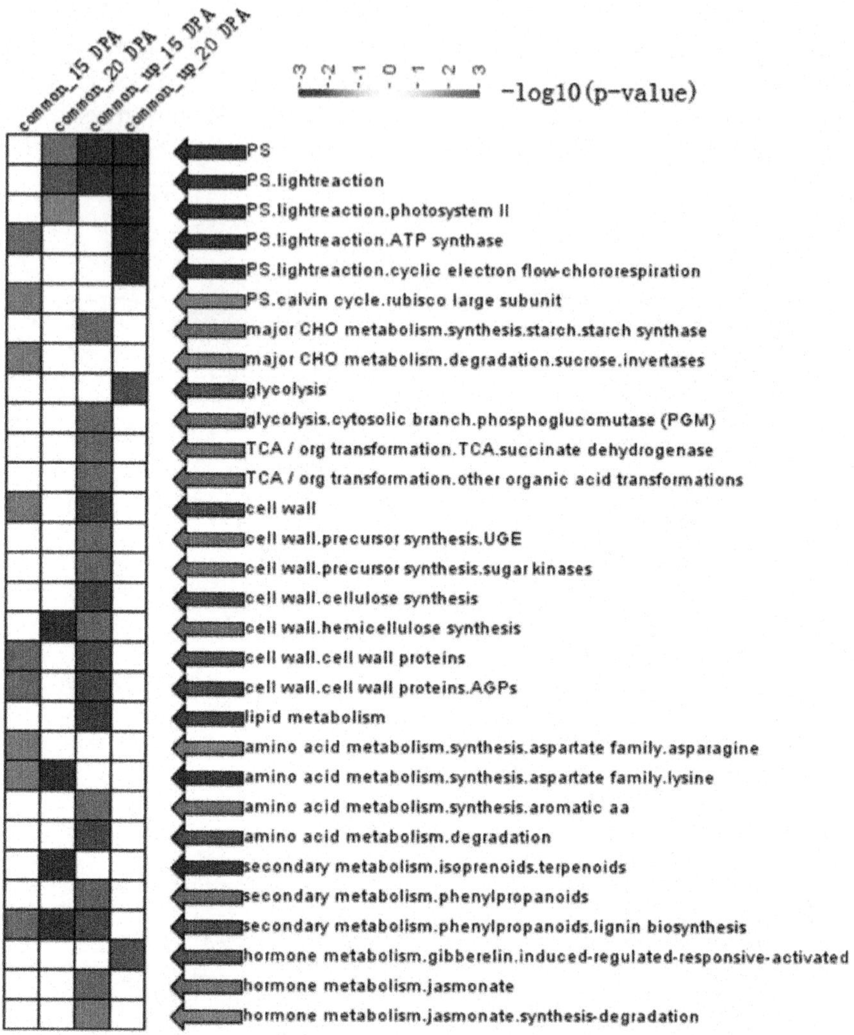

Figure 3. Analysis of common and common upregulated DEGs among three stronger fiber CSILs. (A, B, C, D) Common and common upregulated DEGs among three stronger fiber CSILs at 15 and 20 DPA. Common_up, common regulated DEGs. (B) Functional enrichment analysis of these DEGs using mapman software (Summary statistic type, wlcoxon). Colors from blue to red indicated that functions were enriched more significantly with smaller p-values.

To visualize the expression patterns of DEGs, we performed cluster analysis of 18,288 genes that were differentally expressed between CSIL-35431 and CSIL-31010 (Figure 4). These DEGs could be grouped into six clusters, designated G1–G6, based on their expression patterns. From 15 DPA to 20 DPA, the stages of fast fiber elongation and secondary cell wall deposition overlap, with the latter reaching a peak at around 20–25 DPA. We focused on clusters G1, G4, and G6 to conduct data analysis in order to identify genes that were either up-regulated or down-regulated during the secondary cell wall synthesis stage. Compared to the TM-1 control, 3,658 genes in cluster G1 were highly expressed at 15 and 20 DPA, 4,487 genes in G4 were highly expressed at 15 DPA, 20 DPA, and 25 DPA, 3,033 genes in G6 were highly expressed only at 25 DPA, and the other three groups showed various different expression patterns. Clustering results for 19,742 DEGs from the four CSILs showed five groups, indicating that the gene expression pattern in CSIL-31134 was distinct from the others at 15 DPA and 20 DPA, and that CSIL-35368 was similar to CSIL35431 and CSIL-31010 (Figure S1).

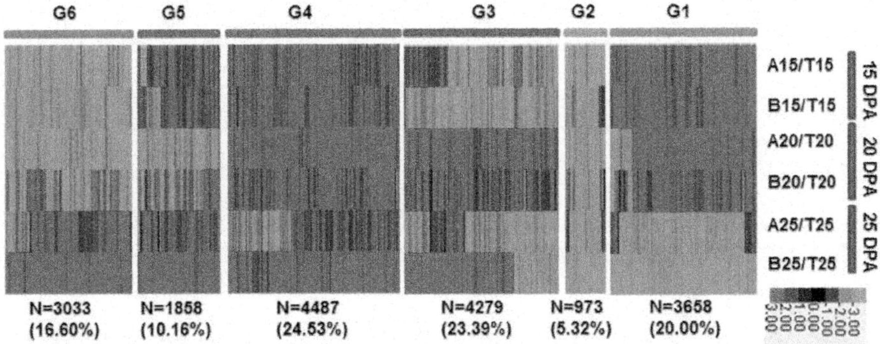

Figure 4. Heat map analysis of the expression of DEGs between CSILs and TM-1. A, B and T indicated CSIL-35431, CSIL-31010 and TM-1, respectively. 15, 15DPA; 20, 20DPA; 25, 25DPA. Red color indicated up-regulated genes and green color indicated down-regulated genes. N=number of DEGs in different group.

Functional Annotation by GO Enrichment and KEGG Analysis

To understand the mechanisms behind the changes in fiber strength observed in the CSILs, we analyzed DEG enrichment in the major functional GO categories of biological process, molecular function, and cellular component between CSIL-35431 and CSIL-31010. Based on the clustering results shown in Figure 4, G1 was enriched in genes for secondary cell wall biogenesis, glucuronoxylan biosynthesis, microtubule-based movement, and cellulose biosynthesis, G4 was enriched in genes for protein phosphorylation, response to chitin, and sugar-mediated signaling pathways, and G6 was enriched in fatty acid biosynthesis genes (Table 2). These data suggest that in the developmental stage of secondary cell wall deposition, DEGs were enriched for carbohydrate synthesis and cell wall formation.

Table 2. Enrichment analysis of gene ontologies from 15 to 25 DPA.

Cluster	GO-ID	GO Ontology (Biological process)
G1	GO:0010417	glucuronoxylan biosynthetic process
15DPA up-regulated	GO:0009834	secondary cell wall biogenesis
20DPA up-regulated	GO:0007018	microtubule-based movement
25DPA down-regulated	GO:0030244	cellulose biosynthetic process
	GO:0009753	response to jasmonic acid stimulus
G2	GO:0015031	protein transport
15DPA down-regulated	GO:0015991	ATP hydrolysis
20DPA down-regulated	GO:0032544	plastid translation
25DPA down-regulated	GO:0016075	rRNA catabolic process
	GO:0006511	ubiquitin catabolic process
G3	GO:0009734	auxin mediated signaling pathway
15DPA down-regulated	GO:0030259	lipid glycosylation
20DPA up-regulated	GO:0018106	peptidyl-histidine phosphorylation
25DPA down-regulated	GO:0009722	detection of cytokinin stimulus
G4	GO:0006468	protein phosphorylation
15DPA up-regulated	GO:0010200	response to chitin
20DPA up-regulated	GO:0006096	glycolysis
25DPA up-regulated	GO:0010182	sugar mediated signaling pathway
	GO:0009966	regulation of signal transduction
G5	GO:0015884	protein folding
15DPA up-regulated	GO:0006886	intracellular protein transport
20DPA down-regulated	GO:0006122	mitochondrial electron transport
25DPA up-regulated	GO:0015914	phospholipid transport
G6	GO:0007267	cell-cell signaling
15DPA down-regulated	GO:0010025	wax biosynthetic process
20DPA down-regulated	GO:0006633	fatty acid biosynthetic process
25DPA up-regulated	GO:0006723	hydrocarbon biosynthetic process

G1–G6 according to Figure 3.
doi:10.1371/journal.pone.0094642.t002

We applied the same GO analysis to the common DEGs at 15 DPA and 20 DPA in CSIL-35431 and CSIL-31010, respectively. These DEGs were enriched in genes for similar functional categories, such as cellular metabolic processes and carbohydrate metabolism, etc. We also found genes for some processes that were enriched only in CSIL35431 or CSIL-31010 (Figure S2).

Further GO analysis for CSIL-35368 and CSIL-31134 indicated that the DEGs in CSIL-35368 at 15 and 20 DPA were enriched in genes for lignin biosynthesis, secondary cell wall biogenesis, and response to chitin, which was similar to the enrichment found in CSIL-35431 and CSIL-31010. But at 15 and 20 DPA in the stronger fiber line CSIL-31134, GO enrichments were different from the other three lines, mainly in genes for ATP synthesis, proton transport, copper ion export, and oxidoreductase activity, but not in cell wall biosynthesis (Table S4).

Based on the results of GO analysis, we know that the secondary cell wall related biological process were impacted in the CSILs, but it is still not very clear how secondary cell wall biosynthesis was affected in the CSILs. Therefore, we performed pathway analysis on 18,288 DEGs in CSIL-35431 and CSIL-31010. The most highly enriched pathways found are listed in Table 3. KEGG analysis showed that the genes were enriched in pathways for starch and sucrose metabolism (328 genes), glycolysis/gluconeogenesis (122 genes), phenylpropanoid biosynthesis (101 genes), and oxidative phosphorylation (87 genes) (Table 3 and Figure S3). The regulation of some enzymes that catalyze sucrose, starch, and cellulose biosynthesis may have a direct or indirect impact on fiber quality. This could be especially true for sucrose and pectin metabolism, and many genes in these pathways were up-regulated. We also found that genes involved in phenylpropanoid and flavonoid biosynthetic processes were enriched in the CSILs.

Table 3. KEGG analysis of DEGs in CSIL-35431 and CSIL-31010.

Pathway	DEGs with pathway annotation (2576)	References genes with pathway annotation(4601)	Ratio	DEGs distribution in each groups					
				G1	G2	G3	G4	G5	G6
Starch and sucrose metabolism	328	563	58.26%	87	15	54	89	31	52
Purine metabolism	251	459	54.68%	35	20	53	65	30	48
Phenylalanine metabolism	140	261	53.64%	29	5	26	47	12	21
Amino sugar and nucleotide sugar metabolism	137	211	64.93%	47	5	23	36	12	14
Pyrimidine metabolism	128	227	56.39%	15	13	26	37	15	22
Glycolysis/Gluconeogenesis	122	188	64.89%	28	4	23	35	13	19
T cell receptor signaling pathway	115	212	54.25%	28	4	23	34	11	15
Pentose and glucuronate interconversions	109	227	48.02%	24	8	20	27	7	23
Glycerolipid metabolism	102	174	58.62%	18	4	29	21	15	15
Pyruvate metabolism	102	182	56.04%	16	8	18	29	12	19
Phenylpropanoid biosynthesis	101	220	45.91%	26	6	17	28	8	16
Galactose metabolism	94	160	58.75%	23	2	20	24	7	18
Cysteine and methionine metabolism	94	145	64.83%	21	2	21	28	10	12
Glycerophospholipid metabolism	94	167	56.29%	14	8	29	21	10	12
Arginine and proline metabolism	93	144	64.58%	16	5	22	19	12	19
Oxidative phosphorylation	87	184	47.28%	19	13	12	11	13	19
Carbon fixation in photosynthetic organisms	84	151	55.63%	19	3	11	34	8	9
Fatty acid degradation	83	133	62.41%	17	5	18	21	7	15

G1–G6 according to Figure 3.
doi:10.1371/journal.pone.0094642.t003

Based on the cluster analysis of the weaker fiber line CSIL-35368, we hypothesized that changes in other biochemical pathways led to reduced fiber strength (Figure S1). Considering only those that were down-regulated in CSIL-35368, we found genes that participated in fatty acid degradation and biosynthesis, and also in carbohydrate metabolic pathways (Figure 2B and Figure S4).

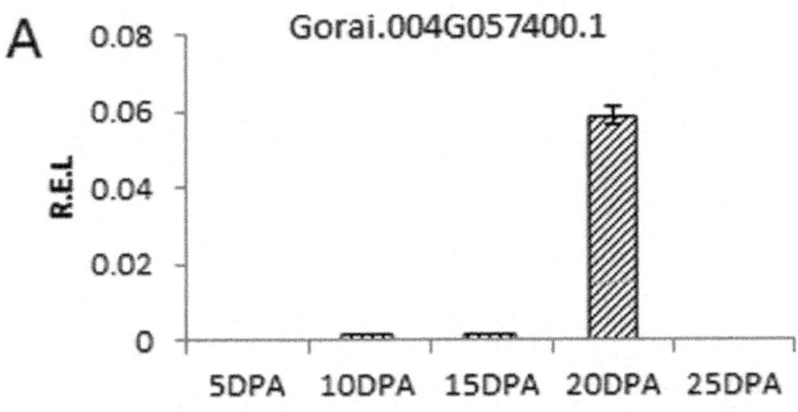

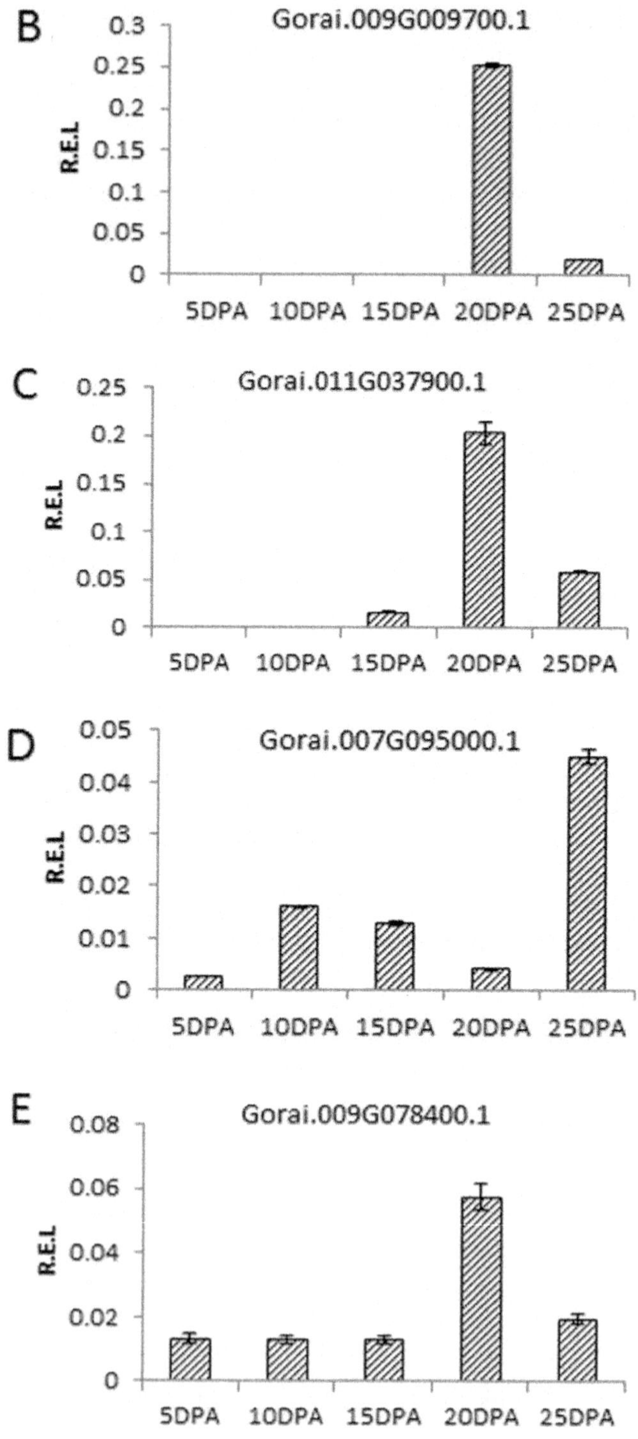
B
Gorai.009G009700.1
R.E.L
0.3
0.25
0.2
0.15
0.1
0.05
0
5DPA 10DPA 15DPA 20DPA 25DPA
C
Gorai.011G037900.1
R.E.L
0.25
0.2
0.15
0.1
0.05
0
5DPA 10DPA 15DPA 20DPA 25DPA
D
Gorai.007G095000.1
R.E.L
0.05
0.04
0.03
0.02
0.01
0
5DPA 10DPA 15DPA 20DPA 25DPA
E
Gorai.009G078400.1
R.E.L
0.08
0.06
0.04
0.02
0
5DPA 10DPA 15DPA 20DPA 25DPA

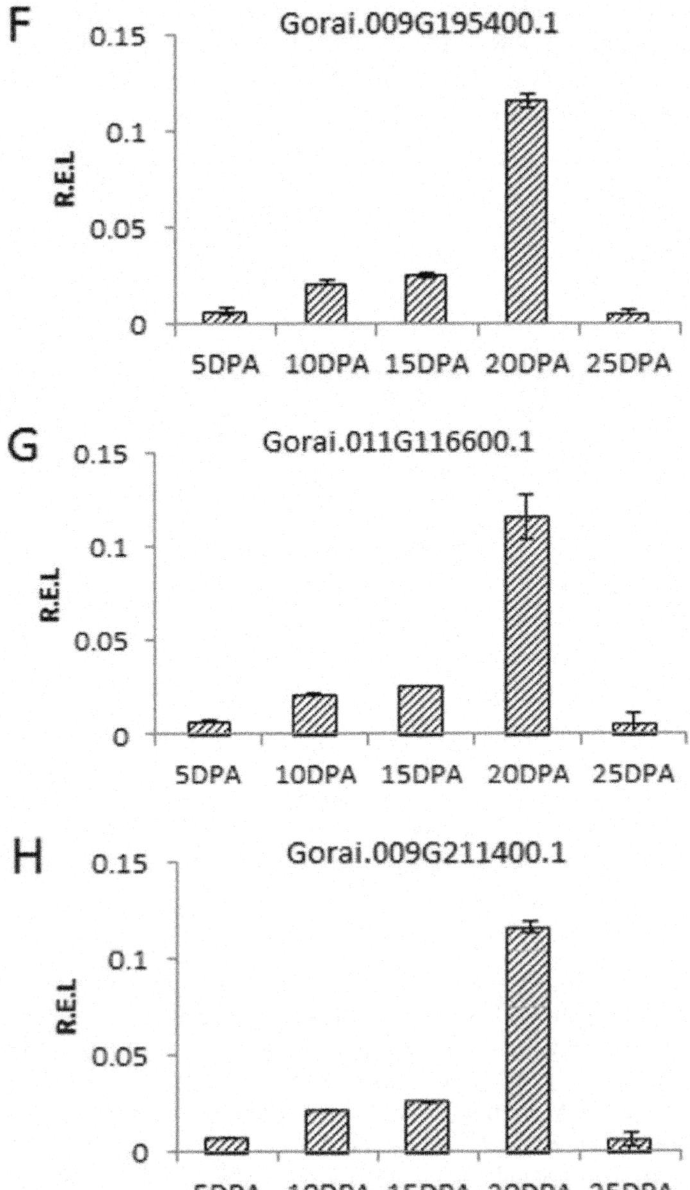

Figure 5. Quantitative RT–PCR validation of tag-mapped genes in TM-1. These genes have been reported before, including 3 CesA genes (A,B,C) (homologous with AtCESA4, AtCESA7, AtCESA8, respectively), xyloglucan endotransglucosylase (D), beta -galactosidase (E), glycosyl hydrolase 9B7 (F), xylan alpha-glucuronosyltransferase 1, GUX1 (G), xylan alpha-glucuronosyltransferase 2, GUX2 (H).

Eight genes previously reported in the carbohydrate pathway were selected for quantitative RT-PCR. The expression patterns of these genes were consistent with the DEG data in TM-1 (Figure 5) and in the CSILs as well (Figure 6B and Figure 7B).

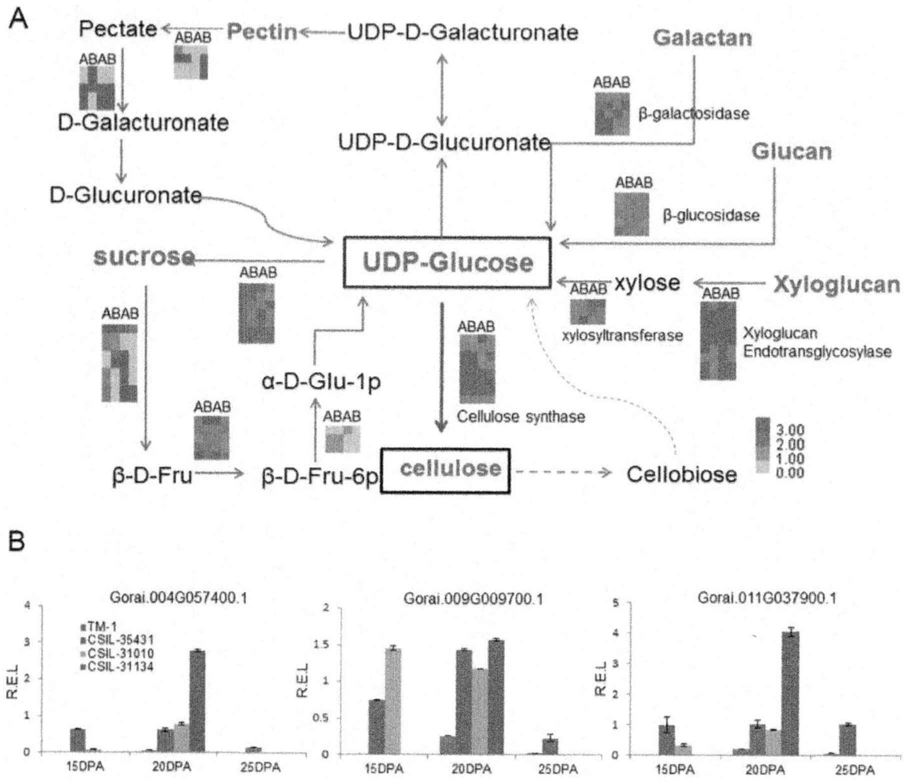

Figure 6. Carbohydrate pathways that are differentially regulated during the secondary cell wall synthesis stage. (A) Carbohydrate pathways. Genes up-regulated in CSIL-315431 and CSIL-31010 were selected to do heat map. ABAB indicated DEGs in CSIL-35431 at 15 DPA, CSIL-35431 at 20DPA, CSIL-3010 at 15DPA and CSIL-31010 at 20DPA, from left to right. Every square stand for one gene and every line stand for the same gene. Genes with red color expressed higher in CSILs than TM-1 and gray color stand for no difference. β-D-Fru, β-D-Fructose; α-D-Glu-1p, α-D-Glucose-1-phosphate; β-D-Fru-6p, β-D-Fructose-6-phosphate. (B) Quantitative RT–PCR validation of four CesA genes in CSILs and TM-1, Gorai.004G057400.1, Gorai.009G009700.1 and Gorai.011G037900.1 homologous with *AtCESA4*, *AtCESA7* and *AtCESA8*, respectively.

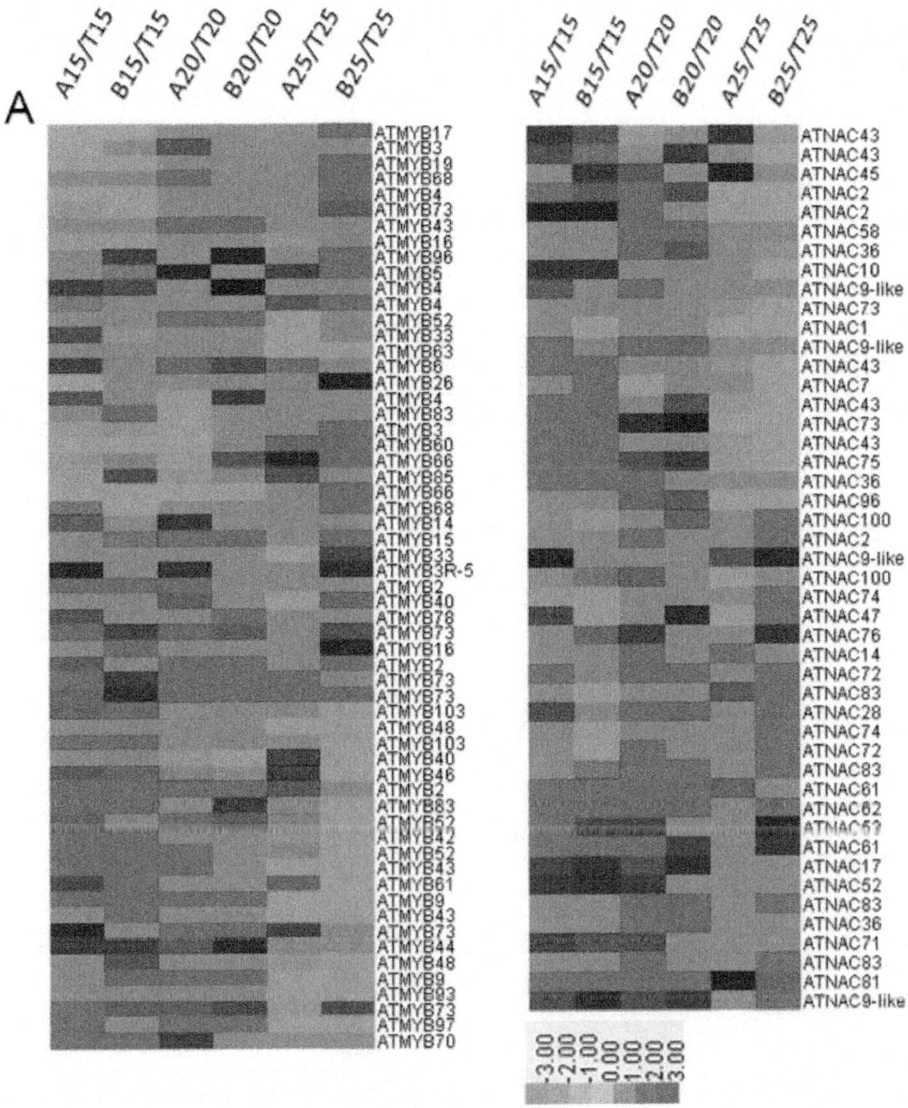
A
A15/T15
B15/T15
A20/T20
B20/T20
A25/T25
B25/T25
ATMYB17
ATMYB3
ATMYB19
ATMYB68
ATMYB4
ATMYB73
ATMYB43
ATMYB16
ATMYB96
ATMYB5
ATMYB4
ATMYB4
ATMYB52
ATMYB33
ATMYB63
ATMYB6
ATMYB26
ATMYB4
ATMYB83
ATMYB3
ATMYB60
ATMYB66
ATMYB85
ATMYB66
ATMYB68
ATMYB14
ATMYB15
ATMYB33
ATMYB3R-5
ATMYB2
ATMYB40
ATMYB78
ATMYB73
ATMYB16
ATMYB2
ATMYB73
ATMYB73
ATMYB103
ATMYB48
ATMYB103
ATMYB40
ATMYB46
ATMYB2
ATMYB83
ATMYB52
ATMYB42
ATMYB52
ATMYB43
ATMYB61
ATMYB9
ATMYB43
ATMYB73
ATMYB44
ATMYB48
ATMYB9
ATMYB93
ATMYB73
ATMYB97
ATMYB70
ATNAC43
ATNAC43
ATNAC45
ATNAC2
ATNAC2
ATNAC58
ATNAC36
ATNAC10
ATNAC9-like
ATNAC73
ATNAC1
ATNAC9-like
ATNAC43
ATNAC7
ATNAC43
ATNAC73
ATNAC43
ATNAC75
ATNAC36
ATNAC96
ATNAC100
ATNAC2
ATNAC9-like
ATNAC100
ATNAC74
ATNAC47
ATNAC76
ATNAC14
ATNAC72
ATNAC83
ATNAC28
ATNAC74
ATNAC72
ATNAC83
ATNAC61
ATNAC62
ATNAC52
ATNAC61
ATNAC17
ATNAC52
ATNAC83
ATNAC36
ATNAC71
ATNAC83
ATNAC81
ATNAC9-like
-3.00
-2.00
-1.00
0.00
1.00
2.00
3.00

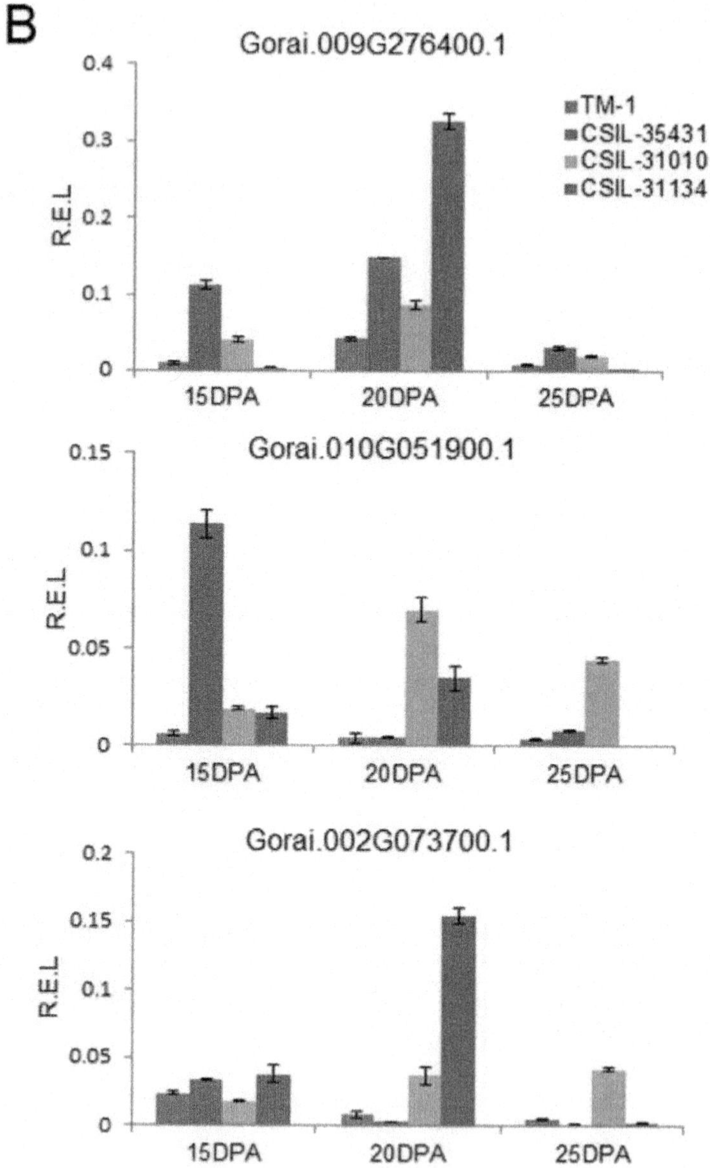

Figure 7. NAC and MYB family genes involved in the regulation of secondary wall biosynthesis. (A) 59 MYB family genes and 47 NAC family genes showed different expression level between CSILs and TM-1 at 15DPA, 20DPA and 25DPA. |Ratio|>2 and FDR<0.001. A, B, T indicated CSIL-35431, CSIL-31010 and TM-1. 15, 15DPA; 20, 20DPA; 25, 25DPA. (B) Quantitative RT–PCR validation of three transcription factors.

Carbohydrate Metabolism in the Secondary Cell Wall Synthesis Stage

Following the start of secondary cell wall formation, protein and carbohydrate metabolism genes involved in cell wall biosynthesis will be up-regulated [27]. We selected 72 DEGs associated with carbohydrate metabolism to investigate the mechanism of fiber development. These genes were related to pectin, sucrose, galactan, glucan, xyloglucan, and cellulose biosynthesis. We were interested in genes that are up-regulated in fiber cells at 15 DPA and 20 DPA, at the start of secondary cell wall formation. A heat map showing the different expression levels for these genes including cellulose synthase, sucrose synthase, pectin lyase, and other polysaccharides degradation in CSIL-35431 and CSIL-31010 is shown in Figure 6A. We found that the cellulose synthase genes were up-regulated in the CSILs at 15 DPA-25 DPA. It has been reported that cellulose biosynthesis predominates, and that many other metabolic pathways are down-regulated during secondary cell wall synthesis [27]. Moreover, we confirmed the expression patterns of cellulose synthase genes, annotated with the *Arabidopsis* genes *AtCESA4*, *AtCESA7* and *AtCESA8*, using quantitative RT-PCR (Figure 4B). Proteins encoded by *AtCESA4*, *7*, and *8* are specifically required to form a functional cellulose synthase complex (CSC) that is essential for secondary cell wall formation [40]–[42].

Transcription Factors Associated with Secondary Cell Wall Synthesis

Recent molecular and genetic studies have identified transcription factors that are involved in regulating secondary cell wall synthesis in *Arabidopsis* [43]–[45]. In our study, 97 MYB-type and 68 NAC-type transcription factors showed changes in expression between the CSILs and TM-1 (Table S5, Table S6). It was interesting that some NACs and MYBs were up-regulated in CSIL-35431 and CSIL-31010 during the secondary cell wall synthesis stage, especially at 15 DPA and 20 DPA. Defined as $|\log_2 \text{Ratio}| \geq 2$, 59 MYB and 47 NAC transcription factors were selected for heat-map analysis (Figure 7A). Among these transcription factors, genes homologous with *ATMYB2*, *ATMYB43*, *ATMYB73*, *ATNAC52*, and *ATNAC61* were expressed at higher levels in the CSILs. We confirmed that three transcription factors were up-regulated in CSILs from 15 DPA to 25 DPA (Figure 7B). In the MYB family, it has been reported that the expression of genes for MYB85, MYB52, MYB54, MYB69, MYB42, and MYB43 are developmentally associated with cells undergoing secondary wall thickening [45].

Different Metabolic Pathways Associated with Altered Fiber Strength

In order to investigate the mechanisms underlying changes in fiber strength, we analyzed several metabolic pathways including cell wall, lipids, minor CHO (carbohydrate) metabolism, and two secondary metabolite pathways. It is interesting that DEGs involved in cell wall proteins, cell wall pectin esterase, cell wall modification, cell wall cellulose synthesis, cell wall degradation/pectate lyases, lipid metabolism/FA synthesis, and lipid degradation showed distinct expression patterns or differential up/down-regulation at 20 DPA (Figure 8A). We found that up-regulated DEGs were similar to down-regulated DEGs both in CSIL-35431 and CSIL-35368. However, most of DEGs in CSIL-31010 were up-regulated at 20 DPA, while the opposite was true for DEGs in CSIL-31134, especially those genes involved in cell wall modification. In CSIL-31134, we also found a few genes in these metabolic pathways that were changed at 15 DPA except in cell wall modification, and in CSIL-31010, we found DEGs enriched in these metabolic pathways at 25 DPA (Figure S5). From the secondary metabolism results, we identified a few DEGs involved in flavonoid biosynthesis in CSIL-35431 and CSIL-31010 at 15 DPA. In contrast, more genes were up-regulated or down-regulated in CSIL-35368 at 15 DPA. It was obvious that DEGs from the phenylpropanoid pathways at 25 DPA were different from one another, and the expression pattern of DEGs in CSIL-31010 changed dramatically.

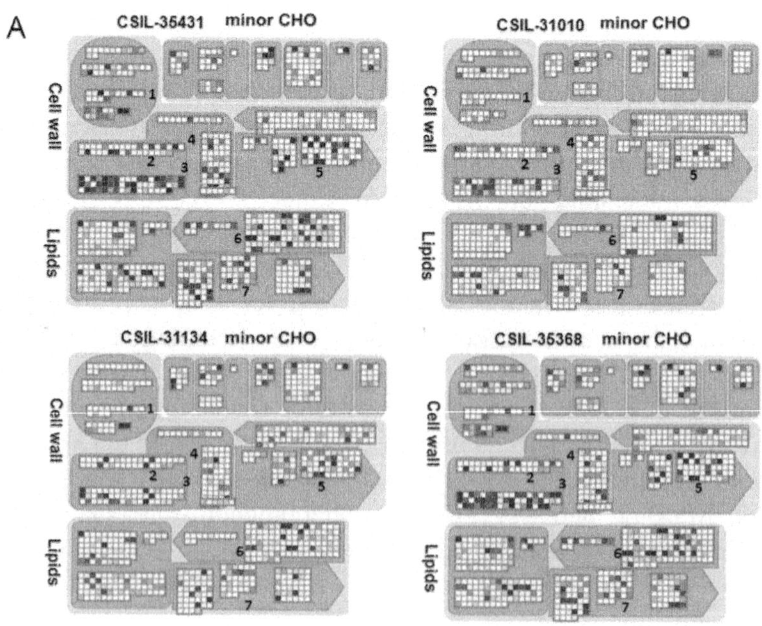

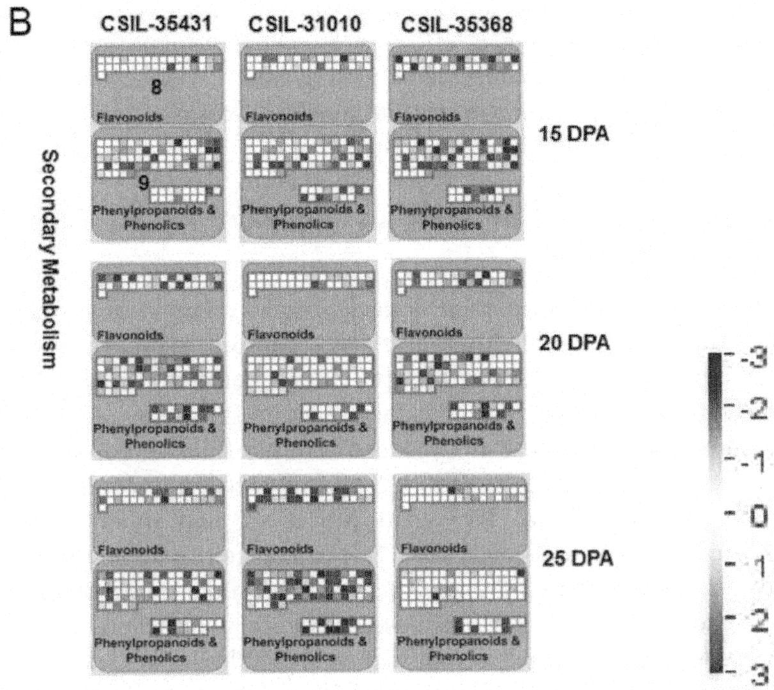

Figure 8. Metabolism analysis of DEGs in CSILs during the secondary cell wall biosynthesis stage. (A) Motabolism overview in four CSILs at 20 DPA. (B) Secondary motabolism analysis in three CSILs at 15 DPA, 20 DPA and 25 DPA. 1, cell wall protein; 2, cell wall pectin esterases; 3, cell wall modification; 4, cell wall cellulose synthesis; 5, cell wall degradation/pectate lyases; 6, lipid metabolism/FA synthesis; 7, lipid degradation; 8, flavonoids; 9, phenylpropanoids/lignin biosynthesis. Blue square, down-regulated genes; Red square, up-regulated genes.

Moreover, there were few genes that were up-regulated or down-regulated in CSIL-35368 at 25 DPA (Figure 7B). We assume that metabolic pathways in the CSILs at different developmental stages were changed in various ways as a result of the introgressed chromosmal segments from *G. barbadense*.

DISCUSSION

G. hirsutum produces a high yield of cotton with moderate fiber strength. *G. barbadense* is characterized by a low yield, but with increased fiber fineness and strength. As a breeding target, we tried to combine the high yield of *G. hirsutum* with the superior fiber qualities of *G. barbadense*, and we also wanted to elucidate the molecular mechanism behind the formation of superior quality fibers. Fiber strength is an important indicator of the cotton fiber quality, and depends on the formation of the secondary cell wall. Genome-wide transcriptome profiling is effective at revealing significant genes and pathways involved in secondary cell wall formation. Transcriptome analysis showed that gene expression patterns and functional distribution were different during secondary cell wall biosynthesis.

Carbohydrate Metabolism Plays an Important Role in Secondary Cell Wall Synthesis

It is well known that the mature cotton fiber is composed of nearly pure cellulose, and that such a high level of cellulose synthesis requires an abundant supply of UDP-glucose [46],[47]. This means that a large amount of cellulose is required during the secondary cell wall synthesis stage. Functional classification and enrichment analysis showed that following the initiation of secondary cell wall synthesis, DEGs were enriched for secondary cell wall biogenesis, glucuronoxylan biological processes, and other carbohydrate metabolic pathways in the CSILs (Table 2). Focusing on carbohydrate metabolic pathways, it is obvious that the key intermediate in the multiple pathways is UDP-glucose, a substrate for cellulose synthesis. Our results showed that several CesA genes are expressed at higher levels during secondary cell wall synthesis than they are at earlier stages (Figure 6B). Ten*AtCESA* genes have been reported in *Arabidopsis*, and *AtCESA4*, *7*, and *8* are specifically required to form the cellulose synthase complex (CSC) that is essential for secondary cell wall synthesis [40]–[42]. Similarly, three CESA isoforms have been identified during secondary cell wall synthesis in rice, maize, and *Populus* [10], [48], [49]. Also, many genes that participate in the degradation of poly- and oligo-saccharides were found to be up-regulated at 15 and 20 DPA, in order to produce more UDP-glucose for cellulose biosynthesis. Similarly, it has also been reported that during the secondary cell wall synthesis stage, certain metabolic pathways, including hydrolysis of fatty acids and non-cellulose poly- and oligo-saccharides, would be up-regulated [27]. Sucrose synthase (SuSy) has long been studied as a cytoplasmic enzyme in plant cells, where it serves to degrade sucrose and provide carbon for respiration and synthesis of cell wall polysaccharides and starch [50]. It has also been shown that genes associated with secondary cell wall biosynthesis are involved in sugar metabolism [51].

Multiple Mechanisms Affect Fiber Strength Development

Except for carbohydrate metabolism, recent research has shown that transcription factors also affect fiber development during secondary cell wall biosynthesis. Several NAC- and MYB-type transcription factors were up-regulated in the CSILs compared to TM-1 from 15 DPA to 25 DPA, and these included cotton homologs of *AtMYB2*, *AtMYB43*, and *AtNAC52*etc. (Figure 7A). The NAC-mediated transcriptional regulation of secondary wall biosynthesis is a conserved mechanism throughout vascular plants [44], [52]. *SND2*, a NAC transcription factor gene, regulates genes involved in secondary cell wall development in*Arabidopsis* fibers and increases fiber cell area in *Eucalyptus* [53]. A MYB75-associated protein complex is likely to be involved in modulating secondary cell wall biosynthesis in both the *Arabidopsis* inflorescence and stem [54]. It has also been found that the rice and maize MYB transcription factors, OsMYB46 and ZmMYB46, are functional orthologs of*Arabidopsis* MYB46/MYB83 and, when overexpressed in *Arabidopsis*, are able to activate the entire secondary wall biosynthetic program [55].

Several metabolic pathways were examined to determine the mechanism behind changes in fiber strength; these included cell wall, lipids, minor CHO metabolism, and two secondary metabolic pathways. Although results of the GO and KEGG analyses showed that CSIL-35431, CSIL-31010, and CSIL-35368 had similar patterns, fiber strength in these three lines were different. Our results support the hypothesis that different metabolic pathways can affect fiber strength, and the same pathway in the CSILs can be altered differentially at various times in development. DEGs in CSIL-31010 were up-regulated at 20 DPA, while the opposite was found for DEGs in CSIL-31134, especially those genes involved in cell wall modification. The expression levels of genes involved in flavonoid biosynthesis in the weak fiber line CSIL-35368 were changed dramatically at 15 DPA, but there were few genes changed at 25 DPA; this patter was the opposite of that in CSIL-35431 and CSIL-31010, lines with high quality fiber. We hypothesize that phenylpropanoid and flavonoid metabolism generally affected the fiber strength of CSIL-35368. Genes for phenylpropanoid and flavonoid biosynthesis showed significant enrichment and temporal differences in gene expression patterns which are associated with xylem formation [56]. It has been reported that expression levels of phenylpropanoid genes showed high correlations with specific fiber properties, supporting a role in determining fiber strength [57].

In conclusion, upland cotton CSILs carrying distinct *G. barbadense* chromosomal segments provide valuable material for research into fiber development. The *G. barbadense*chromosome segments resulted in different patterns of differentially expressed genes, and altered different metabolic pathways, mainly in carbohydrate metabolism. In addition, several transcription factor genes were found to be specifically up-regulated in the CSILs. Metabolic pathways involved in cell wall, lipid, phenylpropanoid, and flavonoid biosynthesis play a significant role during secondary cell wall formation, and are associated with the development of cotton fiber strength.

SUPPORTING INFORMATION

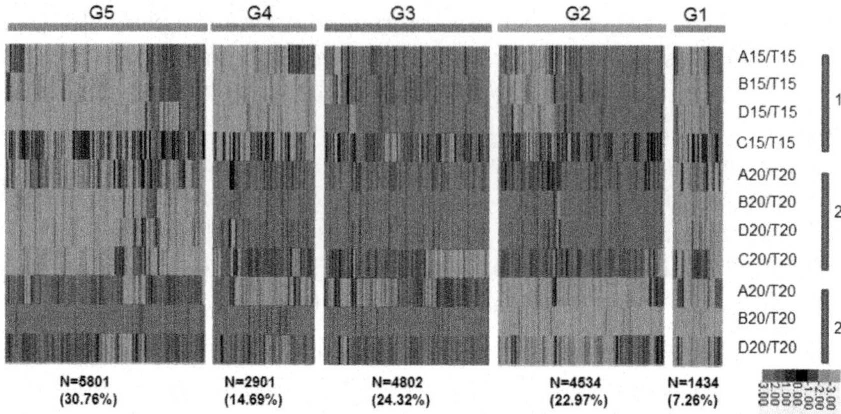

Figure S1. Heat map of the expression of DEGs between 4 CSILs at 15–25 DPA.

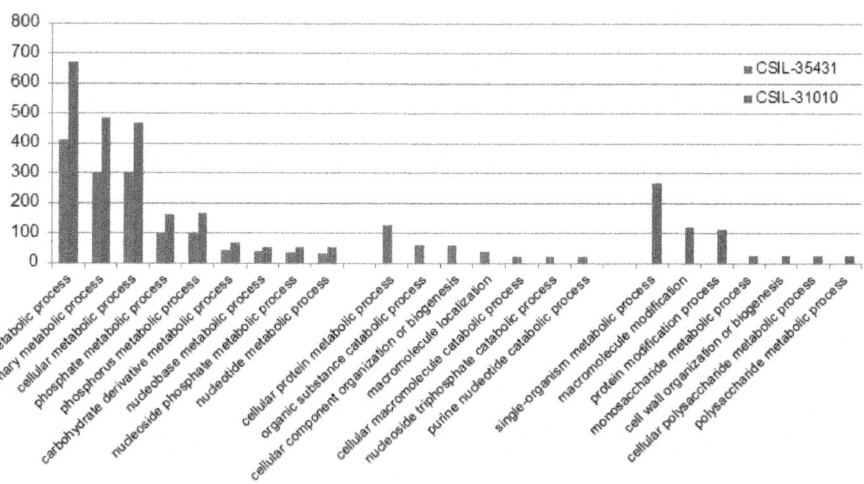

Figure S2. Enrichment analysis of common DEGs at 15DPA and 20DPA in CSIL-35431 and CSIL-31010.

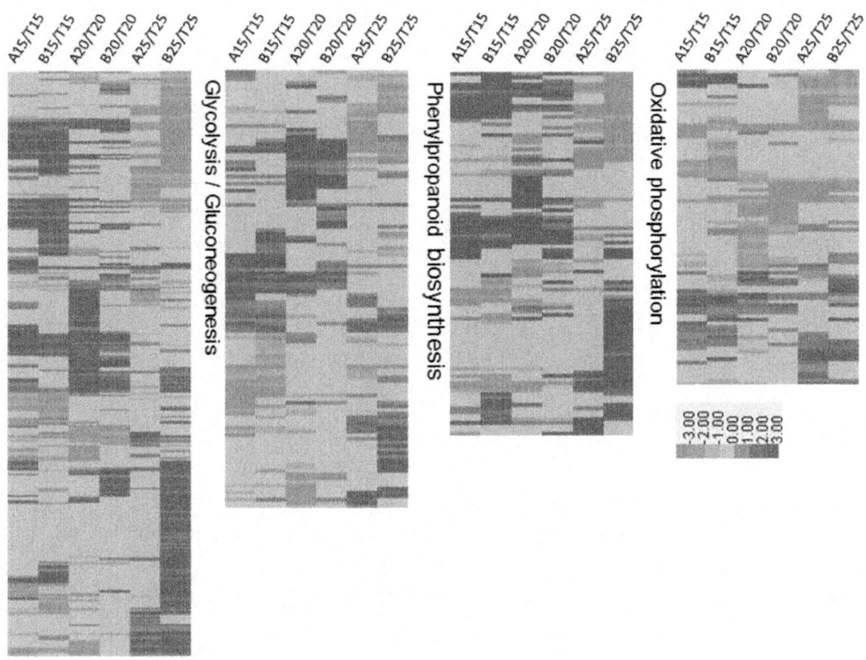

Figure S3. Heat map of DEGs participated in four metabolic pathways from 15 DPA to 25 DPA.

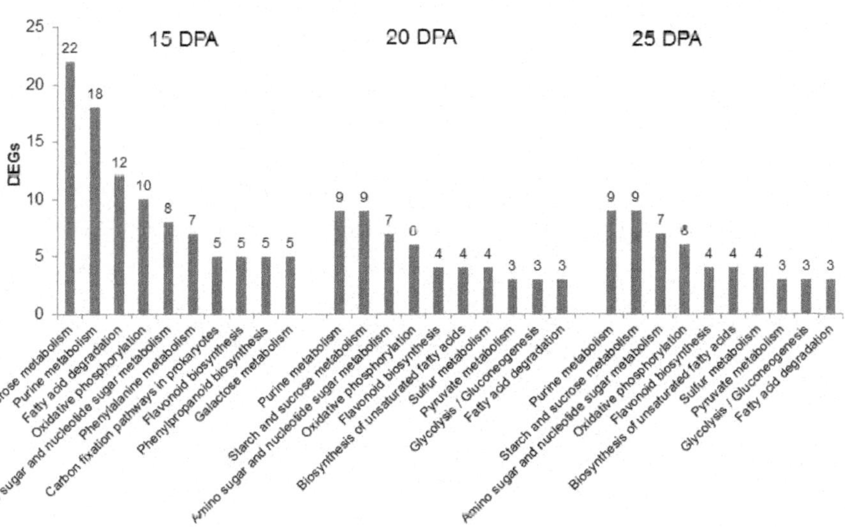

Figure S4. Pathway analysis of genes only down-regulated in CSIL-35368 from 15 DPA to 25 DPA.

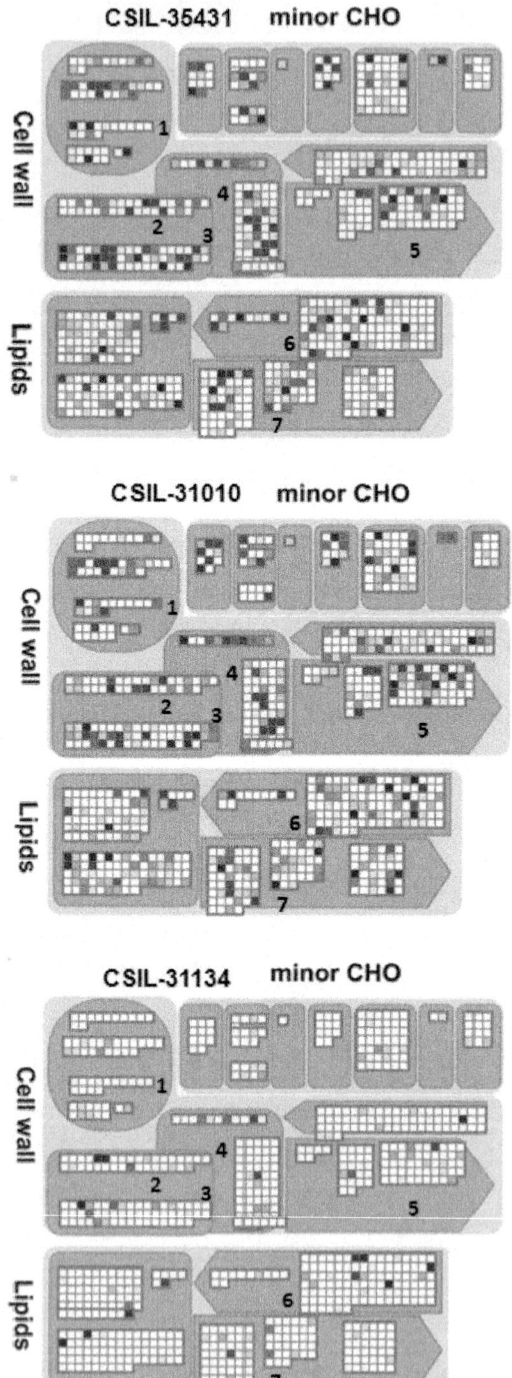
CSIL-35431 minor CHO
Cell wall
Lipids
1
2
3
4
5
6
7
CSIL-31010 minor CHO
Cell wall
Lipids
1
2
3
4
5
6
7
CSIL-31134 minor CHO
Cell wall
Lipids
1
2
3
4
5
6
7

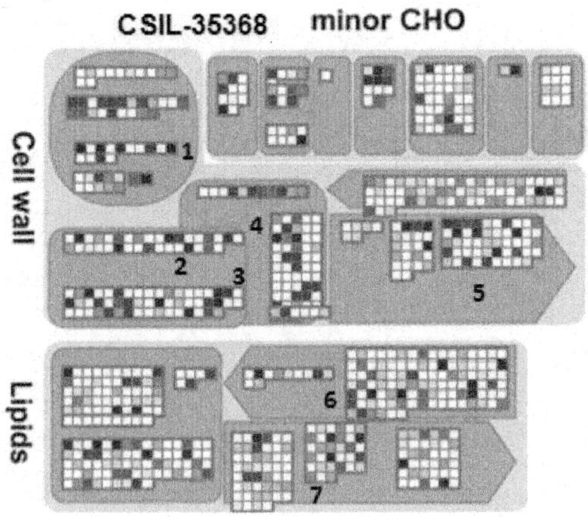
CSIL-35368 minor CHO
Cell wall
Lipids
1
2
3
4
5
6
7

(A)

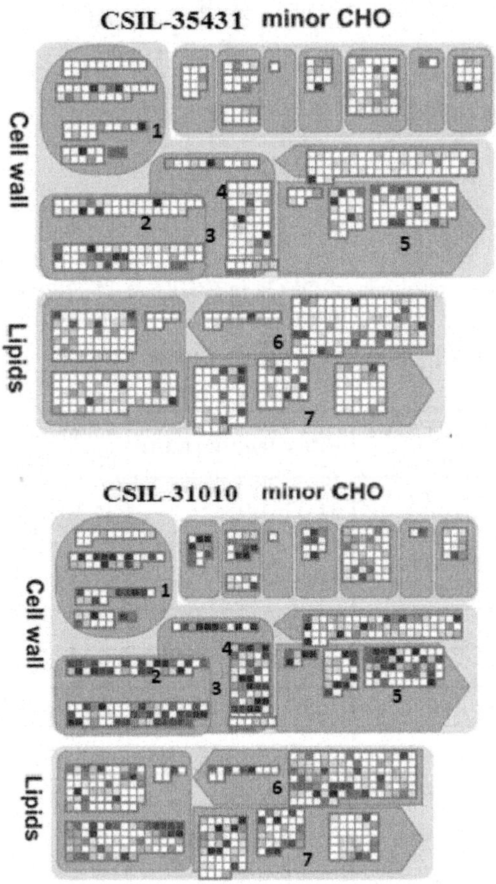
CSIL-35431 minor CHO
Cell wall
Lipids
1
2
3
4
5
6
7
CSIL-31010 minor CHO
Cell wall
Lipids
1
2
3
4
5
6
7

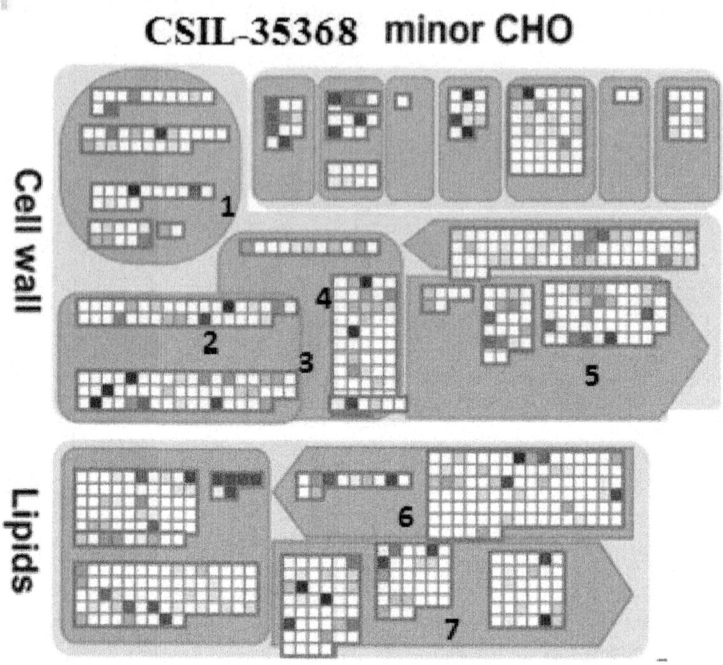

(B)

Figure S5. Metabolism analysis of DEGs in CSILs at 15 DPA and 25 DPA.

Table S1. Average fiber quality of 4 CSILs and TM-1.

CSILs	**Chomosome segment**	**Fiber length (mm)**	**Fiber strength (cN/tex)**	**Mrc**
CSIL-35431	A8(Chr.8) D10(Chr.20)	31.33±1.95**	35.10±2.12**	4.84±0.64
CSIL-31134	A8(Chr.8) D1(Chr.15)	30.63±1.14**	34.73±2.03**	4.59±1.09
CSIL-31010	D2(Chr.14)	30.01±1.05**	34.28±1.10**	5.05±0.94
TM-1		28.89±0.53	30.11±0.43	4.81±0.75
CSIL-35368	D11(Chr. 20)	27.67±1.18**	28.71±3.19	4.47±0.45
**: CSILs have significant difference with TM-1 at 1% level				

Table S2. Primer for quantitative RT-PCR.

Reference Gene	Forward Primer 5'-3'	Reverse Primer 5'-3'
Gorai.004G057400.1	GTGCTGAGAATGCTCTGGTT	GCCTTACTATTGTTGATG-TAATGAT
Gorai.009G009700.1	TGTCCATCTCTATCCATTCCTTAA	GAAGCTAATAGCACTGAC-CATAT
Gorai.011G037900.1	TGCTCGTTACGCCATAAGTC	TTCCACTCATTGCCTATT-GTTCA
Gorai.007G095000.1	ACCCATCAAAGAAATAGC-CATAGAC	AGGAAGCGAGTGGTGGAA
Gorai.009G078400.1	CGCACCCAATAATCCATTCTA	CGATGTATCAAC-CAAGACTTTGG
Gorai.009G195400.1	GTAGCCATAGCCTGTAACTGATT	CGGGCATGCCTCAAA-GATACG
Gorai.011G116600.1	TAAGGACAGGCGAAGGAAGAA	CGAGAGTGAAGCAGTGTT-GATT
Gorai.009G211400.1	TTGCCGAGGAGAGTGAACTT	TGGAGTA-AACTTTGGGTGGGT
Gorai.009G276400.1	CTTCAACTCAAACAGCATCCT	ATACGGATCCTAAGT-CAGTGA
Gorai.010G051900.1	GTCGGAGTCGGTGCCGAGA	ATTGGTGGTGGTGGTGTTG
Gorai.002G073700.1	GGGAGTCTTGGTGGGCTTA	AAGTGTAGGCGGCATTGTG

Table S3. Categorization and abundance of tags.

CSILs	Raw Data	Distinct Tag	Clean Tag	Distinct Tag	Unambiguous Tag Mapping to Gene	Unambiguous Tag Mapping to Gene	Unambiguous Distinct Tag Mapping to Gene	Unambiguous Distinct Tag Mapping to Gene	Unambiguous Tag-mapped Genes	Unambiguous Tag-mapped Genes
35431-15	7461562	375999	7264426	182629	3961228	54.53%	53187	29.12%	21781	58.07%
35431-20	7389820	371748	7213791	199257	3411249	47.29%	48453	24.32%	22325	59.53%
35431-25	7298224	317147	7124575	148121	3463314	48.61%	41065	27.72%	20590	54.90%
TM-1-15	8374304	389162	8144920	169700	3983215	48.90%	45380	26.74%	21498	57.32%
TM-1-20	7231305	394430	7013147	177887	3206328	45.72%	40475	22.75%	20901	55.73%
TM-1-25	14267931	494345	14002534	251173	6829327	48.77%	56172	22.36%	22594	60.24%
35368-15	6938486	282172	6835546	182227	3268867	47.82%	44091	24.20%	21527	57.40%
35368-20	7165879	336637	7020147	194219	3429087	48.85%	51112	26.32%	22434	59.82%
35368-25	7414675	364911	7236602	190038	3469505	47.94%	47697	25.10%	22287	59.42%
31134-15	8710599	404122	8472498	175897	4055377	47.87%	43632	24.81%	21888	58.36%
31134-20	8265105	342671	8070435	158013	3964336	49.12%	42184	26.70%	20962	55.89%

31010-15	7471212	283130	7369012	182184	3671014	49.82%	46720	25.64%	21900	58.39%
31010-20	7007447	337525	6840101	172775	3381955	49.44%	45762	26.49%	22457	59.88%
31010-25	7203054	386302	7002766	188639	3552566	50.73%	48147	25.52%	22811	60.82%

15,15 DPA; 20, 20 DPA; 25, 25 DPA

Clean tags: tags after filtering dirty tags (low quality tags) from raw data.

Distinct tags: different kinds of tags.

Unambiguous tags: the clean tags after removing tags mapped to reference sequences from multiple genes.

Table S4. Enrichment analysis of gene ontologies in CSIL-35368 and CSIL-31010 at 15 DPA and 20 DPA.

CSILs	GO-ID	GO Ontology (Biologic process)
35368/TM-1 common DEGs at 15DPA and 20DPA	GO:0009809	lignin biosynthetic process
	GO:0010200	response to chitin
	GO:0055069	zinc ion homeostasis
	GO:0005985	sucrose metabolic process
	GO:0009834	secondary cell wall biogenesis
	GO:0046686	response to cadmium ion
	GO:0006200	ATP catabolic process
	GO:0009737	response to abscisic acid stimulus
	GO:0007155	cell adhesion
31134/TM-1 common DEGs at 15DPA and 20DPA	GO:0015977	carbon fixation
	GO:0042777	ATP synthesis proton transport
	GO:0060003	copper ion export
	GO:0019318	hexose metabolic process
	GO:0019761	glucosinolate biosynthetic process
	GO:0010252	auxin homeostasis
	GO:0006107	oxaloacetate metabolic process
	GO:0019318	hexose metabolic process
	GO:0016645	oxidoreductase activity

Table S5. Different expression level of 97 MYB transcription factors.

Gene NO.	Annota-tion from Tail10	35431-15/T15	31010-15/T15	35431-20/T20	31010-20/T20	35431-25/T25	31010-25/T25
Gorai.010G016800.1	AT-MYB44	1.6639	1.6660	2.3719	1.3344		
Gorai.009G215900.1	AT-MYB44			7.8806			
Gorai.007G312700.1	AT-MYB97	7.9529		11.7818	8.9573	-10.4945	-4.6585
Gorai.007G037100.1	AT-MYB96		-1.0492		1.2649		4.2777
Gorai.012G054500.1	AT-MYB96		-2.6621	-2.3626			
Gorai.008G192900.1	AT-MYB94	-1.4062	-1.3368	-1.1857			1.3236
Gorai.009G236900.1	AT-MYB93					-10.2625	-10.2625
Gorai.007G362200.1	ATMYB9	9.4834	8.4636	11.3167	9.8471	-4.1267	-10.5227
Gorai.007G348600.1	ATMYB9		8.9726	12.3967	10.7153	-10.6880	-10.6880
Gorai.013G083300.1	AT-MYB88						1.6653
Gorai.006G195700.1	AT-MYB85	-2.8122	-1.1596	-2.2030	-1.9336	2.1863	
Gorai.013G251100.1	AT-MYB83	3.6724	4.4733		1.7719		-13.0322
Gorai.004G172700.1	AT-MYB83		7.8506	-8.2261			
Gorai.004G276500.1	AT-MYB82					6.8111	
Gorai.006G254200.1	AT-MYB79						5.9539
Gorai.011G203900.1	AT-MYB78	2.2646			7.5703		
Gorai.005G068900.1	AT-MYB78			7.7931			2.9997
Gorai.006G080800.1	AT-MYB78						7.6433
Gorai.012G149000.1	AT-MYB73	-2.3107	-3.0086				2.4147
Gorai.010G023500.1	AT-MYB73	1.3119		3.9593	3.5027	-1.0340	
Gorai.007G055600.1	AT-MYB73	1.5588					3.5697

Gorai.009G208900.1	AT-MYB73	2.5760	1.4407	12.4513	9.8698	2.7684	3.5357
Gorai.005G234900.1	AT-MYB73	3.2182	1.7775	4.7444	3.7365	2.7822	
Gorai.009G276400.1	AT-MYB73	3.2822	2.4328	2.4046	1.9634	-2.6031	-1.2363
Gorai.007G218100.1	AT-MYB73	3.6542	1.8614	2.6741	2.4313		2.1775
Gorai.009G275100.1	AT-MYB73			8.1854			
Gorai.012G078200.1	AT-MYB70	8.9529	9.6458	1.6720			-2.0003
Gorai.004G130900.1	ATMYB7						7.2954
Gorai.001G020500.1	ATMYB7						7.1579
Gorai.007G227500.1	AT-MYB68	-1.6581	-9.1415				7.0059
Gorai.001G024300.1	AT-MYB68			7.7000			4.8823
Gorai.012G061800.1	AT-MYB66	-5.7176	-11.5006	-9.5342	-3.6644		9.4506
Gorai.004G196800.1	AT-MYB66	-2.2944	-1.8403	-2.6801	-1.3038	1.2572	5.2921
Gorai.010G096800.1	AT-MYB66	2.3026				2.5598	
Gorai.007G001800.1	AT-MYB66						2.9997
Gorai.012G011700.1	AT-MYB66			-8.0817			4.5847
Gorai.004G269400.1	AT-MYB63			7.3781			
Gorai.002G106300.1	AT-MYB63					-3.4402	-3.0003
Gorai.007G118200.1	AT-MYB61	-1.8896	-1.6247	-1.0407	-1.0764		1.3525
Gorai.001G148500.1	AT-MYB61	1.9420	2.9537			-1.4791	-3.9736
Gorai.004G147600.1	AT-MYB61			-1.4768		1.2469	
Gorai.013G196800.1	AT-MYB60	-3.2670	-3.8229	-3.5643		2.7297	6.4260
Gorai.007G252700.1	AT-MYB60						3.7001
Gorai.013G113000.1	AT-MYB60		7.4636				
Gorai.007G179300.1	ATMYB6	-1.0264	-1.9337	2.6963	2.0599	-1.5731	

Gorai.008G133700.1	AT-MYB55		-1.4871	-1.1369		-1.5187	
Gorai.007G002200.1	AT-MYB52	2.4870		3.4829	2.4420	-1.8997	-10.1030
Gorai.008G293100.1	AT-MYB52	8.8705	8.4636	3.6147			-8.2625
Gorai.009G251700.1	AT-MYB52			7.4935	7.4548	-3.9321	
Gorai.007G260600.1	ATMYB5	1.0038		1.4404			
Gorai.007G350500.1	ATMYB5						-1.4234
Gorai.008G126700.1	ATMYB5		-1.7555	1.2693		-1.1056	3.6709
Gorai.008G288900.1	ATMY-B4R1	2.4870					
Gorai.013G265300.1	AT-MYB48		2.1958			-2.1749	-3.1501
Gorai.003G136400.1	AT-MYB48					-7.5911	-7.5911
Gorai.006G129100.1	AT-MYB46	2.2805	2.6191	-1.8359		-1.0252	-9.7431
Gorai.001G205300.1	AT-MYB43	7.2424	7.2226	4.2072		-5.1647	-9.9758
Gorai.004G064100.1	AT-MYB43		3.3151			-4.3106	-9.1216
Gorai.008G060100.1	AT-MYB43			7.7000	7.9573		
Gorai.004G201900.1	AT-MYB42	8.0309	9.3779			-2.2069	-10.7729
Gorai.008G238800.1	AT-MYB42		8.7632				-7.5911
Gorai.007G074500.1	AT-MYB40			-2.4063	-10.8999	1.8863	-8.2286
Gorai.006G037500.1	AT-MYB40			2.4829		-7.8806	2.6695
Gorai.002G180300.1	ATMYB4	-8.5722	-8.5722				6.6433
Gorai.002G196700.1	ATMYB4	-1.7368				-1.3388	-1.5853
Gorai.007G192900.1	ATMYB4	-1.2735		-2.8398	-1.0319		
Gorai.005G162600.1	ATMYB4	-1.0494	-1.1648		1.0879		
Gorai.008G035700.1	ATMYB4			1.1273			-2.8281
Gorai.006G172800.1	ATMY-B3R-5	1.2906		1.5287			1.4880

Gorai.001G249400.1	ATMY-B3R-3			8.1854			
Gorai.009G301100.3	AT-MYB33	-1.2335	-2.2533			-3.0608	
Gorai.012G022000.1	AT-MYB33					-3.9321	1.9130
Gorai.001G239000.1	AT-MYB30						7.7428
Gorai.004G150800.1	ATMYB3	-7.6180		2.2323			
Gorai.006G192400.1	ATMYB3	-2.7418	-2.3022	-7.2932			9.8803
Gorai.009G288900.1	ATMYB3		1.5237	-1.3384			1.2011
Gorai.004G044000.1	ATMYB3						7.6433
Gorai.011G063900.1	AT-MYB26	-4.1925				-1.9321	1.2425
Gorai.009G142200.1	AT-MYB26	1.1839	1.3067				
Gorai.007G121000.1	ATMYB2	2.4662		4.5241	3.6605		5.1856
Gorai.004G146000.1	ATMYB2	8.4269	9.5445	11.3630			
Gorai.004G145800.1	ATMYB2	9.4834	11.5641	7.2401	5.0804	2.3374	
Gorai.008G277900.1	AT-MYB19	-3.0049	-9.5952				3.0701
Gorai.001G169700.1	AT-MYB17	-2.9779	-9.3459				2.8070
Gorai.004G157600.1	AT-MYB17						7.2954
Gorai.011G122800.1	AT-MYB16	-2.7290	-2.7015			-2.2079	
Gorai.013G088300.1	AT-MYB16						7.7428
Gorai.008G179600.1	AT-MYB16		3.3151				1.3533
Gorai.005G206800.1	AT-MYB15			7.7000			
Gorai.011G173900.1	AT-MYB15			8.6511			
Gorai.001G177100.1	AT-MYB15		-1.6621	4.8419	4.5596		2.9566
Gorai.012G132400.1	AT-MYB15			8.1854			
Gorai.011G021300.2	AT-MYB14	2.3874		1.5760			

Gorai.006G060300.1	AT-MYB116		-8.3728				9.1213
Gorai.007G346400.1	AT-MYB105	-8.0103					2.2627
Gorai.003G155400.1	AT-MYB103	2.5658	3.2282	-2.0357			-13.2135
Gorai.004G138300.1	AT-MYB103	4.3473	4.2035				-14.8496
15, 25DPA; 20, 20DPA; 25, 25DPA; T, TM-1. value=log2Ratio							

Table S6. Different expression level of 68 NAC transcription factors.

Gene NO.	**Annotation from Tail10**	**35431-15/T15**	**31010-15/T15**	**35431-20/T20**	**31010-20/T20**	**35431-25/T25**	**31010-25/T25**
Gorai.003G073300.1	ATNAC9-like	1.0776				-1.2476	1.2798
Gorai.008G261400.1	ATNAC9-like	1.9460	1.2992	1.9988	1.5716		2.6674
Gorai.001G150000.1	ATNAC9-like	2.2525		3.4187			
Gorai.007G017500.1	ATNAC9-like	2.8127		3.1723	2.7059		
Gorai.004G186700.1	ATNAC96			2.5760	-1.3683	-2.8446	-2.8197
Gorai.009G399900.1	ATNAC90						3.0872
Gorai.009G083500.1	ATNAC89			7.4935			
Gorai.007G043900.1	ATNAC83						2.5847
Gorai.005G195400.1	ATNAC83			8.3167			8.8803
Gorai.005G195300.1	ATNAC83			4.2373	2.6914		3.8410
Gorai.011G087700.1	ATNAC83		-3.3783			2.3671	6.6756
Gorai.001G274500.1	ATNAC83			8.3781			
Gorai.012G083700.1	ATNAC83			10.4935			
Gorai.012G125500.1	ATNAC83		-1.7173	3.7096	3.2731		4.7301
Gorai.007G065300.1	ATNAC82		-1.6621				-1.5325
Gorai.009G214000.2	ATNAC76		-1.5069	1.7248			1.5729
Gorai.009G028900.1	ATNAC75	5.9979	4.4671	2.0932	1.6210	-2.5141	-5.9487
Gorai.010G078900.1	ATNAC75			7.3781			
Gorai.004G221300.1	ATNAC74	-2.3851	-8.9755				8.2954
Gorai.007G147300.1	ATNAC74		-8.2618				11.6939

Gorai.001G178800.1	ATNAC74						11.9857
Gorai.008G155200.1	ATNAC73	11.1317	11.6640	1.5183	1.0742	-3.5418	-13.7105
Gorai.010G157600.2	ATNAC73					-2.0794	-6.4238
Gorai.008G227800.1	ATNAC71	1.6741	2.1452	2.0684			
Gorai.003G180600.1	ATNAC71			7.3781			
Gorai.004G149000.1	ATNAC69					-8.0457	
Gorai.006G203800.1	ATNAC62	4.3676	3.3615	5.2406	3.8203		-1.5792
Gorai.001G122800.1	ATNAC61	4.4505	2.8457	2.9228	1.8016		1.5205
Gorai.006G113000.1	ATNAC61	11.0959	10.0486	13.1897	11.3045	2.4984	
Gorai.006G060900.1	ATNAC58			5.5036	3.5804		
Gorai.010G012000.3	ATNAC53	3.7011	1.9526	1.8634			1.2292
Gorai.011G234200.1	ATNAC52	1.4546	1.4127	1.6244			
Gorai.009G309300.1	ATNAC52			1.8073	1.2991		
Gorai.012G007000.1	ATNAC52						2.1737
Gorai.007G267900.1	ATNAC47	-1.0674	-3.9708		1.4146	-4.4232	4.2911
Gorai.007G114500.1	ATNAC45		1.5478	2.3635		1.2254	
Gorai.006G034600.1	ATNAC44			2.3378			
Gorai.002G001900.1	ATNAC36	8.2424	7.7632	4.7142			
Gorai.005G257800.1	ATNAC36			4.6963	2.6990		
Gorai.008G038800.1	ATNAC36			3.9974	2.3162		-2.0998
Gorai.009G204700.1	ATNAC28	2.0263		3.5552	2.6851		4.5543
Gorai.009G309000.1	ATNAC2			1.7927			
Gorai.009G260000.1	ATNAC17	1.7885	1.5132	2.1034	1.3369		
Gorai.005G076200.1	ATNAC17			1.5302			1.6497
Gorai.007G038100.1	ATNAC100						-2.1158
Gorai.007G079900.1	ATNAC100		-2.6921		2.5172		6.1738
Gorai.008G194400.1	ATNAC100		-1.7059	2.4923		-3.7519	
Gorai.009G166300.1	ATNAC10	1.4507	1.5550				-7.0997
Gorai.009G354900.1	ATNAC1		-3.1027			-3.4346	-4.4097
Gorai.013G167100.1	ATNAC2	1.1408	1.1452	3.2667		-3.0493	-3.4769
Gorai.009G433100.1	ATNAC2	3.1110	2.3308	2.9001	2.1553	-3.2263	-5.1823
Gorai.002G073800.1	ATNAC2					-1.1030	

Gorai.005G013300.2	ATNAC14			3.9891		4.3671	
Gorai.003G114100.1	ATNAC7		3.2495	-3.6257			-8.3598
Gorai.003G077700.1	ATNAC43	1.4651	2.1013	-2.5183		1.5167	-2.2509
Gorai.008G259700.1	ATNAC43	2.0395	2.9696		1.7098		-9.6176
Gorai.002G073700.1	ATNAC72	2.9464		6.1881	3.5596		8.4798
Gorai.012G037600.1	ATNAC81	3.8558	4.1055	4.0000	2.8711	1.0013	2.4927
Gorai.007G112500.1	ATNAC43	5.5459	6.2397			-2.1291	-12.1100
Gorai.008G130300.1	ATNAC43	8.8274	10.5181		2.1829	-2.5727	-10.0842
Gorai.004G129200.1	ATNAC43	9.8054	7.2226			-2.5295	-10.8641
Gorai.010G051900.1	ATNAC2			8.4370			4.5233
Gorai.009G433200.1	ATNAC72		-3.8288	4.1131			5.6435
Gorai.009G186000.1	ATNAC31						9.2622
Gorai.002G113300.1	ATNAC31			8.3781			
Gorai.005G088800.1	ATNAC2			10.4513			
Gorai.008G186300.2	ATNAC8						1.8332
Gorai.001G150800.1	ATNAC43			2.5838			
15, 25DPA; 20, 20DPA; 25, 25DPA; T, TM-1. value=log2Ratio							

AUTHOR CONTRIBUTIONS

Conceived and designed the experiments: TZ. Performed the experiments: LF RT SW XL PW. Analyzed the data: LF JC. Wrote the paper: LF TZ.

REFERENCES

1. Basara AS, Malik CP (1984) Development of cotton fiber. Inter Rev Cyto 65–113.
2. Haigler TA, Jernstedt JA (1999) Molecular genetics of developing cotton fibers. In: AM Basra (Ed), Cotton Fibers. Hawthorne Press, New York, 231–267.
3. Kim HJ, Triplett BA (2001) Cotton fiber growth in planta and in vitro. Models for plant cell elongation and cell wall biogenesis. Plant Physiol 127: 1361–1366. doi: 10.1104/pp.010724

4. Lee JJ, Hassan OS, Gao W, Wei NE, Kohel RJ, et al. (2006) Developmental and gene expression analyses of a cotton naked seed mutant. Planta 223: 418–432. doi: 10.1007/s00425-005-0098-7

5. Lee JJ, Woodward AW, Chen ZJ (2007) Gene expression changes and early events in cotton fibre development. Ann Bot 100: 1391–1401.

6. Wilkins TA, Arpat AB (2005) The cotton fiber transcriptome. Physiol Plant 124: 295–300. doi: 10.1111/j.1399-3054.2005.00514.x

7. Meinert MC, Delmer DP (1977) Changes in biochemical composition of the cell wall of the cotton fiber during development. Plant Physiol 59: 1088–1097. doi: 10.1104/pp.59.6.1088

8. Bolton JJ, Soliman KM, Wilkins TA, Jenkins JN (2009) Aberrant Expression of Critical Genes during Secondary Cell Wall Biogenesis in a Cotton Mutant, Ligon Lintless-1 (Li-1). Comp Funct Genom 659301. doi: 10.1155/2009/659301

9. Richmond TA, Somerville CR (2000) The cellulose synthase superfamily. Plant Physiol 124: 495–498. doi: 10.1104/pp.124.2.495

10. Tanaka K, Murata K, Yamazaki M, Onosato K, Miyao A, et al. (2003) Three distinct rice cellulose synthase catalytic subunit genes required for cellulose synthesis in the secondary wall. Plant Physiol 133: 73–83. doi: 10.1104/pp.103.022442

11. Paterson AH, Wendel JF, Gundlach H, Guo H, Jenkins J, et al. (2012) Repeated polyploidization of Gossypium genomes and the evolution of spinnable cotton fibres. Nature 492: 423–427. doi: 10.1038/nature11798

12. Persson S, Wei H, Milne J, Page GP, Somerville CR (2005) Identification of genes required for cellulose synthesis by regression analysis of public microarray data sets. Proc Natl Acad Sci USA 102: 8633–8638. doi: 10.1073/pnas.0503392102

13. Ruan YL, Chourey PS, Delmer DP, Perez-Grau L (1997) The Differential Expression of Sucrose Synthase in Relation to Diverse Patterns of Carbon Partitioning in Developing Cotton Seed. Plant Physiol 115: 375–385.

14. Brill E, van Thournout M, White RG, Llewellyn D, Campbell PM, et al. (2011) A novel isoform of sucrose synthase is targeted to the cell wall during secondary cell wall synthesis in cotton fiber. Plant Physiol 157: 40–54. doi: 10.1104/pp.111.178574

15. Potikha TS, Collins CC, Johnson DI, Delmer DP, Levine A (1999) The involvement of hydrogen peroxide in the differentiation of secondary walls in cotton fibers. Plant Physiol 119: 849–858. doi: 10.1104/pp.119.3.849

16. Yang YM, Xu CN, Wang BM, Jia JZ (2001) Effects of plant growth regulators on secondary wall thickening of cotton fibres. Plant Growth Regul 35: 233–237.

17. Arpat AB, Waugh M, Sullivan JP, Gonzales M, Frisch D, et al. (2004) Functional genomics of cell elongation in developing cotton fibers. Plant Mol Biol 54: 911–929. doi: 10.1007/s11103-004-0392-y

18. Wang QQ, Liu F, Chen XS, Ma XJ, Zeng HQ, et al. (2010) Transcriptome profiling of early developing cotton fiber by deep-sequencing reveals significantly differential expression of genes in a fuzzless/lintless mutant. Genomics 96: 369–376. doi: 10.1016/j.ygeno.2010.08.009

19. Padmalatha KV, Dhandapani G, Kanakachari M, Kumar S, Dass A, et al. (2012) Genome-wide transcriptomic analysis of cotton under drought stress reveal significant down-regulation of genes and pathways involved in fibre elongation and up-regulation of defense responsive genes. Plant Mol Biol 78: 223–246. doi: 10.1007/s11103-011-9857-y

20. Chaudhary B, Hovav R, Rapp R, Verma N, Udall JA, et al. (2008) Global analysis of gene expression in cotton fibers from wild and domesticated *Gossypium barbadense*. Evol Dev 10: 567–582. doi: 10.1111/j.1525-142x.2008.00272.x

21. Hovav R, Udall JA, Chaudhary B, Hovav E, Flagel L, et al. (2008) The evolution of spinnable cotton fiber entailed prolonged development and a novel metabolism. PLoS Genet 4: e25. doi: 10.1371/journal.pgen.0040025

22. Hovav R, Udall JA, Hovav E, Rapp R, Flagel L, et al. (2008) A majority of cotton genes are expressed in single-celled fiber. Planta 227: 319–329. doi: 10.1007/s00425-007-0619-7

23. Ji SJ, Lu YC, Feng JX, Wei G, Li J, et al. (2003) Isolation and analyses of genes preferentially expressed during early cotton fiber development by subtractive PCR and cDNA array. Nucleic Acids Res 31: 2534–2543. doi: 10.1093/nar/gkg358

24. Udall JA, Flagel LE, Cheung F, Woodward AW, Hovav R, et al. (2007) Spotted cotton oligonucleotide microarrays for gene expression analysis. BMC Genomics 8: 81. doi: 10.1186/1471-2164-8-81

25. Qin YM, Hu CY, Pang Y, Kastaniotis AJ, Hiltunen JK, et al. (2007) Saturated very-long-chain fatty acids promote cotton fiber and *Arabidopsis* cell elongation by activating ethylene biosynthesis. Plant Cell 19: 3692–3704. doi: 10.1105/tpc.107.054437

26. Shi YH, Zhu SW, Mao XZ, Feng JX, Qin YM, et al. (2006) Transcriptome profiling, molecular biological, and physiological studies reveal a major

role for ethylene in cotton fiber cell elongation. Plant Cell 18: 651–664. doi: 10.1105/tpc.105.040303

27. Gou JY, Wang LJ, Chen SP, Hu WL, Chen XY (2007) Gene expression and metabolite profiles of cotton fiber during cell elongation and secondary cell wall synthesis. Cell Res 17: 422–434. doi: 10.1038/sj.cr.7310150
28. Eshed Y, Zamir D (1995) An introgression line population of *Lycopersicon pennellii*in the cultivated tomato enables the identification and fine mapping of yield-associated QTL. Genetics 141: 1147–1162.
29. Liu S, Zhou R, Dong Y, Li P, Jia J (2006) Development, utilization of introgression lines using a synthetic wheat as donor. Theor Appl Genet 112: 1360–1373. doi: 10.1007/s00122-006-0238-x
30. Takai T, Nonoue Y, Yamamoto SI, Yamanouchi U, Matsubara K, et al. (2007) Development of chromosome segment substitution lines derived from backcross between indica donor rice cultivar '*Nona bokra*' and japonica recipient cultivar '*Koshihikari*'. Breeding Sci 57: 257–261. doi: 10.1270/jsbbs.57.257
31. Wang P, Ding YZ, Lu QX, Guo WZ, Zhang TZ (2008) Development of *Gossypium barbadense* chromosome segment substitution lines in the genetic standard line TM-1 of *Gossypium hirsutum*. Chi Sci Bull 53: 1512–1517. doi: 10.1007/s11434-008-0220-x
32. Kohel R, Richmond T, Lewis C (1970) Texas marker-1. Description of a genetic standard for *Gossypium hirsutum* L. Crop Sci 10: 670–671. doi: 10.2135/cropsci1970.0011183x001000060019x
33. Pan J, Zhang T, Kuai B (1994) Studies on the inheritance of resistance to *Verticillium*dahliae in cotton. J Nanj Agric Univ 17.
34. Yang C, Guo W, Li G, Gao F, Lin S, et al. (2008) QTLs mapping for *Verticillium* wilt resistance at seedling and maturity stages in *Gossypium barbadense* L. Plant Sci 174: 290–298. doi: 10.1016/j.plantsci.2007.11.016
35. Wang P, Zhu Y, Song X, Cao Z, Ding Y, et al. (2012) Inheritance of long staple fiber quality traits of *Gossypium barbadense* in *G. hirsutum* background using CSILs. Theor Appl Genet 124: 1415–1428. doi: 10.1007/s00122-012-1797-7
36. Jiang JX, Zhang TZ (2003) Extraction of total RNA in cotton tissues with CTAB-acidic phenolic method. Cott Sci 15: 166–167.
37. Benjamini Y, Yekutieli D (2001) The control of the false discovery rate in multiple testing under dependency. Ann Stat 1165–1188. doi: 10.1214/aos/1013699998

38. Herrero J, Valencia A, Dopazo J (2001) A hierarchical unsupervised growing neural network for clustering gene expression patterns. Bioinformatics 17 (2) 126–136. doi: 10.1093/bioinformatics/17.2.126
39. Kanehisa M, Araki M, Goto S, Hattori M, Hirakawa M, et al. (2008) KEGG for linking genomes to life and the environment. Nucleic Acids Res 36: D480–D484. doi: 10.1093/nar/gkm882
40. Taylor NG, Howells RM, Huttly AK, Vickers K, Turner SR (2003) Interactions among three distinct CesA proteins essential for cellulose synthesis. Proc Natl Acad Sci USA 100: 1450–1455. doi: 10.1073/pnas.0337628100
41. Taylor NG, Laurie S, Turner SR (2000) Multiple cellulose synthase catalytic subunits are required for cellulose synthesis in *Arabidopsis*. Plant Cell 12: 2529–2540. doi: 10.1105/tpc.12.12.2529
42. Taylor NG, Scheible WR, Cutler S, Somerville CR, Turner SR (1999) The irregular xylem3 locus of Arabidopsis encodes a cellulose synthase required for secondary cell wall synthesis. Plant Cell 11: 769–780. doi: 10.2307/3870813
43. Olsen AN, Ernst HA, Leggio LL, Skriver K (2005) NAC transcription factors: structurally distinct, functionally diverse. Trends Plant Sci 10: 79–87. doi: 10.1016/j.tplants.2004.12.010
44. Zhong R, Lee C, Ye ZH (2010) Functional characterization of poplar wood-associated NAC domain transcription factors. Plant Physiol 152: 1044–1055. doi: 10.1104/pp.109.148270
45. Zhong R, Lee C, Zhou J, McCarthy RL, Ye ZH (2008) A battery of transcription factors involved in the regulation of secondary cell wall biosynthesis in *Arabidopsis*. Plant Cell 20: 2763–2782. doi: 10.1105/tpc.108.061325
46. Delmer DP, Amor Y (1995) Cellulose biosynthesis. Plant Cell 7: 987–1000. doi: 10.1105/tpc.7.7.987
47. Delmer DP, Haigler CH (2002) The regulation of metabolic flux to cellulose, a major sink for carbon in plants. Metab Eng 4: 22–28. doi: 10.1006/mben.2001.0206
48. Appenzeller L, Doblin M, Barreiro R, Wang HY, Niu XM, et al. (2004) Cellulose synthesis in maize: isolation and expression analysis of the cellulose synthase (CesA) gene family. Cellulose 11: 287–299. doi: 10.1023/b:cell.0000046417.84715.27
49. Song DL, Shen JH, Li LG (2010) Characterization of cellulose synthase complexes in *Populus* xylem differentiation. New Phytol 187: 777–790. doi: 10.1111/j.1469-8137.2010.03315.x

50. Amor Y, Haigler CH, Johnson S, Wainscott M, Delmer DP (1995) A membrane-associated form of sucrose synthase and its potential role in synthesis of cellulose and callose in plants. Proc Natl Acad Sci USA 92: 9353–9357. doi: 10.1073/pnas.92.20.9353

51. Hinchliffe DJ, Meredith WR, Yeater KM, Kim HJ, Woodward AW, et al. (2010) Near-isogenic cotton germplasm lines that differ in fiber-bundle strength have temporal differences in fiber gene expression patterns as revealed by comparative high-throughput profiling. Theor Appl Genet 120: 1347–1366. doi: 10.1007/s00122-010-1260-6

52. Zhong R, Lee C, Ye ZH (2010) Evolutionary conservation of the transcriptional network regulating secondary cell wall biosynthesis. Trends Plant Sci 15: 625–632. doi: 10.1016/j.tplants.2010.08.007

53. Hussey SG, Mizrachi E, Spokevicius AV, Bossinger G, Berger DK, et al. (2011) SND2, a NAC transcription factor gene, regulates genes involved in secondary cell wall development in *Arabidopsis* fibres and increases fibre cell area in Eucalyptus. BMC Plant Biol 11: 173. doi: 10.1186/1471-2229-11-173

54. Bhargava A, Ahad A, Wang S, Mansfield SD, Haughn GW, et al. (2013) The interacting MYB75 and KNAT7 transcription factors modulate secondary cell wall deposition both in stems and seed coat in *Arabidopsis*. Planta 237: 1199–1211. doi: 10.1007/s00425-012-1821-9

55. Zhong R, Lee C, McCarthy RL, Reeves CK, Jones EG, et al. (2011) Transcriptional activation of secondary wall biosynthesis by rice and maize NAC and MYB transcription factors. Plant Cell Physiol 52: 1856–1871. doi: 10.1093/pcp/pcr123

56. Brown DM, Zeef LA, Ellis J, Goodacre R, Turner SR (2005) Identification of novel genes in *Arabidopsis* involved in secondary cell wall formation using expression profiling and reverse genetics. Plant Cell 17: 2281–2295. doi: 10.1105/tpc.105.031542

57. Al-Ghazi Y, Bourot S, Arioli T, Dennis ES, Llewellyn DJ (2009) Transcript profiling during fiber development identifies pathways in secondary metabolism and cell wall structure that may contribute to cotton fiber quality. Plant Cell Physiol 50: 1364–1381. doi: 10.1093/pcp/pcp084

CITATION

CHAPTER 1

Teresa Cuellar, Eric Belhassen, Begoña Fernández-Calvín, Juan Orellana, and Jose L Bella (1996). Chromosomal differentiation in *Helianthus annuus* var.*macrocarpus*: heterochromatin characterization and rDNA location, *Heredity* (1996) 76, 586–591; doi:10.1038/hdy.1996.84

CHAPTER 3

Nagaki K, Tanaka K, Yamaji N, Kobayashi H and Murata M (2015). Sunflower centromeres consist of a centromere-specific LINE and a chromosome-specific tandem repeat. *Front. Plant Sci.* 6:912. doi: 10.3389/fpls.2015.00912

CHAPTER 4

Zamariola L, Tiang CL, De Storme N, Pawlowski W and Geelen D (2014) Chromosome segregation in plant meiosis. *Front. Plant Sci.* **5**:279. doi: 10.3389/fpls.2014.00279

CHAPTER 5

Cristina Maria Pinto de Paula and Vânia Helena Techio (2014). Immunolocalization of chromosome-associated proteins in plants – principles and applications. Botanical Studies 2014 55:63. Doi:10.1186/s40529-014-0063-5

CHAPTER 6

Katsuyuki Ichitani, Satoru Taura, Takahiro Tezuka, Yuuya Okiyama, and Tsutomu Kuboyama (2011). Chromosomal Location of *HWA1* and *HWA2*, Complementary Hybrid Weakness Genes in Rice, Rice (2011) 4: 29–38. DOI 10.1007/s12284-011-9062-2

CHAPTER 7

Koo D-H, Sehgal SK, Friebe B, Gill BS (2015) Structure and Stability of Telocentric Chromosomes in Wheat. PLoS ONE 10(9): e0137747. doi:10.1371/journal.pone.0137747

CHAPTER 8

Genc Y, Taylor J, Rongala J, Oldach K (2014) A Major Locus for Chloride Accumulation on Chromosome 5A in Bread Wheat. PLoS ONE 9(6): e98845. doi:10.1371/journal.pone.0098845

CHAPTER 9

Fang L, Tian R, Chen J, Wang S, Li X, Wang P, et al. (2014) Transcriptomic Analysis of Fiber Strength in Upland Cotton Chromosome Introgression Lines Carrying Different *Gossypium barbadense* Chromosomal Segments. PLoS ONE 9(4): e94642. doi:10.1371/journal.pone.0094642

INDEX

Encyclopaedia of GENETIC RESOURCES, CHROMOSOME ENGINEERING, AND CROP IMPROVEMENT

Encyclopaedia of GENETIC RESOURCES, CHROMOSOME ENGINEERING, AND CROP IMPROVEMENT

Volume II: Fungal Pathogenesis in Plants and Crops

Contributors

Teresa Cuellar,Chia-Lin Chung,and Luciano Lourenço Nass et al.

AURIS
Reference

www.aurisreference.com

Encyclopaedia of Genetic Resources, Chromosome Engineering, and Crop Improvement

Volume II: Fungal Pathogenesis in Plants and Crops

Contributors: Teresa Cuellar,Chia-Lin Chung,and Luciano Lourenço Nass et al.

Published by Auris Reference Limited

www.aurisreference.com

United Kingdom

Printed in 2017 for Sale in the Indian Subcontinent

Fungal Pathogenesis in Plants and Crops

ISBN: 978-1-78154-957-5 (Set of 3 Volumes)

British Library Cataloguing in Publication Data
A CIP record for this book is available from the British Library

Printed in the United Kingdom
Exclusively distributed by CBS Publishers & Distributors Pvt. Ltd.

Contents

List of Abbreviations

AUDPC	Area under the Disease Progress Curve
AWF	Apoplastic Washing Fluid
BCA	Biological Control Agent
CWDE	Cell Wall Degrading Enzymes
DD	Differential Display
DLA	Diseased Leaf Area
DTA	Days to Anthesis
EST	Expressed Sequence Tags
FWER	Family Wise Error Rate
GO	Gene Ontology
GS	Growth Stages
HILIC	Hydrophilic Interaction Liquid Chromatography
IP	Incubation Period
ISR	Induced Systemic Resistance
JA	Jasmonic Acid
LC	Liquid Chromatography
LE	Lesion Expansion
LSD	Least Significant Difference
LSP	Leaderless Secreted Proteins
MAPU	Max Planck Unified Database
MDH	Malate Dehydrogenase
MS	Mass Spectrometry
NIL	Near Isogenic Lines
NLB	Northern Leaf Blight
OPD	Open Proteomic Database
PCA	Principle Component Analysis
PCER	Percomparison Error Rate
PDA	Potato Dextrose Agar
PGPR	Plant Growth Promoting Rhizobacteria
PRIDE	Proteomics Identifications Database
QTL	Quantitative Trait Loci
RFLP	Restriction Fragment Length Polymorphism
ROS	Reactive Oxygen Species
ROS	Reactive Oxygen Species
SA	Salicylic Acid
SAGE	Serial Analyze of Gene Expression
SAR	Systemic Acquired Resistance
SAXS	Small Angle X-ray Scattering
SSR	Simple Sequence Repeat
TCA	Trichloroacetic Acid
TE	Transposable Elements
TILLING	Targeting Induced Local Lesions in Genomes
YRCPDR	Yeast Resource Center Public Data Repository

List of Contributors

Chia-Lin Chung
Dept. of Plant Pathology and Plant-Microbe Biology, Cornell University, Ithaca, NY 14853, USA

Joy M Longfellow
Dept. of Plant Pathology and Plant-Microbe Biology, Cornell University, Ithaca, NY 14853, USA

Ellie K Walsh
Dept. of Plant Pathology and Plant-Microbe Biology, Cornell University, Ithaca, NY 14853, USA

Zura Kerdieh
Dept. of Biology, West Virginia State University, Institute, WV 25112, USA

George Van Esbroeck
Dept. of Crop Science, North Carolina State University, Raleigh, NC 27695, USA

Peter Balint-Kurti
USDA-ARS, Plant Science Research Unit; Dept. of Plant Pathology, North Carolina State University, Raleigh, NC 27695, USA

Rebecca J Nelson
Dept. of Plant Pathology and Plant-Microbe Biology, Cornell University, Ithaca, NY 14853, USA
Dept. of Plant Breeding and Genetics, Cornell University, Ithaca, NY 14853, USA

Raquel González-Fernández
Agricultural and Plant Biochemistry and Proteomics Research Group, Department of Biochemistry and Molecular Biology, University of Córdoba, 14071 Córdoba, Spain

Elena Prats
CSIC, Institute of Sustainable Agriculture, 14080 Córdoba, Spain

Jesús V. Jorrín-Novo
1Agricultural and Plant Biochemistry and Proteomics Research Group, Department of Biochemistry and Molecular Biology, University of Córdoba, 14071 Córdoba, Spain

Richard J O'Connell
Department of Plant Microbe Interactions, Max Planck Institute for Plant Breeding Research, Cologne, Germany
These authors contributed equally to this work. Correspondence should be addressed to R.J.O.

Michael R Thon
Centro Hispano-Luso de Investigaciones Agrarias, Departamento de Microbiología y Genética, Universidad de Salamanca, Villamayor, Spain
These authors contributed equally to this work. Correspondence should be addressed to R.J.O.

Stéphane Hacquard
Department of Plant Microbe Interactions, Max Planck Institute for Plant Breeding Research, Cologne, Germany

Stefan G Amyotte
Department of Plant Pathology, University of Kentucky, Lexington, Kentucky, USA

Jochen Kleemann
Department of Plant Microbe Interactions, Max Planck Institute for Plant Breeding Research, Cologne, Germany

Maria F Torres
Department of Plant Pathology, University of Kentucky, Lexington, Kentucky, USA

Ulrike Damm
Centraalbureau voor Schimmelcultures, Koninklijke Nederlandse Akademie van Wetenschappen, Fungal Biodiversity Centre, Utrecht, The Netherlands

Ester A Buiate
Department of Plant Pathology, University of Kentucky, Lexington, Kentucky, USA

Lynn Epstein
Department of Plant Pathology, University of California, Davis, California, USA

Noam Alkan
Department of Postharvest Science of Fresh Produce, Agricultural Research Organization, The Volcani Center, Bet Dagan, Israel

Janine Altmüller
Cologne Center for Genomics, University of Cologne, Cologne, Germany

Lucia Alvarado-Balderrama
Broad Institute, Cambridge, Massachusetts, USA

Christopher A Bauser
GATC Biotech AG, Konstanz, Germany

Christian Becker
Cologne Center for Genomics, University of Cologne, Cologne, Germany

Bruce W Birren
Broad Institute, Cambridge, Massachusetts, USA

Zehua Chen
Broad Institute, Cambridge, Massachusetts, USA

Jaeyoung Choi
Department of Agricultural Biotechnology, Center for Fungal Genetic Resources, Seoul National University, Seoul, Korea

Jo Anne Crouch
Systematic Mycology and Microbiology Laboratory, US Department of Agriculture, Agricultural Research Service, Beltsville, Maryland, USA

Jonathan P Duvick
Pioneer Hi-Bred International, DuPont Agricultural Biotechnology, Wilmington, Delaware, USA
Department of Genetics, Development and Cell Biology, Iowa State University, Ames, Iowa, USA (J.P.D.), Department of Bioproduction Science, Faculty of Bioresources and Environmental Sciences, Ishikawa Prefectural University, Ishikawa, Japan (H.T.), Department of Plant Pathology, Swammerdam Institute for Life Sciences, University of Amsterdam, Amsterdam, The Netherlands (H.C.v.d.D.) and The College of Natural Sciences, University of Massachu-

setts Amherst, Amherst, Massachusetts, USA (L.-J.M.)
Frances Trail28

Mark A Farman
Department of Plant Pathology, University of Kentucky, Lexington, Kentucky, USA

Pamela Gan
Plant Immunity Research Group, RIKEN Plant Science Center, Yokohama, Japan

David Heiman
Broad Institute, Cambridge, Massachusetts, USA

Bernard Henrissat
Laboratoire Architecture et Fonction des Macromolécules Biologiques, Centre National de la Recherche Scientifique, Unité Mixte de Recherche 7257, Université Aix-Marseille, Marseille, France

Richard J Howard
Pioneer Hi-Bred International, DuPont Agricultural Biotechnology, Wilmington, Delaware, USA

Mehdi Kabbage
Department of Plant Pathology and Microbiology, Institute for Plant Genomics and Biotechnology, Borlaug Genomics and Bioinformatics Center, Texas A&M University, College Station, Texas, USA

Christian Koch
Department of Biology, Division of Biochemistry, Friedrich-Alexander-University ErlangenNuremberg, Erlangen, Germany

Barbara Kracher
Department of Plant Microbe Interactions, Max Planck Institute for Plant Breeding Research, Cologne, Germany

Yasuyuki Kubo
Laboratory of Plant Pathology, Graduate School of Life and Environmental Sciences, Kyoto Prefectural University, Kyoto, Japan

Audrey D Law
Department of Plant Pathology, University of Kentucky, Lexington, Kentucky, USA

Marc-Henri Lebrun
Institut National de la Recherche Agronomique, Biologie et Gestion des Risques en Agriculture—Champignons Pathogènes des Plantes, Thiverval-Grignon, France

Yong-Hwan Lee
Department of Agricultural Biotechnology, Center for Fungal Genetic Resources, Seoul National University, Seoul, Korea

Itay Miyara
Department of Postharvest Science of Fresh Produce, Agricultural Research Organization, The Volcani Center, Bet Dagan, Israel

Neil Moore
Department of Computer Science, University of Kentucky, Lexington, Kentucky, USA

Ulla Neumann
Central Microscopy, Max Planck Institute for Plant Breeding Research, Cologne, Germany

Karl Nordström
Department of Plant Developmental Biology, Max Planck Institute for Plant Breeding Research, Cologne, Germany

Daniel G Panaccione
Division of Plant and Soil Sciences, West Virginia University, Morgantown, West Virginia, USA

Ralph Panstruga
Department of Plant Microbe Interactions, Max Planck Institute for Plant Breeding Research, Cologne, Germany
Institute for Biology I, Unit of Plant Molecular Cell Biology, Rheinisch-Westfälische Technische Hochschule Aachen University, Aachen, Germany

Michael Place
Laboratory for Molecular and Computational Genomics, University of Wisconsin-Madison, Madison, Wisconsin, USA

Robert H Proctor
US Department of Agriculture, Agriculture Research Service, National Center for Agricultural Utilization Research, Peoria, Illinois, USA

Dov Prusky
Department of Postharvest Science of Fresh Produce, Agricultural Research Organization, The Volcani Center, Bet Dagan, Israel

Gabriel Rech
Centro Hispano-Luso de Investigaciones Agrarias, Departamento de Microbiología y Genética, Universidad de Salamanca, Villamayor, Spain

Richard Reinhardt
Max Planck Genome Centre Cologne, Cologne, Germany

Jeffrey A Rollins
Department of Plant Pathology, University of Florida, Gainesville, Florida, USA

Steve Rounsley
Broad Institute, Cambridge, Massachusetts, USA

Christopher L Schard
Plant Immunity Research Group, RIKEN Plant Science Center, Yokohama, Japan

David C Schwartz
Laboratory for Molecular and Computational Genomics, University of Wisconsin-Madison, Madison, Wisconsin, USA

Narmada Shenoy
Broad Institute, Cambridge, Massachusetts, USA

Ken Shirasu
Plant Immunity Research Group, RIKEN Plant Science Center, Yokohama, Japan

Usha R Sikhakolli
Department of Plant Biology, Michigan State University, East Lansing, Michigan, USA

Kurt Stüber
Max Planck Genome Centre Cologne, Cologne, Germany

Serenella A Sukno
Centro Hispano-Luso de Investigaciones Agrarias, Departamento de Microbiología y Genética, Universidad de Salamanca, Villamayor, Spain

James A Sweigard
Pioneer Hi-Bred International, DuPont Agricultural Biotechnology, Wilmington, Delaware, USA

Yoshitaka Takano
Laboratory of Plant Pathology, Graduate School of Agriculture, Kyoto University, Kyoto, Japan

Hiroyuki Takahara
Department of Plant Microbe Interactions, Max Planck Institute for Plant Breeding Research, Cologne, Germany
Department of Genetics, Development and Cell Biology, Iowa State University, Ames, Iowa, USA (J.P.D.), Department of Bioproduction Science, Faculty of Bioresources and Environmental Sciences, Ishikawa Prefectural University, Ishikawa, Japan (H.T.), Department of Plant Pathology, Swammerdam Institute for Life Sciences, University of Amsterdam, Amsterdam, The Netherlands (H.C.v.d.D.) and The College of Natural Sciences, University of Massachusetts Amherst, Amherst, Massachusetts, USA (L.-J.M.)
Frances Trail28

H Charlotte van der Does
Department of Plant Microbe Interactions, Max Planck Institute for Plant Breeding Research, Cologne, Germany
Department of Genetics, Development and Cell Biology, Iowa State University, Ames, Iowa, USA (J.P.D.), Department of Bioproduction Science, Faculty of Bioresources and Environmental Sciences, Ishikawa Prefectural University, Ishikawa, Japan (H.T.), Department of Plant Pathology, Swammerdam Institute for Life Sciences, University of Amsterdam, Amsterdam, The Netherlands (H.C.v.d.D.) and The College of Natural Sciences, University of Massachusetts Amherst, Amherst, Massachusetts, USA (L.-J.M.)

Lars M Voll
Department of Biology, Division of Biochemistry, Friedrich-Alexander-University ErlangenNuremberg, Erlangen, Germany

Isa Will
Department of Plant Microbe Interactions, Max Planck Institute for Plant Breeding Research, Cologne, Germany

Sarah Young
Broad Institute, Cambridge, Massachusetts, USA

Qiandong Zeng
Broad Institute, Cambridge, Massachusetts, USA

Jingze Zhang
Broad Institute, Cambridge, Massachusetts, USA

Shiguo Zhou
Laboratory for Molecular and Computational Genomics, University of Wisconsin-Madison, Madison, Wisconsin, USA

Martin B Dickman
Department of Plant Pathology and Microbiology, Institute for Plant Genomics and Biotechnology, Borlaug Genomics and Bioinformatics Center, Texas A&M University, College Station, Texas, USA

Paul Schulze-Lefert
Department of Plant Microbe Interactions, Max Planck Institute for Plant Breeding Research, Cologne, Germany

Emiel Ver Loren van Themaat
Department of Plant Microbe Interactions, Max Planck Institute for Plant Breeding Research, Cologne, Germany

Li-Jun Ma
Broad Institute, Cambridge, Massachusetts, USA
Department of Genetics, Development and Cell Biology, Iowa State University, Ames, Iowa, USA (J.P.D.), Department of Bioproduction Science, Faculty of Bioresources and Environmental Sciences, Ishikawa Prefectural University, Ishikawa, Japan (H.T.), Department of Plant Pathology, Swammerdam Institute for Life Sciences, University of Amsterdam, Amsterdam, The Netherlands (H.C.v.d.D.) and The College of Natural Sciences, University of Massachusetts Amherst, Amherst, Massachusetts, USA (L.-J.M.)

Lisa J Vaillancourt
Department of Plant Pathology, University of Kentucky, Lexington, Kentucky, USA

Dale R. Walters
Crop and Soil Systems Research Group, Scotland's Rural College, Edinburgh, UK

Neil D. Havis
Crop and Soil Systems Research Group, Scotland's Rural College, Edinburgh,

UK

Linda Paterson
Crop and Soil Systems Research Group, Scotland's Rural College, Edinburgh, UK

Jeanette Taylor
Crop and Soil Systems Research Group, Scotland's Rural College, Edinburgh, UK

David J. Walsh
Engineering, Science and Technology Department, Scotland's Rural College, Edinburgh, UK

Cecile Sablou
Crop and Soil Systems Research Group, Scotland's Rural College, Edinburgh, UK

Asghar Heydari
Department of Plant Pathology, Iranian Research Institute of Plant Protection, Tehran, Iran

Mohammad Pessarakli
Department of Plant Sciences, University of Arizona, Tucson, AZ, 85721, USA

Shaowu Meng
Fungal Genomics Laboratory, Center for Integrated Fungal Research, North Carolina State University, Raleigh, NC 27695, USA
Hayes Laboratory, Lineberger Comprehensive Cancer Center, School of Medicine, CB# 7295, University of North Carolina at Chapel Hill, Chapel Hill, NC 27599-7295, USA

Trudy Torto-Alalibo
Virginia Bioinformatics Institute, Virginia Polytechnic and State University, Blacksburg, VA 24061, USA

Marcus C Chibucos
Virginia Bioinformatics Institute, Virginia Polytechnic and State University, Blacksburg, VA 24061, USA
Institute for Genome Sciences, University of Maryland School of Medicine, Baltimore, MD 21201, USA

Brett M Tyler

Virginia Bioinformatics Institute, Virginia Polytechnic and State University, Blacksburg, VA 24061, USA

Ralph A Dean
Fungal Genomics Laboratory, Center for Integrated Fungal Research, North Carolina State University, Raleigh, NC 27695, USA

Raviraj M. Kalunke
Dipartimento di Scienze e Tecnologie per l'Agricoltura, le Foreste, la Natura e l'Energia, Università della Tuscia, Viterbo, Italy

Silvio Tundo
Dipartimento di Scienze e Tecnologie per l'Agricoltura, le Foreste, la Natura e l'Energia, Università della Tuscia, Viterbo, Italy

Manuel Benedetti
Dipartimento di Biologia e Biotecnologie "Charles Darwin", Sapienza Università di Roma, Roma, Italy

Felice Cervone
Dipartimento di Biologia e Biotecnologie "Charles Darwin", Sapienza Università di Roma, Roma, Italy

Giulia De Lorenzo
Dipartimento di Biologia e Biotecnologie "Charles Darwin", Sapienza Università di Roma, Roma, Italy

Renato D'Ovidio
Dipartimento di Scienze e Tecnologie per l'Agricoltura, le Foreste, la Natura e l'Energia, Università della Tuscia, Viterbo, Italy

Fen Yang
Department of Plant and Environmental Sciences, University of Copenhagen, 1871 Frederiksberg C, Denmark

Wanshun Li
BGI-tech, BGI, 518083 Shenzhen, China

Mark Derbyshire
Department of Plant Biology and Crop Science, Rothamsted Research, Harpenden, Hertfordshire AL5 2JQ, United Kingdom

Martin R Larsen

Department of Biochemistry and Molecular Biology, University of Southern Denmark, 5230 Odense M, Denmark

Jason J Rudd
Department of Plant Biology and Crop Science, Rothamsted Research, Harpenden, Hertfordshire AL5 2JQ, United Kingdom

Giuseppe Palmisano
Department of Biochemistry and Molecular Biology, University of Southern Denmark, 5230 Odense M, Denmark
Institute of Biomedical Science, Department of Parasitology, University of São Paulo, 05508-900 São Paulo, Brazil.

Preface

Encyclopaedia of Genetic Resources, Chromosome Engineering, and Crop Improvement summarizes landmark research and describes medicinal plants as nature's pharmacy. Genetic resources (GRs) refer to genetic material of actual or potential value. Genetic material is any material of plant, animal, microbial or other origin containing functional units of heredity. Fungal pathogenesis is the process by which fungi infect and cause disease in a host. From the molecular basis of host defense mechanisms and molecular events leading to the suppression of defense mechanisms by fungal pathogens to fungal infection processes, the text Fungal Pathogenesis in Plants and Crops covers various aspects of molecular plant pathology. First chapter discusses the use of introgression lines for quantitative trait loci (QTL) mapping and characterization in the maize. Second chapter deals with proteomics of plant pathogenic. Lifestyle transitions in plant pathogenic colletotrichum fungi deciphered by genome and transcriptome analysis have been focused in third chapter. In fourth chapter, we report the results of field experiments over 3 consecutive years, undertaken to determine the potential for use of an elicitor combination to control foliar pathogens of spring barley. The objective of fifth chapter is to present an advanced survey of the nature and practice of biological control as it is applied to the suppression of plant diseases. In sixth chapter, we summarize common mechanisms of pathogenesis displayed by oomycetes and fungi. An update on polygalacturonase-inhibiting protein (PGIP), a leucine-rich repeat protein that protects crop plants against pathogens has been presented in seventh chapter. Last chapter reveals a similarity in fungal protein profiles between two interactions, possibly due to the fact that analysis of whole inoculated leaves resulted in the dominance of most abundant plant and fungal proteins, which largely diluted the information about low abundant fungal proteins likely essential for pathogenicity.

Chapter 1

RESISTANCE LOCI AFFECTING DISTINCT STAGES OF FUNGAL PATHOGENESIS: USE OF INTROGRESSION LINES FOR QTL MAPPING AND CHARACTERIZATION IN THE MAIZE - SETOSPHAERIA TURCICA PATHOSYSTEM

Chia-Lin Chung[1], Joy M Longfellow[1], Ellie K Walsh[1], Zura Kerdieh[3], George Van Esbroeck[4], Peter Balint-Kurti[5] and Rebecca J Nelson[1,2]

[1]Dept. of Plant Pathology and Plant-Microbe Biology, Cornell University, Ithaca, NY 14853, USA

[2]Dept. of Plant Breeding and Genetics, Cornell University, Ithaca, NY 14853, USA

[3]Dept. of Biology, West Virginia State University, Institute, WV 25112, USA

[4]Dept. of Crop Science, North Carolina State University, Raleigh, NC 27695, USA and

[5]USDA-ARS, Plant Science Research Unit; Dept. of Plant Pathology, North Carolina State University, Raleigh, NC 27695, USA

ABSTRACT

Background

Studies on host-pathogen interactions in a range of pathosystems have revealed an array of mechanisms by which plants reduce the efficiency of pathogenesis. While R-gene mediated resistance confers highly effective defense responses against pathogen invasion, quantitative resistance is associated with intermediate levels of resistance that reduces disease progress. To test the hypothesis that specific loci affect distinct stages of fungal pathogenesis, a set of maize introgression lines was used for mapping and characterization of quantitative trait loci (QTL) conditioning resistance to *Setosphaeria turcica*, the causal agent of northern leaf blight (NLB). To better understand the nature of quantitative resistance, the identified QTL were further tested for three secondary hypotheses: (1) that disease QTL differ by host developmental stage; (2) that their performance changes across environments; and (3) that they condition broad-spectrum resistance.

Results

Among a set of 82 introgression lines, seven lines were confirmed as more resistant or susceptible than B73. Two NLB QTL were validated in BC_4F_2 segregating populations and advanced introgression lines. These loci, designated *qNLB1.02* and*qNLB1.06*, were investigated in detail by comparing the introgression lines with B73 for a series of macroscopic and microscopic disease components targeting different stages of NLB development. Repeated greenhouse and field trials revealed that *qNLB1.06* $_{Tx303}$ (the Tx303 allele at bin 1.06) reduces the efficiency of fungal penetration, while *qNLB1.02* $_{B73}$(the B73 allele at bin 1.02) enhances the accumulation of callose and phenolics surrounding infection sites, reduces hyphal growth into the vascular bundle and impairs the subsequent necrotrophic colonization in the leaves. The QTL were equally effective in both juvenile and adult plants; *qNLB1.06* $_{Tx303}$ showed greater effectiveness in the field than in the greenhouse. In addition to NLB resistance, *qNLB1.02* $_{B73}$ was associated with resistance to Stewart›s wilt and common rust, while*qNLB1.06* $_{Tx303}$ conferred resistance to Stewart›s wilt. The non-specific resistance may be attributed to pleiotropy or linkage.

Conclusions

Our research has led to successful identification of two reliably-expressed QTL that can potentially be utilized to protect maize from *S. turcica* in different environments. This approach to identifying and dissecting quantitative resistance in plants will facilitate the application of quantitative resistance in crop protection.

BACKGROUND

Pathogenesis is the series of events that occurs in a host-pathogen interaction, including infection and colonization of the host, and reproduction and dissemination of the pathogen. Genetic variation in host and/or pathogen can have quantitative or qualitative effects on the extent of disease. Many plant genetic factors that modulate pathogenesis have been discovered. The best known group is the R-genes, which provide high levels of resistance or even complete immunity. R-gene mediated resistance is initiated through a gene-for-gene interaction; the recognition of a pathogen effector by a host protein encoded by the R-gene leads to the induction of the hypersensitive response

(HR), the production of antimicrobial metabolites such as phytoalexins, and the expression of pathogenesis-related (PR) proteins [1]. This type of interaction, typically resulting in a highly effective but race-specific defense response against pathogenic invasion, is sometimes known as qualitative resistance. Quantitative resistance, on the other hand, confers intermediate levels of resistance and is believed to be controlled by a set of genes distinct from, or partially overlapping with, those involved in qualitative resistance [2–7].

Although each quantitative resistance locus conditions a relatively small effect on pathogenesis, this type of resistance is of agricultural interest because qualitative resistance tends to be ephemeral in many pathosystems and is unavailable in others. Quantitative resistance is presumably more durable because multiple genes with minor effects lead to lower selection pressure and greater complexity to overcome [8]. A large number of quantitative trait loci (QTL) for disease resistance have been mapped in plants [6, 9], but little is known about the underlying genetic basis or defense mechanisms involved. A range of genetic mechanisms controlling basal resistance, defense signalling pathways, detoxification, morphology, and development in the plant host, is hypothesized to be associated with reducing disease progress [6]. A small number of quantitative resistance genes have recently been cloned [3–5, 7, 10], implicating diverse host functions in quantitative resistance.

Given that diverse host functions affect quantitative resistance, it is likely that QTL act at different stages of pathogenesis. The ways in which quantitative resistance affects different stages of pathogenesis has been addressed, to a limited extent, by comparing trait values obtained using distinct (usually macroscopic) disease components. In most (or probably all) of the phytopathosystems analyzed to date, differences in various disease parameters can be observed among plant genotypes. Previous QTL studies for foliar diseases have mapped distinct loci associated with incubation period, lesion number, lesion size, or diseased leaf area, with results suggesting that defense genes affecting lesion formation and lesion expansion may not be the same. In breeding programs, selection for decreased lesion length or lesion numbers can have insignificant effects on incubation period or disease severity (eg. [11]). These observations suggest that distinct resistance mechanisms govern different macroscopic components of resistance.

More insights into the role of a given disease QTL in limiting pathogenesis can be gained through histopathological analysis. While biochemical and microscopic analyses have been applied to investigate major gene resistance and fungal pathogenicity factors (reviewed by Vidhyasekaran [12]), few studies have reported the effect of individual QTL on distinct stages of pathogenesis from a microscopic view (exceptions include [5, 13]). If QTL effective at specific stages of pathogenesis can be identified, combining favorable alleles for complementary QTL (eg. for infection and colonization) will likely provide greater levels of resistance.

Northern leaf blight (NLB; also known as turcicum blight and northern corn leaf blight) of maize was used as a model system to identify and characterize disease QTL at the macroscopic and microscopic levels. NLB, caused by *Setosphaeria turcica* (anamorph *Exserohilum turcicum,* syn. *Helminthosporium turcicum*), is one of the most prevalent foliar diseases in most maize-growing regions of the world. The disease causes periodic epidemics associated with significant yield losses [14–17], particularly under conditions of moderate temperature and high humidity [18]. Qualitative and quantitative forms of resistance against *S. turcica* are available in maize germplasm [19, 20], and have been widely utilized alone or in combination in resistance breeding programs [21]. A few histological studies have revealed the pathogenesis of *S. turcica*on maize leaves by staining, whole mount and serial dissection [22–25]. Marked phenotypic variation in symptom development has been observed among diverse maize lines in our multiple field and greenhouse trials. How macroscopic and microscopic phenotypes relate to specific QTL remains to be determined.

To answer questions concerning individual QTL effects, such as testing the hypothesis that distinct QTL act at different stages of pathogenesis, well-defined genetic stocks that differ only at specific loci are required. Introgression lines have been successfully used to study QTL in maize [26, 27], rice [28], barley [29, 30], tomato [31], and Arabidopsis [32]. While QTL analysis using recombinant inbred lines (RILs) provides greater statistical power in detecting QTL [33], RIL-based approaches have limitations in estimating QTL effects [31, 32, 34]. Introgression lines can be efficiently used to produce near-isogenic lines (NILs), which permit careful analysis of phenotypic effects associated with introgressed segments [27,31].

NILs allow many long-standing questions about quantitative disease resistance to be addressed, such as the relationship between disease QTL and plant maturity, the interaction of QTL and environmental factors, and the specificity of resistance conditioned by QTL. The interplay between disease resistance and plant development has been widely recognized [35] yet remains poorly understood. In general, the resistance in adult plants or older leaves is greater than in juvenile plants or younger leaves (eg. [36–38]), and a correlation between resistance and flowering time has been found [[39], R. Wisser, J. Kolkman, and P. Balint-Kurti, unpublished]. Some QTL effects may thus be specific to certain plant developmental stages. In addition, the expression and effectiveness of many genes/QTL have been observed to be regulated by environmental conditions [9, 40, 41]. Another issue of fundamental and practical interest is whether a disease QTL confers specific or broad-spectrum resistance. A single locus can condition resistance to more than one disease, if it encompasses linked QTL effective against different diseases, or its underlying genes are involved in broad-spectrum resistance pathways.

Here, we describe the use of introgression lines and derived NILs for QTL mapping and macro-/microscopic characterization in the maize - *S. turcica* pathosystem. We used an available population of introgression lines named TBBC3 [27]. This population is composed of introgression lines, each of which carries one or a few chromosomal segments of the donor genotype Tx303 in the genetic background of the recurrent parent B73. To better understand the nature of quantitative resistance, we assembled a panel of conventional and novel disease components targeting different stages of disease development. These were used to demonstrate that two QTL affect different stages of pathogenesis. The QTL were further characterized to shed light on three secondary hypotheses: (1) that disease QTL differ by host developmental stage (young versus adult plants); (2) that their performance changes across environments (field versus greenhouse); and (3) that they condition broad-spectrum resistance. This approach to identifying and dissecting quantitative resistance in plants will facilitate more effective and efficient application of quantitative resistance in crop protection.

METHODS

Plant Materials

A set of 82 TBBC3 introgression lines was provided by J. Holland of the USDA-ARS unit at North Carolina State University. The TBBC3 (for Tx303 by B73 Backcross 3) population, originally developed by C. Stuber at North Carolina State University [26], was the most extensively developed set of introgression lines available at the time for public use in maize. The population was derived from an initial cross of Tx303 and B73, followed by backcrossing to B73 for three generations. Each line was then selfed for several generations to attain homozygosity. Genotypic information was publicly available for each line, consisting of 14 restriction fragment length polymorphism (RFLP) and 116 simple sequence repeat (SSR) markers across the genome. Each line was known to carry one or more Tx303 introgressions, covering on average 2.5% of the genome, in the background of the sequenced reference maize line B73. Taken together, the set of introgression lines collectively carries ~89% of Tx303 genome [27]. To validate and characterize the effects of Tx303 introgressions, several BC_4F_2 populations were developed by crossing selected TBBC3 lines to B73. Sets of BC_4F_3 and BC_4F_4 lines carrying different introgression(s) were subsequently derived by single-seed descent. After four generations of marker-assisted backcrossing, the BC_4F_3 and BC_4F_4 lines were designated as NILs.

Assessments of Northern Leaf Blight

A single isolate of *S. turcica* (NY001, race 1) was used in the experiments carried out in New York (NY), and a mixture of isolates representing race 1, race 23, and race 23N of *S. turcica* was used in North Carolina (NC). For preparation of liquid inoculum, *S. turcica* was cultured for two to three weeks on lactose - casein hydrolysate agar (LCA) plates under a 12 hr/12 hr normal light-dark cycle at room temperature. The conidia were then dislodged with sterile ddH_2O and a glass rod, filtered through four layers of cheesecloth, and adjusted to the final concentration with the aid of a haemocytometer. Solid inoculum was prepared by culturing *S. turcica* on sorghum grains in plastic milk jugs for two to three weeks under the same condition. For each jug, 900 ml of sorghum grains were soaked overnight in 600 ml of water in a milk jug. The jug was then autoclaved twice at 121°C, 15 lb/cm^2, for 25 minutes per run. The jugs were inoculated by dividing the spore suspension produced from one heavily colonized LCA plate (10 cm diameter) among five jugs. Jugs were shaken daily to prevent caking and accelerate fungal colonization.

In this study, the "juvenile" phase refers to earlier vegetative stages, and the "adult" phase refers to later vegetative stages and reproductive stages. Juvenile plants (at the five- to six-leaf stage) and adult plants (late vegetative stage, around two weeks before tasselling) were used for inoculation. The inoculation technique utilized depended on the specific objectives of the experiment. In the field trials in NY, plants were inoculated with both liquid (0.5 ml of spore suspension, 4×10^3conidia per ml, 0.02% Tween 20) and solid inoculum (1/4 teaspoon, ~1.25 ml of colonized sorghum grains) placed in the whorl. This was done to ensure the viability of inoculum across a range of conditions (under optimal conditions, the liquid inoculum was considered most effective, while the solid inoculum was considered to perform more effectively under dry conditions). In the field trials in NC, ~20 grains of sorghum colonized with *S. turcica* were placed in the whorl.

In greenhouse trials, whorl inoculation with aforementioned liquid inoculum was carried out for assessing individual plants in the segregating populations. Spray inoculation was performed for detailed QTL characterization using NILs, as it provides significantly better differentiation for NLB evaluation (data not shown). The spray method was preferred for microscopic examination and real-time PCR quantification. A higher number of spores could be evenly distributed on leaf surface with spraying, making the subsequent sampling more effective and accurate. About 0.5 ml of concentrated spore suspension (5 × 10^4 conidia per ml, 0.02% Tween 20) was evenly sprayed on the first fully expanded leaf with an airbrush (Badger® Model 150) at 20 psi. After inoculation, the plants were kept overnight in a mist chamber at > 85% RH, then maintained at 22°C day/18°C night temperature with a 14 hr-light/10 hr-dark cycle.

Phenotypic Characterization of Resistance to NLB

Field experiments were conducted at Cornell's Robert Musgrave Research Farm in Aurora, NY and Central Crops Field Station in Clayton, NC. Plants were evaluated for different disease parameters and for days to anthesis (DTA). DTA, which was only assessed for field-grown plants, was scored on a row basis when > 50% of the plants in a row started to shed pollen. An overview of various disease components used in this study and their corresponding stages during NLB development is summarized in Table 1. The evaluation method for each parameter is illustrated as below.

Table 1: Overview of disease components used to target different stages of northern leaf blight (NLB) development

Disease component	Description (unit)	Targeted disease development stage(s)	Evaluation	Literature
Incidence of multiple appressoria[a]	The incidence of > 1 appressorium developed from each germinated conidium (%)	Pre-penetration	Trypan blue staining and microscopy	[25, 42]
Infection efficiency	The incidence of successful infection per germinated conidium (%)	Penetration into the epidermal cell	Trypan blue staining and microscopy, KOH-aniline blue fluorescence microscopy	[25, 42, 94]

Accumulation of callose and phenolics[a]	Diameter of enhanced fluorescing area surrounding the infection site (μm)	Intercellular and intracellular hyphal growth from primary infected cell to surrounding mesophyll cells	KOH-aniline blue fluorescence microscopy	[43]
Vascular invasion efficiency[a]	The incidence of hyphae entering vascular bundles per infection site (%)	Hyphal growth into the vasculature	KOH-aniline blue fluorescence microscopy	[43]
Fungal biomass ratio[a]	The percentage of fungal DNA divided by the total DNA in the infected leaf tissues (%)	Overall fungal growth in leaves before the appearance of necrotic lesions	DNA-based real-time quantitative PCR	[44]
Incubation period (IP)	The number of days from inoculation to the appearance of the first lesion on a plant (days)	Xylem plugging due to extensive hyphal growth in the vascular veins	Visual examination	[73, 74, 76–78,95]
Lesion expansion (LE)	The longitudinal expansion of a lesion per day (mm)	Destructive hyphal growth in primary inoculated leaves	Digital caliper measurement	[78, 96]

Diseased leaf area (DLA)	The percentage of infected leaf area of the entire plant, disregarding decayed bottom leaves (%)	Destructive hyphal growth on the leaves of a entire plant, caused by primary and secondary inoculum	Visual examination	[73, 76, 77, 94,97]
Disease severity[b]	The severity of infected leaf area of the entire plant (scale 1-10, 1: little diseased area)	Destructive hyphal growth on the leaves of a entire plant, caused by primary and secondary inoculum	Visual examination	[98]
Area under the disease progress curve (AUDPC)	Total area under the graph of DLA or other disease severity rating (area unit)	An overall destructive hyphal growth on a plant throughout the season	Calculated from visual examination scores	[76, 77]

[a] To authors' knowledge, not previously utilized as a disease component for evaluating NLB resistance.

[b] Only applied in the 2006 trial in North Carolina.

Microscopic Analysis

Two microscopy techniques were applied to investigate differential development of *S. turcica* in the NILs: trypan blue staining and KOH-aniline blue staining. In greenhouse trials, infected leaf samples were harvested from individual plants. In the field trial, for the purpose of obtaining a sufficient number of infection sites for examination, samples (per genotype per block) were collected from four plants in a row and pooled for subsequent treatments.

Trypan blue staining was performed as previously described [25, 42] with some modifications. Infected tissues were collected at two days post inoculation (dpi) from plants in the greenhouse, and at 3 dpi from plants in the field trials. Leaf samples were cut into 1 × 1 cm^2 segments, cleared first in an acetic acid: ethanol (1:3, v/v) solution overnight, then in an acetic acid: ethanol: glycerol (1:5:1, v/v/v) solution for at least 3 hours. The samples were subsequently incubated overnight in a staining solution of 0.01% (w/v) trypan blue in lactophenol, and rinsed then stored in 60% glycerol until examination. Specimens were transferred onto microscopic slides and examined under a compound microscope. Fifty to 60 germinated conidia were assessed per individual plant (greenhouse) or per row (field).

A modified KOH-aniline blue fluorescence technique [43] was used to visualize the growth of fungal hyphae inside the infected leaves and the accumulation of (plant-produced) callose and phenotypic compounds around the infection sites. Infected leaves were sampled at 4 dpi and 7 dpi in greenhouse trials, and at 6 dpi in the field trial. The samples were cut into $1 \times 1\ cm^2$ segments, incubated in 1 M KOH at room temperature for 24 hours, then autoclaved at 121°C, 15 lb/cm^2 for 2-5 min. Autoclaving time was adjusted according to the rigidity of leaves, which varied with plant genotype and maturity. The autoclaved specimens were rinsed in ddH_2O three times, then stored in autoclaved ddH_2O until examination. Specimens were carefully placed on microscopic slides and mounted in a staining solution of 0.05% aniline blue in 0.067 M K_2HPO_4 (prepared at least 2 hrs prior to use). Thirty five to 40 germinated conidia were checked per individual plant (greenhouse) or per row (field) under a Zeiss fluorescence microscope with a G365 excitation filter, a FT395 dichromatic beam splitter, and an LP420 barrier filter.

Quantitative Real-Time PCR for Quantifying Fungal Colonization

DNA-based quantitative PCR (qPCR) was performed as described by Qi and Yang [44] with some modifications. The specific pair of primers for *S. turcica*: forward: 5'-TCTTTTGCGCACTTGTTGTT and reverse: 5'-CGATGCCAGAACCAAGAGAT, was designed based on the internal transcribed spacer 1 (ITS1) of ribosomal DNA gene in *S. turcica*. The ITS1 sequence (GenBank: AF163067.1) was obtained from the nucleotide database of the National Center for Biotechnology Information (NCBI). PCR amplification resulted in a specific fragment of 170 base pairs.

Inoculation experiments were performed three times in the greenhouse. Five plants per genotype were spray-inoculated. The same amounts of infected tissue (0.12 g per plant) were collected at 9 dpi from the middle part of each leaf. Leaf samples were ground with liquid nitrogen and DNA was extracted following the protocol described later. The extracted DNA from each individual plant was dissolved in 100 µl TE buffer. Total DNA concentration was determined using the PicoGreen® dsDNA quantitation assay kit (Invitrogen,

Eugene, Oregon, USA). Fungal DNA was quantified using qPCR. The ratio of fungal biomass in maize leaves was computed from the amount of fungal DNA divided by total DNA. Each qPCR reaction was performed in a total volume of 25 μl, containing 12.5 μl of iTaq SYBR® Green Supermix with ROX (Bio-Rad Laboratories, Hercules, CA, USA), 3 μl of 7.5-fold diluted DNA from an infected plant and 300 nM of each forward and reverse primer. PCR samples were incubated in an ABI Prism 7000 Sequence Detection System (Applied Biosystems, Foster City, CA, USA) with thermal cycling parameters of 95°C for 2 min followed by 40 cycles of 15 sec at 95°C and 30 sec at 56°C. Two standard curves were constructed by mixing a series of *S. turcica* DNA (0, 1, 10, 50, 100, 500 and 1000 pg) with 50 ng of maize DNA extracted from non-inoculated B73 and Tx303 plants, respectively. The quantification of fungal biomass in infected B73 and derived NILs was based on the first standard curve (created from mixing with a constant amount of B73 DNA), while the quantification for DNA from infected Tx303 was based on the second standard curve. Three technical replicates were carried out in individual plates for both Picogreen quantification and iTaq SYBR Green PCR, with the samples for standard curves repeated twice in the same plates.

Incubation Period (IP)

Individual plants were checked every day after 7 dpi for the appearance of the first wilted lesions. The number of dpi when the first lesions were observed was scored as the IP. In the trials at Aurora NY, IP scores were rated for individual plants, then averaged for the rows. In the trial at Clayton NC, IP was recorded on a row basis when > 50% of the plants in a row started showing lesions.

Lesion Expansion (LE)

LE was scored in the trials in NY but not NC. Around two to three weeks after inoculation, three lesions per plant were randomly chosen for measurement. Lesion margins were marked and then measured 10-14 days later for the longitudinal extension with a digital caliper. The expansion measurements taken from three lesions were averaged, and divided by the number of days from the marking until measurement of the lesions.

Primary Diseased Leaf Area (PrimDLA)

Primary DLA was rated as the percentage of infected leaf area of the inoculated leaves in the 2006 trial in NY. It was scored once on a row basis at around three to four weeks after inoculation.

Diseased Leaf Area (DLA)

DLA was rated as the percentage of infected leaf area of the entire plant, disregarding decayed bottom leaves. DLA was rated on a row basis for TBBC3 lines and derived NILs, and on individual plants for testing trait-marker association in segregating populations. DLA was rated three to four times per season at 10-14 day intervals. The first DLA score was recorded at one to two weeks after observing the onset of secondary infection.

Disease Severity

Disease severity was rated on a row basis four times through the season in the 2006 trial in NC. The severity score was based on a 1 to 9 scale corresponding to the percentage of infected leaf area on primarily the ear leaf as well as the leaves above and below the ear leaf (severity 1: 0%, 2: 12.5%, 3: 25%, ..., 9: 100%).

Area under the Disease Progress Curve (AUDPC)

The AUDPC was calculated as

$$\sum_{i=1}^{n-1} \frac{(yi + y_{i+1})(t_{i+1} - ti)}{2},$$

where y_i = DLA or disease severity at time i, t_{i+1} - t_i = day interval between two ratings, n = number of ratings [45].

Evaluation for Multiple Disease Resistance

Stewart's Wilt

Pantoea stewartii (syn. *Erwinia stewartii*) strain PsNY003, originally collected in NY in 1991, was obtained from Helene Dillard of Cornell University. Inoculum was prepared as previously described [46] with the use of nutrient agar plates and nutrient broth as media. Plants at the five- to six-leaf stage were inoculated with *P. stewartii* following the pinprick method [47, 48] with a modified inoculator. The multiple-pin inoculator was made with 30 T-pins (1.5 inch long), pieces of 5.5 cm × 6.5 cm sponge and cork board (3/8 inch thick) fastened on two arms of a tong with rubber bands. Primary diseased leaf area was rated as the percentage of infected area of the inoculated leaves. It was scored twice (two and three weeks after inoculation) on a row basis and the scores were averaged.

Anthracnose Stalk Rot

A New York isolate of *Colletotrichum graminicola* (teleomorph: *Glomerella graminicola*) (isolate Cg151) was obtained from Gary Bergstrom of Cornell University. Each plant was inoculated with 1 ml of 10^6 conidia per ml (0.02% Tween 20) when more than 50% of the plants in every row were tasseling [49]. Inoculum preparation and inoculation were conducted as described by Muimba-Kankolongo and Bergstrom [50], with the replacement of plastic straw with 1 ml pipette tips. Four weeks after inoculation, stalks were split longitudinally and the percentages of discolored area of individual internodes were visually rated [49] and summed for analysis. In 2007, eight consecutive internodes were scored from four plants per row (inoculation was conducted in NLB plot); in 2008, six consecutive internodes were scored from eight plants per row.

Common Smut

In the 2007 trial, plants in NLB plots were evaluated for the development of ear galls and stalk galls resulting from natural infection. Artificial inoculation was conducted in the 2008 trial using six compatible strains of *Ustilago maydis* (UmNY001, UmNY002, UmNY003, UmNY004, UmNY008 and UmNY009), which were isolated from naturally infected smut galls collected at Aurora NY in 2007. *U. maydis* strains were cultured as described by du Toit and Pataky [51]. Inoculum was prepared by adjusting sporidial suspension (in potato dextrose broth) to a final concentration of 10^6 sporidia per ml (0.02% Tween 20) with sterilized distilled water. Equal amounts of sporidia from the six compatible strains were mixed prior to inoculation. Non-pollinated (shoot-bagged) primary ear of each plant was injected with 2 ml of mixed sporidial suspension at the time that the silks of most primary ears had emerged 1-5 cm. Every plant in the row was rated from four weeks after inoculation for ear galls on a 0-10 scale, corresponding to the number and size of galls and the disease severity of the entire plant (0 = no smut galls and 10 = a dead smut-infected plant). Stalk galls resulting from natural infection were also scored.

Common Rust

Urediniospores of *Puccinia sorghi* were collected from naturally infected leaves at Aurora NY in 2007. Inoculum was increased on three- to four-leaf stage seedlings of susceptible sweet corn in the greenhouse. About 300 mg of stock urediniospores (preserved at -80°C) were suspended in 100 ml of Sortrol oil (Chevron Phillips Chemical Company, Phillips, TX, USA) [52] and evenly applied on leaves with a spray gun (Preval, Yonkers, NY, USA). Plants were kept overnight in a mist chamber at > 85% RH, then grown for two more weeks until *Puccinia sorghi* sporulated vigorously. The urediniospores were collected by agitating infected leaves with mature rust pustules in distilled water and filtering through four layers of cheesecloth. Spore suspension was adjusted to a final concentration of 2×10^5 urediniospores per ml with the aid of a haemocytometer. Field plants were inoculated at the six- to eight-leaf stage by adding 1 ml of spore suspension (0.02% Tween 20) in the whorl [53]. Evaluation for disease severity was based on a 0-10 scale with 0.5 increments, corresponding to the percentage of infected leaf area of the entire plant (0 = no disease, 1 = 10%, ..., 10 = 100%). Disease severity was scored three times at 9-day intervals from four weeks after inoculation in 2008, and at 15-20 day intervals from 10 days after inoculation in 2009. AUDPC was calculated from the three severity scores as described above.

DNA Extraction and Genotyping

About 0.1 g of fresh or lyophilized leaf tissue and a stainless steel ball (5/32 inch diameter, OPS Diagnostics, NJ, USA) were loaded in each well of a 96-well plate (Corning® Costar 96 Well Polypropylene Cluster Tubes). Tissues were frozen at -80°C and pulverized at 450 strokes/min for 50-120 sec using Genogrinder 2000 (SPEX CertiPrep Inc., Metuchen, NJ, USA). Genomic DNA was extracted following a standard CTAB extraction protocol [54, 55] with 500 µl of CTAB extraction buffer, 400 µl of chlorophorm/isoamyl alcohol (24:1, v/v), and 300 µl of isopropanol. Fungal DNA was extracted as described, except that *S. turcica* mycelium was ground in liquid nitrogen with a pestle and mortar, and transferred to a 1.5 ml eppendorf tube for later steps.

Simple sequence repeat (SSR) markers were used for genotypic analysis following a single-reaction nested PCR method [56]. Each PCR reaction was performed as described by Wisser *et al.* [57] in a total volume of 13 µl, with the same thermal cycling parameters as described by Schuelke [56]. Amplicons labeled with different fluorescent dyes were multiplexed (combining up to four PCR reactions, 0.7 µl PCR product per specific primer pair), mixed with 9 µl formamide and 0.05-0.1 µl GeneScan-500 LIZ size standard (Applied Biosystems), and analyzed on the Applied BioSystems 3730xl DNA Analyzer at Biotechnology Resource Center at Cornell University. The sizes of amplicons were scored using GeneMapper v. 3.0 (Applied Biosystems).

Experimental Design

The full set of 82 TBBC3 lines, a subset of 15 TBBC3 lines, and derived sets of BC_4F_3 and BC_4F_4 NILs were evaluated in the field following the resolvable incomplete block design (also known as an alpha design [58]). The 82 TBBC3 lines were evaluated at Aurora NY for IP, PrimDLA, DLA, and DTA. To precisely estimate the effects of Tx303 introgressions, every experimental row was grown next to a row of B73 (B73 in every third row). The whole design had two replicates with 14 blocks per replicate, and six experimental rows plus three B73 rows per block. A parallel experiment was conducted on the 82 TBBC3 lines at Clayton NC with two replicates, nine blocks per replicate and 10 rows per block. The B73 rows were included in every other block for the control (five B73 rows per replicate). Disease severity and DTA were evaluated. The field trials on the 15 selected TBBC3 lines were conducted at Aurora NY and Clayton NC following the alpha design with B73 in every third row as described above. The 15 lines were selected from those significantly different from B73 in NLB resistance (based on the first-year results), and/or from the lines carrying Tx303 introgression(s) corresponding to previously identified NLB QTL. In each location, there were four replicates with four blocks per replicate, and four experimental rows plus two B73 rows per block. The IP, DLA, and DTA were evaluated. An additional disease component, lesion expansion (LE), was evaluated at Aurora NY as well. During the same field season, to understand the effectiveness of identified QTL in adult plants, the same 15 selected TBBC3 lines were grown in a separate field plot at Aurora NY, following identical arrangement and design with three replicates. The plants were inoculated two weeks before flowering (rather than the usual five- to six-leaf stage), and evaluated for IP and DLA. To associate resistance with specific introgressions, several BC_4F_2 populations were genotyped and phenotyped for IP, LE, and DLA in either greenhouse or in the field at Aurora NY. Individual plants of each population were grown within a single block with B73 rows as the border. For QTL confirmation, derived BC_4F_3 and BC_4F_4 NILs were evaluated at Aurora NY for IP, LE, DLA, and DTA, following the previously described alpha design with B73 rows in every third row. The experiment was planted in two replicates, with five blocks per replicate, and four experimental rows plus two B73 rows per block. To test whether the identified NLB QTL confer resistance to other important maize diseases, derived BC_4F_3 and BC_4F_4 NILs were evaluated at Aurora NY for anthracnose stalk rot (ASR) and common smut in 2007 and 2008, and for Stewart's wilt and common rust in 2008 and 2009. In 2007, the assessments of artificially inoculated ASR and naturally occurring common smut were conducted on the plants in NLB trial. In 2008 and 2009, derived NILs were grown in separate field plots for different disease evaluations. Trials for Stewart's wilt and common rust were planted following the previously described alpha design, with two replicates, five blocks per replicate, and four experimental rows plus two B73 rows per block. Trials for ASR were planted following the same alpha design, with two replicates, two blocks per replicate, and seven experimental rows plus four B73 rows per block.

Plants in the trials of common smut were randomized in two replicates. Four genotypes were used for microscopic analysis and DNA-based real time PCR quantification: B73, Tx303, TBBC3-38-05F (the NIL with $qNLB1.06_{Tx303}$), and TBBC3-42-10E-02 (the NIL with $qNLB1.02_{Tx303}$). Three greenhouse trials (each with five plants per genotype) and one field trial (five blocks, one row per genotype per block) were conducted from December 2007 to July 2008. Real-time PCR quantification was not performed in the 2008 field trial. In the greenhouse, plants subject to the same treatment were randomized within a block in order to eliminate the variance due to environmental factors.

Data Analysis

Mixed model analyses were performed in JMP 7.0 for the field trails following an alpha design. For trials conducted in one field location, "maize lines" was specified as a fixed factor, whereas "replicates" and "blocks within replicates" were specified as random. Data from the 2006 trials in NY and NC were analyzed separately because of the different field arrangements and disease rating scales. A combined analysis across two locations (NY and NC) was conducted for the 2007 trials by fitting a mixed model with "maize lines", "locations" and "maize lines by locations" as fixed factors, and "replicates within locations" and "blocks within replicates within locations" as random effects. Phenotypic differences between TBBC3 lines/NILs and B73, or between any two NILs, were determined by pair-wise comparisons of least squares means using two-tailed Student's t-test at $P < 0.05$. Correction for multiple comparisons was not made since all the tests were independent. The TBBC3 lines and derived NILs were each compared to the recurrent parent B73, and pairs of NILs were only compared to each other when sharing identical genotypes except for the target introgression(s) under testing. Overall comparisons of TBBC3 lines or NILs were not intended in this analysis. The QTL effect of each marker locus was investigated in the full TBBC3 population. Considering the presence of unlinked, potentially unrecognized introgressed segments in the same lines, a series of statistical tests was conducted as described by Szalma *et al.* [27] with some modification. Markers likely associated with resistance ($P < 0.05$) were identified by: 1) comparing the least squares mean of TBBC3 lines homozygous for Tx303 alleles at each locus to the least squares mean of B73 rows, and 2) performing one-way analysis of variance (ANOVA) on an individual marker - trait basis in the TBBC3 population. Significant markers were grouped into linked introgression blocks, and the markers with the greatest significance in each linked block were considered to reflect the most likely QTL position. Correlations between the most significant markers were checked. For each pair of unlinked but correlated markers, TBBC3 lines fixed for one marker were grouped and analyzed for the co-segregation of the second marker and disease traits, using two-tailed Student's t test at $P< 0.05$. A putative QTL was declared if the mean of all the lines with Tx303 alleles at this locus was significantly different from B73, and significant difference was detected for lines contrasting for this locus under the condition of another potential QTL locus being fixed.

In segregating populations, the phenotypic and genotypic data were first analyzed by mixed stepwise regression, with a significance probability of $P < 0.05$ for each parameter to enter/leave the model [59]. Once the markers significantly associated with the traits were identified, ANOVA was conducted on a single marker basis at $P < 0.05$ to estimate QTL effects. For the populations with more than one marker showing significant effect on traits, the effective markers were further tested for epistatic interactions by examining marker pairs by two-way ANOVA. Allele effects were determined by pair-wise two-tailed Student›s t test according to the least significant difference (LSD) at $P = 0.05$. The proportion data from microscopic analysis and qPCR were transformed to arcsine of the square roots prior to statistical analysis. Arcsine transformed data were analyzed by fitting a linear least squares model with «genotype» and «replicate» as two independent variables, and with the number in the denominator of each proportion as a weighting factor. For qPCR, considering the biological and technical replicates, the fungal biomass ratios were also arcsine square-root transformed, but were analyzed with a different linear least squares model of $y_{ijm} = \beta_0 + \beta_1 (genotype_i) + \beta_2 (rep_j) + \beta_3(plate_m (rep_j)) + e_{ijm}$ (genotype i = B73, Tx303 and two derived NILs; rep j = 1, 2, 3; plate m = 1, 2, 3; e_{ijm}: random experimental error). Least squares means and 95% confidence intervals were back-transformed to percentages. Confidence intervals were larger than significance levels due to asymmetry resulting from back transformation of arcsine scale. The significant QTL effects were estimated by pair-wise comparisons of least squares means of introgression lines/NILs and B73 using two-tailed Student's t-test at $P < 0.05$. All the statistical analyses in the study were conducted using JMP 7.0.

RESULTS

A stepwise strategy (Figure 1) was conducted to map and characterize QTL using the TBBC3 introgression lines and derived NILs.

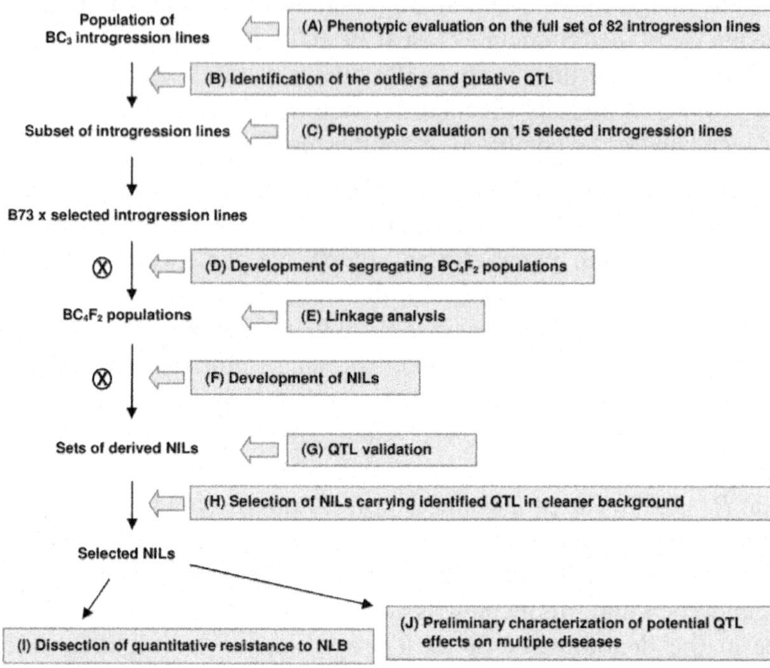

Figure 1: Strategy for mapping and characterizing QTL using introgression lines. (A) Initial screening was conducted at Aurora NY and Clayton NC. Conventional disease components, including incubation period (IP), diseased leaf area (DLA), and disease severity were evaluated. (B) Lines that differed from B73, and lines showed extreme phenotypes in the population were determined. Putative QTL were also identified. (C) Lines carrying putative NLB QTL were evaluated to confirm their differential resistance/susceptibility relative to B73. Plants were tested at the juvenile and adult stages for IP, LE, and DLA. (D) Selected lines were backcrossed to B73, then selfed to generate F_2 populations segregating for the introgressed regions within the populations. (E) Individual plants were tested for markers targeting introgressed regions, and co-segregation was assessed with the disease components measured in the greenhouse or field in NY. Candidate NLB QTL were determined if trait-marker association was detected in more than one F_2population. (F) The BC_3F_3 and/or BC_3F_4 near-isogenic lines (NILs) carrying different Tx303 introgressions were derived from F_2populations. (G) The effects of different introgressions on conventional disease components were further tested. The NLB QTL were declared if the QTL effects were validated in the NILs. (H) Selected NILs carrying identified QTL were used for detailed QTL characterization. (I) Selected NILs were evaluated with a series of disease components targeting different stages of disease development (Table 1) in the greenhouse and field. (J) Selected NILs were also evaluated for anthracnose stalk rot, common rust, common smut and Stewart's wilt at Aurora NY.

Identification of Outliers and Putative NLB QTL in the TBBC3 Population

The full set of 82 TBBC3 lines was screened for NLB resistance at field sites in New York and North Carolina in 2006. As shown in Additional file 1, the TBBC3 population was generally more susceptible than B73 at both locations. In the trial at Aurora NY, three lines were scored as significantly more resistant and 20 lines were scored as significantly more susceptible than B73 for primary DLA, while no lines showed significantly more resistance and 39 lines showed more susceptibility than B73 based on AUDPC. In the trial at Clayton NC, eight lines were significantly more resistant and 13 lines were significantly more susceptible than B73 for AUDPC. The only line that showed contrasting phenotypes in the two environments was TBBC3-25, suggesting experimental error or environmental influence on the expression of NLB QTL.

A combination of analytic approaches was used to associate the resistance or susceptibility effects with specific chromosomal segments [27]. Based on this analysis, 13 putative NLB QTL were identified in the TBBC3 population (Additional file 2). Except for bins 1.01 and 5.08-5.09, the putative QTL co-localized with NLB QTL previously detected in four maize mapping populations [20].

Phenotypic Evaluation on the Subset of 15 TBBC3 Lines

A subset of 15 TBBC3 lines was selected for further evaluation based on the degree of difference in NLB resistance between each individual TBBC3 line and B73 at the two field sites (Additional file 1 and Table 2). Lines significantly or marginally different ($P < 0.15$) from B73 in either of the two locations were taken into consideration. Line selection was also influenced by the identity of the Tx303 introgressions carried by the lines; those carrying introgressions corresponding both to putative QTL identified in the TBBC3 population (Additional file 2), and to previously reported NLB QTL (specifically those at bins 1.07-1.08, 5.02, and 6.04-6.05 [20]) were given priority because of our interest in QTL validation and characterization rather than in QTL discovery.

Table 2: NLB resistance of the subset of 15 TBBC3 introgression lines

Maize line	Introgressed region (Bin)[a]	Days to anthesis	Trait (juvenile plants being inoculated; NY and NC)[b]			Trait (adult plants being inoculated; NY)	
			Relative IP	Relative LE	Relative AUDPC	Relative IP	Relatvie AvgDLA

Tx303		8.7***	**2.1*** **	-0.3	**-164.1*** **	**7.5*** **	**-8.7*** **
TBBC3-38	1.03, 1.06, 5.00, 5.03, 5.07-5.09	4.0***	**2.1*** **	-0.1	**-110.2*** **	**2.6*** **	**-4.5*** **
TBBC3-39	1.03, 1.06, 5.00, 5.07	2.4***	**0.9* **	0.0	**-67.2** **	**1.3* **	**-4.3*** **
TBBC3-18	1.11, 7.05, 7.06, 10.03, 10.04	4.2***	**0.8* **	-0.1	-35.8	**1.3* **	**-2.0** **
TBBC3-19[c]	7.04, 7.06, 9.03, 10.04	1.4*	**1.2** **	0.1	-29.0	1.2	3.2***
TBBC3-26	7.06	1.0	**1.4** **	0.0	-1.8	-0.1	1.0
TBBC3-75[d]	6.04-6.05	1.2	0.4	0.1	12.2	**2.6*** **	**-2.3** **
TBBC3-02	1.07-1.08, 5.01-5.02, 8.03-8.05	3.2***	-0.7	0.2	35.7	0.6	1.5*
TBBC3-14[c]	1.09-1.10, 4.07	3.1***	-0.9*	0.0	67.0**	0.3	6.7***
TBBC3-03	1.01, 1.04, 7.03, 8.03-8.05	2.8***	-0.1	-0.1	117.0***	-1.4*	5.7***
TBBC3-61[c]	1.10	1.4*	-0.5	0.6***	126.7***	0.2	7.0***
TBBC3-77	1.01-1.03, 6.04-6.05	4.3***	-1.4**	0.9***	139.4***	-1.4*	4.0***
TBBC3-30	4.02-4.03, 5.04	0.2	-0.9*	0.6***	195.4***	-0.6	4.7***
TBBC3-21	4,07, 9.01-9.04	0.6	-1.7***	0.8***	275.5***	-1.1	7.8***
TBBC3-36	1.01, 1.03-1.05, 4.01, 8.08	1.3	-0.8	0.7***	305.6***	-1.8**	8.2***
TBBC3-42	1.01-1.02, 4.07-4.08, 5.01-5.02, 8.02-8.05	2.2**	-2.1***	0.7**	310.8***	-1.3*	8.7***

In 2007, the same set of plant materials was inoculated at the five- to six-leaf stage in New York (NY) and North Carolina (NC), and were grown in a separate field plot in NY and inoculated 1-2 weeks before anthesis. Trait values shown are the least squares means of maize lines relative to the recurrent parent B73. Phenotypes that were significantly more resistant than B73 are highlighted in bold, while the phenotypes significantly more susceptible than B73 are underscored. The significance level was determined by pair-wise two-tailed Student's t test on the least square difference (denoted as * 0.01 $<P<$ 0.05; ** 0.001 $<P<0.01$; *** $P<0.001$). Maize lines in bold are the ones that were consistently more resistant or more susceptible than B73 across two field environments.

[a] Introgressed genome regions from Tx303 (heterozygous regions not shown). Bin position was based on genetic map of intermated B73 × Mo17 population (version IBM 2008 neighbors).

[b] Relative least squares means of incubation period (IP) and area under the disease progress curve (AUDPC) shown were from the combined analysis across the trials in NY and NC. Lesion expansion (LE) was only scored in NY.

[c] Individual TBBC3 lines in each field environment were inspected. TBBC3-19 was significantly more resistant than B73 for IP (P = 0.03), DLA1 (P = 0.001), DLA2 (P = 0.035) and AUDPC (P = 0.010) only in NC. TBBC3-14 and TBBC3-61 showed greater susceptibility than B73 for three DLA ratings ($P<$ 0.003) and AUDPC ($P<0.0001$) only in NY. The same differential degrees of resistance or susceptibility in two field environments were also observed in 2006 (Additional file 1).

[d] Resistance to NLB was more effective in adult plants.

In 2007, the 15 selected lines were evaluated at the same field locations in NY and NC to confirm their disease phenotypes. Significant «line» effects ($P<0.0001$) were found for all disease parameters in each of NY and NC, and across two locations. In the combined analysis across environments, significant effects of «location» ($P<0.002$ for IP, DLA and AUDPC) and "line-by-location" ($P<0.0001$ for DLA and AUDPC) were also detected. The variations were attributed to dissimilar pathogen strains, environmental conditions and disease ratings conducted by different scorers at two sites. Inspection of individual TBBC3 lines in NY and NC respectively suggested that most of the lines performed similarly across the two locations. Pair-wise comparisons validated the phenotypic differences for seven out of the 15 TBBC3 lines and B73 (Table 2). Consistently in NY and NC, TBBC3-38 and TBBC3-39 were more resistant, whereas TBBC3-42, TBBC3-36, TBBC3-21, TBBC3-30, and TBBC3-77 were more susceptible than B73.

Ten selected TBBC3 lines showed significantly greater DTA than B73 (about 1-4 days of difference, Table 2), indicating the presence of flowering time QTL in those lines. To test the effect of plant development on QTL expression, the 15 selected lines were inoculated at juvenile and adult stages (Table 2) and their disease traits were compared. In all the lines, QTL effects in reducing or increasing resistance to NLB were generally consistent for juvenile and adult plant stages. No lines showed contrasting phenotypes (more resistant and more susceptible relative to B73) when inoculated at the five- to six-leaf stage or just before flowering, which suggested that allele effects of NLB QTL were not altered by developmental stage. There was, however, evidence suggesting that the effectiveness of QTL might change over plant development. The resistance in TBBC3-38, TBBC3-39, TBBC3-18 and TBBC3-75 was more effective in adult than in juvenile individuals. The formation of NLB lesions was delayed by around 0.5-2 days on mature plants of the same genotypes.

Validation of QTL Effects by Linkage Analysis

Each of the TBBC3 lines had more than one identified introgression from Tx303. To determine which of the introgressed regions in a given line was associated with resistance, several segregating BC_4F_2 populations were developed from crosses between selected TBBC3 lines (TBBC3-02, TBBC3-38, TBBC3-39, TBBC3-42, and TBBC3-77) and the recurrent parent B73. The BC_4F_2 individuals from each cross were evaluated for NLB resistance and genotyped using markers for the respective introgressed regions. Significant trait-marker associations were identified in all the BC_4F_2 populations under investigation (Additional file 3).

The QTL at bin 1.02 was found effective in two segregating populations: those derived from TBBC3-42 and TBBC3-77. Although the introgressed region extended from bin 1.01-1.02, higher significance was detected at the marker located at bin 1.02 so the QTL was designated *qNLB1.02*. In the BC_4F_2 derived from TBBC3-42, ~14% of the IP variation could be attributed to *qNLB1.02*. The B73 allele at *qNLB1.02* was associated with slower disease development. Individuals homozygous for Tx303 alleles in bin 1.02 (replacement of $qNLB1.02_{B73}$) were significantly more susceptible than B73. The Tx303 homozygotes showed lesions ~0.6 days earlier (lower IP; $P < 0.0001$) and lesions expanded more rapidly by ~0.35 mm/day (greater LE; $P < 0.05$) in the B73 × TBBC3-42 population. The same QTL acted differently in the B73 × TBBC3-77 population. While *qNLB1.02* was effective in reducing DLA and AUDPC, no effect on IP was observed in the B73 × TBBC3-77 population. It is worth noting that TBBC3-42 is a much more susceptible genotype than TBBC3-77, which provided a better background for evaluating minor disease QTL.

The QTL at bin 1.06 was not identified based on previously reported genotype information. Although clear phenotypic variation was observed on

IP and DLA in the BC_4F_2 populations from TBBC3-38 and TBBC3-39, no significant association was found between IP and the known introgressions at bins 1.03, 5.00, 5.01-5.03, and 5.07-5.09. The IP on BC_4F_2 individuals of TBBC3-38 and TBBC3-39 ranged from 11 to 18 dpi, while IP on B73 was generally between 12 and 15 dpi. Since these two populations displayed a greater variation in resistance, we inferred that a disease QTL was present in the population but not associated with a recognized introgression. To search for unrecognized introgressions, an additional 68 SSR markers across the genome were chosen to target chromosomal regions that were not well-covered by the original 130 RFLP and SSR markers on TBBC3 map. Pooled DNA samples from BC_4F_2 individuals of TBBC3-38 and TBBC3-39, along with B73 DNA as a control, were tested with the additional SSR markers for heterozygosity. Among the 68 SSR markers tested, *umc1754*and *umc2234* (both in bin 1.06) were the only polymorphic markers segregating in both BC_4F_2 populations. This indicated the presence of a Tx303 introgression at bin 1.06 in both TBBC3-38 and TBC3-39. Individuals in each population were then genotyped with *umc1754* and *umc2234*, and analyzed for the effects on various disease components. A highly significant association was detected for various parameters in both populations (Additional file 3). In the B73 × TBBC3-38 and B73 × TBBC3-39 populations, around 33% and 53% (respectively) of the variation in AUDPC was explained by this QTL. Compared to the B73 allele, the Tx303 allele at *qNLB1.06* increased the IP by ~2 days, decreased LE by ~0.3 mm per day, and reduced DLA by about 5-10%. Another QTL at bins 5.08-5.09 (Tx303 allele for resistance) was found in the B73 × TBBC3-38 population. The effect was relatively minor ($R^2 \approx 0.05$) compared to the *qNLB1.06* $_{Tx303}$ identified in the same population.

QTL were also identified in bins 5.00, 5.01-5.02, and 8.03-8.05. Although contributing significant levels of resistance for IP or DLA in the B73 × TBBC3-77 and B73 × TBBC3-02 populations, the same introgressions/markers were not found effective in the populations derived from TBBC3-38, TBBC3-39 and TBBC3-42. The ambiguous results called the association into question, suggesting that the effectiveness of QTL at bins 5.00, 5.01-5.02, and 8.03-8.05 was affected by genetic background and/or environmental conditions.

Validation of *qNLB1.02* $_{B73}$ and *qNLB1.06* $_{Tx303}$ in Selected NIL Sets

To confirm the QTL effects detected in segregating populations, NILs were generated by selfing selected BC_4F_2 lines from the populations B73 × TBBC3-38, B73 × TBBC3-39 and B73 × TBBC3-42. Selected lines were chosen to represent different introgressed regions in the original TBBC3 lines. The

QTL effects were determined by pair-wise comparisons between individual introgression lines and B73, and between the NILs developed from a TBBC3 line. Within each set of NILs, lines contrasting for each marker locus were grouped and analyzed for their phenotypic differences. The NILs derived from different TBBC3 lines were not compared to each other since the phenotypic difference can be attributed to not only introgressed regions but also to minor differences in genetic backgrounds.

Evaluations of the BC_4F_3 NILs validated the effects of $qNLB1.06_{Tx303}$. Five NILs derived from B73 × TBBC3-38 and four NILs derived from B73 × TBBC3-39 were phenotyped in 2008. Individual trait-locus analysis in the two NIL sets revealed that bin 1.06 was the only locus that showed significant effects on resistance. The QTL effect at bin 5.08-5.09 was detected in the B73 × TBBC3-38 population, but was not confirmed in the NILs. As shown in Additional file 4, most of the NILs carrying Tx303 alleles at bin 1.06 were significantly more resistant than B73. The only exception was the line TBBC3-39-11A, which might have lost the resistance gene(s) due to recombination. Significant differences between lines with $qNLB1.06_{Tx303}$ and B73 were observed for IP, DLA and AUDPC, which was in agreement with the results from linkage analysis in BC_4F_2populations (Additional file 3). The effects of $qNLB1.06_{Tx303}$ relative to $qNLB1.06_{B73}$ on IP (1.5-2.5 days) and DLA (2-10% through the season) were determined by comparing TBBC3-38-05F with B73 (Figure 2A) and TBBC3-39-19E with B73. The significant effect of reducing lesion expansion was not confirmed in the NILs. Lines with $qNLB1.06_{Tx303}$ even showed significantly higher LE scores, indicating some variations from different genetic backgrounds and environments.

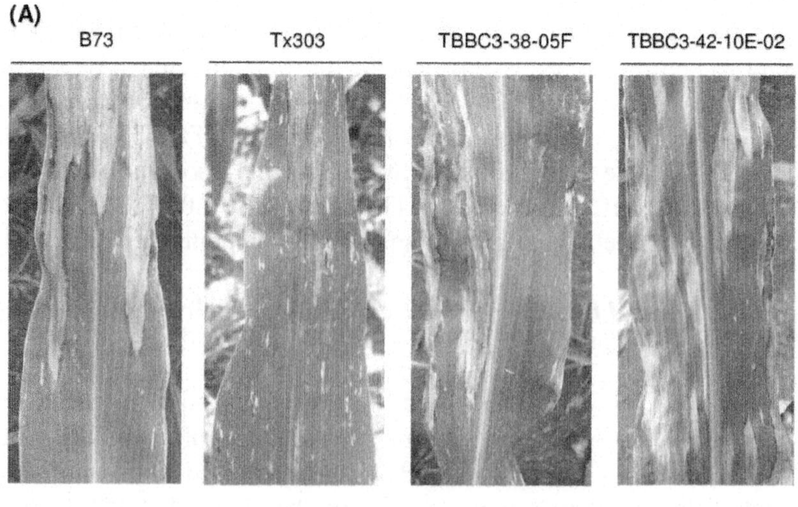

Figure 2: Effects of*qNLB1.06* $_{Tx303}$ and*qNLB1.02* $_{B73}$ in the field. (A) Different levels of NLB resistance in B73, Tx303, TBBC3-38-05F (the NIL carrying *qNLB1.06* $_{Tx303}$), and TBBC3-42-10E-02 (the NIL carrying *qNLB1.02* $_{Tx303}$, which is essentially «replacement of *qNLB1.02* $_{B73}$"). The photographs were taken in Aurora NY, 2008, on the 7th leaves at 28 days after inoculation. (B) Replacement of the B73 allele with the Tx303 allele at bin 1.02 largely increased the susceptibility to NLB. The photograph was taken in Aurora NY, 2008, at 68 days after inoculation. The comparison between B73 and TBBC3-38-05F is not shown because their difference in DLA (< 10%) is not differentiable in photographs.

qNLB1.02 was validated by analyzing six BC_4F_3 NILs and seven BC_4F_4 NILs derived from a B73 × TBBC3-42 cross. Evaluations conducted in 2007 and 2008 led to similar results. Since BC_4F_4 NILs had cleaner genetic backgrounds, the data from the 2008 trial was shown to represent the overall result. Individual trait-locus analysis in this NIL set suggested that bin 1.01-1.02 was the only locus affecting resistance. The NILs carrying Tx303 allele(s) at bin 1.01-1.02 were all significantly more susceptible than the ones carrying B73 allele(s) at the same region (Additional file 5), confirming the resistance effect of*qNLB1.02* $_{B73}$. Highly significant differences between TBBC3-42-10E-02 (the NIL with *qNLB1.02* $_{Tx303}$ in the cleaner background) and B73 were observed across all the disease components under investigation (Figure 2). The B73 allele(s) at*qNLB1.02* delayed lesion formation by 1.6 days ($P = 0.004$), inhibited lesion expansion by 0.84 mm per day ($P < 0.0001$), and reduced DLA 18-38% through the season ($P < 0.0001$). Overall, the relative allele effects detected in the NILs were much greater than in the segregating populations, and the resistance of *qNLB1.02* $_{B73}$ was more effective in the field than in the greenhouse.

Characterization of *qNLB1.02* $_{B73}$ and *qNLB1.06* $_{Tx303}$ using Derived NILs

Two derived NILs carrying identified NLB QTL in "cleaner" B73 backgrounds (minimal amount of introgression from Tx303 not associated with NLB resistance) were chosen for detailed QTL characterization. TBBC3-38-05F was a BC_4F_3 NIL with a Tx303 introgression at bin 1.06 (designated as the NIL with *qNLB1.06* $_{Tx303}$), and TBBC3-42-10E-02 was a BC_4F_4 NIL with a Tx303 introgression at bin 1.02 (designated as the NIL with *qNLB1.02* $_{Tx303}$). To investigate the effect of QTL during NLB development, a series of disease components (Table 1) was selected or developed and applied on the two derived NILs in three replicated experiments in the greenhouse during December 2007 - April 2008, and at Aurora, NY during summer 2008.

Microscopic Investigation on the Pathogenesis of *S. turcica*

Trypan blue staining and KOH-aniline blue fluorescence microscopy techniques were used for histological examination of NLB development *in planta*. To understand typical resistant and susceptible interaction patterns in the NLB pathosystem, preliminary microscopic analysis was conducted on the following maize lines challenged with *S. turcica*: B73, Tx303, TBBC3-38, TBBC3-42, the resistant maize inbred line CML52 and CML103, and a susceptible recombinant inbred line IBM262. As illustrated in Figures 3 and 4, pathogenesis was successfully visualized and can be summarized as follows. After landing and attaching on the leaf surface of a susceptible maize plant, the conidium of *S. turcica* germinates. It forms an appressorium, which produces a penetration peg that punches through the cell wall and into the epidermal cell. From this cell, infective hyphae are produced, grow towards the vascular bundle, enter the vasculature and ramify within it. Aggressive hyphal growth and extension in the vascular bundles can be seen. The hyphae can grow for several days inside the vascular bundles without causing visible symptoms (incubation period was approximately seven days on susceptible maize lines). After extensive growth within the vascular veins, hyphae grow out to colonize the neighboring bundle sheath cells and then branch out to colonize the rest of leaves. This progression is consistent with several previous studies on *S. turcica* pathogenesis on maize [23–25]. The limited damage *S. turcica* causes to host tissues at early phase of pathogenesis and the extended incubation period suggests the possibility that this is a hemibiotrophic interaction.

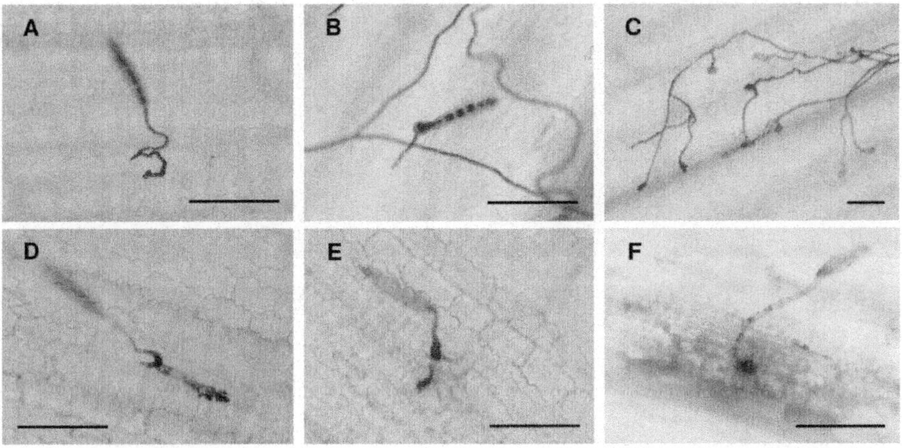

Figure 3: Light micrographs of the infection and early colonization of *Setosphaeria turcica*in corn leaves. Samples from greenhouse trials were stained with trypan blue. (A) A conidium germinated and formed an appressorium. (B) The penetration peg from the appressorium punched through the cuticle and epidermal cell wall. (C) The conidium could continue to produce new germ tubes and appressoria until exhausting its reserves or ultimately gaining entry into a plant cell. This phenomenon was occasionally observed on a resistant inbred line CML52. (D) A subcuticular palm-shaped structure occasionally developed from the penetration peg (likely before hyphae infected the epidermal cell). (E and F) Cytoplasmic depletion of the conidium was seen after the infection process. Infective hyphae spread from primary infected cell to surrounding area, causing host cell death. (Infected leaf samples of A, D and E: CML52, 2dpi; B: B73, 2 dpi; C: CML52, 3 dpi; F: IBM262, 5 dpi) (Scale bars, 100 μm).

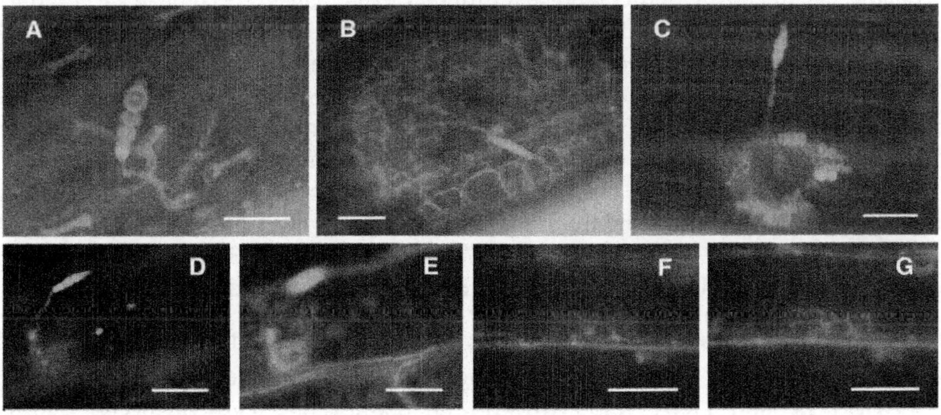

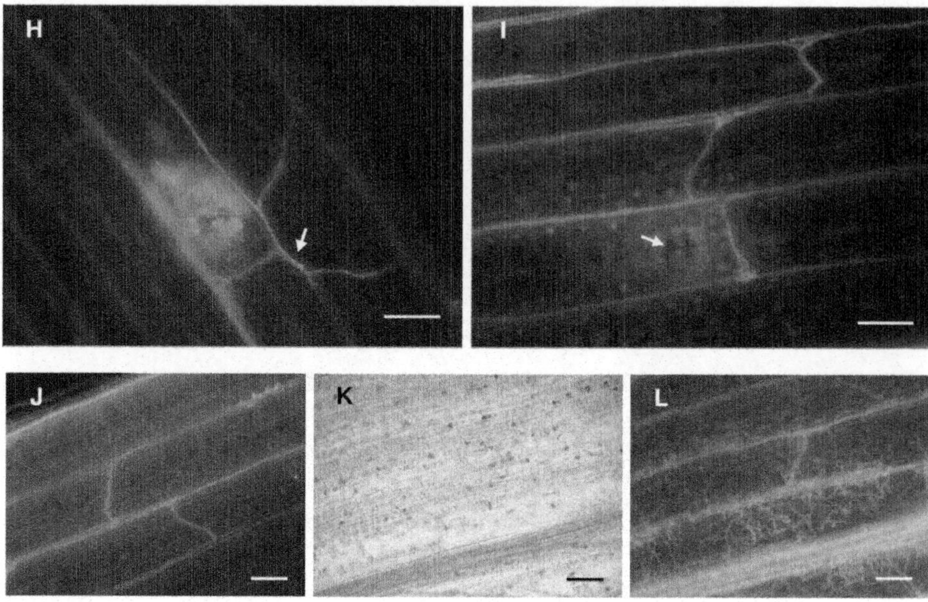

Figure 4: Fluorescence micrographs of the pathogenesis of *Setosphaeria turcica*in corn leaves. Samples from greenhouse trials were treated with KOH-aniline blue. (A) A germinated conidium on leaf surface. (B) Infective hyphae grew into contact with mesophyll cells. (C and H) Defense responses induced around the infection site. The brightly fluorescing area was presumably caused by callose deposition and the accumulation of autofluorescent phenolic compounds. The fluorescing vascular bundles (arrow), possibly due to lignification, could be differentiated from the hyphae growing in the vasculature by the lack of distinguished hyphal coils. (D and E) Infective hyphae grew towards the vascular bundle and invaded it. D and E represent different focal planes. (F and G) Hyphae grew out the vascular vessel to colonize the neighboring bundle sheath cells. F and G represent different focal planes. (I) Hyphae successfully spread through the vascular system. The weak fluorescence and collapsed cells surrounding the infection site (arrow) were typical symptoms of the compatible interaction. (J) Movement of infective hyphae through vascular bundles and cross veins. Leaf tissue remained non-wilted at this stage. (K and L) Colonization of necrotic hyphae in the wilted lesion. Vascular bundles were plugged with aggressively growing hyphae. Hyphae branched out from the colonized vasculature to the rest of the leaf. K and L were viewed in the light- and fluorescent-field, respectively. (Infected leaf samples of A: Tx303, 4 dpi; B: CML52, 4 dpi; C: B73, 4 dpi; D and E: TBBC3-42, 4 dpi; F and G: CML103, 7 dpi; H: B73, 4 dpi; I to L: B73, 10 dpi) (Scale bars, 100 μm).

Among diverse maize genotypes (B73, Tx303, CML52, CML103 and IBM262) and on genotypes carrying different QTL for resistance (B73, TBBC3-38, and TBBC3-42), differential phenotypes were observed for the timing and extent of certain steps in the processes of infection and colonization.

To document these differences, a series of microscopic parameters was chosen for characterizing NLB QTL efficacy at different stages during pathogenesis. The microscopic disease components used in the study included the number of appressoria, infection efficiency, vascular invasion efficiency, the size of strongly fluorescent area surrounding the infection site, and the appearance of necrotic or fortified vascular bundles. The microscopic evaluations on B73, Tx303 and the derived NILs were performed on plants grown in controlled greenhouse conditions as well as in the field.

QTL Effect on the Infection of *S. turcica*

Infection efficiency was examined by trypan blue and KOH-aniline blue staining, followed by scoring the number of conidia that had successfully penetrated the epidermal cell walls. The infection efficiency was defined as the ratio of successful infection sites over the total number of germinated conidia. Non-germinated conidia were excluded because of the difficulty of determining whether a non-germinated conidium was viable, given the fact that a small proportion of damaged conidia was always present in the spore suspension.

(A)

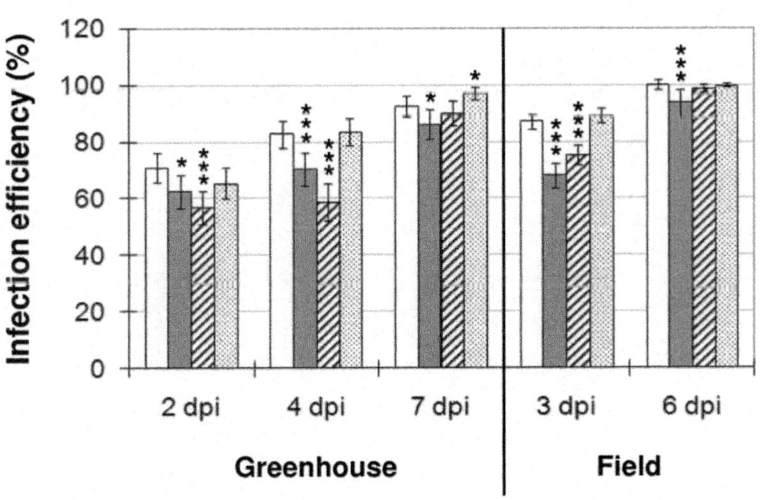

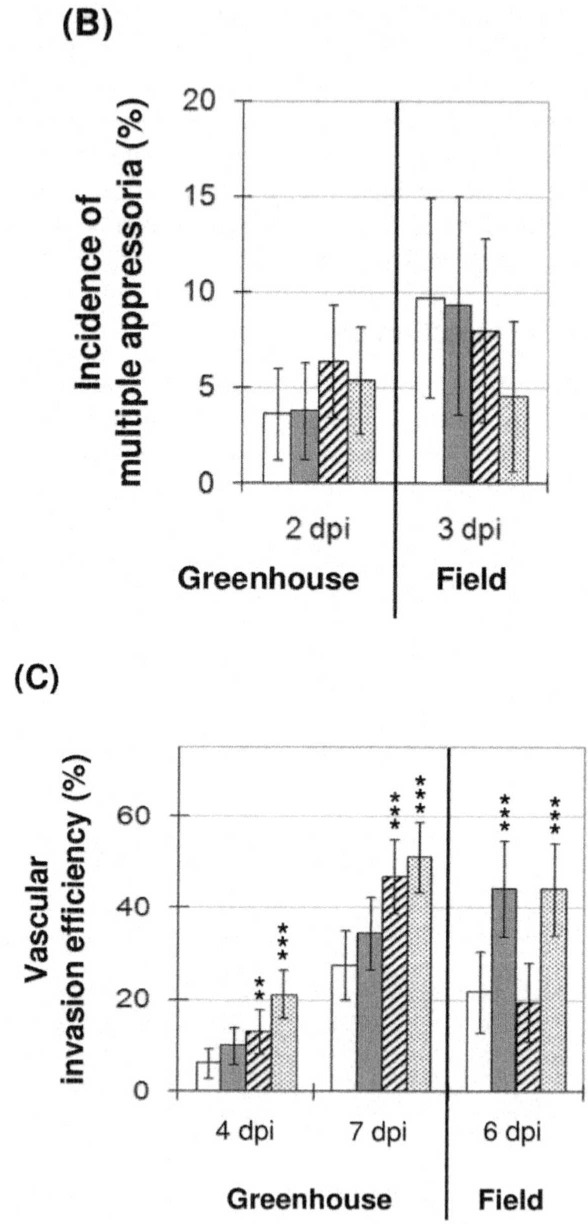
(B)
Incidence of multiple appressoria (%)
20
15
10
5
0
2 dpi
3 dpi
Greenhouse
Field
(C)
Vascular invasion efficiency (%)
60
40
20
0
4 dpi
7 dpi
6 dpi
Greenhouse
Field

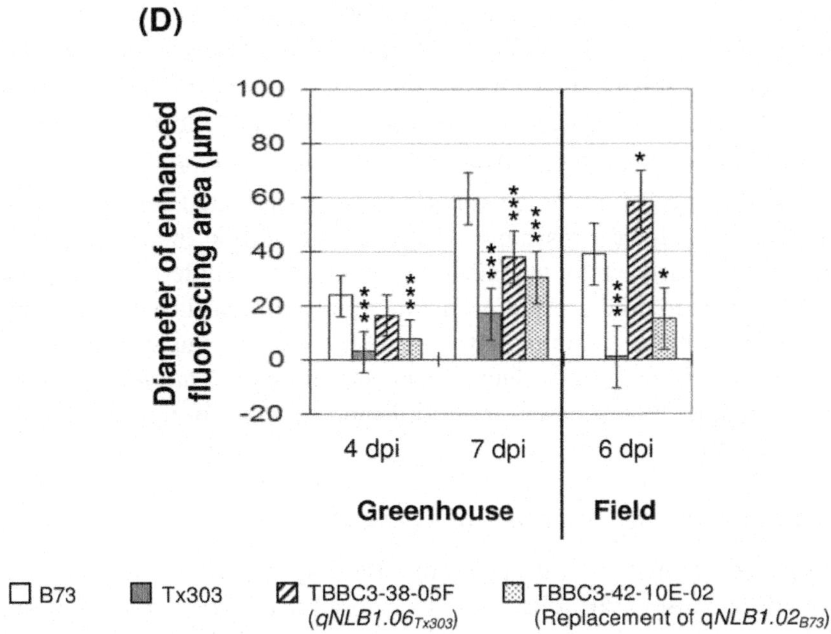

Figure 5: Investigation of QTL effects on microscopic disease components. Microscopic disease components including (A) Infection efficiency, (B) incidence of multiple appressoria, (C) vascular invasion efficiency, and (D) size of strongly fluorescing area surrounding the infection site, were used to assess QTL effects in controlled greenhouse condition and field. Infected leaf samples were collected from maize genotypes B73 (open), Tx303 (gray), TBBC3-38-05F (hatched; the NIL carrying $qNLB1.06_{Tx303}$) and TBBC3-42-10E-02 (dotted; the NIL carrying $qNLB1.02_{Tx303}$, which is essentially «replacement of $qNLB1.02_{B73}$«). Samples collected 2 days post inoculation (dpi) from greenhouse and 3 dpi from field were stained with trypan blue, while the samples collected 4 dpi and 7 dpi from greenhouse and 6 dpi from field were treated with KOH-aniline blue fluorescence technique. Differences between least square means of different genotypes relative to B73 were determined by two-tailed Student›s t test (significance level: * $0.01 < P < 0.05$; ** $0.001 < P < 0.01$; *** $P < 0.001$). In graphs A - C, the proportion data were arcsine transformed for statistical analysis, and the corresponding least squares means and 95% confidence intervals were back transformed to original scale before plotting. The confidence intervals are bigger than significance levels due to asymmetry resulting from back transformation.

In both the greenhouse and field, lower infection efficiency was consistently observed on the NIL with $qNLB1.06_{Tx303}$ at earlier time points examined, as well as on Tx303 at all time points examined (Figure 5A), compared with B73. In the greenhouse, $qNLB1.06_{Tx303}$ reduced infection efficiency by ~14% at 2 dpi and by ~24% at 4 dpi ($P < 0.0001$). Similarly, in the field, infection efficiency was reduced by ~12% at 3 dpi ($P < 0.0001$) relative to B73. The resistant parental genotype Tx303 showed a stronger effect on reducing the infection of *S. turcica* in the field (decreased by ~19% relative to B73, $P < 0.0001$) than in the greenhouse (decreased by 6.5-12.5% relative to B73, $P < 0.05$). No significant effect on pathogen penetration was seen from *qNLB1.02* $_{Tx303}$.

The incidence of multiple appressoria per germinated conidium was used as another indicator for resistance to fungal penetration. It was observed that when a germinated conidium failed to penetrate the epidermis, the conidium would continue to produce new germ tubes and appressoria until exhausting its reserves or ultimately gaining entry into a plant cell (Figure 5B). Despite the effect of *qNLB1.06* $_{Tx303}$ on reducing the penetration of *S. turcica*, no difference was detected on the formation of multiple appressoria among B73, Tx303, and the derived NILs.

QTL Effect on the Vascular Invasion of *S. turcica*

Intracellular hyphal growth from the initially infected epidermal cell to surrounding mesophyll cells, and the subsequent invasion of the vascular bundle, were investigated by KOH-aniline blue fluorescence microscopy. Vascular invasion efficiency was defined as the incidence of hyphal growth into the vasculature per infection site. Differences in vascular invasion efficiency were observed between B73 and the other genotypes (Figure 5C). The NIL with *qNLB1.02* $_{Tx303}$ had greater vascular invasion efficiency, indicating the superiority of the B73 allele(s) at this locus. Replacement of the B73 allele with the Tx303 allele at *qNLB1.02* (in lines carrying an introgression from Tx303 at this locus) led to 15-24% of increase in the incidence of vascular invasion in both greenhouse and field ($P < 0.0001$). The *qNLB1.06* $_{Tx303}$ allele(s) was not effective for inhibiting hyphal growth into the vasculature. Under greenhouse conditions, the NIL with *qNLB1.06* $_{Tx303}$ was observed to have a 7-19% higher susceptibility for vascular invasion ($P < 0.0001$), which is a further indication that this QTL functions at the stage of penetration but not at later stages of pathogenesis. The susceptibility to vascular invasion was also seen in Tx303. In the greenhouse, higher vascular invasion efficiencies were observed in Tx303 than in B73, although with marginal significance levels. In the field, the invasion efficiency in Tx303 was ~22% greater than in B73 ($P < 0.0001$).

QTL Effects on the Accumulation of Defense Compounds around the Infection Site

Infected tissues stained with aniline blue were also used for revealing defense responses induced upon pathogen challenge. The fluorescence emitted from the cells surrounding the infection site is presumably due to callose deposition [60, 61] and the accumulation of autofluorescent phenolic compounds [62–64]. Callose and phenolics are antimicrobial compounds that help restrict the establishment and spread of *S. turcica* in the leaf. In susceptible genotypes, weak fluorescence and collapsed cells were generally observed around the infected sites (Figure 4I), suggesting the lack of induced defense responses. The more resistant genotypes showed enhanced fluorescence from localized cell regions (Figure 4C and 4H).

To assess the degree of defense response around the primary infected cell, the diameter of brightly fluorescing area was measured under a fluorescence microscope with a micrometer (Figure 5D). The average diameter of the fluorescing area in the NIL with $qNLB1.02_{Tx303}$ was consistently smaller than that seen for B73 ($P < 0.0001$ in greenhouse, $P = 0.02$ in the field), indicating that the B73 allele(s) at *qNLB1.02* contribute to the regulation, production or accumulation of callose, lignin or other phenolic compounds. Interestingly, compared to B73, the fluorescing area in the NIL with $qNLB1.06_{Tx303}$ was slightly larger ($P = 0.052$) in the field, but was significantly smaller ($P < 0.0001$) at 7 dpi in the greenhouse, suggesting that the phenotype conferred by $qNLB1.06_{Tx303}$ is vulnerable to environmental influences.

Considering the potential role of vascular resistance in the NLB pathosystem, the incidence of brightly fluorescing vascular bundles (likely caused by lignification) was also evaluated. Although this response was pronounced in CML52 (Figure 4H), this type of resistance was rarely observed in B73, Tx303 and the derived NILs. The induced defense reaction around the infection site was fairly weak at all times examined in either greenhouse or field conditions. It was thus not considered a relevant parameter for testing the effect of $qNLB1.06_{Tx303}$ and $qNLB1.02_{B73}$.

QTL Effect on Mycelial Growth of *S. turcica* in Planta

DNA-based qPCR was developed to precisely quantify mycelial growth of *S. turcica* in maize leaves. The *R*-square values for the two standard curves based on B73 and Tx303 respectively were both higher than 0.99 (Figure 6A), indicating that the designed ITS primer pair had good sensitivity and specificity for reliably amplifying fungal DNA. To assess the qPCR technique as a tool for analyzing the NLB pathosystem, several time-course experiments were performed on seven maize genotypes with a wide range of differential levels of NLB resistance (C. Chung, unpublished). Preliminary results showed that the levels of fungal DNA ratios determined by qPCR approximately conformed to the performance of resistance in the field. For maize genotypes with low to intermediate levels of resistance, fungal DNA was detected as early as three days after inoculation. The detectable fungal DNA ratios increased over time until the end of the incubation period (when visible lesions formed), and thereafter decreased. This reduction in pathogen DNA was unexpected, and apparently corresponded to a loss of DNA integrity. DNA samples extracted from tissues showing lesions were usually brownish and of poor quality. The decrease in detectable fungal DNA ratio in highly diseased leaves is contradictory to our microscopic observation of abundant mycelial growth in NLB lesions. It is likely that in tissues with actively developing lesions, more DNA-degrading enzymes involving in cell death are present and can digest fungal DNA during the extraction. To better differentiate resistant and susceptible responses, it is critical to apply qPCR on infected tissues taken prior to the appearance of necrotic lesions on the most susceptible genotype in an experimental set.

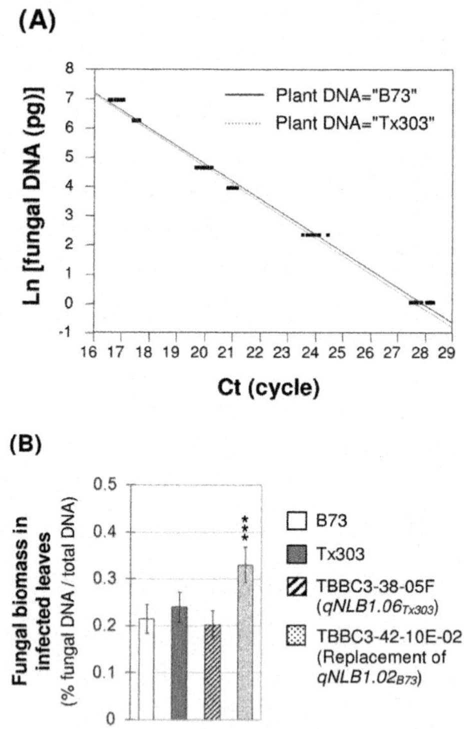

Figure 6: Quantifying QTL effect on mycelial growth of*Setosphaeria turcica in planta*using DNA-based real-time PCR. (A) Two standard curves were constructed by mixing a series of *S. turcica* DNA and 50 ng of maize DNA from non-inoculated B73 and Tx303 plants, respectively. With B73 DNA, Ln[fungal DNA] = 16.912 - 0.606*Ct (R^2= 0.99); with Tx303 DNA, Ln[fungal DNA] = 16.906 - 0.610*Ct (R^2= 0.99). Ln: natural logarithm. Ct: threshold cycle, the number of PCR amplification cycle at which the exponential increase of the product was detected. (B) Measure-

ment of *S. turcica* DNA ratio in infected leaves collected 9 dpi from maize genotypes B73 (open), Tx303 (gray), TBBC3-38-05F (hatched; the NIL carrying *qNLB1.06* $_{Tx303}$) and TBBC3-42-10E-02 (dotted; the NIL carrying *qNLB1.02* $_{Tx303}$, which is essentially «replacement of *qNLB1.02* $_{B73}$«). Differences between least square means of different genotypes relative to B73 were determined by two-tailed Student›s t test (significance level: * 0.01 $<P<0.05$; ** 0.001 $<P<0.01$; *** $P<0.001$). The proportion data were arcsine transformed for statistical analysis, and the corresponding least squares means and 95% confidence intervals were back-transformed to original scale before plotting. The confidence intervals are bigger than significance levels due to asymmetry resulting from back transformation.

Infected leaves from B73, Tx303, and two derived NILs were collected nine days after inoculation (according to the IP of the susceptible NIL), and measured for their fungal DNA content using qPCR. As shown in Figure 6B, neither Tx303 nor*qNLB1.06* $_{Tx303}$ showed significant effect on reducing the growth of *S. turcica* in leaves. However, the fungal biomass ratio (% of fungal DNA in the infected leaf) in the NIL with *qNLB1.02* $_{Tx303}$ was 24% higher than in B73 ($P<0.0001$), indicating that the *in planta* development of *S. turcica* is more extensive on plants without *qNLB1.02* $_{B73}$.

Preliminary Characterization Of Potential QTL Effects for Multiple Disease Resistance using Derived NILs

Significant differences were observed between Tx303 and B73 for resistance to Stewart's wilt, anthracnose stalk rot (ASR), common smut and common rust. In the field, Tx303 was more resistant to Stewart's wilt and ASR, while B73 was more resistant to common smut and common rust. To investigate the resistance spectrum of *qNLB1.02* and *qNLB1.06*, several NILs derived from B73 × TBBC3-38, B73 × TBBC3-39 and B3 × TBBC3-42 were evaluated for the four diseases (Additional files 4 and 5). The choice of NILs was constrained by the availability of seeds.

The evidence suggested that both *qNLB1.06* $_{Tx303}$ and *qNLB1.02* $_{B73}$ were effective for resistance to Stewart›s wilt. As shown in Additional file 4, all the NILs carrying Tx303 introgression at bin 1.06 were significantly more resistant than B73. Based on comparing TBBC3-39-19E to B73, *qNLB1.06* $_{Tx303}$ was inferred to be effective for reducing lesions of Stewart›s wilt on inoculated leaves by as much as 39% ($P<0.0001$). Likewise, in Additional file 5, lines with Tx303 alleles at bin 1.01-1.02 were significantly more susceptible than B73. The effect of *qNLB1.02* $_{B73}$ on Stewart›s wilt was estimated as reducing 19% of primary DLA based on a comparison of TBBC3-42-10E-02 and B73 ($P<0.0001$). Significant differences were also observed among the lines fixed (homozygous for Tx303 or B73 alleles) at bins 1.01-1.02, indicating that there are QTL other than*qNLB1.02* $_{B73}$ segregating in the NIL set. However, none of the rest of the introgressed regions were unambiguously associated with resistance.

qNLB1.06 $_{Tx303}$ was ineffective for resistance to ASR, smut and rust. No phenotypic difference in the development of rust and smut galls was detected among B73, TBBC3-38, TBBC3-39 and their derived NILs. Significant difference in discolored internode area was observed between TBBC3-38 and B73 (~32%, $P = 0.001$), but the variation was not associated with*qNLB1.06*, as TBBC3-39 and the NIL carrying a single introgression at bin 1.06 (TBBC3-39-19E) showed the same resistance level as B73 (Additional file 4).

qNLB1.02 $_{B73}$ was ineffective for resistance to ASR and common smut, but potentially effective for resistance to common rust (Additional file 5). For ASR, some phenotypic difference was found among B73, TBBC3-42 and derived NILs, but none of the identified introgressed regions were associated with resistance. B73 was more resistant than Tx303 for common smut and common rust. For common smut, there were almost no ear galls or stalk galls observed on B73, TBBC3-42 and TBBC3-42-10E-04, suggesting that introgressed segments in TBBC3-42 did not affect smut development. (More ear galls were scored on TBBC3-42-10E-04 in the 2008 trial, but the significance was marginal ($P = 0.04$), and the finding was not consistent with the 2007 result). For common rust, except for TBBC3-42-06F-2, lines with Tx303 alleles at bin 1.01-1.02 were consistently more susceptible than B73 for all three severity ratings (differed by 1.1-3.1 scales in severity, $P < 0.025$) and AUDPC (differed by about 43-53 scale-day, $P < 0.001$). The result indicated the possibility of rust QTL at bins 1.01-1.02 and/or unidentified introgressed regions in TBBC3-42.

DISCUSSION

Identification of NLB QTL using TBBC3 Introgression Lines

Two QTL for resistance to NLB, *qNLB1.02* $_{B73}$ and *qNLB1.06* $_{Tx303}$, were successfully identified, validated and characterized using a population of introgression lines and derived NILs. We applied a stepwise strategy that allowed phenotyping of informative NILs over a series of generations. A special field design, in which each row of CSSL/NIL arranged next to a row of B73, allowed accurate visual comparisons and a relatively uniform epidemic across the field. We detected and sequentially validated QTL using multiple disease components, in a full set of 82 lines, a subset of 15 selected lines, five selected BC_4F_2 populations, and three sets of derived BC_4F_3/BC_4F_4 NILs. Our primary hypothesis that individual QTL affect distinct stage(s) of NLB development was ultimately confirmed using two selected NILs (compared to B73) in repeated greenhouse and field trials, with a panel of conventional and novel components. Our results revealed that *qNLB1.06* $_{Tx303}$ and *qNLB1.02* $_{B73}$ are mainly effective against the infection and colonization of *S. turcica*, respectively. We found that the QTL were both effective in both juvenile and adult plants, and that both chromosomal segments were associated with resistance to more than one disease.

The mechanism underlying resistance conferred by *qNLB1.02* $_{B73}$ and *qNLB1.06* $_{Tx303}$ is unlikely to be the same as the mechanism underlying qualitative resistance conferred by currently known major genes. Qualitative resistance to NLB is generally characterized by chlorotic-necrotic lesions [65–69], chlorotic halo lesions [70], or extremely prolonged IP [71]. Considerable levels of resistance were seen for *qNLB1.02* $_{B73}$ and *qNLB1.06* $_{Tx303}$ in the two field sites, without observing distinct lesion type or much greater IP. Moreover, although co-localized NLB QTL at bins 1.02 and 1.06 have been reported in other populations [72, 73], no major genes have been mapped to these chromosomal regions.

Conventional and Newly-Developed Components of Resistance Targeting Different Stages of NLB Development

QTL effects were analyzed using five conventional macroscopic disease parameters and four microscopic parameters. Each of the macroscopic components reflected different phase(s) of disease development. Incubation period quantifies the time to appearance of wilted lesions, which reflects the speed of xylem plugging due to extensive hyphal growth in the veins. Lesion expansion quantifies the rate of expansion of wilted and necrotic lesions, which reflects the speed of destructive hyphal growth in the leaves. Ratings for diseased leaf area, disease severity and area under the disease progress curve, on the other hand, involve visual quantification of overall disease progress on the entire plant. In our greenhouse and field trials, IP showed a better correlation than LE with DLA, disease severity and AUDPC. This indicates that IP is a more discriminating parameter than LE for measuring NLB severity. This confirms previous reports that IP is a convenient target trait in selection and breeding for resistance to NLB [11, 73–78]. In a recurrent selection study for NLB resistance, selection for prolonged IP resulted in ~7.5% more gain in reducing AUDPC per selection cycle compared to selection for lesion length [11].

QTL effects were also microscopically investigated at the pre-penetration, infection, and colonization phases. In most fungal pathosystems, pre-penetration resistance is associated with specialized physical and chemical features of plant surface which help reduce the incidence of landing, adhesion, germination, appressorium formation and penetration of pathogenic fungi [79]. Post-penetration resistance to microbial attack, on the other hand, is characterized by programmed necrosis, along with the induction of callose, lignin, phenolic compounds and other pathogenesis-related proteins around the primary-infected cell. Four microscopic components, including the incidence of multiple appressoria, infection efficiency, accumulation of defense materials surrounding the infection site, and vascular invasion efficiency, were developed to complement the conventional macroscopic components. Modified trypan blue staining [25, 42] and KOH-aniline blue fluorescence [43] methods were useful in determining the degree and timing of allelic contribution to quantitative resistance to NLB. Previous studies utilizing sectioning and electron microscopy have revealed that xylem plugging is a key stage for the formation of NLB lesions (Jennings and Ullstrup, 1957). Our microscopic examination using KOH-aniline blue fluorescence technique confirmed the finding and allowed the observation of *in planta* colonization of *S. turcica* from a different angle. In all the maize genotypes examined, the mycelium appeared to grow preferentially towards vascular bundle, initially invading the vasculature, then extending through the xylem vessel, and eventually, aggressively growing out to the neighboring mesophyll tissues. To our knowledge, the mechanisms of the post-infection directional growth of vascular fungal pathogens remain to be elucidated.

qPCR was employed for the first time for the measurement of *in planta* biomass of *S. turcica*. Although it has been widely applied in other pathosystems, qPCR can only be used for NLB quantification at early stages of pathogenesis. The accuracy and differentiating power are highly dependent on uniform inoculation and precise sampling at the early stages of pathogenesis, so the method is best suited to work conducted under controlled conditions. The biggest constraint of this technique comes from its not being able to measure the destructive mycelial growth in the blighted leaf areas, possibly due to the poor quality of DNA extracted from necrotic tissues. The cost of qPCR reagents and the required workload make it unfeasible for regular phenotypic screening. However, qPCR serves as a good tool for quantifying the degree of earlier mycelial colonization of *S. turcica* in maize leaves.

Hypothesis #1: Individual QTL Affect Distinct Stages of Disease Development

Detailed macroscopic and microscopic evaluations revealed the distinct features of $qNLB1.06_{Tx303}$ and $qNLB1.02_{B73}$ in NLB development. *qNLB1.06* $_{Tx303}$ conditions resistance mainly against the penetration of *S. turcica*. The anti-penetration effect was observed at earlier but not later time points examined, indicating that *qNLB1.06* $_{Tx303}$ acts in a quantitative manner, which delays rather than prevents the occurrence of infection. In contrast, *qNLB1.02* $_{B73}$ appears to condition resistance not to infection, but rather by enhancing the induction of defense reactions surrounding infection sites, as well as by inhibiting hyphal growth into the vascular bundle, and the subsequent necrotrophic colonization in the leaves. It is difficult to infer whether the resistance is associated with defense reactions in mesophyll, parenchyma, bundle sheath and/or other cells.

Delaying the invasion and extension of *S. turcica* in the vascular system was shown to be critical for quantitative resistance to NLB. This is consistent with previous microscopic analyses [23, 24] suggested that xylem invasion/ plugging is an important pathogenetic phase in NLB progress. Polygenic resistance (in an inbred line CI90A) has been associated with reduced hyphal spread into the xylem [22]. The significance of protecting the vasculature from hyphal invasion is supported by the evidence that the replacement of superior B73 allele with Tx303 allele at *qNLB1.02* led to an 18-38% higher DLA (relative to B73). Incorporating a QTL effective in slowing down the invasion of the vasculature would thus considerably increase overall resistance.

We speculate that preventing the attachment of conidia and the subsequent infection plays a role in quantitative resistance to NLB in the donor line Tx303. Although this was not quantified, it was visually obvious that the number of spores per leaf segment of Tx303 was lower than the spore numbers on B73 or the NILs carrying *qNLB1.06* $_{Tx303}$ and *qNLB1.02* $_{B73}$. Lower infection efficiency of *S. turcica* was observed for Tx303 relative to B73. Interestingly, Tx303 showed a greater susceptibility than B73 for the parameters of vascular invasion efficiency, enhanced fluoresced area, and lesion expansion. Despite the lack of resistance in delaying vascular invasion and extension of *S. turcica* in Tx303, its higher effectiveness in reducing spore attachment and infection may contribute to the overall moderate resistance to NLB.

Hypothesis #2: the Effectiveness of Disease QTL is Affected by Plant Maturity

A number of maize diseases caused by hemibiotrophic fungi (eg. gray leaf spot, anthracnose leaf blight, anthracnose stalk rot) and necrotrophic fungi (eg. southern leaf blight) are known to progress more rapidly on the plants after anthesis [18,49, 80, 81]. In maize, a significant correlation has been detected between disease resistance and flowering time in a panel of 253 diverse lines (R. Wisser, J. Kolkman, P. Balint-Kurti, and R. Nelson, unpublished). The fact that flowering time may account for a considerable proportion of resistance variation leads to the hypotheses that the effects of disease QTL reflect indirect expression of flowering time QTL, and/or that the effects of disease QTL are modulated differently at varied plant developmental stages. In fact, several QTL identified in biparental populations are associated with both NLB resistance and flowering time (J. Poland, pers. comm.).

The association between NLB QTL and plant maturity was investigated in the subset of 15 selected TBBC3 lines. To separate resistance effects from maturity effects, plants in two different fields were inoculated at juvenile and adult stage (two weeks before tasselling) respectively, and evaluated for NLB resistance and flowering time. Generally consistent expression of resistance or susceptibility was observed at the two stages, implying that the effectiveness of these QTL was not altered substantially by plant maturity. While several of the introgression lines (10 out of the 15 TBBC3 lines with differential resistance relative to B73) showed significantly different flowering times relative to B73, the interactions between NLB QTL and flowering time QTL were relatively minor. For the $qNLB1.02_{B73}$ and $qNLB1.06_{Tx303}$, the analysis of advanced NIL sets further provided strong evidence of their independence from flowering time. This agrees with a previous report of flowering-time QTL in TBBC3 introgression lines [27]: bins 1.01-1.02 and 1.06 were not among the QTL affecting days to anthesis, days to silking, or anthesis-silking interval.

Hypothesis #3: the Effectiveness of QTL is Affected by Environmental Conditions

A major limitation for QTL applications is the lack of consistency of QTL effects across environments. The inconsistent detection of QTL has been associated with experimental errors and differential gene expression affected by environmental factors [82]. In view of the widely reported inconsistency, the present study revealed that the introgressions/QTL with relatively large effects (carried in seven introgression lines: TBBC3-38, TBBC3-39, TBBC3-42, TBBC3-36, TBBC3-21, TBBC3-30, and TBBC3-77) were consistent in their performance across field sites and years. Reliable expression of these QTL, including $qNLB1.02_{B73}$ and $qNLB1.06_{Tx303}$, suggests their applicability to resistance breeding. Disease resistance can be expressed differentially, not only in different field environments but also under field versus greenhouse conditions (eg. [80, 83, 84]). Differential efficacy of $qNLB1.02_{B73}$ and $qNLB1.06_{Tx303}$ in controlled greenhouse and field conditions was evaluated using macroscopic and microscopic phenotypes. It should be noted that due to the different inoculation and sampling treatments necessarily employed in different environments, the comparison between the effectiveness of disease QTL in the greenhouse and field should be addressed based on relative observation rather than the absolute values. Our overall results suggest that the resistance of *qNLB1.02* $_{B73}$ was stably expressed in the greenhouse and field, whereas *qNLB1.06* $_{Tx303}$ conferred a higher level of resistance in the field than in the greenhouse. Macroscopically, the effects of *qNLB1.06* $_{Tx303}$ on IP and primary DLA in greenhouse-grown plants were occasionally insignificant. Microscopically, the positive effect of *qNLB1.06* $_{Tx303}$ on triggering defense response surrounding the initially infected cell was only detected in the field. Greenhouse-grown plants carrying *qNLB1.06* $_{Tx303}$, however, displayed an opposite effect (smaller fluorescent area), which suggests that this type of resistance is highly affected by environmental factors. The observation that *qNLB1.06* $_{Tx303}$ increased the efficiency of *S. turcica* invasion into vasculature in the greenhouse but not field, may be an indirect effect of decreased accumulation of host defensive materials at the post-infection stage.

Hypothesis #4: QTL for Resistance to NLB Comprises Genes or Gene Clusters Involving in Broad-Spectrum Resistance

The phenomenon of multiple disease resistance (MDR) has been inferred based on the detection of QTL clusters affecting different diseases [2, 19, 20, 85], genetic correlations in populations [86], non-specific defense mechanisms (eg. SAR [87–89]), and genes (eg. *npr1* [90] and *mlo* [91]). Maize bins 1.02 and 1.06 have been previously associated with a number of disease QTL mapped in diverse maize populations [20]. Our evaluations of resistance to common rust, Stewart›s wilt, anthracnose stalk rot, common smut, and common rust in sets of NILs suggest that both *qNLB1.02* $_{B73}$ and *qNLB1.06* $_{Tx303}$ encompass gene(s) contributing nonspecific defense effects. *qNLB1.02* $_{B73}$ was effective in decreasing NLB, common rust and Stewart›s wilt, while *qNLB1.06* $_{Tx303}$ was effective in decreasing NLB and Stewart›s wilt. Nevertheless, due to the low resolution of QTL localization in this study, it is unclear whether the nonspecific resistance of *qNLB1.02* $_{B73}$ and *qNLB1.06* $_{Tx303}$ is caused by pleiotropy or linkage, or whether their effects are mongenic or polygenic. Although the underlying mechanisms are unknown, the feature of broad-spectrum resistance makes the two identified disease QTL appealing in practical applications.

It is worth noting that the NILs carrying *qNLB1.06* $_{Tx303}$ showed remarkable resistance against Stewart›s wilt, suggesting the involvement of major gene effect. In fact, a dominant major gene locus for Stewart›s wilt, *Sw1*, has been mapped to bin 1.05-1.06 in inbred line Ki14 [92, 93]. The co-localization of QTL for resistance to NLB and Stewart's wilt has also been observed in the NILs derived from the inbred line CML52 crossed to B73 (CML52 allele for resistance; C. Chung, unpublished).

ELECTRONIC SUPPLEMENTARY MATERIAL

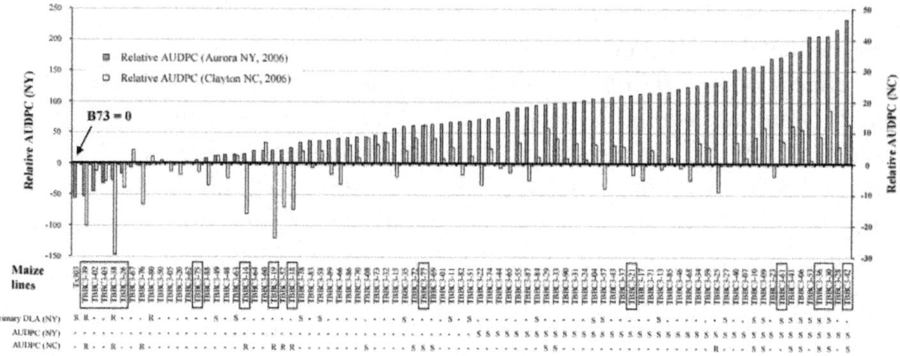

Additional file 1: NLB resistance of the full set of 82 TBBC3 introgression lines.

Additional file 2: Putative NLB QTL identified in the TBBC3 population.

Chr. Bin	Marker	Map Position [a]	Resistance allele	Aurora NY, 2006						Clayton NC, 2006				Previously reported NLB QTL at the locus	
				IP (days)	PrimDLA (%)	DLA1 (%)	DLA2 (%)	DLA3 (%)	AUDPC (%-day)	Severity1 (scale)	Severity2 (scale)	Severity3 (scale)	AUDPC (scale-day)	Mapping population [b]	Reference
Correlated loci															
1.01	*umc1071*	85.2	B73		6.6*		5.5***	8.0***	110.1***	0.7**				None	
1.02	*bnlg1429*	143.5	B73		6.9*	6.2***	7.1***	7.4**	134.7***					B52 x Mo17	[1]
	bnlg1953	170.0	B73		7.9**		6.3***	9.3***	120.0***	0.7**					
4.07	*bnlg1621*	349.6	B73		5.0*	3.0**	4.9***	9.4***	91.8***					D32 x D145	[2]
5.02	*bnlg565*	150.9	B73		4.6*	2.7***	4.9***	10.0***	90.7***					B52 x Mo17	[1, 3]
7.03	*bnlg434*	323.3	B73		6.6*	3.8***	4.7***	4.1*	84.0***					B52 x Mo17	[1, 3]
8.03 [d]	*UMC32B*	199.1	B73			8.8***	12.9***	23.5***	246.0***	1.3**			13.2**	None	
Correlated loci															
1.06 [c]	*umc2234*	529.0	Tx303							-1.1*			-12.2*	D32 x D145	[2]
5.08 [d]	*umc1225*	641.4	Tx303	1.6**	-18.6**					-2.0**	-1.9***	-1.2**	-31.9***	None	
	bnlg1695	664.2	Tx303	1.6**	-18.6**					-2.0**	-1.9***	-1.2**	-31.9***	None	
5.09 [d]	*umc1829*	671.5	Tx303	1.6**	-18.6**					-2.0**	-1.9***	-1.2**	-31.9***	None	
Correlated loci															
4.03	*umc2082*	141.6	B73		11.6***		7.4***	12.4***	132.9***	0.6*	0.4*		7.4*	D32 x D145	[2]
	umc2176	174.6	B73		11.6***		7.4***	12.4***	132.9***	0.6*	0.4*		7.4*		
5.04	*bnlg1208*	323.1	B73	-0.6*	15.4***	3.3**	6.3***	7.4**	103.8***	0.8*	0.5*	0.4*	10.2**	Lo951 x CML202	[4, 5]
														D32 x D145	[2]
Correlated loci															
1.11 [d]	*bnlg131*	1065.6	Tx303							-1.5***	-0.9*		-16.5**	IL731a x W6786	[6]
														Lo951 x CML202	[4, 5]
3.06 [d]	*bnlg2241*	452.7	Tx303							-1.5***	-0.9*		-16.5**	IL731a x W6786	[6]
														B52 x Mo17	[1]
														Lo951 x CML202	[4, 5]

[a] The marker position was based on genetic map of the intermated B73 x Mo17 population (version IBM 2008 neighbors).
[b] The mapping population in which the same locus was detected for resistance to NLB. The resistance donor was underlined.
[c] This marker was first found highly associated with NLB resistance in segregating populations B73 x TBBC3-38 and B73 x TBBC3-39. It was then tested in 82 TBBC3 lines.
[d] Introgressed segment at the locus was only present in a single TBBC line (bin 8.03 in TBBC3-42; bin 5.08-5.09 in TBBC3-38; bin 1.11 and 3.06 in TBBC3-30).

Relative area under the disease progress curve (Relative AUDPC) values shown are the differences of least squares means (from mixed models) between TBBC3 lines and B73 recurrent parent. AUDPC was calculated from three diseased leaf area (DLA) scores in the 2006 trial in NY (solid bars), or three disease severity scores in the 2006 trial in NC (open bars). In NY, primary DLA was also rated for diseased leaf area on inoculated leaves. The letters "R" and "S" below the graph indicate the lines significantly more resistant and more susceptible than B73 at $P < 0.05$, respectively, based on primary DLA and AUDPC. The 15 TBBC3 lines selected for subsequent phenotypic validation are indicated by rectangles highlighting the maize line designation.

Putative QTL for northern leaf blight (NLB QTL) affecting incubation period (IP), primary diseased leaf area (PrimDLA), diseased leaf area (DLA), disease severity (severity), and AUDPC (area under the disease progress curve calculated from DLA or disease severity) were identified using 82 TBBC3 introgression lines. QTL effects for each marker locus are the significant differences of least squares means of Tx303 homozygous genotypes at the locus relative to B73 recurrent parent line (* $0.01 < P < 0.05$, ** $0.001 < P < 0.01$, *** $P < 0.001$). Putative QTL are reported as correlated groups because of the high dependencies among those introgressed segments in TBBC3 lines.

Additional file 3: Validation of NLB QTL in the BC_4F_2 segregating populations.

Introgressed region (Chr. bin)[a]	Map interval (cM)[b]	SSR marker (map position, cM)[b]	Parental donor lines of the BC_4F_2 population[c]	Environment (location, year)	Resistance allele	R^2	Relative allele effect ($LSMean_{Tx303} - LSMean_{B73}$)[d]					
							IP (dpi)	LE (mm/day)	DLA1 (%)	DLA2 (%)	DLA3 (%)	AUDPC (%-day)
1.06[§]	487.5-590.5	*umc2234* (529.0)	TBBC3-38	Aurora NY, 2006	Tx303	0.330[f]	1.96**	-0.29*	-6.36***	-6.69***	-5.31***	-119.27***
1.06[§]	487.5-590.5	*umc2234* (529.0)	TBBC3-39	Aurora NY, 2006	Tx303	0.526[f]	2.38***	-0.33**	-5.79***	-8.39***	-9.54***	-151.70***
1.01-1.02[§]	0-184.2	*bnlg1953* (170.0)	TBBC3-42	Greenhouse, 2006	B73	0.139[e]	-0.60***	0.35*	na	na	na	na
1.01-1.02[§]	0-184.2	*bnlg1953* (170.0)	TBBC3-77	Aurora NY, 2006	B73	0.074[f]	-0.58	0.11	6.05***	3.36*	1.42	70.95*
5.00	0-30.0	*mmc0151* (20.5)	TBBC3-77	Aurora NY, 2006	B73	0.059[f]	-0.71	0.26	4.30**	4.37**	3.04*	80.37**
5.01-5.02	115.1-153.9	*bnlg565* (150.9)	TBBC3-02	Aurora NY, 2006	B73	0.053[e]	-1.30**	0.03	1.98	0.87	0.52	21.13
5.08-5.09	603.0-676.1	*bnlg1829* (313.3)	TBBC3-38	Aurora NY, 2006	Tx303	0.046[f]	0.71	0.01	-2.02	-2.45*	-3.67*	-49.85**
8.03-8.05	265.2-344.0	*umc1130* (330.1)	TBBC3-02	Aurora NY, 2006	Tx303	0.114[f]	0.41	-0.04	-2.97**	-3.90***	-3.02***	-68.90***

[§] The QTL effects on NLB resistance were further confirmed in derived BC_4F_3 and/or BC_4F_4 NILs.

[a] Introgressed regions that were significantly associated with traits in segregating populations. The genome regions are shown as bin positions (eg. bin 1.06 is the sixth segment in maize chromosome 1).

[b] SSR marker with the most significant effects on traits. The marker position was based on genetic map of the intermated B73 x Mo17 population (version IBM 2008 neighbors). The map interval of each introgression is assumed to extend halfway between two markers around the border of each end of the introgression.

[c] The selected TBBC3 lines used for developing the segregating BC_4F_2 populations. The lines were backcrossed to B73 and selfed. Phenotyping and genotyping were conducted on individual plants in the populations.

[d] The significance level was determined by pair-wise two-tailed Student's t test on the least squares difference (denoted as * $0.01 < P < 0.05$; ** $0.001 < P < 0.01$; *** $P < 0.001$).

[e] *R*-square from the ANOVA using IP as the dependent variable, referring to the proportion of the variation in IP accounted for by the marker in the third column.

[f] *R*-square from the ANOVA using AUDPC as the dependent variable, referring to the proportion of the variation in AUDPC accounted for by the marker in the third column.

The BC_4F_2 populations were genotyped and phenotyped for incubation period (IP), lesion expansion (LE), diseased leaf area (DLA), and area under the disease progress curve (AUDPC). The trait-marker association was tested by ANOVA at $P < 0.05$. Introgressions/markers significantly associated with NLB resistance are listed. The relative allele effects are the differences on the least squares means (LSMean) between Tx303 homozygous genotypes and B73 homozygous genotypes at the locus.

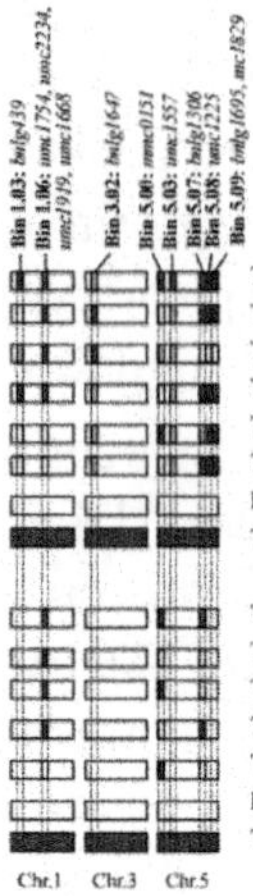

	Northern leaf blight							Stewart's wilt	Anthracnose stalk rot	Common smut		Common rust	Anthesis
	IP	LE	DLA1	DLA2	DLA3	DLA4	AUDPC	PrimDLA	Discolored internode area	Ear gall	Stalk gall	AUDPC	Days to anthesis
TBBC3-38	**17 A**	1.8 DEFG	**10 H**	**13 I**	**18 IJ**	**20 GH**	**462 J**	**0 N**	174 A	0 BC	0 B	53 H	***78 B***
TBBC3-38-11E	**15 ABC**	1.5 GH	15 DEFG	**14 HI**	**19 I**	**20 GH**	**501 IJ**	**16 JKL**				64 DEFGH	***77 BCDEFGH***
TBBC3-38-05F	**15 BCD**	1.4 H	**10 H**	**13 I**	**19 IJ**	**25 GH**	**504 IJ**	**16 KL**				70 DEFGH	75 EFGHI
TBBC3-38-18A	**14 CDEF**	1.6 FGH	**11 GH**	18 FGH	25 EFGH	**28 F**	**637 H**	**20 IJK**				55 GH	***78 BCD***
TBBC3-38-07H	**15 BCD**	1.6 FGH	13 DEFGH	22 DEFG	28 DEFG	38 CD	748 EFG	**30 HIJ**				59 FGH	***76 CDEFGH***
TBBC3-38-16G	13 DEFG	2.0 BCD	21 BC	28 C	29 DEF	34 DE	840 DE	48 DEF	141 CD	1.1 BC	0 B	74 DEF	***77 BC***
B73	13 F	1.5 H	15 DEF	21 EF	26 GH	33 E	710 G	44 F	142 D	0 C	0 B	61 GH	74 I
Tx303	**14 BCDE**	2.0 BCDE	12 EFGH	**14 HI**	**15 J**	**16 H**	**427 J**	**13 KLM**	**97 E**	2.5 A	1.6 A	83 CD	***87 A***
TBBC3-39	**15 BCD**	1.9 CDEF	12 FGH	**18 GHI**	**24 H**	**29 F**	**616 H**	**3 MN**	148 CD	0.1 BC	0 B	57 GH	***78 B***
TBBC3-39-19E	**16 AB**	1.6 FGH	13 EFGH	18 FG	**19 IJ**	**23 G**	**558 HI**	**5 LMN**	141 CD	0.1 BC	0 B	67 DEFGH	***75 FGH***
TBBC3-39-08F	**14 BCD**	1.6 FGH	13 DEFGH	**18 GHI**	25 FGH	**29 F**	**634 H**	**16 JKL**				61 FGH	***76 CDEFGH***
TBBC3-39-11A	12 FGHIJ	1.7 DEFGH	17 CD	26 CD	29 DE	35 CDE	813 DEF	**11 KLMN**				61 FGH	***76 CDEFGH***
TBBC3-39-09F	12 GHIJ	1.8 DEFG	16 DE	27 C	35 C	38 CD	885 D	**35 GH**	155 ABCD			64 EFGH	75 H
B73	13 F	1.5 H	15 DEF	21 EF	26 GH	33 E	710 G	44 F	142 D	0 C	0 B	61 GH	74 I
Tx303	**14 BCDE**	2.0 BCDE	12 EFGH	**14 HI**	**15 J**	**16 H**	**427 J**	**13 KLM**	**97 E**	2.5 A	1.6 A	83 CD	***87 A***

Additional file 4: Genotypes and disease phenotypes for Tx303, B73 and the NIL sets derived from B73 × TBBC3-38 and B73 × TBBC3-39.

Among the target introgressions at bins 1.03, 1.06, 3.02, 5.00, 5.02-5.03 and 5.07-5.09, only *qNLB1.06* $_{Tx303}$ (Tx303 allele at bin 1.06) was validated for association with resistance to NLB. The open bars and solid bars represent the loci homozygous for B73 alleles and Tx303 alleles, respectively. The gray bars represent heterozygous loci or missing genotypic data. Only the chromosomes with introgressed regions in the two NIL sets are shown. The rest of the genome was assumed fixed for B73 alleles. Trait values are least squares means calculated from the mixed model. Pair-wise Student›s t tests were performed to analyze the differences between each NIL and B73, and between every pair of NILs in each set. Trait values with different letters are significantly different from each other. Disease phenotypes that were significantly more resistant than B73 are highlighted in bold and shaded, while the phenotypes significant more susceptible than B73 are underscored. Lines that showed significantly different days to anthesis are in bold italic. *qNLB1.06* $_{Tx303}$ was also effective for resistance to Stewart›s wilt. (IP: incubation period; LE: lesion expansion; DLA: diseased leaf area; PrimDLA: primary DLA; AUDPC: area under the disease progress curve) (PDF 91 KB)

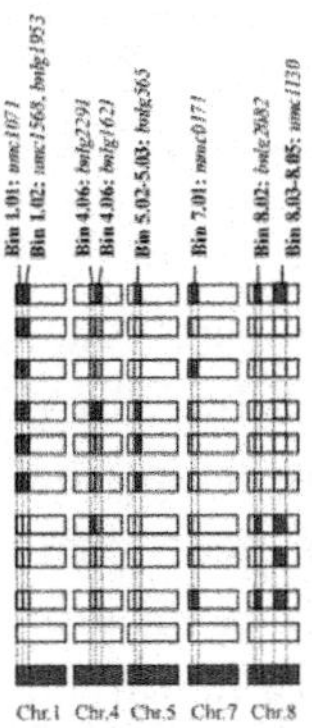

	Northern leaf blight							Stewart's wilt	Anthracnose stalk rot	Common smut		Common rust	Anthesis
	IP	LE	DLA1	DLA2	DLA3	DLA4	AUDPC	PrimDLA	Discolored internode area	Ear gall	Stalk gall	AUDPC	Days to anthesis
TBBC3-42	11 HIJ	2.5 A	30 A	45 A	57 A	67 A	1503 A	73 A	154 ABCD	0.5 BC	0 B	106 A	***76 CDEFG***
TBBC3-42-10E-02	11 HIJ	2.3 AB	33 A	46 A	55 A	71 A	1525 A	63 BC	139 CD			104 AB	***77 BCDE***
TBBC3-42-02E-01	11 IJ	2.2 ABC	31 A	45 A	56 A	71 A	1524 A	71 AB	167 ABC			115 A	***76 CDEFG***
TBBC3-42-06F-02	11 J	2.5 A	25 B	39 B	55 A	69 A	1406 B	58 CD	172 AB			84 BCDE	***76 CDEFGH***
TBBC3-42-12F-01	12 HIJ	2.1 ABC	25 B	38 B	49 B	60 B	1287 C	52 DE	148 ABCD			100 ABC	***76 CDEFGH***
TBBC3-42-10E-04										1.2 B	0.1 B	111 A	***77 BCDEF***
TBBC3-42-05H-04	12 GHIJ	1.5 GH	15 DEFG	24 CDE	30 D	40 C	814 DEF	48 DEF	145 BCD			65 DEFGH	***75 FGH***
TBBC3-42-05H-01	13 FGHI	1.5 FGH	15 DEFG	21 EFG	31 CD	37 CD	781 EF	**30 HI**	146 CD			74 DEFG	***76 DEFGH***
TBBC3-42-09E-01	13 EFGH	1.6 EFGH	15 DEFG	22 DEFG	27 DEFGH	36 CDE	743 FG	45 EFG	142 CD			65 EFGH	***75 GH***
B73	13 F	1.5 H	15 DEF	21 EF	26 GH	33 E	710 G	44 F	142 D	0 C	0 B	61 GH	74 I
Tx303	**14 BCDE**	2.0 BCDE	12 EFGH	**14 HI**	**15 J**	**16 H**	**427 J**	**13 KLM**	**97 E**	2.5 A	1.6 A	83 CD	***87 A***

Additional file 5: Genotypes and disease phenotypes for Tx303, B73 and the NIL set derived from B73 × TBBC3-42.

Among all target introgressions at bins 1.01-1.02, 4.06-4.07, 5.02-5.03, 7.01, 8.02 and 8.03-8.05, only *qNLB1.02* $_{B73}$ (B73 allele at bin 1.02) was validated for association with resistance to NLB. The open bars and solid bars represent the loci homozygous for B73 alleles and Tx303 alleles, respectively. The gray bars represent heterozygous loci or missing genotypic data. Only the chromosomes with introgressed regions in the NIL sets are shown. The rest of the genome was assumed fixed for B73 alleles. Trait values are least squares means calculated from the mixed model. Pair-wise Student›s t tests were performed to analyze the differences between each NIL and B73, and between every pair of NILs. Trait values with different letters are significantly different from each other. Disease phenotypes that were significantly more resistant than B73 are highlighted in bold and shaded, while the phenotypes significant more susceptible than B73 are underscored. Lines that showed significantly different days to anthesis are in bold italic. Preliminary evidence also suggested that *qNLB1.02* $_{B73}$ was effective for resistance to Stewart›s wilt, and QTL at bin 1.01-1.02 and/or 5.02-5.03 were associated with resistance to common rust. (PDF 85 KB)

CONCLUSIONS

Our research has led to successful identification of two reliably-expressed QTL that can potentially be utilized to protect maize from *S. turcica* in different environments. Map-based cloning will reveal more about the genes and mechanisms underlying the distinct features of *qNLB1.02* $_{B73}$ and *qNLB1.06* $_{Tx303}$ in the pre-penetration, penetration and post-penetration phases of pathogenesis. Large mapping populations have been generated from the NILs and fine-mapping of *qNLB1.02* $_{B73}$ and *qNLB1.06* $_{Tx303}$ is in progress.

ACKNOWLEDGEMENTS

We thank Jim Holland for providing the seeds of TBBC3 lines, Jesse Poland for help with rust inoculation, and Pioneer Hi-Bred International Inc. for the supply of sorghum grains used in production of inoculum. We appreciate help from Judith Kolkman, Oliver Ott and Kristen Kennedy with aspects in the research. The work was funded by The CGIAR Generation Challenge Program, The McKnight Foundation, USDA-ARS, and Ministry of Education, Taiwan.

AUTHORS' CONTRIBUTIONS

CC and RN conceived of the study. CC designed and carried out all the experiments in New York, conducted statistical analysis, and drafted the manuscript. JL and EW contributed to the design of microscopic analysis, participated in phenotypic data collection in the greenhouse and field trials in NY during 2007-2008. ZK participated in the inoculation and phenotypic data collection in the field trials in NY in 2008. GVE carried out the field trials in NC. PBK designed and carried out the field trials in NC, participated in coordination of the study, and contributed to its revision. RN participated in the coordination of the work and the experimental design, helped to draft the manuscript and contributed to its revision. All authors read and approved the final manuscript.

REFERENCES

1. Jones JDG, Dangl JL: The plant immune system. *Nature* 2006,444(7117):323–329.
2. Wisser RJ, Sun Q, Hulbert SH, Kresovich S, Nelson RJ: Identification and characterization of regions of the rice genome associated with broad-spectrum, quantitative disease resistance. *Genetics* 2005,169(4):2277–2293.
3. Manosalva PM, Davidson RM, Liu B, Zhu X, Hulbert SH, Leung H, Leach JE: A germin-Like protein gene family functions as a complex quantitative trait locus conferring broad-spectrum disease resistance in rice. *Plant Physiol* 2009,149(1):286–296.
4. Krattinger SG, Lagudah ES, Spielmeyer W, Singh RP, Huerta-Espino J, McFadden H, Bossolini E, Selter LL, Keller B: A putative ABC transporter confers durable resistance to multiple fungal pathogens in wheat. *Science* 2009,323(5919):1360–1363.

5. Fukuoka S, Saka N, Koga H, Ono K, Shimizu T, Ebana K, Hayashi N, Takahashi A, Hirochika H, Okuno K, Yano M: Loss of function of a proline-containing protein confers durable disease resistance in rice. *Science* 2009, 325:998–1001.

6. Poland JA, Balint-Kurti PJ, Wisser RJ, Pratt RC, Nelson RJ: Shades of gray: the world of quantitative disease resistance. *Trends Plant Sci*2009,14(1):21–29.

7. Fu D, Uauy C, Distelfeld A, Blechl A, Epstein L, Chen X, Sela H, Fahima T, Dubcovsky J: A Kinase-START gene confers temperature-dependent resistance to wheat stripe rust. *Science* 2009,323(5919):1357–1360.

8. Parlevliet J: Durability of resistance against fungal, bacterial and viral pathogens; present situation. *Euphytica* 2002, 124:147–156.

9. Young ND: QTL mapping and quantitative disease resistance in plants. *Annu Rev Phytopathol* 1996, 479–501.

10. Broglie KE, Butler KH, Butruille MG, da Silva Conceicao A, Frey TJ, Hawk JA, Jaqueth JS, Jones ES, Multani DS, Wolters PJCC, E.I. du Pont de Nemours and Company, Pioneer Hi-Bred International, Inc., University of Delaware: Polynucleotides and methods for making plants resistant to fungal pathogens. United States Patent 20060223102. *United States Patent and Trademark Office* 2006.

11. Carson ML: Response of a maize synthetic to selection for components of partial resistance to *Exserohilum turcicum* . *Plant Dis*2006,90(7):910–914.

12. Vidhyasekaran P: *Fungal Pathogenesis in Plants and Crops: Molecular Biology and Host Defense Mechanisms.* 2nd edition. Boca Raton, FL: CRC Press; 2007.

13. Moldenhauer J, Pretorius ZA, Moerschbacher BM, Prins R, Van Der Westhuizen AJ: Histopathology and PR-protein markers provide insight into adult plant resistance to stripe rust of wheat. *Mol Plant Pathol* 2008,9(2):137–145.

14. Ullstrup AJ, Miles SR: The effects of some leaf blights of corn on grain yield. *Phytopathology* 1957,47(6):331–336.

15. Raymundo AD, Hooker AL: Measuring the relationship between northern corn leaf blight and yield losses. *Plant Dis* 1981,65(4):325–327.

16. Perkins JM, Pedersen WL: Disease development and yield losses associated with northern leaf blight on corn. *Plant Dis*1987,71(10):940–943.

17. Pingali PL, Pandey S: Meeting world maize needs: technological

opportunities and priorities for the public sector. In *CIMMYT 1999/2000 World Maize Facts and Trends*. Edited by: Pingali PL, Mexico DF. CIMMYT; 2001.

18. Carson ML: Helminthosporium leaf spots and blights. In *Compendium of Corn Diseases*. 3rd edition. Edited by: White DG. St. Paul, Minnesota: The American Phytopathology Society; 1999:15–24.
19. Welz HG, Geiger HH: Genes for resistance to northern corn leaf blight in diverse maize populations. *Plant Breed* 2000,119(1):1–14.
20. Wisser RJ, Balint-Kurti PJ, Nelson RJ: The genetic architecture of disease resistance in maize: a synthesis of published studies. *Phytopathology* 2006,96(2):120–129.
21. Pratt RC, Gordon SG: Breeding for resistance to maize foliar pathogens. *Plant Breed Rev* 2006, 27:119–173.
22. Hilu HM, Hooker AL: Host-pathogen relationship of *Helminthosporium turcicum* in resistant and susceptible corn seedlings. *Phytopathology* 1964,54(5):570–575.
23. Hilu HM, Hooker AL: Localized infection by *Helminthosporium turcicum* on corn leaves. *Phytopathology* 1965,55(2):189–192.
24. Jennings PR, Ullstrup AJ: A histological study of three Helminthosporium leaf blights of corn. *Phytopathology* 1957,47(12):707–714.
25. Knox-Davies PS: Penetration of maize leaves by *Helminthosporium turcicum* . *Phytopathology* 1974,64(11):1468–1470.
26. Stuber CW, Polacco M, Senior ML: Synergy of empirical breeding, marker-assisted selection, and genomics to increase crop yield potential. *Crop Sci* 1999,39(6):1571–1583.
27. Szalma SJ, Hostert BM, LeDeaux JR, Stuber CW, Holland JB: QTL mapping with near-isogenic lines in maize. *Theor Appl Genet*2007,114(7):1211–1228.
28. Li Z-K, Fu B-Y, Gao Y-M, Xu J-L, Ali J, Lafitte HR, Jiang Y-Z, Rey JD, Vijayakumar CHM, Maghirang R, Zheng T-Q, Zhu L-H: Genome-wide introgression lines and their use in genetic and molecular dissection of complex phenotypes in rice (*Oryza sativa* L.). *Plant Mol Biol*2005,59(1):33–52.
29. Toojinda T, Baird E, Booth A, Broers L, Hayes P, Powell W, Thomas W, Vivar H, Young G: Introgression of quantitative trait loci (QTLs) determining stripe rust resistance in barley: an example of marker-assisted line development. *Theor Appl Genet* 1998,96(1):123–131.
30. Brown AHD, Munday J, Oram RN: Use of isozyme-marked segments

from wild barley (*Hordeum spontaneum*) in barley breeding.*Plant Breed* 1988,100(4):280–288.

31. Eshed Y, Zamir D: An introgression line population of *Lycopersicon pennellii* in the cultivated tomato enables the identification and fine mapping of yield-associated QTL. *Genetics* 1995,141(3):1147–1162.
32. Keurentjes JJB, Bentsink L, Alonso-Blanco C, Hanhart CJ, Vries HB-D, Effgen S, Vreugdenhil D, Koornneef M: Development of a near-isogenic line population of*Arabidopsis thaliana* and comparison of mapping power with a recombinant inbred line population.*Genetics* 2007,175(2):891–905.
33. Kaeppler SM: Quantitative trait locus mapping using sets of near-isogenic lines: relative power comparisons and technical considerations. *Theor Appl Genet* 1997,95(3):384–392.
34. Remington DL, Purugganan MD: Candidate genes, quantitative trait loci, and functional trait evolution in plants. *Int J Plant Sci*2003,164(3 Supplement):S7-S20.
35. Whalen MC: Host defence in a developmental context. *Mol Plant Pathol* 2005,6(3):347–360.
36. Kim KD, Hwang BK, Koh YJ: Evaluation of rice cultivars under greenhouse conditions for adult-plant resistance to *Pyricularia oryzae*. *J Phytopathol* 1987,120(4):310–316.
37. Century KS, Lagman RA, Adkisson M, Morlan J, Tobias R, Schwartz K, Smith A, Love J, Ronald PC, Whalen MC: Developmental control of*Xa21* -mediated disease resistance in rice. *Plant J* 1999,20(2):231–236.
38. Kus JV, Zaton K, Sarkar R, Cameron RK: Age-related resistance in *Arabidopsis* is a developmentally regulated defense response to*Pseudomonas syringae* . *Plant Cell* 2002,14(2):479–490.
39. Collins A, Milbourne D, Ramsay L, Meyer R, Chatot-Balandras C, Oberhagemann P, De Jong W, Gebhardt C, Bonnel E, Waugh R: QTL for field resistance to late blight in potato are strongly correlated with maturity and vigour. *Mol Breed* 1999,5(5):387–398.
40. Kumar JR, Kumar BT: Quantitative trait loci (QTL) mapping for crop improvement. *Res J Biotech* 2009,4(2):67–79.
41. Leonard KJ, Levy Y, Smith DR: Proposed nomenclature for pathogenic races of *Exserohilum turcicum* on corn. *Plant Dis*1989,73(9):776 777.
42. Vélez H: *Alternaria alternata* mannitol metabolism in plant-pathogen interactions. *PhD thesis*. North Carolina State University, Department of Plant Pathology; 2005.

43. Hood ME, Shew HD: Applications of KOH-aniline blue fluorescence in the study of plant-fungal interactions. *Phytopathology* 1996,86(7):704–708.

44. Qi M, Yang Y: Quantification of *Magnaporthe grisea* during infection of rice plants using real-time polymerase chain reaction and northern blot/phosphoimaging analyses. *Phytopathology* 2002,92(8):870.

45. Gaurilcikiene I, Deveikyte I, Petraitiene E: Epidemic progress of *Cercospora beticola* Sacc. in *Beta vulgaris* L. under different conditions and cultivar resistance. *Biologija* 2006, 4:54–59.

46. Pataky JK: Influence of host resistance and growth stage at the time of inoculation on Stewart›s wilt and Goss›s wilt development and sweet corn hybrid yield. *Plant Dis* 1989,73(4):339–345.

47. Blanco MH, Johnson MG, Colbert TR, Zuber MS: An inoculation technique for Stewart›s wilt disease of corn. *Plant Dis Rep* 1977,61(5):413–416.

48. Chang C-M, Hooker AL, Lim SM: An inoculation technique for determining Stewart's bacterial leaf blight reaction in corn. *Plant Dis Rep* 1977,61(12):1077–1079.

49. Keller NP, Bergstrom GC: Development predisposition of maize to anthracnose stalk rot. *Plant Dis* 1988,72(11):977–980.

50. Muimba-Kankolongo A, Bergstrom GC: Transitory wound predisposition of maize to Anthracnose stalk rot. *Can J Plant Pathol* 1990,12(1):1–10.

51. du Toit LJ, Pataky JK: Variation associated with silk channel inoculation for common smut of sweet corn. *Plant Dis* 1999,83(8):727–732.

52. Webb CA, Richter TE, Collins NC, Nicolas M, Trick HN, Pryor T, Hulbert SH: Genetic and molecular characterization of the maize *rp3* rust resistance locus. *Genetics* 2002,162(1):381–394.

53. Pataky JK, Campana MA: Reduction in common rust severity conferred by the *Rp1 D* gene in sweet corn hybrids infected by mixtures of *Rp1D* -virulent and avirulent *Puccinia sorghi* . *Plant Dis* 2007,91(11):1484–1488.

54. Doyle JJ, Doyle JL: A rapid DNA isolation procedure for small quantities of fresh leaf tissue. *Phytochem Bull* 1987, 19:11–15.

55. Qiu F, Wang H, Chen J, Zhuang J, Hei L, Cheng S, Wu J: A rapid DNA mini-prep method for large-scale rice mutant screening. *Rice Sci* 2006,13(4):299–302.

56. Schuelke M: An economic method for the fluorescent labeling of PCR fragments. *Nat Biotechnol* 2000,18(2):233–234.

57. Wisser RJ, Murray SC, Kolkman JM, Ceballos H, Nelson RJ: Selection mapping of loci for quantitative disease resistance in a diverse maize

population. *Genetics* 2008,180(1):583–599.

58. Patterson HD, Williams ER: A new class of resolvable incomplete block designs. *Biometrika* 1976,63(1):83–92.
59. van Dam J, Levin I, Struik PC, Levy D: Identification of epistatic interaction affecting glycoalkaloid content in tubers of tetraploid potato (*Solanum tuberosum* L.). *Euphytica* 2003,134(3):353–360.
60. Gonzalez J, Reyes F, Salas C, Santiago M, Codriansky Y, Coliheuque N, Silva H: *Arabidopsis thaliana* : A model host plant to study plant-pathogen interaction using Chilean field isolates of *Botrytis cinerea* . *Biol Res* 2006,39(2):221–228.
61. Johnston PR, Sutherland PW, Joshee S: Visualising endophytic fungi within leaves by detection of (1->3)-beta-D-glucans in fungal cell walls. *Mycologist* 2006,20(4):159–162.
62. Soylu S: Accumulation of cell-wall bound phenolic compounds and phytoalexin in *Arabidopsis thaliana* leaves following inoculation with pathovars of *Pseudomonas syringae* . *Plant Sci* 2006,170(5):942–952.
63. Ficke A, Gadoury DM, Seem RC: Ontogenic resistance and plant disease management: A case study of grape powdery mildew. *Phytopathology* 2002,92(6):671–675.
64. Bennett M, Gallagher M, Fagg J, Bestwick C, Paul T, Beale M, Mansfield J: The hypersensitive reaction, membrane damage and accumulation of autofluorescent phenolics in lettuce cells challenged by *Bremia lactucae* . *Plant J* 1996,9(6):851–865.
65. Hooker AL: Inheritance of chlorotic-lesion resistance to *Helminthosporium turcicum* in seedling corn. *Phytopathology* 1963,53(6):660–662.
66. Hooker AL: A second major gene locus in corn for chlorotic lesion resistance to *Helminthosporium turcicum* . *Crop Sci* 1977,17(1):132–135.
67. Hooker AL: Resistance to *Helminthosporium turcicum* from *Tripsacum floridanum* incorporated into corn. *Maize Genet Coop Newsl* 1981, 55:87–88.
68. Ogliari JB, Guimaraes MA, Geraldi IO, Aranha Camargo LE: New resistance genes in the *Zea mays - Exserohilum turcicum* pathosystem. *Genet Mol Biol* 2005,28(3):435–439.
69. Ogliari JB, Guirnaraes MA, Aranha Carnargo LE: Chromosomal locations of the maize (*Zea mays* L.) *HtP* and *rt* genes that confer resistance to *Exserohilum turcicum* . *Genet Mol Biol* 2007,30(3):630–634.
70. Carson ML: A new gene in maize conferring the «chlorotic halo» reaction to infection by *Exserohilum turcicum* . *Plant Dis* 1995,79(7):717–720.

71. Gevers HO: A new major gene for resistance to *Helminthosporium turcicum* leaf blight of maize. *Plant Dis Rep* 1975,59(4):296–299.

72. Freymark PJ, Lee M, Woodman WL, Martinson CA: Quantitative and qualitative trait loci affecting host-plant response to*Exserohilum turcicum* in maize (*Zea mays* L.). *Theor Appl Genet* 1993,87(5):537–544.

73. Welz HG, Xia XC, Bassetti P, Melchinger AE, Luebberstedt T: QTLs for resistance to *Setosphaeria turcica* in an early maturing Dent × Flint maize population. *Theor Appl Genet* 1999,99(3–4):649–655.

74. Carson ML: Inheritance of latent period length in maize infected with *Exserohilum turcicum* . *Plant Dis* 1995,79(6):581–585.

75. SmithDR,KinseyJG:Latentperiod-apossibleselectiontoolfor*Exserohilum turcicum* resistance in corn (*Zea mays* L.). *Maydica*1993, 38:205–208.

76. Welz HG, Schechert AW, Geiger HH: Dynamic gene action at QTLs for resistance to *Setosphaeria turcica* in maize. *Theor Appl Genet*1999,98(6–7):1036–1045.

77. Schechert AW, Welz HG, Geiger HH: QTL for resistance to *Setosphaeria turcica* in tropical African maize. *Crop Sci* 1999,39(2):514–523.

78. Brewster VA, Carson ML, Wicks ZW III: Mapping components of partial resistance to northern leaf blight of maize using reciprocal translocation. *Phytopathology* 1992,82(2):225–229.

79. Tucker SL, Talbot NJ: Surface attachment and pre-penetration stage development by plant pathogenic fungi. *Annu Rev Phytopathol*2001, 385–417.

80. Bubeck DM, Goodman MM, Beavis WD, Grant D: Quantitative trait loci controlling resistance to gray leaf spot in maize. *Crop Sci*1993,33(4):838–847.

81. Leonard KJ, Thompson DL: Effects of temperature and host maturity on lesion development of *Colletotrichum graminicola* on corn. *Phytopathology* 1976,66(5):635–639.

82. Sofi P, Rather AG: QTL analysis in rice improvement: concept, methodology and application. *Biotechnology* 2007,6(1):1–13.

83. Trognitz BR, Orrillo M, Portal L, Roman C, Ramon P, Perez S, Chacon G: Evaluation and analysis of reduction of late blight disease in a diploid potato progeny. *Plant Pathol* 2001,50(3):281–291.

84. Dinh SQ, Joyce DC, Irving DE, Wearing AH: Effects of combined methyl jasmonate and ethylene-inhibitor treatments against *Botrytis*

cinerea infecting Geraldton waxflower. *Acta Hort (ISHS)* 2007, 755:527–533.

85. Balint-Kurti PJ, Wisser R, Zwonitzer JC: Use of an advanced intercross line population for precise mapping of quantitative trait loci for gray leaf spot resistance in maize. *Crop Sci* 2008,48(5):1696–1704.

86. Mitchell-Olds T, James RV, Palmer MJ, Williams PH: Genetics of*Brassica rapa* (syn. *campestris*). 2. Multiple disease resistance to three fungal pathogens. *Peronospora parasitica* , *Albugo candida* and *Leptosphaeria maculans* . *Heredity* 1995,75(4):362–369.

87. Asai T, Tena G, Plotnikova J, Willmann MR, Chiu W-L, Gomez-Gomez L, Boller T, Ausubel FM, Sheen J: MAP kinase signalling cascade in*Arabidopsis* innate immunity. *Nature* 2002,415(6875):977–983.

88. Durrant WE, Dong X: Systemic acquired resistance. *Annu Rev Phytopathol* 2004, 42:185–209.

89. Yang Y, Shah J, Klessig DF: Signal perception and transduction in plant defense responses. *Gene Dev* 1997,11(13):1621–1639.

90. Mou Z, Fan W, Dong X: Inducers of plant systemic acquired resistance regulate NPR1 function through redox changes. *Cell*2003,113(7):935–944.

91. Consonni C, Humphry ME, Hartmann HA, Livaja M, Durner J, Westphal L, Vogel J, Lipka V, Kemmerling B, Schulze-Lefert P, Somerville SC, Panstruga R: Conserved requirement for a plant host cell protein in powdery mildew pathogenesis. *Nat Genet* 2006,38(6):716–720.

92. Pataky JK, Bohn MO, Lutz JD, Richter PM: Selection for quantitative trait loci associated with resistance to Stewart›s wilt in sweet corn. *Phytopathology* 2008,98(4):469–474.

93. Ming R, Brewbaker JL, Moon HG, Musket TA, Holley RN, Pataky JK, McMullen MD: Identification of RFLP makers linked to a major gene, *sw1* , conferring resistance to Stewart›s wilt in maize. *Maydica* 1999, 44:319–323.

94. Freymark PJ, Lee M, Martinson CA, Woodman WL: Molecular-marker-facilitated investigation of host-plant response to *Exserohilum turcicum* in maize (*Zea mays* L.): components of resistance. *Theor Appl Genet* 1994,88(3–1):305–313.

95. Carson ML, Van Dyke CG: Effect of light and temperature on expression of partial resistance of maize to *Exserohilum turcicum* .*Plant Dis* 1994,78(5):519–522.

96. Sigulas KM, Hill RRJ, Ayers JE: Genetic analysis of *Exserohilum turcicum* lesion expansion on corn. *Phytopathology* 1988,78(2):149–153.

97. Brown AF, Juvik JA, Pataky JK: Quantitative trait loci in sweet corn associated with partial resistance to Stewart›s wilt, northern corn leaf blight, and common rust. *Phytopathology* 2001,91(3):293–300.

98. Balint-Kurti PJ, Zwonitzer JC, Wisser RJ, Carson ML, Oropeza-Rosas MA, Holland JB, Szalma SJ: Precise mapping of quantitative trait loci for resistance to southern leaf blight, caused by *Cochliobolus heterostrophus* race O, and flowering time using advanced intercross maize lines. *Genetics* 2007,176(1):645–657.

Chapter 2

PROTEOMICS OF PLANT PATHOGENIC FUNGI

Raquel González-Fernández,[1] Elena Prats,[2] and Jesús V. Jorrín-Novo[1]

[1]Agricultural and Plant Biochemistry and Proteomics Research Group, Department of Biochemistry and Molecular Biology, University of Córdoba, 14071 Córdoba, Spain

[2]CSIC, Institute of Sustainable Agriculture, 14080 Córdoba, Spain

ABSTRACT

Plant pathogenic fungi cause important yield losses in crops. In order to develop efficient and environmental friendly crop protection strategies, molecular studies of the fungal biological cycle, virulence factors, and interaction with its host are necessary. For that reason, several approaches have been performed using both classical genetic, cell biology, and biochemistry and the modern, holistic, and high-throughput, omic techniques. This work briefly overviews the tools available for studying Plant Pathogenic Fungi and is amply focused on MS-based Proteomics analysis, based on original papers published up to December 2009. At a methodological level, different steps in a proteomic workflow experiment are discussed. Separate sections are devoted to fungal descriptive (intracellular, subcellular, extracellular) and differential expression proteomics and interactomics. From the work published we can conclude that Proteomics, in combination with other techniques, constitutes a powerful tool for providing important information about pathogenicity and virulence factors, thus opening up new possibilities for crop disease diagnosis and crop protection.

INTRODUCTION: PLANT PARASITIC FUNGI

Fungi form a large and heterogeneous eukaryotic group of living organisms characterized by their lack of photosynthetic pigment and their chitinous cell wall. It has been estimated that the fungal kingdom contains more than 1.5 million species, but only around 100,000 have so far been described, with yeast, mold, and mushroom being the most familiar [1]. Although the majority of fungal species are saprophytes, a number of them are parasitics, in order to complete their biological cycle, animals or plants, with around 15,000 of them causing disease in plants, the majority belonging to the Ascomycetes and Basidiomycetes [2] (Table 1). Within a fungal plant pathogen species, for example, in Fusarium oxysporum, up to 120 different formae specialis can be found based on specificity to host species belonging to a wide range of plant families [3].

Table 1: Main plant pathogenic fungi causing disease in plants

Phylum	Genus	Anamorphic stage	Hosts	Disease	Example
Chytridiomycota	Olpidium		cabbage	root diseases	O. brassicae
	Physoderma		corn	brown spot	P. maydis
			alfalfa	crown wart	P. (= Urophlyctis) alfalfae
	Synchytrium		potato	potato wart	S. endobioticum
Zygomycota	Rhizopus		fruits and vegetables	bread molds and soft rot	R. oligosporus
	Choanephora		squash	soft rot	C. cucurbitarum
	Mucor		fruits and vegetables	bread mold and storage rots	M. indicus
Ascomycota	Taphrina		peach plum oak	leaf curl leaf blister and so forth	T. deformans
	Galactomyces		citrus	sour rot	G. candidum
	Blumeria		cereals and grasses	powdery mildew	B. graminis [1]
	Erysiphe	Oidium	many herbaceous plants	powdery mildew	E. pisi
	Leveillula		tomato	powdery mildew	L. taurica
	Microsphaera		lilac	powdery mildew	M. penicullata
	Oidium		tomato	powdery mildew	O. neolycopersici
	Podosphaera		apple	powdery mildew	P. leucotricha
	Sphaerotheca		roses and peach	powdery mildew	S. pannosa
	Uncinula		grape	powdery mildew	U. necator

	Nectria		trees	twig and stem cankers	N. galligena
	Gibberella		corn and small grains	foot or stalk rot	F. graminearum [1]
		Fusarium	several plants	vascular wilts root rots stem rots seed infections	F. oxysporum[1]
	Claviceps		grain crops	ergot	C. purpurea
	Ceratocystis	Chalara	oak	oak wilt	C. fagacearum
			stone fruit and sweet potato	cankers and root rot	C. fimbriata
			pineapple	butt rot	C. paradoxa
	Monosporascus		cucurbits	root rot and collapse	M. cannonballus
	Glomerella		apple	anthracnoses and bitter rot	G. cingulata
		Colletotrichum	many plants	anthracnoses	C. lindemuthianum
	Phyllachora		grasses	leaf spots	P. graminis
	Ophiostoma	Sporothrix and Graphium	elm	Dutch elm disease	O. novo-ulm
	Diaporthe			citrus melanose	D. citri
				eggplant fruit rot	D. vexans
				soybean pod and stem rot	D. phaseolorum
	Gaeumannomyces		grain crops and grasses	take-all disease	G. graminis
	Magnaporthe		rice	rice blast	M. grisea [1]
	Cryphonectria		chestnut	blight disease	C. parasitica
	Leucostoma		peach and other trees	canker diseases	L. persoonii
	Hypoxylon		poplars	canker disease	H. mammatum
	Rosellinia		fruit trees and vines	root diseases	R. necatrix
	Xylaria		trees	tree cankers and wood decay	X. longipes
	Eutypa		fruit trees and vines	canker	E. armeniacae

	Mycosphaerella	Cercospora	Banana	Sigatoka disease	M. musicola and M. fijiensis
		Septoria	cereals and grasses	leaf spots	M. graminicola
			strawberry	leaf spot	M. fragariae
	Elsinoë		citrus trees	citrus scab	E. fawcetti
			grape	anthracnose	E. ampelina
			raspberry	anthracnose	E. veneta
	Capnodium		most plants	sooty molds	C. elaeophilum
	Cochliobolus	Bipolaris	grain crops and grasses	leaf spots and root rots	C. carbonum and B. maydis
		Curvularia	grasses	leaf spots	C. lunata [1]
	Pyrenophora	Drechslera	cereals and grasses	leaf spots	P. graminea
	Setosphaera		cereals and grasses	leaf spots	S. turcica
	Pleospora	Stemphylium	tomato	black mold rot	P. lycopersici and S. solani
	Leptosphaeria		cabbage	black leg and foot rot	L. maculans [1]
	Venturia		apple	apple scab	V. inaequalis
			pear	pear scab	V. pyrina
		Cladosporium	tomato	leaf mold	C. fulvum
			peach and almond	scab	C. carpophilum
	Guignardia	Phyllosticta	grapes	black rot	G. bidwellii
	Apiosporina		cherries and plums	black knot	A. morbosa
	Hypoderma		pines	needle cast	H. desmazierii
	Lophodermium		pines	needle cast	L. pinastri
	Rhabdocline		pines	Douglas fir needle cast	R. weirii
	Rhytisma		maple	tar spot of leaves	R. acerinum
	Monilinia		stone fruit	brown rot disease	M. fruticola
	Sclerotinia		vegetables	white mold	S. sclerotiorum [1]
	Stromatinia		gladiolus	corm rot	S. gladioli
	Pseudopeziza		alfalfa	leaf spot	P. trifolii

	Diplocarpon		quince and pear	black spot	D. maculatum
	Talaromyces	Penicillium	fruits	blue mold rot	P. digitatum
		Aspergillus	seeds	bread mold and seed decays	A. niger [1]
	Hypocrea	Verticillium	many plants	vascular wilts	V. dahliae [1]
	Lewia	Alternaria	many plants	leaf spots and blights	A. alternata
	Setosphaera	Exserohi-lum	grasses	leaf spots	E. longirostra-tum
	Botryosphaeria	Sphaeropsis	apple	black rot	S. pyriputre-scens
	Botryotinia	Botrytis	many plants	gray mold rots	B. cinerea [1]
	Monilinia	Monilia	stone fruits	brown rot	M. fruticola
	Diplocarpon	Entomo-sporium	pear	leaf and fruit spot	E. mespili
	Greeneria	Melanco-nium	grape	bitter rot	M. fuligineum
Basidio-mycota	Ustilago		corn	smut	U. maydis [1]
			oats	loose smuts	U. avenae
			barley	loose smuts	U. nuda
			wheat	loose smuts	U. tritici
	Tilletia		wheat	covered smut or bunt	T. caries
			wheat	Karnal bunt	T. indica
	Urocystis		onion	smut	U. cepulae
	Sporisorium		sorghum	covered kernel smut	S. sorghi
			sorghum	loose sorghum smut	S. cruentum
	Sphacelotheca		sorghum	head smut	S. reiliana
	Cronartium		pines	blister rust	C. ribicola
	Gymnosporan-gium		apple	cedar-apple rust	G. juniperi-virginianae
	Hemileia		coffee	rust	H. vastatrix
	Melampsora		flax	rust	M. lini
	Phakopsora		soybeans	rust	P. pachyrrhizi
	Puccinia		cereals	rust	P. recondita
	Uromyces		beans	rust	U. appendicu-latus [1]

	Exobasidium		ornamentals	leaf flower and stem galls	E. japonicum
	Athelia		many plants	Southern blight	A. rolfsii
		Sclerotium	onions	white rot	S. cepivorum
	Thanatephorus	Rhizoctonia	many plants	root and stem rots damping-off and fruit rots	T. cucumeris and R.solani
	Typhula		turf grasses	snow mold	T. incarnata
	Armillaria		trees	root rots	A. mellea
	Crinipellis		cacao	witches'-broom	C. perniciosus
	Marasmius		turf grasses	fairy ring disease	M. oreades
	Pleurotus		trees	white rot on logs tree stumps and living trees	P. ostreatus [1]
	Pholiota		trees	brown wood rot	P. squarrosa
	Chondrostereum		trees	silver leaf disease	C. purpureum
	Corticium		turf grasses	red thread disease	C. fuciforme
	Heterobasidion		trees	root and butt rot	H. annosum
	Ganoderma		trees	root and basal stem rots	G. boninense
	Inonotus		trees	heart rot	I. hispidus
	Polyporus		trees	heart rot	P. glomeratus
	Postia		trees	wood and root rot	P. fragilis

[1]These phytopathogenic fungi are named in the text.

According to the type of parasitism and infection strategy, fungi are classified as biotrophic (e.g., Blumeria graminis), necrotophic (e.g., Botrytis cinerea), or hemibiotrophic (e.g., Colletotrichum destructivum). While the former derives nutrients from dead cells, the latter takes nutrients from the plant but does not kill it [4]. Hemibiotrophes sequentially deploy a biotrophic and then a necrotrophic mode of nutrition. Necrotrophic species tend to attack a broad range of plant species; on the contrary, biotrophes usually exhibit a high degree of specialization for individual plant species. Most biotrophic fungi are obligatory parasites, surviving only limited saprophytic phases. Differently from necrotrophes, the cultivation of biotrophic fungi succeeds only in a few exceptions, for example, Podosphaera fusuca [5] or B. graminis (M. M. Corbitt, personal communication, adapted from [5]). Fungal diseases are, in nature, more the exception than the rule. Thus, only a limited number of fungal species are able to penetrate and invade host tissues, avoiding recognition and plant defence responses, in order to obtain nutrients from them, causing disease and sometimes host death. In agriculture, annual crop losses due to pre- and post harvest fungal diseases exceed 200 billion euros, and, in the United Stated alone, over $600 million are annually spent on fungicides [6].

Fungal pathogens have complicated life cycles, with both asexual and sexual reproduction, and stages involving the formation of different infective, vegetative, and reproductive structures [7]. The primary events in a disease cycle are establishment of infection, colonization (invasion), growth and reproduction, of the pathogen, dissemination of the pathogen, and survival of the pathogen in the absence of the host, that is, overwintering or oversummering (overseasoning) of the pathogen (Figure 1). However, the execution of each stage largely differs depending on the pathogen [8]. In polycyclic diseases there are several infection cycles within one, the so-called secondary cycles [3] (Figure 1).

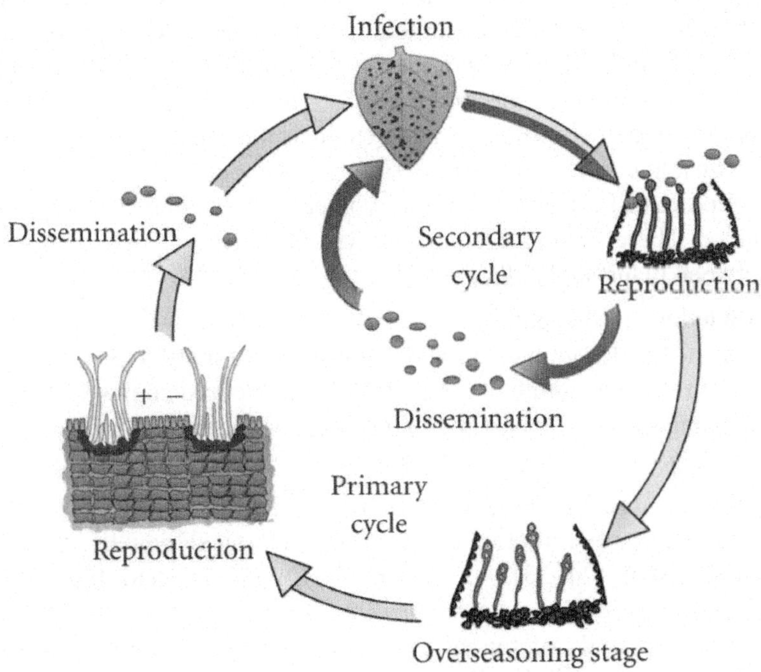

Figure 1: Diagram of monocyclic (yellow) and polycyclic (yellow and blue) fungi. In monocyclic diseases the fungus produces spores at the end of the season that serve as primary and only inoculum for the following year. The primary inoculum infects plants during the growth season and, at the end of the growth season, produces new spores in the infected tissues. These spores remain in the soil (overseasoning stage) and serve as the primary inoculum the following season. In polycyclic fungal patho-

gens, the primary inoculum often consists of the sexual (perfect) spore or, in fungi that lack the sexual stage, some other structures such as sclerotia, pseudosclerotia, or mycelium in infected tissue. This inoculum causes the primary infection and then large numbers of asexual spores (secondary inoculum) are produced at each infection site and these spores can themselves cause new (secondary) infections that produce more asexual spores for more infections.

The fungal plant interplay depends on mutual recognition, signalling, and the expression of pathogenicity and virulence factors, from the fungal side, and the existence of passive, preformed, or inducible defence mechanisms in the plant, resulting in compatible (susceptibility) or incompatible (nonhost, basal or host specific resistance) interactions. From a genetic point of view, and according to the gene-for-gene interaction hypothesis, proposed by Flor while studying flax rust [9], resistance results from the combination of a dominant avirulence (Avr) gene in the pathogen and a cognate resistance (R) gene in the host; the interaction of both gene products leads to the activation of host defence responses, such as the hypersensitive response, that arrests the growth of fungi. This hypothesis has been experimentally demonstrated for a number of pathosystems, mainly involving biotrophic fungi, with a number of avirulent genes identified [10]. A number of fungal mechanisms and molecules have been shown to contribute to fungal pathogenicity or virulence, understood as the capacity to cause damage in a host, in absolute or relative terms. Among them, cell wall degrading proteins, inhibitory proteins [11], and enzymes involved in the synthesis of toxins [12–15] are included. These virulence factors are typically involved in evolutionary arms races between plants and pathogens [16, 17].

Knowledge of the pathogenic cycle and that of virulence factors [18, 19] is crucial for designing effective crop protection strategies, including the development of resistant plant genotypes through classical plant breeding [20] or genetic engineering [21], fungicides [22], or the use of biological control strategies [23].

Studies of fungal pathogens and their interactions with plants have been performed using several approaches, from classical genetic, cell biology, and biochemistry [24–33], to the modern, holistic, and high-throughput omic techniques [34, 35] accompanied by proper bioinformatic tools [36]. In recent years, the study of fungal plant pathogens has been greatly promoted by the availability of their genomic sequences and resources for functional genomic analysis, including transcriptomics, proteomics, and metabolomics [37], which, in combination with targeted mutagenesis or transgenic studies, are unravelling molecular host-pathogen crosstalk, the complex mechanisms involving pathogenesis and host avoidance [38]. This review work makes an overview and summarizes the contribution of the most recent molecular techniques to the knowledge of phytopathogenic fungi biology and is mainly focused on the MS-based Proteomics approach.

FROM STRUCTURAL TO FUNCTIONAL GENOMICS

The importance of plant pathogenic fungi studies is underlined by the increasing number of fungal genome sequencing projects. Currently, over 40 fungal genomes have been sequenced, 16 of which are phytopathogenic (Table 2), with more than 300 sequencing projects being in progress (Genomes Online database,http://www.genomesonline.org/). Sequence information, while valuable and a necessary starting point, is insufficient alone to answer questions concerning gene function, regulatory networks, and the biochemical pathways activated during pathogenesis. Based on the accumulation of a wealth of fungal genomic sequences, the traditional pursuit of a gene starting with a phenotype (forward genetics) has given way to the opposite situation where the gene sequences are known but not their functions. Thus, the challenge is now to decipher the function of the thousands of genes identified by genome projects, and reverse genetics methodologies are key tools in this endeavour [39].

Table 2: Publicly available plant pathogenic fungal Genome sequences

Phytopathogen Speciesa	URL
Ascomycota	
Dothydeomycetes	
Mycosphaerella fijensis (Banana black leaf streak)	http://genome.jgi-psf.org/Mycfi1/Mycfi1.home.html
Mycosphaerella graminicola (Wheat leaf blotch)	http://genome.jgi-psf.org/Mycgr1/Mycgr1.home.html
Pyrenophora tritici-repentis (Wheat disease)	http://www.broad.mit.edu/annotation/genome/pyrenophora_tritici_repentis/Home-html
Stagonospora nodorum (Wheat glume blotch)	http://www.broad.mit.edu/annotation/fungi/stagonospora_nodorum
	http://www.acnfp.murdoch.edu.au/Mission.htm
Eurotiomycetes	
Aspergillus flavus	http://www.aspergillusflavus.org/
Leotiomycetes	
Botrytis cinerea (Grape/other host grey rot) BO5.10	http://www.broad.mit.edu/annotation/fungi/botrytis_cinerea

T4	http://urgi.versailles.inra.fr/proyects/Botrytis/index.php
Sclerotinia sclerotium (Multi-host rot diseases)	http://www.broad.mit.edu/annotation/fungi/sclerotinia_sclerotium
Saccharomycetes	
Ashbya gossypii (Cotton/citrus fruits disease)	http://agd.vital-it.ch/index.html
Sordariomycetes	
Fusarium graminearum (Wheat/barley head blight)	http://www.broad.mit.edu/annotation/genome/fusarium_group
Fusarium oxysporum (Multi-host wilt disease)	http://www.broad.mit.edu/annotation/genome/fusarium_group
Fusarium verticillioides (Maize seed rot)	http://www.broad.mit.edu/annotation/genome/fusarium_group
Magnaporthe grisea (Rice blast disease)	http://www.broad.mit.edu/annotation/genome/magnaporthe_grisea/MultiHome.htlm
Nectria haematococca (Pea wilt)	http://genome.jgi-psf.org/Necha2/ Necha2.home.html
Verticillium dahliae VdLs.17 (Multi-host wilt)	http://www.broad.mit.edu/annotation/genome/verticillium_dahliae/MultiHome.html
Basidiomycota	
Pucciniomycetes	
Puccinia graminis (Cereal rusts)	http://www.broad.mit.edu/annotation/genome/puccinia_graminis
Ustilaginomycetes	
Ustilago maydis (Corn smut disease)	http://www.broad.mit.edu/annotation/fungi/ustilago_maydis

[a]Species are grouped by phylum and class. In parenthesis below the species's name and associated with each species, the most common or most widely recognized diseases are listed.

The study of gene function in filamentous fungi has made great advances in recent years [69]. Some of the techniques used in high-throughput reverse genetics approaches are targeted gene disruption/replacement (knock-out) [70], gene silencing (knock-down) [71], insertional mutagenesis [72], or targeting induced local lesions in genomes (TILLING) [73] (for review and examples see [39]). Thus, a number of pathogenicity factors have been targeted [74–76], and among them, several signaling pathways such as the cAMP and a mitogen activated protein kinase (MAPK) pathways have been shown to be crucial to virulence in several phytopathogenic fungi [77–82]. Random insertional mutagenesis is an excellent approach for dissecting complex biological traits, such as pathogenicity, because it does not require any prior information or assumptions on gene function. Recently, transposable elements (TEs) have been used for insertional mutagenesis and large-scale transposon mutagenesis has been developed as a tool for the genome-wide identification of virulence determinants in F. oxysporum [83].

Otherwise, transcriptomics, the global analysis of gene expression at the mRNA level, is also an attractive method for analyzing the molecular basis of fungal-plant interactions and pathogenesis [84–87]. For understanding the transcriptional activation or repression of genes during the infection process tools such as Differential Display (DD) [88], cDNA-Amplified Fragment-Length Polymorphism (cDNA-AFLP) [89], Suppression Subtractive Hybridization (SSH) [90], Serial Analyse of Gene Expression (SAGE) [91], expressed sequence tags (ESTs) [92], or DNA microarrays [91] have been developed in addition to older techniques such as Northern blotting, and they are reviewed in [84, 93].

PROTEOMICS

Until the early 1990s most biological research was focused on the in vitro studies of individual components in which genes and proteins were investigated one at a time. This strategy shifted in the early and mid 1990s to in vivo and molecular large-scale research, starting with structural genomics and transcriptomics research projects, then moving to proteomics, and recently to metabolomics. All these together constitute the methodological bases of the Modern Systems Biology [37, 94]. Since no single approach can fully clear up the complexity of living organisms, each approach does contribute and must be validated, this being considered as part of a multidisciplinary integrative analysis at different levels, extending from the gene to the phenotype through proteins and metabolites.

Within the "-omics" techniques, Proteomics constitutes, nowadays, priority research for any organism and configures a fundamental discipline in the postgenomic era. It is true that, at 2010, the realities are below the expectations originally generated and that the results gained over the last 15 years have shown that the dynamism, variability, and behaviour of proteins are more complex than had ever been imagined, especially as refers to a number of protein species per gene as a result of alternative splicing, reading frame, and posttranslational modifications, trafficking, and interactions, and considering that protein complexes, rather than individual proteins, are the functional units of the biological machines. However, and differently from other biological systems, mainly yeast [95] and humans [96], the full potential of proteomics is far from being fully exploited in fungal pathogen research, as refers to the low number of fungal pathogen species under investigation at the proteomic level, the low proteome coverage in those species investigated, and the almost unique use of classical, first generation techniques, those based on 2-DE coupled to MS.

The term proteomics was coined by Marc Wilkins, back during the 1994 Siena Meeting, to simply refer to the "PROTein complement of a genOME" [97]. Fifteen years later proteomics has become more than just an appendix of genomics or an experimental approach but a complex scientific discipline dealing with the study of the cell proteome. In the broadest sense, the proteome can be defined as being the total set of protein species present in a biological unit (organule, cell, tissue, organ, individual, species, ecosystem) at any developmental stage and under specific environmental conditions. By using proteomics we aim to know how, where, when, and what for are the several hundred thousands of individual protein species produced in a living organism, how they interact with one another and with other molecules to construct the cellular building, and how they work with each other to fit in with programmed growth and development, and to interact with their biotic and abiotic environment. In the last ten years, excellent reviews and monographs on the fundamentals, concepts, applications, power, and limitations of proteomics have appeared [95, 98–104], some of them dealing with fungal pathogens [37, 105, 106]. It is not the objective of this review to comment or discuss every aspect but instead to show which one has been its contribution to the knowledge of fungal pathogens.

In Proteomics, several areas can be defined:

- Descriptive Proteomics, including Intracellular and Subcellular Proteomics;
- Differential Expression Proteomics,
- Post-translational Modifications;
- Interactomics; and
- Proteinomics (targeted or hypothesis-driven Proteomics).

In the case of fungi, a new area can also be defined as Secretomics (the secretome is defined as being the combination of native proteins and cell machinery involved in their secretion), since many fungi secrete a vast number of proteins to accommodate their saprotrophic life-style; this would be the case of proteins implicated in the adhesion to the plant surface [107], host-tissue penetration and invasion effectors, [11, 108, 109], and other virulence factors [110].

Proteomics is constantly being renewed to respond to the question of the role of the proteins expressed in a living organism, experiencing, in the last ten years, an explosion of new protocols, and platforms with continuous improvements made at all workflow stages, starting from the laboratory (tissue and cell fractionation, protein extraction, depletion, purification, separation, MS analysis) and ending at the computer (algorithms for protein identification and bioinformatics tools for data analysis, databases, and repositories). These techniques will be briefly introduced and discussed in the next section.

Despite the technological achievements in proteomics, only a tiny fraction of the cell proteome has been characterized so far, and only for a few biological systems (human, fruit fly, Arabidopsis, rice). Even for these organisms, the function of quite a number of proteins remains to be investigated [99]. Proteomics techniques have a number of limitations, such as sensitivity, resolution, and speed of data capture. They also face a number of challenges, such as deeper proteome coverage, proteomics of unsequenced "orphan" organisms, top-down Proteomics [100], protein quantification [98], PTMs [105], and Interactomics [55, 57]. Most of these limitations and challenges reflect the difficulty of working with the biological diversity of proteins and their range of physicochemical properties.

The relevance of proteomics in plant fungal pathogens research is very well illustrated by the pioneer work on the Cladosporium fulvum-tomato interaction carried out by the Pierre de Wit research group back in 1985 [111] that allowed the characterization of the first avirulence gene product (Avr9) after purification from tomato apoplastic fluids by preparative polyacrylamide gel electrophoresis followed by reverse-phase HPLC and EDMAN N-terminal sequencing [112]. Later on, a number of avirulence gene product effectors have been discovered, mainly by genomic approaches [113]. Curiously, this pioneer work followed the typical proteomics strategy even before MS-based powerful techniques were developed. Another good example is the tomato-F. oxysporum pathosystem, in which the first effector of root invading fungi was identified and sequenced, in this case by MS, the Six1, corresponding to a 12 kDa cysteine-rich protein [114]. Other further protein effectors have been characterized in different fungi [115].

Next, we describe the state of the art in the methodology of fungal plant pathogen proteomics and summarize the works published in this field up to December 2009, which so far are 30 (Table 3) out of a total of over 14000 original papers in Proteomics in the last ten years.

Table 3: Original proteomics papers and reviews published on plant pathogenic fungi

Fungus	Proteomic approach (reference)
Aspergillus ssp.	1-DE, MALDI-TOF-MS [40]
Aspergillus flavus	1-DE/2-DE, nanoLC-MS/MS [41]
	1-DE/2-DE, MALDI-TOF-MS [42]
Blumeria graminis f.sp. hordei	2-DE, MALDI-TOF/TOF-MS/MS [43]
	nanoLC-MS/MS [44]
Botrytis cinerea	2-DE, MALDI-TOF/TOF-MS/MS, nanoLC-MS/MS, ESI-IT-MS/MS [45]
	2-DE, MALDI-TOF/TOF-MS/MS, ESI-IT-MS/MS [46]
	2-DE, MALDI-TOF/TOF-MS/MS [47]
	1-DE, nanoLC MS/MS [48]
	1-DE, nanoLC MS/MS [49]
Curvularia lunata	2-DE, MALDI-TOF/TOF-MS/MS [50]
Fusarium graminearum	1-DE/2-DE, nanoLC-Q-TOF-MS/MS [51]
	1-DE, CID-LTQ-MS [52]
	2-DE, IT-MS/MS, iTRAQ-MS/MS [53]
	2-DE, ESI-MS/MS [54]
	Interactome [55]
Leptosphaeria maculans	1-DE, liquid-phase IEF, 2-DE [56]
Magnaporthe grisea	Interactome [57]
Phytophthora infestans	2-DE, MALDI-TOF-MS [58]
	2-DE, nanoLC-MS/MS [59]
Phytophthora palmivora	2-DE, MALDI-TOF-MS [58]
Phytophthora ramorum	HPLC-ESI-Q-TOF-nanoLC-MS/MS [60]
Pyrenophora tritici-repentis	2-DE, ESI-Q-TOF-MS/MS [61]
Pleorotus ostreatus	1-DE, MALDI-TOF-MS, ESI-Q-TOF-MS/MS/de novo sequencing [62]
Rhizoctonia solani	2-DE, MALDI-TOF-MS [63]
Sclerotinia sclerotiorum	2-DE, ESI-Q-TOF-nanoLC-MS/MS [64]
Stagnospora nodorum	2 -DE, LC-MS/MS [65]
	2-D LC-MALDI-MS/MS [66]
Uromyces appendiculatus	MudPit-MS/MS [67]
Ustilago maydis	2-DE, MALDI-TOF-MS, nanoLC-Q-TOF-MS/MS [68]

Methodology

The workflow of a standard MS-based proteomics experiment includes all or most of the following steps: experimental design, sampling, tissue/cell or organelle preparation, protein extraction/fractionation/purification, labeling/ modification, separation, MS analysis, protein identification, statistical analysis of data, validation of identification, protein inference, quantification, and data analysis, management, and storage (Figure 2) [102, 116]. The most appropriate protocol to be used depends on and must be optimized for the biological system (i.e., fungal species, plant species, organ, tissue, cells), as well as the objectives of the research (descriptive, comparative, PTMs, interactions, targeted Proteomics) [102].

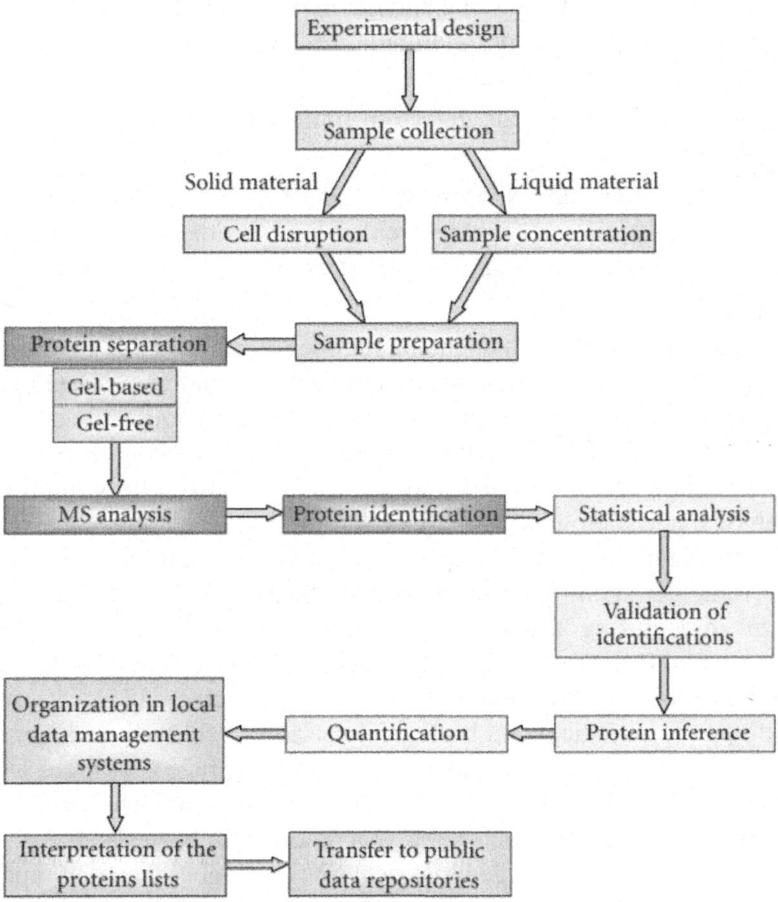

Figure 2: Schematic overview of the work flow in a fungal proteomatic approach (adapted from Deutsch et al., 2008 [116]).

Experimental Design

A good experimental design is crucial to the success of any proteomics experiment. Eriksson and Fenyö [117] have developed a simulation tool for evaluating the success of current designs and for predicting the performance of future, better-designed proteomics experiments. The simulation gives a holistic view of a general analytical experiment and attempts to identify the factors that affect the success rate. It has been used to predict the success of proteome analyses of Human tissue and body fluids that use various experimental design principles. Several parameters, are required to simulate the steps of a proteome analysis:

- the distribution of protein amounts in the sample analyzed,
- the loss of analyte material and the maximal limit of the amount loaded at each step of sample manipulation (e.g., separation, digestion, and chemical modification),
- the dynamic range, the detection limit, and the losses associated with MS analysis.

Depending on what experiment is being modelled, the detection limit used in a simulation can represent either protein identification only (lower identification limit) or protein identification with quantification (lower quantification limit) [102].

The establishment of an adequate number of replicates is essential to any differential expression proteomics experiment. This number should be set up while taking into account the dynamic nature of the proteome, and a good number will allow a correct interpretation of the results and the confident assignment of any protein to the group of variable ones [102, 118]. Furthermore, a study of analytical and biological variability should be carried out. Thus, analytical variability examines both the experiment procedures (protein extraction, IEF, SDS-PAGE, gel staining-destaining) and the accuracy of the hardware and software in acquiring and analysing images, and biological variability tests look at several different samples. For example, in the studies to characterize the protein profile of the fungal phytopathogen B. cinerea [45], it previously determined the analytical and biological variability. Sixty-four major spots in three 2-DE replicate gels were analyzed and the average CVs were 16.1% for the analytical variance and 37.5% for the biological variance for this fungus. The analytical variance was similar to that reported for bacteria, plants, mice, or human extract and the average biological CV was higher than that reported for other organisms grown under controlled conditions [45]. While characterizing the Pinus radiata needle proteome [119], we found differences in the standard error of mean spot quantity, depending on the number of replicates; the error ranged from 111 and 115 ng for two analytical and biological replicates to 58 and 59 ng for 10 analytical and 12 biological gel replicates.

Using more than six biological replicates did not significantly reduce the standard error; so this figure could be optimal for comparative proteomic experiments. Since normally this is not feasible, most papers in our literature review used only three biological replicates. Given the susceptibility of the data to variation, in 2-DE comparative proteomics it is necessary to be restrictive when deciding whether a spot showed variation. First, all the spots considered had to be consistent, that is, present or absent in all the biological replicates of the particular stage in question; second, when not qualitative (presence vs. absence), differences had to be statistically significant (ANOVA); finally, the variance with respect to a control had to be higher than the average biological coefficient of variance determined for a representative set of at least 150 spots.

Cell Disruption and Sample Preparation

The importance of the extraction protocol in a proteomic experiment can be summarized in the following statement: only if you can extract and solubilize a protein, you have a chance of detecting and identifying it. This sentence sums up the importance of the preparation of protein samples in a proteomic experiment [102]. This is more important in the case of plant tissue or fungal material. In the case of filamentous fungi, the protein extraction is also influenced by the presence of a cell wall which makes up the majority of the cell mass. This cell wall is exceptionally robust [120], and the effective extraction of proteins is a critical step and essential for reproducible results in fungal proteomic studies. Therefore, cell breakdown is an important element in sample preparation for fungal proteomics.

Several early studies were performed to overcome this challenge by providing an effective means of cell lysis for an adequate release of intracellular proteins. For example, mechanical lysis was used to release the cytoplasmic protein via glass beads [59, 121–125], using a cell mill [68], or by sonication [40, 126, 127], these being more efficient than either chemical or enzyme extraction methods [128]. Shimizu and Wariishi [129] utilized an alternative approach to avoid the difficulty of lysing the fungal cell wall by generating protoplasts of Tyromyces palustris, since 2-DE patterns from protoplast were better visualized than protein obtained from disrupting the fungal cell wall using SDS extraction. However, the most widely used method for cell disruption is pulverizing the fungus material in liquid nitrogen using a mortar and pestle [45–47, 64, 126, 130, 131].

The extraction protocol most amply utilized implicates the use of protein precipitation media containing organic solvents, such as trichloroacetic acid

(TCA)/acetone, followed by solubilization of the precipitate in an appropriate buffer. It allows an increase in the protein concentration and removal of contaminants (salts, lipids, polysaccarides, phenols, and nucleic acids) which can be a problem during IEF [132] and prevents protease activities. This method has often been applied to the preparation of extracts from plants [133–135] and fungi [123, 136]. TCA-treatment makes subsequent protein solubilization for IEF difficult, especially with hydrophobic proteins. These problems have been partly overcome by the use of new zwitterionic detergents [137–142] and thiourea [143], or by a brief treatment with sodium hydroxide [123], which led to an increase in the resolution and capacity of 2-DE gels. Other protein extraction methods have reported an improvement when using acidic extraction solution to reduce streaking of fungal samples caused by their cell wall [144], as well as the use of a phosphate buffer solubilization before the precipitation [45, 46]. In a study to develop an optimized protein extraction protocol for Rhizoctonia solani, Lakshman et al. [63] compared TCA/acetone precipitation and phosphate solubilization before TCA/acetone precipitation. Both protocols worked well forR. solani protein extraction, although selective enrichment of some proteins was noted with either method. Finally, the combined use of TCA precipitation and phenol extraction gave a better spot definition, because it reduced streaking and led to a higher number of detected spots [47, 145].

In B. cinerea, our group has optimized a protein precipitation protocol from mycelium based on a combination of TCA/acetone and phenol/methanol protein extraction methods described by Maldonado et al.[135] with some modifications (Gonzalez-Fernandez et al., unpublished data). Mycelium is lyophilized, pulverized with liquid nitrogen using a mortar and pestle, sonicated in 10% (w/v) TCA/acetone solution with a sonic probe, washed sucessively with methanol and acetone, and finally a protein extraction is released with phenol/methanol precipitation method. We have used a similar protocol from conidia of this fungus by sonicating directly the spores collected in 10% (w/v) TCA/acetone solution (Gonzalez-Fernandez et al., unpublished results). In this sense, specific protocols for protein extraction from spores were optimized inAspergillus ssp. using acidic conditions, step organic gradient, and variable sonication treatment (sonic probe and water bath) [40]. In this study, the use of a sonic probe was the best method to break the robust cell wall of conidia and obtain more proteins.

Special protocols are required in the case of secreted proteins, facing problems such as the very low protein concentration, sometimes below the detection limit of colorimetric methods for determining protein concentration such as Bradford, Lowry, or BCA, and the presence of polysaccharides, mucilaginous material, salts, and secreted metabolites (low-molecular organic acids, fatty acids, phenols, quinines, and other aromatic compounds). Moreover, the presence of these extracellular compounds may impair standard methods for protein quantification and can result in a strong overestimation of total protein [146]. This determination can also be affected by the high concentration of reagents from the solubilisation buffer (such as urea, thiourea or DTT) that may interfere in the spectrophotometric measurement, producing an overestimation of the total amount of protein. Fragner et al., showed that, depending on the method, the differences varied in the order of two magnitudes, indicating that only the Bradford assay does not lead to an overestimation of the proteins [147].

Francisco and colleagues provided pioneering contributions to this field, establishing a sample preparation protocol for fungal secretome [148], including steps of lyophilization or ultrafiltration plus dialysis, precipitation (TCA/ethanol or chloroform/methanol), deglycosylation, and solubilization for SDS-PAGE or 2-DE.

Comparison of different standard methods for protein precipitation has demonstrated their limited applicability to analyzing the whole fungal secretome [149–152]. Usually, the fungal liquid culture is clarified by filtering and centrifuging, then dialyzed and concentrated up by lyophilizing. Recently, a new optimized protocol has been developed to obtain extracellular proteins from several higher basidiomycetes (Coprinopsis cinerea, Pleorotus ostreatus, Phanerochaete chrysosporium, Polyporus brumalis, and Schizophyllum commune) [147]. In this work, several precipitation methods, (TCA/acetone, phenol/methanol, other precipitation methods and an optimized protocol by high-speed centrifugation/TCA precipitation/Tris-acetone washes) were compared from liquid cultures of these fungi. The best method was the use of high-speed centrifugation, since it removed a considerable gelatinous material from the liquid culture supernatants. Then fungal proteins were effectively enriched by TCA precipitation and coprecipitated metabolites hampering 2-DE were removed through the application of Tris/acetone washes [147].

Vincent et al. [56], using the plant pathogenic fungus Leptosphaeria maculans and symbiont Laccaria bicolorgrown in culture, have established a proteomic protocol for the analysis of the secretome. They evaluate different protocols, including ultrafiltration followed by TCA/acetone precipitation or phenol/ammonium acetate-phase partition, successive TCA/acetone precipitations without ultrafiltration, phenol/ether extraction without ultrafiltration, lyophilization, and prefractionation of secretome samples using liquid-phase IEF. Liquid-phase IEF followed by dialysis and lyophilization as a prefractionation prior to IPG-IEF produced the best results, with up to 2000 spots well resolved on 2-DE. This protocol can be applicable to a reduced number of samples and be very useful for descriptive proteomics but not for comparative proteomics experiments because of the excessive manipulation of the sample. Thus, in our work with B. cinerea, we aim to compare the proteome of wild-type and a high number of mutant strains (close to 100) affected in pathogenicity, and therefore we have optimized a protocol including the following steps: liquid media filtering, dialyzing, lyophilizing, and TCA/acetone-phenol/methanol protein precipitation. It is true that the number of resolved spots is much lower, but still enough for our purposes (Gonzalez-Fernandez et al., unpublished data).

A new field of study has been opened up with the analysis of infection structures such as appressorium and haustoria. In this case, specific protocols for the isolation of such structures are required [85] and the main problem is the large amount of material required for proteomic analysis if compared to transcriptomics. Godfrey et al. have developed a procedure for isolating haustoria from the barley powdery mildew fungus based on filtration and the use of differential and gradient centrifugation [44].

Up to now we have made reference to experiments with in vitro grown fungi; studies in planta are much more complicated due to the presence of both proteomes than of the plant and the pathogen. Bindschedler et al.[153] undertook a systematic shotgun proteomics analysis of the obligate biotroph B. graminis f. sp. hordei at different stages of development in the host: ungerminated spores, epiphytic sporulating hyphae, and haustoria, this being, as far as we know, the only large-scale comparative study of proteomes of phytopathogenic fungi during in planta colonization in addition to those analyses of whole extracts from host infected tissue [154,155] or intercellular washing fluids [156].

In short, since no single protein extraction protocol can capture the full proteome, the chosen protocol should be optimized for the research objective. The ideal method should be highly reproducible and should extract the greatest number of protein species, while at the same time reduce the level of contaminants and minimize artifactual protein degradation and modifications [148, 157, 158]. Fortunately, each protocol takes us to a specific fraction of the proteome, thus being complementary [135]. Another issue to consider is the extreme complexity of the proteome and the wide dynamic range in protein abundance, which exceeds the capability of all currently available analytical platforms. Sample prefractionation is a good approach to reducing the complexity of the proteome sample and decreasing the dynamic range [159], with EQUALIZER being the last developed technology [160].

Protein Separation, Mass Spectrometry and Protein Identification

In fungal plant pathogen research electrophoresis, including denaturing 1-DE SDS-PAGE and 2-DE, with IEF as first dimension, and SDS-PAGE as the second, is almost the only protein separation technique employed, with both crude total protein extracts and protein fractions obtained from various prefractionation procedures [161]. Despite its simplicity, 1-DE can still be quite a valid technique providing relevant information, especially in the case of comparative proteomics with large numbers of samples to be compared. Thus, using this technique, it is possible to distinguish between genotypes of different wild-type strains of B. cinerea and identify proteins involved in the pathogenicity mechanisms (Figure 3) (Gonzalez-Fernandez et al., unpublished results). With appropriate software, 1-DE is a simple and reliable technique for finger-printing crude extracts and it is especially useful in the case of hydrophobic and low-molecular-weight proteins [162]. The combination of 1-DE, band cutting, trypsin digestion, and LC separation of the resulting peptides remains the proteomic technique capable of providing the greatest protein coverage [163, 164]. Therefore, the 1-DE is a good approach to obtaining preliminary results before the study by 2-DE [41, 42, 50, 165]. For example, the use of 1-DE in combination with MS/MS analysis has allowed the detection of both known [166] or new proteins [62] of interest in fungal pathogenicity.

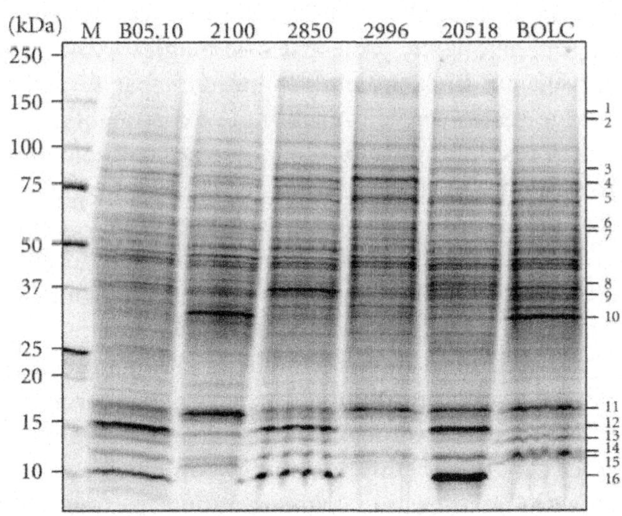

Figure 3: One-DE of 15 g of mycelium protein extract of six different strains of B. cinerea (B05.10, CECT 2100, CECT 2850, CECT 2996, CECT 20518, BOLC (isolated from infected lentil plants)). This approach allowed us to observe differences in the

protein band patterns among strains. The bands were cut out and the protein identification was made using MALDI-TOF/TOF MS/MS, and PMF search and a combined search (+MS/MS) were performed in nrNCBI database of proteins using MASCOT. Some of these proteins identified have been reported to be involved in pathogenicity in B. cinereaor in other phytopathogenic fungi, such as malate dehydrogenase (10), woronin body major protein (11), peptidyl-prolyl cis-trans isomerase (14) and PIC5 protein (15), or implicated in fungal growth and differentiation, such as nucleoside diphosphate kinase (12). The abundance of these proteins was different among isolated (Gonzalez-Fernandez et al., unpublished results).

In fungal proteomics, 2-DE separation techniques [132, 167, 168] are widely used (Table 3), these having the advantage of separating the proteins at the protein species level with a high resolution of up to 10,000 spots [169]. Briefly, the 2-DE consists of a tandem pair of electrophoretic separations: in the first dimension, proteins are resolved according to their isoelectric points (pIs) normally using IEF, and in the second dimension, proteins are separated according to their approximate molecular weight using SDS-PAGE. These proteins can then be detected by a variety of staining techniques:

- organic dyes, such as colloidal Coomassie blue staining,
- zinc-imidazole staining,
- silver staining, and
- fluorescence-based detection.

Excellent reviews describe and discuss the features and protocols of electrophoretic separations in proteomics strategies [132, 170]. The main advantages of 2-DE are high protein separation capacity and the possibility of making large-scale protein-profiling experiments. However, the reproducibility and resolution of this technique are still remaining challenges. Moreover, this method was reported to under-represent proteins with extreme physicochemical properties (size, isoelectric point, transmembrane domains) and those of a low abundance [171]. These limitations to analytical protein profiling have led to the more recent development of techniques based on LC separation of proteins or peptides, including two-dimensional liquid-phase chromatography (based on a high-perfomance chromatofocusing in the first dimension followed by high-resolution reversed-phase chromatography in the second one) [172], and one-dimensional 1-DE-nanoscale capillary LC-MS/MS, namely, GeLC-MS/MS (this technique combines a size-based protein separation with an in-gel digestion of the resulting fractions). Recorbert et al. [173] explored the efficiency of GeLC-MS/MS to identify proteins from the mycelium of Glomus intraradices developed on root organ cultures, reporting on the identification of 92 different proteins. Overall, this GeLC-MS/MS strategy paves the way towards analysing on a large-scale fungal response environmental cues on the basis of quantitative shotgun protein profiling experiments. Despite the existence of quite a number of different methods developed for protein extraction and separation, it is clear that, all in all, it is not enough to allow for the analysis of entire proteomes (organelle, cell, tissue, or organ). Some methodologies have proven to be more powerful and decisive than others, with regard to the number of proteins identified. This is the case of Multidimensional Protein Identification Technology (MudPIT), an LC-based strategy, which allows the detection of a much larger number of proteins compared to gel-based methods, its drawback being the lack of quantitative data [174]. MudPIT was first applied to the fungal proteome of the S. cerevisiae and yielded the largest proteome analysis to date, in which a total of 1484 proteins were detected and identified [175]. The categorization of these hits demonstrated the ability of this technology to detect and identify proteins rarely seen in proteome analysis, including low abundance proteins like transcription factors and protein kinases [175]. It has been reported that a set of proteins can only be detected by a specific technology [176, 177], which is in agreement with the idea that a combination of different methodologies is still needed to characterize entire proteomes [131]. MudPIT has been used to compare the proteome from germinating and ungerminated asexual uredospores of the biotrophic fungal pathogen Uromyces appendiculatus, which is the causal agent of rust disease in beans [67].

Mass spectrometry is the basic technique for global proteomic analysis due to its accuracy, resolution, and sensitivity, small amounts of sample (femtomole to attomole concentrations), and having the capacity for a high throughput. It allows not only to profile a proteome from a qualitative and quantitative point of view but also, and more important, to identify protein species and characterize postranslational modifications. Proteins are identified from mass spectra of intact proteins (top-down proteomics) or peptide fragments obtained after enzymatic (mostly tryptic peptides) or chemical treatment (bottom-up proteomics). Protein species are identified by comparison of the experimental spectra and the theoretical one obtained in silico from protein, genomic, ESTs sequence, or MS spectra databases. For that purpose, different instrumentation, algorithms, databases, and repositories are available [178, 179].

Different strategies and algorithms can be used for protein identification (i) including peptide mass fingerprinting (PMF, only valid if the protein sequence is present in the database of interest, generally used if the genome of the organism of study is fully sequenced); (ii) tandem mass spectrometry (where peptide sequences are identified by correlating acquired fragment ion spectra with theoretical spectra predicted for each peptide contained in a protein sequence database, or by correlating acquired fragment ion spectra with libraries of experimental MS/MS spectra identified in previous experiments); (iii) de novo sequencing, where peptide sequences are explicitly read out directly from fragment ion spectra; and (iv) hybrid approaches, such as those based on the extraction of short sequence tags of 3–5 residues in length, followed by "error-tolerant" database searching [179]. In a genomic-based proteomics strategy the percentage of proteins identified from MS data is dependent on the availability of genomic DNA or EST sequences. The construction of protein repositories with signature peptides and derived MS spectra will open up new possibilities for protein identification. The availability of ESTs from unsequenced "orphan" organisms as is the case of most plant fugal pathogens of interest will increase the percentage of identified proteins [93]. There are relatively few and a slow accumulation of EST data derived from a number of plant fungal pathogens and related species in public databases (e.g., dbEST at the National Centre for Biotechnology Information (NCBI),http://www.ncbi.nlm.nih.gov/dbEST/; and COGEME, Phytopathogenic Fungi and Oomycete EST Database [180, 181] at Exeter University, UK, http://cogeme.ex.ac.uk/). In Table 4, the number of EST entries for a number of fungi at the NCBI and Dana Faber databases is listed.

Table 4: Number of ESTs entries for some fungi of interest up to December 2009

Fungus	**Dana Faber/NBCIa**
Aspergillus flavus	20372/22452
Aspergillus niger	No entries/47082
Blumeria graminis hordei	No entries/17142
Botrytis cinerea	No entries/28531
Claviceps purpurea	No entries/8789
Curvularia lunata	No entries/1488
Fusarium graminearum	No entries/58011
Fusarium oxysporum	No entries/17478
Fusarium verticillioides	86908/87134
Leptosphaeria maculans	No entries/20034
Magnaporthe grisea	87403/110613
Puccinia graminis	No entries/209
Phytophthora infestans	90287/164143
Phytophthora palmivora	No entries/14824
Pyrenophora tritici-repentis	No entries/1
Nectria haematococca mpVI	No entries/33122

Sclerotinia sclerotiorum	No entries/2578
Stagonospora nodorum	No entries/16447
Ustilago maydis	No entries/39717

[a]Data taken from The Gene Index Proyect at the Dana Farber Cancer Institute

Second-Generation Techniques for Quantitative Proteomics

By using proteomics it is not only aimed to identify protein species (main objective of descriptive proteomics) but also quantify them, at least in relative terms, by comparing two biological units (genotypes, cells, organules) under different spatiotemporal parameters and environmental conditions. Absolute, rather than relative, protein quantitation remains one of the main challenges in proteomics [102]. There are different methods to dissect the proteome in a quantitative manner:

- methods based on 2-DE with poststaining, such as colloidal Comassie blue staining [182] and fluorescence staining [183], or prelabeling such as two-dimensional fluorescence difference gel electrophoresis (2-DIGE) [184, 185]; and
- gel-free methods based on in vitro or in vivo protein targetting with a stable isotope, such as isotope-coded affinity tags (ICATs) [186,187], ^{18}O labeling [188], or stable isotope labelling in cell culture (SILAC) [189], or isobaric tags, such as isobaric tag for relative and absolute quantitation (iTRAQ) [190].

Novel, label-free approaches, such as spectral counting, are being developed [191, 192]. The use of second generation proteomic techniques based on protein labelling and those label-free ones are far from being fully exploited in fungal pathogen research.

In 2-DIGE, proteins in two samples are labeled in vitro through cysteine or lysine residues using two different fluorescent cyanine dyes differing in their excitation and emission wavelengths, but with an identical relative mass. Labeled samples are then mixed and subjected to 2-DE on the same gel. After consecutive excitation with both wavelengths, the images are overlaid and normalized, whereby only differences between the two samples are visualized [193]. 2-DIGE enables to perform high-throughput, differential protein expression analysis to compare directly, on a single gel, the differences in protein expression levels between different complex protein samples. The main advantage of 2-DIGE on 2-DE is its unrivalled performance, attributable to a unique experiment, in which each protein spot on the gel is represented by its own internal standard [105].

The classical proteomic quantification electrophoretic methods utilizing dyes, fluorophores, or radioactivity have provided very good sensitivity, linearity, and dynamic ranges, but they suffer from two important shortcomings: first, they require high-resolution protein separation typically provided by 2-DE gels, which limits their applicability to abundant and soluble proteins; and second, they do not reveal the identity of the underlying protein [194]. Both of these problems are overcome by modern LC-MS/MS techniques [95,195–197]. However, the MS-based techniques are not inherently quantitative because proteolytic peptides exhibit a wide range of physicochemical properties (size, charge, hydrophobicity) that lead to large differences in mass spectrometric response. Therefore, in MS-based gel-free quantification it is necessary to use isotopic labeling. Observed peak ratios for isotopic analogs are highly accurate, because there are no chemical differences between the species, and they are analyzed in the same experiment. Mass spectrometry can recognize the mass difference between the labeled and unlabeled forms of a peptide and the quantification is achieved by comparing their respective signal intensities [194].

A number of isotopic labeling techniques have recently been proposed that share the requirement of the chemical modification of the peptides or proteins. One of these strategies is the ICAT method for relative quantitation of protein abundance [186]. In this approach, an isotopically labeled affinity reagent is attached to particular amino acids in all proteins in the population. After digestion of the protein to peptides, as a necessary step in all mainstream proteomic protocols, the labeled peptides are affinity-purified using the newly incorporated affinity tag, thereby achieving a simplification of the peptide mixture at the same time as incorporating the isotopic label. This method addresses many of the above limitations and leads to a larger number of identifications of cysteine-containing peptides. However, the method is performed by cross-linking peptides to beads via their cysteine groups and photo-releasing them afterwards, which may compromise low-level analysis. The iTRAQ is used to identify and quantify proteins from different sources in one experiment. The method is based on the covalent labeling of the N-terminus and side-chain amines of peptides from protein digestions with tags of varying mass. The fragmentation of the attached tag generates a low molecular mass reporter ion that can be used to relatively quantify the peptides and the proteins from which they originated [198]. In SILAC (stable isotope labeling by amino acids in cell culture), labeled, essential amino acids are added to amino acid-deficient cell culture media and are therefore incorporated into all proteins as they are synthesized, "encoded into the proteome" [189]. No chemical labeling or affinity purification steps are performed, and the method is compatible with virtually all cell culture conditions. Finally, label-free protein quantification methods are promising alternatives. It is based on precursor signal intensity, which is, in most cases, applied to data acquired on high mass precision spectrometers. The mass spectral peak intensities of the peptide ion correlate well with protein abundances in complex samples [199–201]. Another label-free method is spectral counting, which simply counts the number of MS/MS spectra identified for a given peptide and then integrates the results for all measured peptides of proteins that are quantified [202]. An advantage of this technique is that relative abundances of different proteins can in principle be measured. These new quantification techniques have become powerful tools to overcome the inherent problems of the 2-DE including identification of proteins of a low abundance, high hydrophobicity, extreme pl or high MW.

These second-generation MS technologies for Quantitative Proteomics have not begun to be applied to Fungal Proteomics yet. The only one nongel-based quantitative proteomics example is the use of iTRAQ to study the profile protein expression differences on F. graminearum which allowed the identification of numerous candidate pathogenicity proteins [53].

Although this review is focused on MS-based proteomics, we want to make a brief mention of protein microarrays because they are powerful tools for individual studies as well as systematic characterization of proteins and their biochemical activities and regulation [203]. The arrays can be used to screen nearly the entire proteome in an unbiased fashion and have an enormous utility for a variety of applications. These include protein-protein interactions, identification of novel lipid- and nucleic acid-binding proteins, and finding targets of small molecules, protein kinases, and other modification enzymes.

In short, all these technologies have a great potential in protein separation and remain a challenge for future research works in Fungal Proteomics.

Data Analysis and Statistical Validation

Proteomics tools generate an important amount of data, because a single proteomics experiment reveals the expression information for hundreds or thousands of proteins. Therefore, data analysis and bioinformatics are essential for this type of research and in many cases take more time than the actual experiment and require special skills and tools (for review see [116, 118]). All 2-DE software permits a fast and reliable gel comparison, and multiple gel analyses, including filtering of 2-DE images, automatic spot detection, normalization of the volume of each protein spot, and differential and statistical analyses [204–206]. A great resource for finding software tools for proteomics can be found in the website https://proteomecommons.org/.

Before the protein identification, the remaining challenge is to determine whether the putative identification is, in fact, correct. Statistical tools help us to validate information. Although postsearch statistical validation still does not enjoy universal application, its importance has been recognized by most researchers and codified in the editorial policies of some leading journal [207]. In this decade, an important number of commercial software involving even more powerful algorithms and statistical tools than those of the previous generations have been designed to help researchers deal with the sheer quantity of data produced [194, 195, 208].

Statistical data analyses can be classified as univariated or multivariated [209]. The univariated methods, such as the Student's -Test, are used to detect significant changes in the expression of individual proteins. They are the simplest to interpret conceptually and the most common ones used. The multivariated methods, such as principle component analysis (PCA), look for patterns in expression changes and utilize all the data simultaneously. Early expression studies compared one sample with another, generally by the calculation of a ratio, and the analyses were restricted to looking for changes above a threshold determined by the system experimental noise. This method of analysis limits the sensitivity of the system, as biologically relevant changes smaller than the threshold cannot be detected. Using a threshold is a rather simplistic approach and does not take into account the variability of each protein, running the risk of selecting variable proteins due to sample selection. The use of a fold change has a potential role in preliminary experiments but the limitation of this method must always be considered when interpreting the data [209].

Hypothesis tests, for example, Student's -Test, assess whether the differences between groups are an effect of chance arising from a sampling effect or reflect a real statistical significant difference between groups [209]. Hypothesis tests are usually stated in terms of both a condition that is in doubt (the null hypothesis) and a condition that is believed to exist (the alternative hypothesis). The tests calculate a -score, which is the probability of obtaining these results assuming that the null hypothesis is correct. Hypothesis tests can be divided into two groups: parametric and nonparametric. Parametric tests assume that the distribution of the variables being assessed belong to known probability distributions. For example, the Student's -Test assumes that the variable comes from a normal distribution. Nonparametric tests, also called distribution-free methods, do not rely on estimation of parameters such as the mean. An example of nonparametric tests is the Mann-Whitney -Test, which ranks all values from low to high and then compares the mean ranking the two groups. Otherwise, Student's -Test or Mann-Whitney -Test compares two groups and ANOVA or Krustal-Wallis compares more than two groups. These tests analyze individual spots instead of the complete set, omitting information about correlated variables. Biron et al. recommend assessing the normality for each protein species and then selecting either a parametric or nonparametric test [208].

In expression studies, many thousands of statistical tests are conducted, one for each protein species. A substantial number of false positives may accumulate which is termed the multiple testing problem and is a general property of a confidence-based statistical test when applied many times [209]. One approach to addressing the multiple testing problem is to control the family wise error rate (FWER), which control the probability of one or more false rejections among all the tests conducted. The simplest and most conservative approach is the Bonferroni correction, which adjusts the threshold of significance by dividing the percomparison error rate (PCER) by the number of comparisons being completed [210]. This has led to the application of methodologies to control the false discovery rate (FDR), where the focus is on achieving an acceptable ratio of true- and false-positives. The FDR is a proportion of changes identified as significant that are false [211]. An extension to the FDR calculates a -value for each tested feature and is the expected proportion of false positives incurred when making a call that this feature has a significant change in the expression [212]. For each -value, a -value will be reported on an overall estimation for the proportion of species changing in the study.

Multiple testing correction methods, such as the Bonferroni correction and testing for the false discovery rate (FDR) [213], fit the Student -test or ANOVA values for each protein spot to keep the overall error rate as low as possible. Multivariate data analysis methods, such as PCA, are now used to pinpoint spots that differ between samples. These multivariate methods focus not only on differences in individual spots but also on the covariance structure between proteins [214]. However, the results of these methods are sensitive to data scaling, and they may fail to produce valid multivariate models due to the large number of spots in the gels that do not contribute to the discrimination process [215]. One of the limitations of PCA analysis is that it does not allow to miss values, a problem that can be avoided by imputing them when possible (if enough replicates are available) [216].

Databases and Repositories

The huge amount of data generated are being deposited and organized in several databases available to the scientific community: the UniProt knowledgebase reported by Schneider et al. [217] and other Proteome Databases mentioned in Table 5. After 20 years of Proteomics research, it is possible to look back at previous research and publications, identifying errors from the experimental design, the analysis, and the interpretation of the data [179]. In addition, data validation is done in a purely descriptive or speculative manner, as well as it is common to find low-confidence protein identification in the literature, especially in the case of unsequenced organisms and inappropriate statistical analyses of results have often been performed. It is interesting to see how many manuscripts contain the term "proteome" when probably only a tiny fraction of the total proteome has been analysed. About this problem, HUPO's Proteomic Standard Initiative has developed guidance modules [218] that have been translated into Minimal Information about a Proteomic Experiment (MIAPE) documents. The MIAPE documents recommend proteomic techniques that should be considered and followed when conducting a proteomic experiment. Proteomics journals should be, and in fact are, extremely strict when recommending that investigators follow the MIAPE standards for publishing a proteomic experiment. On the other hand, many journals recommend or require the original data generated in a proteomic experiment to be submitted to public repositories [207, 219]. A shift in the protein identification paradigm is currently underway, moving from sequencing and database searching to spectrum searching in spectral libraries. This underscores the importance of repositories for Proteomics [220–222]. The main public peptide and protein identification repositories are GPMDB (Global Proteome Machine database) [223], PeptideAtlas [224], and Proteomics Identifications database (PRIDE) [225]. Other emerging and smaller systems include Genome Annotating Proteomic Pipeline database (GAPP database) [226], Tranche (Falkner, J. A., Andrews, P. C., HUPO Conference 2006, Long Beach, USA, Poster presentation), PepSeeker [227], Max-Planck Unified database (MAPU) [228], the Open Proteomic Database (OPD) [229], and the Yeast Resource Center Public Data Repository (YRC PDR) [230].

Table 5: Useful online resources and Fungal Genome and Proteome Databases

Name/description	URL
Genome Databases	
National Center for Biotechnology Information (NCBI).	http://www.ncbi.nlm.nih.gov/
NIH genetic sequences database.	http://www.ncbi.nlm.nih.gov/Genbank/
Fungal Genomes Central, information and resources pertaining to fungi and fungal sequencing projects.	http://www.ncbi.nlm.nih.gov/projects/genome/guide/fungi/

The Gene Index proyect (GI). The Computational Biology and Functional Genomics Laboratory, and the Dana-Faber Institute and Public School of Public Health.	http://compbio.dfci.harvard.edu/tgi/fungi.html
Fungal Genome Initiative of The Broad Institute (FGI).	http://www.broadinstitute.org/science/projects/fungal-genome-initiative/fungal-genome-initiative
Genoscope, Sequencing National Centre.	http://www.genoscope.cns.fr/spip/Fungi-sequenced-at-Genoscope.html
Joint Genome Institute (JGI).	http://www.jgi.doe.gov/
The Genome Center at Washington University (WU-GSC).	http://genome.wustl.edu/genomes/list/plant_fungi
The Sanger Institute fungal sequencing.	http://www.sanger.ac.uk/Projects/Fungi/
Genome projects.	http://genomesonline.org/
The MIPS F. Graminearum Genome Database.	http://mips.gsf.de/projects/fungi/Fgraminearum/
The MIPS U. Maydis Database.	http://mips.gsf.de/genre/proj/ustilago
The MIPS Neurospora crassa Genome Database.	http://mips.helmholtz-muenchen.de/genre/proj/ncrassa/
COGEME, Phytopathogenic Fungi and Oomycete EST Database (v1.6), constructed and maintained by Darren Soanes (University of Exeter, UK).	http://cogeme.ex.ac.uk/
SGD, Saccaromyces Genome Database, scientific database of the molecular biology and genetics of the yeast Saccharomyces cerevisiae.	http://www.yeastgenome.org/
e-Fungi, warehouse which integrates sequence data (genomic data, EST data, Gene Ontology annotation, KEGG pathways and results of the following analyses performed on the genomic data) from multiple fungal sequences in a way that facilitates the systematic comparative study of those genomes (School of Computer Science and the Faculty of Life Sciences at the University of Manchester and the Departments of Computer Science and Biological Sciences at the University of Exeter).	http://beaconw.cs.manchester.ac.uk/efungi/execute/welcome

CADRE, Central Aspergillus Database Repository, resource for viewing assemblies and annotated genes arising from various Aspergillus sequencing and annotation projects.	http://www.cadre-genomes.org.uk/
FungalGenome, website with several links and references for the currently available fungal genomes sequences or proposed fungal genomes.	http://fungalgenomes.org/wiki/Fungal_Genomes_Links
Proteome Databases	
The Expasy (Expert Protein Analysis System) proteomics server of Swiss Institute of Bioinformatics (SIB). Analysis of protein sequences, structures and 2-D-PAGE.	http://ca.expaxy.org/
MIPS, Munich Information Center for Protein Sequences.	http://mips.gsf.de/http://ca.expaxy.org/
The PRIDE, Proteinomics IDEntifications Database. EMBL-EBI (European Bioinformatic Institute).	http://www.ebi.ac.uk/pride/
Integr8, Integrated information about deciphered genomes and their corresponding proteomes. EMBL-EBI.	http://www.ebi.ac.uk/integr8/EBI-Integr8-HomePage.do
SNAPPVIEW (Structure, iNterfaces and Aligments for Protein-Protein Interactions).	http://www.compbio.dundee.ac.uk/SNAPPI/download.jsp
Phospho3. Database of three-dimensional structures of phosphorylation sites.	http://cbm.bio.uniroma2.it/phospho3d/
Proteome Analyst PA-GOSUB 2.5. Sequences, predicted GO molecular functions and subcellular localisations.	http://www.cs.ualberta.ca/~bioinfo/PA/GOSUB/
RCSB, The Research Collaboratory for Structural Bioinformatics. Protein Database (PDB).	http://www.rcbs.org/pdb/home/home.do
PDB-Site. Comprehensive structural and functional information on PTMs, catalytic active sites, ligand binding (protein-protein, protein-DNA, protein-RNA interacions) in the Protein Data Bank (PDB).	http://wwwmgs.bionet.nsc.ru/mgs/gnw/pbbsite/
WoLF PSORT, Protein Subcellular Localization Prediction.	http://wolfpsort.org/

NMPdb, Nuclear Matrix Associated Proteins.	http://cubic.bioc.columbia.edu/db/NMPdb/
TargetP, predicts the subcellular location of eukaryotic proteins, based on the predicted presence of the N-terminal presequences.	http://www.cbs.dtu.dk/services/TargetP/
MitoP2, Mitochondrial Database. This database provides a comprehensive list of mitochondrial proteins of yeast, mouse, Arabidopsis thaliana, neurospora and human.	http://www.mitop.de:8080/mitop2/
The SecretomeP, Prediction of protein secretion and information on various PTMs and localisational aspect of the protein.	http://www.cbs.dtu.dk/services/SecretomeP/
MASCOT, a powerful search engine that uses MS data to identify proteins from primary sequence databases.	http://www.matrixscience.com/
VEMS, Virtual Expert Mass Spectrometrist. Program for integrated proteome analysis.	http://www.yass.sdu.dk/
The NetPhos server produces neural network predictions for serine, threonine and tyrosine phosphorylation sites in eukaryotic proteins.	http://www.cbs.dtu.dk/services/NetPhos/
ProPrInt, Protein-Protein Interaction Predictor. Compilation of web-resources in the field of Protein-Protein Interaction (PPI).	http://www.imtech.res.in/raghava/proprint/resources.htm
ProteomeCommons, public proteomics database for annotations and other information linked to the Tranche data repository and to other resources. It provides public access to free, open-source proteomics tools and data.	https://proteomecommons.org
Fungal Proteome Specialized Databases	
MPID, Protein-protein interaction Database of M. grisea.	http://bioinformatics.cau.edu.cn/cgi-bin/zzd-cgi/ppi/mpid.pl
FPPI, Protein-protein interaction database of F. graminearum.	http://csb.shu.edu.cn/fppi

Proteomics of Plant Pathogenic Fungi

Several proteomic studies have been carried out in order to understand fungal pathogenicity or plant-fungus interactions (for reviews see [266–268]), although the plant-fungus association has been the one most studied by Proteomics approaches (Table 6), which is outside the scope of this review. On the other hand, some fungal species have attracted an increasing interest in the biotechnological industry, in food science, or in agronomy as biocontrol agents (Table 7), which is also beyond the objectives of this work. At this point, this review describes studies published up to December 2009 in plant pathogenic fungi in descriptive proteomics (intracellular proteomics, subproteomics, and secretomics), differential expression proteomics, as well as some basic knowledge about the Interactomics in fungi (Table 3).

Table 6: Original proteomics papers published in plant-pathogenic fungi interactions

Pathogen-Host	Description of study (References)
Alternaria bassicicola-Arabidopsis	Study of change in the Arabidopsis secretome in response to salicylic acid and identifying of several proteins involved in pathogen response such as GDSL LIPASE1 (GLIP1) [231].
Aphanomyces euticher-Medicago truncatula	Identification of several proteins which play a major role during root adaptation to various stress conditions [232], and study of parasitic plant-pathogen interactions formed between legumes and this oomycete [233, 234].
Black point disease-Barley	Identification of a novel late embryogenesis abundant (LEA) protein and a barley grain peroxidase 1 (BP1) that were specifically more abundant in healthy grain and black pointed grain, respectively [235].
B. graminis hordei-Barley	Systematic shotgun proteomics analysis at different stages of development of powdery mildew in the host to gain further understanding of the biology during infection of this fungus [153].
Cladosporium fulvum-Tomato	Identified 3 novel fungal secretory proteins [236].
Cronartium ribicola-Pinus strobus	Study of molecular basis of white pine blister rust resistance [237].
Diploidia scrobiculata-Pinus nigra Sphaeropsis sapinea-Pinus nigra	Study about defense protein responses in phloem of Austrian pine inoculated with D. scrobiculata and S. sapinea[238].
Erysiphe pisi-Pea	Identification of proteins implicated in powdery mildew resistance [239].

F. graminearum-Barley	Identification of proteins associated with resistance to Fusarium head blight in barley [240].
F. graminearum-Wheat	Identification of proteins associated with resistance to Fusarium head blight in wheat [241] and which have a role in interaction between F. graminearum and T. aestivum [155].
F. graminearum-Wheat	Identification of proteins associated with resistance to scab in wheat spikes [238].
F. moniliforme-Arabidopsis	Study of changes in the extracellular matrix of A. thaliana cell suspension cultures with fungal pathogen elicitors of F. moniliforme [242].
F. oxysporum-Sugar beet	Study of resistance to F. oxysporum disease [243].
F. oxysporum-Tomato	Identification of 21 tomato and 7 fungal proteins in the xylem sap of tomato plants infected by F. oxysporum [156].
F. verticillioides-Maize	Identification of protein change patterns in germinating maize embryos in response to infection with F. verticillioides[244].
Fusarium-Arabidopsis	Identification of differentially expressed proteins in response to treatments with pathogen-derived elicitors to identify pivotal genes' role in pathogen defence systems [245].
Fusarium-Maize	Study of the role of the extracellular matrix in signal modulation during pathogen-induced defence responses [246].
Gossypium hirsutum-Cotton	Identified pathogen-induced cotton proteins implicated in post-invasion defence reponses (PR-proteins related to oxidative burst), nitrogen metabolism, amino acid synthesis and isoprenoid synthesis [247].
Hyaloperonospora parasitica-Arabidopsis B. cinerea-Arabidopsis	Study of pathogenic resistance of Arabidopsis wild-type and CaHIR1-overexpressing transgenic plants inoculated with these fungi, among other pathogens [248].
Leptosphaeria maculans-Brassica	Identification of Brassica proteins involved in resistance to this fungus [249].
L. maculans-Brassica carinata L. maculans-Brassica napus	Study of changes in the leaf protein profiles of Brassica napus (highly susceptible) and Brassica carinata (highly resistant) in order to understand the biochemical basis for the observed resistance to L. maculans [250].

M. grisea-Rice	Change protein analysis during blast fungus infection of rice leaves with different levels of nitrogen nutrient [251]. Analysis of differentially expressed proteins induced by blast fungus in suspension-cultured cells [252] in leaves [253] and during appressorium formation [254]. Proteomic approach of differentially expressed proteins in rice plant leaves at 12 h and 24 h after treatment with the glycoprotein elicitor CSB I, purified from ZC(13), a race of the rice blast fungus M. grisea [255].
Marsonina brunnea f. sp. Multigermtub-Populus euramericana	Identification of proteins related to black spot disease resistance in poplar leaves [256].
Moliniophtora perniciosa-Cocoa	Optimization of protein extraction for cocoa leaves and meristemes infected by this fungus that causes witches' broom disease [257].
Nectria haematococca-Pea	Study of extracellular proteins in pea roots inoculated with N. haematococca [258].
Penincillium exposum-Piccia membranefaciens-Peach fruit	Peach fruit inoculated with P. exposum and treated with SA and P. membranefaciens [259].
Phellinus sulphurasces-Pseudotsuga menzeii	Comparative proteomic study to explore the molecular mechanisms that underlie the defense response of Douglas-fir to laminated root rot disease caused by P. sulphurascens [260].
Peronospora viciae-Pea	Catalogued host (pea) leaf proteins, which showed alternation in their abundance levels during a compatible interaction with P. viciae [261].
Plasmodiophora brassicae-Brassica napus	Study of changes in the root protein profile of canola with clubroot disease [262].
Puccinia triticina-Wheat	Change analysis in the proteomes of both host and pathogen during development of wheat leaf rust disease [154].
Rhizoctonia solani-Rice	Identification of proteins and DNA markers in rice associated with response to infection by R. solani [263].
Rust-Phaseolus vulgaris	Study of basal and R-gene-mediated plant defense in bean leaves against this pathogen [264].
S. sclerotiorum-Brassica napus	Study of changes in the leaf proteome of B. napus accompanying infection by S. sclerotiorum [265].

Table 7: Original proteomics papers published on fungi for biotechnological or agricultural applications

Fungus	**Interest (reference)**
Boletus edilus	Study of salinity stress of this ectomycorrhizal fungus for its importance in reforestation in saline areas [269].
Coprinopsis cinerea	Optimization of a protocol for 2-DE of extracellular proteins from these wood-degrading fungi [147].
Pleorotus ostreatus	
Phanerochaete chryso-sporium	
Polyporus brumalis	
Schizophyllum com-mune	
Glomus intraradices	Study of arbuscular mycorrhiza symbiosis [173].
Metarhizium aniso-pliae	Study of bioinsecticidal activity of this fungus to develop novel compounds or produce genetically modified plants resistant to insect pests [270].
Monascus pilosus	Study of the influence of nitrogen limitation for industrial production of many poliketide secondary metabolites [271].
Phanerochaete chryso-sporium	Several studies of ligninolytic processes for wood biodelignification in cellulose pulp industries [130, 131, 149, 150, 272].
Pleurotus sapidus	Study of secretome for wood biodelignification for peanuts industry applications [152].
Trichoderma atro-viride	Several studies in these fungus for their biocontrol properties [126, 273–275]
T. harzianum	
T. reesei	Study of cell-wall-degrading enzymes in A. bisporigera (an ectomycorrhizal basidiomyctous fungus) comparing with a MS/MS-based shotgun proteomics of the secretome of T. reesei [276].
Amanita bisporigera	

Descriptive, Subcellular, and Differential Expression Proteomics

Within this section, papers devoted to establishing reference proteome maps of fungal cells and structures and subcellular fractions, and to study changes in the protein profile between species, races, populations, mutants, growth and developmental stages, as well as growth conditions, are discussed, paying special attention to proteins related to pathogenicity and virulence.

Most of the reported work mainly uses mycelia from in vitro grown fungi, and 2-DE coupled to MS as proteomic strategy. Thus, a partial proteome map has been reported for the ascomycete B. cinerea Pers. Fr. (teleomorf Botryotinia fuckeliana (de Bary) Whetzel), a phytopathogenic necrotroph pathogen causing significant yield losses in a number of crops, by Fernández-Acero et al. They have reported the detection of 400 spots in Coomassie-stained 2-DE gels, covering the 5.4–7.7 pH and 14–85 kDa ranges. Out of 60 spots subjected to MS analysis, twenty-two proteins were identified by MALDI-TOF or ESI IT MS/MS, with some of them corresponding to forms of malate dehydrogenase, glyceraldehyde-3-phosphate dehydrogenase, and a cyclophilin, proteins that have been related to virulence [45]. In a second study, comparative proteomic analysis of two B. cinerea strains differing in virulence and toxin production revealed the existence of qualitative and quatitative differences in the 2-DE protein profile. Some of them were the same proteins mentioned above and they appeared overexpressed or exclusively in the most virulent strain [46]. A third and more exhaustive work tried to establish a proteomic map of B. cinerea during cellulose degradation [47]. Using 2-DE and MALDI-TOF/TOF MS/MS, 306 proteins were identified, mostly representing unannoted proteins. The authors conclude that since cellulose is one of the major components of the plant cell wall, many of the identified proteins may have a crucial role in the pathogenicity process, be involved in the infection cycle, and be potential antifungal targets.

A close relative to B. cinerea is the soil-borne Sclerotinia sclerotiorum. Yajima and Kav [64] performed the first comprehensive proteome-level study in this important phytopathogenic fungus, in order to gain a better understanding of its life cycle and its ability to infect susceptible plants. For the high-throughput identification of secreted as well as mycelial proteins, they employed 2-DE and MS/MS. Eighteen secreted and 95 mycelial proteins were identified. Many of the annotated secreted proteins were cell wall-degrading enzymes that had been previously identified as pathogenicity or virulence factors of S. Sclerotiorum. Furthermore, this study has allowed the annotation of a number of proteins that were unnamed, predicted, or hypothetical proteins with undetermined functions in the available databases.

Xu et al. [50] analyzed the proteome profile of six different isolates of Curvularia lunata, a maize phytopatogenic fungus, by means of 1-DE and 2-DE, in an attempt to correlate the band or spot pattern with virulence. According to the 1-DE band pattern, isolates were clustered into three groups consisting of different virulent types. By 2-DE 423 spots were resolved with 29 of them being isolate-specific, and 39 showed quantitative differences. Twenty proteins were identified by MALDI-TOF-TOF, most of them associated with virulence differentiation, metabolisms, stress response, and signal transduction. One of them was identified as Brn1 protein which has been reported to be related to melanin biosynthesis and the virulence differentiation in fungi.

The fungal pathogen F. graminearum (teleomorph Gibberella zeae) is the causal agent of Fusarium head blight in wheat, barley, and oats and Gibberella ear rot in maize in temperate climates worldwide. It synthesizes trichothecene mycotoxins during plant host attack to facilitate spread within the host. In order to study proteins and pathways that are important for successful host invasion, Taylor et al. [53] conducted experiments in which F. graminearum cells were grown in aseptic liquid culture conditions conducive to trichothecene and butenolide production in the absence of host plant tissue. Protein samples were extracted from three biological replicates of a time course study and subjected to iTRAQ (isobaric tags for relative and absolute quantification) analysis. Statistical analysis of a filtered dataset of 435 proteins revealed 130 F. graminearum proteins that exhibited significant changes in expression, 72 of which were upaccumulated relative to their level at the initial phase of the time course. There was good agreement between upaccumulated proteins identified by 2-DE-MS/MS and iTRAQ. RT-PCR and northern hybridization confirmed that genes encoding proteins that were upregulated based on iTRAQ were also transcriptionally active under mycotoxin-producing conditions. Numerous candidate pathogenicity proteins were identified using this technique, including many predicted secreted proteins. Curiously, enzymes catalyzing reactions in the mevalonate pathway leading to trichothecene precursors were either not identified or only identified in one replicate, indicating that proteomics approaches cannot always probe biological characteristics. Two-DE with MS has been used to compare the proteome of virus-free and virus-(FgV-DK21-) infected F. graminearum cultures [54]. The virus perturbs fungal developmental processes such as sporulation, morphology, and pigmentation and attenuates its virulence. A total of 148 spots showing differences in abundance were identified. Among these spots, 33 spots were subjected to ESI-MS/MS, with 23 identified. Seven proteins including sporulation-specific gene SPS2, triose phosphate isomerase, nucleoside diphosphate kinase, and woronin body major protein precursor were upaccumulated while 16, including enolase, saccharopine dehydrogenase, flavohemoglobin, mannitol dehydrogenase, and malate dehydrogenase, were downaccumulated. Variations in protein abundance were investigated at the mRNA level by real-time RT-PCR analysis, which confirmed the proteomic data for 9 out of the representative 11 selected proteins.

There are a few proteomics studies on fungal spores published. The results in the characterization of Penicillium spores by MALDI-TOF MS with different matrices demonstrated its ability for the classification of fungal spores [277]. Recently, Sulc et al. [40] have reported protein profiling of intact Aspergillus ssp. spores, including some plant pathogenic species, by MALDI-TOF MS, and they built up a mass spectral database with twenty-four Aspergillus strains. Thus, these mass finger-printing generated by MS can be used for typing and characterizing different fungal strains and finding new biomarkers in host-pathogen interactions. Another study on B. graminis f.sp. hordei (Bgh) spores concluded with the first proteome of Bgh, using a combination of 2-DE and MS analyses and matched to NCBInr EST on Bgh-translated genome databases [43]. The identity of 123 distinct fungal gene products was determined, most of them with a predicted function in carbohydrate, lipid, or protein metabolism indicating that the conidiospore is geared for the breakdown of storage compounds and protein metabolites during germination correlating with previously reported transcriptomic data [278, 279]. These results allowed a functionally annotated reference proteome for Bgh conidia.

Holzmuller et al. [118] have reported a technique to isolate the fungal haustorium (specialised structures, existing in intimate contact with the host cell, are required by the pathogen to acquire nutrients from the host cell) from infected plants, using the barley powdery mildew as an experimental system. The technique is of relevance in the study of the molecular bases of biotrophy considering that biotrophic fungi, including downy mildews (Oomycota), powdery mildews (Ascomycota), and rust fungi (Basidiomycota), are some of the most destructive pathogens on many plants. Extracted proteins were separated and analyzed by LC-MS/MS. The searches were made against a custom Bgh EST sequence database and the NCBInr fungal protein database, using the MS/MS data, and 204 haustorium proteins were identified. The majority of the proteins appeared to have roles in protein metabolic pathways and biological energy production. Surprisingly, pyruvate decarboxylase (PDC), involved in alcoholic fermentation and commonly abundant in fungi and plants, was absent in both their Bgh proteome data set and in their EST sequence database. Significantly, BLAST searches of the recently available Bgh genome sequence data also failed to identify a sequence encoding this enzyme, strongly indicating that Bgh does not have a gene for PDC [44].

In order to overcome the low proteome coverage of most of the proteomic platforms available, this being related to the physicochemical and biological complexity and high dynamism range of proteins, different strategies directed at subfractionating the whole proteome have been developed, most of them involving cell fractionation. The analysis of the subcellular proteomes [280] not only allows a deeper proteome coverage but also provides relevant informaction on the biology of the different organules, protein location, and trafficking. The number of intracellular subproteomic studies carried out with fungal plant pathogens is minimum. Next we introduce a couple of papers appearing in the literature. The number of them devoted to the cell wall and extracellular fraction is much higher, and because of that a specific section is devoted to them. Hernández-Macedo et al. [131] have reported differences in the patterns of cellular and membrane proteins obtained from iron-sufficient and iron-deficient mycelia from P. chrysosporium and L. edodes by using SDS-PAGE and 2-DE. Mitochondria have also received attention. Grinyer et al. [121] were the first to publish a mitochondrial subproteome, describing a successful sample preparation protocol and mitochondrial proteome map for T. harzianum. Based on protein databases of N. crassa, A. nidulans, A. oryzae, S. cerevisiae, andSchizosaccharomyces pombe, they identified 25 unique mitochondrial proteins involved in the tricarboxylic acid cycle, chaperones, binding-proteins and transport proteins, as well as mitochondrial integral membrane proteins. More recently, the same researchers separated and identified 13 of the 14 subunits of the T. reesei 20S proteasome [281], providing the first filamentous fungal proteasome proteomics and paving the way for future differential display studies addressing intracellular degradation of endogenous and foreign proteins in filamentous fungi.

Relevant information on biological systems and processes comes from comparative studies in which genotypes, including mutants, developmental stages, or environmental conditions supply the knowledge inferred from the observed differences. Fungal pathogenicity requires the coordinated regulation of multiple genes (and their protein products) involved in host recognition, spore germination, hyphal penetration, appressorium formation, toxin production, and secretion. To study the infection cycle and to identify virulence factors, proteomics provides us with a powerful tool for analyzing changes in protein expression between races and stages. However, most of these studies are made in planta after the plant inoculation, which is outside the scope of this review. In the case of plant fungal pathogens at least four papers have reported changes in the protcome at different developmental stages or strains. The dimorphic phytopathogenic fungusUstilago maydis has been established as a valuable model system to study fungal dimorphism and pathogenicity. In its haploid stage, the fungus is unicellular and multiplies vegetatively by budding and undergoing a dimorphic transition infective filamentous growth. This process is coordinately regulated by the bW/bE transcription factor. Böhmer et al. [68] reported the first proteome reference map of U. maydis cells, in which proteins were identified combining 2-DE with MALDI-TOF MS and ESI-MS/MS analyses. The authors observed 13 proteins spots accumulated in greater abundance in the bW/bE-induced filamentous form than in the budding state. The majority of the identified proteins might have putative roles in energy and general metabolism. Comparison of Rac1- and -b-regulated protein sets supports the hypothesis that filament formation during pathogenic development occurs via stimulation of a Rac1-containing signalling module. The proteins identified in this study might prove to be potential targets for antibiotic substances specifically targeted at dimorphic fungal pathogens. Detailed information on the proteins can be found in an interactive map accessible at the MIPS Ustilago Maydis Database (MUMDB;http://mips.gsf.de/genre/proj/ustilago/Maps/2D/). The reference map generated from U. maydis had a coverage of 4% of all annotated genes, indicating the low proteome coverage encompassed by standard proteomic techniques.

By using 2-DE and MALDI-TOF MS, specific proteins of asexual life stages from Phytophthora palmivora, a pathogen of cocoa and other economically important tropical crops, were analyzed [58]. From 400 (cyst and germinated cyst) to 800 (sporangial) could be resolved. Approximately 1% of proteins appeared to be specific for each of the mycelial, sporangial, zoospore, cyst, and germinated cyst stages of the life cycle. Moreover, they made the protein profiles of parallel samples of P. palmivora and P. infestans and demonstrated that precisely 30% of proteins comigrated suggesting that proteomics could be used to proteptype Phytophthora spp. In this work, only three identified proteins were reported, corresponding to actin isoforms. More recently, Ebstrup et al. [59] performed a proteomic study of proteins from cysts, germinated cysts, and appressoria on P. infestansgrown in vitro, identifying significant changes in the amount of several proteins. These identified proteins were most likely important for disease establishment and some of the proteins could therefore be putative targets for disease control. For example, downregulation of the crinkling- and necrosis-inducing (CRN2) protein in appressoria compared to germinated cysts and the discovery of upregulation of a putative elongation factor (EF-3) are of great interest. On the one hand, CRN2 protein might have an important function in the interaction with the host-plant before and after penetration into the leaf, this being a putative target for disease control. Since plants presumably do not contain EF-3, it could represent a putative antioomycete as well as a putative antifungal target. Furthermore, several representatives of housekeeping systems were upaccumulated, and these changes are most likely involved in the runup to the establishment of the infection of the host plant. The biotrophic fungal pathogen U. appendiculatus is the causal agent of rust disease of beans. Cooper et al. [67] surveyed the proteome from germinating and ungerminated asexual uredospores of this pathogen, using MudPIT MS/MS. The proteins identified revealed that uredospores require high energy and structural proteins during germination, indicating a metabolic transition from dormancy to germination.

The role of signal transduction in the pathogenecity of Stagonospora nodorum is well established and the inactivation of heterotrimeric G protein signaling caused developmental defects and reduced pathogenicity [282]. In a follow-up study, the S. nodorum wild-type and Galpha-defective mutant (gna1) proteomes were compared via 2-DE coupled to LC-MS/MS. By matching the protein mass spectra to the translate S. nodorumgenome, the study identified several Gna1-regulated proteins, including a positively regulated short-chain dehydrogenase (Sch1) [65].

Cao et al. [61] released an evaluation of pathogenic ability of Pyrenophora tritici-repentis and the possible adaptation to a saprophytic habit of an avirulent race. This fungus causes tan spot, an important foliar disease of wheat, and produces multiple host-specific toxins, including Ptr ToxB, which is also found in avirulent isolates of the fungus. In order to improve the understanding of the role of this homolog and evaluate the general pathogenic ability of P. tritici-repentis, the authors compared both full mycelial and secreted proteomes of avirulent and virulent isolates of the pathogen, by 2-DE and ESI-q-TOF MS/MS. The proteomic analysis revealed a number of the proteins found to be upregulated in a virulent race, which has been implicated in microbial virulence in other pathosystems, such as the secreted enzymes a-mannosidase and exo-b-1,3-glucanase, heat-shock and bip proteins, and various metabolic enzymes, which suggests a reduced general pathogenic ability in avirulent race of P. tritici-repentis, irrespective of toxin production.

Extracellular and Cell Wall Proteins: The Secretome

Most eukaryotic plant pathogens initially invade the space between host cell walls (the apoplastic space), and much of the initial host defence and pathogen counter defence happens in the apoplast and commonly involves secreted pathogen and host-derived proteins and metabolites [283]. While some pathogens remain exclusively in the apoplast, such as Cladosporium fulvum, others, including mildews, rusts smuts, Phytophthora, and Magnaporthe species, breach host cell walls but remain external to and separated from the host cytoplasm by host and pathogen cell membranes. Some host wall-breaching pathogens, like rusts, mildews, and oomycetes, form specialised expanded hyphal protuberances called haustoria whereas others, like maize smut and the rice blast fungi, use unexpanded but probably specialised intrahost cell wall hyphae [284]. The role of these structures was initially thought to be primarily nutrient acquisition, but recently their additional role in secretion of effectors, some of which are translocated to the host cytoplasm, has become more apparent. These issues have been recently reviewed by Ellis et al. [113].

The secretome has been defined as being the combination of native secreted proteins and the cell machinery involved in their secretion [285]. A defining characteristic of plant pathogenic fungi is the secretion of a large number of degradative enzymes and other proteins, which have diverse functions in nutrient acquisition, substrate colonization, and ecological interactions [286–288]. Several extracellular fungal enzymes, such as polygalacturonase, pectate lyase, xylanase, and lipase, have been shown or postulated to be required for virulence in at least one host-pathogen interaction [289–295]. Proteomics is the right approach to study the interaction between plants and microbes mediated by excreted molecules, the role of the cell wall and the interface, and to identify fungal protein effectors facilitating either infection (virulence factors, enzymes of the toxin biosynthesis pathways) or trigger defence responses (avirulence factors). In the light of this, it has been said that, unlike animals, "fungi digest their food and [then] eat it" [296], illustrating the large number of extracellular hydrolytic enzymes necessary to digest a plethora of potential substrates. Therefore, many of these proteins are of special interest in the study of plant pathogens [46, 64]. This might also be owing to the fact that secretome sample preparation is much faster and simpler than extraction and preparation of intracellular proteins. Next, a number of papers covering this topic are presented, including those dealing with the secretome of Trichoderma spp, a study directed at identifying proteins related to its biofungicidal activity.

Pioneering work on this field comes before the arrival of proteomics during the 1990s, with typical studies focused on the identification, purification, and characterization of single secreted proteins, under the influence of the biotechnology industry for the production of enzymes for commercial and industrial use [297]. The first complete proteomic study of secreted proteins was released on the filamentous fungus A. flavus [41, 42]. The interest of this study was the ability of both A. flavus and A. parasiticus to degrade the flavonoids that plants produce as typical secondary metabolites against invading microorganisms. The secreted proteins were analyzed by 2-DE and MALDI-TOF mass spectrometry, with 15 rutin-induced proteins and 7 noninduced proteins identified, among them enzymes of routine catabolism pathway and glycosidases.

In F. graminearum, a devastating pathogen of wheat, maize, and other cereals, Phalip et al. [51] investigated the exoproteome of this fungus grown on glucose and on plant cell wall (Humulus lupulus, L.). The culture medium was found to contain a larger amount of proteins and these were more diverse when the fungus grew on the cell wall. Using both 1-DE and 2-DE coupled to LC-MS/MS analysis and protein identification based on similarity searches, 84 unique proteins were identified in the cell wall-grown fungal exoproteome and 45% were implicated in plant cell wall degradation. These cell wall-degradating enzymes were predominantly matches to putative carbohydrate active enzymes implicated in cellulose, hemicelluloses, and pectin, catabolism. As expected,

F. graminearum grown on glucose produced relatively few cell wall-degrading enzymes. These results indicated that fungal metabolism becomes oriented towards the synthesis and secretion of a whole arsenal of enzymes able to digest almost the complete plant cell wall.

The secretome has also been analyzed in S. sclerotium as commented above [64]. In this study, 52 secreted proteins were identified and many of the annotated secreted proteins were cell wall-degrading enzymes that had been identified previously as pathogenic or virulence factors of S. sclerotium. However, one of them, α-L-arabinofuranosidase, which is involved in the initiation or progression of plant diseases, was not detected by previous EST studies, clearly demonstrating the merit of performing proteomic research.

Two studies have been published reporting the B. cinerea (B05.10) secreted proteins analysis [48, 49]. First, secretions were collected from fungus grown on a solid substrate of cellophane membrane while mock infecting media supplemented with the extract of full red tomato, ripened strawberry, or Arabidopsis leaf extract. Overall, 89 B. cinerea proteins were identified by high-throughput LC-MS/MS from all growth conditions. Sixty of these proteins were predicted to contain a SignalP motif indicating the extracellular location of the proteins. The proteins identified were transport proteins, proteins well-characterized for carbohydrate metabolism, peptidases, oxidation/reduction, and pathogenicity factors that could provide important insights into how B. cinerea might use secreted proteins for plant infection and colonization [48]. In the second work, the impact of degree of esterification of pectin on secreted enzyme of B. cinerea was studied, because changes during the ripening process of fruits appear to play an important role in the activation of the dormant infection. All the major components of the fruit cell wall (pectin, cellulose, hemicellulose) undergo these changes. By 1-DE and LC-MS/MS, 126 proteins were identified and 87 proteins were predicted secreted by SignalP, some of them being pectinases. The results showed that the growth of B. cinerea and the secretion of proteins were similar in cultures containing differently esterified pectins, and therefore it is likely that the activation of this fungi from dormant state is not solely dependent on changes in the degree of esterification of the pectin component of the plant cell wall [49]. Therefore, future studies of the B. cinerea secretome in infections of ripe and unripe fruits will provide important information for describing the mechanisms that the fungus employs to access nutrients and decompose tissues.

Using the plant pathogenic fungus L. maculans and symbiont Laccaria bicolor grown in culture, Vincent et al.[56] established a proteomic protocol for extraction, concentration, and resolution of the fungal secretome. These authors used both broad and narrow acidic and basic pH range in IEF. The quality of protein extracts was assessed by both 1-DE and 2-DE and MS identification. Compared with the previously published protocols for which only dozens of 2-DE spots were recovered from fungal secretome samples, in this study, up to approximately 2000 2-DE spots were resolved. This high resolution was confirmed with the identification of proteins along several pH gradients as well as the presence of major secretome markers such as endopolygalacturonases, beta-glucanosyltransferases, pectate lyases, and endoglucanases. Thus, shotgun proteomic experiments evidenced the enrichment of secreted protein within the liquid medium.

One of the earliest works was released on Trichoderma reesei mycelium cell wall, one of the most powerful producers of extracellular proteins, this study being justified in order to find out the protein secretory pathways and the effect of the fungal genus, strain, and media condition on the excretion through the cell wall [298]. A total of 220 cell envelope-associated proteins were successfully extracted and separated by 2-DE fromTrichoderma reesei mycelia actively secreting proteins and from mycelia in which the secretion of proteins is low. Out of the 52 2-DE spots subjected to ESI-TOF MS, 20 were identified, with HEX1, the major protein in Woronin body, a structure unique to filamentous fungi, being the most abundant one. Suárez et al. [299] studied the secretome of T. harzianum grown using either chitin (a key cell wall component) or cell wall of other fungi (R. solani, B. cinerea, or Pythium ultimum) as a nutrient source. For each different substrate, they found significant differences in 2-DE maps of extracellular proteins. However, despite these differences, the most abundant protein under all conditions was a novel aspartic protease (P6281), which showed a strong homology with polyporopepsin from Irpex lacteus. This led to speculation that this protein plays a fundamental role in the parasitic activity of Trichoderma spp. Marra et al. [273] have studied interactions between T. atroviride, two different fungal phytopathogens (B. cinerea and R. solani), and plants (bean). Two-DE was used to analyze separately collected proteomes from each single, two- or three-partner interaction. Then, differential proteins were subjected to MALDI-TOF MS and in silico analysis to search homologies with known proteins. Thus, a large number of protein factors associated with the multiplayer interactions examined were identified, including protein kinases, cyclophilines, chitine synthase, and ABC transporters. Recently, another similar study was released between T. harzianum and R. solani by analysing the secretome to identify the target proteins that are directly related to biocontrol mechanism [274]. Seven cell-wall degrading enzymes, chitinase, cellulase, xylanase, β-1,3-glucanase, β-1,6-glucanase, mannanase, and protease, were revealed by activity assay, in-gel activity stain, 2-DE, and LC-MS/MS analysis, these being increased in response to R. solani.

A cell wall proteome has been proposed for the oomycete Phytophthora ramorum, the causal agent of sudden oak death, in order to study its pathogenic factors [60]. This study showed an inventory of cell wall-associated proteins based on MS sequence analysis. Seventeen secreted proteins were identified by homology searches. The functional classification revealed several cell wall-associated proteins, thus suggesting that cell wall proteins may also be important for fungal pathogenicity.

The filamentous fungus Neurospora crassa is a model laboratory organism but in nature is commonly found growing on dead plant material, particularly grasses. Using functional genomics resources available for N. crassa, which include a near-full genome deletion strain set and whole genome microarrays, Tian et al. [300] undertook a system-wide analysis of plant cell wall and cellulose degradation, identifying approximately 770 genes that showed expression differences when N. crassa was cultured on ground Miscanthus stems as a sole carbon source. An overlap set of 114 genes was identified from expression analysis of N. crassa grown on pure cellulose. Functional annotation of upregulated genes showed enrichment for proteins predicted to be involved in plant cell wall degradation, but also many genes encoding proteins of an unknown function. As a complement to expression data, the secretome associated with N. crassa growth on Miscanthus and cellulose was determined using a shotgun MudPIT proteomic strategy. Over 50 proteins were identified, including 10 of the 23 predicted N. crassa cellulases. Strains containing deletions in genes encoding 16 proteins detected in both the microarray and mass spectrometry experiments were analyzed for phenotypic changes during growth on crystalline cellulose and for cellulase activity. While growth of some of the deletion strains on cellulose was severely diminished, other deletion strains produced higher levels of extracellular proteins that showed increased cellulase activity. These results show that proteomics in combination with other powerful tools available in model systems such as N. crassa allow for a comprehensive system level understanding of fungal biology.

Interactomics

The biological organization in living cells can be regarded as being part of a complex network [301–303]. Traditional approaches studied a single gene or unique protein and therefore did not provide a complete knowledge of the biological processes. Proteins release their functional roles through their interactions with one another in vivo. Thus, developing a protein-protein interaction (PPI) network can lead to a more comprehensive understanding of the cell processes [304]. Interactomics is a discipline at the intersection of bioinformatics and biology that deals with studying both the interactions and the consequences of those interactions between and among proteins and other molecules within a cell [305]. The network of all such interactions is called the interactome. Interactomics thus aims to compare these interaction networks (i.e., interactomes) between and within species in order to find how the traits of such networks are either preserved or varied. Interactomics is an example of top-down systems biology, which takes an overhead, as well as overall, view of a biosystem or organism.

In recent years, high-throughput methods have been implemented to identify PPIs [306–310] and these have recently been reviewed in [311, 312]. Using these experimental methods, such as yeast two-hybrid screens, PPI networks for a series of model organisms were determined and allowed us to better understand the function of proteins at the level of system biology. Two-hybrid screening (also known as yeast two hybrid system or Y2H) is a powerful tool for identifying PPI. The premise behind the test is the activation of dowstream reporter gene(s) by the binding of a transcription factor onto an upstream activating sequence (UAS). For the purposes of two-hybrid screening, the transcription factor is split into two separate fragments, called the binding domain (BD) and activating domain (AD). The BD is the domain responsible for binding to the UAS and the AD is the domain responsible for activation of transcription [313].

In parallel with the large-scale experimental determination of PPI, many PPI prediction methods were also developed. These methods are based on diverse attributes, concepts, or data types, such as interolog [314], gene expression profiles [315], gene ontology (GO) annotations [316], domain interactions [317], coevolution [318], and structural information [319]. Some machine learning methods, such as support vector machines (SVMs) have also been used to predict PPIs [320, 321]. Among the above-mentioned computational methods the interolog approach has been widely implemented [322] and has proved to be reliable for predicting PPI from model organisms [323]. The core idea of the interolog approach is that many PPIs are conserved in different organisms [324]. Accumulated PPI data from model organisms as well as advances in detecting orthologous proteins in different organisms [280] have continuously made the interolog method an increasingly powerful tool for constructing PPI maps for entire proteomes.

Using the interolog method, He et al. [57] constructed the first PPI network for M. grisea. Thus, 11674 PPIs among 3017 M. grisea proteins were deduced from the experimental PPI data in different organisms, although the predicted PPI network covered approximately only one-fourth of the fungal proteome and may still contain many false-positives. Moreover, they built two subnets called pathogenicity and secreted proteins networks, which may be helpful in constructing an interactome between the rice blast fungus and rice (MPID website, http://bioinformatics.cau.edu.cn/zzd_lab/MPID.html).

A F. graminearum protein-protein interaction database providing comprehensive information on protein-protein interactions based on both interologs from several protein-protein interaction of seven species and domain-domain interactions experimentally determined based on protein is available athttp://csb.shu.edu.cn/fppi [55]. It contains 223 166 interactions among 7406 proteins for F. graminearum, covering 52% of the whole F. graminearum proteome.

CONCLUDING REMARKS

In the current scientific scenario, Proteomics should be understood to be part of a multidisciplinary approach. A combination of high-throughput "Omics" (Genomics, Transcriptomics, Proteomics, and Metabolomics) and classical biochemistry and cell biology techniques should be used for data validation and to deepen the knowledge of living organisms. Proteomic techniques are used to characterize a specific protein or a structural or functional group of proteins. This is what we can call "Hypothesis-driven Proteomics", "Targeted Proteomics", or "Proteinomics". This type of study will provide relevant information on protein structure and function, isoforms, organs, cells, and subcellular location and trafficking, processing, signal peptides, PTMs, expression kinetics, and correlation with RNA and metabolites. At the same time, it is a method for validating data obtained using one specific approach [102].

Despite the continuous development and improvement of powerful proteomic techniques, protocols, equipments, and bioinformatic tools, just a minimal fraction of the cell proteome, and for only a few organisms, has been characterized so far. This is mainly due to the enormous diversity and complexity of proteomes, and to technical limitations in quantification, sensitivity, resolution, speed of data capture, and analysis.

In the field of Fungal MS-based Proteomics, great progress has been made in past years. This is because of the increasing number of fungal genomes available and the developments in sample preparation, high-resolution protein separation techniques, MS, MS software for effective protein identification and characterization, and bioinformatics technology. The tremendous diversity and genome flexibility in fungi, however, will make this task a difficult one. Thus, a key step in sequence analysis is the annotation. The existing programs for automated gene prediction are not perfect and need to be improved or trained better. Follow-up manual annotation is also necessary to improve the accuracy of automated annotation, but this is time-consuming and labor-intensive.

Ultimately, a comprehensive genome database similar to YPD (http://www.yeastgenome.org/) will be desirable for fungal pathogens.

To date, most proteomic studies in plant pathogenic fungi have been limited to 1- and 2-DE analysis. However, various powerful proteomic methods have been developed for genomewide analysis of protein expression, protein localization, and protein-protein interaction in fungi. Whole-genome protein arrays and systematic yeast two-hybrid assays have been used to characterize the yeast proteome and interactome. Integration of large-scale genomics and proteomics data enables the elucidation of global networks and system biology studies in yeast. It is necessary for similar advanced proteomic resources to be soon available for some fungal plant pathogens. Moreover, second-generation MS technologies for Quantitative Proteomics such as 2-DIGE, stable isotope labeling, (ICAT, iTRAQ and SILAC), or label-free methods (peak integration, spectral counting) have not yet begun or are beginning to be applied to Fungal Proteomics research.

Otherwise, the major challenge is the analysis and significance of PTMs because proteins have properties arising from their folded structure and so generic methods are difficult to design and apply.

In conclusion, since plant pathogenic fungi cause important losses in a number of crops, it is necessary to make high-throughput studies on these organisms to identify pathogenicity factors. Although genomics-based investigation of host-pathogen interactions can provide valuable information on the changes in gene expression, the investigation into changes in protein abundance is also important, in order to identify those proteins that are essential during such interactions. This is because there is often a poor correlation between transcript and protein abundance [325]. Proteomics analysis is an excellent tool that can give us a great deal of information about fungal pathogenicity by high-throughput studies. This approach has allowed the identification of new fungal virulence factors, characterizing signal transduction or biochemical pathways, studying the fungal life cycle and their life-style. We can use this information to provide new targets for disease crop diagnosis focused on fungicide design. Otherwise, the secretome analysis is especially important because fungi secrete an arsenal of extracellular enzymes to break down the plant cell wall for pathogen penetration and nutrient consumption. In this sense, Proteomics allows us to identify numerous differential proteins involved in multiple-player cross-talk normally occurring in nature between plants and pathogens, the so-called “interaction proteomes”. Finally, MS-based Proteomics can help us to characterize fungal strains and find new biomarkers in host-pathogen interactions. In short, Fungal Proteomics is in the first step. Therefore, we still have a long way to go in the “Omics” of Plant Pathogenic Fungi compared to studies made in Humans, Bacteria, Yeast, or Plants. The important investment made by both the public and private sector in recent years augurs good prospects in fungal proteomics research in the future.

ABBREVIATIONS

1-DE:	One-dimentional electrophoresis
2-DE:	Two-dimentional electrophoresis
ESI:	Electrospray ionization
HPLC:	High performance liquid chromatograohy
IEF:	Isoelectrofocusing
LC:	Liquid chromatography
MALDI:	Matrix-assisted laser desorption/ionization
MS:	Mass spectrometry
SDS-PAGE:	Sodium dodecyl sulphate polyacrylamide gel electrophoresis
TOF:	Time of flight.

ACKNOWLEDGMENTS

This work was supported by the Spanish “Ministerio de Ciencia e Innovación” (Project BOTBANK EUI2008-03686), the “Junta de Andalucía”, and the “Universidad de Córdoba” (AGR 164: Agricultural and Plant Biochemistry and Proteomics Research Group).

REFERENCES

1. D. L. Hawksworth, “The fungal dimension of biodiversity: magnitude, significance, and conservation,”Mycological Research, vol. 95, no. 6, pp. 641–655, 1991. ·
2. D. L. Hawksworth, P. M. Kirk, B. C. Sutton, and D. N. Pegler, Ainsworth and Bisby›s Dictionary of the Fungi, International Mycological Institute, Surrey, UK, 2001.
3. G. N. Agrios, Plant Pathology, Elsevier Academic Press, San Diego, Calif, USA, 2005.
4. A. M. Green, U. G. Mueller, and R. M. Adams, “Extensive exchange of fungal cultivars between sympatric species of fungus-growing ants,” Molecular Ecology, vol. 11, no. 2, pp. 191–195, 2002.
5. A. Pérez-García, E. Mingorance, M. E. Rivera, et al., “Long-term

preservation of Podosphaera fusca using silica gel," Journal of Phytopathology, vol. 154, no. 3, pp. 190–192, 2006.

6. D. K. Arora, P. D. Bridge, and D. Bhatnagar, Fungal Biotechnology in Agricultural, Food and Environmental Applications, Marcel Dekker, New York, NY, USA, 2004.
7. D. A. Glawe, "The powdery mildews: a review of the world›s most familiar (yet poorly known) plant pathogens," Annual Review of Phytopathology, vol. 46, pp. 27–51, 2008.
8. S. Isaac, Fungal-Plant Interactions, Chapman & Hall, London, UK, 1992.
9. H. H. Flor, "Inheritance of pathogenicity in Melampsora lini," Phytopathology, vol. 32, pp. 653–669, 1942.
10. P. J. De Wit, R. Mehrabi, H. A. Van den Burg, and I. Stergiopoulos, "Fungal effector proteins: past, present and future," Molecular Plant Pathology, vol. 10, no. 6, pp. 735–747, 2009.
11. H. P. van Esse, J. W. Van›t Klooster, M. D. Bolton, et al., "The Cladosporium fulvum virulence protein Avr2 inhibits host proteases required for basal defense," Plant Cell, vol. 20, no. 7, pp. 1948–1963, 2008. · ·
12. T. L. Friesen, J. D. Faris, P. S. Solomon, and R. P. Oliver, "Host-specific toxins: effectors of necrotrophic pathogenicity," Cellular Microbiology, vol. 10, no. 7, pp. 1421–1428, 2008.
13. X. Gao and M. V. Kolomiets, "Host-derived lipids and oxylipins are crucial signals in modulating mycotoxin production by fungi," Toxin Reviews, vol. 28, no. 2-3, pp. 79–88, 2009.
14. K. S. Kim, J. Y. Min, and M. B. Dickman, "Oxalic acid is an elicitor of plant programmed cell death during Sclerotinia sclerotiorum disease development," Molecular Plant-Microbe Interactions, vol. 21, no. 5, pp. 605–612, 2008.
15. C. B. Lawrence, T. K. Mitchell, K. D. Craven, Y. Cho, R. A. Cramer, and K. H. Kim, "At death›s door: alternaria pathogenicity mechanisms," Plant Pathology Journal, vol. 24, no. 2, pp. 101–111, 2008.
16. H. C. van der Does and M. Rep, "Virulence genes and the evolution of host specificity in plant-pathogenic fungi," Molecular Plant-Microbe Interactions, vol. 20, no. 10, pp. 1175–1182, 2007.
17. Z. Zhang, T. L. Friesen, K. J. Simons, S. S. Xu, and J. D. Faris, "Development, identification, and validation of markers for marker-assisted selection against the Stagonospora nodorum toxin sensitivity genes Tsn1 and Snn2 in wheat," Molecular Breeding, vol. 23, no. 1, pp.

35–49, 2009.

18. B. Pariaud, V. Ravigné, F. Halkett, H. Goyeau, J. Carlier, and C. Lannou, "Aggressiveness and its role in the adaptation of plant pathogens," Plant Pathology, vol. 58, no. 3, pp. 409–424, 2009.
19. D. Parker, M. Beckmann, H. Zubair, et al., "Metabolomic analysis reveals a common pattern of metabolic re-programming during invasion of three host plant species by Magnaporthe grisea," Plant Journal, vol. 59, no. 5, pp. 723–737, 2009.
20. S. McCouch, "Diversifying selection in plant breeding," PLoS Biology, vol. 2, no. 10, article e347, 2004. · ·
21. Y. Yang, H. Zhang, G. Li, W. Li, X. Wang, and F. Song, "Ectopic expression of MgSM1, a Cerato-platanin family protein from Magnaporthe grisea, confers broad-spectrum disease resistance inArabidopsis," Plant Biotechnology Journal, vol. 7, no. 8, pp. 763–777, 2009.
22. B. S. Kim and B. K. Hwang, "Microbial fungicides in the control of plant diseases," Journal of Phytopathology, vol. 155, no. 11-12, pp. 641–653, 2007. · ·
23. G. Berg, "Plant-microbe interactions promoting plant growth and health: perspectives for controlled use of microorganisms in agriculture," Applied Microbiology and Biotechnology, vol. 84, no. 1, pp. 11–18, 2009.
24. G. Aguileta, M. E. Hood, G. Refrégier, and T. Giraud, "Chapter 3 genome evolution in plant pathogenic and symbiotic fungi," Advances in Botanical Research, vol. 49, pp. 151–193, 2009.
25. G. Aguileta, G. Refrégier, R. Yockteng, E. Fournier, and T. Giraud, "Rapidly evolving genes in pathogens: methods for detecting positive selection and examples among fungi, bacteria, viruses and protists," Infection, Genetics and Evolution, vol. 9, no. 4, pp. 656–670, 2009.
26. M. D. C. Alves and E. A. Pozza, "Scanning electron microscopy applied to seed-borne fungi examination," Microscopy Research and Technique, vol. 72, no. 7, pp. 482–488, 2009.
27. V. Farkaš, "Structure and biosynthesis of fungal cell walls: methodological approaches," Folia Microbiologica, vol. 48, no. 4, pp. 469–478, 2003.
28. A. R. Hardham and H. J. Mitchell, "Use of molecular cytology to study the structure and biology of phytopathogenic and mycorrhizal fungi," Fungal Genetics and Biology, vol. 24, no. 1-2, pp. 252–284, 1998.
29. R. J. Howard, "Cytology of fungal pathogens and plant-host interactions," Current Opinion in Microbiology, vol. 4, no. 4, pp. 365–373, 2001.

30. S. Koh and S. Somerville, "Show and tell: cell biology of pathogen invasion," Current Opinion in Plant Biology, vol. 9, no. 4, pp. 406–413, 2006.

31. P. Pérez and J. C. Ribas, "Cell wall analysis," Methods, vol. 33, no. 3, pp. 245–251, 2004.

32. J. A. Poland, P. J. Balint-Kurti, R. J. Wisser, R. C. Pratt, and R. J. Nelson, "Shades of gray: the world of quantitative disease resistance," Trends in Plant Science, vol. 14, no. 1, pp. 21–29, 2009.

33. J. Xu, "Fundamentals of fungal molecular population genetic analyses," Current Issues in Molecular Biology, vol. 8, no. 2, pp. 75–89, 2006. ·

34. M. Choquer, E. Fournier, C. Kunz, et al., "Botrytis cinerea virulence factors: new insights into a necrotrophic and polyphageous pathogen," FEMS Microbiology Letters, vol. 277, no. 1, pp. 1–10, 2007. · ·

35. M. J. Egan and N. J. Talbot, "Genomes, free radicals and plant cell invasion: recent developments in plant pathogenic fungi," Current Opinion in Plant Biology, vol. 11, no. 4, pp. 367–372, 2008.

36. L. A. Hadwiger, "Localization predictions for gene products involved in non-host resistance responses in a model plant/fungal pathogen interaction," Plant Science, vol. 177, no. 4, pp. 257–265, 2009.

37. K.-C. Tan, S. V. Ipcho, R. D. Trengove, R. P. Oliver, and P. S. Solomon, "Assessing the impact of transcriptomics, proteomics and metabolomics on fungal phytopathology," Molecular Plant Pathology, vol. 10, no. 5, pp. 703–715, 2009.

38. S. Walter, P. Nicholson, and F. M. Doohan, "Action and reaction of host and pathogen during Fusariumhead blight disease," New Phytologist, vol. 185, no. 1, pp. 54–66, 2010.

39. V. Bhadauria, S. Banniza, Y. Wei, and Y.-L. Peng, "Reverse genetics for functional genomics of phytopathogenic fungi and oomycetes," Comparative and Functional Genomics, vol. 2009, Article ID 380719, 11 pages, 2009.

40. M. Sulc, K. Peslova, M. Zabka, M. Hajduch, and V. Havlicek, "Biomarkers of Aspergillus spores: strain typing and protein identification," International Journal of Mass Spectrometry, vol. 280, no. 1–3, pp. 162–168, 2009.

41. M. L. Medina, P. A. Haynes, L. Breci, and W. A. Francisco, "Analysis of secreted proteins fromAspergillus flavus," Proteomics, vol. 5, no. 12, pp. 3153–3161, 2005.

42. M. L. Medina, U. A. Kiernan, and W. A. Francisco, "Proteomic analysis

of rutin-induced secreted proteins from Aspergillus flavus," Fungal Genetics and Biology, vol. 41, no. 3, pp. 327–335, 2004.

43. S. Noir, T. Colby, A. Harzen, J. Schmidt, and R. Panstruga, "A proteomic analysis of powdery mildew (Blumeria graminis f.sp. hordei) conidiospores," Molecular Plant Pathology, vol. 10, no. 2, pp. 223–236, 2009.

44. D. Godfrey, Z. Zhang, G. Saalbach, and H. Thordal-Christensen, "A proteomics study of barley powdery mildew haustoria," Proteomics, vol. 9, no. 12, pp. 3222–3232, 2009.

45. F. J. Fernández-Acero, I. Jorge, E. Calvo, et al., "Two-dimensional electrophoresis protein profile of the phytopathogenic fungus Botrytis cinerea," Proteomics, vol. 6, supplement 1, pp. S88–S96, 2006.

46. F. J. Fernández-Acero, I. Jorge, E. Calvo, et al., "Proteomic analysis of phytopathogenic fungus Botrytis cinerea as a potential tool for identifying pathogenicity factors, therapeutic targets and for basic research," Archives of Microbiology, vol. 187, no. 3, pp. 207–215, 2007.

47. F. J. Fernández-Acero, T. Colby, A. Harzen, J. M. Cantoral, and J. Schmidt, "Proteomic analysis of the phytopathogenic fungus Botrytis cinerea during cellulose degradation," Proteomics, vol. 9, no. 10, pp. 2892–2902, 2009.

48. P. Shah, J. A. Atwood, R. Orlando, H. E. Mubarek, G. K. Podila, and M. R. Davis, "Comparative proteomic analysis of Botrytis cinerea secretome," Journal of Proteome Research, vol. 8, no. 3, pp. 1123–1130, 2009.

49. P. Shah, G. Gutierrez-Sanchez, R. Orlando, and C. Bergmann, "A proteomic study of pectin-degrading enzymes secreted by Botrytis cinerea grown in liquid culture," Proteomics, vol. 9, no. 11, pp. 3126–3135, 2009.

50. S. Xu, J. I. E. Chen, L. Liu, X. Wang, X. Huang, and Y. Zhai, "Proteomics associated with virulence differentiation of Curvularia iunata in maize in China," Journal of Integrative Plant Biology, vol. 49, no. 4, pp. 487–496, 2007.

51. V. Phalip, F. Delalande, C. Carapito, et al., "Diversity of the exoproteome of Fusarium graminearumgrown on plant cell wall," Current Genetics, vol. 48, no. 6, pp. 366–379, 2005.

52. J. M. Paper, J. S. Scott-Craig, N. D. Adhikari, C. A. Cuomo, and J. D. Walton, "Comparative proteomics of extracellular proteins in vitro and in planta from the pathogenic fungus Fusarium graminearum,"Proteomics, vol. 7, no. 17, pp. 3171–3183, 2007.

53. R. D. Taylor, A. Saparno, B. Blackwell, et al., "Proteomic analyses of Fusarium graminearum grown under mycotoxin-inducing conditions," Proteomics, vol. 8, no. 11, pp. 2256–2265, 2008.

54. S.-J. Kwon, S.-Y. Cho, K.-M. Lee, J. Yu, M. Son, and K.-H. Kim, "Proteomic analysis of fungal host factors differentially expressed by Fusarium graminearum infected with Fusarium graminearum virus-DK21," Virus Research, vol. 144, no. 1-2, pp. 96–106, 2009. · ·

55. X.-M. Zhao, X.-W. Zhang, W.-H. Tang, and L. Chen, "FPPI: Fusarium graminearum protein-protein interaction database," Journal of Proteome Research, vol. 8, no. 10, pp. 4714–4721, 2009.

56. D. Vincent, M.-H. Balesdent, J. Gibon, et al., "Hunting down fungal secretomes using liquid-phase IEF prior to high resolution 2-DE," Electrophoresis, vol. 30, no. 23, pp. 4118–4136, 2009.

57. F. He, Y. Zhang, H. Chen, Z. Zhang, and Y. L. Peng, "The prediction of protein-protein interaction networks in rice blast fungus," BMC Genomics, vol. 9, article 519, 2008.

58. S. J. Shepherd, P. van West, and N. A. R. Gow, "Proteomic analysis of asexual development ofPhytophthora palmivora," Mycological Research, vol. 107, no. 4, pp. 395–400, 2003.

59. T. Ebstrup, G. Saalbach, and H. Egsgaard, "A proteomics study of in vitro cyst germination and appressoria formation in Phytophthora infestans," Proteomics, vol. 5, no. 11, pp. 2839–2848, 2005.

60. H. J. G. Meijer, P. J. I. van de Vondervoort, Q. Y. Yin, et al., "Identification of cell wall-associated proteins from Phytophthora ramorum," Molecular Plant-Microbe Interactions, vol. 19, no. 12, pp. 1348–1358, 2006.

61. T. Cao, Y. M. Kim, N. N. V. Kav, and S. E. Strelkov, "A proteomic evaluation of Pyrenophora tritici-repentis, causal agent of tan spot of wheat, reveals major differences between virulent and avirulent isolates," Proteomics, vol. 9, no. 5, pp. 1177–1196, 2009. · ·

62. M. Matis, M. Žakelj-Mavrič, and J. Peter-Katalinić, "Mass spectrometry and database search in the analysis of proteins from the fungus Pleurotus ostreatus," Proteomics, vol. 5, no. 1, pp. 67–75, 2005.

63. D. K. Lakshman, S. S. Natarajan, S. Lakshman, W. M. Garrett, and A. K. Dhar, "Optimized protein extraction methods for proteomic analysis of Rhizoctonia solani," Mycologia, vol. 100, no. 6, pp. 867–875, 2008.

64. W. Yajima and N. N. V. Kav, "The proteome of the phytopathogenic fungus Sclerotinia sclerotiorum,"Proteomics, vol. 6, no. 22, pp. 5995–6007, 2006.

65. K. A. R. C. Tan, J. L. Heazlewood, A. H. Millar, G. Thomson, R. P. Oliver, and P. S. Solomon, "A signaling-regulated, short-chain dehydrogenase of Stagonospora nodorum regulates asexual development," Eukaryotic Cell, vol. 7, no. 11, pp. 1916–1929, 2008.

66. S. Bringans, J. K. Hane, T. Casey, et al., "Deep proteogenomics; high throughput gene validation by multidimensional liquid chromatography and mass spectrometry of proteins from the fungal wheat pathogen Stagonospora nodorum," BMC Bioinformatics, vol. 10, article 301, 2009.

67. B. Cooper, A. Neelam, K. B. Campbell, et al., "Protein accumulation in the germinating Uromyces appendiculatus uredospore," Molecular Plant-Microbe Interactions, vol. 20, no. 7, pp. 857–866, 2007. · ·

68. M. Böhmer, T. Colby, C. Böhmer, A. Bräutigam, J. Schmidt, and M. Bölker, "Proteomic analysis of dimorphic transition in the phytopathogenic fungus Ustilago maydis," Proteomics, vol. 7, no. 5, pp. 675–685, 2007.

69. R. J. Weld, K. M. Plummer, M. A. Carpenter, and H. J. Ridgway, "Approaches to functional genomics in filamentous fungi," Cell Research, vol. 16, no. 1, pp. 31–44, 2006.

70. J. Schumacher, I. F. de Larrinoa, and B. Tudzynski, "Calcineurin-responsive zinc finger transcription factor CRZ1 of Botrytis cinerea is required for growth, development, and full virulence on bean plants,"Eukaryotic Cell, vol. 7, no. 4, pp. 584–601, 2008.

71. N. Ajiro, Y. Miyamoto, A. Masunaka, et al., "Role of the host-selective ACT-toxin synthesis gene ACTTS2 encoding an enoyl-reductase in pathogenicity of the tangerine pathotype of alternaria alternata," Phytopathology, vol. 100, no. 2, pp. 120–126, 2010.

72. M.-H. Chi, S.-Y. Park, S. Kim, and Y.-H. Lee, "A novel pathogenicity gene is required in the rice blast fungus to suppress the basal defenses of the host," PLoS Pathogens, vol. 5, no. 4, Article ID e1000401, 2009.

73. K. H. Lamour, L. Finley, O. Hurtado-Gonzales, D. Gobena, M. Tierney, and H. J. Meijer, "Targeted gene mutation in Phytophthora spp.," Molecular Plant-Microbe Interactions, vol. 19, no. 12, pp. 1359–1367, 2006.

74. A. Di Pietro, M. P. Madrid, Z. Caracuel, J. Delgado-Jarana, and M. I. G. Roncero, "Fusarium oxysporum: exploring the molecular arsenal of a vascular wilt fungus," Molecular Plant Pathology, vol. 4, no. 5, pp. 315–325, 2003.

75. S. R. Herron, J. A. Benen, R. D. Scavetta, J. Visser, and F. Jurnak, "Structure and function of pectic enzymes: virulence factors of plant pathogens," Proceedings of the National Academy of Sciences of the

United States of America, vol. 97, no. 16, pp. 8762–8769, 2000.

76. S. Meng, T. Torto-Alalibo, M. C. Chibucos, B. M. Tyler, and R. A. Dean, "Common processes in pathogenesis by fungal and oomycete plant pathogens, described with Gene Ontology terms," BMC Microbiology, vol. 9, supplement 1, p. S7, 2009.

77. J. Delgado-Jarana, A. L. Martínez-Rocha, R. Roldán-Rodriguez, M. I. Roncero, and A. D. Pietro, "Fusarium oxysporum G-protein β subunit Fgb1 regulates hyphal growth, development, and virulence through multiple signalling pathways," Fungal Genetics and Biology, vol. 42, no. 1, pp. 61–72, 2005.

78. N. Lee, C. A. D›Souza, and J. W. Kronstad, "Of smuts, blasts, mildews, and blights: cAMP signaling in phytopathogenic fungi," Annual Review of Phytopathology, vol. 41, pp. 399–427, 2003. · ·

79. N. Rispail and A. Di Pietro, "Fusarium oxysporum Ste12 controls invasive growth and virulence downstream of the Fmk1 MAPK cascade," Molecular Plant-Microbe Interactions, vol. 22, no. 7, pp. 830–839, 2009.

80. J. R. Xu and J. E. Hamer, "MAP kinase and cAMP signaling regulate infection structure formation and pathogenic growth in the rice blast fungus Magnaporthe grisea," Genes and Development, vol. 10, no. 21, pp. 2696–2706, 1996. ·

81. L. Zheng, M. Campbell, J. Murphy, S. Lam, and J. R. Xu, "The BMP1 gene is essential for pathogenicity in the gray mold fungus Botrytis cinerea," Molecular Plant-Microbe Interactions, vol. 13, no. 7, pp. 724–732, 2000. ·

82. N. Rispail, D. M. Soanes, C. Ant, et al., "Comparative genomics of MAP kinase and calcium-calcineurin signalling components in plant and human pathogenic fungi," Fungal Genetics and Biology, vol. 46, no. 4, pp. 287–298, 2009.

83. M. S. López-Berges, A. Di Pietro, M. J. Daboussi, et al., "Identification of virulence genes in Fusarium oxysporum f. sp. lycopersici by large-scale transposon tagging," Molecular Plant Pathology, vol. 10, no. 1, pp. 95–107, 2009.

84. V. Bhadauria, L. Popescu, W. S. Zhao, and Y. L. Peng, "Fungal transcriptomics," Microbiological Research, vol. 162, no. 4, pp. 285–298, 2007.

85. Y. Oh, N. Donofrio, H. Pan, et al., "Transcriptome analysis reveals new insight into appressorium formation and function in the rice blast fungus Magnaporthe oryzae," Genome Biology, vol. 9, no. 5, R85, 2008.

86. H. Takahara, A. Dolf, E. Endl, and R. O›Connell, "Flow cytometric purification of Colletotrichum higginsianum biotrophic hyphae from Arabidopsis leaves for stage-specific transcriptome analysis,"Plant Journal, vol. 59, no. 4, pp. 672–683, 2009.

87. R. P. Wise, M. J. Moscou, A. J. Bogdanove, and S. A. Whitham, "Transcript profiling in host-pathogen interactions," Annual Review of Phytopathology, vol. 45, pp. 329–369, 2008.

88. B. Venkatesh, U. Hettwer, B. Koopmann, and P. Karlovsky, "Conversion of cDNA differential display results (DDRT-PCR) into quantitative transcription profiles," BMC Genomics, vol. 6, article 51, 2005.

89. X. Wang, C. Tang, G. Zhang, et al., "cDNA-AFLP analysis reveals differential gene expression in compatible interaction of wheat challenged with Puccinia striiformis f. sp. tritici," BMC Genomics, vol. 10, article 289, 2009.

90. C. Fekete, R. W. Fung, Z. Szabó, et al., "Up-regulated transcripts in a compatible powdery mildew-grapevine interaction," Plant Physiology and Biochemistry, vol. 47, no. 8, pp. 732–738, 2009.

91. R. C. Venu, Y. Jia, M. Gowda, et al., "RL-SAGE and microarray analysis of the rice transcriptome after Rhizoctonia solani infection," Molecular Genetics and Genomics, vol. 278, no. 4, pp. 421–431, 2007.

92. C. Yin, X. Chen, X. Wang, Q. Han, Z. Kang, and S. Hulbert, "Generation and analysis of expression sequence tags from haustoria of the wheat stripe rust fungus Puccinia striiformis f. sp. Tritici," BMC Genomics, vol. 10, article 626, 2009.

93. W. Skinner, J. Keon, and J. Hargreaves, "Gene information for fungal plant pathogens from expressed sequences," Current Opinion in Microbiology, vol. 4, no. 4, pp. 381–386, 2001.

94. F. J. Bruggeman and H. V. Westerhoff, "The nature of systems biology," Trends in Microbiology, vol. 15, no. 1, pp. 45–50, 2007.

95. P. Picotti, B. Bodenmiller, L. N. Mueller, B. Domon, and R. Aebersold, "Full dynamic range proteome analysis of S. cerevisiae by targeted proteomics," Cell, vol. 138, no. 4, pp. 795–806, 2009.

96. N. L. Anderson, N. G. Anderson, T. W. Pearson, et al., "A human proteome detection and quantitation project," Molecular and Cellular Proteomics, vol. 8, no. 5, pp. 883–886, 2009.

97. M. R. Wilkins, J. C. Sanchez, A. A. Gooley, et al., "Progress with proteome projects: why all proteins expressed by a genome should be identified and how to do it," Biotechnology and Genetic Engineering Reviews, vol. 13,

pp. 19–50, 1996.

98. J. Cox and M. Mann, "Is proteomics the new genomics?" Cell, vol. 130, no. 3, pp. 395–398, 2007.

99. B. F. Cravatt, G. M. Simon, and J. R. Yates III, "The biological impact of mass-spectrometry-based proteomics," Nature, vol. 450, no. 7172, pp. 991–1000, 2007.

100. X. Han, M. Jin, K. Breuker, and F. W. McLafferty, "Extending top-down mass spectrometry to proteins with masses great than 200 kilodaltons," Science, vol. 314, no. 5796, pp. 109–112, 2006.

101. X. Han, A. Aslanian, and J. R. Yates III, "Mass spectrometry for proteomics," Current Opinion in Chemical Biology, vol. 12, no. 5, pp. 483–490, 2008.

102. J. V. Jorrín-Novo, A. M. Maldonado, S. Echevarría-Zomeño, et al., "Plant proteomics update (2007-2008): second-generation proteomic techniques, an appropriate experimental design, and data analysis to fulfill MIAPE standards, increase plant proteome coverage and expand biological knowledge," Journal of Proteomics, vol. 72, no. 3, pp. 285–314, 2009.

103. M. Mann, "Comparative analysis to guide quality improvements in proteomics," Nature Methods, vol. 6, no. 10, pp. 717–719, 2009.

104. A. Schmidt, M. Claassen, and R. Aebersold, "Directed mass spectrometry: towards hypothesis-driven proteomics," Current Opinion in Chemical Biology, vol. 13, no. 5-6, pp. 510–517, 2009.

105. V. Bhadauria, W. S. Zhao, L. X. Wang, et al., "Advances in fungal proteomics," Microbiological Research, vol. 162, no. 3, pp. 193–200, 2007.

106. Y. Kim, M. P. Nandakumar, and M. R. Marten, "Proteomics of filamentous fungi," Trends in Biotechnology, vol. 25, no. 9, pp. 395–400, 2007.

107. L. J. Newey, C. E. Caten, and J. R. Green, "Rapid adhesion of Stagonospora nodorum spores to a hydrophobic surface requires pre-formed cell surface glycoproteins," Mycological Research, vol. 111, no. 11, pp. 1255–1267, 2007.

108. P. N. Dodds, M. Rafiqi, P. H. P. Gan, A. R. Hardham, D. A. Jones, and J. G. Ellis, "Effectors of biotrophic fungi and oomycetes: pathogenicity factors and triggers of host resistance," New Phytologist, vol. 183, no. 4, pp. 993–1000, 2009.

109. P. M. Houterman, B. J. Cornelissen, and M. Rep, "Suppression of plant resistance gene-based immunity by a fungal effector," PLoS Pathogens,

vol. 4, no. 5, Article ID e1000061, 2008.

110. J. Song, J. Win, M. Tian, et al., "Apoplastic effectors secreted by two unrelated eukaryotic plant pathogens target the tomato defense protease Rcr3," Proceedings of the National Academy of Sciences of the United States of America, vol. 106, no. 5, pp. 1654–1659, 2009.

111. P. J. G. M. De Wit, M. B. Buurlage, and K. E. Hammond, "The occurrence of host-, pathogen- and interaction-specific proteins in the apoplast of Cladosporium fulvum (syn. Fulvia fulva) infected tomato leaves," Physiological and Molecular Plant Pathology, vol. 29, no. 2, pp. 159–172, 1986.

112. I. M. J. Schottens-Toma and P. J. G. M. De Wit, "Purification and primary structure of a necrosis-inducing peptide from the apoplastic fluids of tomato infected with Cladosporium fulvum (syn. Fulvia fulva)," Physiological and Molecular Plant Pathology, vol. 33, no. 1, pp. 59–67, 1988.

113. J. G. Ellis, M. Rafiqi, P. Gan, A. Chakrabarti, and P. N. Dodds, "Recent progress in discovery and functional analysis of effector proteins of fungal and oomycete plant pathogens," Current Opinion in Plant Biology, vol. 12, no. 4, pp. 399–405, 2009.

114. M. Rep, H. C. van der Does, M. Meijer, et al., "A small, cysteine-rich protein secreted by Fusarium oxysporum during colonization of xylem vessels is required for I-3-mediated resistance in tomato,"Molecular Microbiology, vol. 53, no. 5, pp. 1373–1383, 2004.

115. M. Rep, "Small proteins of plant-pathogenic fungi secreted during host colonization," FEMS Microbiology Letters, vol. 253, no. 1, pp. 19–27, 2005.

116. E. W. Deutsch, H. Lam, and R. Aebersold, "Data analysis and bioinformatics tools for tandem mass spectrometry in proteomics," Physiological Genomics, vol. 33, no. 1, pp. 18–25, 2008.

117. J. Eriksson and D. Fenyö, "Improving the success rate of proteome analysis by modeling protein-abundance distributions and experimental designs," Nature Biotechnology, vol. 25, no. 6, pp. 651–655, 2007.

118. P. Holzmuller, P. Grébaut, J. P. Brizard, et al., "Pathogeno-proteomics," Annals of the New York Academy of Sciences, vol. 1149, pp. 66–70, 2008.

119. L. Valledor, M. A. Castillejo, C. Lenz, R. Rodríguez, M. J. Cañal, and J. Jorrín, "Proteomic analysis of Pinus radiata needles: 2-DE map and protein identification by LC/MS/MS and substitution-tolerant database searching," Journal of Proteome Research, vol. 7, no. 7, pp. 2616–2631,

2008.

120. J. Ruiz-Herrera, Fungal Cell Wall: Structure, Synthesis and Assembly, CRC Press, Boca Raton, Fla, USA, 1992.

121. J. Grinyer, M. McKay, B. Herbert, and H. Nevalainen, "Fungal proteomics: mapping the mitochodrial proteins of a Trichoderma harzianum strain applied for biological control," Current Genetics, vol. 45, no. 3, pp. 170–175, 2004.

122. P. Melin, J. Schnürer, and E. G. Wagner, "Proteome analysis of Aspergillus nidulans reveals proteins associated with the response to the antibiotic concanamycin A, produced by Streptomyces species,"Molecular Genetics and Genomics, vol. 267, no. 6, pp. 695–702, 2002.

123. M. P. Nandakumar, J. Shen, B. Raman, and M. R. Marten, "Solubilization of trichloroacetic acid (TCA) precipitated microbial proteins via NaOH for two-dimensional electrophoresis," Journal of Proteome Research, vol. 2, no. 1, pp. 89–93, 2003.

124. K. Ström, J. Schnürer, and P. Melin, "Co-cultivation of antifungal Lactobacillus plantarum MiLAB 393 and Aspergillus nidulans, evaluation of effects on fungal growth and protein expression," FEMS Microbiology Letters, vol. 246, no. 1, pp. 119–124, 2005. · ·

125. J. M. Bruneau, T. Magnin, E. Tagat, et al., "Proteome analysis of Aspergillus fumigatus identifies glycosylphosphatidylinositol-anchored proteins associated to the cell wall biosynthesis," Electrophoresis, vol. 22, no. 13, pp. 2812–2823, 2001.

126. J. Grinyer, S. Hunt, M. McKay, B. E. N. R. Herbert, and H. Nevalainen, "Proteomic response of the biological control fungus Trichoderma atroviride to growth on the cell walls of Rhizoctonia solani,"Current Genetics, vol. 47, no. 6, pp. 381–388, 2005.

127. J. Grinyer, M. McKay, H. Nevalainen, and B. E. N. R. Herbert, "Fungal proteomics: initial mapping of biological control strain Trichoderma harzianum," Current Genetics, vol. 45, no. 3, pp. 163–169, 2004. · ·

128. M. P. Nandakumar and M. R. Marten, "Comparison of lysis methods and preparation protocols for one- and two-dimensional electrophoresis of Aspergillus oryzae intracellular proteins," Electrophoresis, vol. 23, no. 14, pp. 2216–2222, 2002.

129. M. Shimizu and H. Wariishi, "Development of a sample preparation method for fungal proteomics,"FEMS Microbiology Letters, vol. 247, no. 1, pp. 17–22, 2005. · ·

130. M. Shimizu, N. Yuda, T. Nakamura, H. Tanaka, and H. Wariishi,

"Metabolic regulation at the tricarboxylic acid and glyoxylate cycles of the lignin-degrading basidiomycete Phanerochaete chrysosporium against exogenous addition of vanillin," Proteomics, vol. 5, no. 15, pp. 3919–3931, 2005. · ·

131. M. L. Hernández-Macedo, A. Ferraz, J. Rodrguez, L. M. M. Ottoboni, and M. P. De Mello, "Iron-regulated proteins in Phanerochaete chrysosporium and Lentinula edodes: differential analysis by sodium dodecyl sulfate polyacrylamide gel electrophoresis and two-dimensional polyacrylamide gel electrophoresis profiles," Electrophoresis, vol. 23, no. 4, pp. 655–661, 2002.
132. A. Görg, W. Weiss, and M. J. Dunn, "Current two-dimensional electrophoresis technology for proteomics," Proteomics, vol. 4, no. 12, pp. 3665–3685, 2004.
133. S. C. Carpentier, E. Witters, K. Laukens, P. Deckers, R. Swennen, and B. Panis, "Preparation of protein extracts from recalcitrant plant tissues: an evaluation of different methods for two-dimensional gel electrophoresis analysis," Proteomics, vol. 5, no. 10, pp. 2497–2507, 2005.
134. C. Damerval, D. de Vienne, M. Zivy, and H. Thiellement, "Technical improvements in two-dimensional electrophoresis increase the level of genetic variation detected in wheat-seedling proteins,"Electrophoresis, vol. 7, pp. 52–54, 1986.
135. A. N. A. M. Maldonado, S. Echevarría-Zomeño, S. Jean-Baptiste, M. Hernández, and J. V. Jorrín-Novo, "Evaluation of three different protocols of protein extraction for Arabidopsis thaliana leaf proteome analysis by two-dimensional electrophoresis," Journal of Proteomics, vol. 71, no. 4, pp. 461–472, 2008. · ·
136. R. Wildgruber, G. Reil, O. Drews, H. Parlar, and A. Görg, "Web-based two-dimensional database ofSaccharomyces cerevisiae proteins using immobilized pH gradients from pH 6 to pH 12 and matrix-assisted laser desorption/ionization-time of flight mass spectrometry," Proteomics, vol. 2, no. 6, pp. 727–732, 2002. ·
137. H. Everberg, N. Gustavasson, and F. Tjerned, "Enrichment of membrane proteins by partitioning in detergent/polymer aqueous two-phase systems," Methods in Molecular Biology, vol. 424, pp. 403–412, 2008. ·
138. S. Luche, V. Santoni, and T. Rabilloud, "Evaluation of nonionic and zwitterionic detergents as membrane protein solubilizers in two-dimensional electrophoresis," Proteomics, vol. 3, no. 3, pp. 249–253, 2003.
139. M. P. Molloy, B. R. Herbert, K. L. Williams, and A. A. Gooley,

"Extraction of Escherichia coli proteins with organic solvents prior to two-dimensional electrophoresis," Electrophoresis, vol. 20, no. 4-5, pp. 701–704, 1999. ·

140. T. Rabilloud, "Solubilization of proteins for electrophoretic analyses," Electrophoresis, vol. 17, no. 5, pp. 813–829, 1996.
141. T. Rabilloud, C. Adessi, A. Giraudel, and J. Lunardi, "Improvement of the solubilization of proteins in two-dimensional electrophoresis with immobilized pH gradients," Electrophoresis, vol. 18, no. 3-4, pp. 307–316, 1997.
142. O. Kniemeyer, F. Lessing, O. Scheibner, C. Hertweck, and A. A. Brakhage, "Optimisation of a 2-D gel electrophoresis protocol for the human-pathogenic fungus Aspergillus fumigatus," Current Genetics, vol. 49, no. 3, pp. 178–189, 2006.
143. T. Rabilloud, "Use of thiourea to increase the solubility of membrane proteins in two-dimensional electrophoresis," Electrophoresis, vol. 19, no. 5, pp. 758–760, 1998.
144. B. E. N. R. Herbert, J. Grinyer, J. T. McCarthy, et al., "Improved 2-DE of microorganisms after acidic extraction," Electrophoresis, vol. 27, no. 8, pp. 1630–1640, 2006.
145. O. Guais, G. Borderies, C. Pichereaux, et al., "Proteomics analysis of "Rovabiot Excel", a secreted protein cocktail from the filamentous fungus Penicillium funiculosum grown under industrial process fermentation," Journal of Industrial Microbiology & Biotechnology, pp. 1659–1668, 2008.
146. S. H. Kao, H. I. N. K. Wong, C. Y. Chiang, and H. A. N. M. Chen, "Evaluating the compatibility of three colorimetric protein assays for two-dimensional electrophoresis experiments," Proteomics, vol. 8, no. 11, pp. 2178–2184, 2008.
147. D. Fragner, M. Zomorrodi, U. Kües, and A. Majcherczyk, "Optimized protocol for the 2-DE of extracellular proteins from higher basidiomycetes inhabiting lignocellulose," Electrophoresis, vol. 30, no. 14, pp. 2431–2441, 2009.
148. M. L. Medina and W. A. Francisco, "Isolation and enrichment of secreted proteins from filamentous fungi," in 2D PAGE: Sample Preparation and Fractionation, pp. 275–285, Springer, New York, NY, USA, 2008.
149. A. Abbas, H. Koc, F. Liu, and M. Tien, "Fungal degradation of wood: initial proteomic analysis of extracellular proteins of Phanerochaete chrysosporium grown on oak substrate," Current Genetics, vol. 47, no. 1, pp. 49–56, 2005.

150. H. Ravalason, G. Jan, D. Mollé, et al., “Secretome analysis of Phanerochaete chrysosporium strain CIRM-BRFM41 grown on softwood,” Applied Microbiology and Biotechnology, vol. 80, no. 4, pp. 719–733, 2008.

151. A. V. Wymelenberg, P. Minges, G. Sabat, et al., “Computational analysis of the Phanerochaete chrysosporium v2.0 genome database and mass spectrometry identification of peptides in ligninolytic cultures reveal complex mixtures of secreted proteins,” Fungal Genetics and Biology, vol. 43, no. 5, pp. 343–356, 2006.

152. H. Zorn, T. Peters, M. Nimtz, and R. G. Berger, “The secretome of Pleurotus sapidus,” Proteomics, vol. 5, no. 18, pp. 4832–4838, 2005.

153. L. V. Bindschedler, T. A. Burgis, D. J. S. Mills, J. T. C. Ho, R. Cramer, and P. D. Spanu, “In planta proteomics and proteogenomics of the biotrophic Barley fungal pathogen Blumeria graminis f. sp.hordei,” Molecular and Cellular Proteomics, vol. 8, no. 10, pp. 2368–2381, 2009.

154. C. Rampitsch, N. V. Bykova, B. McCallum, E. V. A. Beimcik, and W. Ens, “Analysis of the wheat and Puccinia triticina (leaf rust) proteomes during a susceptible host-pathogen interaction,” Proteomics, vol. 6, no. 6, pp. 1897–1907, 2006.

155. W. Zhou, F. Eudes, and A. Laroche, “Identification of differentially regulated proteins in response to a compatible interaction between the pathogen Fusarium graminearum and its host, Triticum aestivum,”Proteomics, vol. 6, no. 16, pp. 4599–4609, 2006.

156. P. M. Houterman, D. Speijer, H. L. Dekker, C. G. de Koster, B. J. C. Cornelissen, and M. Rep, “The mixed xylem sap proteome of Fusarium oxysporum-infected tomato plants,” Molecular Plant Pathology, vol. 8, no. 2, pp. 215–221, 2007.

157. A. Harder, “Sample preparation procedure for cellular fungi,” Methods in Molecular Biology, vol. 425, pp. 265–273, 2008. ·

158. A. Pitarch, C. Nombela, and C. Gil, “Cell wall fractionation for yeast and fungal proteomics,” Methods in Molecular Biology, vol. 425, pp. 217–239, 2008. ·

159. X. Jiang, M. Ye, and H. Zou, “Technologies and methods for sample pretreatment in efficient proteome and peptidome analysis,” Proteomics, vol. 8, no. 4, pp. 686–705, 2008.

160. E. Boschetti and P. G. Righetti, “The art of observing rare protein species in proteomes with peptide ligand libraries,” Proteomics, vol. 9, no. 6, pp. 1492–1510, 2009.

161. D. Robertson, G. P. Mitchell, J. S. Gilroy, C. Gerrish, G. P. Bolwell, and A. R. Slabas, "Differential extraction and protein sequencing reveals major differences in patterns of primary cell wall proteins from plants," Journal of Biological Chemistry, vol. 272, no. 25, pp. 15841–15848, 1997. · ·

162. F. Supek, P. Peharec, M. Krsnik-Rasol, and T. Šmuc, "Enhanced analytical power of SDS-PAGE using machine learning algorithms," Proteomics, vol. 8, no. 1, pp. 28–31, 2008.

163. L. M. F. de Godoy, J. V. Olsen, G. A. de Souza, G. Li, P. Mortensen, and M. Mann, "Status of complete proteome analysis by mass spectrometry: SILAC labeled yeast as a model system," Genome Biology, vol. 7, no. 6, article R50, 2006.

164. F. Tribl, C. Lohaus, T. Dombert, et al., "Towards multidimensional liquid chromatography separation of proteins using fluorescence and isotope-coded protein labelling for quantitative proteomics,"Proteomics, vol. 8, no. 6, pp. 1204–1211, 2008.

165. B. G. Fryksdale, P. T. Jedrzejewski, D. L. Wong, A. L. Gaertner, and B. S. Miller, "Impact of deglycosylation methods on two-dimensional gel electrophoresis and matrix assisted laser desorption/ionization-time of flight-mass spectrometry for proteomic analysis," Electrophoresis, vol. 23, no. 14, pp. 2184–2193, 2002. ·

166. A. V. Wymelenberg, G. Sabat, M. Mozuch, P. J. Kersten, D. A. N. Cullen, and R. A. Blanchette, "Structure, organization, and transcriptional regulation of a family of copper radical oxidase genes in the lignin-degrading basidiomycete Phanerochaete chrysosporium," Applied and Environmental Microbiology, vol. 72, no. 7, pp. 4871–4877, 2006.

167. P. Jungblut and B. Thiede, "Protein identification from 2-DE gels by MALDI mass spectrometry," Mass Spectrometry Reviews, vol. 16, no. 3, pp. 145–162, 1997. ·

168. P. H. O›Farrell, "High resolution two dimensional electrophoresis of proteins," Journal of Biological Chemistry, vol. 250, no. 10, pp. 4007–4021, 1975. ·

169. J. Klose and U. Kobalz, "Two-dimensional electrophoresis of proteins: an updated protocol and implications for a functional analysis of the genome," Electrophoresis, vol. 16, no. 6, pp. 1034–1059, 1995. ·

170. T. Rabilloud, A. R. Vaezzadeh, N. Potier, C. Lelong, E. Leize-Wagner, and M. Chevallet, "Power and limitations of electrophoretic separations in proteomics strategies," Mass Spectrometry Reviews, vol. 28, no. 5, pp. 816–843, 2009.

171. P. A. Haynes and T. H. Roberts, "Subcellular shotgun proteomics in

plants: looking beyond the usual suspects," Proteomics, vol. 7, no. 16, pp. 2963–2975, 2007. · ·

172. A. Pirondini, G. Visioli, A. Malcevschi, and N. Marmiroli, "A 2-D liquid-phase chromatography for proteomic analysis in plant tissues," Journal of Chromatography B, vol. 833, no. 1, pp. 91–100, 2006. · ·

173. G. Recorbet, H. Rogniaux, V. Gianinazzi-Pearson, and E. DumasGaudot, "Fungal proteins in the extra-radical phase of arbuscular mycorrhiza: a shotgun proteomic picture," New Phytologist, vol. 181, no. 2, pp. 248–260, 2009.

174. M. Ye, X. Jiang, S. Feng, R. Tian, and H. Zou, "Advances in chromatographic techniques and methods in shotgun proteome analysis," Trends in Analytical Chemistry, vol. 26, no. 1, pp. 80–84, 2007.

175. M. P. Washburn, D. Wolters, and J. R. Yates III, "Large-scale analysis of the yeast proteome by multidimensional protein identification technology," Nature Biotechnology, vol. 19, no. 3, pp. 242–247, 2001.

176. T. McDonald, S. Sheng, B. Stanley, et al., "Expanding the subproteome of the inner mitochondria using protein separation technologies: one- and two-dimensional liquid chromatography and two-dimensional gel electrophoresis," Molecular and Cellular Proteomics, vol. 5, no. 12, pp. 2392–2411, 2006.

177. W. W. Wu, G. Wang, S. J. Baek, and R. F. Shen, "Comparative study of three proteomic quantitative methods, DIGE, cICAT, and iTRAQ, using 2D gel- or LC-MALDI TOF/TOF," Journal of Proteome Research, vol. 5, no. 3, pp. 651–658, 2006.

178. B. Domon and R. Aebersold, "Mass spectrometry and protein analysis," Science, vol. 312, no. 5771, pp. 212–217, 2006.

179. A. I. Nesvizhskii, O. Vitek, and R. Aebersold, "Analysis and validation of proteomic data generated by tandem mass spectrometry," Nature Methods, vol. 4, no. 10, pp. 787–797, 2007.

180. D. M. Soanes, W. Skinner, J. Keon, J. Hargreaves, and N. J. Talbot, "Genomics of phytopathogenic fungi and the development of bioinformatic resources," Molecular Plant-Microbe Interactions, vol. 15, no. 5, pp. 421–427, 2002. ·

181. D. M. Soanes and N. J. Talbot, "Comparative genomic analysis of phytopathogenic fungi using expressed sequence tag (EST) collections," Molecular Plant Pathology, vol. 7, no. 1, pp. 61–70, 2006. · ·

182. J. Fiévet, C. Dillmann, G. Lagniel, et al., "Assessing factors for reliable quantitative proteomics based on two-dimensional gel electrophoresis,"

Proteomics, vol. 4, no. 7, pp. 1939–1949, 2004.

183. G. B. Smejkal, M. H. Robinson, and A. Lazarev, "Comparison of fluorescent stains: relative photostability and differential staining of proteins in two-dimensional gels," Electrophoresis, vol. 25, no. 15, pp. 2511–2519, 2004.

184. Y. I. Hu, G. Wang, G. Y. J. Chen, X. I. N. Fu, and S. Q. Yao, "Proteome analysis of Saccharomyces cerevisiae under metal stress by two-dimensional differential gel electrophoresis," Electrophoresis, vol. 24, no. 9, pp. 1458–1470, 2003.

185. J. U. N. X. Yan, A. T. Devenish, R. Wait, T. I. M. Stone, S. Lewis, and S. U. E. Fowler, "Fluorescence two-dimensional difference gel electrophoresis and mass spectrometry based proteomic analysis ofEscherichia coli," Proteomics, vol. 2, no. 12, pp. 1682–1698, 2002.

186. S. P. Gygi, B. Rist, S. A. Gerber, F. Turecek, M. H. Gelb, and R. Aebersold, "Quantitative analysis of complex protein mixtures using isotope-coded affinity tags," Nature Biotechnology, vol. 17, no. 10, pp. 994–999, 1999.

187. H. Zhou, J. A. Ranish, J. D. Watts, and R. Aebersold, "Quantitative proteome analysis by solid-phase isotope tagging and mass spectrometry," Nature Biotechnology, vol. 20, no. 5, pp. 512–515, 2002.

188. X. Yao, A. Freas, J. Ramirez, P. A. Demirev, and C. Fenselau, "Proteolytic 18O labeling for comparative proteomics: model studies with two serotypes of adenovirus," Analytical Chemistry, vol. 73, no. 13, pp. 2836–2842, 2001.

189. P. A. Everley, J. Krijgsveld, B. R. Zetter, and S. P. Gygi, "Quantitative cancer proteomics: stable isotope labeling with amino acids in cell culture (SILAC) as a tool for prostate cancer research," Molecular and Cellular Proteomics, vol. 3, no. 7, pp. 729–735, 2004.

190. L. R. Zieske, "A perspective on the use of iTRAQTM reagent technology for protein complex and profiling studies," Journal of Experimental Botany, vol. 57, no. 7, pp. 1501–1508, 2006. · ·

191. W. M. Old, K. Meyer-Arendt, L. Aveline-Wolf, et al., "Comparison of label-free methods for quantifying human proteins by shotgun proteomics," Molecular and Cellular Proteomics, vol. 4, no. 10, pp. 1487–1502, 2005.

192. B. Zhang, N. C. VerBerkmoes, M. A. Langston, E. Uberbacher, R. L. Hettich, and N. F. Samatova, "Detecting differential and correlated protein expression in label-free shotgun proteomics," Journal ofProteome Research, vol. 5, no. 11, pp. 2909–2918, 2006. · ·

193. M. Ünlü, M. E. Morgan, and J. S. Minden, "Difference gel electrophoresis: a single gel method for detecting changes in protein extracts," Electrophoresis, vol. 18, no. 11, pp. 2071–2077, 1997.

194. M. Bantscheff, M. Schirle, G. Sweetman, J. Rick, and B. Kuster, "Quantitative mass spectrometry in proteomics: a critical review," Analytical and Bioanalytical Chemistry, vol. 389, no. 4, pp. 1017–1031, 2007.

195. L. N. Mueller, M. I. Y. Brusniak, D. R. Mani, and R. Aebersold, "An assessment of software solutions for the analysis of mass spectrometry based quantitative proteomics data," Journal of Proteome Research, vol. 7, no. 1, pp. 51–61, 2008.

196. V. J. Patel, K. Thalassinos, S. E. Slade, et al., "A comparison of labeling and label-free mass spectrometry-based proteomics approaches," Journal of Proteome Research, vol. 8, no. 7, pp. 3752–3759, 2009.

197. S. Xie, C. Moya, B. Bilgin, A. Jayaraman, and S. P. Walton, "Emerging affinity-based techniques in proteomics," Expert Review of Proteomics, vol. 6, no. 5, pp. 573–583, 2009.

198. I. A. N. P. Shadforth, T. P. J. Dunkley, K. S. Lilley, and C. Bessant, "i-Tracker: for quantitative proteomics using iTRAQTM," BMC Genomics, vol. 6, article 145, 2005.

199. P. V. Bondarenko, D. Chelius, and T. A. Shaler, "Identification and relative quantitation of protein mixtures by enzymatic digestion followed by capillary reversed-phase liquid chromatography-tandem mass spectrometry," Analytical Chemistry, vol. 74, no. 18, pp. 4741–4749, 2002.

200. D. Chelius and P. V. Bondarenko, "Quantitative profiling of proteins in complex mixtures using liquid chromatography and mass spectrometry," Journal of Proteome Research, vol. 1, no. 4, pp. 317–323, 2002.

201. W. Wang, H. Zhou, H. U. A. Lin, et al., "Quantification of proteins and metabolites by mass spectrometry without isotopic labeling or spiked standards," Analytical Chemistry, vol. 75, no. 18, pp. 4818–4826, 2003.

202. H. Liu, R. G. Sadygov, and J. R. Yates III, "A model for random sampling and estimation of relative protein abundance in shotgun proteomics," Analytical Chemistry, vol. 76, no. 14, pp. 4193–4201, 2004.

203. J. Fasolo and M. Snyder, "Protein microarrays," Methods in Molecular Biology, vol. 548, pp. 209–222, 2009.

204. J. Barrett, P. M. Brophy, and J. V. Hamilton, "Analysing proteomic data," International Journal for Parasitology, vol. 35, no. 5, pp. 543–553, 2005.

205. E. Marengo, E. Robotti, F. Antonucci, D. Cecconi, N. Campostrini, and P. G. Righetti, "Numerical approaches for quantitative analysis of two-dimensional maps: a review of commercial software and home-made systems," Proteomics, vol. 5, no. 3, pp. 654–666, 2005.

206. Å. M. Wheelock and S. Goto, "Effects of post-electrophoretic analysis on variance in gel-based proteomics," Expert Review of Proteomics, vol. 3, no. 1, pp. 129–142, 2006.

207. E. W. Deutsch, H. Lam, and R. Aebersold, "PeptideAtlas: a resource for target selection for emerging targeted proteomics workflows," EMBO Reports, vol. 9, no. 5, pp. 429–434, 2008.

208. D. G. Biron, C. Brun, T. Lefevre, et al., "The pitfalls of proteomics experiments without the correct use of bioinformatics tools," Proteomics, vol. 6, no. 20, pp. 5577–5596, 2006.

209. N. A. Karp and K. S. Lilley, "Design and analysis issues in quantitative proteomics studies," Proteomics, vol. 2, no. 1, pp. 42–50, 2007.

210. J. M. Bland and D. G. Altman, "Multiple significance tests: the Bonferroni method," British Medical Journal, vol. 310, no. 6973, article 170, 1995.

211. Y. Benjamini and Y. Hochberg, "Controlling the false discovery rate: a practical and powerful approach to multiple testing," Journal of the Royal Statistical Society. Series B, vol. 57, pp. 289–300, 1995.

212. J. D. Storey and R. Tibshirani, "Statistical significance for genomewide studies," Proceedings of the National Academy of Sciences of the United States of America, vol. 100, no. 16, pp. 9440–9445, 2003.

213. K. Strimmer, "fdrtool: a versatile R package for estimating local and tail area-based false discovery rates,"Bioinformatics, vol. 24, no. 12, pp. 1461–1462, 2008.

214. S. Jacobsen, H. Grove, K. N. Jensen, et al., "Multivariate analysis of 2-DE protein patterns—practical approaches," Electrophoresis, vol. 28, no. 8, pp. 1289–1299, 2007.

215. K. N. Jensen, F. Jessen, and B. O. M. Jørgensen, "Multivariate data analysis of two-dimensional gel electrophoresis protein patterns from few samples," Journal of Proteome Research, vol. 7, no. 3, pp. 1288–1296, 2008.

216. R. Pedreschi, M. L. A. T. M. Hertog, S. C. Carpentier, et al., "Treatment of missing values for multivariate statistical analysis of gel-based proteomics data," Proteomics, vol. 8, no. 7, pp. 1371–1383, 2008.

217. M. Schneider, L. Lane, E. Boutet, et al., "The UniProtKB/Swiss-Prot knowledgebase and its Plant Proteome Annotation Program," Journal of

Proteomics, vol. 72, no. 3, pp. 567–573, 2009.

218. C. F. Taylor, N. W. Paton, K. S. Lilley, et al., “The minimum information about a proteomics experiment (MIAPE),” Nature Biotechnology, vol. 25, no. 8, pp. 887–893, 2007.

219. J. A. Mead, I. A. N. P. Shadforth, and C. Bessant, “Public proteomic MS repositories and pipelines: available tools and biological applications,” Proteomics, vol. 7, no. 16, pp. 2769–2786, 2007.

220. M. Fälth, M. M. Savitski, M. L. Nielsen, F. Kjeldsen, P. E. R. E. Andren, and R. A. Zubarev, “SwedCAD, a database of annotated high-mass accuracy MS/MS spectra of tryptic peptides,” Journal of Proteome Research, vol. 6, no. 10, pp. 4063–4067, 2007.

221. J. A. N. Hummel, M. Niemann, S. Wienkoop, et al., “ProMEX: a mass spectral reference database for proteins and protein phosphorylation sites,” BMC Bioinformatics, vol. 8, article 216, 2007.

222. H. Lam, E. W. Deutsch, J. S. Eddes, J. K. Eng, S. E. Stein, and R. Aebersold, “Building consensus spectral libraries for peptide identification in proteomics,” Nature Methods, vol. 5, no. 10, pp. 873–875, 2008. · ·

223. R. Craig, J. P. Cortens, and R. C. Beavis, “Open source system for analyzing, validating, and storing protein identification data,” Journal of Proteome Research, vol. 3, no. 6, pp. 1234–1242, 2004.

224. F. Desiere, E. W. Deutsch, N. L. King, et al., “The peptide atlas project,” Nucleic Acids Research, vol. 34, pp. D655–658, 2006. ·

225. L. Martens, H. Hermjakob, P. Jones, et al., “PRIDE: the proteomics identifications database,” Proteomics, vol. 5, no. 13, pp. 3537–3545, 2005.

226. I. A. N. Shadforth, W. Xu, D. Crowther, and C. Bessant, “GAPP: a fully automated software for the confident identification of human peptides from tandem mass spectra,” Journal of Proteome Research, vol. 5, no. 10, pp. 2849–2852, 2006.

227. T. McLaughlin, J. A. Siepen, J. Selley, et al., “PepSeeker: a database of proteome peptide identifications for investigating fragmentation patterns,” Nucleic Acids Research, vol. 34, pp. D649–D654, 2006. ·

228. Y. Zhang, Y. Zhang, J. U. N. Adachi, et al., “MAPU: Max-Planck Unified database of organellar, cellular, tissue and body fluid proteomes,” Nucleic Acids Research, vol. 35, database issue, pp. D771–D779, 2007. · ·

229. J. T. Prince, M. W. Carlson, R. Wang, P. Lu, and E. M. Marcotte, “The need for a public proteomics repository,” Nature Biotechnology, vol. 22, no. 4, pp. 471–472, 2004.

230. M. Riffle, L. Malmström, and T. N. Davis, "The yeast resource center public data repository," Nucleic Acids Research, vol. 33, pp. D378–D382, 2005.

231. I. L. S. Oh, A. E. R. A. N. Park, M. I. N. S. Bae, et al., "Secretome analysis reveals an Arabidopsis lipase involved in defense against Altemaria brassicicola," Plant Cell, vol. 17, no. 10, pp. 2832–2847, 2005.

232. F. Colditz, O. Nyamsuren, K. Niehaus, H. Eubel, H. P. Braun, and F. Krajinski, "Proteomic approach: identification of Medicago truncatula proteins induced in roots after infection with the pathogenic oomycete Aphanomyces euteiches," Plant Molecular Biology, vol. 55, no. 1, pp. 109–120, 2004.

233. F. Colditz, H. P. Braun, C. Jacquet, K. Niehaus, and F. Krajinski, "Proteomic profiling unravels insights into the molecular background underlying increased Aphanomyces euteiches-tolerance of Medicago truncatula," Plant Molecular Biology, vol. 59, no. 3, pp. 387–406, 2005.

234. F. Colditz, K. Niehaus, and F. Krajinski, "Silencing of PR-10-like proteins in Medicago truncatula results in an antagonistic induction of other PR proteins and in an increased tolerance upon infection with the oomycete Aphanomyces euteiches," Planta, vol. 226, no. 1, pp. 57–71, 2007.

235. T. J. March, J. A. Able, C. J. Schultz, and A. J. Able, "A novel late embryogenesis abundant protein and peroxidase associated with black point in barley grains," Proteomics, vol. 7, no. 20, pp. 3800–3808, 2007.

236. M. D. Bolton, H. P. van Esse, J. H. Vossen, et al., "The novel Cladosporium fulvum lysin motif effector Ecp6 is a virulence factor with orthologues in other fungal species," Molecular Microbiology, vol. 69, no. 1, pp. 119–136, 2008.

237. J. A. Smith, R. A. Blanchette, T. A. Burnes, et al., "Proteomic comparison of needles from blister rust-resistant and susceptible Pinus strobus seedlings reveals upregulation of putative disease resistance proteins," Molecular Plant-Microbe Interactions, vol. 19, no. 2, pp. 150–160, 2006.

238. D. Wang, A. Eyles, D. Mandich, and P. Bonello, "Systemic aspects of host-pathogen interactions in Austrian pine (Pinus nigra): a proteomics approach," Physiological and Molecular Plant Pathology, vol. 68, no. 4–6, pp. 149–157, 2006.

239. M. Curto, E. Camafeita, J. A. Lopez, A. N. A. M. Maldonado, D. Rubiales, and J. V. Jorrín, "A proteomic approach to study pea (Pisum sativum) responses to powdery mildew (Erysiphe pisi)," Proteomics, vol. 6, supplement 1, pp. S163–S174, 2006.

240. J. Geddes, F. Eudes, A. Laroche, and L. B. Selinger, "Differential expression of proteins in response to the interaction between the pathogen Fusarium graminearum and its host, Hordeum vulgare," Proteomics, vol. 8, no. 3, pp. 545–554, 2008.

241. W. Zhou, F. L. Kolb, and D. E. Riechers, "Identification of proteins induced or upregulated by Fusarium head blight infection in the spikes of hexaploid wheat (Triticum aestivum)," Genome, vol. 48, no. 5, pp. 770–780, 2005.

242. B. K. Ndimba, S. Chivasa, J. M. Hamilton, W. J. Simon, and A. R. Slabas, "Proteomic analysis of changes in the extracellular matrix of Arabidopsis cell suspension cultures induced by fungal elicitors,"Proteomics, vol. 3, no. 6, pp. 1047–1059, 2003.

243. R. L. Larson, A. M. Y. L. Hill, and A. Nuñez, "Characterization of protein changes associated with sugar beet (Beta vulgaris) resistance and susceptibility to Fusarium oxysporum," Journal of Agricultural and Food Chemistry, vol. 55, no. 19, pp. 7905–7915, 2007.

244. S. Campo, M. Carrascal, M. Coca, J. Abián, and B. San Segundo, "The defense response of germinating maize embryos against fungal infection: a proteomics approach," Proteomics, vol. 4, no. 2, pp. 383–396, 2004.

245. S. Chivasa, J. M. Hamilton, R. S. Pringle, et al., "Proteomic analysis of differentially expressed proteins in fungal elicitor-treated Arabidopsis cell cultures," Journal of Experimental Botany, vol. 57, no. 7, pp. 1553–1562, 2006.

246. S. Chivasa, W. J. Simon, X. L. Yu, N. Yalpani, and A. R. Slabas, "Pathogen elicitor-induced changes in the maize extracellular matrix proteome," Proteomics, vol. 5, no. 18, pp. 4894–4904, 2005.

247. J. V. F. Coumans, A. Poljak, M. J. Raftery, D. Backhouse, and L. Pereg-Gerk, "Analysis of cotton (Gossypium hirsutum) root proteomes during a compatible interaction with the black root rot fungusThielaviopsis basicola," Proteomics, vol. 9, no. 2, pp. 335–349, 2009.

248. H. O. W. O. N. Jung, C. W. O. O. Lim, S. C. Lee, H. W. O. O. Choi, C. H. Hwang, and B. K. Hwang, "Distinct roles of the pepper hypersensitive induced reaction protein gene CaHIR1 in disease and osmotic stress, as determined by comparative transcriptome and proteome analyses," Planta, vol. 227, no. 2, pp. 409–425, 2008.

249. N. Sharma, N. Hotte, M. H. Rahman, M. Mohammadi, M. K. Deyholos, and N. N. V. Kav, "Towards identifying Brassica proteins involved in mediating resistance to Leptosphaeria maculans: a proteomics-based approach," Proteomics, vol. 8, no. 17, pp. 3516–3535, 2008.

250. B. Subramanian, V. K. Bansal, and N. N. V. Kav, "Proteome-level investigation of Brassica carinata-derived resistance to Leptosphaeria maculans," Journal of Agricultural and Food Chemistry, vol. 53, no. 2, pp. 313–324, 2005.

251. H. Konishi, K. Ishiguro, and S. Komatsu, "A proteomics approach towards understanding blast fungus infection of rice grown under different levels of nitrogen fertilization," Proteomics, vol. 1, no. 9, pp. 1162–1171, 2001.

252. S. U. N. T. A. E. Kim, K. Y. U. S. Cho, S. Yu, et al., "Proteomic analysis of differentially expressed proteins induced by rice blast fungus and elicitor in suspension-cultured rice cells," Proteomics, vol. 3, no. 12, pp. 2368–2378, 2003.

253. S. T. Kim, S. G. Kim, D. H. Hwang, et al., "Proteomic analysis of pathogen-responsive proteins from rice leaves induced by rice blast fungus, Magnaporthe grisea," Proteomics, vol. 4, no. 11, pp. 3569–3578, 2004. · ·

254. S. T. Kim, S. Yu, S. G. Kim, et al., "Proteome analysis of rice blast fungus (Magnaporthe grisea) proteome during appressorium formation," Proteomics, vol. 4, no. 11, pp. 3579–3587, 2004.

255. M. Liao, Y. Li, and Z. Wang, "Identification of elicitor-responsive proteins in rice leaves by a proteomic approach," Proteomics, vol. 9, no. 10, pp. 2809–2819, 2009. · ·

256. K. U. N. Yuan, B. O. Zhang, Y. Zhang, Q. Cheng, M. Wang, and M. Huang, "Identification of differentially expressed proteins in poplar leaves induced by Marssonina brunnea f. sp. Multigermtubi,"Journal of Genetics and Genomics, vol. 35, no. 1, pp. 49–60, 2008.

257. C. P. Pirovani, H. A. S. Carvalho, R. C. R. Machado, et al., "Protein extraction for proteome analysis from cacao leaves and meristems, organs infected by Moniliophthora perniciosa, the causal agent of the witches› broom disease," Electrophoresis, vol. 29, no. 11, pp. 2391–2401, 2008.

258. F. Wen, H. D. VanEtten, G. Tsaprailis, and M. C. Hawes, "Extracellular proteins in pea root tip and border cell exudates," Plant Physiology, vol. 143, no. 2, pp. 773–783, 2007.

259. Z. Chan, G. Qin, X. Xu, B. Li, and S. Tian, "Proteome approach to characterize proteins induced by antagonist yeast and salicylic acid in peach fruit," Journal of Proteome Research, vol. 6, no. 5, pp. 1677–1688, 2007.

260. M. A. Islam, R. N. Sturrock, and A. K. M. Ekramoddoullah, "A proteomics approach to identify proteins differentially expressed in Douglas-fir seedlings infected by Phellinus sulphurascens," Journal of Proteomics,

vol. 71, no. 4, pp. 425–438, 2008.

261. R. C. Amey, T. Schleicher, J. Slinn, et al., "Proteomic analysis of a compatible interaction between Pisum sativum (pea) and the downy mildew pathogen Peronospora viciae," European Journal of Plant Pathology, vol. 122, no. 1, pp. 41–55, 2008.

262. T. Cao, S. Srivastava, M. H. Rahman, et al., "Proteome-level changes in the roots of Brassica napus as a result of Plasmodiophora brassicae infection," Plant Science, vol. 174, no. 1, pp. 97–115, 2008.

263. J. Lee, T. M. Bricker, M. Lefevre, S. R. M. Pinson, and J. H. Oard, "Proteomic and genetic approaches to identifying defence-related proteins in rice challenged with the fungal pathogen Rhizoctonia solani,"Molecular Plant Pathology, vol. 7, no. 5, pp. 405–416, 2006. · ·

264. J. Lee, J. Feng, K. B. Campbell, et al., "Quantitative proteomic analysis of bean plants infected by a virulent and avirulent obligate rust fungus," Molecular and Cellular Proteomics, vol. 8, no. 1, pp. 19–31, 2009.

265. Y. U. Liang, S. Srivastava, M. H. Rahman, S. E. Strelkov, and N. N. V. Kav, "Proteome changes in leaves of Brassica napus L. as a result of Sclerotinia sclerotiorum challenge," Journal of Agricultural and Food Chemistry, vol. 56, no. 6, pp. 1963–1976, 2008.

266. V. Bhadauria, S. Banniza, L.-X. Wang, Y.-D. Wei, and Y.-L. Peng, "Proteomic studies of phytopathogenic fungi, oomycetes and their interactions with hosts," European Journal of Plant Pathology, vol. 126, no. 1, pp. 81–95, 2010.

267. U. Mathesius, "Comparative proteomic studies of root-microbe interactions," Journal of Proteomics, vol. 72, no. 3, pp. 353–366, 2009.

268. A. Mehta, A. C. M. Brasileiro, D. S. L. Souza, et al., "Plant-pathogen interactions: what is proteomics telling us?" FEBS Journal, vol. 275, no. 15, pp. 3731–3746, 2008.

269. Y. U. Liang, H. U. I. Chen, M. Tang, and S. Shen, "Proteome analysis of an ectomycorrhizal fungusBoletus edulis under salt shock," Mycological Research, vol. 111, no. 8, pp. 939–946, 2007.

270. A. M. Murad, E. F. Noronha, R. N. Miller, et al., "Proteomic analysis of Metarhizium anisopliae secretion in the presence of the insect pest Callosobruchus maculatus," Microbiology, vol. 154, no. 12, pp. 3766–3774, 2008.

271. W. U. N. Y. Lin, J. U. I. Y. Chang, C. H. Hish, and T. Z. U. M. Pan, "Profiling the Monascus pilosusproteome during nitrogen limitation," Journal of Agricultural and Food Chemistry, vol. 56, no. 2, pp. 433–441,

2008.

272. A. V. Wymelenberg, G. Sabat, D. Martinez, et al., "The Phanerochaete chrysosporium secretome: database predictions and initial mass spectrometry peptide identifications in cellulose-grown medium,"Journal of Biotechnology, vol. 118, no. 1, pp. 17–34, 2005. · ·

273. R. Marra, P. Ambrosino, V. Carbone, et al., "Study of the three-way interaction between Trichoderma atroviride, plant and fungal pathogens by using a proteomic approach," Current Genetics, vol. 50, no. 5, pp. 307–321, 2006.

274. S. C. Tseng, S. H. U. Y. Liu, H. H. Yang, C. T. Lo, and K. O. U. C. Peng, "Proteomic study of biocontrol mechanisms of Trichoderma harzianum ETS 323 in response to Rhizoctonia solani," Journal of Agricultural and Food Chemistry, vol. 56, no. 16, pp. 6914–6922, 2008.

275. V. Seidl, I. S. Druzhinina, and C. P. Kubicek, "A screening system for carbon sources enhancing β-N-acetylglucosaminidase formation in Hypocrea atroviridis (Trichoderma atroviride)," Microbiology, vol. 152, no. 7, pp. 2003–2012, 2006.

276. S. Nagendran, H. E. Hallen-Adams, J. M. Paper, N. Aslam, and J. D. Walton, "Reduced genomic potential for secreted plant cell-wall-degrading enzymes in the ectomycorrhizal fungus Amanita bisporigera, based on the secretome of Trichoderma reesei," Fungal Genetics and Biology, vol. 46, no. 5, pp. 427–435, 2009.

277. K. J. Welham, M. A. Domin, K. Johnson, L. Jones, and D. S. Ashton, "Characterization of fungal spores by laser desorption/ionization time-of-flight mass spectrometry," Rapid Communications in Mass Spectrometry, vol. 14, no. 5, pp. 307–310, 2000. ·

278. M. Both, M. Csukai, M. P. H. Stumpf, and P. D. Spanu, "Gene expression profiles of Blumeria graminisindicate dynamic changes to primary metabolism during development of an obligate biotrophic pathogen," Plant Cell, vol. 17, no. 7, pp. 2107–2122, 2005. · ·

279. M. Both, S. E. Eckert, M. Csukai, E. Muller, G. Dimopoulos, and P. D. Spanu, "Transcript profiles ofBlumeria graminis development during infection reveal a cluster of genes that are potential virulence determinants," Molecular Plant-Microbe Interactions, vol. 18, no. 2, pp. 125–133, 2005.

280. M. R. Remm, C. E. V. Storm, and E. L. L. Sonnhammer, "Automatic clustering of orthologs and in-paralogs from pairwise species comparisons," Journal of Molecular Biology, vol. 314, no. 5, pp. 1041–1052, 2001.

281. J. Grinyer, L. Kautto, M. Traini, et al., "Proteome mapping of the Trichoderma reesei 20S proteasome,"Current Genetics, vol. 51, no. 2, pp. 79–88, 2007.

282. P. S. Solomon, K. A. R. C. Tan, P. Sanchez, R. M. Cooper, and R. P. Oliver, "The disruption of a Gαsubunit sheds new light on the pathogenicity of Stagonospora nodorum on wheat," Molecular Plant-Microbe Interactions, vol. 17, no. 5, pp. 456–466, 2004. ·

283. J. C. Misas-Villamil and R. A. van der Hoorn, "Enzyme-inhibitor interactions at the plant-pathogen interface," Current Opinion in Plant Biology, vol. 11, no. 4, pp. 380–388, 2008.

284. P. Kankanala, K. Czymmek, and B. Valent, "Roles for rice membrane dynamics and plasmodesmata during biotrophic invasion by the blast fungus," Plant Cell, vol. 19, no. 2, pp. 706–724, 2007.

285. H. Tjalsma, A. Bolhuis, J. D. H. Jongbloed, S. Bron, and J. M. van Dijl, "Signal peptide-dependent protein transport in Bacillus subtilis: a genome-based survey of the secretome," Microbiology and Molecular Biology Reviews, vol. 64, no. 3, pp. 515–547, 2000. ·

286. R. P. De Vries, "Regulation of Aspergillus genes encoding plant cell wall polysaccharide-degrading enzymes; relevance for industrial production," Applied Microbiology and Biotechnology, vol. 61, no. 1, pp. 10–20, 2003. ·

287. F. M. Freimoser, S. Screen, G. Hu, and R. St. Leger, "EST analysis of genes expressed by the zygomycete pathogen Conidiobolus coronatus during growth on insect cuticle," Microbiology, vol. 149, no. 7, pp. 1893–1900, 2003. ·

288. J. D. Walton, "Deconstructing the cell wall," Plant Physiology, vol. 104, no. 4, pp. 1113–1118, 1994. ·

289. N. Brito, J. J. Espino, and C. González, "The endo-β-1,4-xylanase Xyn11A is required for virulence inBotrytis cinerea," Molecular Plant-Microbe Interactions, vol. 19, no. 1, pp. 25–32, 2006. · ·

290. H. Deising, R. L. Nicholson, M. Haug, R. J. Howard, and K. Mendgen, "Adhesion pad formation and the involvement of cutinase and esterases in the attachment of uredospores to the host cuticle," Plant Cell, vol. 4, no. 9, pp. 1101–1111, 1992. ·

291. A. Isshiki, K. Akimitsu, M. Yamamoto, and H. Yamamoto, "Endopolygalacturonase is essential for citrus black rot caused by Alternaria citri but not brown spot caused by Alternaria alternata," Molecular Plant-Microbe Interactions, vol. 14, no. 6, pp. 749–757, 2001. ·

292. B. Oeser, P. M. Heidrich, U. Müller, P. Tudzynski, and K. B. Tenberge, "Polygalacturonase is a pathogenicity factor in the Claviceps purpurea/rye interaction," Fungal Genetics and Biology, vol. 36, no. 3, pp. 176–186, 2002.

293. A. ten Have, W. Mulder, J. Visser, and J. A. L. van Kan, "The endopolygalacturonase gene Bcpg1 is required to full virulence of Botrytis cinerea," Molecular Plant-Microbe Interactions, vol. 11, no. 10, pp. 1009–1016, 1998.

294. C. A. Voigt, W. Schäfer, and S. Salomon, "A secreted lipase of Fusarium graminearum is a virulence factor required for infection of cereals," Plant Journal, vol. 42, no. 3, pp. 364–375, 2005.

295. N. Yakoby, D. Beno-Moualem, N. T. Keen, A. Dinoor, O. Pines, and D. Prusky, "Colletotrichum gloeosporioides pelB is an important virulence factor in avocado fruit-fungus interaction," Molecular Plant-Microbe Interactions, vol. 14, no. 8, pp. 988–995, 2001.

296. J. W. Bennett, "The molds of Katrina. Update," New York Academy of Sciences Magazine, pp. 6–9, 2006.

297. G. L. F. Wallis, R. J. Swift, F. W. Hemming, A. P. J. Trinci, and J. F. Peberdy, "Glucoamylase overexpression and secretion in Aspergillus niger: analysis of glycosylation," Biochimica et Biophysica Acta, vol. 1472, no. 3, pp. 576–586, 1999.

298. D. Lim, P. Hains, B. Walsh, P. Bergquist, and H. Nevalainen, "Proteins associated with the cell envelope of Trichoderma reesei: a proteomic approach," Proteomics, vol. 1, no. 7, pp. 899–910, 2001.

299. M. B. Suárez, L. Sanz, M. I. Chamorro, et al., "Proteomic analysis of secreted proteins from Trichoderma harzianum: identification of a fungal cell wall-induced aspartic protease," Fungal Genetics and Biology, vol. 42, no. 11, pp. 924–934, 2005.

300. C. Tian, W. T. Beeson, A. T. Iavarone, et al., "Systems analysis of plant cell wall degradation by the model filamentous fungus Neurospora crassa," Proceedings of the National Academy of Sciences of the United States of America, vol. 106, no. 52, pp. 22157–22162, 2009.

301. J. M. Cork and M. D. Purugganan, "The evolution of molecular genetic pathways and networks,"BioEssays, vol. 26, no. 5, pp. 479–484, 2004.

302. T. Köcher and G. Superti-Furga, "Mass spectrometry-based functional proteomics: from molecular machines to protein networks," Nature Methods, vol. 4, no. 10, pp. 807–815, 2007.

303. W. F. Loomis and P. W. Sternberg, "Genetic networks," Science, vol.

269, no. 5224, p. 649, 1995. ·

304. M. E. Cusick, N. Klitgord, M. Vidal, and D. E. Hill, "Interactome: gateway into systems biology," Human Molecular Genetics, vol. 14, no. 2, pp. R171–R181, 2005. · ·
305. L. Kiemer and G. Cesareni, "Comparative interactomics: comparing apples and pears?" Trends in Biotechnology, vol. 25, no. 10, pp. 448–454, 2007.
306. L. Giot, J. S. Bader, C. Brouwer, et al., "A protein interaction map of Drosophila melanogaster," Science, vol. 302, no. 5651, pp. 1727–1736, 2003.
307. D. J. LaCount, M. Vignali, R. Chettier, et al., "A protein interaction network of the malaria parasitePlasmodium falciparum," Nature, vol. 438, no. 7064, pp. 103–107, 2005.
308. S. Li, C. M. Armstrong, N. Bertin, et al., "A map of the interactome network of the metazoan C. elegans,"Science, vol. 303, no. 5657, pp. 540–543, 2004.
309. J. R. Parrish, J. Yu, G. Liu, et al., "A proteome-wide protein interaction map for Campylobacter jejuni,"Genome Biology, vol. 8, no. 7, p. R130, 2007.
310. P. Uetz, L. Glot, G. Cagney, et al., "A comprehensive analysis of protein-protein interactions inSaccharomyces cerevisiae," Nature, vol. 403, no. 6770, pp. 623–627, 2000.
311. T. Berggård, S. Linse, and P. James, "Methods for the detection and analysis of protein-protein interactions," Proteomics, vol. 7, no. 16, pp. 2833–2842, 2007.
312. J. A. N. A. Miernyk and J. A. Y. J. Thelen, "Biochemical approaches for discovering protein-protein interactions," Plant Journal, vol. 53, no. 4, pp. 597–609, 2008.
313. B. Causier and B. Davies, "Analysing protein-protein interactions with the yeast two-hybrid system,"Plant Molecular Biology, vol. 50, no. 6, pp. 855–870, 2002. · ·
314. L. R. Matthews, P. Vaglio, J. Reboul, et al., "Identification of potential interaction networks using sequence-based searches for conserved protein-protein interactions or "interologs"," Genome Research, vol. 11, no. 12, pp. 2120–2126, 2001.
315. T. Ideker, O. Ozier, B. Schwikowski, and A. F. Siegel, "Discovering regulatory and signalling circuits in molecular interaction networks," Bioinformatics, vol. 18, supplement 1, pp. S233–S240, 2002. ·

316. X. Wu, L. E. I. Zhu, J. I. E. Guo, D. A. Y. Zhang, and K. U. I. Lin, "Prediction of yeast protein-protein interaction network: insights from the Gene Ontology and annotations," Nucleic Acids Research, vol. 34, no. 7, pp. 2137–2150, 2006.

317. S. E. E. K. Ng, Z. Zhang, and S. H. Tan, "Integrative approach for computationally inferring protein domain interactions," Bioinformatics, vol. 19, no. 8, pp. 923–929, 2003.

318. R. Jothi, M. G. Kann, and T. M. Przytycka, "Predicting protein-protein interaction by searching evolutionary tree automorphism space," Bioinformatics, vol. 21, supplement 1, pp. i241–i250, 2005. · ·

319. U. Ogmen, O. Keskin, A. S. Aytuna, R. Nussinov, and A. Gursoy, "PRISM: protein interactions by structural matching," Nucleic Acids Research, vol. 33, no. 2, pp. W331–W336, 2005.

320. Y. Qi, J. Klein-Seetharaman, and Z. I. V. Bar-Joseph, "A mixture of feature experts approach for protein-protein interaction prediction," BMC Bioinformatics, vol. 8, supplement 10, p. S6, 2007.

321. J. Shen, J. Zhang, X. Luo, et al., "Predicting protein-protein interactions based only on sequences information," Proceedings of the National Academy of Sciences of the United States of America, vol. 104, no. 11, pp. 4337–4341, 2007.

322. C. von Mering, L. J. Jensen, M. Kuhn, et al., "STRING 7—recent developments in the integration and prediction of protein interactions," Nucleic Acids Research, vol. 35, no. 1, pp. D358–D362, 2007.

323. D. R. Rhodes, S. A. Tomlins, S. Varambally, et al., "Probabilistic model of the human protein-protein interaction network," Nature Biotechnology, vol. 23, no. 8, pp. 951–959, 2005.

324. E. Hirsh and R. Sharan, "Identification of conserved protein complexes based on a model of protein network evolution," Bioinformatics, vol. 23, no. 2, pp. e170–e176, 2007.

325. S. P. Gygi, Y. Rochon, B. R. Franza, and R. Aebersold, "Correlation between protein and mRNA abundance in yeast," Molecular and Cellular Biology, vol. 19, no. 3, pp. 1720–1730, 1999.

Chapter 3

LIFESTYLE TRANSITIONS IN PLANT PATHOGENIC COLLETOTRICHUM FUNGI DECIPHERED BY GENOME AND TRANSCRIPTOME ANALYSES

Richard J O'Connell[1], Michael R Thon[2], Stéphane Hacquard[1], Stefan G Amyotte[3], Jochen Kleemann[1], Maria F Torres[3], Ulrike Damm[4], Ester A Buiate[3], Lynn Epstein[5], Noam Alkan[6], Janine Altmüller[7], Lucia Alvarado-Balderrama[8], Christopher A Bauser[9], Christian Becker[7], Bruce W Birren[8], Zehua Chen[8], Jaeyoung Choi[10], Jo Anne Crouch[11], Jonathan P Duvick[12,30], Mark A Farman[3], Pamela Gan[13], David Heiman[8], Bernard Henrissat[14], Richard J Howard[12], Mehdi Kabbage[15], Christian Koch[16], Barbara Kracher[1], Yasuyuki Kubo[17], Audrey D Law[3], Marc-Henri Lebrun[18], Yong-Hwan Lee[10], Itay Miyara[6], Neil Moore[19], Ulla Neumann[20], Karl Nordström[21], Daniel G Panaccione[22], Ralph Panstruga[1,23], Michael Place[24], Robert H Proctor[25], Dov Prusky[6], Gabriel Rech[2], Richard Reinhardt[26], Jeffrey A Rollins[27], Steve Rounsley[8], Christopher L Schardl[3], David C Schwartz[24], Narmada Shenoy[8], Ken Shirasu[13], Usha R Sikhakolli[28], Kurt Stüber[26], Serenella A Sukno[2], James A Sweigard[12], Yoshitaka Takano[29], Hiroyuki Takahara[1,30], Frances Trail[28], H Charlotte van der Does[1,30], Lars M Voll[16], Isa Will1, Sarah Young[8], Qiandong Zeng[8], Jingze Zhang[8], Shiguo Zhou[24], Martin B Dickman[15], Paul Schulze-Lefert[1], Emiel Ver Loren van Themaat[1], Li-Jun Ma[8,30] & Lisa J Vaillancourt[3]

[1]Department of Plant Microbe Interactions, Max Planck Institute for Plant Breeding Research, Cologne, Germany

[2]Centro Hispano-Luso de Investigaciones Agrarias, Departamento de Microbiología y Genética, Universidad de Salamanca, Villamayor, Spain

[3]Department of Plant Pathology, University of Kentucky, Lexington, Kentucky, USA

[4]Centraalbureau voor Schimmelcultures, Koninklijke Nederlandse Akademie van Wetenschappen, Fungal Biodiversity Centre, Utrecht, The Netherlands

[5]Department of Plant Pathology, University of California, Davis, California, USA

[6]Department of Postharvest Science of Fresh Produce, Agricultural Research Organization, The Volcani Center, Bet Dagan, Israel

[7]Cologne Center for Genomics, University of Cologne, Cologne, Germany

[8]Broad Institute, Cambridge, Massachusetts, USA

[9]GATC Biotech AG, Konstanz, Germany

[10]Department of Agricultural Biotechnology, Center for Fungal Genetic Resources, Seoul National University, Seoul, Korea

[11]Systematic Mycology and Microbiology Laboratory, US Department of Agriculture, Agricultural Research Service, Beltsville, Maryland, USA

[12]Pioneer Hi-Bred International, DuPont Agricultural Biotechnology, Wilmington, Delaware, USA

[13]Plant Immunity Research Group, RIKEN Plant Science Center, Yokohama, Japan

[14]Laboratoire Architecture et Fonction des Macromolécules Biologiques, Centre National de la Recherche Scientifique, Unité Mixte de Recherche 7257, Université Aix-Marseille, Marseille, France

[15]Department of Plant Pathology and Microbiology, Institute for Plant Genomics and Biotechnology, Borlaug Genomics and Bioinformatics Center, Texas A&M University, College Station, Texas, USA

[16]Department of Biology, Division of Biochemistry, Friedrich-Alexander-University ErlangenNuremberg, Erlangen, Germany

[17]Laboratory of Plant Pathology, Graduate School of Life and Environmental Sciences, Kyoto Prefectural University, Kyoto, Japan

[18]Institut National de la Recherche Agronomique, Biologie et Gestion des Risques en Agriculture—Champignons Pathogènes des Plantes, Thiverval-Grignon, France

[19]Department of Computer Science, University of Kentucky, Lexington, Kentucky, USA

[20]Central Microscopy, Max Planck Institute for Plant Breeding Research, Cologne, Germany

[21]Department of Plant Developmental Biology, Max Planck Institute for Plant Breeding Research, Cologne, Germany

[22]Division of Plant and Soil Sciences, West Virginia University, Morgantown, West Virginia, USA

[23]Institute for Biology I, Unit of Plant Molecular Cell Biology, Rheinisch-Westfälische Technische Hochschule Aachen University, Aachen, Germany

[24]Laboratory for Molecular and Computational Genomics, University of Wisconsin-Madison, Madison, Wisconsin, USA

[25]US Department of Agriculture, Agriculture Research Service, National Center for Agricultural Utilization Research, Peoria, Illinois, USA

[26]Max Planck Genome Centre Cologne, Cologne, Germany

[27]Department of Plant Pathology, University of Florida, Gainesville, Florida, USA

[28]Department of Plant Biology, Michigan State University, East Lansing, Michigan, USA

[29]Laboratory of Plant Pathology, Graduate School of Agriculture, Kyoto University, Kyoto, Japan

[30]Department of Genetics, Development and Cell Biology, Iowa State University, Ames, Iowa, USA (J.P.D.), Department of Bioproduction Science, Faculty of Bioresources and Environmental Sciences, Ishikawa Prefectural University, Ishikawa, Japan (H.T.), Department of Plant Pathology, Swammerdam Institute for Life Sciences, University of Amsterdam, Amsterdam, The Netherlands (H.C.v.d.D.) and The College of Natural Sciences, University of Massachusetts Amherst, Amherst, Massachusetts, USA (L.-J.M.)

Colletotrichum **species are fungal pathogens that devastate crop plants worldwide. Host infection involves the differentiation of specialized cell types that are associated with penetration, growth inside living host cells** (biotrophy) and tissue destruction (necrotrophy). We report here genome and transcriptome analyses of *Colletotrichum higginsianum* **infecting** *Arabidopsis thaliana* **and** *Colletotrichum graminicola* **infecting maize. Comparative genomics showed that both fungi have large sets of pathogenicity-related genes, but families of genes encoding secreted effectors, pectin-degrading enzymes, secondary metabolism enzymes, transporters and peptidases are expanded in** *C. higginsianum*. Genome-wide expression profiling revealed that these genes are transcribed in successive waves that are linked to pathogenic transitions: effectors and secondary metabolism enzymes are induced before penetration and during biotrophy, whereas most hydrolases and transporters are upregulated later, at the switch to necrotrophy. Our findings show that preinvasion perception of plant-derived signals substantially reprograms fungal gene expression and indicate previously unknown functions for particular fungal cell types.

The genus *Colletotrichum* (Sordariomycetes, Ascomycota; Fig. 1a) comprises ~600 species[1]attacking over 3,200 species of monocot and dicot plants. These pathogens use a multistage hemibiotrophic infection strategy[2]: dome-shaped appressoria first puncture host surfaces using a combination of mechanical force and enzymatic degradation, bulbous biotrophic hyphae enveloped by an intact host plasma membrane then develop inside living epidermal cells, and finally, the fungus switches to necrotrophy and differentiates thin, fast-growing hyphae that kill and destroy host tissues (Fig. 1b,c). We sequenced two *Colletotrichum* species with different host specificities and infection strategies: *C. higginsianum* attacks several members of *Brassicaceae*, including *Arabidopsis*, and has emerged as a tractable model for studying fungal pathogenicity and plant immune responses[3, 4, 5]. Biotrophy in this fungus is confined to the first invaded host cell and is followed by a complete switch to necrotrophy[5]. In contrast, *C. graminicola* primarily infects maize (*Zea mays*), causing annual losses of approximately 1 billion dollars in the United States alone[6]. In this species, biotrophy extends into many host cells and persists at the advancing colony margin while the center of the colony becomes necrotrophic[7].

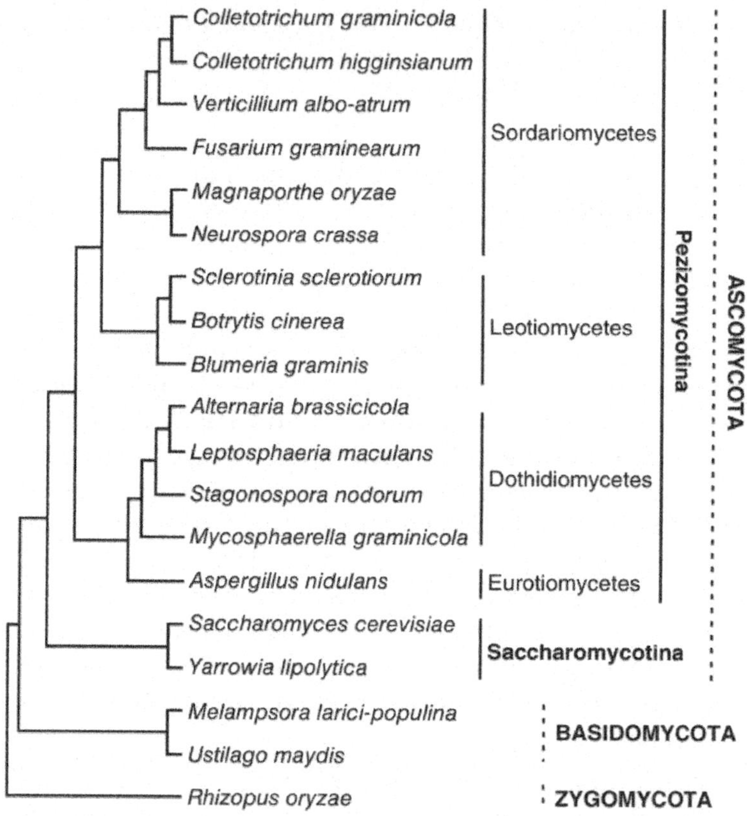

(a)

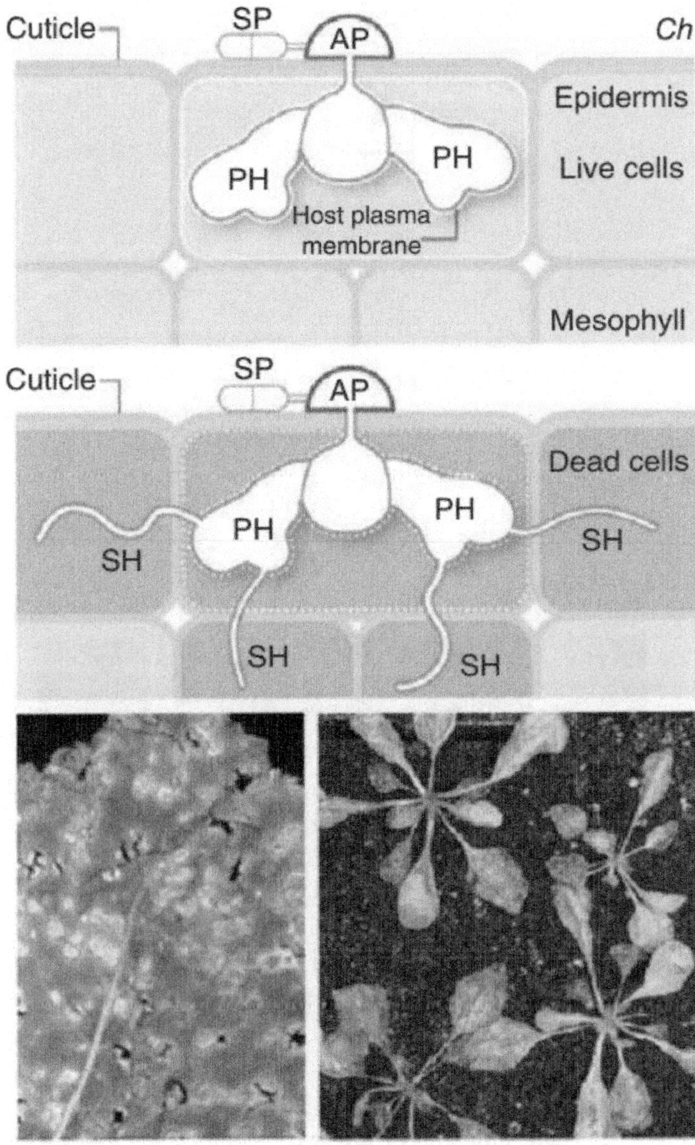
Cuticle
SP
AP
Ch
Epidermis
PH
PH
Live cells
Host plasma membrane
Mesophyll
Cuticle
SP
AP
Dead cells
PH
PH
SH
SH
SH
SH

(b)

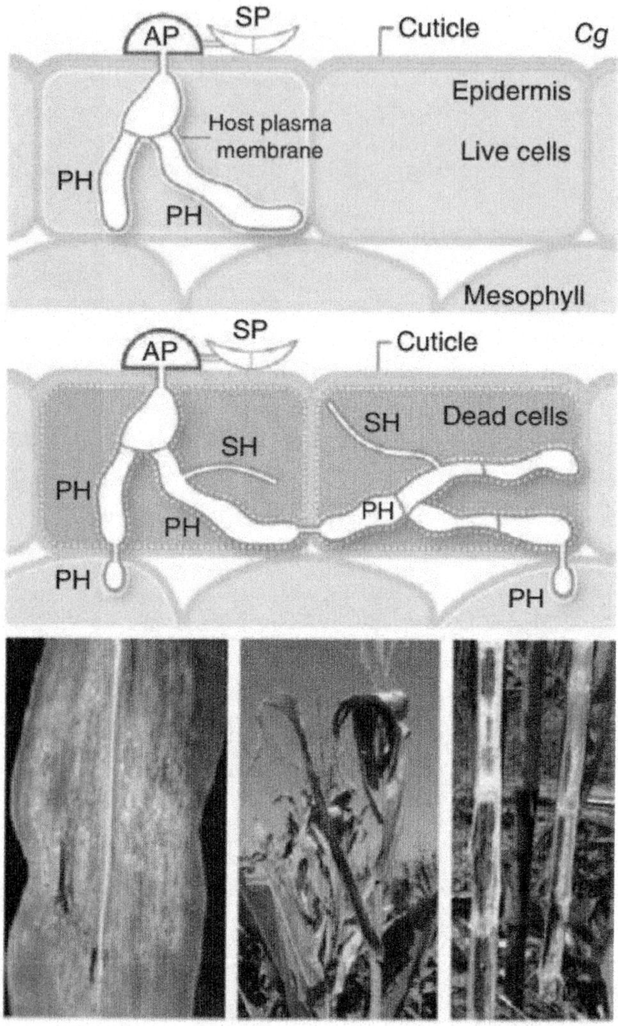

(c)

Figure 1: Phylogeny and infection of the two *Colletotrichum* **species analyzed in this study.** (a) Cladogram showing the phylogenetic relationship of *Colletotrichum* to other sequenced fungi, including 13 species used for comparative analyses (see Fig. 3). The unscaled tree was constructed using CVTree[34] with *Rhizopus oryzae* as the out-group. (b) Infection process of *C. higginsianum* (*Ch*) and leaf anthracnose symptoms on *Brassica* and *Arabidopsis*. The *Brassica* image is reproduced with permission of University of Georgia Plant Pathology Archive (bugwood.org/). (**c**) Infection process of *C. graminicola*(*Cg*), and leaf-blight, top die-back and stalk-rot symptoms on maize. SP, spore; AP, appressorium; PH, biotrophic primary hyphae; SH, necrotrophic secondary hyphae.

Optical mapping showed that the genomes of the two species are similar in size and structure. *C. graminicola* has a 57.4-Mb genome that is distributed among 13 chromosomes, including three minichromosomes less than 1 Mb in size, whereas *C. higginsianum* has a 53.4-Mb genome comprising 12 chromosomes, including two minichromosomes. We sequenced the *C. graminicola* genome using Sanger and 454 platforms, which provided a high-quality reference assembly of 50.9 Mb. We sequenced the *C. higginsianum* genome using 454 and Illumina platforms, yielding an assembly of 49.3 Mb. Repetitive DNA comprises 12.2% of the *C. graminicola* genome assembly and 1.2% of the *C. higginsianum* assembly. The repeats clustered in genomic regions with low GC content in *C. graminicola*, similar to the AT-rich isochores found in *Leptosphaeria maculans*[8]. Including unassembled genomic regions (mostly repeats, such as ribosomal DNA, telomeres, centromeres and transposons), repetitive DNA was estimated to total 22.3% of the *C. graminicola* genome and 9.1% of the *C. higginsianum* genome. The two*Colletotrichum* species diverged relatively recently (~47 million years ago), after the separation of monocots and dicots 140–150 million years ago[9].

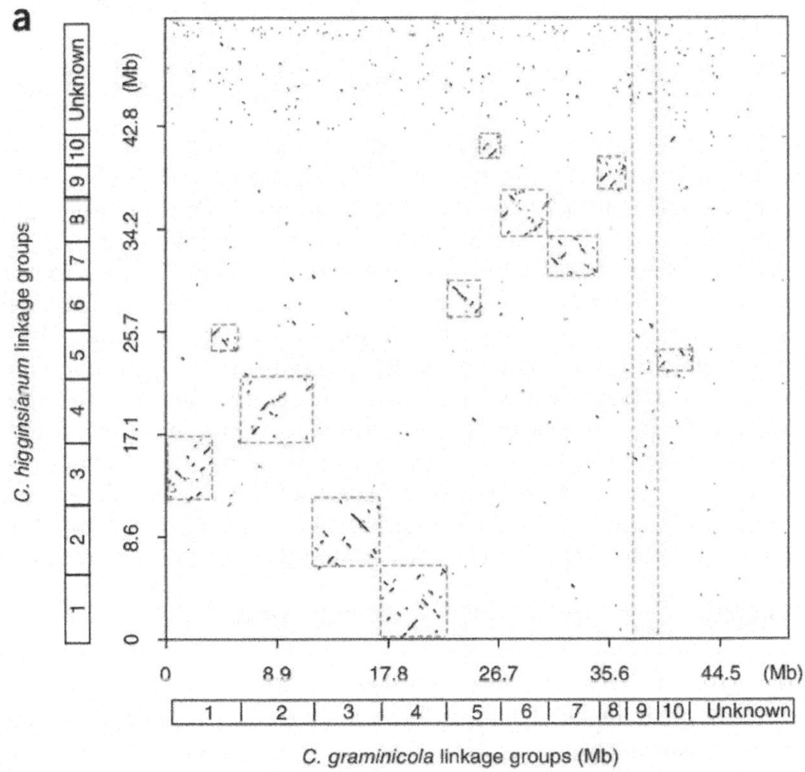

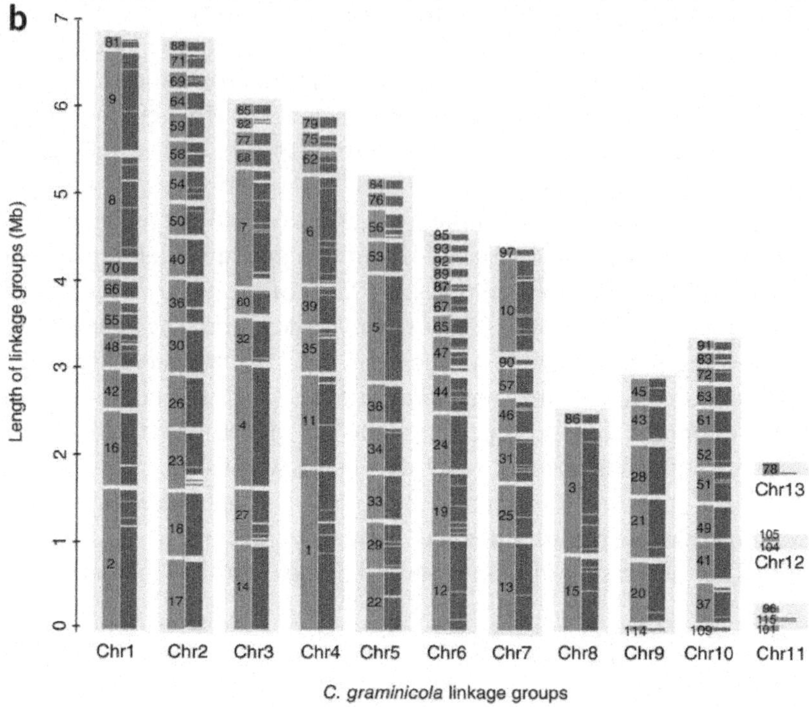

Figure 2: Conservation of synteny between the genomes of *C. graminicola* **and** *C. higginsianum*. (a) Dot plot showing the syntenic blocks between the 13 chromosomes (optical linkage groups) of *C. graminicola* (horizontal axis) and the 12 chromosomes of *C. higginsianum* (vertical axis). Homologies between chromosomes of each species are highlighted in red dashed boxes. Homologous sequences of*C. graminicola* chromosome 9, indicated between the blue dashed lines, are dispersed among many *C. higginsianum* chromosomes. (**b**) Global view of syntenic alignments between the genomes of *C. graminicola* and *C. higginsianum.* Linkage groups of *C. graminicola* are shown as the reference, with linkage group lengths defined by the *C. graminicola* optical map. For each chromosome, numbered genomic scaffolds (dark gray) positioned on the optical linkage groups are separated by scaffold breaks. The magenta blocks show syntenic mapping of the *C. higginsianum* sequences; notably, there is a near absence of homologous sequences among the minichromosomes.

Although *C. graminicola* and *C. higginsianum* belong to sister clades within the genus, only 35% of the two genomes are syntenic, which is less than the synteny between *Botrytis cinerea* and *Sclerotinia sclerotiorum*[10]. Nevertheless, an analysis of synteny between the two *Colletotrichum* genomes identified homologous chromosomes and revealed that major intrachromosomal rearrangements have occurred in one or both species (Fig.

2a). The minichromosomes do not contain homologous sequences (Fig. 2b), suggesting that they are lineage-specific innovations, and in *C. graminicola*, the minichromosomes are enriched with repetitive DNA (averaging 23%) compared to the core genome (averaging 5.5%).

We predicted the existence of 12,006 protein-coding genes in *C. graminicola* compared to 16,172 in *C. higginsianum* . Having been compiled from short-read data only, the *C. higginsianum* assembly is more fragmented than that of *C. graminicola*, resulting in some genes (5.2%) being split into two or more gene models, whereas others (4%) are truncated versions of the complete gene. After correcting for this fragmentation, the estimated gene content of *C. higginsianum* (15,331) is still markedly larger than that of *C. graminicola*. The two species share 9,795 orthologous genes. Using Markov clustering (MCL)[11]to analyze the proteomes, we found that 10,077 *C. higginsianum* genes belong to multicopy gene clusters, compared to 5,342 genes in *C. graminicola*, suggesting that the greater gene content of*C. higginsianum* results partly from gene duplication. The MCL analysis also revealed that gene clusters encoding serine proteases, methyl transferases, polyketide synthases, cytochrome P450 enzymes and small-molecule efflux pumps are expanded in *C. higginsianum* compared to *C. graminicola*, which we verified by manual inspection. Clusters that are expanded in *C. graminicola* relative to *C. higginsianum* include a family of genes encoding atypical cellulases (glycoside hydrolase GH61, described below) and another encoding secreted histidine acid phosphatases, which probably mobilize phytic acid, the main form of stored phosphorus in plants[12].

C. higginsianum and *C. graminicola* are particularly well equipped with genes encoding carbohydrate-active enzymes (CAZymes)[13] that potentially degrade the plant cell wall[14] (Fig. 3a) and modify the fungal cell wall. Both species encode more CAZymes than 13 other fungal genomes we examined. These expanded CAZyme arsenals are more similar to those of other hemibiotrophic and necrotrophic pathogens than to the highly reduced set found in biotrophs such as *Melampsora* and *Blumeria* (Fig. 3a). The exceptionally large and diverse inventory of CAZymes encoded by both *Colletotrichum* genomes provides a rich source of enzymes for potential commercial exploitation[15]. *C. higginsianum* encodes over twice as many pectin-degrading enzymes as does *C. graminicola* (Fig. 3b), the majority (62%) of which are activated during necrotrophy. Conversely, although both species encode similar numbers of cellulases and hemicellulases, *C. graminicola* activates many more of these genes during necrotrophy (48%) than does *C. higginsianum* (26%), including 22 GH61 copper-dependent oxygenases, which act in concert with classical cellulases to enhance lignocellulose hydrolysis[16, 17]. Thus, *C.*

graminicola and *C. higginsianum* use very different strategies to deconstruct plant cell walls, reflecting their host preferences: dicot cell walls are enriched with pectin (35% in dicots compared to 10% in maize), whereas the cell walls of grasses contain more hemicellulose (60% in grasses compared to 30% in dicots) and phenolics (up to 5%)[18].

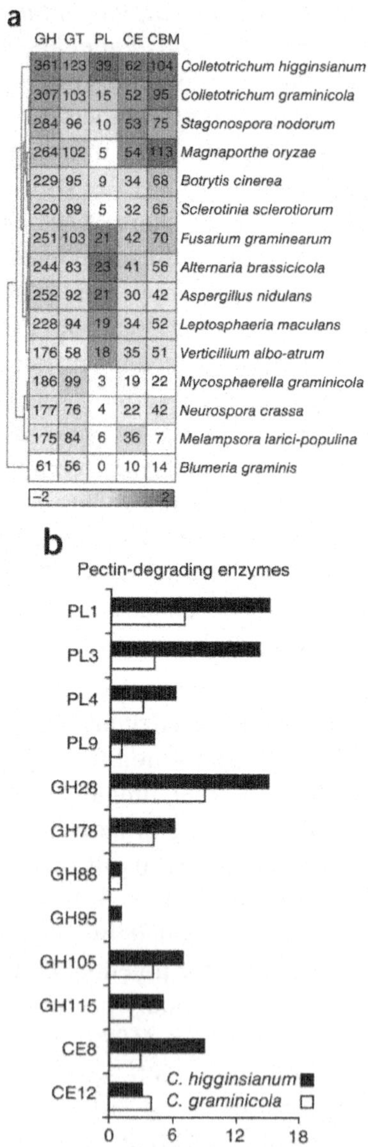

Figure 3: Comparison of fungal carbohydrate-active enzyme (CAZyme) repertoires. (a) Hierarchical clustering of CAZyme classes from *Colletotrichum* and 13 other

fungal genomes. GH, glycoside hydrolase; GT, glycosyltransferase; PL, polysaccharide lyase; CE, carbohydrate esterase; CBM, carbohydrate-binding module. The numbers of enzyme modules in each genome are shown. Overrepresented (orange to red) and underrepresented modules (pale yellow to white) are depicted as fold changes relative to the class mean. (**b**) Comparison of the pectin-degrading enzyme repertoires of *C. higginsianum* and *C. graminicola* shown as the number of modules in each CAZyme family[11]. In total, *C. higginsianum* encodes 86 such modules, whereas *C. graminicola* encodes only 42.

Many phytopathogens secrete proteins known as effectors that facilitate infection by reprogramming host cells and modulating plant immunity[19]. By defining candidate secreted effectors (CSEPs) as predicted extracellular proteins without any homology to proteins outside the genus *Colletotrichum*, we found 177 CSEP-encoding genes in *C. graminicola*, 85 (48%) of which were species specific. In contrast, *C. higginsianum* encodes twice as many CSEPs (365), including more species-specific proteins (264, or 72%). The CSEPs are mostly small proteins (averaging 110 residues and 175 residues in *C. higginsianum* and *C. graminicola*, respectively) and are more cysteine rich than the total proteome. CSEP-encoding genes are randomly distributed across the chromosomes of *C. graminicola*, with no evidence for clustering, enrichment on particular chromosomes or localization near transposable elements or telomeres, as has been reported for some other plant pathogens[8, 20,21, 22] . An MCL analysis revealed that relatively few *Colletotrichum* CSEPs (14% in both species) belong to small multigenic families with two to five members. The larger, more diversified CSEP repertoire of *C. higginsianum* might be an adaptation to invade a broader range of host plants than *C. graminicola*, which is restricted to infection of *Zea* under field conditions[23].

Both *Colletotrichum* species encode markedly more secondary metabolism enzymes (103 in *C. higginsianum* and 74 in *C. graminicola*) than other sequenced fungi (2–58 in ascomycetes[24, 25]). In fungi, secondary metabolism genes are typically located in clusters[26]; we found 42 of these clusters in *C. graminicola* and 39 in *C. higginsianum*, surpassing the numbers found in most other sequenced ascomycetes. Only 11 secondary metabolism gene clusters are shared between the two *Colletotrichum* species, and only 6 of these clusters show limited synteny (Fig. 4). This cluster diversity seems to result from gene duplication or loss and chromosomal rearrangements and may be related to the association of secondary metabolism gene clusters (71% in *C. graminicola*) with repetitive DNA. Because each secondary metabolism gene cluster is probably involved in the biosynthesis of a specific metabolite[24], each *Colletotrichum* species can be expected to produce unusually large and divergent spectra of secondary metabolites, some of which may be previously unknown bioactive molecules.

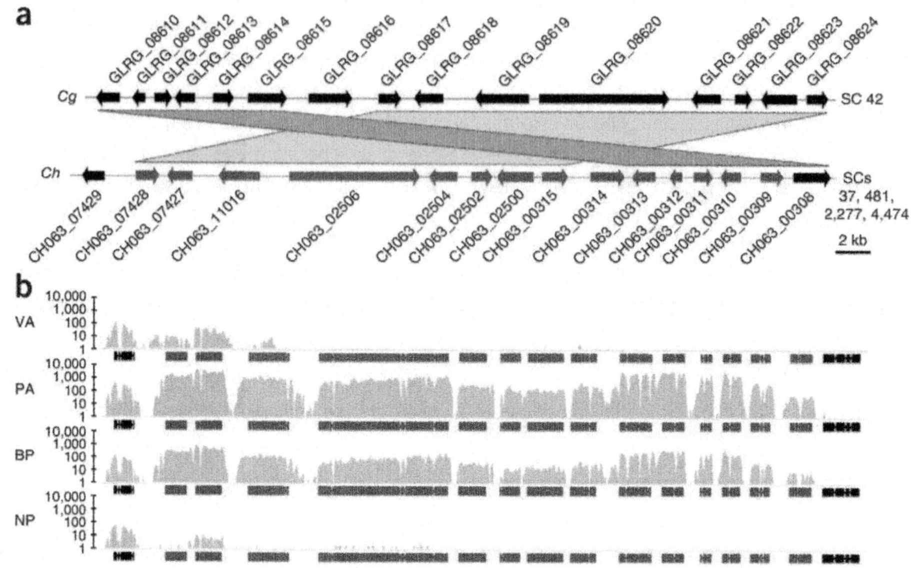

Figure 4: Structure and transcription of a secondary metabolism gene cluster. (a) Gene cluster 18 from *C. graminicola* (*Cg*) is orthologous to cluster 10 from *C. higginsianum* (*Ch*). The latter is split between four small scaffolds in the Broad Institute genome annotation (supercontigs (SCs) 37, 481, 2,277 and 4,474) and was reconstructed based on an improved genome assembly. Microsynteny is indicated by gray bars. The 14 genes highlighted in red in the *C. higginsianum* cluster are co-regulated. (**b**) Visualization of RNA-Seq coverage across the *C. higginsianum* polyketide biosynthesis cluster. The gray curves indicate read coverage (log scale) for the four samples. Co-regulated gene models are highlighted in red. VA, *in vitro* appressoria; PA, *in planta* appressoria; BP, biotrophic phase; NP, necrotrophic phase.

To investigate how the fungal genetic program is deployed during host infection, we applied Illumina RNA sequencing to both pathosystems. We collected samples from infected *Arabidopsis* or maize leaves at intervals corresponding to pre-penetration appressoria, the early biotrophic phase and the transition to necrotrophy and from *C. higginsianum* appressoria formed *in vitro* (Fig. 5a). Almost all the gene models were transcribed *in planta* (14,972 *C. higginsianum* genes, or 92%, and 10,812 *C. graminicola* genes, or 90%). However, this transcription was highly dynamic, particularly in *C. higginsianum*, where 7,162 genes (44%) were differentially regulated (log2 fold change >2, $P < 0.05$) between one or more of the infection stages. Fewer genes (2,619, or 22%) were differentially regulated in *C. graminicola*, which may reflect the contrasting biology of this species, where biotrophic and necrotrophic growth occur simultaneously. The more clearly defined infection stages of *C. higginsianum* provided better temporal and spatial resolution of expression changes, and we therefore highlight our results for this species.

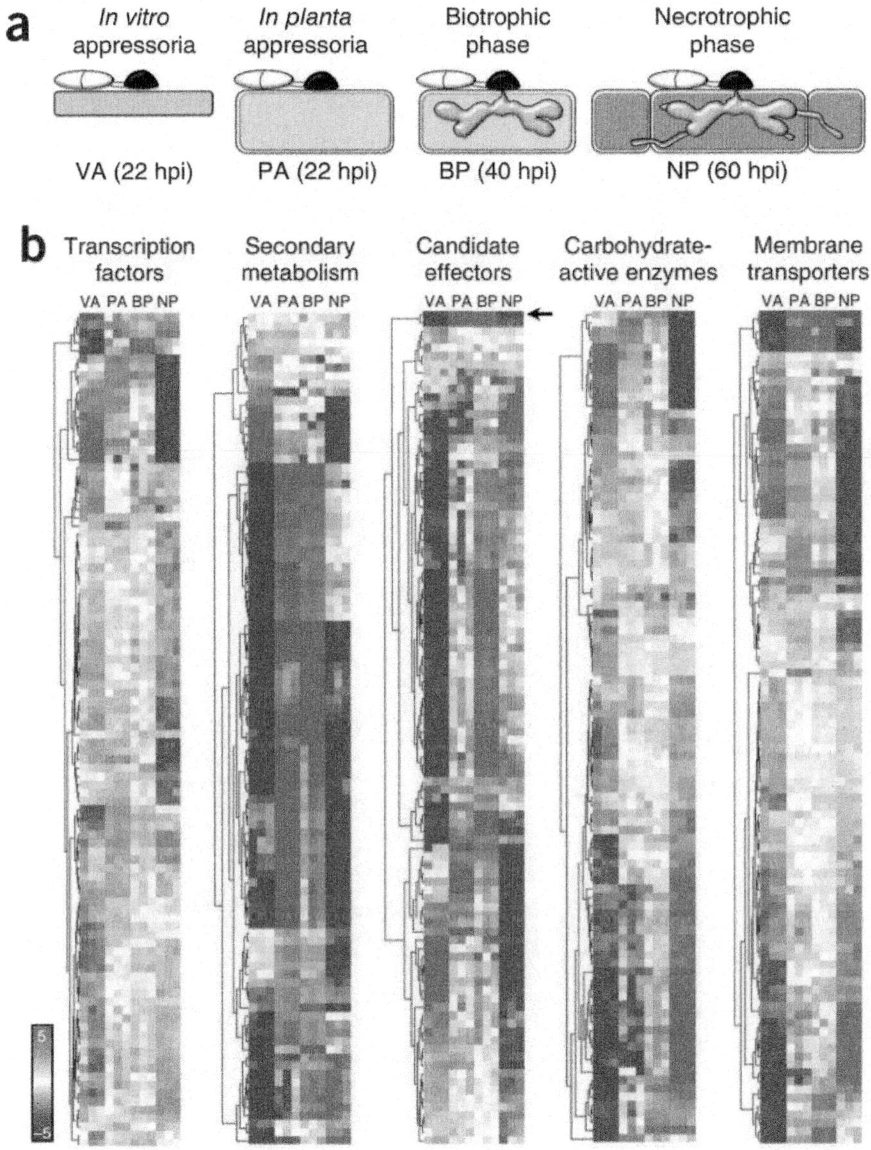

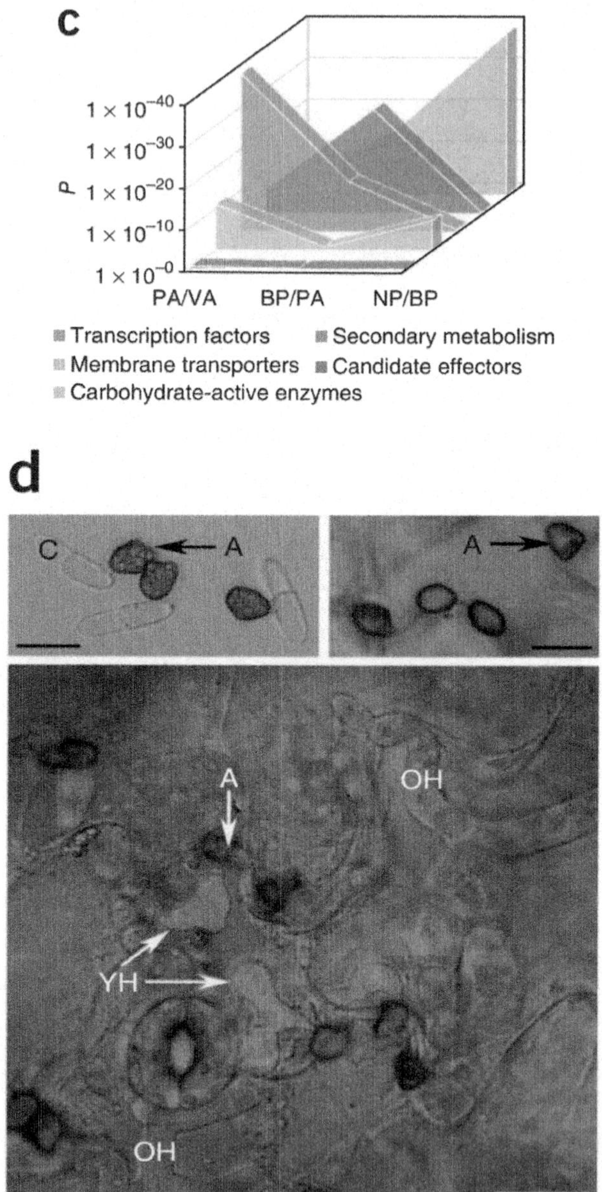

Figure 5: Expression profiling of pathogenicity-related genes in *C. higginsianum*. (**a**) Schematic representation of the four *C. higginsianum* developmental stages selected for RNA sequencing. Gray indicates polystyrene, green indicates living plant cell, and brown indicates dead plant cell. Hpi, hours post-inoculation. (**b**) Heatmaps of gene expression showing the 100 most highly expressed and significantly regulated genes (log2 fold change >2, $P < 0.05$) in five functional categories. Overrepresented (pale

red to dark red) and underrepresented transcripts (pale blue to dark blue) are shown as log2 fold changes relative to the mean expression measured across all four stages. The arrow indicates the CSEP-encoding gene *ChEC6* (CH063_01084). (**c**) The statistical significance of gene induction (*y* axis) in five functional categories during fungal developmental transitions (*x* axis). The *P*values were calculated using a one-sided Fisher's exact test and represent the probability of observing the number of significantly induced genes for a specific category during a transition given the total number of significantly induced genes during that transition (log2 fold change >2, $P < 0.05$) and the total number of genes in the category. (**d**) Transcriptional regulation of the effector gene *ChEC6* by plant-derived signals. Confocal micrographs showing *C. higginsianum* expressing the *mCherry* reporter gene under the native*ChEC6* promoter (overlays of bright-field and fluorescence channels). Appressoria (A) formed on polystyrene are unlabeled (top left), whereas those on the leaf surface (top right) have fluorescent cytoplasm. After host penetration, labeling is visible in young biotrophic hyphae (YH) but not older biotrophic hyphae (OH) (bottom). Scale bars, 10 µm. C, conidium.

Five gene categories relevant to pathogenicity (encoding transcription factors, secondary metabolism enzymes, CSEPs, CAZymes and transporters) had markedly different expression patterns during infection (Fig. 5b). We distinguished three waves of gene activation corresponding to pathogenic transitions (Fig. 5c). Among the genes upregulated at the appressorial phase were those encoding CAZymes that are predicted to degrade cutin, cellulose, hemicellulose and pectin, which may contribute to initial host penetration, together with a larger set of enzymes that potentially remodel the fungal cell wall (Fig. 5b). However, early during infection, the transcriptome of *C. higginsianum* was dominated by secondary metabolism genes, with 12 different secondary metabolism gene clusters being induced before penetration and during biotrophy (Fig. 5b,c). This indicates previously unsuspected roles for appressoria and biotrophic hyphae in synthesizing an array of small molecules for delivery to the first infected plant cells. Because these cells initially remain alive, such molecules are probably not toxins and may instead function in host manipulation, similar to protein effectors[27]. Remarkably, the *C. higginsianum* secondary metabolism gene cluster with the strongest activation at this stage was silent in *C. graminicola* at all the infection stages we examined (Fig. 4), suggesting that additional metabolite diversity is generated through transcriptional regulation.

Different sets of CSEP-encoding genes were expressed at each infection stage, but the majority of these genes were strongly induced during biotrophy (Fig. 5b,c). This suggests *Colletotrichum* requires a maximum capacity for host manipulation during intracellular colonization and that biotrophic hyphae provide a major interface for effector delivery to host cells. These specialized hyphae morphologically resemble the haustoria of obligate biotrophs, which function both as platforms for effector secretion and feeding structures for the uptake of sugars and amino acids[28, 29]. However, we found no evidence for specific transcriptional reprogramming of nutrient transporters in *C. higginsianum* during biotrophy (Fig. 5b,c), suggesting that the biotrophic hyphae of this pathogen function primarily to deliver protein effectors and secondary metabolites to the plant cell.

Transcripts encoding a vast array of lytic enzymes are induced at the transition to necrotrophy, when the pathogen uses dead and dying host cells as a nutrient source to support rapid colonization and sporulation (Fig. 5b,c). These enzymes include 44 putative secreted proteases and 146 CAZymes that potentially cleave all major polysaccharides in the host wall. Concomitantly, numerous genes encoding plasma membrane transporters that may be required for assimilating the products of this degradative activity, for example, oligopeptides, amino acids and sugars, are also induced (Fig. 5b,c). In fungi, genes encoding secreted proteases, CAZymes and permeases are often subject to pH regulation[30]. Consistent with this, we found evidence that necrotrophy in *C. higginsianum* is associated with local alkalinization of*Arabidopsis* tissue, probably resulting from fungal ammonia secretion[31], but tissue alkalinization was less pronounced in maize colonized by *C. graminicola*at this stage.

Notably, although appressoria *in vitro* are morphologically indistinguishable from those *in planta*, their transcriptomes are substantially different, with 1,515 genes significantly induced by host contact. One of these, the CSEP-encoding gene called *ChEC6*(CH063_01084)[32], is the most highly and significantly induced of all *C. higginsianum* genes (>50,000-fold) compared to appressoria *in vitro*. To experimentally verify this expression pattern at the cellular level, we generated transgenic *C. higginsianum* strains expressing a reporter gene under the control of the *ChEC6* promoter (Fig. 5d). Using this method, we confirmed that the transcription of *ChEC6* was plant specific, starting in the appressorium before penetration and continuing in young biotrophic hyphae, but it was switched off before the hyphae were fully expanded, indicating that its expression is transient and tightly regulated. The large-scale reprogramming of appressorial gene expression *in planta* shows that these specialized cells are highly responsive to host-derived cues that are perceived before penetration. Long regarded as organs of attachment and penetration[33], our findings assign a previously unsuspected sensory function to fungal appressoria, enabling the pathogen to prepare for the subsequent invasion of living host cells.

Major hemibiotrophic plant pathogens such as *Colletotrichum* and the rice blast fungus*Magnaporthe oryzae* undergo major transformations in cell morphology and infection mode when switching from growth on the plant surface to intracellular biotrophy and from biotrophy to necrotrophy. Genome sequencing combined with high-throughput transcriptome sequencing revealed the transcriptional dynamics underlying these transitions and led us to redefine the functions of appressoria and intracellular hyphae. Despite their similar morphologies, a genomic comparison of *C. higginsianum* and *C. graminicola* uncovered major differences in their gene content. We propose that the diversification of functions required for host interaction, notably, the secretion of small-molecule and protein effectors and the degradation of plant polymers, allows *C. higginsianum* to colonize a wider range of plant species. In contrast, *C. graminicola*, a pathogen that is adapted to a narrow range of hosts, has maintained a more targeted arsenal of virulence factors.

METHODS

Sequencing and Assembly

C. graminicola strain M1.001 (M2) was collected in Missouri from infected maize (Fungal Genetics Stock Center culture 10212). *C. higginsianum* strain IMI349063 was isolated from *Brassica campestris* in Trinidad and Tobago (CABI culture collection, Wallingford, UK). The genome assemblies of *C. graminicola* were generated at the Broad Institute by combining data from Sanger and 454 pyrosequencing using a Newbler hybrid approach. Paired-end reads from 468,734 plasmids and 67,151 fosmids improved the continuity of the assembly. In the assembled genome, more than 98.5% of the sequence bases had quality scores >40. The *C. higginsianum* genome assembly was generated by GATC Biotech AG (Konstanz, Germany) by combining 454 GS-FLX shotgun reads and Illumina GAII mate-pair reads. Additionally, 864 fosmids were end-sequenced with Sanger technology. After removing dinucleotide repeats, the 454 reads and the fosmid end sequences were coassembled using the SeqMan NGen assembler (DNAStar Inc., USA). Contigs were then sorted into scaffolds using the paired-end information derived from an Illumina 3-kb–insert mate-pair library (2 × 36 bp reads). Scaffolds were manually edited to correct falsely joined contigs and falsely arranged scaffolds. To correct homopolymer sequencing errors in the 454 data, the Illumina GA data (76-fold coverage) were mapped to the scaffolded contigs, and the depth of coverage was used to create a final corrected consensus sequence.

Gene Annotation

A total of 28,424 expressed sequence tags (ESTs) from two *C. graminicola* complementary DNA libraries and 828,592 ESTs from six *C. higginsianum* libraries were used to provide a training set for the gene-calling pipeline and for validating the gene models. Protein-coding genes were annotated in *C. graminicola* using multiple lines of evidence from BLAST, PFAM searches and EST alignments, as described previously[20]. Gene structures were predicted using the Broad Institute automated gene-calling pipeline[35] based on a combination of gene models predicted by the programs FGENESH (Softberry Inc., USA), GENEID[36], GeneMark[37], SNAP[38] and Augustus[39] together with EST-based and manually curated gene models. GENEID, FGENESH, SNAP and Augustus were trained using a set of high-confidence EST-based gene models generated by clustering Blat-aligned species-specific ESTs. By combining BLAST, EST and *ab initio* predictions, annotators manually built additional gene models that were otherwise missed by the automated annotation. *C. graminicola* was predicted to have 12,006 gene models, 39% of which were verified by the alignment of 13,600 Sanger EST reads. The *C. higginsianum*gene set was created similarly and was filtered using TBLASTN alignments from 10,661 of the *C. graminicola* gene models ($<1 \times 10^{-10}$). Another 1,564 gene models were based on evidence from*C. higginsianum* ESTs, and 600 were based on EVidenceModeler (EVM) models having BLAST hits to proteins in the UniRef90 database. *C. higginsianum* was predicted to have 16,172 protein-coding genes, 89% of which were validated by the alignment of 135,923 ESTs from 454 sequencing.

Optical Mapping

C. graminicola and *C. higginsianum* protoplasts[40] were lysed and prepared for optical mapping[41]using *Mlu*I (with an average fragment size for both genomes of 9.2 kb). Raw datasets comprising single DNA molecule maps (Rmaps; 300× coverage per genome) were assembled into genome-wide contig maps spanning each chromosome using divide-and-conquer[41] and iterative assembly strategies[42]. PROmer from the MUMmer package[43] was used to conduct pairwise comparisons between the *C. graminicola* and *C. higginsianum* genomes (Fig. 2a). The synteny map (Fig. 2b) was generated using the Argo browser[44].

Transposable Element Analysis

Repetitive DNA elements were identified by performing a self BLASTN of each genome and processing the output with a custom Perl script (available on request), which identified multicopy sequences and organized them into nonredundant families. Consensus sequences of these families were then used to generate a custom library for RepeatMasker to scan both genome assemblies. The distributions of the genes, the transposable elements and the GC content were examined within a 100-kb window, sliding 10 kb across each chromosome.

Orthology and Multigene Families

To identify differences in gene family size between *C. graminicola* and *C. higginsianum*, we clustered their proteomes using the Markov clustering program MCL[11]. An all-versus-all BLASTP search was performed using default parameters, followed by clustering with MCL using an inflation value of 2.0. We also included the proteomes of 13 additional fungal species (Fig. 3). Sequences were aligned using MAFFT[45], and phylogenetic trees were constructed using the neighbor-joining method, followed by a bootstrap test with 100 replications. Sequence editing and alignment and phylogenetic analyses were performed using Geneious Pro (version 5.5; see URLs).

Annotation of Specific Gene Categories

Secretomes of both *Colletotrichum* species were predicted using WoLF-PSORT[46]. CSEPs were defined as extracellular proteins with no significant BLAST homology (expect value $<1 \times 10^{-3}$) to sequences in the UniProt database (SwissProt and TrEMBL components). Homologs of proteins from outside the genus *Colletotrichum* were excluded. Genes encoding putative carbohydrate-active enzymes were identified using the CAZy annotation pipeline[13]. To identify secreted peptidase genes, sequences of predicted extracellular proteins were subjected to a MEROPS Batch BLAST analysis[47]. Membrane transporters were identified from BLAST searches against the Transporter Collection Database (see URLs) and *Saccharomyces* Genome Database (see URLs). Secondary metabolism genes were initially identified using MCL and gene family searches using the Broad Institute *Colletotrichum* database, BLAST searches against GenBank and InterproScan analysis (see URLs). The Secondary Metabolite Unknown Region Finder (SMURF)[26] was used to predict secondary metabolism gene clusters. SMURF was applied to the Velvet assembly of *C. higginsianum* . Candidate genes identified using automated searches were inspected manually, including protein sequence alignments to known enzymes and searches against the NCBI Conserved Domain Database. The Fungal Cytochrome P450 Database (see URLs) and Fungal Transcription Factor Database were used to annotate cytochrome P450 enzymes and transcription factors, respectively.

Whole-Genome Transcriptome Profiling

Arabidopsis leaves infected by *C. higginsianum* were obtained as described previously[48]. Sampling and RNA isolation of the pre-penetration stage (22 h hpi), the early biotrophic stage (40 hpi), the switch between biotrophy and necrotrophy (60 hpi) and *in vitro* appressoria (22 hpi) have been described previously[32, 49]. Each experimental repetition of the *in planta* stages was based on RNA extracted from ~300 leaves. Maize leaf sheaths infected by *C. graminicola* were obtained as described previously[50]. Sheaths from the maize inbred line Mo940 at the V3 stage were cut into 5-cm–long segments and inoculated with two 10-μl drops of spore suspension (5×10^5spores per ml). Sheaths containing mature pre-penetration appressoria (24 hpi), intracellular biotrophic hyphae (36 hpi) and necrotrophic hyphae with water-soaked lesions (60 hpi) were sampled. Each leaf sheath was trimmed to include only the inoculated area, and total RNA was extracted as described previously[51] (15 maize sheaths per experimental repetition). The RNA integrity of all samples was verified on an Agilent 2100 Bioanalyzer.

Twelve *C. higginsianum* libraries (four developmental stages and three biological replicates) and nine *C. graminicola* libraries (three developmental stages and three biological replicates) were prepared with the Illumina TruSeq RNA Sample Preparation Kit and sequenced using the Illumina Genome Analyzer IIx (single reads, 100 bp for *C. higginsianum* and 76 bp for *C. graminicola*). The RNA-Seq reads were mapped to the annotated genomes with TopHat ($a = 10$, $g = 5$)[52] and transformed into counts per annotated gene per sample with the 'coverageBed› function from the BEDtools suite[53] and custom R scripts. Differentially expressed genes between two developmental stages were detected using the 'exactTest› function from the R package EdgeR[54]. To calculate fold changes, the number of reads for each gene in each library was normalized by the total number of mapped reads for the library, and direct ratios (log2) were calculated between the different developmental stages. Transcripts with a significant P value (<0.05) and more than a twofold change (log2) in transcript level were considered to be differentially expressed. All P values were corrected for false discoveries resulting from multiple hypothesis testing using the Benjamini-Hochberg procedure. Heatmaps of gene expression profiles were generated with the Genesis expression analysis package[55]. All codes for the RNA-Seq processing are available upon request. The *C. higginsianum* RNA-Seq data were also mapped onto the unannotated Velvet genome assembly using bowtie[56] and visualized with SAMtools (see URLs) and the IGV browser. RNA-Seq expression profiles were validated by quantitative RT-PCR.

Molecular Phylogeny and Evolutionary Divergence Date Estimation

A whole-genome cladogram showing the phylogenetic relationships of *C. graminicola* and *C. higginsianum* to 17 other sequenced fungi was constructed with CVTree[34] (Fig. 1). A phylogeny was generated for the genus *Colletotrichum* based on sequencing five genes in 28 selected isolates. To estimate the evolutionary divergence date for *C. graminicola* and *C. higginsianum*, a phylogenetic analysis was performed using the 13 species. The proteomes were clustered using MCL, and proteins in each cluster were aligned using MUSCLE. Sixty-four clusters containing only one protein from each species and having at least 80% average pairwise nucleotide identity were used for further analyses. Sequence alignments were concatenated, and a phylogenetic tree was constructed with MrBayes[57] using the WAG amino acid substitution model. Date estimates were computed using the program r8s[58] with the nonparametric rate smoothing (NPRS) method using date estimates by Lücking *et al.*[59].

Fluorescent Reporter Gene Assay

The promoter of the CSEP-encoding gene *ChEC6* (CH063_01084) was fused to *mCherry*[60] and a transcriptional terminator by overlap fusion PCR[61] using the primer pairs shown. The genomic region between the *ChEC6* start codon and the stop codon of its upstream gene (1,198 bp) was amplified with primer pair 1. The *mCherry* gene was amplified with primer pair 2. The transcriptional terminator of *Aspergillus nidulans trpC* was amplified from the plasmid pBin-GFP-hph[5] with primer pair 3. After fusion, the insert was subcloned into the plasmid pENTR/D-TOPO (Invitrogen) and verified by sequencing. The insert was cut out with BamHI and EcoRI and ligated into the plasmid pBIGDR1, providing direct repeat recombination-mediated gene targeting[62]. A *ku70* mutant of *C. higginsianum* strain IMI349063 (ref. 62) was used for *Agrobacterium*-mediated transformation[3]. Confocal images of transformants were obtained using a Leica TCS SP2 confocal laser scanning microscope. Excitation for imaging mCherry fluorescence was at 563 nm, and emission was detected at 566–620 nm.

Host Tissue Alkalinization

The pH of the host cells during infection was measured using the cell-permeant pH-sensitive dye 2′,7′-bis(carboxyethyl)-5(6)-carboxyfluorescein (BCECF) for analysis by epifluorescence microscopy[31]. Fluorescence intensity values were correlated with direct pH determinations obtained with a piercing-tip pH electrode (Eutech, Singapore). Ammonia concentrations in infected maize and *Arabidopsis* leaf tissues were measured using a photometric ammonium assay kit (Merck, Germany).

ACCESSION CODES

The *C. graminicola* and *C. higginsianum* genome assemblies have been deposited in NCBI's Whole-Genome Shotgun Project with accession numbers ACOD0100000000 and CACQ0200000000, respectively. The RNA-Seq data for *C. graminicola* and *C. higginsianum*have been deposited in the NCBI Gene Expression Omnibus under GEO Series accession numbers GSE34632 and GSE33683, respectively.

Primary Accessions

Gene Expression Omnibus

- GSE34632
- GSE33683

ACKNOWLEDGMENTS

This manuscript is dedicated to the memory of Robert Hanau. This work was primarily supported by US Department of Agriculture (USDA)–Cooperative State Research, Education and Extension Service (CSREES) grant 2007-35600-17829 (L.J.V., M.B.D., S.R. and M.R.T.), the Max Planck Society (R.J.O.) and Deutsche Forschungsgemeinschaft (SPP1212) grant OC 104/1-3 (R.J.O.). Other funding sources included USDA-CSREES grant 2009-34457-20125 (L.J.V.), the University of Kentucky College of Agriculture Research Office (L.J.V.), Ministerio de Ciencia e Innovación (MICINN) of Spain grants AGL2008-03177/AGR and AGL2011-29446/AGR (M.R.T.) and the Programme for Promotion of Basic and Applied Research for Innovations in Bio-oriented Industry (P.G. and K. Shirasu). This is manuscript number 12-12-011, published with the approval of the Director of the University of Kentucky Agricultural Experiment Station.

REFERENCES

1. Crous, P.W., Gams, W., Stalpers, J.A., Robert, V. & Stegehuis, G. MycoBank: an online initiative to launch mycology into the 21st century. *Stud. Mycol.* **50**, 19–22 (2004).
2. Perfect, S.E., Hughes, H.B., O'Connell, R.J. & Green, J.R. *Colletotrichum*—a model genus for studies on pathology and fungal-plant interactions. *Fungal Genet. Biol.* **27**, 186–198(1999).
3. Huser, A., Takahara, H., Schmalenbach, W. & O'Connell, R. Discovery of pathogenicity genes in the crucifer anthracnose fungus *Colletotrichum higginsianum*, using random insertional mutagenesis. *Mol. Plant Microbe*

Interact. **22**, 143–156 (2009).

4. Narusaka, M. *et al. RRS1* and *RPS4* provide a dual resistance-gene system against fungal and bacterial pathogens. *Plant J.* **60**, 218–226 (2009).
5. O'Connell, R. *et al.* A novel *Arabidopsis-Colletotrichum* pathosystem for the molecular dissection of plant-fungal interactions. *Mol. Plant Microbe Interact.* **17**, 272–282 (2004).
6. Frey, T.J., Weldekidan, T., Colbert, T., Wolters, P.J.C.C. & Hawk, J.A. Fitness evaluation of*Rcg1*, a locus that confers resistance to *Colletotrichum graminicola* (Ces.) G.W. Wils. Using Near-Isogenic Maize Hybrids. *Crop Sci.* **51**, 1551–1563 (2011).
7. Mims, C.W. & Vaillancourt, L.J. Ultrastructural characterization of infection and colonization of maize leaves by *Colletotrichum graminicola*, and by a *C. graminicola* pathogenicity mutant. *Phytopathology* **92**, 803–812 (2002).
8. Rouxel, T. *et al.* Effector diversification within compartments of the *Leptosphaeria maculans*genome affected by repeat-induced point mutations. *Nat. Commun.* **2**, 202 (2011).
9. Chaw, S.M., Chang, C.C., Chen, H.L. & Li, W.H. Dating the monocot-dicot divergence and the origin of core Eudicots using whole chloroplast genomes. *J. Mol. Evol.* **58**, 424–441(2004).
10. Amselem, J. *et al.* Genomic analysis of the necrotrophic fungal pathogens *Sclerotinia sclerotiorum* and *Botrytis cinerea*. *PLoS Genet.* **7**, e1002230 (2011).
11. Enright, A.J., Van Dongen, S. & Ouzounis, C.A. An efficient algorithm for large-scale detection of protein families. *Nucleic Acids Res.* **30**, 1575–1584 (2002).
12. Oh, B.C., Choi, W.C., Park, S., Kim, Y.O. & Oh, T.K. Biochemical properties and substrate specificities of alkaline and histidine acid phytases. *Appl. Microbiol. Biotechnol.* **63**,362–372 (2004).
13. Cantarel, B.L. *et al.* The Carbohydrate-Active EnZymes database (CAZy): an expert resource for glycogenomics. *Nucleic Acids Res.* **37**, D233–D238 (2009).
14. van den Brink, J. & de Vries, R.P. Fungal enzyme sets for plant polysaccharide degradation. *Appl. Microbiol. Biotechnol.* **91**, 1477–1492 (2011).
15. King, B.C. *et al.* Arsenal of plant cell wall degrading enzymes reflects host preference among plant pathogenic fungi. *Biotechnol. Biofuels* **4**, 4 (2011).

16. Quinlan, R.J. *et al.* Insights into the oxidative degradation of cellulose by a copper metalloenzyme that exploits biomass components. *Proc. Natl. Acad. Sci. USA* **108**,15079–15084 (2011).
17. Beeson, W.T., Phillips, C.M., Cate, J.H. & Marletta, M.A. Oxidative cleavage of cellulose by fungal copper-dependent polysaccharide monooxygenases. *J. Am. Chem. Soc.* **134**,890–892 (2012).
18. Vogel, J. Unique aspects of the grass cell wall. *Curr. Opin. Plant Biol.* **11**, 301–307 (2008).
19. Stergiopoulos, I. & de Wit, P.J.G.M. Fungal effector proteins. *Annu. Rev. Phytopathol.* **47**,233–263 (2009).
20. Ma, L.-J. *et al.* Comparative genomics reveals mobile pathogenicity chromosomes in*Fusarium*. *Nature* **464**, 367–373 (2010).
21. Farman, M.L. Telomeres in the rice blast fungus *Magnaporthe oryzae*: the world of the end as we know it. *FEMS Microbiol. Lett.* **273**, 125–132 (2007).
22. Kämper, J. *et al.* Insights from the genome of the biotrophic fungal plant pathogen *Ustilago maydis*. *Nature* **444**, 97–101 (2006).
23. Crouch, J.A. & Beirn, L.A. Anthracnose of cereals and grasses. *Fungal Divers.* **39**, 19–44(2009).
24. Collemare, J., Billard, A., Bohnert, H.U. & Lebrun, M.H. Biosynthesis of secondary metabolites in the rice blast fungus *Magnaporthe grisea*: the role of hybrid PKS-NRPS in pathogenicity. *Mycol. Res.* **112**, 207–215 (2008).
25. Spanu, P.D. *et al.* Genome expansion and gene loss in powdery mildew fungi reveal functional trade-offs in extreme parasitism. *Science* **330**, 1543–1546 (2010).
26. Khaldi, N. *et al.* SMURF: genomic mapping of fungal secondary metabolite clusters. *Fungal Genet. Biol.* **47**, 736–741 (2010).
27. Collemare, J. & Lebrun, M.-H. Fungal secondary metabolites: ancient toxins and novel effectors in plant-microbe interactions. in *Effectors in Plant-Microbe Interactions* (eds. Martin, F. & Kamoun, S.) 379–402 (Wiley-Blackwell, Oxford, 2011).
28. Voegele, R.T. & Mendgen, K.W. Nutrient uptake in rust fungi: how sweet is parasitic life?*Euphytica* **179**, 41–55 (2011).
29. Catanzariti, A.-M., Dodds, P.N., Lawrence, G.J., Ayliffe, M.A. & Ellis, J.G. Haustorially expressed secreted proteins from flax rust are highly enriched for avirulence elicitors. *Plant Cell* **18**, 243–256 (2006).
30. Peñalva, M.A. & Arst, H.N. Regulation of gene expression by ambient

pH in filamentous fungi and yeasts. *Microbiol. Mol. Biol. Rev.* **66**, 426–446 (2002).

31. Alkan, N., Fluhr, R., Sherman, A. & Prusky, D. Role of ammonia secretion and pH modulation on pathogenicity of *Colletotrichum coccodes* on tomato fruit. *Mol. Plant Microbe Interact.* **21**, 1058–1066 (2008).
32. Kleemann, J. *et al.* Sequential delivery of host-induced virulence effectors by appressoria and intracellular hyphae of the phytopathogen. *Colletotrichum higginsianum. PLoS Pathog.* **8**, e1002643 (2012).
33. Frank, B. Uebereinige neue und weniger bekannte Pflanzenkrankheiten. *Ber. Deutsch. Bot. Gesell.* **1**, 29–34 (1883).
34. Wang, H., Xu, Z., Gao, L. & Hao, B. A fungal phylogeny based on 82 complete genomes using the composition vector method. *BMC Evol. Biol.* **9**, 195 (2009).
35. Stanke, M., Steinkamp, R., Waack, S. & Morgenstern, B. AUGUSTUS: a web server for gene finding in eukaryotes. *Nucleic Acids Res.* **32**, W309–W312 (2004).
36. Haas, B.J., Zeng, Q., Pearson, M.D., Cuomo, C.A. & Wortman, J.R. Approaches to fungal genome annotation. *Mycology* **2**, 118–141 (2011).
37. Parra, G., Blanco, E. & Guigó, R. GeneID in *Drosophila*. *Genome Res.* **10**, 511–515 (2000).
38. Borodovsky, M. & McIninch, J. GeneMark: parallel gene recognition for both DNA strands. *Comput. Chem.* **17**, 123–133 (1993).
39. Korf, I. Gene finding in novel genomes. *BMC Bioinformatics* **5**, 59 (2004).
40. Thon, M.R., Nuckles, E.M. & Vaillancourt, L.J. Restriction enzyme-mediated integration used to produce pathogenicity mutants of *Colletotrichum graminicola*. *Mol. Plant Microbe Interact.* **13**, 1356–1365 (2000).
41. Zhou, S. *et al.* Validation of rice genome sequence by optical mapping. *BMC Genomics* **8**,278 (2007).
42. Zhou, S. *et al.* A single molecule scaffold for the maize genome. *PLoS Genet.* **5**, e1000711(2009).
43. Kurtz, S. *et al.* Versatile and open software for comparing large genomes. *Genome Biol.* **5**,R12 (2004).
44. Engels, R. *et al.* Combo: a whole genome comparative browser. *Bioinformatics* **22**,1782–1783 (2006).
45. Katoh, K., Misawa, K., Kuma, K. & Miyata, T. MAFFT: a novel method for rapid multiple sequence alignment based on fast Fourier

transform. *Nucleic Acids Res.* **30**, 3059–3066(2002).

46. Horton, P. *et al.* WoLF PSORT: protein localization predictor. *Nucleic Acids Res.* **35**,W585–W587 (2007).
47. Rawlings, N.D., Barrett, A.J. & Bateman, A. MEROPS: the peptidase database. *Nucleic Acids Res.* **38**, D227–D233 (2010).
48. Takahara, H., Dolf, A., Endl, E. & O'Connell, R. Flow cytometric purification of*Colletotrichum higginsianum* biotrophic hyphae from *Arabidopsis* leaves for stage-specific transcriptome analysis. *Plant J.* **59**, 672–683 (2009).
49. Kleemann, J., Takahara, H., Stüber, K. & O'Connell, R. Identification of soluble secreted proteins from appressoria of *Colletotrichum higginsianum* by analysis of expressed sequence tags. *Microbiology* **154**, 1204–1217 (2008).
50. Kankanala, P., Czymmek, K. & Valent, B. Roles for rice membrane dynamics and plasmodesmata during biotrophic invasion by the blast fungus. *Plant Cell* **19**, 706–724(2007).
51. Metz, R.P., Kwak, H.-I., Gustafson, T., Laffin, B. & Porter, W.W. Differential transcriptional regulation by mouse single-minded 2s. *J. Biol. Chem.* **281**, 10839–10848 (2006).
52. Trapnell, C., Pachter, L. & Salzberg, S.L. TopHat: discovering splice junctions with RNA-Seq. *Bioinformatics* **25**, 1105–1111 (2009).
53. Quinlan, A.R. & Hall, I.M. BEDTools: a flexible suite of utilities for comparing genomic features. *Bioinformatics* **26**, 841–842 (2010).
54. Robinson, M.D., McCarthy, D.J. & Smyth, G.K. EdgeR: a Bioconductor package for differential expression analysis of digital gene expression data. *Bioinformatics* **26**, 139–140(2010).
55. Sturn, A., Quackenbush, J. & Trajanoski, Z. Genesis: cluster analysis of microarray data.*Bioinformatics* **18**, 207–208 (2002).
56. Langmead, B., Trapnell, C., Pop, M. & Salzberg, S.L. Ultrafast and memory-efficient alignment of short DNA sequences to the human genome. *Genome Biol.* **10**, R25 (2009).
57. Ronquist, F. & Huelsenbeck, J.P. MrBayes 3: Bayesian phylogenetic inference under mixed models. *Bioinformatics* **19**, 1572–1574 (2003).
58. Sanderson, M.J. r8s: inferring absolute rates of molecular evolution and divergence times in the absence of a molecular clock. *Bioinformatics* **19**, 301–302 (2003).
59. Lücking, R., Huhndorf, S., Pfister, D.H., Plata, E.R. & Lumbsch,

H.T. Fungi evolved right on track. *Mycologia* **101**, 810–822 (2009).

60. Shaner, N.C. *et al.* Improved monomeric red, orange and yellow fluorescent proteins derived from *Discosoma* sp. red fluorescent protein. *Nat. Biotechnol.* **22**, 1567–1572 (2004).
61. Szewczyk, E. *et al.* Fusion PCR and gene targeting in *Aspergillus nidulans*. *Nat. Protoc.* **1**,3111–3120 (2006).
62. Ushimaru, T. *et al.* Development of an efficient gene targeting system in *Colletotrichum higginsianum* using a non-homologous end-joining mutant and *Agrobacterium tumefaciens*–mediated gene transfer. *Mol. Genet. Genomics* **284**, 357–371 (2010).

Chapter 4

CONTROL OF FOLIAR PATHOGENS OF SPRING BARLEY USING A COMBINATION OF RESISTANCE ELICITORS

Dale R. Walters[1], Neil D. Havis[1], Linda Paterson[1], Jeanette Taylor[1], David J. Walsh[2] and Cecile Sablou[1]

[1]Crop and Soil Systems Research Group, Scotland's Rural College, Edinburgh, UK

[2]Engineering, Science and Technology Department, Scotland's Rural College, Edinburgh, UK

The ability of the resistance elicitors acibenzolar-S-methyl (ASM), β-aminobutyric acid (BABA), *cis*-jasmone (CJ), and a combination of the three products, to control infection of spring barley by *Rhynchosporium commune* was examined under glasshouse conditions. Significant control of *R. commune* was provided by ASM and CJ, but the largest reduction in infection was obtained with the combination of the three elicitors. This elicitor combination was found to up-regulate the expression of *PR-1b*, which is used as a molecular marker for systemic acquired resistance (SAR). However, the elicitor combination also down-regulated the expression of *LOX2*, a gene involved in the biosynthesis of jasmonic acid (JA). In field experiments over 3 consecutive years, the effects of the elicitor combination were influenced greatly by crop variety and by year. For example, the elicitor combination applied on its own provided significant control of powdery mildew (*Blumeria graminis* f.sp. *hordei*) and *R. commune* in 2009, whereas no control on either variety was observed in 2007. In contrast, treatments involving both the elicitor combination and fungicides provided disease control and yield increases which were equal to, and in some cases better than that provided by the best fungicide-only treatment. The prospects for the use of elicitor plus fungicide treatments to control foliar pathogens of spring barley in practice are discussed.

INTRODUCTION

Application of various agents to plants can lead to the induction of resistance to subsequent pathogen attack, both locally, and systemically (Walters et al., 2013). Such induced resistance can be split into systemic acquired resistance (SAR) and induced systemic resistance (ISR). SAR is characterized by a restriction of pathogen growth and a suppression of disease symptom development compared to non-induced plants infected with the same pathogen. Its onset is associated with an accumulation of salicylic acid (SA) at sites of infection and systemically, and with the coordinated activation of a specific set of genes encoding pathogenesis-related (PR) proteins. Treatment of plants with SA or one of its functional analogs e.g., acibenzolar-S-methyl (ASM; marketed in Europe as Bion® and in North America as Actigard®), induces SAR and activates the same set of*PR* genes. ISR develops as a result of colonization of plant roots by plant growth-promoting rhizobacteria (PGPR) and has been shown to function independently of SA and activation of *PR* genes, requiring instead, jasmonic acid (JA), and ethylene (ET) (Pieterse et al., 2012; Spoel and Dong, 2012).

Research over the past decade suggests that SA induces defenses against biotrophic pathogens, while JA mediates defenses against necrotrophic pathogens (Glazebrook, 2005). It is thought that cross-talk between the two signaling pathways might help to fine-tune defense responses against a particular pathogen according to its mode of infection (Beckers and Spoel, 2006). Interestingly, Spoel et al. (2007) found that infection with the biotrophic pathogen *Pseudomonas syringae*, which induces SA-mediated defense, rendered *Arabidopsis thaliana* more susceptible to the necrotrophic pathogen *Alternaria brassicicola*, although this trade-off was restricted to plant tissue adjacent to the initial infection site. In terms of plant-insect interactions, JA is known to play a major role in mediating defenses against insect herbivory (Bostock, 2005), although the situation is rather more complex than was once thought, since both SA- and JA-responsive gene expression can be elicited by aphids and whiteflies. In terms of insect herbivory, although there are examples of negative cross-talk i.e., SA-mediated suppression of JA-inducible gene expression, there are also examples of no trade-offs, and even of positive effects (see Walters et al., 2013).

Because induced resistance offers the prospect of broad spectrum disease control using the plant's own resistance mechanisms, there has been great interest in the development of agents which can mimic natural inducers of resistance (Lyon, 2007). These include elicitor molecules released during the early stages of the plant-pathogen interaction and the signaling pathways used to trigger defenses locally and systemically. Examples include ASM, which has been shown to elicit SAR in a wide range of plant-pathogen interactions (Leadbeater and Staub, 2007), the non-protein amino acid β-aminobutyric acid (BABA), and the oxylipin, *cis*-jasmone (CJ) (Walters et al., 2007).

BABA is known to induce resistance against pathogens in various systems, including tomato, potato, grapevine, and pea (Cohen et al., 1999; Jakab et al., 2001). In field experiments, Cohen (2002) found that BABA provided significant control of late blight of potato, while Liljeroth et al. (2010) showed that BABA used together with a reduced fungicide dose gave the same level of late blight control as a full dose of the standard fungicide treatment.

CJ is structurally related to JA and methyl jasmonate (MeJA), both of which are well known to activate plant defenses (Farmer and Ryan, 1990; Thaler et al., 1996), although CJ activates a unique and more limited set of genes than does treatment with MeJA (e.g., Pickett et al., 2007). CJ is released naturally from plants damaged by insects and when applied artificially, can activate defense against insects (Birkett et al., 2000; Bruce et al., 2003).

The efficacy of induced resistance under field conditions is variable, representing a major obstacle to its use in practical crop protection (Reglinski and Walters, 2009). Induced resistance is a complex plant response to pathogen attack and as such, will be modified by many factors including genotype. However, insufficient attention has been paid to investigating the mechanisms underlying variable efficacy and approaches that might be adopted to incorporate elicitors into crop protection practice, such as use of elicitors and fungicides together in the same disease control program, and use of combinations of elicitors. The latter aspect has received little attention to date, probably because of the trade-offs that might be associated with using elicitors which activate different signaling pathways, as mentioned above. In this paper, we report the results of field experiments over 3 consecutive years, undertaken to determine the potential for use of an elicitor combination to control foliar pathogens of spring barley. Some preliminary data from this study have appeared previously in a conference paper (Walters et al., 2010).

MATERIALS AND METHODS

Plant Growth and Pathogen Inoculation under Glasshouse Conditions

The spring barley (*Hordeum vulgare* L.) variety Cellar was used for glasshouse studies. Cellar was chosen since it exhibits moderate susceptibility to *Rhynchosporium commune* (HGCA, 2009). Seeds were sown in pots in Fisons Levington compost and grown in a walk-in growth room at 18°C with a 16 h photoperiod (190 μmol $m^{-2}s^{-1}$ provided by 400 W mercury vapor lamps). The experiment was laid out in a randomized block design, with each of the five treatment groups consisting of 15 plants, with 10 plants used for disease assessment and 5 plants for gene expression analysis. Plants were used for efficacy experiments when the sixth leaf was fully formed and the seventh leaf emerging. Leaves 1–4 were sprayed with elicitors using a hand-held sprayer. Two days later, plants were inoculated with the leaf scald pathogen, *Rhynchosporium commune*, by spraying with a suspension of spores (1×10^5 spores/ml) in distilled water containing 0.01% Tween 20. Inoculated plants were then covered with plastic bags for 48 h (the first 24 h in the dark) and kept at 16°C to provide the conditions necessary for spore germination and early fungal development. Thereafter, the temperature of the growth room was increased to 18°C for the remainder of the experiment. Infection intensity on leaves 5–7 was assessed 21 days after inoculation by determining the % leaf area exhibiting symptoms on each of 10 plants. For gene expression experiments, leaves three and four were treated with elicitor and 2 days later were inoculated with*R. commune*. Leaves were harvested 2 days later and frozen at −80°C for gene expression analysis. Data presented are the means of three replicates.

The elicitors used in these experiments were ASM, BABA, and CJ. ASM (Bion®) was a gift from Syngenta, Basel, Switzerland; BABA was purchased from Sigma, Poole, Dorset, UK; CJ was purchased from Sigma Aldrich, Dorset, UK. ASM (1 mM), BABA (1 mM) and CJ (0.625 g/l) were made up in distilled water containing 0.01% Tween 20.

Field EXPERIMENTS

Field experiments were conducted in 2007 at Tibbermore, Perthshire, Scotland, and in 2008 and 2009 at Drumalbin, Lanark, Scotland. Total rainfall and average air temperatures at these sites during the period 1 June—1 September were:

Perthshire, 2007: rainfall = 219 mm; average air temperature = 17°C

Lanark, 2008: rainfall = 317 mm; average air temperature = 16°C

Lanark, 2009: rainfall = 338 mm; average air temperature = 17°C.

Two spring barley varieties (Cellar and Optic) used in all field experiments reported here. Cellar has a resistance rating (RR) of 9 for powdery mildew and 4 for *Rhynchosporium commune*, and Optic has a RR of 5 for powdery mildew and 4 for *R. commune* (RR scale: 10 = high resistance, 1 = low resistance; HGCA, 2009). Each variety was sown in a randomized block design at a seed rate of 360 seeds m^{-2} and an individual plot size of 10×2 m, using three plots per treatment. For each barley variety, the factor tested was the applied treatment (i.e., elicitor, fungicide, elicitor + fungicide). Plots received standard fertilizer and herbicide regimes and 16 treatment programs were compared (Table 1). Spray dates for treatments were based on plant growth stage as described by Zadoks et al. (1974) and were applied with a knapsack sprayer using an equivalent spray volume of 200 l ha-1. Disease symptoms (% leaf area infected) and % GLA were assessed using 10 plants per plot at spray dates and at 14 day intervals after the final spray. Area under the disease progress curves (AUDPC) were calculated using the formula:

$$\Sigma(yi+y(i+1))/2\times(t(i+1)-ti)\Sigma(yi+y(i+1))/2\times(t(i+1)-ti)$$

where y_i is the disease rating at time t_i.

Table 1. Elicitor and fungicide treatments applied in field experiments in 2007–2009

Treatment	GS24	GS31	GS39
1	Nil	Nil	Nil
2	Elicitors	Nil	Nil
3	Nil	Elicitors	Nil
4	Nil	Nil	Elicitors
5	Elicitors	Elicitors	Nil
6	Elicitors	Nil	Elicitors
7	Fungicide[1]	Nil	Nil
8	Nil	Fungicide[1]	Nil
9	Nil	Nil	Fungicide[2]
10	Nil	Fungicide[1]	Fungicide[2]
11	Fungicide[1] + Elicitors	Nil	Nil
12	Nil	Fungicide[1] + Elicitors	Nil
13	Nil	Nil	Fungicide[2] + Elicitors
14	Elicitors	Fungicide[1]	Fungicide[2]
15	Elicitors	Fungicide[3]	Fungicide[4]
16	Elicitors	Elicitors + Fungicide[3]	Fungicide[4]

Treatments were applied at different growth stages (GS) as described in the Materials and Methods.

GS = growth stage (see Zadoks et al., 1974).

Elicitors = ASM + BABA + C.

Fungicide[1] = prothioconazole at full-rate x 0.4 + cyprodinil and picoxystyrobin at full-rate x 0.5.

Fungicide[2] = prothioconazole at full-rate x 0.4 + chlorothalonil at full-rate.

Fungicide[3] = prothioconazole at full-rate x 0.2 + cyprodinil and picoxystrobin at full-rate x 0.3.

Fungicide[4] = prothioconazole at full-rate x 0.2 + chlorothalonil at full-rate x 0.5 Plots were harvested at the end of the trial and yields expressed as tonnes/hectare at 85% dry matter content.

Gene Expression

Total RNA was extracted from barley leaves using a RNeasy™ kit (Qiagen, West Sussex, UK) and RNA yield determined using a Nanodrop 1000 spectrophotometer (Nanodrop Technologies, Wilmington, DE, USA). In order to remove any remaining trace of DNA likely to interfere with measurements, samples were treated with desoxyribonuclease enzymes using the DNA-*free*™ kit from Applied Biosystems (California, USA). The final quantity and quality of the RNA was tested using a RNA 6000 Nano Chip kit (Agilent Technologies, Santa Clara, CA, USA).

Primer sequences for *PR1-b, LOX2*, and the cyclophilin gene (internal control) are listed in Table 2. All sequences were purchased from Eurofins MWG Operon (Ebersburg, Germany) and all primers were designed using Beacon Designer software (Premier Biosoft International, Palo Alto, California, USA).

Table 2. Primer sequences for genes used in this study

Gene	Primer sequence (F)	Primer sequence (R)
PR1-b	CTACGACTACGGCTCCAACAC	GCATCACGGTTAGTATGGTTTCTG
LOX2	CGGCAGACTCCCTCATCACTAAAG	GGCAGCAACAGGTCGTGGTAG
Cyclophilin	CCTGTCGTGTCGTCGGTCTAAA	ACGCAGATCCAGCAGCCTAAAG

Amplicon lengths: PR1-b = 190 base pairs. LOX2 = 121 base pairs. Cyclophilin = 122 base pairs.

Following RNA extraction, cDNA was generated using a SuperScript™ first-strand cDNA synthesis kit (Invitrogen, USA). Quantitative real-time PCR (qRT-PCR) was then performed with a MX3000P system (Stratagene, CA, USA) using a Brilliant 11SYBR Green QPCR master mix with ROX (Agilent Technologies, Santa Clara, CA, USA). In order to construct standard curves for the genes, six data points were used with a 5-fold dilution series (1:10–1:31,250). A 25 μl reaction for PCR amplification contained 12.5 μl of SYBR Green master mix (see above), 0.75 μl forward primer, 0.75 μl reverse primer, 6 μl sterile distilled water, and 5 μl cDNA. All PCR reactions were performed in duplicate. The cycling conditions were as follows: pre-incubation for 10 min at 95°C, followed by 40 cycles, each consisting of 30 s denaturing at 95°C, 60 s annealing at 57°C, 30 s at 72°C for new strand synthesis. The standard curves were used to calculate the absolute quantity of the product in each sample, Relative expression values were then calculated by normalizing against the cyclophilin gene as an internal control.

Statistical Analysis

All data were subjected to One-Way ANOVA using the GenStat Release 11.1 statistical program. The effect of blocks was considered random and the effect of applied treatments was defined as fixed. % leaf area infected values from glasshouse experiments and % GLA data from field experiments were log-transformed prior to analysis. Comparison of treatment means was performed using Fisher's protected least significant difference (LSD) Test.

RESULTS

Effects of Elicitors under Glasshouse Conditions

Initial experiments were conducted under glasshouse conditions to examine the effects of Bion®, BABA, and CJ, singly and in combination, on infection of the barley variety Cellar by *R. commune*. Although BABA reduced infection compared to the untreated control, this difference was not significant. In contrast, treatment with Bion® and CJ led to significant reductions in *R. commune* infection, with Bion® reducing infection by 70% and CJ by 64% (Figure 1). The largest reduction in infection (96%) was obtained using a combination of Bion®, BABA, and CJ.

Application of the elicitor combination to leaves 1 and 2 of the barley variety Cellar led to changes in the expression of two defense-related genes in leaves 3 and 4. Thus, elicitor treatment resulted in significant increases in expression of *PR1b* in both leaves 3 and 4 (4.3-fold and 3.8-fold, respectively; Figures 2A,B). Expression of *PR1b* was also increased significantly in leaf 3 following inoculation of untreated leaves (2.6-fold increase), but was not affected in leaf 4 (Figures 2A,B). However, the largest increases in *PR1b* expression were obtained when leaves 3 and 4 were first treated with elicitor and subsequently inoculated with *R. commune*. Here, *PR1b* expression was increased 7-fold in leaves 3 and 4 compared to the untreated control (Figures2A,B). This suggests that the elicitor combination primes the plant for enhanced expression of *PR1b*.

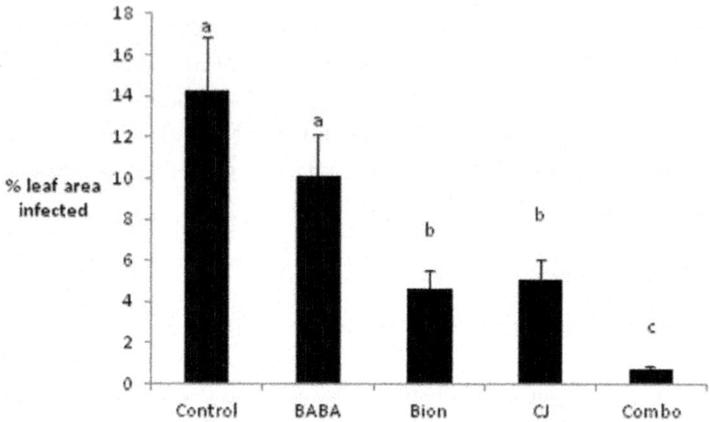

Figure 1. Effects of elicitors, singly and in combination, on infection of barley with *R. commune*. Leaves 1–4 were sprayed with elicitor and inoculated with *R. commune* 2 days later. Infection intensity was assessed 21 days later on leaves 5–7. Bars with a different letter are significantly different at $P < 0.05$ (Fisher's LSD).

A

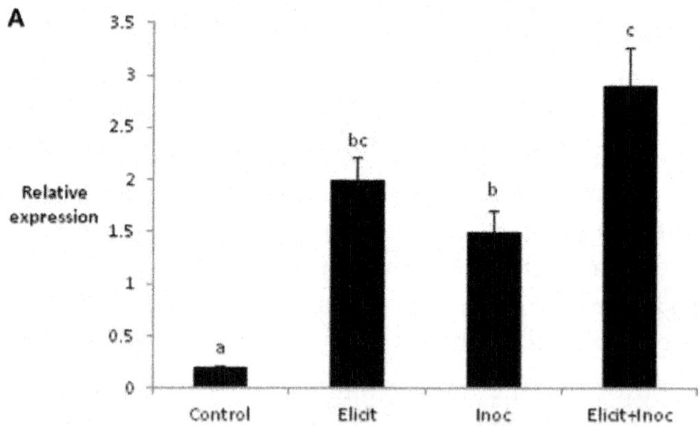

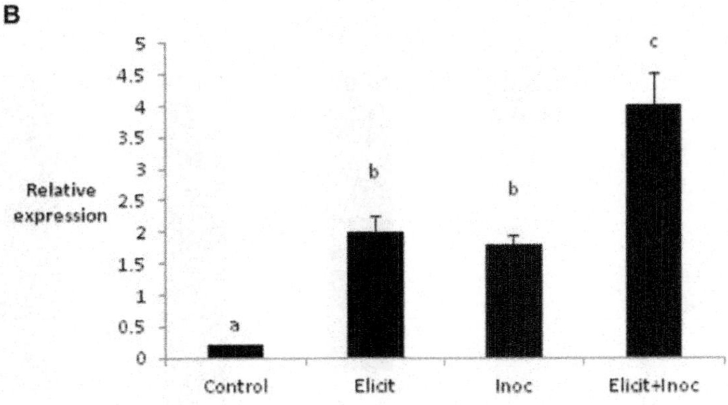

Figure 2. Relative quantity of *PR1-b* **in leaf 3 (A) and leaf 4 (B) of barley plants treated with the elicitor combination**. Leaves three and four were treated with elicitor and inoculated with *R. commune* 2 days later. Leaves were harvested 2 days after inoculation for analysis of gene expression. Bars with a different letter are significantly different at $P < 0.05$ (Fisher's LSD).

In contrast to *PR1b*, expression of *LOX2* was reduced by treatment with the elicitor combination, compared to the untreated control. Indeed, all three treatments led to substantial and significant decreases in expression of*LOX2* in leaves 3 and 4 (Figures 3A,B).

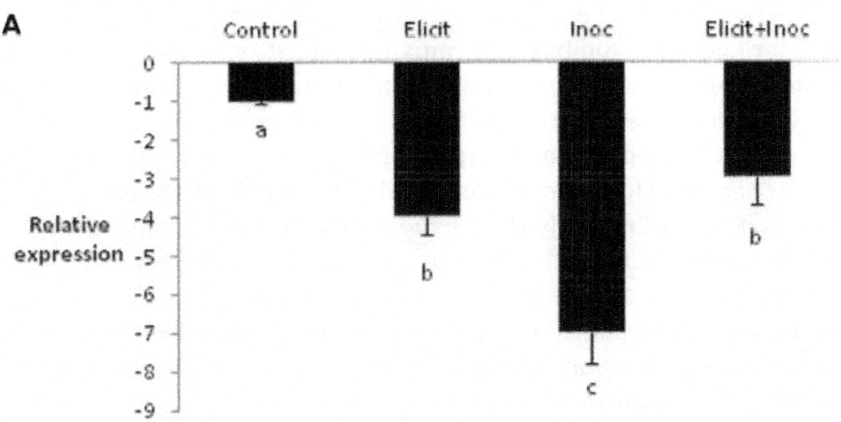

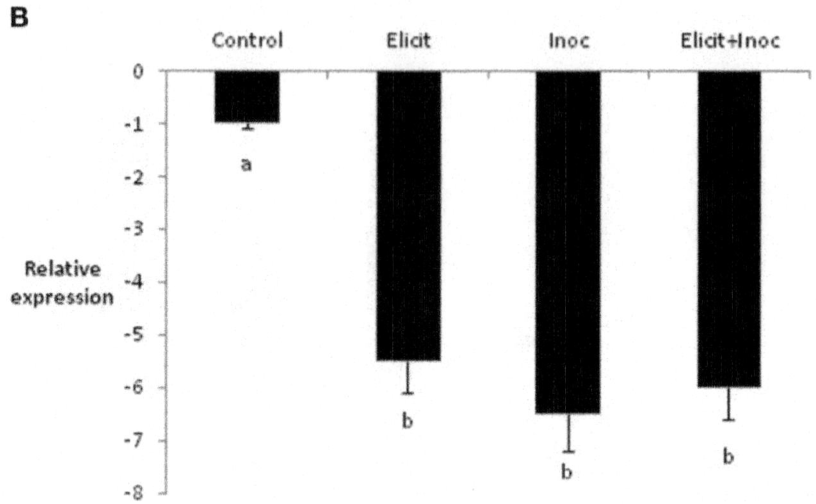

Figure 3. Relative quantity of *LOX2* **in leaf 3 (A) and leaf 4 (B) of barley plants treated with the elicitor combination.** Leaves three and four were treated with elicitor and inoculated with *R. commune* 2 days later. Leaves were harvested 2 days after inoculation for analysis of gene expression. Bars with a different letter are significantly different at $P < 0.05$ (Fisher's LSD).

Field Experiments

Since the elicitor combination provided most effective control of *R. commune* under glasshouse conditions, this treatment was chosen for inclusion in field experiments. Over the 3 years of field experiments, the two major foliar diseases detected on the spring barley crops were powdery mildew and *R. commune*. However, disease levels varied between years and in some years (2008 and 2009), only *R. commune* was observed on the variety Cellar. This probably reflected differences in weather at the different sites, since, for example, the 2007 season at the Perth site was considerably drier than the Lanark sites in 2008 and 2009. The 3 years of field experimentation, involving 17 different treatments on two varieties generated too much data to be shown here. Instead, only data for selected treatments are shown. These treatments are untreated (1), one fungicide-only treatment (10), one elicitor-only treatment (5), and one elicitor + fungicide (reduced rate) treatment (13). The data for all treatments are provided in Supplementary Material.

The efficacy of the elicitor combination was dependent on crop variety and year. Thus, in 2009, the elicitor combination provided significant control of both powdery mildew and *R. commune* on the variety Optic and of *R. commune* on variety Cellar (Figure 4). However, on both varieties, most effective disease control was achieved using a combination of the elicitor combination and fungicide. Indeed, the level of disease control achieved with the elicitor and fungicide treatment was at least as good as that obtained using the fungicide only treatment (Figure 4). Interestingly, although the elicitor combination on its own provided significant disease control on the two varieties, grain yield remained unchanged compared to untreated plants (Figure 4). In contrast, grain yield was increased significantly ($P < 0.05$) in both barley varieties treated with the elicitor plus fungicide treatment (Figure 4).

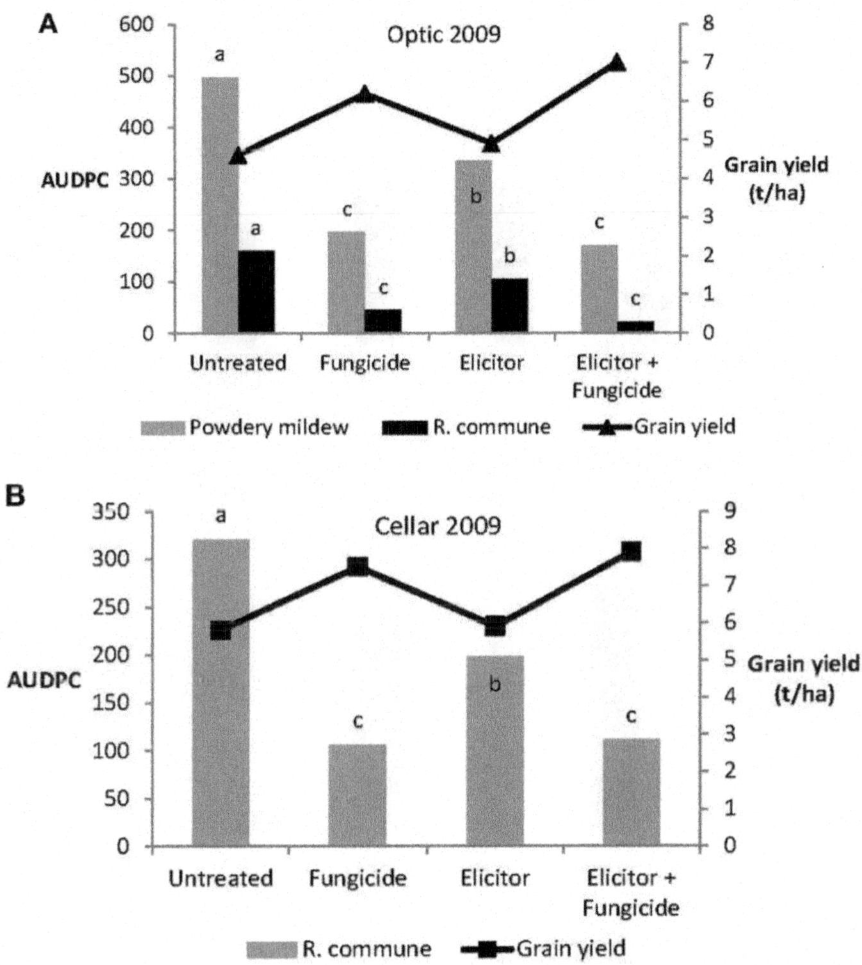

Figure 4. Effects of the elicitor combination and fungicides on (A) AUDPC for powdery mildew and *R. commune*, and grain yield in the spring barley variety Optic, and (B) on AUDPC for *R. commune* **and grain yield in the variety Cellar, in a field**

experiment in 2009. Bars with different letters are significantly different at $P < 0.05$ (Fisher's LSD).

In 2008, the elicitor combination applied on its own reduced levels of powdery mildew significantly on variety Optic, but provided no control of *R. commune* on either Optic or Cellar (Figure 5). Treatment with elicitor plus fungicide provided significant control of both diseases on Optic and of *R. commune* on Cellar. In both varieties, highest grain yields were obtained from plants receiving the elicitor plus fungicide treatment (Figure 5).

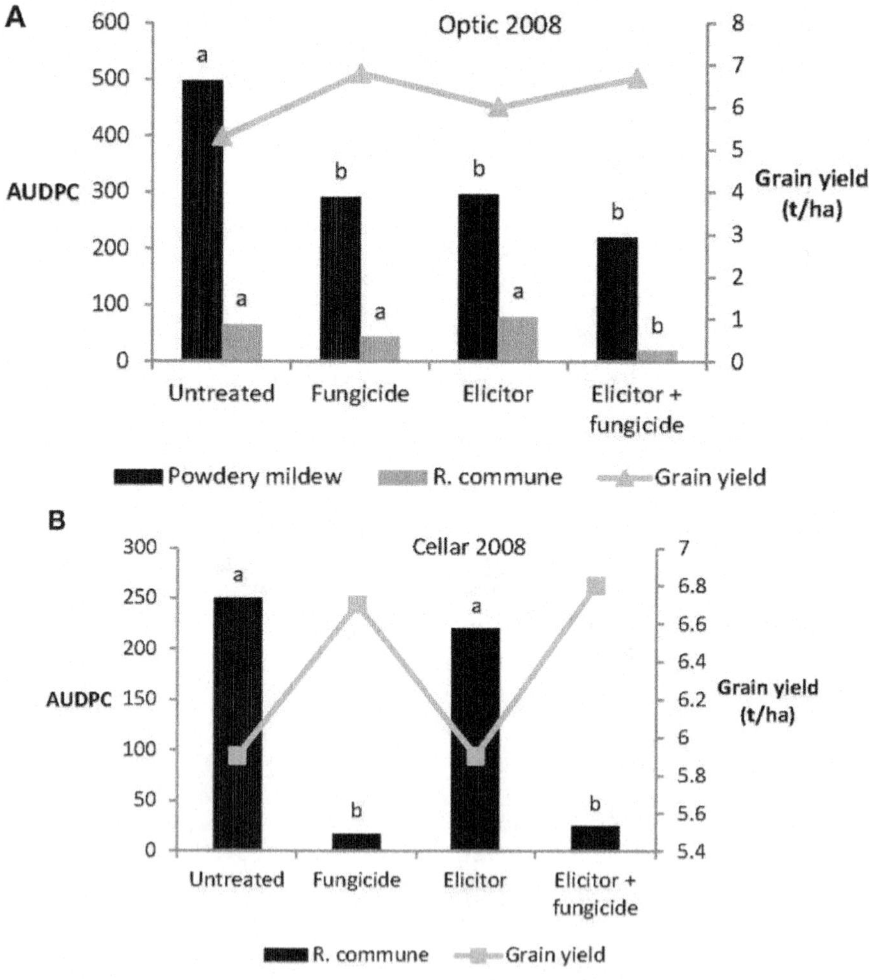

Figure 5. Effects of the elicitor combination and fungicides on (A) AUDPC for powdery mildew and *R. commune*, and grain yield in the spring barley variety Optic, and (B) on AUDPC for *R. commune* **and grain yield in the variety Cellar, in a field**

experiment in 2008. Bars with different letters are significantly different at $P < 0.05$ (Fisher's LSD).

Levels of both powdery mildew and *R. commune* were not significantly affected by treatment of either variety with the elicitor combination on its own in 2007 (Figure 6). In contrast, on both varieties, treatment with elicitor plus fungicide provided significant control of both diseases and in most cases the level of disease control achieved was as good as that obtained using the fungicide only treatment (Figure 6). Grain yields of both varieties were significantly increased in plants receiving the elicitor plus fungicide treatment compared to the other treatments, while in Optic, the elicitor combination on its own actually reduced grain yield (Figure 6).

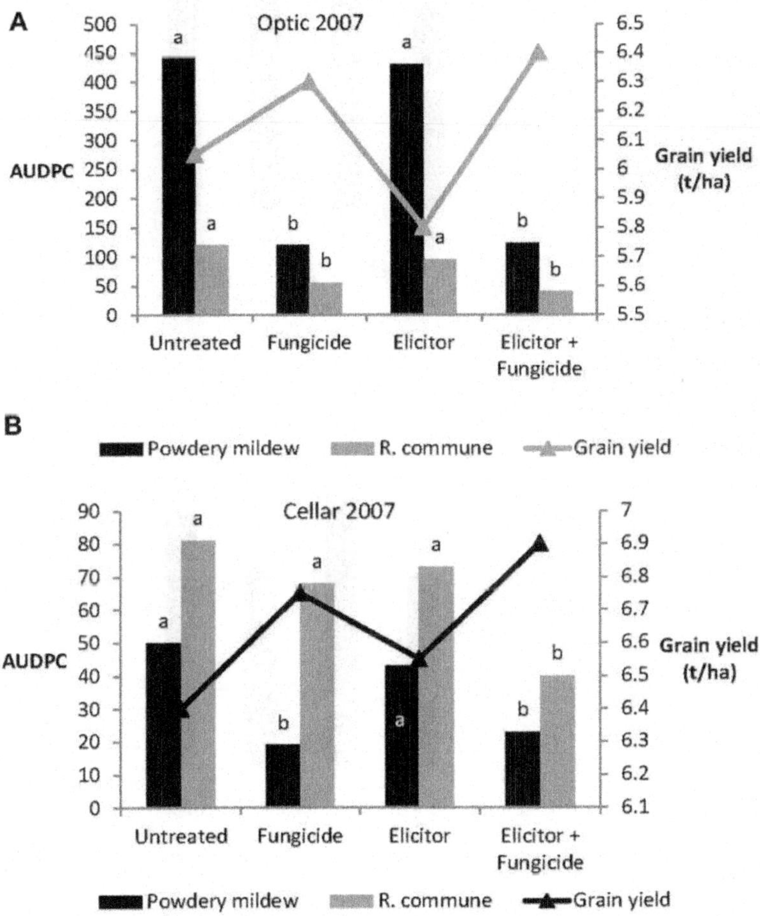

Figure 6. Effects of the elicitor combination and fungicides on (A) AUDPC for powdery mildew and *R. commune*, and grain yield in the spring barley variety Optic, and

(B) in the variety Cellar, in a field experiment in 2007. Bars with different letters are significantly different at $P < 0.05$ (Fisher's LSD).

The absence of any significant effect of the elicitor-only treatment on grain yields in the 2 varieties across the years probably reflects, in part, the lack of a significant effect of this treatment on green leaf area (GLA) (Figure7). This is supported by the fact that the significant reduction in grain yield in the variety Optic in 2007 (Figure6) was associated with a significant reduction in GLA (Figure 7E).

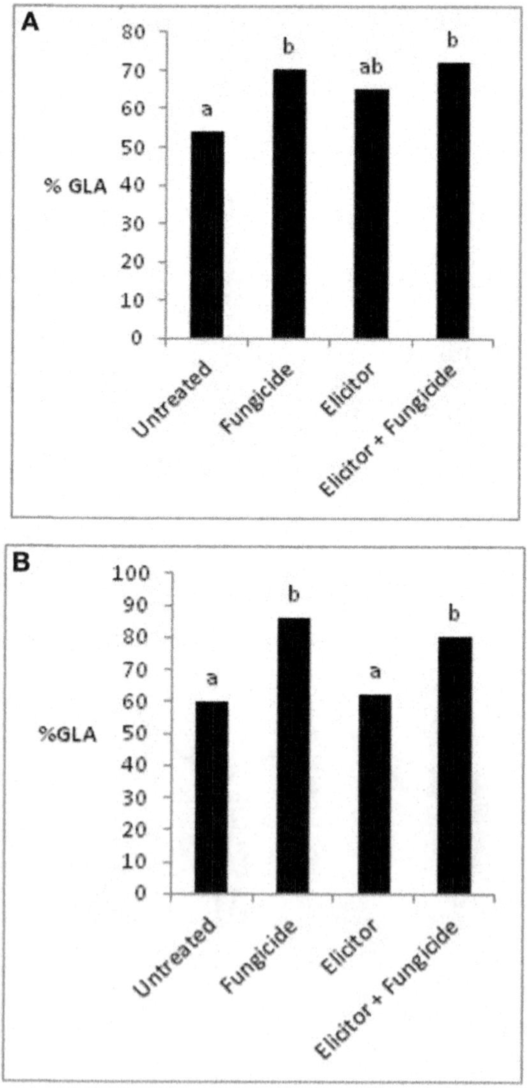

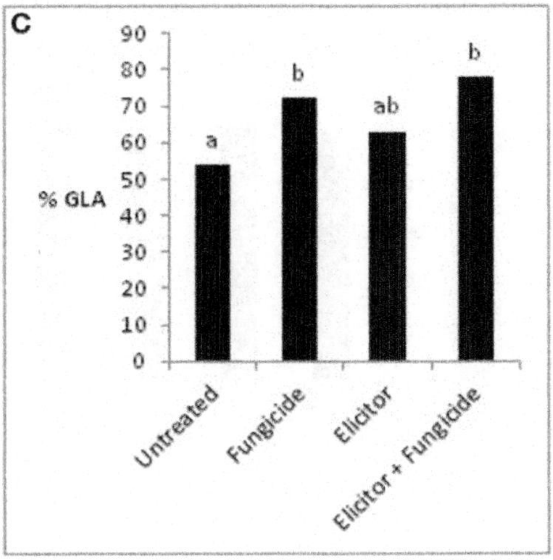
C
90
80
70
60
50
% GLA
40
30
20
10
0
a
b
ab
b
Untreated
Fungicide
Elicitor
Elicitor + Fungicide

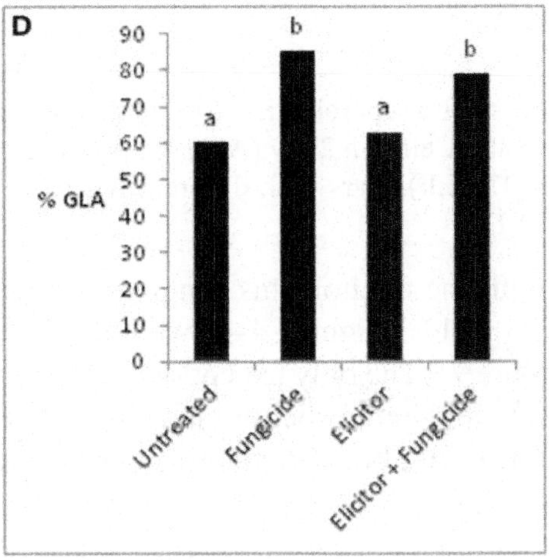
D
90
80
70
60
50
% GLA
40
30
20
10
0
a
b
a
b
Untreated
Fungicide
Elicitor
Elicitor + Fungicide

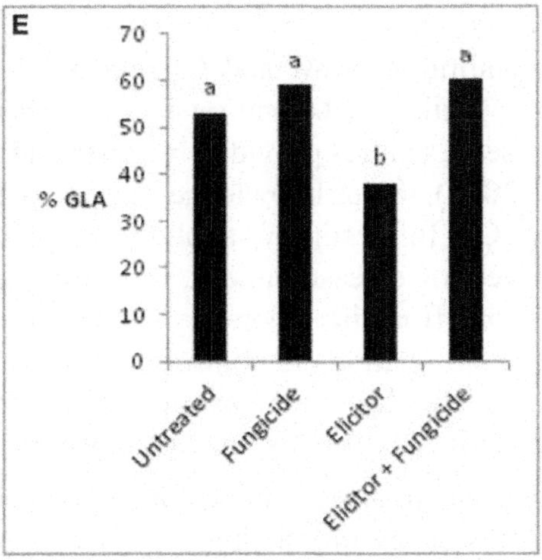
E
70
60
50
40
% GLA
30
20
10
0
a
a
b
a
Untreated
Fungicide
Elicitor
Elicitor + Fungicide

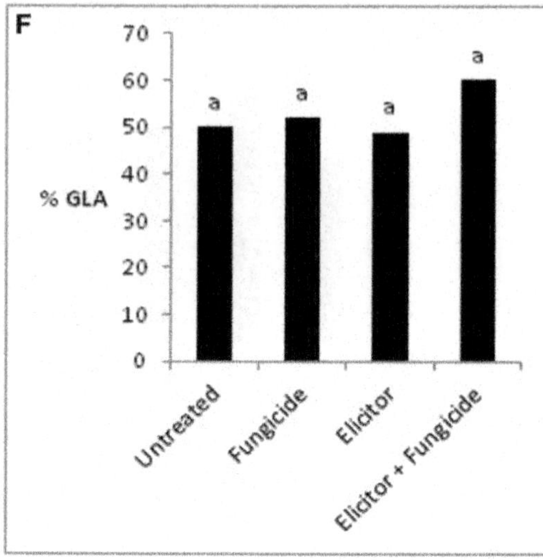

Figure 7. Effects of the elicitor combination and fungicides on percentage green leaf area (% GLA) in Optic and Cellar in 2009 (A,B), Optic and Cellar in 2008 (C,D), and Optic and Cellar in 2007 (E,F). Bars with different letters are significantly different at $P < 0.05$ (Fisher›s LSD).

This contrasts with the elicitor plus fungicide treatment, where, in most cases, increased grain yields (Figures 4–6) were associated with significantly increased GLAs (Figure 7). The only exception was the variety Cellar in 2007, where increased grain yield in the elicitor plus fungicide treatment (Figure 6B) was accompanied by a small, but statistically non-significant increase in GLA (Figure 7F).

DISCUSSION

Under glasshouse conditions, ASM and CJ reduced *R. commune* infection of barley by 64–70%, while BABA had much less effect. Although there are many examples of disease control provided by ASM and BABA (Cohen et al., 2010; Walters et al., 2013), to our knowledge, this is the first report of disease control provided by CJ. Interestingly, applying the three elicitors together provided the best levels of disease control, reducing infection by 96%, and confirms previous reports from this laboratory (Walters et al., 2011a,b). Reports of the use of elicitor combinations to control plant disease are rare, although combinations of ASM and Milsana® (extract of *Reynoutria sachalinensis*) were found to control powdery mildew on cucumber (Bokshi et al., 2008).

Molecular studies on tobacco and *Arabidopsis* have shown that ASM activates the SAR pathway by mimicking the activity of SA (Gaffney et al.,

1993; Friedrich et al., 1996; Lawton et al., 1996). The situation with BABA is more complex and seems to involve SA-dependent, SA-independent, and abscisic acid (ABA)-dependent mechanisms, with the relative importance of the different signaling pathways depending on the particular host-pathogen interaction (Ton et al., 2005). As indicated earlier, CJ is structurally related to JA and MeJA, although it up-regulates a unique set of genes compared to MeJA (Birkett et al., 2000; Pickett et al., 2007). Treatment of leaves one and two of barley with the combination of ASM, BABA, and CJ led to an approximately 6-fold up-regulation of *PR-1b* in leaves three and four, confirming previous reports on the effects of the elicitor combination on defense-related gene expression (Walters et al., 2011a,b). Expression of a *PR-1* gene is usually considered to be a molecular marker for SAR (van Loon et al., 2006) and therefore, these data suggest that the elicitor combination activated SAR in barley. *PR-1b* was also up-regulated by inoculation with *R. commune*(~4-fold increase in leaf 3, for example). However, the largest up-regulation of *PR-1b* was obtained when elicitor-treated plants were subsequently inoculated with *R. commune*. This suggests that the elicitor combination primed *PR-1b* gene expression in these plants. In contrast to *PR-1b*, the expression of *LOX2*exhibited a completely different trend. Treatment of barley plants with the elicitor combination resulted in a substantial down-regulation of *LOX2* (4-fold in leaf 3 and 5.5-fold in leaf 4). Inoculation with *R. commune* led to a greater down-regulation of *LOX2* in the treated leaves, while application of the elicitor combination followed by inoculation had no further effect on gene expression than the elicitor combination only. The *LOX2*gene, which is involved in the octadecanoid pathway, is auto-regulated by JA, thereby controlling a feed-forward loop in JA biosynthesis (Bell et al., 1995). Suppression of *LOX2* in transgenic *Arabidopsis* was shown to block JA biosynthesis during pathogen infection (Spoel et al., 2003), and the present data on *LOX2* in barley suggests that the elicitor combination might suppress activity of the JA pathway in barley. Concomitant activation of SAR and suppression of the JA pathway might be expected to enhance defense against biotrophic pathogens and increase susceptibility to necrotrophic pathogens (Glazebrook, 2005). *R. commune* is a hemibiotrophic pathogen, possessing an initial biotrophic phase followed by a prolonged necrotrophic phase (Walters et al., 2008). Perhaps it is no surprise therefore, that the elicitor combination reduces infection by *R. commune*. It is possible that the elicitor combination might compromise the ability of the plant to defend itself against necrotrophic pathogens and indeed, the elicitor combination was shown to provide control of powdery mildew and *R. commune*, but was associated with increased symptoms of leaf spot caused by the necrotrophic pathogen *Ramularia collo-cygni* on spring barley in the field (Walters et al., 2011b). Plant defense against chewing insects is mediated by

JA signaling (Pieterse et al., 2012), as is establishment of functional arbuscular mycorrhizal (AM) symbioses (Pozo and Azcón-Aguilar, 2007). In view of the down-regulation of *LOX2* in barley treated with the elicitor combination, it would be prudent to examine the effects of treated plants on defense against herbivorous insects and on the establishment of AM symbiosis.

In all 3 years of field experiments, the elicitor combination applied on its own was either partially effective or ineffective at controlling powdery mildew and *R. commune* on the two barley cultivars. In 2009, the elicitor combination provided significant control of both diseases on both spring barley varieties, and also controlled powdery mildew on Optic in 2008. However, the elicitor treatment did not control *R. commune* on either variety in 2008 and provided no disease control in 2007. Perhaps unsurprisingly therefore, given the poor levels of disease control provided by the elicitor-only treatment, grain yield was not significantly affected, apart from 2007, when grain yield of Optic was significantly reduced compared to the untreated control. Here, the reduced grain yield probably reflected the significantly reduced GLA in the elicitor-only treatment. Infection by many pathogens, including biotrophic pathogens such as powdery mildews, results in chlorosis and reduced photosynthetic rates (e.g., Walters and McRoberts, 2007) and the failure of the elicitor-only treatment to control infection would probably have affected rates of photosynthesis. Whether the elicitor combination affects photosynthesis in barley is not known, but should be examined.

The data on disease control presented in this paper highlight two typical problems associated with the practical use of elicitors on certain crops under field conditions, namely inconsistency and poor levels of disease control (Walters and Fountaine, 2009). In contrast to the elicitor-only treatments, the performance of the elicitor plus fungicide treatment was better both in terms of disease control and consistency. The acid test for such combined treatments is whether the performance of the combination is superior to that of the fungicide treatment on its own. Unfortunately, in most cases, although the combined elicitor and fungicide treatment performed as well as the fungicide-only treatment in terms of disease control, only in one case (control of *R. commune* on Optic in 2008) did the elicitor plus fungicide treatment out-perform the fungicide-only treatment. A similar situation was found for grain yield. Here, although grain yield tended to be increased by application of the elicitor plus fungicide treatment, most of these increases were not significantly different from the fungicide-only treatment. The lack of consistency in terms of disease control shown by the elicitor combination in barley, contrasts with the situation in oilseed rape (*Brassica napus*). Here, application of the elicitor combination to winter oilseed rape provided better control of light leaf spot caused by *Pyrenopeziza brassicae*, than standard fungicide treatments (Oxley and Walters, 2012).

Over the 3 years of field experiments, the elicitor plus fungicide combinations providing most consistent disease control were treatments 13, 14, and 16. As indicated earlier, only data for treatment 13 are shown, since this treatment performed most consistently throughout the study. Treatment 13 involved application of a combination of fungicide at reduced rate plus elicitor at GS39, with no control treatments applied at earlier growth stages. It has been suggested that application of elicitors earlier in the season might reduce inoculum levels, thereby requiring less fungicide to be applied later (Walters et al., 2013). On the basis of the results obtained in the present paper, this suggestion does not appear to work for spring barley. This suggests that, at least for spring barley, protecting later stages of crop growth is important in maintaining grain yield.

In some crops, use of elicitors and fungicides (or bactericides) can be effective. For example, the use of ASM (as Actigard®) in combination with fungicides and bactericides was recommended in tomato spray programs in North Carolina, USA (Ivors and Louws, 2007). The rationale here was that the elicitor would increase plant resistance, while the fungicides and bactericides would reduce pathogen inoculum levels. On mandarins (variety Murcott), tank-mixing ASM with azoxystrobin improved the efficacy of the fungicide by more than 50% (Miles et al., 2005), although this effect was clearly variety-specific, since no extra benefit of tank-mixing the elicitor and fungicide was obtained with the mandarin variety Imperial (Miles et al., 2004).

It has been suggested that one of the reasons for the relatively poor performance of elicitors is due to the likelihood that under field conditions, plants are already induced (Walters, 2009). Indeed, Herman et al. (2007) found that in tomato under field conditions, some defense genes were already expressed prior to ASM application. Nevertheless, the expression of these genes was increased further following ASM application. In preliminary work, examination of CAD activity in leaves from the field experiment in 2007 indicated that activity of the enzyme was already high prior to elicitor treatment (Paterson and Walters, unpublished results). Although it is tempting to suggest that CAD activity was increased further following elicitor application, any increases observed were not significant. It is possible that in barley, unlike tomato, prior induction of resistance compromises the ability of the plant to respond effectively to elicitors. Indeed, this was reported for barley treated with the elicitor combination, where prior inoculation with *R. commune* compromised the ability of the plant to respond to subsequent elicitor treatment (Walters et al., 2011a). It was suggested that this might help to explain the relatively poor performance of induced resistance in the field, particularly in cereals, compared to plants grown under controlled conditions (Walters et al., 2011a).

The results presented in this paper indicate quite clearly that use of a combination of elicitors alone does not provide effective disease control in spring barley. In contrast, using the elicitor combination and fungicides, even at half-rate, can provide levels of disease control and yield increases that are equal to the best fungicide-only treatment. From a practical perspective, an elicitor plus fungicide program is only likely to be attractive to a grower if it is cost-effective i.e., it provides levels of disease control and yield increases above that achieved using the fungicide on its own. This suggests that barley growers are unlikely to find the elicitor plus fungicide treatments examined in this work an attractive proposition. This might change however, if fungicide availability is further reduced through legislation. Indeed, elicitor/fungicide combinations could be valuable in reducing fungicide use, and prolonging the useful life of certain fungicides.

ACKNOWLEDGMENTS

We are grateful to Duncan McKenzie of Syngenta, Switzerland, for the gift of Bion®. This research was supported by the Scottish Government (Work Packages 1.4 and 6.4).

SUPPLEMENTARY MATERIAL

The Supplementary Material for this article can be found online at:http://www.frontiersin.org/journal/10.3389/fpls.2014.00241/abstract

REFERENCES

1. Beckers, G. J. M., and Spoel, S. H. (2006). Fine-tuning plant defence signalling: salicylate versus jasmonate. *Plant Biol.* 8, 1–10. doi: 10.1055/s-2005-872705
2. Bell, E., Creelman, R. A., and Mullett, J. E. (1995). A chloroplast lipoxygenase is required for wound-induced accumulation of jasmonic acid in*Arabidopsis*. *Proc. Natl. Acad. Sci. U.S.A.* 92, 8675–8679. doi: 10.1073/pnas.92.19.8675
3. Birkett, M. A., Campbell, C. A., Chamberlain, K., Guerrieri, E., Hick, A. J., Martin, J. L., et al. (2000). New roles for *cis*-jasmone as an insect semiochemical and in plant defense. *Proc. Natl. Acad. Sci. U.S.A.* 97, 9329–9334. doi: 10.1073/pnas.160241697

4. Bokshi, A. I., Jobling, J., and McConchie, R. (2008). A single application of Milsana® followed by Bion® assists in the control of powdery mildew in cucumber and helps overcome yield losses. *J. Hort. Sci. Biotech.* 83, 701–706.
5. Bostock, R. M. (2005). Signal crosstalk and induced resistance: straddling the line between cost and benefit. *Annu. Rev. Phytopathol.* 43, 545–580. doi: 10.1146/annurev.phyto.41.052002.095505
6. Bruce, T. J. A., Pickett, J. A., and Smart, L. E. (2003). *cis*-jasmone switches on plant defence against insects. *Pest. Out.* 14, 96–98. doi: 10.1039/b305499n
7. Cohen, Y. (2002). β-aminobutyric acid-induced resistance against plant pathogens. *Plant Dis.* 86, 448–457. doi: 10.1094/PDIS.2002.86.5.448
8. Cohen, Y., Reuveni, M., and Baider, A. (1999). Local and systemic activity of BABA (DL- 3-aminobutyric acid) against *Plasmopara viticola* in grapevines. *Eur. J. Plant Pathol.* 105, 351–361. doi: 10.1023/A:1008734019040
9. Cohen, Y., Rubin, A. E., and Kilfin, G. (2010). Mechanisms of induced resistance in lettuce against *Bremia lactucae* by DL-β-aminobutyric acid (BABA). *Eur. J. Plant Pathol.* 126, 553–573. doi: 10.1007/s10658-009-9564-6
10. Farmer, E. E., and Ryan, C. A. (1990). Interplant communication: airborne methyl jasmonate induces synthesis of proteinase inhibitors in plant leaves. *Proc. Natl. Acad. Sci. U.S.A.* 87, 7713–7716. doi: 10.1073/pnas.87.19.7713
11. Friedrich, L., Lawton, K., Ruess, W., Masner, P., Specker, N., Gut Rella, M., et al. (1996). A benzothiadiazole derivative induces systemic acquired resistance in tobacco. *Plant J.* 10, 61–70. doi: 10.1046/j.1365-313X.1996.10010061.x
12. Gaffney, T., Friedrich, L., Vernooij, B., Negrotto, D., Nye, G., Uknes, S., et al. (1993). Requirement of salicylic acid for the induction of systemic acquired resistance. *Science* 261, 754–756. doi: 10.1126/science.261.5122.754
13. Glazebrook, J. (2005). Contrasting mechanisms of defense against biotrophic and necrotrophic pathogens. *Annu. Rev. Phytopathol.* 43, 205–227. doi: 10.1146/annurev.phyto.43.040204.135923
14. Herman, M. A. B., Restrepo, S., and Smart, C. D. (2007). Defense gene expression patterns of three SAR-induced tomato cultivars in the field.*Physiol. Mol. Plant Pathol.* 71, 192–200. doi: 10.1016/j.

pmpp.2008.02.002

15. Home-Grown Cereals Authority (HGCA). (2009). *Recommended List for Spring Barley, 2009. © Agriculture and Horticulture Development Board*. Stoneleigh, UK: HGCA publications.
16. Ivors, K., and Louws, F. (2007). *Foliar Fungicide Spray Guide for Tomatoes in NC*. Available online at:www.ces.ncsu.edu/fletcher/programs/plantpath/tomato-spray-guide
17. Jakab, G., Cottier, V., Toquin, V., Rigoli, G., Zimmerli, L., Metraux, J.-P., et al. (2001). β- aminobutyric acid-induced resistance in plants. *Eur. J. Plant Pathol.* 107, 29–37. doi: 10.1023/A:1008730721037
18. Lawton, K. A., Friedrich, L., Hunt, M., Weymann, K., Delaney, T. P., Kessmann, H., et al. (1996). Benzothiadiazole induces disease resistance in*Arabidopsis* by activation of Systemic acquired resistance signal transduction pathway. *Plant J.* 10, 71–82. doi: 10.1046/j.1365-313X.1996.10010071.x
19. Leadbeater, A., and Staub, T. (2007). "Exploitation of induced resistance: a commercial Perspective," in *Induced Resistance For Plant Defence: A Sustainable Approach To Crop Protection*, eds D. Walters, A. Newton, and G. Lyon (Oxford: Blackwell Publishing), 229–242.
20. Liljeroth, E., Bengtsson, T., Wiik, L., and Andreasson, E. (2010). Induced resistance in potato to *Phytophthora infestans* – effects of BABA in greenhouse and field tests with different potato varieties. *Eur. J. Plant Pathol.* 127, 171–183. doi: 10.1007/s10658-010-9582-4
21. Lyon, G. (2007). "Agents that can elicit induced resistance," in *Induced Resistance For Plant Defence: A Sustainable Approach To Crop Protection*, eds D. Walters, A. Newton, and G. Lyon (Oxford: Blackwell Publishing), 9–29.
22. Miles, A. K., Willingham, S. L., and Cooke, A. W. (2004). Field evaluation of strobilurins and a plant activator for the control of citrus black spot. *Aust. Plant Pathol.* 33, 371–378. doi: 10.1071/AP04025
23. Miles, A. K., Willingham, S. L., and Cooke, A. W. (2005). Field evaluation of a plant activator, captan, chlorothalonil, copper hydroxide, iprodione, mancozeb and strobilurins for the control of citrus brown spot of mandarin. *Aust. Plant Pathol.* 34, 63–71. doi: 10.1071/AP04085
24. Oxley, S. J. P., and Walters, D. R. (2012). Control of light leaf spot (*Pyrenopeziza brassicae*) on winter oilseed rape (*Brassica napus*) with resistance elicitors. *Crop Prot.* 40, 59–62. doi: 10.1016/j.cropro.2012.04.028

25. Pickett, J. A., Birkett, M. A., Moraes, M. C. B., Bruce, T. J. A., Chamberlain, K., Gordon-Weeks, R., et al. (2007). *cis*-jasmone as an allelopathic agent in inducing plant defence. *Allelopathy J.* 19, 109–117.
26. Pieterse, C. M. J., van der Does, D., Zamioudis, C., Leon-Reyes, A., and van Wees, S. C. M. (2012). Hormonal modulation of plant immunity. *Annu. Rev. Plant Cell Dev. Biol.* 28, 489–521. doi: 10.1146/annurev-cellbio-092910-154055
27. Pozo, M. J., and Azcón-Aguilar, C. (2007). Unraveling mycorrhiza-induced resistance. *Curr. Opin. Plant Biol.* 10, 393–398. doi: 10.1016/j.pbi.2007.05.004
28. Reglinski, T., and Walters, D. (2009). "Induced resistance for plant disease control," in *Disease Control in Crops: Biological And Environmentally Friendly Approaches*, ed D. Walters (Oxford: Wiley-Blackwell), 62–92.
29. Spoel, S. H., and Dong, X. (2012). How do plants achieve immunity? Defence without specialized immune cells. *Nat. Rev. Immunol.* 12, 89–100. doi: 10.1038/nri3141
30. Spoel, S. H., Johnson, J. S., and Dong, X. (2007). Regulation of tradeoffs between plant defences against pathogens with different lifestyles. *Proc. Natl. Acad. Sci. U.S.A.* 104, 18842–18847. doi: 10.1073/pnas.0708139104
31. Spoel, S. H., Kornneef, A., Claessens, S. M. C., Korzrlius, J. P., van Pelt, J. A., Mueller, J. A., et al. (2003). NPR1 modulates cross-talk between salicylate and jasmonate dependent defense pathways through a novel function in the cytosol. *Plant Cell* 15, 760–770. doi: 10.1105/tpc.009159
32. Thaler, J. S., Stout, M. J., Karban, R., and Duffey, S. S. (1996). Exogenous jasmonates simulate insect wounding in tomato plants (*Lycopersicon esculentum*) in the laboratory and field. *J. Chem. Ecol.* 22, 1767–1781. doi: 10.1007/BF02028503
33. Ton, J., Jakab, G., Toquin, V., Flors, V., Iavicoli, A., Maeder, M. N., et al. (2005). Dissecting the β-aminobutyric acid induced priming phenomenon in *Arabidopsis*. *Plant Cell* 17, 987–999. doi: 10.1105/tpc.104.029728
34. van Loon, L. C., Rep, M., and Pieterse, C. M. J. (2006). Significance of inducible defense-related proteins in infected plants. *Annu. Rev. Phytopathol.* 44, 135–162. doi: 10.1146/annurev.phyto.44.070505.143425
35. Walters, D., Newton, A., and Lyon, G. (2007). *Induced Resistance for Plant Defence: A Sustainable Approach to Crop Protection*. Oxford: Blackwell Publishing. doi: 10.1002/9780470995983
36. Walters, D. R. (2009). Are plants in the field already induced? Implications for practical disease control. *Crop Prot.* 28, 459–465. doi: 10.1016/j.

cropro.2009.01.009

37. Walters, D. R., and Fountaine, J. M. (2009). Practical application of induced resistance to plant diseases: an appraisal of effectiveness under field conditions. *J. Agric. Sci.* 147, 523–535. doi: 10.1017/S0021859609008806
38. Walters, D. R., Havis, N. D., Sablou, C., and Walsh, D. J. (2011b). Possible trade-off associated with use of a combination of resistance elicitors.*Physiol. Mol. Plant Pathol.* 75, 188–192. doi: 10.1016/j.pmpp.2011.02.001
39. Walters, D. R., and McRoberts, N. (2007). Plants and biotrophs: a pivotal role for cytokinins? *Trends Plant Sci.* 11, 581–586. doi: 10.1016/j.tplants.2006.10.003
40. Walters, D. R., McRoberts, N., and Fitt, B. D. L. (2008). Are green islands red herrings? Significance of green islands in plant interactions with pathogens and pests. *Biol. Rev.* 83, 79–102. doi: 10.1111/j.1469-185X.2007.00033.x
41. Walters, D. R., Paterson, L., and Havis, N. D. (2010). "Control of foliar diseases of spring barley using resistance elicitors," in *Proceedings Crop Protection in Northern Britain 2010* (Dundee), 91–96.
42. Walters, D. R., Paterson, L., Sablou, C., and Walsh, D. J. (2011a). Existing infection with *Rhynchosporium secalis* compromises the ability of barley to express induced resistance. *Eur. J. Plant Pathol.* 130, 73–82. doi: 10.1007/s10658-010-9733-7
43. Walters, D. R., Ratsep, J., and Havis, N. D. (2013). Controlling crop diseases using induced resistance: challenges for the future. *J. Exp. Bot.* 64, 1263–1280. doi: 10.1093/jxb/ert026
44. Zadoks, J. C., Chang, T. T., and Konzak, C. F. (1974). A decimal code for the growth stages of cereals. *Weed Res.* 14, 415–421. doi: 10.1111/j.1365-3180.1974.tb01084.x

Chapter 5

A REVIEW ON BIOLOGICAL CONTROL OF FUNGAL PLANT PATHOGENS USING MICROBIAL ANTAGONISTS

Asghar Heydari[1] and Mohammad Pessarakli[2]

[1]Department of Plant Pathology, Iranian Research Institute of Plant Protection, Tehran, Iran

[2]Department of Plant Sciences, University of Arizona, Tucson, AZ, 85721, USA

ABSTRACT

The objective of this study was to review the published research works on biological control of fungal plant diseases during past 50 years. Fungal plant pathogens are among the most important factors that cause serious losses to agricultural products every year. Biological control of plant diseases including fungal pathogens has been considered a viable alternative method to chemical control. In plant pathology, the term biocontrol applies to the use of microbial antagonists to suppress diseases. Throughout their lifecycle, plants and pathogens interact with a wide variety of organisms. These interactions can significantly affect plant health in various ways. Different mode of actions of biocontrol-active microorganisms in controlling fungal plant diseases include hyperparasitism, predation, antibiosis, cross protection, competition for site and nutrient and induced resistance. Successful application of biological control strategies requires more knowledge-intensive management. Various methods for application of biocontrol agents include: application directly to the infection court at a high population level to swamp the pathogen, application at one place in which biocontrol microorganisms are applied at one place (each crop year) but at lower populations which then multiply and spread to other plant parts and give protection against pathogens and one time or occasional application that maintain pathogen populations below threshold levels. Commercial use and application of biological disease control have been slow mainly due to their variable performances under different environmental conditions in the field. To overcome this problem and in order to take the biocontrol technology to the field and improve the commercialization of biocontrol, it is important to develop new formulations of biocontrol microorganisms with higher degree of stability and survival. Majority of biocontrol products are applied against seed borne and soil borne fungal pathogens, including the causal agents of seed rot, damping-off and root rot diseases. These products are mostly used as seed treatment and have been effective in protecting several major crops such as wheat, rice, corn, sugar beet and cotton against fungal pathogens. However, in some cases, biocontrol microorganisms have also been tested as spray application on foliar diseases, including powdery mildew, downy mildew, blights and leaf spots. A few post-harvest fungal diseases have also been controlled by the use of antagonistic fungi and bacteria. Biocontrol microorganisms are also being used as the form of composts in some plants. Research data and observations in nurseries have shown that addition of composted organic matter to potting mixes results in suppression of soil borne diseases. A significant improvement have been made in different aspects of biological control of fungal plant diseases, but this area still need much more development and investigations to solve the existing problems. In order to have more effective biological control strategies in the future, it is critical to carry out more research studies on some less developed aspects of biocontrol, including development of novel formulations, understanding the impact of environmental factors on biocontrol agents, mass production of biocontrol microorganisms and the use of biotechnology and nano-technology in improvement of biocontrol mechanisms and strategies. Future outlooks of biocontrol of plant diseases is bright and promising and with the growing demand for biocontrol products among the growers, it is possible to use the biological control as an effective strategy to manage plant diseases, increase yield, protect the environment and biological resources and approach a sustainable agricultural system.

INTRODUCTION

Plant pests (harmful insects, parasitic weeds and pathogens) are among the most important biotic agents causing serious losses and damages to agricultural products. Plant pests need to be controlled to ensure food, feed and fiber production quantitatively and qualitatively. A number of different strategies are currently being employed to manage and control plant pests (Agrios, 1988; Baker, 1987; Cook, 1993; Bargabus et al., 2002, 2004; Benhamou, 2004; Chisholm et al., 2006; Heydari, 2007; Heydari et al., 2007; Islam et al., 2005; Kloepperet al., 2004). Beyond good agronomic and cultural practices, growers often rely heavily on chemical pesticide application (Agrios, 1988; Baker, 1987). However, the environmental pollution caused by excessive use of agrochemicals has led to considerable changes in people's attitudes towards the use of pesticides in agriculture. Today, there are strict regulations on chemical pesticide use and there is political pressure to remove the most hazardous chemicals from the market. In addition to the above-mentioned issues, the spread of plant diseases in natural ecosystems may preclude successful application of chemicals, because of the scale to which such applications might have to be applied. Consequently, some pest management researchers have focused their efforts on developing alternative inputs to synthetic chemicals for controlling pests and diseases (Baker, 1987; Cook, 1993).

Plant diseases are mostly controlled by the use of chemical pesticides and in some cases by cultural practices (Agrios, 1988; Cook, 1993). However, the widespread use of chemicals in agriculture has been a subject of public concern and scrutiny due to the potential harmful effects on the environment, their undesirable effects on non-target organisms and possible carcinogenicity of some chemicals (Agrios, 1988;Cook, 1993; Heydari, 2007; Heydari et al., 2007). Other problems include development of resistant races of pathogens, a gradual elimination and phasing out of some available pesticides and the reluctance of some chemical companies to develop and test new chemicals due to the problems with registration process and cost (Cook, 1993). The need for the development of non-chemical alternative methods to control plant diseases is therefore clear.

Biological control of plant diseases has been considered a viable alternative method to manage plant diseases (Cook, 1993). Biological control is the inhibition of growth, infection or reproduction of one organism using another organism (Cook, 1993; Baker, 1987). Biocontrol is environmentally safe and in some cases is the only option available to protect plants against pathogens (Cook, 1993). Biological controlemploys natural enemies of pests or pathogens to eradicate or control their population. This can involve the introduction of exotic species, or it can be a matter of harnessing whatever form of biological control exists naturally in the ecosystem. The induction of plant resistance using non-pathogenic or incompatible microorganisms is also a form of biological control (Cook, 1993; Schouten et al., 2004). Fungal plant diseases are considered the most important microbial agents causing serious losses in the agriculture annually (Agrios, 1988). Some fungal diseases that have successfully been controlled using biological agents are pathogens of pruning wounds and other cut surfaces, diseases of leaves and flowers, such as powdery mildew, diseases of fruits and vegetables, such as Botrytis and fungal pathogens in the soil (Agrios, 1988;Baker, 1987; Cook, 1993; Heydari, 2007; Heydari et al., 2004, 2007; Heydari and Misaghi, 1998, 1999, 2003). A variety of biological controls are available for use, but further development and effective adoption will require a greater understanding of the complex interactions among plants, people and the environment. To that end, the objectives of this review chapter is to present an advanced survey of the nature and practice of biological control as it is applied to the suppression of plant diseases. In this review, different aspects of biological control of fungal plant diseases including definitions, modes of action, application strategies, current status and future development and outlooks will be discussed.

TERMINOLOGY

The term biological control and its abbreviated synonym biocontrol have been used in different fields of biology, most notably entomology and plant pathology. In plant pathology, the term applies to the use of microbial antagonists to suppress diseases as well as the use of host-specific pathogens to control weed populations (Cook, 1993). In both fields, the organism that suppresses the pest or pathogen is referred to as the Biological Control Agent (BCA). More broadly, the term biological control also has been applied to the use of the natural products extracted or fermented from various sources (Cook, 1993). These formulations may be very simple mixtures of natural ingredients with specific activities or complex mixtures with multiple effects on the host as well as the target pest or pathogen. While such inputs may mimic the activities of living organisms, non-living inputs should more properly be referred to as biopesticides or biofertilizers, depending on the primary benefit provided to the host plant (Cook, 1993).

The various definitions offered in the scientific literature have sometimes caused confusion and controversy. For example, members of the United States National Research Council took into account modern biotechnological developments and referred to biological control as the use of natural or modified organisms, genes, or gene products, to reduce the effects of undesirable organisms and to favor desirable organisms such as crops, beneficial insects and microorganisms, but this definition spurred much subsequent debates and it was frequently considered too broad by many scientists who worked in the field.

Published definitions of biocontrol differ depending on the target of suppression; number, type and source of biological agents and the degree and timing of human intervention (Cook, 1993). Most broadly, biological control is the suppression of damaging activities of one organism by one or more other organisms, often referred to as natural enemies. With regards to plant diseases, suppression can be accomplished in many ways. If grower's activities are considered relevant, cultural practices such as the use of rotations and planting of disease resistant cultivars (whether naturally selected or genetically engineered) would be included in the definition (Cook, 1993).

Because the plant host responds to numerous biological factors, pathogenic and non-pathogenic, induced host resistance might be considered a form of biological control (Cook, 1993). More narrowly, biological control refers to the purposeful utilization of introduced or resident living organisms, other than disease resistant host plants, to suppress the activities and populations of one or more plant pathogens. This may involve the use of microbial inoculants to suppress a single type or class of plant diseases. This may also involve managing soils to promote the combined activities of native soil and plant-associated organisms that contribute to general suppression (Cook, 1993). Most narrowly, biological control refers to the suppression of a single pathogen by a single antagonist, in a single cropping system. Most specialists in the field would concur with one of the narrower definitions presented above.

INTERACTIONS BETWEEN PLANTS AND BENEFICIAL MICROBES

Throughout their lifecycle, plants and pathogens interact with a wide variety of organisms. These interactions can significantly affect plant health in various ways (Agrios, 1988; Bull et al., 2002; Katska, 1994; Chisholm et al., 2006; Fitter and Garbaye, 1994; McSpadden-Gardener and Weller, 2001). In order to understand the mechanisms of biological control, it is helpful to appreciate the different ways that organisms interact. Note too, in order to interact, organisms must have some form of direct or indirect contact. The types of interactions between plants and microorganisms have been referred to as mutualism, protocooperation, commensalisms, neutralism, competition, amensalism, parasitism and predation (Bankhead et al., 2004; Bull et al., 2002; Katska, 1994; Chisholm et al., 2006; Fitter and Garbaye, 1994; Hoitink and Boehm, 1999). While the terminology has been developed for macroecology, examples of all of these types of interactions can be found in the natural world at both the macroscopic and microscopic level. And, because the development of plant diseases involves both plants and microbes, the interactions that lead to biological control take place at multiple levels of scale (Bull et al., 2002; Katska, 1994; Chisholm et al., 2006; Fitter and Garbaye, 1994).

From the plant's point of view, biological control may be considered a positive result arising from different specific and non-specific interactions (Cook, 1993; Weller et al., 2002). We can begin to classify and functionally delineate the diverse components of ecosystems that contribute to biological control. Mutualism is an association among several species where all of them are benefited from this association (Biermann and Linderman, 1983; Bull et al., 2002; Katska, 1994; Chisholm et al., 2006; Duchesne, 1994; Fitter and

Garbaye, 1994; Garcia-Garrido and Ocampo, 1989; Kerry, 2000). Sometimes, it can be an obligatory relation involving close physical and biochemical contact between two organisms, such as those between plants and mycorrhizal fungi (Bull et al., 2002; Katska, 1994; Chisholm et al., 2006; Fitter and Garbaye, 1994). However, they are generally facultative and opportunistic.

For example, Rhizobium bacteria reproduce either in the soil or, to a much greater degree, through their mutualistic association with legume plants. These types of mutualism can contribute to biological control, by providing plant with improved nutrition and/or by stimulating host defense mechanism and ability (Chisholm et al., 2006; Fitter and Garbaye, 1994). Many of the microorganisms isolated and classified as biocontrol agents (BCA) can be considered facultative mutualists, because host and disease suppression by them will vary depending on the prevailing environmental conditions (Cook, 1993).

Commensalism is also a symbiotic interaction between two living organisms, where one organism benefits and the other is neither harmed nor benefited (Fitter and Garbaye, 1994). Most plant-associated microorganisms are assumed to be commensals with regards to the host plant, because their presence, individually or in total, rarely results in positive or negative consequences to the plant (Katska, 1994; Chisholm et al., 2006). While the presence of these microorganisms may present a variety of challenges to an infecting pathogen, their absence decreases pathogen infection or disease severity and is indicative of commensal interactions (Cook, 1993).

Biological interactions in which the population density of one species has absolutely no effect on the other are called neutralism (Berg et al., 2005; Chisholm et al., 2006). Related to biological control, an inability to associate the population dynamics of pathogen with that of another organism would indicate neutralism (Chisholm et al., 2006). In contrast, antagonism between organisms results in a negative outcome for one or both. Competition within and between species caused a decreased growth, activity, and/or fecundity of the interacting organisms (Cook, 1993). Biocontrol can occur when non-pathogens compete with pathogens for nutrients and sites in host plant. Direct interactions that benefit one population at the expense of another also affect our understanding of biological control (Cook, 1993).

Parasitism is also a symbiotic relation in which two organisms coexist over a prolonged period of time (Cook, 1993; Chisholm et al., 2006; Lo et al., 1997). In this type of interaction, one organism, usually the physically smaller (parasite) benefits and the other (host) is harmed. The activities of various hyperparasites, for example those agents that parasitize plant pathogens, can result in biocontrol (Lo et al., 1997). Another interesting contribution to biocontrol is when host infection and parasitism by relatively avirulent pathogens may lead to biocontrol of more virulent pathogens through the stimulation of host defense systems (Cook, 1993). Finally predation refers to the hunting and killing of one organism by another for consumption and sustenance. While the term predator typically refers to animals that feed at higher trophic levels in the macroscopic world, it has also been applied to the actions of microorganisms such as protists and mesofauna, e.g. fungal feeding nematodes and microarthropods, that consume pathogen biomass for sustenance (Cook, 1993).

Biological control can result in various forms of these types of interactions, depending on the environmental conditions within which they occur. Significant biological control, as was described above, generally arises from manipulating mutualisms between microorganisms and their plant hosts or from manipulating antagonisms between microbes and pathogens (Bull et al., 2002; Katska, 1994; Chisholm et al., 2006;Fitter and Garbaye, 1994).

MECHANISMS OF BIOLOGICAL CONTROL

Since biological control is a result of many different types of interactions among microorganisms, scientists have concentrated on characterization of mechanisms occurring in different experimental situations (Audenaert et al., 2002; De Meyer and Hofte, 1997; Elad and Baker, 1985; Heydari et al., 1997; Homma et al., 1989; Howell et al., 1988; Islam et al., 2005; Meziane et al., 2005; Ryu et al., 2004; Van Dijk and Nelson, 2000). In all cases, pathogens are antagonized by the presence and activities of other microorganisms that they encounter.

Direct antagonism results from physical contact and/or a high-degree of selectivity for the pathogen by the mechanism(s) expressed by the biocontrol active microorganisms. In this type of interaction, Hyperparasitism by obligate parasites of a plant pathogen would be considered the most direct type of mechanism because the activities of no other organism would be required to exert a suppressive effect (Harman et al., 2004; Linderman, 1994). In contrast, indirect antagonism is resulted from the activities that do not involve targeting a pathogen by a biocontrol active microorganism. Improvement and stimulation of plant host defense mechanism by non-pathogenic microorganisms is the most indirect form of antagonism (Kloepper et al., 1980; Lafontaine and Benhamo, 1996; Leeman et al., 1995; Maurhofer et al., 1994; Silva et al., 2004). While many studies have concentrated on the establishment of the importance of specific mechanisms of biocontrol to particular pathosystems, all of the mechanisms described below are likely to be operating to some extent in all natural and managed ecosystems.

The most effective biocontrol active microorganisms studied appear to antagonize plant pathogens employing several modes of actions (Cook, 1993). For example, pseudomonads known to produce the antibiotic 2, 4-diacetylphloroglucinol (DAPG) may also induce host defenses (Kloepper et al., 1980; Lafontaine and Benhamou, 1996; Leeman et al., 1995; Maurhofer et al., 1994; Silva et al., 2004). Additionally, DAPG-producers bacterial antagonists can aggressively colonize roots, a trait that might further contribute to their ability to suppress pathogen activity in the rhizosphere of plant through competition for organic nutrients. However, the most important modes of actions of biocontrol active microorganisms are as follows:

Mycoparasitism

In Hyperparasitism, the pathogen is directly attacked by a specific biocontrol agent (BCA) that kills it or its propagules. Four major groups of hyperparasites have generally been identified which include hypoviruses, facultative parasites, obligate bacterial pathogens and predators. An example of hypoparasites is the virus that infects Cryphonectria parasitica, the fungal causal agent of chestnut blight, which causes hypovirulence, a reduction in pathogenicity of the pathogen. This phenomenon has resulted in the control of chestnut blight in many places (Milgroom and Cortesi, 2004). However, the interaction of virus, fungus, tree and environment determines the success or failure of hypovirulence.

In addition to hypoviruses several fungal hypoparasites have also been identified including those that attack sclerotia (e.g., Coniothyrium minitans) or others that attack fungal hyphae (e.g. Pythium oligandrum). In some cases, a single fungal pathogen can be attacked by multiple hyperparasites. For example, Acremonium alternatum, Acrodontium crateriforme, Ampelomyces quisqualis, Cladosporium oxysporum and Gliocladium virens are just a few of the fungi that have the capacity to parasitize powdery mildew pathogens (Milgroom and Cortesi, 2004).

In contrast to hyperparasitism, microbial predation is more general, non-specific and generally provides less predictable levels of disease control. Some biocontrol agents exhibit predatory behavior under nutrient-limited conditions. Such as Trichoderma, a fungal antagonist that produces a range of enzymes that are directed against cell walls of pathogenic fungi. However, when fresh bark is used in composts,Trichoderma sp. does not directly attack the plant pathogen, Rhizoctonia solani. But, in decomposing bark, the concentration of readily available cellulose decreases and this activates the chitinase genes of Trichoderma sp. Which, in turn, produce chitinase to parasitize R. solani (Benhamou and Chet, 1997).

Antibiosis

Many microbes produce and secrete one or more compounds with antibiotic activity (Homma et al., 1989; Howell and Stipanovic, 1980; Islam et al., 2005; Leclére et al., 2005; Shahraki et al., 2009; Shanahan et al., 1992; Thomashow et al., 1990; Thomashow and Weller, 1988). In a general definition antibiotics are microbial toxins that can, at low concentrations, poison or kill other microorganisms. It has been shown that some antibiotics produced by microorganisms are particularly effective against plant pathogens and the diseases they cause (Homma et al., 1989; Howell and Stipanovic, 1980; Islam et al., 2005; Shanahan et al., 1992; Thomashow et al., 1990, 2002; Thomashow and Weller, 1988). In all cases, the antibiotics have been shown to be particularly effective at suppressing growth of the target pathogen in vitro and/or in situconditions. An effective antibiotic must be produced in sufficient quantities (dose) near the pathogen. In situ production of antibiotics by several different biocontrol agents has been studied (Thomashow et al., 1990). While several procedures have been developed to ascertain when and where biocontrol agents may produce antibiotics detecting expression in the infection court is difficult because of the heterogenous distribution of plant-associated microbes and the potential sites of infection (Thomashow et al., 1990).

However, in some cases, the relative importance of antibiotic production by biocontrol bacteria has been demonstrated. For example, mutant strains incapable of producing phenazines (Thomashow and Weller, 1988) or phloroglucinols (Keel et al., 1989) have been shown to be equally capable of colonizing the rhizosphere, but much less capable of suppressing soil borne root diseases than the corresponding wild-type and complemented mutant strains. Many biocontrol strains have been shown to produce multiple antibiotics which can suppress one or more pathogens (Homma et al., 1989; Howell and Stipanovic, 1980; Islam et al., 2005; Shanahan et al., 1992; Thomashow et al., 1990; Thomashow and Weller, 1988). The ability of production of several antibiotics probably results in suppression of diverse microbial competitors and plant pathogens.

Metabolite Production

Many biocontrol active microorganisms produce other metabolites that can interfere with pathogen growth and activities. Lytic enzymes are among these metabolites that can break down polymeric compounds, including chitin, proteins, cellulose, hemicellulose and DNA (Anderson et al., 2004; Howell et al., 1988; Loper and Buyer, 1991; Ordentlich et al., 1988; Press et al., 2001; Wilhite et al., 2001). Studies have shown that some of these metabolites can sometimes directly result in the suppression of plant pathogens. For example, control of Sclerotium rolfsii by Serratia marcescens appeared to be mediated by chitinase expression (Ordentlich et al., 1988). It seems more likely that antagonistic activities of these metabolites are indicative of the need to degrade complex polymers in order to obtain carbon nutrition. Microorganisms that show a preference in colonizing and suppression of plant pathogens might be classified as biocontrol agents. For example, Lysobacter and Myxobacteria that produce lytic enzymes have been shown to be effective against some plant pathogenic fungi (Bull et al., 2002).

Studies have shown that some products of lytic enzyme activity may have indirect efficacy against plant pathogens. For example, oligosaccharides derived from fungal cell walls have been shown to induce plant host defenses (Howell et al., 1988). It is believed that the effectiveness of the above compounds against plant pathogens is dependent on the composition and carbon and nitrogen sources of the soil and rhizosphere. For example, in post-harvest disease control, addition of chitosan which is a non-toxic and biodegradable polymer of beta-1, 4-glucosamine produced from chitin by alkaline deacylation stimulated microbial degradation of pathogens (Benhamou, 2004). Amendment of plant growth substratum with chitosan suppressed the root rot caused by Fusarium oxysporum f. sp. radicis-lycopersici in tomato (Lafontaine and Benhamou, 1996).

In addition to the above-mentioned metabolites, other microbial byproducts may also play important roles in plant disease biocontrol (Phillipset al., 2004). For example, Hydrogen cyanide (HCN) effectively blocks the cytochrome oxidase pathway and is highly toxic to all aerobic microorganisms at picomolar concentrations (Ramette et al., 2003). The production of HCN by certain fluorescent pseudomonads is believed to be effective against plant pathogens. Results of some research studies in this regard have shown that P. fluorescens CHA0, an antagonistic bacterium, produces antibiotics including siderophores and HCN, but suppression of black rot of tobacco caused by Thielaviopsis basicolaappeared to be due primarily to HCN production. In another study Howell et al. (1988) reported that volatile compounds such as ammonia produced by Enterobacter cloacae were involved in the suppression of cotton seedling damping-off caused by Pythium ultimum.

Competition

The nutrient sources in the soil and rhizosphere are frequently not sufficient for microorganisms. For a successful colonization of phytosphere and rhizosphere a microbe must effectively compete for the available nutrients (Elad and Baker, 1985; Keel et al., 1989; Loper and Buyer, 1991). On plant surfaces, host-supplied nutrients include exudates, leachates, or senesced tissue. In addition to these, nutrients can also be obtained from waste products of other organisms such as insects and the soil. This is a general believe that competition between pathogens and non-pathogens for nutrient resources is an important issue in biocontrol (Elad and Baker, 1985; Keel et al., 1989; Loper and Buyer, 1991). Ii is also believed that competition for nutrients is more critical for soil borne pathogens, including Fusarium and Pythiumspecies that infect through mycelial contact than foliar pathogens that germinate directly on plant surfaces and infect through appressoria and infection pegs (Elad and Baker, 1985; Keel et al., 1989; Loper and Buyer, 1991). Results of a study by Anderson et al. (1988) revealed that production of a particular plant glycoprotein called agglutinin was correlated with potential of Pseudomonas putida to colonize the root system. P. putida mutants deficient in this ability exhibited reduced capacity to colonize the rhizosphere and a corresponding reduction in Fusarium wilt suppression in cucumber (Tari and Anderson, 1988).

It has been shown that non-pathogenic plant-associated microrganisms generally protect the plant by rapid colonization and thereby exhausting the limited available substrates so that none are available for pathogens to grow. For example, effective catabolism of nutrients in the spermosphere has been identified as a mechanism contributing to the suppression of Pythium ultimum by Enterobacter cloacae (Van Dijk and Nelson, 2000; Kageyama and Nelson, 2003). At the same time, these microbes produce metabolites that are effective in suppression of pathogens. These microbes colonize the sites where water and carbon-containing nutrients are most readily available, such as exit points of secondary roots, damaged epidermal cells and nectaries and utilize the root mucilage.

Competition for rare but essential micronutrients, such as iron, has also been shown to be important in biological disease control. Iron is extremely limited in the rhizosphere, depending on soil pH. In highly oxidized and aerated soil, iron is present in ferric form (Kageyama and Nelson, 2003; Shahraki et al., 2009), which is insoluble in water and the concentration may be extremely low. This very low concentration cannot support the growth of microorganisms. To survive in such environment, organisms were found to secrete iron-binding ligands called Siderophores having high ability to obtain iron from the micro-organisms (Shahraki et al., 2009). Almost all microorganisms produce siderophores, of either the catechol type or hydroxamate type (Kageyama and Nelson, 2003).

A direct correlation was established in vitro between siderophore synthesis in fluorescent pseudomonads and their capacity to inhibit germination of chlamydospores of F. oxysporum (Elad and Baker, 1985). It was shown that mutants incapable of producing some siderophores, such as pyoverdine, were reduced in their capacity to suppress different plant pathogens (Keel et al., 1989; Loper and Buyer, 1991). The increased efficiency in iron uptake of the commensal microorganisms is thought to be a critical factor in their root colonization ability which is a major factor in biocontrol performance of bacterial antagonists.

Induction of resistance: Plants actively respond to a variety of environmental stimulating factors, including gravity, light, temperature, physical stress, water and nutrient availability and chemicals produced by soil and plant associated microorganisms (Audenaert et al., 2002; De Meyer and Hofte, 1997; Kloepper et al., 1980; Leeman et al., 1995; Moyne et al., 2000; Vallad and Goodman, 2004; Van Loon et al., 1998; Van Peer and Schippers, 1992; Van Wees et al., 1997). Such stimuli can either induce or condition plant host defenses through biochemical changes that enhance resistance against subsequent infection by a variety of pathogens. Induction of host defenses can be local and/or systemic in nature, depending on the type, source and amount of stimulation agents (Audenaert et al., 2002; De Meyer and Hofte, 1997; Kloepper et al., 1980; Leeman et al., 1995; Moyne et al., 2000; Vallad and Goodman, 2004; Van Loon et al., 1998; Van Peer and Schippers, 1992; Van Wees et al., 1997).

Recently, plant pathologists have begun to characterize the determinants and pathways of induced resistance stimulated by biological control agents and other non-pathogenic microorganisms (Audenaert et al., 2002; Moyne et al., 2000; Vallad and Goodman, 2004). The first pathway called Systemic Acquired Resistance (SAR), is mediated by Salicylic Acid (SA), a chemical compound which is usually produced after pathogen infection and typically leads to the expression of Pathogenesis-related (PR) proteins (Vallad and Goodman, 2004). These PR proteins include a variety of enzymes some of which may act directly to lyse invading cells, reinforce cell wall boundaries to resist infections, or induce localized cell death (Vallad and Goodman, 2004).

Second pathway, called Induced Systemic Resistance (ISR), is mediated by Jasmonic Acid (JA) and/or ethylene, which are produced following applications of some nonpathogenic rhizobacteria (Audenaert et al., 2002; De Meyer and Hofte, 1997; Kloepper et al., 1980; Leeman et al., 1995;Moyne et al., 2000; Van Loon et al., 1998; Van Peer and Schippers, 1992; Van Wees et al., 1997). Interestingly, the SA- and JA- dependent defense pathways can be mutually antagonistic and some bacterial pathogens take advantage of this to overcome the SAR. For example, pathogenic strains of Pseudomonas syringae produce coronatine, which is similar to JA, to overcome the SA-mediated pathway (Vallad and Goodman, 2004). Since the various host-resistance pathways can be activated to variable degrees by different microorganisms and insect feeding, it is therefore possible that multiple stimuli are constantly being received and processed by the plant. Thus, the magnitude and duration of host defense induction will likely vary over time. Only if induction can be controlled, i.e., by overwhelming or synergistically interacting with endogenous signals, will host resistance be increased (Audenaert et al., 2002; De Meyer and Hofte, 1997; Kloepper et al., 1980; Leeman et al., 1995; Moyne et al., 2000).

Some strains of root-colonizing microorganisms have been identified as potential elicitors of plant host defenses. For example, some biocontrol active strains of Pseudomonas sp. and Trichoderma sp. are known to strongly induce plant host defenses (Haas and Defago, 2005;Harman et al., 2004). In other instances, inoculation with Plant Growth Promoting Rhizobacteria (PGPR) have been shown to be effective in controlling multiple diseases caused by different fungal pathogens, including anthracnose (Colletotrichum lagenarium). A number of chemical elicitors of SAR and ISR such as salicylic acid, siderophore, lipopolysaccharides and 2, 3-butanediol may be produced by the PGPR strains upon inoculation (Ryu et al., 2004; Van Loon et al., 1998).

A substantial number of microbial products have been reported to elicit host defenses, indicating that host defenses are likely stimulated continually during the plant's lifecycle (Ryu et al., 2004; Van Loon et al., 1998). These inducers include lipopolysaccharides and flagellin from Gram-negative bacteria; cold shock proteins of diverse bacteria; transglutaminase, elicitins and a-glucans in Oomycetes; invertase in yeast; chitin and ergosterol in all fungi; and xylanase in Trichoderma (Ryu et al., 2004). These findings indicate that plants would detect the composition of their plant-associated microbial communities and respond to changes in the quantity, quality and localization of many different signals. The importance of such interactions is indicated by the fact that further induction of host resistance pathways, by chemical and microbiological inducers, is not always effective in improving plant health or productivity in the field (Vallad and Goodman, 2004).

METHODS OF APPLICATION OF ANTAGONISTS

Overall Application

Successful application of biological control strategies requires more knowledge-intensive management (Baker, 1987; Cook, 1993; Heydari et al., 2004; Shah-Smith and Burns, 1997). Understanding when and where biological control of plant pathogens can be profitable, requires an appreciation of its place within integrated pest management systems (Cook, 1993; Heydari et al., 2004; Shah-Smith and Burns, 1997).

In general, the foundation of a sound pest and disease management program in an annual cropping system begins with cultural practices that alter the farm landscape to promote crop health (Cook, 1993; Heydari et al., 2004; Shah-Smith and Burns, 1997). These include crop rotations that limit the availability of host material used by plant pathogens (Cook, 1993). Proper use of tillage can disrupt pathogen life cycles and prepare seed beds of optimal moisture and bulk density. Careful management of soil fertility and moisture can also limit plant diseases by minimizing plant stress (Cook, 1993). In nurseries and greenhouses environmental control can be more tightly regulated in terms of temperature, light, moisture and soil composition, but the design of such systems cannot wholly eliminate disease problems (Paulitz and Belanger, 2001).

The second layer of defense against pests consists of the quality of crop germplasm. Breeding for pathogen resistance including fungal pathogens contributes substantially to crop success in most regions (Cook, 1993). Newer technologies that directly incorporate genes into crop genomes, commonly referred to as genetic modification or genetic engineering, are bringing new traits into crop. Other technologies, such as seed washing, testing for pathogens and treatments are also used to keep germplasm pathogen-free. In perennial cropping systems, such as orchards and forests, germplasm quality may be more important than cultural practices, because rotation and tillage cannot be used as regularly (Agrios, 1988; Cook, 1993). Upon these two layers, growers can further reduce pathogen pressure by considering both biological and chemical inputs.

Biologically based inputs such as microbial fungicides can be used to interfere with pathogen activities. Registered biofungicides are generally labeled with short reentry intervals and pre-harvest intervals, giving greater flexibility to growers who need to balance their operational requirements and disease management goals. When living microorganisms are introduced, they may also augment natural beneficial populations to further reduce the damage caused by targeted pathogens (Cook, 1993; Heydari et al., 2004; Shah-Smith and Burns, 1997).

Applying to the Infection Site

Application directly to the infection court at a high population level to swamp the pathogen (inundate application), seed coating and treatment with antagonistic fungi and bacteria, e.g., *Trichoderma harzianum* and *Psudomonas fluorescens*(Cook, 1993; Heydari and Misaghi, 2003; Heydari et al., 2004), antagonists applied to fruit for protection in storage, e.g., *Pseudomonas fluorescens* (De Capdeville et al., 2002; El-Ghaouth et al., 2000; Janisiewicz and Korsten, 2002; Janisiewicz and Peterson, 2004) and application to soil at the site of seed placement (Heydari and Misaghi, 2003). These types of applications are the most commonly used procedures which have resulted in the successful control of several fungal plant pathogens.

One place application: in this procedure, biocontrol microorganisms are applied at one place (each crop year), but at lower populations which then multiply and spread to other plant parts and give protection (augmentative application) against fungal pathogens. An Example of this method is Plant Growth Promoting Rhizobacteria (PGPR) and atoxigenic *Aspergillus flavus* on wheat seed scattered on the soil to spread to cotton flowers where they displace aflatoxin producing strains of A. flavus and fungal antagonists added to soil (Islam et al., 2005; Kloepperet al., 2004).

Occasional application: One time or occasional application maintains pathogen populations below threshold levels. In theory, parasites of the pathogen, or hypovirulent (disease carrying) strains of the pathogen, might be used and not require yearly repetition (e.g., hypovirulent strains of the chestnut blight pathogen) in which host plant is inoculated with attenuated strains of pathogenic that protects the host plant against the virulent strains of pathogen (Milgroom and Cortesi, 2004).

BIOCONTROL OF DIFFERENT FUNGAL PATHOGENS

Microorganisms naturally present in the plants ecosystem will help reduce disease potential or disease damage, but only if they are allowed to grow vigorously (Cook, 1993). They accomplish these tasks by competing with the pathogens for food sources, producing metabolites that inhibit the growth of the pathogens and physically eliminating the pathogens from the plant by occupying the space and sites first. Microorganisms not naturally present in plant environment can be introduced in an attempt to control diseases (Cook, 1993). This can be done by application of organic materials that contain natural microbial populations such as composts or natural microbial populations added to them including natural organic fertilizers with microbial supplements. In both cases, the products must be applied prior to disease development as they are preventive and not curative (Baker, 1987; Cook, 1993). Natural organic fertilizers should be used for their nutritional value (nitrogen and potassium) and not for any possible secondary effects.

Fungal plant pathogens are very diverse and cause diseases on different parts of plants such as root, stem, leaf, fruit, etc. In this section, application of biological control strategies for controlling fungal diseases on different parts of plants will be discussed.

The majority of research on biocontrol of fungal diseases have focused on soil borne diseases rather than foliar or post-harvest. According to the results of numerous research projects, several fungal and bacterial biocontrol agents have been used as seed and soil application to reduce the incidence of plant diseases caused by soil borne fungal pathogens (Cook, 1993; Heydari, 2007; Heydari et al., 2004; Heydari and Misaghi, 2003; Lo et al., 1995, 1996, 1997; McSpadden-Gardener, 2001; Naraghi et al., 2004; Ramette et al., 2003; Scheuerell et al., 2005). Since many plant pathogens can spread readily in the foliar parts, control of these diseases requires both suppression of initial plant infection and reduction of the infection rate (Lo et al., 1997). Granular applications of strain 1295-22 of *Trihoderma harzianum* has been shown to significantly inhibit disease severity of some plant diseases during the initial stage of disease development, most likely by reducing levels of the pathogen inoculum in the soil (Lo et al., 1995, 1996, 1997). It is apparent, therefore, that soil applications alone cannot effectively control the foliar phases of this disease.

Additives have been commonly used with fungicides to improve efficacy and they also may enhance the ability of biocontrol agents to reduce plant diseases. For example, it was reported that seed treatment using 10% Pelgel with solid matrix priming markedly enhanced the efficacy of Trichoderma strains to control Pythium sp. on various crops (Lo et al., 1997). Research has indicated that for control of multiple fungal plant diseases, greater control was obtained when Triton X-100 was included than when no additives, Pelgel, or Tween 20 were used (Lo et al., 1997). The use of specific surfactants with Trichoderma strains seems essential to obtain levels of control equivalent to those achieved with chemical fungicides. Detergents such as Triton X-100 may have several functions in biocontrol systems. They may slow the growth of pathogens more than that of the biocontrol agents or they may enhance wetting and adhesion of spores to infection courts (Lo et al., 1997). In preliminary experiments, both Tween 20 and Triton X-100 slowed the growth of both *T. harzianum* and the pathogens, but the ratio of the growth rates of *T. harzianum* and pathogens was greater with Triton X-100 than with Tween 20 (Lo et al., 1997).

Living organisms, in addition to yielding a large quantity of biomass of the bioprotectant fungus, must perform effectively in each application. To examine this, different spore formulations of *Trihoderma harzianum* were compared in a study for controlling plant diseases (Lo et al., 1996). It was found that all formulations provided equivalent levels of control, indicating that the method of spore production may not be a key factor in the efficacy of this fungal biocontrol agent in controlling these diseases (Lo et al., 1996). To predictably and successfully use biological control agents for fungal disease control, it is critical that their biology and ecology be more completely understood. Therefore, effective antagonists must become established in plant ecosystems and remain active against target pathogens during periods favorable for plant infection.

Broadcast application of granules of Trichoderma to control plant diseases has resulted in establishment of stable and effective populations of plants in soils (Lo et al., 1995, 1996, 1997). Similarly, it was shown that the populations of *T. harzianum* in soils treated with spray applications were as high as those in soils treated with granular formulations (Lo et al., 1996). Population levels of strain 1295-22 in about 5x10 5 cfu g-1 of soil significantly reduced Pythium blight, root rot and brown patch diseases (Lo et al., 1997). However, spray applications, even though resulted in numerically similar levels of root colonization, did not provide the same benefit. This may reflect the differences in inoculum potential of granules versus spray applications. Granules are applied as a several-millimeter-diameter particle that is completely colonized by the fungus. Conidial inoculum, on the other hand, is much smaller and would therefore be expected to possess lower inoculum potential than the granular formulation (Lo et al., 1997).

Conversely, in greenhouse and field experiments, it was found that *Trihoderma harzianum* significantly reduced some foliar phases of plant diseases when spray applications of conidial suspensions containing Triton X-100 were used (Lo et al., 1995, 1996, 1997). Weekly spray applications were as effective as the standard (monthly) fungicide applications. These results indicate that the efficacy of *T. harzianum* against plant diseases, especially those involving secondary infections, is very strongly affected by the method of application (Lo et al., 1997).

The ability to survive on the plant phylloplane is also a desirable trait for strains of fungal and bacterial antagonists used as biocontrol agents against foliar diseases (Lo et al., 1997). Spray applications of strain 1295-22 of *T. harzianum* has resulted in disease suppressive population levels on leaf (Lo et al., 1997). These populations were sufficient to suppress Pythium root rot, brown patch and dollar spot over the entire season. Thus, *T. harzianum* 1295-22 may possess a measure of phylloplane competence on the plants. The ideal biocontrol strategy attempts to introduce or promote the activity of biocontrol agents only when and where they are needed or are most effective and minimizes wasteful application of inoculum to non-target habitats (Lo et al., 1997). Thus, for effective delivery, it is necessary to consider plant–pathogen–antagonist interactions in terms of time and space.

Pythium, Rhizoctonia and *Sclerotinia* are important soil borne pathogens of many plant species and their survival structures in soil serve as primary inoculum. Consequently, suppression of the initial inoculum will be the first step in managing these pathogens (Lo et al., 1997). The granular application of biocontrol agents should be followed by monthly spray applications to suppress foliar phases of these diseases. Inhibition of the secondary infection and dissemination of these pathogens is also important for disease management (Lo et al., 1997). Monthly spray applications of *T. harzianum* could provide a second step in protection of plant foliage from attack by preventing these pathogens from initially infecting leaves and by reducing the spread of disease or other methods of inoculum dissemination. Finally, results of Lo et al. (1997) study have indicated that it will be necessary to apply weekly sprays for highly effective control of these pathogens under severe disease conditions.

In addition to Trihoderma and other fungal antagonists, several antagonistic bacterial species including Pseudomonas fluorescens, P. putida,P. aerofaciens, Burkholderia cepacia, Bacillus subtillis, B. Polymyxa and B. cerrues have also been used successfully in biological control of different soil borne fungal diseases (Heydari et al., 1997, 2004, 2007; Heydari and Misaghi, 2003; Kloepper et al., 2004; Leeman et al., 1995;Shahraki et al., 2009; Shishido et al., 2005; Weller and Cook, 1983; Zaki et al., 1998). By application of these bacterial antagonists, various fungal pathogens including Rhizoctonia solani, Fusarium moxysporium, F. solani, Verticillium dahliae, Gaummannomuces graminis and soil borne diseases caused by them such as seed rot, damping-off, root rot, vascular wilt and take-all have been biologically controlled on major agricultural crops including cotton, sugar beet, wheat, rice and different vegetables.

Although the majority of biological control research have been concentrated on soil borne fungal diseases, a number of studies have focused on fungal pathogens causing diseases and disorders in above-ground parts of plants (Kessel et al., 2005; Khodakaramian et al., 2008; Kovach et al., 2000; Milgroom and Cortesi, 2004; Smith et al., 1993). For example, Anderson et al. (2004) studied the possibility of biological control of fungal pathogens in the phylosphere and proposed that it may be possible to reduce the incidence and development of these diseases using fungal and bacterial antagonists.

In another study, biological control of powdery mildew disease on different crops using antagonistic fungi was investigated and it was found that biocontrol-active microorganisms can potentially be applied against this very important foliar diseases. Botrytis cinera which is the causal agent of gray mold on many plants (Agrios, 1988) was successfully controlled by the use of biocontrol-active microorganism on strawberry (Kovach et al., 2000). In another study conducted by Smith et al. (1993) biological control of cotton leak of cucumber caused by a fungal foliar pathogen was studied. It was found that

Bacillus cerrues, a bacterial antagonist was capable of reducing the incidence of the disease significantly (Smith et al., 1993).

Another example of using biocontrol-active microorganisms against foliar fungal pathogen is the study in which chestnut blight was successfully controlled by the virus that infects Cryphonectria parasitica, the fungal causal agent of the disease through the mechanism of hypovirulence, a reduction in pathogenicity of the pathogen. This phenomenon has resulted in control of the chestnut blight in many places (Milgroom and Cortesi, 2004). However, the interactions of virus, fungus, tree and environment play very important role in the success of disease control.

In addition to soil borne and foliar diseases some studies have also tested the efficacy of biocontrol-active microorganisms on post harvest fungal pathogens which cause losses to fruits and vegetables during post harvest and storage periods (Janisiewicz and Korsten, 2002). Spray applications of fungal and bacterial antagonists have resulted in significant reduction in the infection caused by some fungal pathogens in the storage.

THE USE OF COMPOST AS BIOFERTILIZER

Research data and observations in nurseries have shown that addition of composted organic matter to potting mixes results in suppression of soil borne fungal diseases (McKellar and Nelson, 2003; Paulitz and Belanger, 2001). The concentration of suppressive microorganisms in compost amended substrates is very high, but greatly reduced in soils or potting mixes after the amendment (McKellar and Nelson, 2003; Paulitz and Belanger, 2001). As a result, predictive disease suppression models have been developed based on the composition and concentration of microbial biomass.

The effectiveness of composts in suppression of soil borne diseases is dependent on heat kill, organic matters decomposition, recolonization of compost by suppressive microorganisms following heat kill and physical and chemical factors (McKellar and Nelson, 2003). Although previous works have focused on plant soil borne diseases, current research indicates that potting mixes containing composted organic materials which also have been inoculated with Trichoderma hamatum can be effective as a biocontrol alternative to foliar fungicides; however, the mechanism of this systemic type of induced resistance is not yet understood (McKellar and Nelson, 2003). Although the growers have traditionally relied on aged pine bark and composted biosolids to provide the potential for disease suppression, research indicates that composted animal manure have the potential to replace some of these components, but a consistent quantity and quality of these materials will need to be incorporated (McKellar and Nelson, 2003; Paulitz and Belanger, 2001). The maturity (stability) of the composted manure and its salinity largely determine its ability to induce suppression.

COMMERCIALIZATION OF BIOCONTROL

Commercial use and application of biological disease control have been slow mainly due to their variable performances under different environmental conditions in the field (Fravel, 2005; Mercier and Lindow, 2001; Paulitz and Belengar, 2001; Wang et al., 2003). Many biocontrol agents perform well in the laboratory and green house conditions but fail to do so in the field. This problem can only be solved by better understanding of the environmental parameters that affect biocontrol agents (Fravel, 2005; Mercier and Lindow, 2001; Paulitz and Belengar, 2001; Wang et al., 2003). In addition to this problem, there has also been relatively little investment in the development and production of commercial formulation of biocontrol-active microorganisms probably due to the cost of developing, testing, registering and marketing of these products (Heydari et al., 2007; Ardakani et al., 2009).

Biological control agents are generally formulated as wetable powders, dusts, granules and aqueous or oil-based liquid products using different mineral and organic carriers (Ardakani et al., 2009).

Currently in the market, a number of biologically based products are being sold for the control of fungal plant diseases (Ardakani et al., 2009). A growing number of companies are also developing new products that are in the process of registration. Many of these companies are small, privately owned firms with a limited product-line. Others are publicly traded and have substantial capitalization values. In addition, larger companies with more diverse product lines that include a variety of agrochemicals and biotechnological products have played a significant role in the development and marketing of products for the control of plant pathogens (Ardakani et al., 2009).

Biocontrol products are either marketed as stand-alone products or formulated as mixtures with other microbials. Some products with biocontrol properties may not be registered, but are sold instead as plant strengtheners or growth promoters without any specific claims regarding disease control (Ardakani et al., 2009). To help improve the global market perception of biopesticides as effective products, the biopesticide Industry Alliance is establishing a certification process to ensure industry standards for efficacy, quality and consistency. To improve commercial use and application of biological disease control it is extremely important to emphasize and concentrate on several factors including training of growers, formulation of biocontrol microorganisms and studying the role of environmental factors.

FUTURE OUTLOOK

Biological control really developed as an academic discipline during the 1970s and is now a mature science supported in both the public and private sector. Research related to biological control is published in many different scientific journals, particularly those related to plant pathology and entomology. Additionally, there are some academic journals specifically devoted to this disipline. In the United States, research funds for the biological control are provided primarily by several USDA programs. These include the Section 406 programs, regional IPM grants, Integrated Organic Program, IR-4 and several programs funded as part of the National Research Initiative (Bloom et al., 2003). Monies also exist to stimulate the development of commercial ventures through the small business innovation research programs. Such ventures are intended to be conduits for academic research that can be used to develop new companies (Spadaro and Gullino, 2005).

Much has been learned from the biological control research conducted over the past forty years. But, in addition to learning the lessons of the past, biocontrol researchers need to look forward to define new and different questions, the answers to which will help facilitate new biocontrol technologies and applications. Currently, fundamental advances in computing, molecular biology, analytical chemistry and statistics have led to new research aimed at characterizing the structure and functions of biocontrol agents, pathogens and host plants at the molecular, cellular and ecological levels (Spadaro and Gullino, 2005). Some of the research criteria that will advance our understanding of biological control and the conditions under which it can be most fruitfully applied are as follows:

Ecology of Antagonistic Microbes

Ecological factors play very important roles in the performance and activity of biocontrol-active microorganisms. In this regard, the following criteria need to be clarified and studied:

- The distribution of fungal pathogens and their antagonists in the environment
- The optimum conditions in which biocontrol microorganisms exert their suppressive capacities
- The response of native and introduced populations to different management practices
- The determinants factor of successful colonization and expression of biocontrol traits

- The components and dynamics of plant host defense induction

Application Methods

In regard with application strategies still there are some areas which should be investigated and developed for the enhancement of the effectiveness of biocontrol microorganisms. These areas are as follows:

- The search for more effective strains or strain variants for current applications
- The use of genetic engineering of microbes and plants for enhancing biocontrol application methods
- The development of proper formulations to enhance activities of known biocontrol agents

Introducing New Strains and Mechanisms

Since fungal plant pathogens are very diverse and their pathogenicity is different on host plants, it is therefore very important to look for new and novel biocontrol microorganisms with different mechanisms. In this regard, the following criteria need further investigation:

- The use of previously uncharacterized microbes as biological control agents
- Study on the roles of other genes and gene products which are involved in pathogen suppression
- The efficacy of using novel strain combinations in comparison with individual agents
- Study on the signal molecules of plant and microbial origin which regulate the expression of biocontrol traits by different agents

Integrated Pest Management

Since the ultimate goal of biological control of plant diseases is to assist the growers to combat and control plant pathogens in the field which is the real agricultural environment, it is therefore important to practically integrate biocontrol strategies into agricultural system. In this regard, the following criteria should be considered and followed carefully:

- Selection of production systems that can most benefit from biocontrol for disease management
- Application and use of biocontrol strategies which best fit with other IPM system components

- Development of effective biocontrol-cultivar combinations by plant breeders

Research and development: Nowadays, growers are interested in reducing dependence on chemical inputs, so biological controls (defined in the narrow sense) can be expected to play an important role in Integrated Pest Management (IPM) systems (Jacobsen et al., 2004). Good agricultural practices (GAP) including appropriate site selection, crop rotations, tillage, fertility and water management, provide the foundation for successful pest management by providing a fertile growing environment for the crop. The use of disease-resistant varieties, developed through conventional breeding or genetic engineering, provides the next line of defense. However, such measures are not always sufficient to be productive or economically sustainable. In such cases, the next step would be to deploy biorational controls of diseases. These include BCAs, introduced as inoculants or amendments, as well as active ingredients directly derived from natural origins and having a low impact on the environment and non-target microorganisms (Guetsky et al., 2001; Jacobsen et al., 2004).

If these foundational options are not sufficient to ensure plant health and/or economically sustainable production, then less specific and less harmful synthetic chemical toxins can be used to ensure productivity and profitability. With the growing interest in reducing chemical inputs, companies involved in the manufacturing and marketing of BCAs should experience continued growth. However, stringent quality control measures must be adopted so that farmers get quality products. New, more effective and stable formulations also will need to be developed.

Most fungal pathogens are susceptible to one or more biocontrol strategies, but practical implementation on a commercial scale has been constrained by a number of factors. Cost, convenience, efficacy and reliability of biological controls are important considerations, but only in relation to the alternative disease control strategies. Cultural practices (e.g., good sanitation, soil preparation and water management) and host resistance can go a long way towards controlling many diseases, so biocontrol should be applied only when such agronomic practices are insufficient for effective disease control. As long as petroleum is cheap and abundant, the cost and convenience of chemical pesticides will be difficult to surpass. However, if the infection court or target pathogen can be effectively colonized using inoculation, the ability of the living organism to reproduce could greatly reduce application costs.

In general, although, regulatory and cultural concerns about the health and safety of specific classes of pesticides are the primary economic drivers promoting the adoption of biological control strategies in urban and rural landscapes (Timms-Wilson et al., 2004). Self-perpetuating biological controls (e.g., hypovirulence of the chestnut blight pathogen) are also needed for control of diseases in forested and rangeland ecosystems where high application rates over larger land areas are not economically feasible. In terms of efficacy and reliability, the greatest successes in biological control have been achieved in situations where environmental conditions are most controlled or predictable and where biocontrol agents can preemptively colonize the infection court (Fravel, 2005). Monocyclic, soilborne and post-harvest fungal diseases have been controlled effectively by biological control agents that act as bioprotectants (i.e., preventing infections). Specific applications for the high value crops targeting specific diseases (e.g., downy mildew, powdery mildew and several other fungal diseases) have also been adopted (Kessel et al., 2005). As research unravels the various conditions needed for successful biocontrol of different fungal diseases, the adoption of BCAs in IPM systems is bound to increase in the years ahead.

CONCLUSIONS

Due to the serious environmental and health problems that wide spread use of chemical pesticides has created in the world, search for alternative safe methods is unavoidable. Biological control of plant diseases has been the subject of numerous research projects in recent years (Bargabus et al., 2004; Benhamou, 2004; Chisholm et al., 2006; Heydari, 2007; Islam et al., 2005). There is a growing demand for biologically based pest management practices. Recent surveys of both conventional and organic growers indicate an interest in using biocontrol products suggesting that the market potential of biocontrol products will increase in the future (Joshi and Gardener, 2006). Application of different biological control strategies has been successful in the greenhouse industry and continues to increase (Jacobsen et al., 2004). An upswing in commercial interests has also developed in the past few years and prospects for increased growth are positive. The Biopesticide Industry Alliance has formed and it is now actively promoting the value and efficacy of biopesticides (including those that control fungal plant pathogens). Clearly, the future success of the biological control industry will depend on innovative business management, product marketing, extension education and research (Timms-Wilson et al., 2004; Joshi and Gardener, 2006).

Increased demand for organic products in home gardening activities by using non-chemical methods has enlarged the market for biocontrol products. The field of plant pathology will contribute substantially to making the 21st century the ages of biotechnology by the development of innovative biocontrol strategies. A variety of research questions remain to be fully answered about the nature of biological control and the means to most effectively manage it under production conditions. Advanced molecular techniques are now being used to characterize the diversity, abundance and activities of microbes that live in and around plants, including those that significantly impact plant health (Joshi and Gardener, 2006). Still, much remains to be learned about the microbial ecology of both plant pathogens and their microbial antagonists in different agricultural systems. Fundamental work remains to be done on characterizing the different mechanisms by which organic amendments reduce plant disease including those caused by fungal pathogens. More studies on the practical aspects of mass production and formulation need to be undertaken to make new biocontrol products stable, effective, safer and more cost-effective.

Fungal pathogens are among the most important factors that cause serious damages and losses to plants. Harmful impacts of the chemical pesticides on the environment and non-target organisms have clearly been documented. The need for the development of non-chemical alternative strategies to protect plants against plant diseases including fungal pathogens is therefore clear. Biological control using fungal and bacterial antagonists to manage plant diseases seems to be a promising alternative strategy and have successfully been applied to control some diseases on different plants and crops. Biocontrol strategies may also be used to manage other plant diseases including foliar ones. Some of the important factors that affect the efficacy of microbial biocontrol agents in controlling plant diseases which should carefully be considered include method of application, formulation of biocontrol microorganisms and timing of application. Various composts and organic amendments as other means of biological control have also been tested on some plants and proven to be promising.

There are many products composed of living organisms, primarily bacteria and fungi, being sold that claim they will increase plant health. However, for any material to be considered a biological fungicide the Environmental Protection Agencies and Organizations must register it (Bloom et al., 2003). This registration indicates that the safety of the product to humans, non-humans (fish for example) and the environment has been determined. Materials that have not been approved should be used with caution.

Complete elimination of chemical pesticides for controlling plant pests and diseases in modern agriculture may be impossible, but a logical reduction in their application is absolutely feasible. To have a sustainable agricultural system with minimum contamination and risks to the environment, a combination of all available methods should be applied to manage pest problems and this can be achieved by Integrated Pest Management (IPM). Implementation of IPM strategies may be the safest solution for management of pest problems including fungal diseases in every cropping system and with no doubt biological control is one of the most important components of Integrated Pest Management which can lead us toward a sustainable agricultural system in the future.

REFERENCES

1. Agrios, N.A., 1988. Plant Pathology. 3rd Edn., Academic Press, USA., pp: 220-222.
2. Anderson, A.J., P. Habibzadegah-Tari and C.S. Tepper, 1988. Genetic studies on the role of an agglutinin-in root colonization byPseudomonas putida. Applied Environ. Microbiol., 54: 375-380.
3. Anderson, L.M., V.O. Stockwell and J.E. Loper, 2004. An extracellular protease of-Pseudomonas fluorescens inactivates antibiotics of Pantoea agglomerans. Phytopathology, 94: 1228-1234.
4. Ardakani, S., A. Heydari, N. Khorasani, R. Arjmandi and M. Ehteshami, 2009. Preparation of new biofungicides using antagonistic bacteria and mineral compounds for controlling cottn seedling damping-off disease. J. Plant Prot. Res., 49: 49-55.
5. Audenaert, K., T. Pattery, P. Cornelis and M. Hofte, 2002. Induction of systemic resistance to-Botrytis cinerea in tomato by Pseudomonas aeruginosa 7NSK2: role of salicylic acid, pyochelin and pyocyanin. Mol. Plant-Microbe. Interact., 15: 1147-1156.
6. Baker, K.F., 1987. Evolving concepts of biological control of plant pathogens. Annu. Rev. Phytopathol., 25: 67-85.
7. Bankhead, S.B., B.B. Landa, E. Lutton, D.M. Weller and B.B. Gardener, 2004. Minimal changes in rhizosphere population structure following root colonization by wild type and transgenic biocontrol strains. FEMS Microb. Ecol., 49: 307-318.
8. Bargabus, R.L., N.K. Zidack, J.E. Sherwood and B.J. Jacobsen, 2002. Characterization of systemic resistance in sugar beet elicited by a non-pathogenic, phylosphere-colonizing Bacillus mycoides, biological control agent. Physiol. Mol. Plant Pathol., 61: 289-298.
9. Bargabus, R.L., N.K. Zidack, J.E. Sherwood and B.J. Jacobsen, 2004. Screening for the identification of potential biological control agents that induce systemic acquired resistance in sugar beet. Biol. Control, 30: 342-350.
10. Benhamou, N. and I. Chet, 1997. Cellular and molecular mechanisms involved in the interaction between Trichoderma harzianum andPythium ultimum. Applied Environ. Microbiol., 63: 2095-2099.

11. Benhamou, N., 2004. Potential of the mycoparasite, Verticillium lecanii, to protect citrus fruit-against Penicillium digitatum, the causal agent of green mold: A comparison with-the effect of chitosan. Phytopathol., 94: 693-705.

12. Berg, G., A. Krechel, M. Ditz, R.A. Sikora, A. Ulrich and J. Hallmann, 2005. Endophytic and ectophytic potato-associated bacterial communities differ in structure and antagonistic function against plant pathogenic fungi. FEMS Microbiol. Ecol., 51: 215-229.

13. Biermann, B. and R.G. Linderman, 1983. Use of vesicular-arbuscular mycorrhizal roots, intraradical vesicles and extraradical vesicles as inoculum. New Phytol., 95: 97-105.

14. Bloom, B., R. Ehlers, S. Haukeland-Salinas, H. Hoddanen and K. Jung et al., 2003. Biological control agents: Safety and regulatory policy. BioControl, 48: 477-484.

15. Bull, C.T., K.G. Shetty and K.V. Subbarao, 2002. Interactions between Myxobacteria, plant pathogenic fungi and biocontrol agents. Plant Dis., 86: 889-896.

16. Chisholm, S.T., G. Coaker, B. Day and B.J. Staskawicz, 2006. Host-microbe interactions: Shaping the evolution of the plant immune response. Cell, 124: 803-814.

17. Cook, R.J., 1993. Making greater use of introduced microorganisms for biological control of plant pathogens. Annu. Rev. Phytopathol., 31: 53-80.

18. De Capdeville, G., C.L. Wilson, S.V. Beer and J.R. Aist, 2002. Alternative disease control agents induce resistance to blue mold in harvested Red Delicious apple fruit. Phytopathology, 92: 900-908.

19. De Meyer, G. and M. Hofte, 1997. Salicylic acid produced by the rhizobacterium Pseudomonas aeruginosa 7NSK2 induces resistance to leaf infection by Botrytis cinera on bean. Phytopathology, 87: 588-593.

20. Duchesne, L.C., 1994. Role of Ectomycorrhizal Fungi in Biocontrol. In: Mycorrhizae and Plant Health, Pfleger, F.L. and R.G. Linderman (Eds.). APS Press, Paul, MN., pp: 27-45.

21. El-Ghaouth, A., J.L. Smilanick, G.E. Brown, A. Ippolito, M. Wisniewski and C.L. Wilson, 2000. Application of Candida saitoana and glycolchitosan for the control of post harvest diseases of apple and citrus fruit under semi-commercial conditions. Plant Dis., 84: 243-248.

22. Elad, Y. and R. Baker, 1985. Influence of trace amounts of cations and siderophore-producing pseudomonads on chlamydospore germination of

Fusarium oxysporum. Phytopathology, 75: 1047-1052.

23. Fitter, A.H. and J. Grabaye, 1994. Interactions between mycorrhizal fungi and other soil organisms. Plant Soil, 159: 123-132.
24. Fravel, D., 2005. Commercialization and implementation of biocontrol. Annu. Rev. Phytopathol., 43: 337-359.
25. Garcia-Garrido, J.M. and J.A. Ocampo, 1989. Effect of VA mycorrhizal infection of tomato on damage caused by Pseudomonas syringae. Soil Biol. Biochem., 21: 165-167.
26. Guetsky, R., D. Shtienberg, Y. Elad and A. Dinoor, 2001. Combining biocontrol agents to reduce the variability of biological control. Phytopathology, 91: 621-627.
27. Haas, D. and G. Defago, 2005. Biological control of soil-borne pathogens by fluorescent pseudomonads. Nat. Rev. Microbiol., 3: 307-319.
28. Harman, G.E., C.R. Howell, A. Viterbo, I. Chet and M. Lorito, 2004. Trichoderma species-opportunistic, avirulent plant symbionts. Nat. Rev. Microbiol., 2: 43-56.
29. Heydari, A. and I.J. Misaghi, 1998. Biocontrol activity of Burkholderia cepacia against Rhizoctonia solani in herbicide-treated soils. Plant Soil, 202: 109-116.
30. Heydari, A. and I.J. Misaghi, 1999. Herbicide-Mediated Changes in the Populations and Activity of Root Associated Microorganisms: a Potential Cause of Plant Stress. In: Hand Book of Plant and Crop Stress, Pessarakli, M. (Ed.). 2nd Edn., Marcel Dekker Press, New York.
31. Heydari, A. and I.J. Misaghi, 2003. The role of rhizosphere bacteria in herbicide-mediated increase in Rhizoctonia solani-induced cotton seedling damping-off. Plant Soil, 257: 391-396.
32. Heydari, A., 2007. Biological Control of Turfgrass Fungal Diseases. In: Turfgrass Management and Physiology, Pessarakli, M. (Ed.). CRC Press, Florida, USA.
33. Heydari, A., H. Fattahi, H.R. Zamanizadeh, N.H. Zadeh and L. Naraghi, 2004. Investigation on the possibility of using bacterial antagonists for biological control of cotton seedling damping-off in greenhouse. Applied Entomol. Phytopathol., 72: 51-68.
34. Heydari, A., I.J. Misaghi and G.M. Balestra, 2007. Pre-emergence herbicides influence the efficacy of fungicides in controlling cotton seedling damping-off in the field. Int. J. Agric. Res., 2: 1049-1053.
35. Heydari, A., I.J. Misaghi and W.B. McCloskey, 1997. Effects of three soil-applied herbicides on populations of plant disease suppressing

bacteria in the cotton rhizosphere. Plant Soil, 195: 75-81.

36. Hoitink, H.A.J. and M.J. Boehm, 1999. Biocontrol within the context of soil microbial communities: A substrate-dependent phenomenon. Annu. Rev. Phytopathol., 37: 427-446.

37. Howell, C.R. and R.D. Stipanovic, 1980. Suppression of Pythium ultimum induced damping-off of cotton seedlings by Pseudomonas fluorescens and its antibiotic pyoluteorin. Phytopathology, 70: 712-715.

38. Howell, C.R., R.C. Beier and R.D. Stipanovic, 1988. Production of ammonia by Enterobacter cloacae and its possible role in the biological control of Pythium pre-emergence damping-off by the bacterium. Phytopathology, 78: 1075-1078.

39. Islam, M.T., Y. Hashidoko, A. Deora, T. Ito and S. Tahara, 2005. Suppression of damping-off-disease in host plants by the rhizoplane bacterium Lysobacter sp. strain SB-K88 is-linked to plant colonization and antibiosis against soilborne peronosporomycetes. Appl. Environ. Microbiol., 71: 3786-3796.

40. Jacobsen, B.J., N.K. Zidack and B.J. Larson, 2004. The role of Bacillus-based biological control agents in integrated pest management systems: Plant diseases. Phytopathology, 94: 1272-1275.

41. Janisiewicz, W.J. and D.L. Peterson, 2004. Susceptibility of the stem pull area of mechanically harvested apples to blue mold decay and its control with a biocontrol agent. Plant Dis., 88: 662-664.

42. Janisiewicz, W.J. and L. Korsten, 2002. Biological control of post harvest disease of fruits. Ann. Rev. Phytopathol., 40: 411-441.

43. Joshi, R. and B.B.M. Gardener, 2006. Identification and characterization of novel genetic markers associated with biological control activities in Bacillus subtilis. Phytopathology, 96: 145-154.

44. Kageyama, K. and E.B. Nelson, 2003. Differential inactiviation of seed exudates stimulation of-Pythium ultimum sporangium germination byEnterobacter cloacae influences biological control efficacy on different plant species. Applied Environ. Microbiol., 69: 1114-1120.

45. Katska, V., 1994. Interrelationship between vesicular-arbuscular mycorrhiza and rhizosphere microflora in apple replant disease. Biol. Plant, 36: 99-104.

46. Keel, C., C. Voisard, C.H. Berling, G. Kahir and G. Defago, 1989. Iron sufficiency is a prerequisit for suppression of tobacco black root rot byPseudomonas fluorescnes strain CHA0 under gnotobiotic contiditions. Phytopathology, 79: 584-589.

47. Kerry, B.R., 2000. Rhizosphere interactions and the exploitation of microbial agents for the biological control of plant-parasitic nematodes. Ann. Rev. Phytopathol., 38: 423-441.

48. Kessel, G.J.T., J. Kohl, J.A. Powell, R. Rabbinge and W. van der Werf, 2005. Modeling spatial characteristics in the biological control of fungi at the leaf scale: Competitive substrate colonization by Botrytis cinerea and the saprophytic antagonist Ulocladium atrum. Phytopathology, 95: 439-448.

49. Khodakaramian, G., A. Heydari and G.M. Balestra, 2008. Evaluation of pseudomonads bacterial isolates in biological control of citrus bacterial canker disease. Int. J. Agric. Res., 3: 268-272.

50. Kloepper, J.W., C.M. Ryu and S. Zhang, 2004. Induce systemic resistance and promotion of plant growth by Bacillus spp. Phytopathology, 94: 1259-1266.

51. Kloepper, J.W., J. Leong, M. Teintze and M.N. Schroth, 1980. Pseudomonas siderophores: A mechanism explaining disease suppression in soils. Curr. Microbiol., 4: 317-320.

52. Kovach, J., R. Petzoldt and G.E. Harman, 2000. Use of honey and bumble bees to disseminate Trichoderma harzianum 1295-22 to strawberries for Botrytis control. Biol. Control, 18: 235-242.

53. Lafontaine, P.J. and N. Benhamou, 1996. Chitosan treatment: An emerging strategy for enhancing resistance of greenhouse tomato plants to infection by Fusarium oxysporum f.sp. radicis-lycopersici. Biocont. Sci. Technol., 6: 111-124.

54. Leclere, V., M. Bechet, A. Adam, J.S. Guez and B. Wathelet et al., 2005. Mycosubtilin overproduction by Bacillus subtilis BBG100 enhances the organism's antagonistic and biocontrol activities. Applied Environ. Microbiol., 71: 4577-4584.

55. Leeman, M., J.A. van Pelt, F.M. den Ouden, M. Heinsbroek, P.A.H.M. Bakker and B. Schippers, 1995. Induction of systemic resistance byPseudomonas fluorescens in radish cultivars differing in susceptibility to Fusarium wilt, using novel bioassay. Eur. J. Plant Pathol., 101: 655-664.

56. Linderman, R.G., 1994. Role of VAM Fungi in Biocontrol. In: Mycorrhizae and Plant Health. Pfleger, F.L. and R.G. Linderman (Eds.). American Phytological Society, St. Paul. MN, APS Press, USA., ISBN-10: 0890541582.

57. Lo, C.T., E.B. Nelson and G.E. Harman, 1995. Improved biocontrol efficacy of Trichoderma harzianum for foliar phases of turf diseases by

use of spray applications. Plant Dis., 81: 1132-1138.

58. Lo, C.T., E.B. Nelson and G.E. Harman, 1996. Biological control of turfgrass diseases with a rhizosphere component strain of Trichoderma harzianum. Plant Dis., 82: 736-741.

59. Lo, C.T., E.B. Nelson and G.E. Harman, 1997. Biological control of Pythium, Rhizoctonia and Sclerotinia infected diseases of turfgrass withTrichoderma harzianum. Phytopathology, 84: 1372-1379.

60. Loper, J.E. and J.S. Buyer, 1991. Siderophores in microbial interactions of plant surfaces. Mol. Plant-Microbe Interact, 4: 5-13.

61. Maurhoffer, M., C. Hase, J.P. Matraux and G. Defago, 1994. Induction of systemic resistance of tobacco to tobacco necrosis virus by the root-colonizing Pseudomonas fluorescens strains CHAO: Influence of the gacA gene and pyoverdine production. Phytopathology, 84: 139-146.

62. McKellar, M.E. and E.B. Nelson, 2003. Compost-induced suppression of Pythium damping-off is mediated by fatty-acid-metabolizing seed-colonizing microbial communities. Applied Environ. Microbiol., 69: 452-460.

63. McSpadden-Gardener, B.B. and D.M. Weller, 2001. Changes in populations of rhizosphere bacteria associated with take-all disease of wheat. Applied Environ. Microbiol., 67: 4414-4425.

64. Mercier, J. and S.E. Lindow, 2001. Field performance of antagonistic bacteria identified in a novel assay for biological control of firelight. Biol. Control, 22: 66-71.

65. Meziane, H., I. van der Sluis, L.C. van Loon, M. Hofte and P.A.H.M. Bakker, 2005. Determinants of Pseudomonas putida WCS358 involved in inducing systemic resistance in plants. Mol. Plant Pathol., 6: 177-185.

66. Milgroom, M.G. and P. Cortesi, 2004. Biological control of chestnut blight with hypovirulence:a critical analysis. Annu. Rev. Phytopathol., 42: 311-338.

67. Moyne, A.L., R. Shelby, T.E. Cleveland and S. Tuzun, 2001. Bacillomycin D: An iturin with antifungal activity against Aspergillus flavus. J. Applied Microbiol., 90: 622-629.

68. Naraghi, L., A. Heydari, A. Karimi-Roozbehani and D. Ershad, 2004. Isolation of Talaromyces flavus from Golestan cotton fields and its antagonistic effects on Verticillium dahliae the causal agent of cotton verticillium wilt. Iran. J. Plant Pathol., 39: 109-2004.

69. Ordentlich, A., Y. Elad and I. Chet, 1988. The role of chitinase of Serratia marcescens in biocontrol of Sclerotium rolfsii. Phytopathology, 78: 84-

87.

70. Paulitz, T.C. and R.R. Belanger, 2001. Biological control in greenhouse systems. Annu. Rev. Phytopathol., 39: 103-133.

71. Phillips, A.D., T.C. Fox, M.D. King, T.V. Bhuvaneswari and L.R. Teuber, 2004. Microbial products trigger amino acid exudation from plant roots. Plant Physiol., 136: 2887-2894.

72. Press, C.M., J.E. Loper and J.W. Kloepper, 2001. Role of iron in rhizobacteria mediated induced systemic resistance of cucumber. Phytopathology, 91: 593-598.

73. Ramette, A., Y. Moenne-Loccoz and G. Defago, 2003. Prevalence of fluorescent-pseudomonads producing antifungal phloroglucinols and/or hydrogen cyanide in soils naturally suppressive or conducive to tobacco root rot. FEMS Microb. Ecol., 44: 35-43.

74. Ryu, C.M., M.A. Farag, C.H. Hu, M.S. Reddy, J.W. Kloepper and P.W. Pare, 2004. Bacterial volatiles induce systemic resistance in Arabidopsis. Plant Physiol., 134: 1017-1026.

75. Scheuerell, S.J., D.M. Sullivan and W.F. Mahaffee, 2005. Suppression of seedling damping-off caused by Pythium ultimum, P. irregulare andRhizoctonia solani in container media amended with a diverse range of pacific northwest compost sources. Phytopathology, 95: 306-315.

76. Schouten, A, G. van den Berg, V. Edel-Hermann, C. Steinberg and N. Gautheron et al., 2004. Defense responses of Fusarium oxysporum to 2,4-DAPG, a broad spectrum antibiotic produced by Pseudomonas fluorescens. Mol. Plant-Microbe Interact, 17: 1201-1211.

77. Shah-Smith, D.A. and R.G. Burns, 1997. Shelf-life of a biocontrol Pseudomonas putida applied to the sugar beet seeds using commercial coatings. Biocontrol Sci. Technol., 7: 65-74.

78. Shahraki, M., A. Heydari and N. Hassanzadeh, 2009. Investigation of antibiotic, siderophore and volatile metabolites production by Bacillus and Pseudomonas bacteria. Iran. J. Biol., 22: 71-85.

79. Shanahan, P., D.J. O'Sullivan, P. Simpson, J.D. Glennon and F. O`Gara, 1992. Isolation of 2,4-diacetylphloroglucinol from a fluorescent pseudomonad and investigation of physiological parameters influencing its production. Applied Environ. Microbiol., 58: 353-358.

80. Shishido, M., C. Miwa, T. Usami, Y. Amemiya and K.B. Johnson, 2005. Biological control efficiency of fusarium wilt of tomato by nonpathogenic F. oxysporum Fo-B2 in different environments. Phytopathology, 95: 1072-1080.

81. Silva, H.S.A., R.D.S. Romeiro, D. Macagnan, B.D.A. Halfeld-Vieira, M.C.B. Pereira and A. Mounteer, 2004. Rhizobacterial induction of systemic resistance in tomato plants: Non-specific protection and increase in enzyme activities. Biol. Control, 29: 288-295.

82. Smith, K.P., M.J. Havey and J. Handelsman, 1993. Suppression of cottony leak of cucumber with Bacillus cereus strain UW85. Plant Dis., 77: 139-142.

83. Spadaro, D. and M.L. Gullino, 2005. Improving the efficacy of biocontrol agents against soilborne pathogens. Crop Prot., 24: 601-613.

84. Tari, P.H. and A.J. Anderson, 1988. Fusarium wilt suppression and agglutinability of Pseudomonas putida. Applied Environ. Microbiol., 54: 2037-2041.

85. Thomashow, L.S. and D.M. Weller, 1988. Role of a phenazine antibiotic from Pseudomonas fluorescens in biological control ofGaeumannomyces graminis var. Tritici. J. Bacteriol., 170: 3499-3508.

86. Thomashow, L.S., D.M. Weller, R. Bonsall and L.S. Pierson, 1990. Production of de antibiotic phenazine 1-carboxylic acid by fluorescent pseudomonad species in the rhizosphere of wheat. Applied Environ. Microbiol, 56: 908-912.

87. Thomashow, L.S., R.F. Bonsall and D.M. Weller, 2002. Antibiotic Production by Soil and Rhizosphere Microbes in situ. 2nd Edn., ASM Press, Washington DC., pp: 638-647.

88. Timms-Wilson, T.M., K. Kerry and J.B. Mark, 2004. Risk assessment for engineered bacteria used in biocontrol of fungal disease in agricultural crops. Plant Soil, 266: 57-60.

89. Vallad, G.E. and R. Goodman, 2004. Systemic acquired resistance and induced systemic resistance in conventional agriculture. Crop Sci., 44: 1920-1934.

90. Van Dijk, K. and E.B. Nelson, 2000. Fatty acid competition as a mechanism by which Enterobacter cloacae suppresses Pythium ultimumsporangium germination and damping-off. Applied Environ. Microbiol., 66: 5340-5347.

91. Van Loon, L.C., P.A.H.M. Bakker and C.M.J. Pieterse, 1998. Systemic resistance induced by rhizosphere bacteria. Annu. Rev. Phytopathol., 36: 453-483.

92. Van Peer, R. and B. Schippers, 1992. Lipopolysaccharides of plant-growth promoting Pseudomonas spp. strain WCS 417r induce resistance in carnation to Fusarium wilt. Neth. Eur. J. Plant Pathol., 98: 129-139.

93. Van Wees, S.C., C.M. Pieterse, A. Trijssenaar, Y.A. van't Westende, F. Hartog and L.C. van Loon, 1997. Differential induction of systemic resistance in Arabidopsis by biocontrol bacteria. Mol. Plant-Microbe Interact, 10: 716-724.

94. Wang, H., S.F. Hawang, K.F. Chang, G.D. Turnbull and R.J. Howard, 2003. Suppression of important pea diseases by bacterial antagonists. Bio Control, 48: 447-460.

95. Weller, D.M. and R.J. Cook, 1983. Suppression of take-all of wheat by seed treatments with fluorescent pseudomonads. Phytopathology, 73: 463-469.

96. Weller, D.M., J.M. Raaijmakers, B.B.M. Gardener and L.S. Thomashow, 2002. Microbial populations responsible for specific soil suppressiveness to plant pathogens. Annu. Rev. Phytopathol., 40: 309-348.

97. Wilhite, S.E., R.D. Lumsden and D.C. Strancy, 2001. Peptide synthetase gene in Trichoderma virens. Applied Environ. Microbiol., 67: 5055-5062.

98. Yoshihisa, H., S. Zenji, H. Fukushi, K. Katsuhiro, S. Haruhisa and S. Takahito, 1989. Production of antibiotics by Pseudomonas cepacia as an agent for biological control of soilborne Plant pathogens. Soil Biol. Biochem., 21: 723-728.

99. Zaki, K., I.J. Misaghi, A. Heydari and M.N. Shatla, 1998. Control of cotton seedling damping-off in the field by Burkholdria cepacia. Plant Dis., 82: 291-293.

Chapter 6

COMMON PROCESSES IN PATHOGENESIS BY FUNGAL AND OOMYCETE PLANT PATHOGENS, DESCRIBED WITH GENE ONTOLOGY TERMS

Shaowu Meng[1,3], Trudy Torto-Alalibo[2], Marcus C Chibucos[2,4], Brett M Tyler[2] and Ralph A Dean[1]

[1]Fungal Genomics Laboratory, Center for Integrated Fungal Research, North Carolina State University, Raleigh, NC 27695, USA

[2]Virginia Bioinformatics Institute, Virginia Polytechnic and State University, Blacksburg, VA 24061, USA

[3]Hayes Laboratory, Lineberger Comprehensive Cancer Center, School of Medicine, CB# 7295, University of North Carolina at Chapel Hill, Chapel Hill, NC 27599-7295, USA

[4]Institute for Genome Sciences, University of Maryland School of Medicine, Baltimore, MD 21201, USA

ABSTRACT

Plant diseases caused by fungi and oomycetes result in significant economic losses every year. Although phylogenetically distant, the infection processes by these organisms share many common features. These include dispersal of an infectious particle, host adhesion, recognition, penetration, invasive growth, and lesion development. Previously, many of these common processes did not have corresponding Gene Ontology (GO) terms. For example, no GO terms existed to describe processes related to the appressorium, an important structure for infection by many fungi and oomycetes. In this mini-review, we identify common features of the pathogenic processes of fungi and oomycetes and create a pathogenesis model using 256 newly developed and 38 extant GO terms, with an emphasis on the appressorium and signal transduction. This set of standardized GO terms provides a solid base to further compare and contrast the molecular underpinnings of fungal and oomycete pathogenesis.

COMMON PATHOGENESIS PROGRAMS OF FUNGI AND OOMYCETES

Oomycetes, although phylogenetically very distant, share many common morphological and physiological features with the true fungi [1–3]. For example, they have similar filamentous, branching, indeterminate bodies, and they acquire nutrition by secreting digestive enzymes and then absorbing the resultant breakdown products. More importantly, fungi and oomycetes share a unique capability compared with other microbial pathogens, namely that they are able to breach cuticles of host plants and establish infection rapidly [4]. Consequently, both are causal agents of many destructive plant diseases and are responsible for significant economic losses every year.

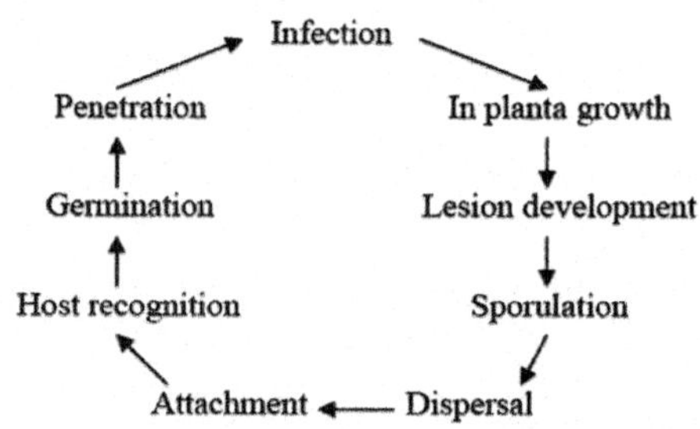

Figure 1: A generalized diagram displaying infection and disease cycle caused by fungi and oomycetes.

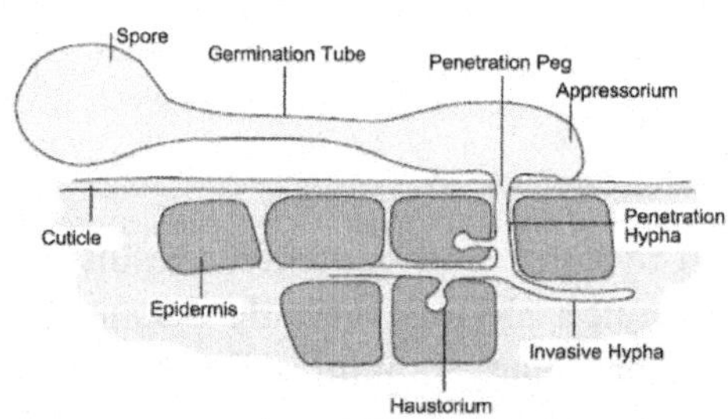

Figure 2: The infection process in fungal and oomycete pathogens. Modified by permission from Schumann, G. L., 1991, Plant diseases: Their biology and social impact, American Phytopathological Society, St. Paul, MN.

In this review, we summarize common mechanisms of pathogenesis displayed by oomycetes and fungi. Pathogenesis by a fungus or oomycete is a complex process. Briefly, it includes the following steps: dispersal and arrival of an infectious particle (usually a spore of some kind) in the vicinity of the host, adhesion to the host, recognition of the host (which may occur prior to adhesion), penetration into the host, invasive growth within the host, lesion development in the host, and finally production of additional infectious particles [5, 6] (see Figures 1, 2).

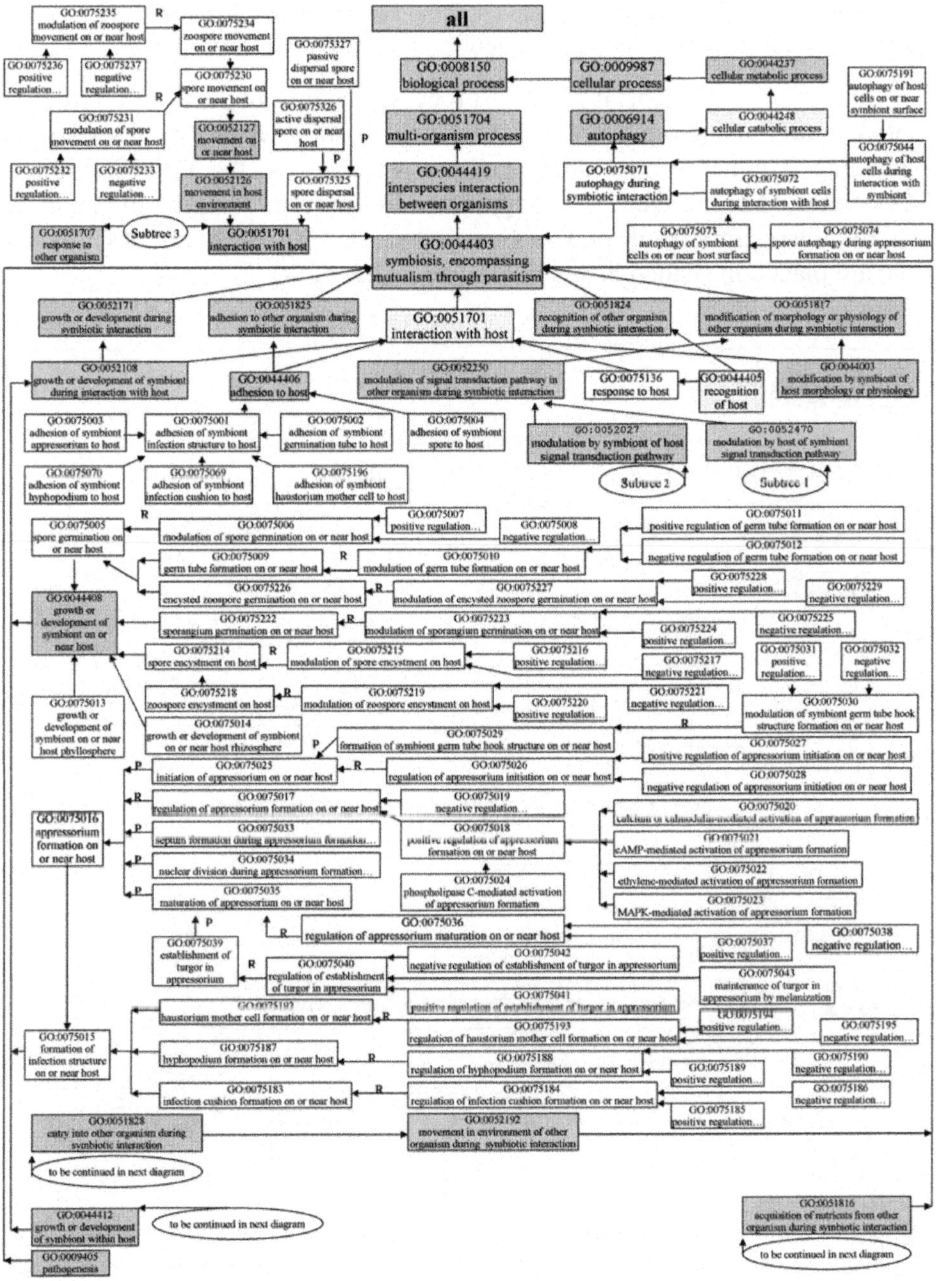

Figure 3: Gene Ontology terms for processes related to infection and disease (Part 1). Subtree 1 and 2 are depictured in Figure 5, and Subtree 3 is depictured in Figure 6. Shaded boxes indicate pre-existing GO terms, and unshaded boxes represent GO terms developed under the PAMGO project. "R" indicates "regulates relationship",

"P" indicates "part of relationship", and null indicates "is a relationship" (see the Gene Ontology website at http://www.geneontology.org for further information).

In order to describe the entire process, we formulate a description of pathogenesis using standardized terms from the Gene Ontology (GO), including 256 new terms developed by members of the PAMGO (Plant-Associated Microbe Gene Ontology) consortium http://pamgo.vbi.vt.edu, an official interest group of the GO Consortium, as well as 38 extant GO terms that are placed in shaded boxes in Figures 3, 4,5, 6.

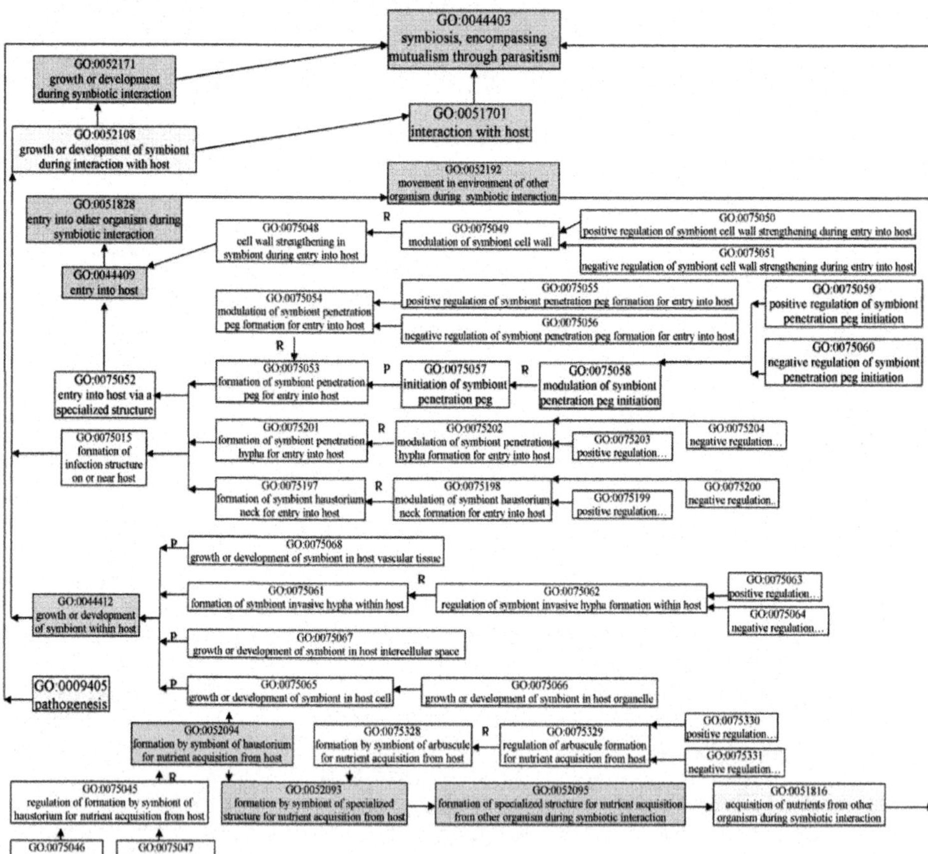

Figure 4: Gene Ontology terms for processes related to infection and disease (Part 2). Shaded boxes indicate pre-existing GO terms, and unshaded boxes represent GO terms developed under the PAMGO project. "R" indicates "regulates relationship", "P" indicates "part of relationship", and null indicates "is a relationship" (see the Gene Ontology website at http://www.geneontology.org for further information).

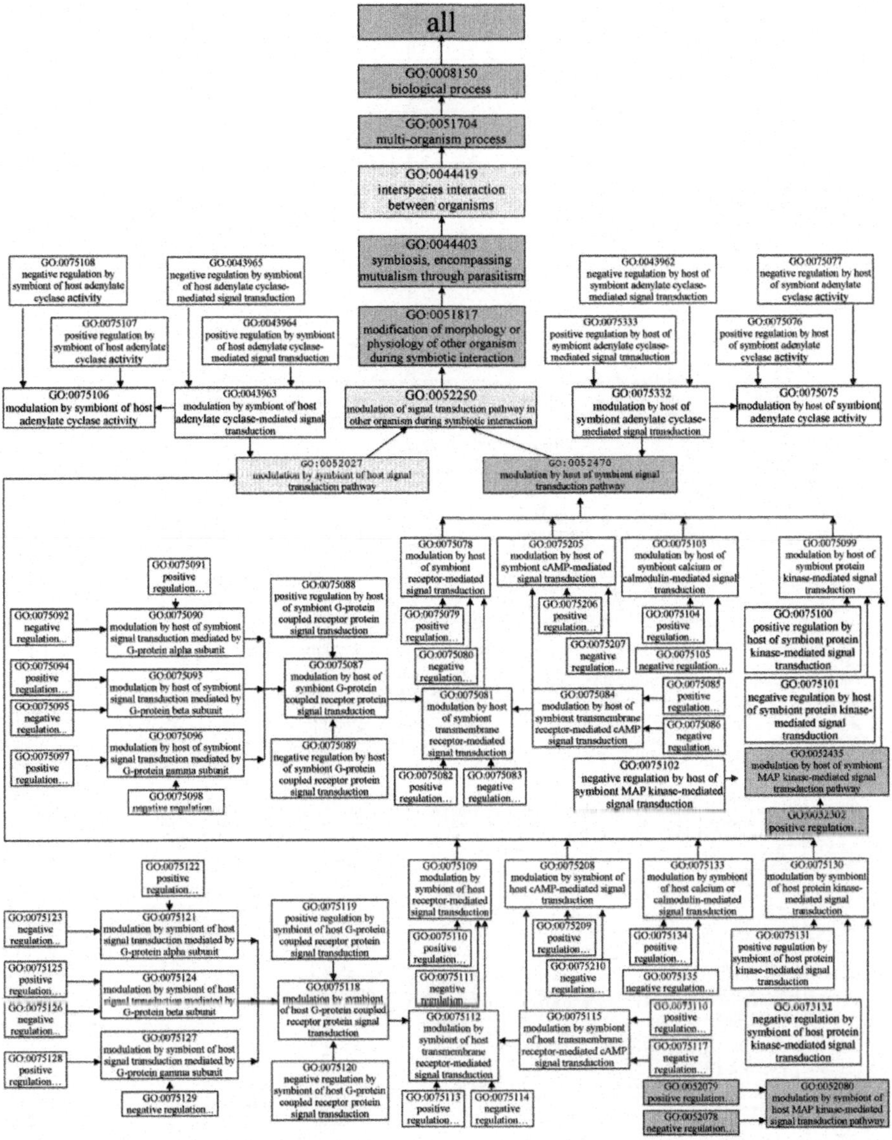

Figure 5: Gene Ontology terms for signal transduction processes related to infection and disease (Part 1). Subtree 1 consists of GO terms intending to annotate host gene products that stimulate signal transduction in symbiont. Subtree 2 represents the opposite perspective of Subtree 1. Shaded boxes indicate pre-existing GO terms, and unshaded boxes represent GO terms developed under the PAMGO project.

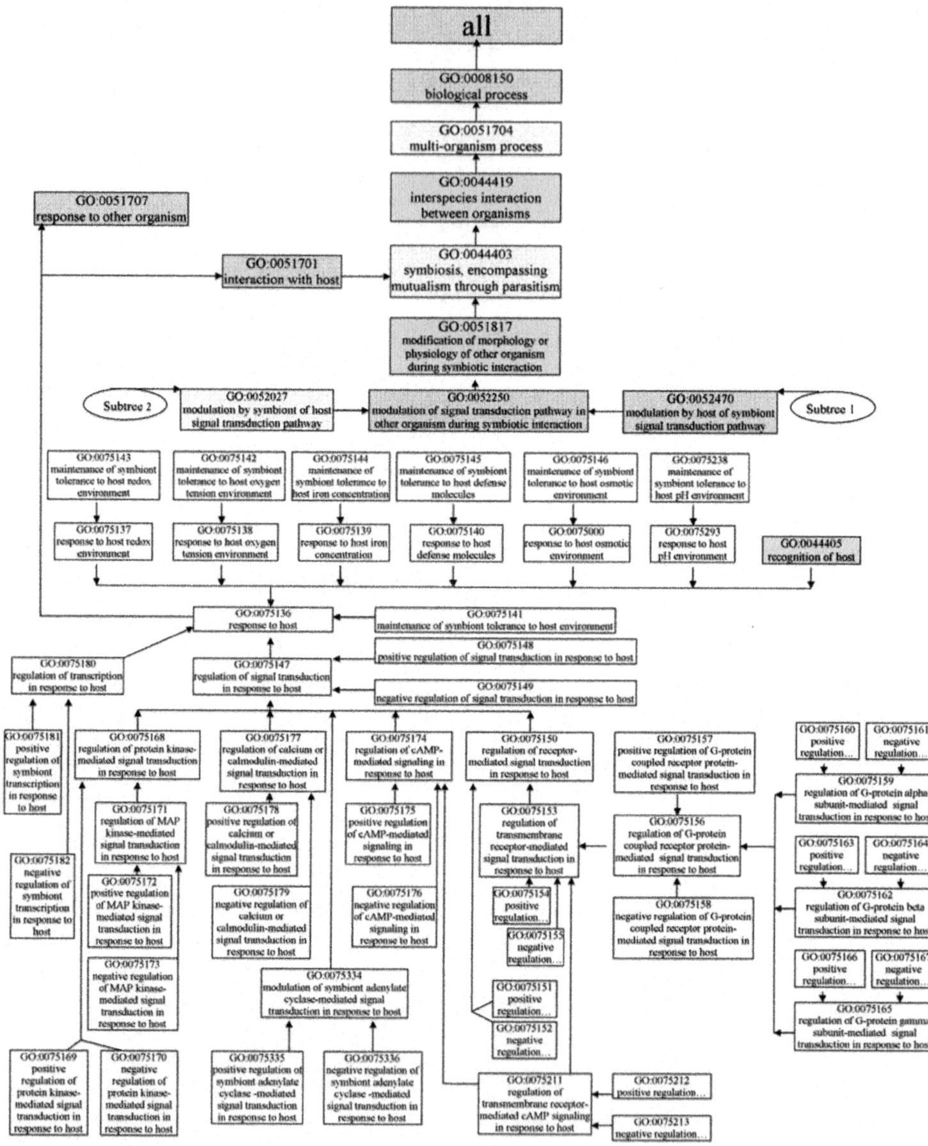

Figure 6: Gene Ontology terms for signal transduction processes related to infection and disease (Part 2). Subtree 3 consists of GO terms intending to annotate symbiont gene products that stimulate signal transduction in symbiont in response to host. Shaded boxes indicate pre-existing GO terms, and unshaded boxes represent GO terms developed under the PAMGO project.

All of the 256 new terms are placed within GO under the node "GO ID 0044403 symbiosis, encompassing mutualism through parasitism" (note the original, broad definition of "symbiosis" used in GO, which specifies use of the words "symbiont" and "host" as the smaller and larger, respectively, of the symbiotic interactants). Among these new terms, every term that starts with "modulation" or "regulation" has two child terms, one is "positive regulation of...", and the other is "negative regulation of..." Note that these child terms are general GO terms; "position regulation," for example, includes induction, upregulation, stimulation, etc. Four diagrams (see Figures 3, 4, 5, 6) encompassing the 256 new and 38 extant GO terms explicitly depict our description of pathogenesis, with an emphasis on appressorium formation and signal transduction. More details about each step are presented in the following sections.

Spore Dispersal

Dispersal of spores is the most common process to initiate new infections [5], though direct infection by hyphae may occur. An example of the latter is the spread of ectomycorrhizal basidiomycetes in forest soils. Dispersal mechanisms can be grouped into two types: one is passive dispersal by wind, water or animal [7], and the other is active dispersal such as shooting ascospores through the boundary layer of air surrounding the fruiting body by forcible discharge [8]. Similarly, spores can be grouped into two types according to their motility. In fungi non-motile spores include sexual spores such as ascospores, rust urediniospores, sclerotia and conidiospores, while non-motile oomycete spores include oospores, sporangiospores and conidia. Motile spores with flagella, called zoospores, are ubiquitous among oomycetes and are also found in chytrid fungi [9]. Additionally, spores vary in requirements for dormancy. Some spores, such as zoospores, must encyst and differentiate to acquire qualities of dormancy before they become true spores [10]. Different spores vary in the length of the dormancy period; for example, some ascospores and oospores show extended dormancy, while others such as zoospores and ascomycete conidia are usually short-lived.

Three new GO terms including the term "GO ID 0075325 spore dispersal on or near host" were developed under the node "GO ID 0051701 ...interaction with host" to describe the mechanisms of spore dispersal. New terms describing active or passive dispersal mechanisms were placed as children of "spore dispersal on or near host" (see Figure 3).

Eight new GO terms describing spore motility were listed under the node "GO ID 0052127 movement on or near host". The term "GO ID 0075230 spore movement on or near host" is central to these eight terms with "GO ID 0075234 zoospore movement on or near host" as the principal child term (see Figure 3).

Similarly, eight new GO terms were added under "GO ID 0044408 growth or development of symbiont on or near host" to describe spore encystment. The term "GO ID 0075214 spore encystment on host" is central to these eight terms, with the term "GO ID 0075218 zoospore encystment on host" as the main child term (see Figure 3).

These 19 new terms are appropriate for annotating gene products known to be involved in spore dispersal. For example, inhibition of the vacuolar H+-ATPase by potassium nitrate causes a reduction in vacuole expulsion in zoospores of the oomycete *Phytophthora nicotianae* and leads to premature encystment [11]. Thus, H+-ATPase negatively regulates zoospore encystment and can be annotated with the new term «GO ID 0075221 negative regulation of zoospore encystment on host».

Adhesion to the Host

Adhesion of spores to the host involves physical and chemical processes [3]. Typically, when spores reach the surface of a host tissue, they attach via adhesion molecules [5]. A germination tube then emerges from the spore or the encysted zoospore (see Figure 2). From the germination tube, a growth hypha or an infection structure such as an appressorium [12–16] develops, which also becomes firmly attached to the host surface via adhesion molecules. A variety of other infection structures such as hyphopodia [17–19], haustorium mother cells [20–23], or infection cushions [24] are generated by fungal pathogens after germinating on the host surface. These all serve a common function of facilitating the pathogen's entry into the host tissue. It should be noted that the sporangia of many oomycetes may germinate directly to form an infection hypha, or else in the presence of abundant water they may differentiate, through specialized cleavage vesicles, into 10–30 zoospores that can individually disperse to initiate sites of infection [25].

Seven new GO terms under the parent, "GO ID 0044406 adhesion to host", were developed to describe in detail the biological process of adhesion to a host. The term "GO ID 0075001 adhesion of symbiont infection structure to host" is central to this section. Among the seven terms, five terms that describe adhesion of a specific infection structure, including appressorium, hyphopodium, haustorium mother cell, infection cushion, or germination tube, are children of "adhesion of symbiont infection structure to host" (see Figure 3).

To describe spore germination on or near host tissue, 16 new terms under the parent, "GO ID 0044408 growth or development of symbiont on or near host", were developed. The 16 terms cover spore germination, sporangium germination, encysted zoospore germination, and germ tube formation. The term "GO ID 0075005 spore germination on or near host" is central to this section. Major relationships among the sixteen terms are shown in Figure 3.

The 23 new GO terms in this section are useful for annotating pathogen gene products involved in adhesion to host tissue. For example, Car (cyst-germination-specific acidic repeat) proteins of the oomycete *Phytophthora infestans* are transiently expressed during germination of cysts (i.e., encysted zoospores) and during formation of appressoria, and they are localized at the surface of germlings. These proteins have considerable sequence homology to mucins and have an internal octapeptide repeat with the mucin consensus sequence TTYAPTEE. Therefore, like mucins, Car proteins should serve as a mucous cover protecting the germling and assisting in adhesion to the leaf surface [26]. Thus, the Car proteins can be annotated with the new terms «GO ID 0075226 encysted zoospore germination on or near host» and «GO ID 0075001 adhesion of symbiont infection structure to host», using the GO evidence code ISS (Inferred from Sequence or Structural Similarity).

Signal Transduction during Recognition of the Host

Signal transduction is an integral component of the host recognition process. Examples include protein kinase-mediated signal transduction [27], receptor-mediated signal transduction [28], G-protein coupled receptor protein signal transduction, G-protein subunit-mediated signal transduction [29], cAMP-mediated signal transduction [30], calcium or calmodulin-mediated signal transduction [31], and adenylate cyclase-mediated signal transduction [12].

In order to adequately describe signal transduction during symbiont interaction with its host, three sets of new terms were developed. Signal transduction pathways involved in the recognition between host and symbiont are generally quite extensively characterized and consequently 127 new terms were developed.

The first set of new terms is intended for annotation of host gene products that stimulate symbiont signal transduction (see Subtree 1, which includes terms under the node "GO ID 0052470 modulation by host of symbiont signal transduction pathway" in Figure 5). This set has 37 new terms. Five of these terms describing different types of signal transduction pathways are children of "GO ID 0052470" (see Subtree 1 in Figure 5).

The second set of new terms is intended for annotation of symbiont gene products that stimulate host signal transduction (see Subtree 2, which includes terms under the node "GO ID 0052027 modulation by symbiont of host signal transduction pathway" in Figure 5). This set has 36 new GO terms and has the same structure as the first set (see Subtree 2 in Figure 5). The terms in the second set are essentially the converse of the terms in the first set. For example, the term "GO ID 0075130 modulation by symbiont of host protein kinase-mediated signal transduction" in the second term set has a complementary term "GO ID 0075099 modulation by host of symbiont protein kinase-mediated signal transduction" in the first term set.

The third set of new terms is intended for annotation of symbiont gene products that stimulate symbiont signal transduction in response to the host (see Subtree 3, which includes terms under the nodes "GO ID 0051701 interaction with host" and "GO ID 0051707 response to other organism" in Figure 6). This set has 56 new GO terms. The new term "GO ID 0075136 response to host" is central to the 56 new terms. This term has 10 child terms that describe 10 symbiont responses to host organisms, such as "GO ID 0075147 regulation of signal transduction in response to host", "GO ID 0075140 response to host defense molecules", and "GO ID 0075180 regulation of transcription in response to host" etc. (see Figure 6). Eight of the 10 terms have their own child and lower level offspring terms, and each of those "response" terms has a child term such as "maintenance of symbiont tolerance to host ..." (see details in Figure 6).

The term "GO ID 0075147 regulation of signal transduction in response to host" has five children to describe different types of signal transduction, similar to the five child terms of "GO ID 0052470 modulation by host of symbiont signal transduction pathway" in the first set. Each of the five terms has child terms for positive regulation and negative regulation.

The three sets of new GO terms can be used to explicitly describe genes of signal transduction pathways involved in host recognition. For instance, the PMK1 gene of the rice blast fungus *Magnaporthe oryzae* encodes a mitogen-activated protein kinase (MAPK), which is a key component in the MAPK signaling cascade and is involved in appressorium formation and infectious growth [32]. Thus, the PMK1 protein can be annotated with the term «GO ID 0075171 regulation of MAP kinase-mediated signal transduction in response to host». Note that this gene product would <u>not</u> be annotated with "GO ID 0052435 modulation by host of symbiont MAP kinase-mediated signal transduction pathway» since this latter GO term is reserved to annotate host gene products. Similarly, this protein should <u>not</u> be annotated with "GO ID 0052080 modulation by symbiont of host MAP kinase-mediated signal transduction pathway" since PMK1 belongs to the symbiont's and not the host's signaling transduction pathway.

In addition, the modulation terms have children that describe more specific kinds of signal transduction. For example, «GO ID 0075168 regulation of protein kinase-mediated signal transduction in response to host» has a child

«GO ID 0075171 regulation of MAP kinase-mediated signal transduction in response to host» (see details in Figure 6).

Penetration into the Host

Pathogens have evolved several mechanisms that include structural and/or enzymatic components in order to enter into their plant hosts [5]. Many fungi, such as *Alternaria alternata, Colletotrichum graminicola, M. oryzae, Pyrenophora teres*, and many oomycetes, such as *P. infestans* and *Phytophthora cinnamomi*, develop appressoria to directly penetrate plant cuticles [13, 33–38]. An appressorium is a highly specialized structure that differentiates from the end of a symbiont germ tube. It is a swollen, dome-shaped or cylindrical organ, from which a narrow penetration peg emerges to rupture the plant cuticle and cell wall [33]. The penetration peg extends and forms a penetration hypha to penetrate through the epidermal cells and emerge into the underlying tissue [34, 35]. In some instances, penetration is driven by astoundingly high turgor pressures within the appressoria [36, 38]. Generation of turgor is due in part to a thick inner cell wall layer of melanin in mature appressoria. In other instances, cell wall degrading enzymes may play a primary role in or may facilitate the penetration process [39–41]. Appressoria produced by some fungi, such as rust fungi, do not penetrate directly through the cuticle, but gain entry through stomata [42].

Sixty-four new GO terms were developed to describe the biological process of penetration into the host, and they form two groups. The first group includes 43 new GO terms related to infection structures established on the outside of the host tissue, such as appressoria, hyphopodia, infection cushions, and haustorium mother cells. The second group has 21 new terms related to specialized structures that directly pierce the surface of the host, for example penetration pegs, penetration hyphae, and haustorium necks.

All of the 43 terms in the first group are children or lower level offspring of "GO ID 0052108 growth or development of symbiont during interaction with host". The core of this group is "GO ID 0075015 formation of infection structure on or near host". Twenty-eight terms in this group are related to appressorium formation. In particular, five of the 28 terms describe in detail the process of appressorium formation, namely "GO ID 0075025 initiation of appressorium on or near host", "GO ID 0075034 nuclear division during appressorium formation on or near host", "GO ID 0075033 septum formation during appressorium formation on or near host", "GO ID 0075035 maturation of appressorium on or near host", and "GO ID 0075017 regulation of appressorium formation on or near host" (see details in Figure 3). Besides the child term "GO ID 0075016 appressorium formation on or near host", the term

"GO ID 0075015 formation of infection structure on or near host" has three more detailed child terms: "GO ID 0075192 haustorium mother cell formation on or near host", "GO ID 0075187 hyphopodium formation on or near host", and "GO ID 0075183 infection cushion formation on or near host" (see details in Figure 3).

All of the 21 terms in the second group are children or lower level offspring of "GO ID 0044409 entry into host". The core of this group is "GO ID 0075052 entry into host via a specialized structure", which has three child terms related to penetration peg, penetration hypha, or haustorium neck for entry into the host (see details in Figure 4).

The 64 new terms can be used to annotate the gene products of penetration-related genes. For example, genes involved in melanin biosynthesis in the rice blast fungus, such as ALB1, RSY1 and BUF1, are required for appressorium function since mutants lacking these genes make appressoria, but are unable to penetrate susceptible rice leaves [43]; these can be annotated with the term "GO ID 0075053 formation of symbiont penetration peg for entry into host".

Invasive Growth within the Host

After successful penetration, invasive hyphae are formed that ramify through the host tissue [44, 45]. In some cases, special structures, such as a haustorium or an arbuscle, are formed in host cells for the symbiont to absorb nutrition [22,23].

To describe invasive growth, 15 new GO terms were developed that are children or lower level offspring of "GO ID 0044412 growth or development of symbiont within host". The term "GO ID 0075065 growth or development of symbiont in host cell" has two children, "GO ID 0052094 formation by symbiont of haustorium for nutrient acquisition from host" and "GO ID 0075066 growth or development of symbiont in host organelle". Additionally, arbuscules produced by mycorrhizal fungi are a type of structure functionally similar to haustoria, and thus "GO ID 0075328 formation by symbiont of arbuscule for nutrient acquisition from host" is a sibling of "GO ID 0052094" (see details in Figure 4).

The 15 new GO terms in this section meet the need to annotate pathogen genes that are involved in invasive growth. For example, the *MST12* gene in the rice blast fungus *M. grisea* was found to regulate infectious growth but not appressorium formation [46]. In particular, no obvious defects in vegetative growth, conidiation, or conidia germination were observed in*MST12* deletion mutants. Also, *MST12* mutants produce typical dome-shaped and melanized appressoria. When inoculated through wound sites, *MST12* mutants fail to cause spreading lesions and appear to be defective in infectious growth. As a result, *MST12* mutants are nonpathogenic [46]. Thus, the *MST12* gene can be annotated with the term "GO ID 0075061 formation of symbiont invasive hypha within host».

Lesion Development in the Host

The eventual result of infection in most cases is lesion development. A lesion can be defined as any abnormality involving any tissue or organ due to any disease or any injury (cited from MedicineNet.com). Not surprisingly, there are many types of lesions including those caused by damage such as cold injury or insects' bites etc. It is difficult to define lesions objectively, as this requires a subjective judgment on what constitutes abnormal or damage and from what perspective, ranging for example from perturbation of a few cells to death of an entire tissue or organ. Similarly, formation of a lesion is not a specific process belonging to either the pathogen or the host and can be highly dependent on the environment. Therefore, at this time only one term, "GO ID 0009405 pathogenesis", is appropriate for genes involved in lesion formation.

Other New GO Terms

Six new terms were placed jointly under the nodes "GO ID 0006914 autophagy" and "GO ID 0044403 symbiosis, encompassing mutualism through parasitism". The term "GO ID 0075071 autophagy during symbiotic interaction" is the core of the six terms, and it has two complementary children, "GO ID 0075044 autophagy of host cells during interaction with symbiont" and "GO ID 0075072 autophagy of symbiont cells during interaction with host". The latter term has a child "GO ID 0075073 autophagy of symbiont cells on or near host surface", which itself has a lower level child "GO ID 0075074 spore autophagy during appressorium formation on or near host" (see details in Figure 3).

The six autophagy-related GO terms are applicable to describe the functions of several genes in fungal pathogens during symbiotic interaction. For example, formation of a functional appressorium in the rice blast fungus requires autophagic cell death of the conidium, which is controlled by the *MgATG8* gene. Deletion of *MgATG8* results in impaired autophagy, arrested conidial cell death, and a nonpathogenic fungus [14]. Thus, *MgATG8* can be annotated with the new term "GO ID 0075074 spore autophagy during appressorium formation on or near host".

CONCLUSION

Two hundred fifty-six new GO terms were developed to annotate genes or gene products involved in common pathogenic processes in fungi and oomycetes, including spore dispersal, host adhesion, recognition, penetration, and invasive growth. These new GO terms provide the opportunity to apply a standard set of terms to annotate gene products of fungi, oomycetes, and their associated hosts, as well as those of other plant-associated pathogens and their hosts. The ability to compare and contrast these annotations for widely different plant-associated microbes and their hosts, using a standardized vocabulary, will greatly facilitate the identification of unique and conserved features of pathogenesis across different kingdoms. In addition, such comparisons should provide insight into the evolution of pathogenic processes.

ACKNOWLEDGEMENTS

All authors read and approved the final manuscript. We thank Candace Collmer, Michelle Gwinn Giglio, and the editor at The Gene Ontology Consortium Jane Lomax for their comments and suggestions in developing these PAMGO terms. This work is a part of PAMGO project, which is supported by the USDA NRI-CSREES (grant number 2005-35600-16370), and the National Science Foundation (grant number EF-0523736).

This article has been published as part of *BMC Microbiology* Volume 9 Supplement 1, 2009: The PAMGO Consortium: Unifying Themes In Microbe-Host Associations Identified Through The Gene Ontology. The full contents of the supplement are available online at http://www.biomedcentral.com/1471-2180/9?issue=S1.

REFERENCES

1. Money NP: Why oomycetes have not stopped being fungi. Mycology Research. 1998, 102 (6): 767-768. 10.1017/S095375629700556X.
2. Latijnhouwers M, Wit PJGMD, Govers F: Oomycetes and fungi: similar weaponry to attack plants. Trends in Microbiology. 2003, 11 (10): 462-469. 10.1016/j.tim.2003.08.002.
3. Epstein L, Nicholson RL: Adhesion and adhesives of fungi and oomycetes. Biological Adhesives. Edited by: Smith AM, Callow JA. 2006, Springer-Verlag Berlin Heidelberg
4. Soanes DM, Richards TA, Talbot NJ: Insights from sequencing fungal and oomycete genomes: what can we learn about plant disease and the evolution of pathogenicity?. The Plant Cell. 2007, 19: 3318-3326. 10.1105/tpc.107.056663.
5. Agrios GN: Plant pathology. 2005, Elsevier Academic Press, London, UK, Fifth

6. Colditz F, Krajinski F, Niehaus K: Plant proteomics upon fungal attack. Plant Proteomics. Edited by: Šamaj J, Thelen J. 2007, Springer

7. Ingold GT: Dispersal in Fungi. 1953, Clarendon Press, Oxford; Oxford University Press, New York

8. Trail F: Fungal cannons: explosive spore discharge in the Ascomycota. FEMS Microbiol Lett. 2007, 276: 12-18. 10.1111/j.1574-6968.2007.00900.x.

9. James TY, Letcher PM, Longcore JE, Mozley-Standridge SE, Porter D, Powell MJ, Griffith GW, Vilgalys R: A molecular phylogeny of the flagellated fungi (Chytridiomycota) and description of a new phylum (Blastocladiomycota). Mycologia. 2006, 98 (6): 860-871. 10.3852/mycologia.98.6.860.

10. Griffin DH: Fungal Physiology. 1996, Published by Wiley_Default

11. Mitchell HJ, Hardham AR: Characterisation of the water expulsion vacuole in *Phytophthora nicotianae* zoospores. Protoplasma. 1999, 206: 118-130. 10.1007/BF01279258.

12. Choi W, Dean RA: The adenylate cyclase gene MAC1 of *Magnaporthe grisea* controls appressorium formation and other aspects of growth and development. Plant Cell. 1997, 9 (11): 1973-1983. 10.1105/tpc.9.11.1973.

13. Hardham AR: The cell biology behind *Phytophthora* pathogenicity. Australasian Plant Pathology. 2001, 30: 91-98. 10.1071/AP01006.

14. Veneault-Fourrey C, Barooah MK, Egan MJ, Talbot NJ: Autophagic fungal cell death is necessary for infection by the rice blast fungus. Science. 2006, 312 (5773): 580-583. 10.1126/science.1124550.

15. Talbot NJ: Fungal genomics goes industrial. Nature Biotechnology. 2007, 25 (5): 542-543. 10.1038/nbt0507-542.

16. Grenville-Briggs LJ, Anderson VL, Fugelstad J, Avrova AO, Bouzenzana J, Williams A, Wawra S, Whisson SC, Birch PRJ, Bulone V, West PV: Cellulose synthesis in *Phytophthora infestans* is required for normal appressorium formation and successful infection of potato. The Plant Cell. 2008, 20: 720-738. 10.1105/tpc.107.052043.

17. Onyile AB, Edwards HH, Gessner RV: Adhesive material of the hyphopodia of *Buergenerula spartinae*. Mycologia. 1982, 74 (5): 777-784. 10.2307/3792864.

18. Du M, Schard CL, Nuckles EM, Vaillancourt LJ: Using mating-type gene sequences for improved phylogenetic resolution of*Collectotrichum* species complexes. Mycologia. 2005, 97 (3): 641-658. 10.3852/mycologia.97.3.641.

19. Sukno SA, García VM, Shaw BD, Thon MR: Root infection and systemic colonization of maize by *Colletotrichum graminicola*. Applied and environmental microbiology. 2008, 74 (3): 823-832. 10.1128/AEM.01165-07.
20. Heath MC, Heath IB: Ultrastructural changes associated with the haustorial mother cell ceptum during haustorium formation in*Uromyces phaseoli* var. *vignae*. Protoplasma. 1975, 84: 297-314. 10.1007/BF01279359.
21. Glidewell DC, Mims CW: Ultrastructure of the haustorial apparatus in the rust fungus *Kunkelia nitens*. Botanical Gazette. 1979, 140 (2): 148-152. 10.1086/337071.
22. Mendgen K, Dressler E: Culturing *Puccinia coronata* on a cell monolayer of the *Avena saliva* coleoptile. Phytopath. 1983, 108: 226-234. 10.1111/j.1439-0434.1983.tb00583.x.
23. Wiethölter N, Horn S, Reisige K, Beike U, Moerschbacher BM: In vitro differentiation of haustorial mother cells of the wheat stem rust fungus, *Puccinia graminis* f. sp. *tritici*, triggered by the synergistic action of chemical and physical signals. Fungal Genetics and Biology. 2003, 38: 320-326. 10.1016/S1087-1845(02)00539-X.
24. Demirci E, Döken MT: Host penetration and infection by the anastomosis groups of *Rhizoctonia solani* Kühn isolated from potatoes. Tr J of Agriculture and Forestry. 1998, 22: 609-613.
25. Birch PRJ, Cooke DEL: Mechanisms of infection: Oomycetes. Encyclopedia of Plant and Crop Science. 2004, 1 (1): 697-700.
26. Görnhardt B, Rouhara I, Schmelzer E: Cyst germination proteins of the potato pathogen *phytophthora infestans* share homology with human mucins. Mol Plant-Micro Interact. 2000, 13 (1): 32-42. 10.1094/MPMI.2000.13.1.32.
27. Zhao X, Kim Y, Park G, Xu J-R: A mitogen-activated protein kinase cascade regulating infection-related morphogenesis in *Magnaporthe grisea*. The Plant Cell. 2005, 17: 1317-1329. 10.1105/tpc.104.029116.
28. Ligterink W, Kroj T, Nieden UZ, Hirt H, Scheel D: Receptor-mediated activation of a MAP kinase in pathogen defense of plants. Science. 1997, 276 (27): 2054-2057. 10.1126/science.276.5321.2054.
29. Miwa T, Takagi Y, Shinozaki M, Yun C-W, Schell WA, Perfect JR, Kumagai H, Tamaki H: Gpr1, a putative G-protein-coupled receptor, regulates morphogenesis and hypha formation in the pathogenic fungus *Candida albicans*. Eukaryotic Cell. 2004, 3 (4): 919-931. 10.1128/EC.3.4.919-931.2004.

30. Lafon A, Han K-H, Seo J-A, Yu J-H, d'Enfert C: G-protein and cAMP-mediated signaling in aspergilli: A genomic perspective. Fungal Genetics and Biology. 2006, 43 (7): 490-502. 10.1016/j.fgb.2006.02.001.

31. Praveen RJ, Reena G, Subramanyam C: Calmodulin-dependent protein phosphorylation during conidial germination and growth of*Neurospora crassa*. Mycol Res. 1997, 101: 1484-1488. 10.1017/S0953756297004255.

32. Xu J-R, Hamer JE: MAP kinase and cAMP signaling regulate infection structure formation and pathogenic growth in the rice blast fungus*Magnaporthe grisea*. Genes & Dev. 1996, 10: 2696-2706. 10.1101/gad.10.21.2696.

33. DeZwaan TM, Carroll AM, Valent B, Sweigard JA: *Magnaporthe grisea* Pth11p is a novel plasma membrane protein that mediates appressorium differentiation in response to inductive substrate cues. The Plant Cell. 1999, 11: 2013-2030. 10.2307/3871094.

34. Clergeot P-H, Gourgues M, Cots J, Laurans F, Latorse M-P, Pépin R, Tharreau D, Notteghem J-L, Lebrun M-H: *PLS1*, a gene encoding a tetraspanin-like protein, is required for penetration of rice leaf by the fungal pathogen *Magnaporthe grisea*. PNAS. 2001, 98 (12): 6963-6968. 10.1073/pnas.111132998.

35. Park G, Bruno KS, Staiger CJ, Talbot NJ, Xu J-R: Independent genetic mechanisms mediate turgor generation and penetration peg formation during plant infection in the rice blast fungus. Molecular Microbiology. 2004, 53 (6): 1695-1707. 10.1111/j.1365-2958.2004.04220.x.

36. Wang ZY, Jenkinson JM, Holcombe LJ, Soanes DM, Veneault-Fourrey C, Bhambra GK, Talbot NJ: The molecular biology of appressorium turgor generation by the rice blast fungus *Magnaporthe grisea*. Biochem Soc Trans. 2005, 33 (Pt 2): 384-388.

37. Blanco FA, Judelson HS: A bZIP transcription factor from *Phytophthora* interacts with a protein kinase and is required for zoospore motility and plant infection. Molecular Microbiology. 2005, 56 (3): 638-648. 10.1111/j.1365-2958.2005.04575.x.

38. Liu XH, Lu JP, Zhang L, Dong B, Min H, Lin FC: Involvement of a *Magnaporthe grisea* serine/threonine kinase gene, MgATG1, in appressorium turgor and pathogenesis. Eukaryot Cell. 2007, 6 (6): 997-1005. 10.1128/EC.00011-07.

39. Tonukari NJ, Scott-Craig JS, Walton JD: The *Cochliobolus carbonum SNF1* gene is required for cell wall-degrading enzyme expression and virulence on maize. Plant Cell. 2000, 12: 237-248. 10.1105/tpc.12.2.237.

40. Li D, Ashby AM, Johnstone K: Molecular evidence that the extracellular

cutinase Pbc1 is required for pathogenicity of *Pyrenopeziza brassicae* on oilseed rape. Mol Plant-Micro Interact. 2003, 16: 545-552. 10.1094/MPMI.2003.16.6.545.

41. Aro N, Pakula T, Penttila M: Transcriptional regulation of plant cell wall degradation by filamentous fungi. FEMS Microbiology Reviews. 2005, 29: 719-739. 10.1016/j.femsre.2004.11.006.
42. Prats E, Llamas MJ, Jorrin J, Rubiales D: Constitutive coumarin accumulation on sunflower leaf surface prevents rust germ tube growth and appressorium differentiation. Crop science. 2007, 47: 1119-1124. 10.2135/cropsci2006.07.0482.
43. Chumley FG, Valent B: Genetic analysis of melanin-deficient, nonpathogenic mutants of *Magnaporthe grisea*. Mol Plant-Micro Interact. 1990, 3: 135-143.
44. Walker SK, Chitcholtan K, Yu YP, Christenhusz GM, Garrill A: Invasive hyphal growth: An F-actin depleted zone is associated with invasive hyphae of the oomycetes *Achlya bisexualis* and *Phytophthora cinnamomi*. Fungal Genetics and Biology. 2006, 43 (5): 357-365. 10.1016/j.fgb.2006.01.004.
45. Bassilana M, Blyth J, Arkowitz RA: Cdc24, the GDP-GTP exchange factor for Cdc42, is required for invasive hyphal growth of *Candida albicans*. Eukaryotic cell. 2003, 2 (1): 9-18. 10.1128/EC.2.1.9-18.2003.
46. Park G, Xue C, Zheng L, Lam S, Xu J-R: *MST12* regulates infectious growth but not appressorium formation in the rice blast fungus *Magnaporthe grisea*. Mol Plant-Micro Interact. 2002, 15 (3): 183-192. 10.1094/MPMI.2002.15.3.183.

Chapter 7

AN UPDATE ON POLYGALACTURONASE-INHIBITING PROTEIN (PGIP), A LEUCINE-RICH REPEAT PROTEIN THAT PROTECTS CROP PLANTS AGAINST PATHOGENS

Raviraj M. Kalunke[1], Silvio Tundo[1], Manuel Benedetti[2], Felice Cervone[2], Giulia De Lorenzo[2*] and Renato D'Ovidio[1]

[1]Dipartimento di Scienze e Tecnologie per l'Agricoltura, le Foreste, la Natura e l'Energia, Università della Tuscia, Viterbo, Italy

[2]Dipartimento di Biologia e Biotecnologie "Charles Darwin", Sapienza Università di Roma, Roma, Italy

Polygalacturonase inhibiting proteins (PGIPs) are cell wall proteins that inhibit the pectin-depolymerizing activity of polygalacturonases secreted by microbial pathogens and insects. These ubiquitous inhibitors have a leucine-rich repeat structure that is strongly conserved in monocot and dicot plants. Previous reviews have summarized the importance of PGIP in plant defense and the structural basis of PG-PGIP interaction; here we update the current knowledge about PGIPs with the recent findings on the composition and evolution of *pgip* gene families, with a special emphasis on legume and cereal crops. We also update the information about the inhibition properties of single *pgip* gene products against microbial PGs and the results, including field tests, showing the capacity of PGIP to protect crop plants against fungal, oomycetes and bacterial pathogens.

INTRODUCTION

Successful colonization of plant tissues by microbial pathogens requires the overcoming of the cell wall. To this end, pathogens produce a wide array of plant cell wall degrading enzymes (CWDEs), among which endo-polygalacturonases (PGs; EC 3.2.1.15) are secreted at very early stages of the infection process (ten Have et al., 1998). PGs cleave the α-(1–4) linkages between the D-galacturonic acid residues of homogalacturonan, the main component of pectin, causing cell separation and maceration of the host tissue. To counteract the activity of PGs, plants deploy the cell wall polygalacturonase inhibiting proteins (PGIPs) that inhibit the pectin-depolymerizing activity of PGs. No plant species or mutants totally lacking PGIP activity have been characterized so far. The structure of PGIPs is typically formed by 10 imperfect leucine-rich repeats (LRRs) of 24 residues each, which are organized to form two β-sheets, one of which (sheet B1) occupies the concave inner side of the molecule and contains residues crucial for the interaction with PGs (Di Matteo et al., 2003). In addition to PG inhibition, the interaction between PGs and PGIPs promotes the formation of oligogalacturonides (OGs), which are elicitors of a variety of defense responses (Cervone et al., 1989; Ridley et al., 2001; Ferrari et al., 2013). Since many aspects of the PGIP biology have been already summarized in previous reviews (De Lorenzo et al., 2001; De Lorenzo and Ferrari, 2002; D'Ovidio et al., 2004a; Gomathi and Gnanamanickam, 2004; Shanmugam, 2005; Di Matteo et al., 2006; Federici et al., 2006; Cantu et al., 2008;Misas-Villamil and van der Hoorn, 2008; Protsenko et al., 2008; Reignault et al., 2008; Lagaert et al., 2009), here we present an overview of the recent findings on genome composition and evolution of *pgip* gene families and on the efficacy of PGIP to limit the development of diseases caused by microbial pathogens in crop plants.

PGIP Genes and their Genomic Organization

Early characterization of a polygalacturonase-inhibiting activity was reported in 1970s (Albersheim and Anderson, 1971) and the first *pgip* gene was isolated 20 years later in French bean (Toubart et al., 1992). Since then, several PGIPs and a large number of *pgip* genes have been characterized. Up to now more than 170 complete or partial *pgip* genes from dicot and monocot plants have been deposited in nucleotide databases (e.g., http://www.ncbi.nlm.nih.gov/). Most of these genes have been identified as *pgip* genes on the basis of sequence identity but only a few of them have been shown to encode proteins with PG-inhibitory activity.

Genome analysis has shown that *pgip* genes did not undergo a large expansion and may exist as single genes, as in diploid wheat species (Di Giovanni et al., 2008), or organized into gene families, the members of which are organized in tandem and can vary from two, as in *Arabidopsis thaliana* (Ferrari et al., 2003), to sixteen, as in *Brassica napus* (Hegedus et al., 2008). The majority of *pgip* genes are intronless, however, some of them can contain a short intron as in *Atpgip1* and *Atpgip2* (Ferrari et al., 2003). Moreover, *pgip* genes can be inactivated by transposon elements as in cultivated and wild wheat where the occurrence of *Copia*-retrotransposon and *Vacuna* transposons has been reported (Di Giovanni et al., 2008). Characterized *pgip* loci are shown in Figure 1. Like other families of defense-related genes, *pgip* families show variation in the expression pattern of the different members, some of which are constitutive, others are tissue-specific and, in most cases, up-regulated following stress stimuli (see reviews indicated above; Table 1).

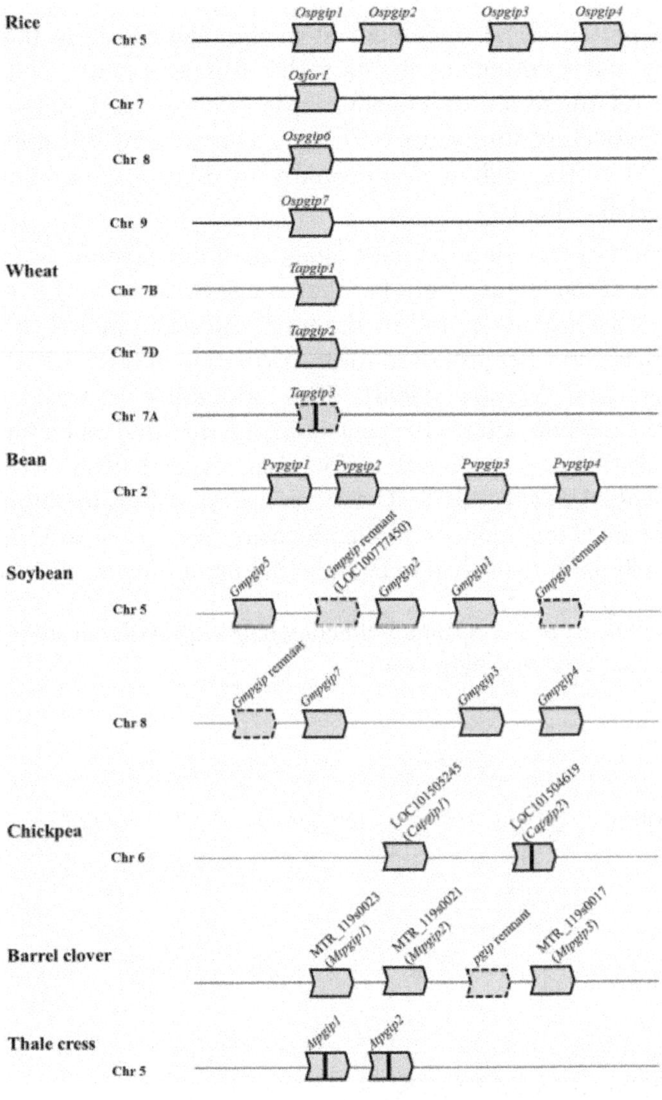

Figure 1. Schematic representation of the genomic organization *pgip* **families in rice, wheat, bean, soybean, chickpea, barrel clover and thale cress**. Each block-arrow with compound-type lines represents a predicted *pgip* gene and a block-arrow with

dash type lines represents a predicted pseudo-gene or remnant gene. Vertical line within block-arrow indicates introns (*Capgip2*, *Atpgip1*, and *Atpgip2*) or a Copia ret-rotransposon (*Tapgip3*). The direction of the arrow indicates ATG to stop codon. The location of*pgip* genes of legume species are based on Kalunke et al. (2014), those of rice and wheat on Janni et al. (2006) and Di Giovanni et al. (2008), and those of thale crest on Ferrari et al. (2003). Chr, chromosome.

At the protein level, members of a *pgip* family show both functional redundancy and sub-functionalization (De Lorenzo et al., 2001;Federici et al., 2006). As suggested previously, these features likely have an adaptive significance for combating more efficiently a broad array of pathogens (Ferrari et al., 2003) or responding more rapidly to diverse environmental stimuli (D›Ovidio et al., 2004b).

In support of this view, a recent analysis of the genomic organization and composition of the legume *pgip* families suggested that the forces driving the evolution of the*pgip* genes follow the birth-and-death model (Kalunke et al., 2014), similarly to what proposed for the evolution of NBS-LRR-type *R* genes (Michelmore and Meyers, 1998). This possibility is based on genomic features that include inferred recent duplications, diversification as well as pseudogenization of *pgip* copies, as found in soybean, bean, barrel clover and chickpea (Kalunke et al., 2014). The organization of the *pgip* families therefore supports the view that tandem duplications are frequent in stress-related genes and are beneficial for survival in challenging environments (Oh et al., 2012).

Table 1. Treatments or stress stimuli affecting *pgip* **expression in some plant species with a well characterized pgip family**

***Pgip* family**	**Treatments or stress stimuli**	**References**
Rice	Abscisic acid (ABA), brassinosteroid, gibberellic acid (GA), 3-indole acetic acid (IAA), jasmonic acid (JA), kinetin, naphthalene acetic acid (NAA), salicylic acid (SA); *Rhizoctonia solani* (necrotrophic fungus)	Janni et al., 2006; Lu et al., 2012
Wheat	*Bipolaris sorokiniana* (necrotrophic fungus) and mechanical wounding	Janni et al., 2013
Bean	Oligogalacturonides (OGs); mechanical wounding; *Botrytis cinerea*, *Sclerotinia sclerotiorum* (necrotrophic fungi); *Colletotrichum lindemuthianum* (hemibiotrophic fungus)	Bergmann et al., 1994; Nuss et al., 1996; Devoto et al., 1997; D'Ovidio et al., 2004b; Oliveira et al., 2010; Kalunke et al., 2011
Soybean	Mechanical wounding; *S. sclerotiorum* (necrotrophic fungus)	D'Ovidio et al., 2006; Kalunke et al., 2014
M. truncatula	JA, SA, ABA; *Colletotrichum trifolii* (hemibiotrophic fungus)	Song and Nam, 2005
Rapeseed	JA, SA, mechanical wounding; *S. sclerotiorum*	Hegedus et al., 2008
Pepper	SA, Methyl jasmonate (Me-JA), ABA, wounding, cold treatment	Wang et al., 2013
Arabidopsis	OGs; JA; *B. cinerea; Stemphylium solani* (necrotrophic fungus); aluminum, low-pH, cold; geminivirus	Ferrari et al., 2003; Ascencio-Ibanez et al., 2008; Sawaki et al., 2009; Di et al., 2012; Kobayashi et al., 2014

Inhibition Activity of PGIPs

A number of papers deals with the inhibition activity of PGIPs purified from several plant tissues. This aspect has been reviewed several years ago (De Lorenzo et al., 2001); here, we present an update of this information (Table 2).

Because purified PGIPs may contain a mix of highly similar PGIP isoforms, the activity detected in a tissue may result from the contribution of the activities of different PGIPs expressed in that tissue. An appropriate approach to study the inhibition activity of individual PGIP isoforms is their expression in a heterologous system. However, only a few of the more than 170 *pgip* genes isolated so far from different plant species have been investigated. As reported in Table 3, individual heterologous expression and analysis of all members of a *pgip* family has been performed only for *Arabidopsis* (Ferrari et al., 2003), common bean (D›Ovidio et al., 2004b), soybean (D›Ovidio et al., 2006; Kalunke et al., 2014) and wheat (Janni et al., 2013). PGIPs have been expressed in prokaryotic systems, as a fusion with the maltose-binding protein (MBP) (Jang et al., 2003; Table 3) or using lower temperature for bacterial growth (Chen et al., 2011), in *Pichia pastoris* and in plants by stable transformation or, transiently, by virus-mediated expression (Table 3). In some cases, the proteins were successfully expressed, but did not show any inhibitory activity *in vitro*, as, for example, in the case of some GmPGIPs (D'Ovidio et al., 2006). GmPGIP3, but not GmPGIP1, GmPGIP2, and GmPGIP7 showed inhibitory activity, whereas no expression of GmPGIP5 was obtained (D'Ovidio et al., 2006; Kalunke et al., 2014). Similarly, TaPGIP1 and TaPGIP2, encoded by the two members of the wheat *pgip* family, were successfully expressed but showed no inhibition activity (Janni et al., 2013).

Table 2. Bulk PGIP purified from plants and tested against microbial PGs. These data update those reported in De Lorenzo et al. (2001)

Plant	Tissue	PGIP preparation	Polygalacturonases		References
			Inhibited	Not inhibited	
Tomato (*Solanum lycopersici* L.)	Stem	Crude extract	*Ralstonia solanacearum*		Schacht et al., 2011
Tobacco (*Nicotiana tabacum* L.)	Nectar		*Botrytis cinerea*		Thornburg et al., 2003
Potato (*Solanum tuberosum* L.)		Gel chromatography	*Aspergillus niger* *Fusarium moniliforme*[§] *Fusarilm solani* isolate 1402	*Fusarium solani* (isolate 3122)	Machinandiarena et al., 2001
Common Bean (*Phaseolus vulgaris* L.)	Leaves	PG-Sepharose chromatography	*Fusarium anthophilum* *Fusarium circinatum* *Fusarium subglutinans.* *Fusarium proliferatum* isolate 1152 *Fusarium proliferatum* PVS-Fu 64 *Fusarium sacchari* *Fusarium fujikuroi* *F. thapsinum*	*Fusarium verticillioides* *Fusarium proliferatum* ISPAVEmc 1189 *Fusarium nygamai*	Raiola et al., 2008
			Fusarium moniliforme[§] FC-10	*Fusarium moniliforme*[§] PD	Sella et al., 2004

Leek (*Allium ampeloprasum* L.)	Basal leaves	Mono-S chromatography		*Fusarium anthophilum* *Fusarium circinatum* *Fusarium subglutinans* *Fusarium proliferatum* *Fusarium sacchari* *Fusarium fujikuroi* *Fusarium verticillioides* *Fusarium proliferatum* ISPAVEmc 1189 *Fusarium nygamai*	Raiola et al., 2008
Asparagus (*Asparagus officinalis* L.)	White spear	Mono-S chromatography		*Fusarium anthophilum* *Fusarium circinatum* *Fusarium subglutinans.* *Fusarium proliferatum* *Fusarium sacchari* *Fusarium fujikuroi* *Fusarium verticillioides* *Fusarium proliferatum* ISPAVEmc 1189 *Fusarium nygamai*	Raiola et al., 2008
Pepper (*Capsicum annuum L.*)	Fruit	Ion-exchange chromatography	*Colletotrichum gleosporoides*, *Colletotrichum capsici*, *Colletotrichum lindemuthianum* *Sclerotium rolfsi* *Fusarium moniliforme*[§]		Shivashankar et al., 2010

Plant	Tissue	PGIP preparation	Polygalacturonases		References
			Inhibited	**Not inhibited**	
Guava (*Psidium guajava* L.)	Fruit	Purified using a Sephadex G-100	*Aspergillus niger*		Deo and Shastri, 2003
"Oroblanco" grapefruit hybrid (*Citrus grandis* × *C. paradisi* Macf.)	Fruit	Anion exchange chromatography	*Penicillium Italicum* *Botrytis cinerea*		D'hallewin et al., 2004
Apple (*Malus domestica* L.)	Fruit		*Colletotrichum acutatum*		Gregori et al., 2008
	Fruit skin	Partial purified	*Botryosphaeria dothidea*	*Glomerella cingulata*	Lee et al., 2006
	Parenchymal tissues	Partial purified	*Monilia fructigena*		Buza et al., 2004
Cantaloupe (*Cucumis melo* L.)	Fruit	Cation exchange chromatography	*Phomopsis cucurbitae* *Aspergillus niger* *Fusarium solani*	*Didymella bryoniae* *Rhizopus* PG *Fusarium verticillioides*	Fish and Davis, 2004
Cotton (*Gossypium hirsutum* L.)	Stem	PG-affinity chromatography	*Aspergilus niger*		James and Dubery, 2001
Pear (*Pyrus communis* L.)	Fruit	Partial purified	*Verticillium dahliae* *Botrytis cinerea* *Venturia nashicola*		Ladu et al., 2012; Faize et al., 2003
Pearl millets (*Pennisetum glaucum* (L) R. Br.)	Seedlings	Crude extract	*Aspergilus niger*		Prabhu et al., 2012

Plant	Tissue	PGIP preparation	Polygalacturonases		References
			Inhibited	Not inhibited	
Grass pea (*Lathyrus sativus* L.)	Seeds	Gel-filtration chromatography	*Aspergillus niger* *Rhizopus spp*		Tamburino et al., 2012
Orange (*Citrus reticulate* L.)	Fruit	Partial purified	*Diaprepes abbreviatus*		Doostdar et al., 1997
Blue mustard (*Chorispora bungeana*)	Leaves, stem, root	Partial purified	*Aspergillus niger* *Stemphylium solani*		Di et al., 2009
Ginseng (*Panax ginseng* L.)		Crude extract	*Colletotrichum gloeosporioides* *Phythium ultimum* *Fusarium oxysporum* *Rhizoctonia solani*		Sathiyaraj et al., 2010
Bread wheat (*Triticum aestivum* L.)	Leaves	Cation exchange chromatography	*Cochliobolus sativus*	*Aspergillus niger* (EPG I and II) *Cryphonectria parasitica* *Postia placenta* *Fusarium moniliforme*[§] *Colletotrichum lindemuthianum* *Aspergillus niger* exopolygalacturonase *Ralstonia solanacearum*	Kemp et al., 2003
Durum wheat (*Triticum turgidum ssp. dicoccoides*	Leaves	Crude extract	*Fusarium graminearum* *Bipolaris sorokiniana* *Stenocarpella maydis*	*Fusarium phyllophylum*	Janni et al., 2013

[§] *Reclassified as Fusarium phyllophilum* (Mariotti et al., 2008).

Table 3. *Pgip* genes individually expressed in plants or in heterologous systems and tested for inhibition activity against microbial PGs

Species	Gene	Heterologous systems	Origin of purified PG		References
			Inhibited	Not inhibited	
Common bean (*Phaseolus vulgaris* L.)	PvPGIP1	Transgenic tomato		*Fusarium oxysporum* *Botrytis cinerea* *Alternaria solani*	Desiderio et al., 1997
			Stenocarpella maydis *Aspergillus niger*		Berger et al., 2000
	PvPGIP1 PvPGIP2 PvPGIP3 PvPGIP4	PVX/*Nicotiana benthamiana*	*Aspergillus niger* *Fusarium moniliforme*[§] *Stenocarpella maydis* *Colletotrichum acutatum* *Botrytis cinerea*	*Lygus rugulipennis* *Adelphocoris lineolatus* *Orthops kalmi* *Closterotomus norwegicus*	D'Ovidio et al., 2006; Frati et al., 2006
	PvPGIP2	Transgenic wheat	*Bipolaris sorokiniana* *F. graminearum*	*Claviceps purpurea*	Janni et al., 2008; Volpi et al., 2013
		Transgenic *Brassica napus*	*Rhizoctonia solani*		Akhgari et al., 2012
		Transgenic sugarbeet	*Fusarium phyllophilum* FC10		Mohammadzadeh et al., 2012
		PVX/*Nicotiana benthamiana*	*Fusarium phyllophilum* FC-10 *Fusarium phyllophilum* 10241 *Fusarium phyllophilum* 25219 *Fusarium phyllophilum* 25218	*Fusarium phyllophilum* 25305 *Fusarium verticillioides* 62264 *Fusarium verticillioides* PD	Mariotti et al., 2008

Species	Gene	Heterologous systems	Origin of purified PG		References
			Inhibited	Not inhibited	
Runner bean (*Phaseolus coccineus* L.)	PcPGIP2	PVX/*Nicotiana benthamiana*	*Fusarium moniliforme*[§] *Aspergillus niger* *Colletotrichum lupini* *Botrytis cinerea*		Farina et al., 2009
Tepary bean (*Phaseolus acutifolius* L.)	PaPGIP2	PVX/*Nicotiana benthamiana*	*Fusarium moniliforme*[§] *Aspergillus niger* *Colletotrichum lupini* *Botrytis cinerea*		Farina et al., 2009
Lima bean (*Phaseolus lunatus* L.)	PlPGIP2	PVX/*Nicotiana benthamiana*	*Fusarium moniliforme*[§] *Aspergillus niger* *Colletotrichum lupini* *Botrytis cinerea*		Farina et al., 2009
Soybean (*Glycine max* L.)	GmPGIP1 GmPGIP2	PVX/*Nicotiana benthamiana*		*Sclerotinia sclerotiorum* PGb *Sclerotinia sclerotiorum* PGa *Fusarium moniliforme*[§] *Botrytis aclada* *Aspergillus niger* *Botrytis cinerea* *Colletotrichum acutatum* *Fusarium graminearum* *Lygus rugulipennis* *Adelphocoris lineolatus* *Orthops kalmi* *Closterotomus norwegicus*	D'Ovidio et al., 2006; Frati et al., 2006
	GmPGIP3	PVX/*Nicotiana benthamiana*	*Sclerotinia sclerotiorum* PGb *Sclerotinia sclerotiorum* PGa *Fusarium moniliforme*[§] *Botrytis aclada* *Aspergillus niger* *Botrytis cinerea* *Colletotrichum acutatum* *Fusarium graminearum*		D'Ovidio et al., 2006; Frati et al., 2006
	GmPGIP4	PVX/*Nicotiana benthamiana*		*Sclerotinia sclerotiorum* PGb *Sclerotinia sclerotiorum* PGa *Fusarium moniliforme*[§] *Botrytis aclada* *Aspergillus niger* *Botrytis cinerea* *Colletotrichum acutatum* *Fusarium graminearum*	D'Ovidio et al., 2006; Frati et al., 2006
	GmPGIP7	PVX/*Nicotiana benthamiana*		*Sclerotinia sclerotiorum* *Fusarium graminearum* *Colletotrichum acutatum* *Aspergillus niger*	Kalunke et al., 2014
Pepper (*Capsicum annum* L.)	CaPGIP1 CaPGIP2	*Escherichia coli*	*Alternaria alternata* *Colletotrichum nicotianae*		Wang et al., 2013

Species	Gene	Heterologous systems	Origin of purified PG		References
			Inhibited	**Not inhibited**	
Rapeseed (*Brassica napus* L.)	BnPGIP1	*Pichia pastoris*	*Sclerotinia sclerotiorum* PG6		Bashi et al., 2013
Chinese cabbage (*Brassica rapa* L.)	BrPGIP2	Transgenic *Brassica rapa*	*Pectobacterium carotovorum* *Botryosphaeria dothidea*		Hwang et al., 2010
	BrPGIP2	*Escherichia coli*	*Sclerotinia sclerotiorum*		HuangFu et al., 2014
Grapevine (*Vitis vinifera* L.)	VvPGIP1	Transgenic tobacco	*Botrytis cinerea* PGI *Botrytis cinerea* PG4 *Botrytis cinerea* PG6 *Aspergillus. niger* PGA *Aspergillus niger* PGB	*Botrytis cinerea* PG3 *Aspergillus niger* PGII	Joubert et al., 2006
			Aspergillus niger PGI	*Botrytis cinerea* PG2	Joubert et al., 2007
Apple (*Malus domestica* Borkh.)	MdPGIP1	Transgenic tobacco	*Colletotrichum lupini* *Botryosphaeria obtusa* *Diaporthe ambigua*	*Aspergillus niger*	Oelofse et al., 2006
		Transgenic potato	*Verticillium dahliae*		Gazendam et al., 2004
Pear (*Pyrus communis* L.)	PpPGIP	Transgenic grape	*Botrytis cinerea*		Agüero et al., 2005
		Transgenic tomato	*Botrytis cinerea*		Powell et al., 2000
		Transgenic persimmon	*Botrytis cinerea*		Tamura et al., 2004
Raspberry (*Rubus idaeus* L.)	RiPGIP	Transgenic pea	*Stenocarpella maydis* *Colletotrichum lupini*		Richter et al., 2006
Wheat (*Triticum aestivum* L.)	TaPGIP1 TaPGIP2	PVX/*Nicotiana benthamiana*		*Fusarium phyllophylu* *Stenocarpella maydis* *Bipolaris sorokiniana* *Fusarium graminearum*	Janni et al., 2013
Rice (*Oryza sativa* L.)	OsPGIP1	PVX/*Nicotiana benthamiana*	*Sclerotinia sclerotiorum* *Fusarium moniliforme*§ *Fusarium graminearum* *Aspergillus niger* *Botrytis cinerea*		Janni et al., 2006
	OsFOR1	*Escherichia coli* BL21	*Aspergillus niger* PG		Jang et al., 2003
Pearl millet [*Pennisetum glaucum* (L.) R. Br.]	PglPGIP1	*Escherichia coli SHuffle*® *T7 Express*	*Aspergillus niger*, AnPGII	*Fusarium moniliforme*, FmPGIII	Prabhu et al., 2014
Arabidopsis thaliana	AtPGIP1 AtPGIP2	Transgenic *Arabidopsis*	*Colletotrichum gloeosporioides* *Stenocarpella maydis* *Botrytis cinerea* *Fusarium graminearum*	*Aspergillus niger* *Fusarium moniliforme*§ *Lygus rugulipennis* *Adelphocoris lineolatus* *Orthops kalmi* *Closterotomus norwegicus*	Frati et al., 2006; Ferrari et al., 2012, 2003

§*Reclassified as Fusarium phyllophilum FC10* (Mariotti et al., 2008).

The absence of inhibition activity *in vitro* may also reflect the possibility that some PGIPs are active only in the *in planta* environment, as suggested by Joubert et al. (2006) in the case of the *Botrytis cinerea* BcPG2 and VvPGIP1 from grapevine (*Vitis vinifera L.)*. These proteins do not interact *in vitro*, although VvPGIP1 reduces symptoms caused by BcPG2 upon co-infiltration in leaves. The number and sources of PGs tested is also limited; only a few studies have been carried out against PGs of bacteria and insects (Doostdar et al., 1997;D'Ovidio et al., 2004b; Frati et al., 2006; Hwang et al., 2010; Schacht et al., 2011; Kirsch et al., 2012). The limitations of data prevents to draw conclusions about correlations between PGIPs of specific plant families and specific pathogens. Notably, PG produced by a highly detrimental pathogen, *Fusarium verticillioides*, is not inhibited by any known PGIP (see Table 2). This PG has been a target of an unsuccessful attempt to render PvPGIP2 an efficient inhibitor against this PG (see below, Benedetti et al., 2011a).

The utilization of *pgip* genes for crop protection relies on the identification of inhibitors with broad specificities against the many PGs produced by phytopathogens and/or the construction of novel PGIPs with stronger and broader inhibitor activity. Many more PGIPs than those reported in Tables 2, 3 exist in nature and are likely to have different specificities against microbial PGs, considering that single amino acid changes are able to change specificity of the inhibitors (Leckie et al., 1999). Searching for PGIPs with novel specificities may allow to count on a much larger reservoir of possible genes for crop protection. A direct and simple strategy to isolate PGIPs with recognition capability against a given PG may be based on affinity chromathography methods, similar to that originally used to purify PGIP from *P. vulgaris* (Cervone et al., 1987), and mass spectrometry. Attempts to drive *in vitro* evolution of PGIPs to generate proteins with improved inhibition properties have not been successful yet (Benedetti et al., 2011a).

The occurrence of PG-inhibiting activity in crude leaf protein extracts of tetraploid wild wheat (*T. dicoccoides*) possessing nonfunctional *pgip* genes (Di Giovanni et al., 2008) suggested the existance of *pgip* genes with a sequence divergent from the classical one. This possibility, which deserves further investigation, is also supported by the finding that the wheat tissue contains PG-inhibiting proteins with N-terminal sequences (Lin and Li, 2002; Kemp et al., 2003) different from TaPGIP1 and TaPGIP2 (Janni et al., 2013) and from the *pgip*sequences reported so far (http://www.ncbi.nlm.nih.gov/nucleotide/). Recently, a wheat gene with some sequence similarity to *pgip* genes has been reported and was shown to be involved in the defense response against *Fusarium graminearum* (Hou et al., 2014).

Structural Studies on the PG-PGIP Interaction

Thus, the possibility of engineering new forms of PGIPs depends on the detailed structural knowledge of the PG-PGIP interaction. Several structural studies have been performed (Mattei et al., 2001; King et al., 2002;Benedetti et al., 2011b, 2013; Gutierrez-Sanchez et al., 2012), but a high resolution 3D-structure of the PG-PGIP complex is still missing. The enzyme-inhibitor combinations that have been more extensively investigated, are those that PGIP2 from *Phaseolus vulgaris* (PvPGIP2) forms with PG from *A. niger* (AnPGII),*F. phyllophilum* (FpPG) and *C. lupini* (ClPG). Site-directed mutagenesis has shown that the residues involved in the interaction are located in the concave surface of the inhibitor (Leckie et al., 1999; Federici et al., 2001;Spinelli et al., 2009; Benedetti et al., 2011b, 2013). Computational methods such as the Codon Substitution Model in combination with the Desolvation Energy Calculation and the Repeat Conservation Mapping (RCM;Helft et al., 2011) have pinpointed several residues of PvPGIP2 responsible for the PG-inhibiting activity (Casasoli et al., 2009).

On the other hand, residues of PG that are critical for the interaction with PGIP have been also studied. FvPG is 92.5% identical to FpPG, but is inhibited by neither PvPGIP2 nor other known PGIPs. By both loss- and gain-of-function site-directed mutations, a single amino acid at position 274 of both FvPG and FpPG was demonstrated to act as a switch for recognition by PvPGIP2 (Raiola et al., 2008; Benedetti et al., 2013). Unfortunately, the lack of high-resolution structural information on the PG-PGIP complex does not allow to precisely identify the contacting residue in PGIP. Moreover, both PGs and PGIPs are glycosylated proteins (Caprari et al., 1993; Lim et al., 2009); however, whether glycosylation plays a role in the PGIP-PG interaction requires further investigation. For example, glycosylation in pearl millet PGIP was found to affect pH and temperature stability of the protein but not its capability of inhibiting AnPGII (Prabhu et al., 2015).

A single PGIP may display different mechanisms of PG inhibition (competitive, non competitive and mixed) suggesting that the protein is highly versatile in recognizing different epitopes of various PGs (Federici et al., 2001; King et al., 2002; Sicilia et al., 2005; Bonivento et al., 2008). Consequently, many 3D-models based on docking predictions have been proposed so far (Sicilia et al., 2005; Maulik et al., 2009; Prabhu et al., 2014). Techniques such as the mass amide exchange mass spectrometry in the case of AnPGII and FpPG and the Small Angle X-ray Scattering (SAXS) in the case of FpPG and ClPG have produced models that, in some cases, are discordant. For example, while the mass amide exchange mass spectrometry predicts that the area of FpPG in contact with PvPGIP2 is located at the N-terminus and predominantly on the underside of the enzyme beta-barrel structures (Gutierrez-Sanchez et al., 2012), the SAXS analysis indicates that the protein region in contact with PvPGIP2 is located at the C-terminus of the enzyme and includes the loops surrounding the active site cleft. A site-directed mutagenesis analysis has been used to validate this second view (Benedetti et al., 2013). In general, low resolution techniques such as SAXS analysis or mass amide exchange mass spectrometry require validation by site-directed mutagenesis to locate the contacting residues in a protein complex. The X-ray crystallography, successfully used to solve several high-resolution structures of PGs (van Santen et al., 1999; Federici et al., 2001; Bonivento et al., 2008) and that of PvPGIP2 (Di Matteo et al., 2003), was so far unsuccessful in the case of the PG-PGIP complex. This is probably due to the intrinsic instability of the PG-PGIP interaction, which only occurs, under apoplastic conditions of pH and ionic strength, through the contact of only a few, sometimes only one, residues (Leckie et al., 1999).

The use of a cross-linker for stabilizing the PG-PGIP complex coupled to techniques that allow the protein analysis directly in solution, such as SAXS and NMR spectroscopy (Wand and Englander, 1996; Nietlispach et al., 2004), may be a valid alternative in order to obtain a detailed map of the contacting residues but this requires a subsequent validation by site-directed mutagenesis.

PGIPs Engineered in Dicot Crops

The important role of PGIP in plant defense has been demonstrated by overexpressing *pgip* genes in several plant species. In these experiments, the source of the used genes was either the same plant species utilized for transformation or a different one (Table 4). The transformation of the model plant *A. thaliana* has been particularly useful to highlight the potentiality of several *pgip* genes, namely the endogenous *Atpgip1* and*Atpgip2*, the bean *Pvpgip2* and the rapeseed (*Brassica napus*) *Bnpgip1* or *Bnpgip2*. Arabidopsis plants overexpressing *Atpgip1* or *Atpgip2* showed a significant reduction of disease symptoms caused by *B. cinerea*(Ferrari et al., 2003) and were less susceptible against the hemibiotrophic fungal pathogen *F. graminearum*(Ferrari et al., 2012), the major causal agent of Fusarium head blight (FHB).

Conversely, silencing of their expression using an antisense *Atpgip*, led to enhanced susceptibility (Ferrari et al., 2006). Arabidopsis plants expressing *Pvpgip2*, encoding an efficient inhibitor of the *B. cinerea* PG (ten Have et al., 1998), showed reduction of disease symptoms caused by *B. cinerea* and those expressing the rapeseed genes *Bnpgip1* and*Bnpgip2* delayed the symptoms caused by *S. sclerotiorum* (Bashi et al., 2013).

The protective potential of *pgip* genes has also been demonstrated in transgenic crops. The first transgenic crop plant obtained by using a*pgip* gene and tested against pathogenic microorganisms were tomatos expressing PvPGIP1 from *P. vulgaris*. These plants, however, did not show any increased resistance against *Fusarium oxysporum* f. sp. *lycopersici*, *B. cinerea*, and *Alternaria solani*. The negative result was due to the inability of PvPGIP1 to inhibit the PGs secreted by these fungi, as shown by *in vitro* inhibition assays and led to discovery of other forms of PGIPs and eventually to the existence of a complex PGIP family in French bean (Desiderio et al., 1997).

Table 4. List of transgenic crops produced using the gene coding for PGIP and their response to fungal, oomycetes or bacterial phytopathogens

Transgenic crops	PGIP gene[c]	Tested against fungal, oomycetes or bacterial phytopathogens	References
Tomato[a] (*Solanum lycopersicum* L.)	PcPGIP PvPGIP1	*Botrytis cinerea** *Fusarium oxysporum* f.sp. *lycopersici*† *Botrytis cinerea*† *Alternaria solani*†	Powell et al., 2000, 1994 Desiderio et al., 1997
Tobacco[a] (*Nicotiana tabacum* L.)	PvPGIP2	*Botrytis cinerea** *Rhizoctonia solani** *Phytophthora parasitica** *Peronospora hyoscyami**	Manfredini et al., 2005 Borras-Hidalgo et al., 2012
	CaPGIP1	*Alternaria alternata** *Colletotrichum nicotianae**	Wang et al., 2013
	VvPGIP1 BrPGIP2	*Botrytis cinerea** *Pectobacterium carotovorum**	Joubert et al., 2006 Hwang et al., 2010
Potato[a] (*Solanum tuberosum* L.)	MdPGIP1 StPGIP	*Verticillium dahliae*† *Verticillium dahliae**	Gazendam et al., 2004; Guo et al., 2014
Brassica rapa[a]	BrPGIP2	*Pectobacterium carotovorum**	Hwang et al., 2010
Rapeseed[a] (*Brassica napus* L.)	BnPGIP2	*Sclerotinia sclerotiorum**	HuangFu et al., 2014
Pea[a] (*Pisum sativum* L.)	RiPGIP	*Glomus intraradices*[Ψ]	Hassan et al., 2012
Grapevine[a] (*Vitis vinifera* L.) Rice[a] (*Oriza sativa* L.)	PcPGIP OsPGIP1	*Botrytis cinerea** *Xylella fastidiosa** *Rhizoctonia solani*	Agüero et al., 2005; Wang et al., 2014b
Wheat[b] (*Triticum aestivum* L., *Triticum durum* Desf.)	PvPGIP2GmPGIP3	*Bipolaris sorokiniana** *Fusarium graminearum** *Claviceps purpurea*† *Bipolaris sorokiniana** *Gaeumannomyces graminis* var. *tritici**	Janni et al., 2008 Ferrari et al., 2012 Volpi et al., 2013; Wang et al., 2014a
Arabidopsis thaliana L.[a]	PvPGIP2 AtPGIP1 AtPGIP2 BnPGIP1 BnPGIP2	*Botrytis cinerea** *Fusarium graminearum** *Sclerotinia sclerotiorum**	Manfredini et al., 2005 Ferrari et al., 2012 Bashi et al., 2013

[a] *The transgenic gene was under control of CaMV 35S promoter.*
[b] *The transgenic gene was under control of Ubiquitin promoter.*
[c] *Pc, Pyrus communis; Pv, Phaseolus vulgaris; Ca, Capsicum annum; Vv, Vitis vinifera; Br, Brassica rapa; Md, Malus domestica; St, Solanum torvum; Ri, Rubus idaeus; Ac, Actinidia deliciosa; At, Arabidopsis thaliana; Bn, Brassica napa.*
* *Showed enhanced resistance.*
† *No evidence of enhanced resistance.*
Ψ *No effect on mycorrhization.*

A few years later, transgenic tomato plants expressing a pear (*Pyrus communis* L.) PGIP (PcPGIP) capable of inhibiting the PGs secreted by *B. cinerea*, showed a reduction of disease lesions caused by this fungus both on ripening fruit (15% reduction) and leaves (about 25% reduction). The initial establishment of infection was not affected in the transgenic plants but the later colonization of the host tissue was significantly reduced (Powell et al., 2000).

Tobacco has been the most used crop plant for testing the effect of PGIP expression on resistance to pathogens. Constitutive and high-level expression of *Pvpgip2* (from *P. vulgaris*), *Vvpgip1* (from *V. vinifera*), *Capgip1*[from pepper (*Capsicum annum*)] and *Brpgip2* (from *B. rapa*) have been obtained in transgenic tobacco. Plants expressing PvPGIP2 showed about 35% reduction of symptoms caused by *B. cinerea* (Manfredini et al., 2005) and, more recently, were shown to display reduced disease symptoms against *Rhizoctonia solani* and two oomycete pathogens, *Phytophthora parasitica* var. *nicotianae* and the blue mold-causing agent*Peronospora hyoscyami* f. sp. *tabacina* (Borras-Hidalgo et al., 2012). Notably, the experiments against *P. hyoscyami* f. sp. *tabacina* were performed in the field during seasonal conditions that favor the pathogen spreading. In agreement with what observed under controlled conditions, resistance of transgenic plants was comparable to that exhibited by *Nicotiana* species (*N. rustica*, *N. debneyi* and *N. megalosiphon*) that are highly resistant to blue mold disease. These transgenic plants expressing PvPGIP2 represented the first example of PGIP-expressing plants subjected to field trails. Recently, transgenic rice expressing OsPGIP1 showed also improved resistance against *Rhizoctonia solani* in field experiments (Wang et al., 2014b).

Transgenic tobacco plants expressing the grapevine *pgip* gene *Vvpgip1* (Joubert et al., 2006) also showed a reduced (from 47 to 69%) disease susceptibility to *B. cinerea* infection. As for plants expressing PvPGIP2, the resistance phenotype correlated with the accumulation of VvPGIP1 as well as with its capability of inhibiting the activity of PG secreted by *B. cinerea*, namely BcPG1, BcPG3, and BcPG6. Several observations, however, suggest that PGIP may improve resistance by mechanisms other than classical PGIP-PG inhibition. For example, non-infected transgenic tobacco plants expressing *Vvpgip1* show modified expression patterns of genes involved in various metabolic pathways (Alexandersson et al., 2011) and an altered cell wall structure (Nguema-Ona et al., 2013). In these plants, lignin accumulation and arabinoxyloglucan-cellulose re-organization leads to a general strengthening/reinforcing of the cell wall that may contribute to an improved resistance against *B. cinerea*.

A reduction of disease symptoms (about 50%) caused by *Alternaria alternata* and *Colletotrichum nicotianae*was also observed in transgenic tobacco lines expressing the pepper CaPGIP1 and, once again, resistance correlated with the inhibition capacity of purified CaPGIP1 against PG activity of both fungal pathogens (Wang et al., 2013).

Within the Solanaceae family, transgenic potato (*Solanum tuberosum*) plants expressing the gene *StPGIP1*from *S. torvum* showed a 50% reduction of wilt disease symptoms caused by *Verticillium dahliae* and a normal plant growth (Guo et al., 2014). Transgenic potato plants overexpressing the apple *pgip1* gene showed protection against the same fungal pathogen but displayed an extended juvenile phase (Gazendam et al., 2004). Transgenic grapevine (*V. vinifera*) plants constitutively expressing the pear PcPGIP gene represent an interesting example of the potential of PGIP for protection against pathogens other than fungi and oomycetes. These plants show a delayed development of the Pierce's disease (PD) caused by bacterial pathogen *Xylella fastidiosa* (Agüero et al., 2005). Not only leaf scorching and *Xylella* titre were reduced but also plants showed a better re-growth after pruning compared to infected untransformed controls. Moreover, an inverse dose-effect relationship was shown between development of PD and levels of PcPGIP activity in the tissues. The improved resistance of the grapevine plants expressing PcPGIP against a bacterial pathogen was unexpected, because until then the PGIP inhibition activity was thought to be limited to fungal and insect PGs (Cervone et al., 1990;Johnston et al., 1993; D›Ovidio et al., 2004b). It was later shown that pear PcPGIP inhibits the PG encoded by*X. fastidiosa* and that PG activity is a virulence factor of this pathogen (Roper et al., 2007; Pérez-Donoso et al., 2010). The observation that PcPGIP is present in xylem exudates of non-transgenic scions grafted on transgenic rootstocks expressing PcPGIP suggests that grafting of non transgenic varieties on transgenic rootstocks represents, in this case, a useful agronomical practice for plant protection (Agüero et al., 2005).

The results obtained with *X. fastidiosa* prompted further investigations on the capability of PGIP of controlling bacterial diseases (summarized in Table 4). Transgenic tobacco plants expressing *B. rapa* BrPGIP2 were resistant against *Pectobacterium carotovorum*, the causal agent of the soft rot disease, with a strong reduction (66–88%) of the symptoms as compared to wild-type plants (Hwang et al., 2010). The resistance correlated with the inhibitory activity against *P. carotovorum* PG activity found in the total protein extracts of the transgenic plants (Hwang et al., 2010). Also chinese cabbage (*B. rapa* ssp. *pekinensis*) plants overexpressing BrPGIP2 showed higher resistance against *P. carotovorum* and produced normal looking pods-like structures with no viable seeds. Combination of crossing with non-transgenic plants did not restore fertility of the transgenic plants, suggesting that mechanisms such as ploidy changes occurring during the tissue culture stage or changes in cell-wall architecture of sexual organs are responsible for the abnormality (Hwang et al., 2010).

No phenotypic abnormalities were, instead, found in transgenic tobacco plants expressing BrPGIP2 (Hwang et al., 2010), nor in rapeseed plants overexpressing the *B. napus Bnpgip2*. The latter plants displayed a significant reduction of rot caused by the necrotrophic fungal pathogen *S. sclerotiorum* (HuangFu et al., 2014).

Additional PGIP-transgenic crops include pea (*Pisum sativum* L.), transformed with *Ripgip* from raspberry (*Rubus idaeus* L.) (Richter et al., 2006), persimmon (*Diospyros kaki* L.) and apple (*Malus domestica* Borkh.) transformed with pear PcPGIP (Szankowski et al., 2003; Tamura et al., 2004), sugarbeet (*Beta vulgaris* L.) transformed with bean *Pvpgip2* (Mohammadzadeh et al., 2012), chickpea transformed with either *Ripgip* or a*pgip* gene from kiwi fruit (Senthil et al., 2004), tobacco transformed with PpPGIP gene from *Pyrus pyrifolia*Nakai (Liu et al., 2013) and maize (*Zea mays* L.) transformed with bean *Pvpgip1* (O›Kennedy et al., 2001). The response of these plants to pathogens has not been reported yet. Transgenic pea plants expressing RiPGIP were instead evaluated for their response to beneficial microorganisms. *Glomus intraradices*, an arbuscular mycorrhizal fungus, colonized roots of transgenic plants at an extend comparable to that observed in control non transgenic plants, indicating that the expression of RiPGIP does not affect mycorrhization (Hassan et al., 2012).

PGIPs Engineered in Monocot Crops

Although the low pectin content of cereal species like wheat and rice indicates that this cell wall component may have a marginal role during infection, results show that the expression of PGIP in transgenic plants limits some diseases caused by fungal pathogens (Janni et al., 2008; Ferrari et al., 2012; Wang et al., 2014a,b). In our labs, the bean *Pvpgip2* gene was used under the constitutive promoter of the maize unbiquitin gene (*Ubi-1*) to transform both durum and bread wheat by particle bombardment. PvPGIP2 was correctly targeted to the apoplast and the transgenic plants did not show any major morphological and growth defects. Transgenic wheat showed a significant reduction (46–50%) of foliar spot blotch symptoms caused by the hemibiotrophic fungal pathogen *Bipolaris sorokiniana* and improved resistance (25–30%) against the hemibiotrophic fungal pathogen *F. graminearum* (Ferrari et al., 2012), the major causal agent of FHB in wheat. A reduced degradability of the transgenic tissue by PG treatments correlated with the capacity of PvPGIP2 to inhibit PG activity of *B. sorokiniana* and less strongly PG of *F. graminearum* (Janni et al., 2008; Ferrari et al., 2012). An interesting aspect of the wheat plants expressing PvPGIP2 is that, under moderate infection with *F. graminearum*, the reduced FHB symptoms are concomitant with a greater amount of total starch in the grains as compared to control plants (D'Ovidio et al., 2012). On the other hand, wheat plants expressing PvPGIP2 were susceptible to the biotrophic fungal pathogen *Claviceps purpurea*, the causal agent of ergot disease probably because PvPGIP2 is not able to inhibit the activity of *C. purpurea* CpPG1 and CpPG2 (Volpi et al., 2013). Recently, transgenic wheat expressing the soybean GmPGIP3 was shown to be resistant to both take-all and common root rot diseases caused by the fungal pathogen *Gaeumannomyces graminis* var. *tritici* and *B. sorokiniana*, respectively; symptoms were reduced of about 47–83% and 42–60%, respectively (Wang et al., 2014a). Similarly, the expression of OsPGIP1 in transgenic rice enhanced resistence against *Rhizoctonia solani*in field tests and resistance was related with the expression levels of OsPGIP1 (Wang et al., 2014b).

CONCLUDING REMARKS AND FUTURE CHALLENGES

The results reported in this review clearly indicate that PGIP is useful to improve resistance in different crop species. High-level expression of PGIP does not prevent infection but limits significantly the colonization of the host tissue with a consequent positive impact on crop yield and product quality. The efficacy of PGIP to control diseases has been demonstrated against fungi, oomycetes and bacteria and is equally efficient against necrorophic and hemibiotrophic pathogens. The experiments performed with biotrophs do not allow to draw any clear conclusion since the only fungal biotrophic pathogen analyzed, *C. purpures*, produced PG activity that was not inhibited by the PGIP expressed in the transgenic plants (Volpi et al., 2013). The identification and development of PGIPs with stronger and broader inhibitory capacities may be useful to utilize these proteins in crop protection. Germplasm analysis to identify novel PGIPs is still limited (Farina et al., 2009) and the initial attempts to drive *in vitro* evolution of PGIP to generate proteins with improved inhibition properties have not been particularly successful (Benedetti et al., 2011a). Structural studies should be implemented in order to obtain a detailed map of the contacts between various PGs and PGIPs. This is necessary not only for constructing novel inhibitors with stronger activities but also for future programs of genome editing in which the existing genes of a plant species may be ameliorated to better adapt to new virulent strains of microorganisms evolving in nature.

The available results support the notion that inhibition of the microbial PG by PGIP is a prerequisite of the inhibitors to confer resistance to transgenic plants against microbes. The delay of symptoms is often related to the capacity of PGIP to inhibit the PG activity secreted by the pathogens and, consequently, to reduce both tissue maceration and favor the release of OGs, as summarized in Figure 2. However, this aspect of the PGIP›s biology needs further investigation. In some cases PGIP has been reported to confer resistance without any evidence of PG-inhibition *in vitro* (Joubert et al., 2006). Moreover, some evidence suggests that the capability of reducing tissue maceration is associated with the property of PGIP to bind pectin, likely shielding this component of the cell wall from PG activity (Spadoni et al., 2006). In this regard the observation that transgenic plants expressing PGIPs exhibit an altered gene expression and cell wall composition is also intriguing. It is not yet clear the mechanism that links the ectopic expression of PGIP to alteration of gene expression and whether this contributes to disease resistance (Alexandersson et al., 2011; Nguema-Ona et al., 2013).

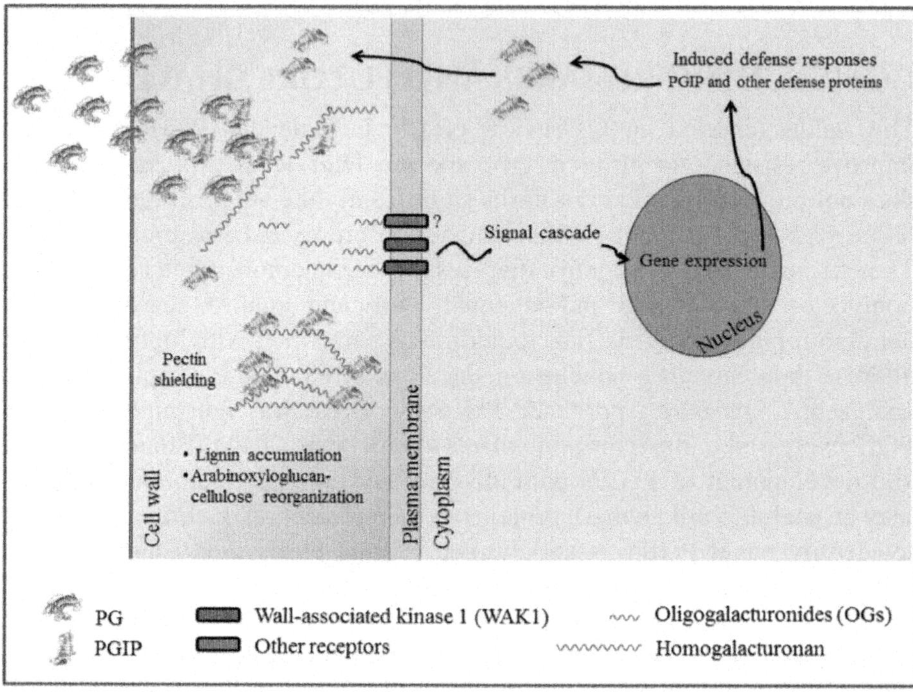

Figure 2. A model for the role of PGIP in the defense response against pathogens. Delay of symptoms is related to the inhibitory activity of PGIP toward PGs secreted by the pathogens and likely to the accumulation of oligogalacturonide (OG) elicitors, which are recognized by WAK1 and likely other receptors not yet characterized. Cell wall modification and pectin shielding could also play a role. Signaling cascades activated by OGs are described in Ferrari et al. (2013).

An important but very little explored aspect of the PGIP biology is its possible role in processes of growth and development. Although plants overexpressing PGIPs do not show obvious morphological alterations, indeed several reports point to PGIP as a player in development. PGIP are induced, not only by phosphate deficiency, but also by auxin treatment and in mutants defective in SIZ1, a SUMO (small ubiquitin-related modifier) E3 ligase that is involved in several stress responses, including Pi starvation, and flowering (Sato and Miura, 2011). Suppression of PGIPs under the control ABA insensitive 5

(ABI5) transcription factor accompanies promotion of seed germination by the peroxisomal ABC transporter PED3 (Kanai et al., 2010). Upregulation of *PGIP2*correlates with the acquisition of competence to form green callus in an auxin-rich callus induction medium (Che et al., 2007) and occurs in Arabidopsis tissue culture lines in which the expression of the peroxidases PRX33 and PRX34 is knocked down by antisense expression (O'Brien et al., 2012), whereas PGIP1 was identified in a proteomic study performed on Arabidopsis etiolated hypocotyls used as a model of cells undergoing elongation followed by growth arrest within a short time (Irshad et al., 2008). Finally, both PGIP1 and PGIP2 are associated with cell wall stabilization at low pH under the control of the zinc-finger protein STOP1 (Sensitive to Proton Rhizotoxicity 1) and STOP2 (Kobayashi et al., 2014). A role of PGIP not only in defense but also in growth and development implies that the inhibitor may affect one or more of the many endogenous PGs expressed by plants. This is also an unexplored aspect of the PGIP biology and, at the moment, only one very old evidence is available showing that PGIP may have a plant-derived PG partner (Cervone et al., 1990).

ACKNOWLEDGMENTS

This research was supported by the Italian Ministry of University and Research (PRIN 2010–2011) to RD.

REFERENCES

1. Agüero, C. B., Uratsu, S. L., Greve, C., Powell, A. L. T., Labavitch, J. M., Meredith, C. P., et al. (2005). Evaluation of tolerance to Pierce's disease and Botrytis in transgenic plants of *Vitis vinifera* L. expressing the pear PGIP gene. *Mol. Plant Pathol.* 6, 43–51. doi: 10.1111/j.1364-3703.2004.00262.x
2. Akhgari, A., Motallebi, B., and Zamani, M. (2012). Bean polygalacturonase-inhibiting protein expressed in transgenic *Brassica napus* inhibits polygalacturonase from its fungal pathogen *Rhizoctonia solani*. *Plant Prot. Sci.* 48, 1–9.
3. Albersheim, P., and Anderson, A. J. (1971). Proteins from plant cell walls inhibit polygalacturonases secreted by plant pathogens. *Proc. Natl. Acad. Sci. U.S.A.* 68, 1815–1819.
4. Alexandersson, E., Becker, J. V. W., Jacobson, D., Nguema-Ona, E., Steyn, C., Denby, K. J., et al. (2011). Constitutive expression of a grapevine polygalacturonase-inhibiting protein affects gene expression and cell wall properties in uninfected tobacco. *BMC Res. Notes* 4:493.

doi: 10.1186/1756-0500-4-493

5. Ascencio-Ibanez, J. T., Sozzani, R., Lee, T.-J., Chu, T.-M., Wolfinger, R. D., Cella, R., et al. (2008). Global analysis of Arabidopsis gene expression uncovers a complex array of changes impacting pathogen response and cell cycle during geminivirus infection. *Plant Physiol.* 148, 436–454. doi: 10.1104/pp.108.121038
6. Bashi, Z. D., Rimmer, S. R., Khachatourians, G. G., and Hegedus, D. D. (2013). *Brassica napus* polygalacturonase inhibitor proteins inhibit*Sclerotinia sclerotiorum* polygalacturonase enzymatic and necrotizing activities and delay symptoms in transgenic plants. *Can. J. Microbiol.* 59, 79–86. doi: 10.1139/cjm-2012-0352
7. Benedetti, M., Andreani, F., Leggio, C., Galantini, L., Di Matteo, A., Pavel, N. V., et al. (2013). A single amino-acid substitution allows endo-polygalacturonase of *Fusarium verticillioides* to acquire recognition by PGIP2 from *Phaseolus vulgaris*. *PLoS ONE* 8:e80610. doi: 10.1371/journal.pone.0080610
8. Benedetti, M., Bastianelli, E., Salvi, G., De Lorenzo, G., and Caprari, C. (2011a). Artificial evolution corrects a repulsive amino acid in polygalacturonase inhibiting proteins (PGIPs). *J. Plant Pathol.* 93, 89–95. doi: 10.4454/jpp.v93i1.277
9. Benedetti, M., Leggio, C., Federici, L., De Lorenzo, G., Pavel, N. V., and Cervone, F. (2011b). Structural resolution of the complex between a fungal polygalacturonase and a plant polygalacturonase-inhibiting protein by small-angle X-ray scattering. *Plant Physiol.* 157, 599–607. doi: 10.1104/pp.111.181057
10. Berger, D. K., Oelofse, D., Arendse, M. S., Du Plessis, E., and Dubery, I. A. (2000). Bean polygalacturonase inhibitor protein-1 (PGIP-1) inhibits polygalacturonases from *Stenocarpella maydis*. *Physiol. Mol. Plant Pathol.* 57, 5–14. doi: 10.1006/pmpp.2000.0274
11. Bergmann, C. W., Ito, Y., Singer, D., Albersheim, P., Darvill, A. G., Benhamou, N., et al. (1994). Polygalacturonase-inhibiting protein accumulates in *Phaseolus vulgaris* L. *in response to wounding, elicitors and fungal infection*. *Plant J.* 5, 625–634. doi: 10.1111/j.1365-313X.1994.00625.x
12. Bonivento, D., Pontiggia, D., Di Matteo, A., Fernandez-Recio, J., Salvi, G., Tsernoglou, D., et al. (2008). Crystal structure of the endopolygalacturonase from the phytopathogenic fungus *Colletotrichum lupini* and its interaction with polygalacturonase inhibiting proteins. *Proteins* 70, 294–299. doi: 10.1002/prot.21610

13. Borras-Hidalgo, O., Caprari, C., Hernandez-Estevez, I., De Lorenzo, G., and Cervone, F. (2012). A gene for plant protection: expression of a bean polygalacturonase inhibitor in tobacco confers a strong resistance against *Rhizoctonia solani* and two oomycetes. *Front. Plant Sci.* 3:268. doi: 10.3389/fpls.2012.00268

14. Buza, N. L., Krinitsyna, A. A., Protsenko, M. A., and Vartapetyan, V. V. (2004). Role of the polygalacturonidase inhibitor Protein in the ripening of apples and their resistance to *Monilia fructigena*, a causative agent of fruit rot. *Appl. Biochem. Microbiol.* 40, 89–92. doi: 10.1023/B:ABIM.0000010361.48129.6e

15. Cantu, D., Vicente, A. R., Labavitch, J. M., Bennett, A. B., and Powell, A. L. (2008). Strangers in the matrix: plant cell walls and pathogen susceptibility. *Trends Plant Sci.* 13, 610–617. doi: 10.1016/j.tplants.2008.09.002

16. Caprari, C., Bergmann, C., Migheli, Q., Salvi, G., Albersheim, P., Darvill, A., et al. (1993). *Fusarium moniliforme* secretes four endopolygalacturonases derived from a single gene product. *Physiol. Mol. Plant Pathol.* 43, 453–462. doi: 10.1006/pmpp.1993.1073

17. Casasoli, M., Federici, L., Spinelli, F., Di Matteo, A., Vella, N., Scaloni, F., et al. (2009). Integration of evolutionary and desolvation energy analysis identifies functional sites in a plant immunity protein. *Proc. Natl. Acad. Sci. U.S.A.* 106, 7666–7671. doi: 10.1073/pnas.0812625106

18. Cervone, F., De Lorenzo, G., Degrà, L., Salvi, G., and Bergami, M. (1987). Purification and characterization of a polygalacturonase-inhibiting protein from *Phaseolus vulgaris* L. *Plant Physiol.* 85, 631–637. doi: 10.1104/pp.85.3.631

19. Cervone, F., De Lorenzo, G., Pressey, R., Darvill, A. G., and Albersheim, P. (1990). Can *Phaseolus* PGIP inhibit pectic enzymes from microbes and plants? *Phytochemistry* 29, 447–449. doi: 10.1016/0031-9422(90)85094-V

20. Cervone, F., Hahn, M. G., De Lorenzo, G., Darvill, A., and Albersheim, P. (1989). Host-pathogen interactions. XXXIII. A plant protein converts a fungal pathogenesis factor into an elicitor of plant defense responses. *Plant Physiol.* 90, 542–548. doi: 10.1104/pp.90.2.542

21. Che, P., Lall, S., and Howell, S. H. (2007). Developmental steps in acquiring competence for shoot development in Arabidopsis tissue culture.*Planta* 226, 1183–1194. doi: 10.1007/s00425-007-0565-4

22. Chen, X., Liu, X., Zuo, S., Ma, Y., Tong, Y., Pan, X., et al. (2011). Prokaryotic expression of rice *Ospgip1* gene and bioinformatic analysis

of encoded product. *Rice Sci.* 18, 250–256. doi: 10.1016/S1672-6308(12)60002-X

23. D'hallewin, G., Schirra, M., Powell, A. L., Greve, L. C., and Labavitch, J. M. (2004). Properties of a polygalacturonase-inhibiting protein isolated from 'Oroblanco' grapefruit. *Physiol. Plant.* 120, 395–404. doi: 10.1111/j.0031-9317.2004.00264.x
24. D'Ovidio, R., Laino, P., Janni, M., Botticella, E., Di Carli, M. S., Benvenuto, E., et al. (2012). *"Proteomic analysis of mature kernels of Fusarium graminearum-infected transgenic bread wheat expressing PGIP,"* in *Proceedings of 11th International Gluten Workshop*, eds Z. He and D. Wang (Beijing), 18.
25. D'Ovidio, R., Mattei, B., Roberti, S., and Bellincampi, D. (2004a). Polygalacturonases, polygalacturonase-inhibiting proteins and pectic oligomers in plant-pathogen interactions. *Biochim. Biophys. Acta* 1696, 237–244. doi: 10.1016/j.bbapap.2003.08.012
26. D›Ovidio, R., Raiola, A., Capodicasa, C., Devoto, A., Pontiggia, D., Roberti, S., et al. (2004b). Characterization of the complex locus of bean encoding polygalacturonase-inhibiting proteins reveals subfunctionalization for defense against fungi and insects. *Plant Physiol.* 135, 2424–2435. doi: 10.1104/pp.104.044644
27. D'Ovidio, R., Roberti, S., Giovanni, M. D., Capodicasa, C., Melaragni, M., Sella, L., et al. (2006). The characterization of the soybean polygalacturonase-inhibiting proteins (Pgip) gene family reveals that a single member is responsible for the activity detected in soybean tissues. *Planta* 224, 633–645. doi: 10.1007/s00425-006-0235-y
28. De Lorenzo, G., D›Ovidio, R., and Cervone, F. (2001). The role of polygalacturonase-inhibiting proteins (PGIPs) in defense against pathogenic fungi. *Annu. Rev. Phytopathol.* 39, 313–335. doi: 10.1146/annurev.phyto.39.1.313
29. De Lorenzo, G., and Ferrari, S. (2002). Polygalacturonase-inhibiting proteins in defense against phytopathogenic fungi. *Curr. Opin. Plant Biol.* 5, 295–299. doi: 10.1016/S1369-5266(02)00271-6
30. Deo, A., and Shastri, N. V. (2003). Purification and characterization of polygalacturonase-inhibitory proteins from *Psidium guajava* Linn. (guava) fruit. *Plant Sci.* 164, 147–156. doi: 10.1016/S0168-9452(02)00337-0
31. Desiderio, A., Aracri, B., Leckie, F., Mattei, B., Salvi, G., Tigelaar, H., et al. (1997). Polygalacturonase-inhibiting proteins (PGIPs) with different specificities are expressed in *Phaseolus vulgaris*. *Mol. Plant Microbe Interact.* 10, 852–860. doi: 10.1094/MPMI.1997.10.7.852

32. Devoto, A., Clark, A. J., Nuss, L., Cervone, F., and De Lorenzo, G. (1997). Developmental and pathogen-induced accumulation of transcripts of polygalacturonase-inhibiting protein in *Phaseolus vulgaris* L. *Planta* 202, 284–292. doi: 10.1007/s004250050130

33. Di, C., Li, M., Long, F., Bai, M., Zheng, X., Xu, S., et al. (2009). Molecular cloning, functional analysis and localization of a novel gene encoding polygalacturonase-inhibiting protein in *Chorispora bungeana*. *Planta* 231, 169–178. doi: 10.1007/s00425-009-1039-7

34. Di, C. X., Zhang, H., Sun, Z. L., Jia, H. L., Yang, L. N., Si, J., et al. (2012). Spatial distribution of polygalacturonase-inhibiting proteins in*Arabidopsis* and their expression induced by *Stemphylium solani* infection. *Gene* 506, 150–155. doi: 10.1016/j.gene.2012.06.085

35. Di Giovanni, M., Cenci, A., Janni, M., and D›Ovidio, R. (2008). A LTR copia retrotransposon and Mutator transposons interrupt *pgip* genes in cultivated and wild wheats. *Theor. Appl. Genet.* 116, 859–867. doi: 10.1007/s00122-008-0719-1

36. Di Matteo, A., Bonivento, D., Tsernoglou, D., Federici, L., and Cervone, F. (2006). Polygalacturonase-inhibiting protein (PGIP) in plant defence: a structural view. *Phytochemistry* 67, 528–533. doi: 10.1016/j.phytochem.2005.12.025

37. Di Matteo, A., Federici, L., Mattei, B., Salvi, G., Johnson, K. A., Savino, C., et al. (2003). The crystal structure of polygalacturonase-inhibiting protein (PGIP), a leucine-rich repeat protein involved in plant defense. *Proc. Natl. Acad. Sci. U.S.A.* 100, 10124–10128. doi: 10.1073/pnas.1733690100

38. Doostdar, H., McCollum, T. G., and Mayer, R. T. (1997). Purification and characterization of an endo-polygalacturonase from the gut of West Indies sugarcane rootstalk borer weevil (*Diaprepes abbreviatus* L.) larvae. *Comp. Biochem. Physiol. Part B Biochem. Mol. Bio.* 118, 861–867.

39. Faize, M., Sugiyama, T., Faize, L., and Ishii, H. (2003). Polygalacturonase-inhibiting protein (PGIP) from Japanese pear: possible involvement in resistance against scab. *Physiol. Mol. Plant Pathol.* 63, 319–327. doi: 10.1016/j.pmpp.2004.03.006

40. Farina, A., Rocchi, V., Janni, M., Benedettelli, S., De Lorenzo, G., and D'Ovidio, R. (2009). The bean polygalacturonase-inhibiting protein 2 (PvPGIP2) is highly conserved in common bean (*Phaseolus vulgaris* L.) germplasm and related species. *Theor. Appl. Genet.* 118, 1371–1379. doi: 10.1007/s00122-009-0987-4

41. Federici, L., Caprari, C., Mattei, B., Savino, C., Di Matteo, A., De Lorenzo, G., et al. (2001). Structural requirements of endopolygalacturonase for the interaction with PGIP (polygalacturonase-inhibiting protein). *Proc. Natl. Acad. Sci. U.S.A.* 98, 13425–13430. doi: 10.1073/pnas.231473698

42. Federici, L., Di Matteo, A., Fernandez-Recio, J., Tsernoglou, D., and Cervone, F. (2006). Polygalacturonase inhibiting proteins: players in plant innate immunity? *Trends Plant Sci.* 11, 65–70. doi: 10.1016/j.tplants.2005.12.005

43. Ferrari, S., Galletti, R., Vairo, D., Cervone, F., and De Lorenzo, G. (2006). Antisense expression of the *Arabidopsis thaliana AtPGIP1* gene reduces polygalacturonase-inhibiting protein accumulation and enhances susceptibility to *Botrytis cinerea*. *Mol. Plant Microbe Interact.* 19, 931–936. doi: 10.1094/MPMI-19-0931

44. Ferrari, S., Savatin, D. V., Sicilia, F., Gramegna, G., Cervone, F., and De Lorenzo, G. (2013). Oligogalacturonides: plant damage-associated molecular patterns and regulators of growth and development. *Front. Plant Sci.* 4:49. doi: 10.3389/fpls.2013.00049

45. Ferrari, S., Sella, L., Janni, M., De Lorenzo, G., Favaron, F., and D'Ovidio, R. (2012). Transgenic expression of polygalacturonase-inhibiting proteins in *Arabidopsis* and wheat increases resistance to the flower pathogen *Fusarium graminearum*. *Plant Biol.* 14, 31–38. doi: 10.1111/j.1438-8677.2011.00449.x

46. Ferrari, S., Vairo, D., Ausubel, F. M., Cervone, F., and De Lorenzo, G. (2003). Tandemly duplicated *Arabidopsis* genes that encode polygalacturonase-inhibiting proteins are regulated coordinately by different signal transduction pathways in response to fungal infection. *Plant Cell Online* 15, 93–106. doi: 10.1105/tpc.005165

47. Fish, W. W., and Davis, A. R. (2004). The purification, physical/chemical characterization, and cDNA sequence of cantaloupe fruit polygalacturonase-inhibiting protein. *Phytopathology* 94, 337–344. doi: 10.1094/PHYTO.2004.94.4.337

48. Frati, F., Galletti, R., De Lorenzo, G., Salarino, G., and Conti, E. (2006). Activity of endo-polygalacturonases in mirid bugs (Heteroptera: Miridae) and their inhibition by plant cell wall proteins (PGIPs). *Eur. J. Entomol.* 103, 515–522. doi: 10.14411/eje.2006.067

49. Gazendam, I., Oelofse, D., and Berger, D. K. (2004). High-level expression of apple PGIP1 is not sufficient to protect transgenic potato against*Verticillium dahliae*. *Physiol. Mol. Plant Pathol.* 65, 145–155. doi: 10.1016/j.pmpp.2005.01.002

50. Gomathi, V., and Gnanamanickam, S. S. (2004). Polygalacturonase-inhibiting proteins in plant defence. *Curr. Sci.* 87, 1211–1217.

51. Gregori, R., Mari, M., Bertolini, P., Barajas, J. A., Tian, J. B., and Labavitch, J. M. (2008). Reduction of *Colletotrichum acutatum* infection by a polygalacturonase inhibitor protein extracted from apple. *Postharvest. Biol. Technol.* 48, 309–313. doi: 10.1016/j.postharvbio.2007.10.006

52. Guo, J.-L., Zhu, Y.-P., Shi, K., Jue, D.-W., Liu, S.-P., Hong, Y.-B., et al. (2014). Over-expression of *Solanum torvum* PGIP enhances resistance to*Verticillium dahliae* in transgenic potato plants. *Bothalia Pretoria.* 44, 392–404.

53. Gutierrez-Sanchez, G., King, D., Kemp, G., and Bergmann, C. (2012). SPR and differential proteolysis/MS provide further insight into the interaction between PGIP2 and EPGs. *Fungal Biol.* 116, 737–746. doi: 10.1016/j.funbio.2012.04.010

54. Hassan, F., Noorian, M. S., and Jacobsen, H. J. (2012). Effect of antifungal genes expressed in transgenic pea (*Pisum sativum* L.) on root colonization with *Glomus intraradices*. *GM Crops Food* 3, 301–309. doi: 10.4161/gmcr.21897

55. Hegedus, D. D., Li, R., Buchwaldt, L., Parkin, I., Whitwill, S., Coutu, C., et al. (2008). *Brassica napus* possesses an expanded set of polygalacturonase inhibitor protein genes that are differentially regulated in response to *Sclerotinia sclerotiorum* infection, wounding and defense hormone treatment. *Planta* 228, 241–253. doi: 10.1007/s00425-008-0733-1

56. Helft, L., Reddy, V., Chen, X., Koller, T., Federici, L., Recio, J. F., et al. (2011). LRR Conservation mapping to predict functional sites within protein leucine-rich repeat domains. *PLoS ONE* 6:e21614. doi: 10.1371/journal.pone.0021614

57. Hou, W., Mu, J., Li, A., Wang, H., and Kong, L. (2014). Identification of a wheat polygalacturonase-inhibiting protein involved in Fusarium head blight resistance. *Eur. J. Plant Pathol.* 141, 731–745. doi: 10.1007/s10658-014-0574-7

58. HuangFu, H., Guan, C., Jin, F., and Yin, C. (2014). Prokaryotic expression and protein function of *Brassica napus* PGIP2 and its genetic transformation. *Plant Biotechnol. Rep.* 8, 171–181. doi: 10.1007/s11816-013-0307-y

59. Hwang, B. H., Bae, H., Lim, H. S., Kim, K. B., Kim, S. J., Im, et al. (2010). Overexpression of polygalacturonase-inhibiting protein 2 (PGIP2) of Chinese cabbage (*Brassica rapa* ssp. pekinensis) increased

resistance to the bacterial pathogen *Pectobacterium carotovorum* ssp. carotovorum.*Plant Cell Tiss. Organ Cult.* 103, 293–305. doi: 10.1007/s11240-010-9779-4

60. Irshad, M., Canut, H., Borderies, G., Pont-Lezica, R., and Jamet, E. (2008). A new picture of cell wall protein dynamics in elongating cells of*Arabidopsis thaliana*: Confirmed actors and newcomers. *BMC Plant Biol.* 8:94. doi: 10.1186/1471-2229-8-94
61. James, J. T., and Dubery, I. A. (2001). Inhibition of polygalacturonase from *Verticillium dahliae* by a polygalacturonase inhibiting protein from cotton. *Phytochemistry* 57, 149–156. doi: 10.1016/S0031-9422(01)00024-3
62. Jang, S., Lee, B., Kim, C., Kim, S. J., Yim, J., Han, J. J., et al. (2003). The *OsFOR1* gene encodes a polygalacturonase-inhibiting protein (PGIP) that regulates floral organ number in rice. *Plant Mol. Biol.* 53, 357–372. doi: 10.1023/B:PLAN.0000006940.89955.f1
63. Janni, M., Bozzini, T., Moscetti, I., Volpi, C., and D'Ovidio, R. (2013). Functional characterisation of wheat Pgip genes reveals their involvement in the local response to wounding. *Plant Biol. Stuttg. Ger.* 15, 1019–1024. doi: 10.1111/plb.12002
64. Janni, M., Giovanni, M., Roberti, S., Capodicasa, C., and D'Ovidio, R. (2006). Characterization of expressed Pgip genes in rice and wheat reveals similar extent of sequence variation to dicot PGIPs and identifies an active PGIP lacking an entire LRR repeat. *Theor. Appl. Genet.* 113, 1233–1245. doi: 10.1007/s00122-006-0378-z
65. Janni, M., Sella, L., Favaron, F., Blechl, A. E., De Lorenzo, G., and D'Ovidio, R. (2008). The expression of a bean PGIP in transgenic wheat confers increased resistance to the fungal pathogen *Bipolaris sorokiniana. Mol. Plant Microbe Interact.* 21, 171–177. doi: 10.1094/MPMI-21-2-0171
66. Johnston, D. J., Ramanathan, V., and Williamson, B. (1993). A protein from immature raspberry fruits which inhibits endopolygalacturonases from *Botrytis cinerea* and other micro-organisms. *J. Exp. Bot.* 44, 971–976. doi: 10.1093/jxb/44.5.971
67. Joubert, D. A., Kars, I., Wagemakers, L., Bergmann, C., Kemp, G., Vivier, M. A., et al. (2007). A polygalacturonase-inhibiting protein from grapevine reduces the symptoms of the endopolygalacturonase BcPG2 from *Botrytis cinerea* in *Nicotiana benthamiana* leaves without any evidence for *in vitro* interaction. *Mol. Plant Microbe Interact.* 20, 392–402. doi: 10.1094/MPMI-20-4-0392

68. Joubert, D. A., Slaughter, A. R., Kemp, G., Becker, J. V. W., Krooshof, G. H., Bergmann, C., et al. (2006). The grapevine polygalacturonase-inhibiting protein (*VvPGIP1*) reduces *Botrytis cinerea* susceptibility in transgenic tobacco and differentially inhibits fungal polygalacturonases. *Transgenic Res.* 15, 687–702. doi: 10.1007/s11248-006-9019-1

69. Kalunke, R. M., Cenci, A., Volpi, C., O'Sullivan, D. M., Sella, L., Favaron, F., et al. (2014). The pgip family in soybean and three other legume species: evidence for a birth-and-death model of evolution. *BMC Plant Biol.* 14:189. doi: 10.1186/s12870-014-0189-3

70. Kalunke, R. M., Janni, M., Sella, L., David, P., Geffroy, V., Favaron, F., et al. (2011). Transcript analysis of the bean polygalacturonase inhibiting protein gene family reveals that Pvpgip2 is expressed in the whole plant and is strongly induced by pathogen infection. *J. Plant Pathol.* 93, 141–148. doi: 10.4454/jpp.v93i1.284

71. Kanai, M., Nishimura, M., and Hayashi, M. (2010). A peroxisomal ABC transporter promotes seed germination by inducing pectin degradation under the control of ABI5. *Plant J.* 62, 936–947. doi: 10.1111/j.1365-313X.2010.04205.x

72. Kemp, G., Bergmann, C. W., Clay, R., Van der Westhuizen, A. J., and Pretorius, Z. A. (2003). Isolation of a polygalacturonase-inhibiting protein (PGIP) from wheat. *Mol. Plant Microbe Interact.* 16, 955–961. doi: 10.1094/MPMI.2003.16.11.955

73. King, D., Bergmann, C., Orlando, R., Benen, J. A., Kester, H. C., and Visser, J. (2002). Use of amide exchange mass spectrometry to study conformational changes within the endopolygalacturonase II-homogalacturonan-polygalacturonase inhibiting protein system. *Biochemistry* 41, 10225–10233. doi: 10.1021/bi020119f

74. Kirsch, R., Wielsch, N., Vogel, H., Svatoš, A., Heckel, D. G., and Pauchet, Y. (2012). Combining proteomics and transcriptome sequencing to identify active plant-cell-wall-degrading enzymes in a leaf beetle. *BMC Genomics* 13:587. doi: 10.1186/1471-2164-13-587

75. Kobayashi, Y., Ohyama, Y., Kobayashi, Y., Ito, H., Iuchi, S., Fujita, M., et al. (2014). STOP2 activates transcription of several genes for Al- and low pH-tolerance that are regulated by STOP1 in Arabidopsis. *Mol. Plant* 7, 311–322. doi: 10.1093/mp/sst116

76. Ladu, G., Pani, G., Venditti, T., Dore, A., Molinu, M. G., and D›Hallewin, G. (2012). Natural resistance against pre- and post-harvest pathogens in Sardinian pears germoplasm. *Commun. Agric. Appl. Biol. Sci.* 77, 163–171.

77. Lagaert, S., Beliën, T., and Volckaert, G. (2009). "Plant cell walls: protecting the barrier from degradation by microbial enzymes" *Semin. Cell Dev. Biol.* 20, 1064–1073. doi: 10.1016/j.semcdb.2009.05.008

78. Leckie, F., Mattei, B., Capodicasa, C., Hemmings, A., Nuss, L., Aracri, B., et al. (1999). The specificity of polygalacturonaseinhibiting protein (PGIP): a single amino acid substitution in the solvent-exposed b-strand/ b-turn region of the leucine-rich repeats (LRRs) confers a new recognition capability. *EMBO J.* 18: 2352–2363 doi: 10.1093/emboj/18.9.2352

79. Lee, H. D., Bae, H., Kang, I. K., Byun, J. K., and Kang, S. G. (2006). Characterization of an apple polygalacturonase-inhibiting protein (PGIP) that specifically inhibits an endopolygalacturonase (PG) purified from apple fruits infected with *Botryosphaeria dothidea*. *J. Microbio. Biotech.* 16, 1192–1200.

80. Lim, J. M., Aoki, K., Angel, P., Garrison, D., King, D., Tiemeyer, M., et al. (2009). Mapping glycans onto specific N-linked glycosylation sites of*Pyrus communis* PGIP redefines the interface for EPG-PGIP interactions. *J. Proteome Res.* 8, 673–680. doi: 10.1021/pr800855f

81. Lin, W., and Li, Z. (2002). The partial structure of wheat polygalacturonase inhibiting protein. *Chin. J. Biochem. Mol. Bio.* 18, 197–201.

82. Liu, D., Li, W., He, X., Ding, Y., Chen, C., and Ge, F. (2013). Characterization and functional analysis of a novel PGIP gene from *Pyrus pyrifolia*Nakai cv Huobali. *Acta Physiol. Plant.* 35, 1247–1256. doi: 10.1007/s11738-012-1164-y

83. Lu, L., Zhou, F., Zhou, Y., Fan, X., Ye, S., Wang, L., et al. (2012). Expression profile analysis of the polygalacturonase-inhibiting protein genes in rice and their responses to phytohormones and fungal infection. *Plant Cell Rep.* 31, 1173–1187. doi: 10.1007/s00299-012-1239-7

84. Machinandiarena, M. F., Olivieri, F. P., Daleo, G. R., and Oliva, C. R. (2001). Isolation and characterization of a polygalacturonase-inhibiting protein from potato leaves. Accumulation in response tosalicylic acid, wounding and infection. *Plant Physiol. Biochem.* 39, 129–136. doi: 10.1016/S0981-9428(00)01228-6

85. Manfredini, C., Sicilia, F., Ferrari, S., Pontiggia, D., Salvi, G., Caprari, C., et al. (2005). Polygalacturonase-inhibiting protein 2 of *Phaseolus vulgaris* inhibits BcPG1, a polygalacturonase of *Botrytis cinerea* important for pathogenicity, and protects transgenic plants from infection.*Physiol. Mol. Plant Pathol.* 67, 108–115. doi: 10.1016/j.pmpp.2005.10.002

86. Mariotti, L., Casasoli, M., Migheli, Q., Balmas, V., Caprari, C., and De Lorenzo, G. (2008). Reclassification of *Fusarium verticillioides* (syn. *F.*

moniliforme) strain FC-10 as *F. phyllophilum*. *Mycol. Res.* 112, 1010–1011. doi: 10.1016/j.mycres.2008.07.004

87. Mattei, B., Bernalda, M. S., Federici, L., Roepstorff, P., Cervone, F., and Boffi, A. (2001). Secondary structure and post-translational modifications of the leucine-rich repeat protein PGIP (polygalacturonase-inhibiting protein) from *Phaseolus vulgaris*. *Biochemistry* 40, 569–576. doi: 10.1021/bi0017632

88. Maulik, A., Ghosh, H., and Basu, S. (2009). Comparative study of protein-protein interaction observed in Polygalacturonase-inhibiting proteins from *Phaseolus vulgaris* and *Glycine max* and Polygalacturonase from *Fusarium moniliforme*. *BMC Genomics* 10:S19. doi: 10.1186/1471-2164-10-S3-S19

89. Michelmore, R. W., and Meyers, B. C. (1998). Clusters of resistance genes in plants evolve by divergent selection and a birth-and-death process.*Genome Res.* 8, 1113–1130.

90. Misas-Villamil, J. C., and van der Hoorn, R. A. (2008). Enzyme-inhibitor interactions at the plant-pathogen interface. *Curr. Opin. Plant Biol.* 11, 380–388. doi: 10.1016/j.pbi.2008.04.007

91. Mohammadzadeh, R., Zamani, M., Motallebi, M., Norouzi, P., Jourabchi, E., Benedetti, M., et al. (2012). Agrobacterium tumefaciens-mediated introduction of polygalacturonase inhibiting protein 2 gene (PvPGIP2) from Phaseolus vulgaris into sugar beet (Beta vulgaris L.). *Aust. J. Crop Sci.* 6, 1290–1297.

92. Nguema-Ona, E., Moore, J. P., Fagerström, A. D., Fangel, J. U., Willats, W. G., Hugo, A., et al. (2013). Overexpression of the grapevine PGIP1 in tobacco results in compositional changes in the leaf arabinoxyloglucan network in the absence of fungal infection. *BMC Plant Biol.* 13:46. doi: 10.1186/1471-2229-13-46

93. Nietlispach, D., Mott, H. R., Stott, K. M., Nielsen, P. R., Thiru, A., and Laue, E. D. (2004). Structure determination of protein complexes by NMR.*Methods Mol. Biol.* 278, 255–288. doi: 10.1385/1-59259-809-9:255

94. Nuss, L., Mahe, A., Clark, A. J., Grisvard, J., Dron, M., Cervone, F., et al. (1996). Differential accumulation of PGIP (polygalacturonase inhibiting protein) mRNA in two near-isogenic lines of *Phaseolus vulgaris* L. upon infection with *Colletotrichum lindemuthianum*. *Physiol. Mol. Plant Pathol.* 48, 83–89.

95. O'Kennedy, M., Burger, J., and Berger, D. (2001). Transformation of elite white maize using the particle inflow gun and detailed analysis of a

low-copy integration event. *Plant Cell Rep*. 20, 721–730. doi: 10.1007/s002990100383

96. O'Brien, J. A., Daudi, A., Finch, P., Butt, V. S., Whitelegge, J. P., Souda, P., et al. (2012). A peroxidase-dependent apoplastic oxidative burst in cultured arabidopsis cells functions in MAMP-elicited defence. *Plant Physiol*. 158, 2013–2027. doi: 10.1104/pp.111.190140
97. Oelofse, D., Dubery, I. A., Meyer, R., Arendse, M. S., Gazendam, I., and Berger, D. K. (2006). Apple polygalacturonase inhibiting protein1 expressed in transgenic tobacco inhibits polygalacturonases from fungal pathogens of apple and the anthracnose pathogen of lupins. *Phytochemistry* 67, 255–263. doi: 10.1016/j.phytochem.2005.10.029
98. Oh, D.-H., Dassanayake, M., Bohnert, H. J., and Cheeseman, J. M. (2012). Life at the extreme: lessons from the genome. *Genome Biol*. 13, 241. doi: 10.1186/gb-2012-13-3-241
99. Oliveira, M. B., Nascimento, L. B., Junior, M. L., and Petrofeza, S. (2010). Characterization of the dry bean polygalacturonase-inhibiting protein (PGIP) gene family during *Sclerotinia sclerotiorum* (*Sclerotiniaceae*) infection. *Genet. Mol. Res*. 9, 994–1004. doi: 10.4238/vol9-2gmr776
100. Pérez-Donoso, A. G., Sun, Q., Roper, M. C., Greve, L. C., Kirkpatrick, B., and Labavitch, J. M. (2010). Cell wall-degrading enzymes enlarge the pore size of intervessel pit membranes in healthy and *Xylella fastidiosa*-infected grapevines. *Plant Physiol*. 152, 1748–1759. doi: 10.1104/pp.109.148791
101. Powell, A. L. T., Stotz, H. U., Labavitch, J. M., and Bennett, A. B. (1994). *"Glycoprotein inhibitors of fungal polygalacturonases,"* in *Advances in Molecular Genetics of Plant-Microbe Interactions, Current Plant Science and Biotechnology in Agriculture*, eds M. J. Daniels, J. A. Downie, and A. E. Osbourn (Dordrecht: Springer), 399–402.
102. Powell, A. L. T., van Kan, J., ten Have, A., Visser, J., Greve, L. C., Bennett, A. B., et al. (2000). Transgenic expression of pear PGIP in tomato limits fungal colonization. *Mol. Plant Microbe Interact*. 13, 942–950. doi: 10.1094/MPMI.2000.13.9.942
103. Prabhu, S. A., Kini, K. R., Raj, S. N., Moerschbacher, B. M., and Shetty, H. S. (2012). Polygalacturonase-inhibitor proteins in pearl millet: possible involvement in resistance against downy mildew. *Acta Biochim. Biophys. Sinica*. 44, 415–423. doi: 10.1093/abbs/gms015
104. Prabhu, S. A., Singh, R., Kolkenbrock, S., Sujeeth, N., El Gueddari, N. E., Moerschbacher, B. M., et al. (2014). Experimental and bioinformatic

characterization of recombinant polygalacturonase-inhibitor protein from pearl millet and its interaction with fungal polygalacturonases. *J. Exp. Bot.* 65, 5033–5047. doi: 10.1093/jxb/eru266

105. Prabhu, S. A., Wagenknecht, M., Melvin, P., Gnanesh Kumar, B. S., Veena, M., Shailasree, S., et al. (2015). Immuno-affinity purification of PglPGIP1, a polygalacturonase-inhibitor protein from pearl millet: studies on its inhibition of fungal polygalacturonases and role in resistance against the downy mildew pathogen. *Mol. Biol. Rep.* doi: 10.1007/s11033-015-3850-5. [Epub ahead of print].
106. Protsenko, M. A., Buza, N. L., Krinitsyna, A. A., Bulantseva, E. A., and Korableva, N. P. (2008). Polygalacturonase-inhibiting protein is a structural component of plant cell wall. *Biochem. (Moscow)* 73, 1053–1062. doi: 10.1134/S0006297908100015
107. Raiola, A., Sella, L., Castiglioni, C., Balmas, V., and Favaron, F. (2008). A single amino acid substitution in highly similar endo-PGs from*Fusarium verticillioides* and related *Fusarium* species affects PGIP inhibition. *Fungal Genet. Biol.* 45, 776–789. doi: 10.1016/j.fgb.2007.11.003
108. Reignault, P., Valette-Collet, O., and Boccara, M. (2008). The importance of fungal pectinolytic enzymes in plant invasion, host adaptability and symptom type. *Eur. J. Plant Pathol.* 120, 1–11. doi: 10.1007/s10658-007-9184-y
109. Richter, A., Jacobsen, H. J., De Kathen, A., De Lorenzo, G., Briviba, K., Hain, R., et al. (2006). Transgenic peas (*Pisum sativum*) expressing polygalacturonase inhibiting protein from raspberry (*Rubus idaeus*) and stilbene synthase from grape (*Vitis vinifera*). *Plant Cell Rep.* 25, 1166–1173. doi: 10.1007/s00299-006-0172-z
110. Ridley, B. L., O'Neill, M. A., and Mohnen, D. (2001). Pectins: structure, biosynthesis, and oligogalacturonide-related signaling. *Phytochemistry*57, 929–967. doi: 10.1016/S0031-9422(01)00113-3
111. Roper, M. C., Greve, L. C., Warren, J. G., Labavitch, J. M., and Kirkpatrick, B. C. (2007). *Xylella fastidiosa* requires polygalacturonase for colonization and pathogenicity in *Vitis vinifera* grapevines. *Mol. Plant Microbe Interact.* 20, 411–419. doi: 10.1094/MPMI-20-4-0411
112. Sathiyaraj, G., Srinivasan, S., Subramanium, S., Kim, Y. J., Kim, Y. J., Kwon, W. S., et al. (2010). Polygalacturonase inhibiting protein: isolation, developmental regulation and pathogen related expression in *Panax ginseng* C.A. Meyer. *Mol. Biol. Rep.* 37, 3445–3454. doi: 10.1007/s11033-009-9936-1

113. Sato, A., and Miura, K. (2011). Root architecture remodeling induced by phosphate starvation. *Plant Signal. Behav.* 6, 1122–1126. doi: 10.4161/psb.6.8.15752

114. Sawaki, Y., Iuchi, S., Kobayashi, Y., Kobayashi, Y., Ikka, T., Sakurai, N., et al. (2009). STOP1 regulates multiple genes that protect Arabidopsis from proton and aluminum toxicities. *Plant Physiol.* 150, 281–294. doi: 10.1104/pp.108.134700

115. Schacht, T., Unger, C., Pich, A., and Wydra, K. (2011). Endo- and exopolygalacturonases of *Ralstonia solanacearum* are inhibited by polygalacturonase-inhibiting protein (PGIP) activity in tomato stem extracts. *Plant Physiol. Biochem.* 49, 377–387. doi: 10.1016/j.plaphy.2011.02.001

116. Sella, L., Castiglioni, C., Roberti, S., D'Ovidio, R., and Favaron, F. (2004). An endo-polygalacturonase (PG) of *Fusarium moniliforme* escaping inhibition by plant polygalacturonase-inhibiting proteins (PGIPs) provides new insights into the PG-PGIP interaction. *FEMS Microbiol. Lett.* 240, 117–124. doi: 10.1016/j.femsle.2004.09.019

117. Senthil, G., Williamson, B., Dinkins, R. D., and Ramsay, G. (2004). An efficient transformation system for chickpea (*Cicer arietinum* L.). *Plant Cell Rep.* 23, 297–303. doi: 10.1007/s00299-004-0854-3

118. Shanmugam, V. (2005). Role of extracytoplasmic leucine rich repeat proteins in plant defence mechanisms. *Microbiol. Res.* 160, 83–94. doi: 10.1016/j.micres.2004.09.014

119. Shivashankar, S., Thimmareddy, C., and Roy, T. K. (2010). Polygalacturonase inhibitor protein from fruits of anthracnose resistant and susceptible varieties of Chilli (*Capsicum annuum* L). *Indian J. Biochem. Biophys.* 47, 243–248.

120. Sicilia, F., Fernandez-Recio, J., Caprari, C., De Lorenzo, G., Tsernoglou, D., Cervone, F., et al. (2005). The polygalacturonase-inhibiting protein PGIP2 of *Phaseolus vulgaris* has evolved a mixed mode of inhibition of endopolygalacturonase PG1 of *Botrytis cinerea*. *Plant Physiol.* 139, 1380–1388. doi: 10.1104/pp.105.067546

121. Song, K. H., and Nam, Y. W. (2005). Genomic organization and differential expression of two polygalacturonase-inhibiting protein genes from*Medicago truncatula*. *J. Plant Biol.* 48, 467–478. doi: 10.1007/BF03030589

122. Spadoni, S., Zabotina, O., Di Matteo, A., Mikkelsen, J. D., Cervone, F., De Lorenzo, G., et al. (2006). Polygalacturonase-inhibiting protein interacts with pectin through a binding site formed by four clustered

residues of arginine and lysine. *Plant Physiol.* 141, 557–564. doi: 10.1104/pp.106.076950

123. Spinelli, F., Mariotti, L., Mattei, B., Salvi, G., Cervone, F., and Caprari, C. (2009). Three aspartic acid residues of polygalacturonase-inhibiting protein (PGIP) from *Phaseolus vulgaris* are critical for inhibition of *Fusarium phyllophilum* PG. *Plant Biol.* 11, 738–743. doi: 10.1111/j.1438-8677.2008.00175.x

124. Szankowski, I., Briviba, K., Fleschhut, J., Schönherr, J., Jacobsen, H. J., and Kiesecker, H. (2003). Transformation of apple (*Malus domestica*Borkh.) with the stilbene synthase gene from grapevine (*Vitis vinifera* L.) and a PGIP gene from kiwi (*Actinidia deliciosa*). *Plant Cell Rep.* 22, 141–149. doi: 10.1007/s00299-003-0668-8

125. Tamburino, R., Chambery, A., Parente, A., and Di Maro, A. (2012). A Novel Polygalacturonase-Inhibiting Protein (PGIP) from *Lathyrus sativus*L. seeds. *Protein Pept. Lett.* 19, 820–825. doi: 10.2174/092986612801619561

126. Tamura, M., Gao, M., Tao, R., Labavitch, J. M., and Dandekar, A. M. (2004). Transformation of persimmon with a pear fruit polygalacturonase inhibiting protein (PGIP) gene. *Sci. Hortic.* 103, 19–30. doi: 10.1016/j.scienta.2004.04.006

127. ten Have, A., Mulder, W., Visser, J., and van Kan, J. A. (1998). The endopolygalacturonase gene *Bcpg1* is required for full virulence of *Botrytis cinerea*. *Mol. Plant Microbe Interact.* 11, 1009–1016. doi: 10.1094/MPMI.1998.11.10.1009

128. Thornburg, R. W., Carter, C., Powell, A. L., Mittler, R., Rizhsky, R., and Horner, H. T. (2003). A major function of the tobacco floral nectary is defense against microbial attack. *Plant Syst. Evol.* 238, 211–218. doi: 10.1007/s00606-003-0282-9

129. Toubart, P., Desiderio, A., Salvi, G., Cervone, F., Daroda, L., De Lorenzo, G., et al. (1992). Cloning and characterization of the gene encoding the endopolygalacturonase-inhibiting protein (PGIP) of *Phaseolus vulgaris* L. *Plant J.* 2, 367–373. doi: 10.1046/j.1365-313X.1992.t01-35-00999.x

130. van Santen, Y., Benen, J. A., Schroter, K. H., Kalk, K. H., Armand, S., and Visser, J. (1999). 1.68-Å crystal structure of endopolygalacturonase II from *Aspergillus niger* and identification of active site residues by site-directed mutagenesis. *J. Biol. Chem.* 274, 30474–30480. doi: 10.1074/jbc.274.43.30474

131. Volpi, C., Raiola, A., Janni, M., Gordon, A., O'Sullivan, D. M., Favaron,

F., et al. (2013). *Claviceps purpurea* expressing polygalacturonases escaping PGIP inhibition fully infects PvPGIP2 wheat transgenic plants but its infection is delayed in wheat transgenic plants with increased level of pectin methyl esterification. *Plant Physiol. Biochem.* 73, 294–301. doi: 10.1016/j.plaphy.2013.10.011

132. Wand, A. J., and Englander, S. W. (1996). Protein complexes studied by NMR spectroscopy. *Curr. Opin. Biotechnol.* 7, 403–408.
133. Wang, A., Wei, X., Rong, W., Dang, L., Du, L. P., Qi, L., et al. (2014a). GmPGIP3 enhanced resistance to both take-all and common root rot diseases in transgenic wheat. *Funct. Integr. Genomics*. doi: 10.1007/s10142-014-0428-6. [Epub ahead of print].
134. Wang, R., Lu, L., Pan, X., Hu, Z., Ling, F., Yan, Y., et al. (2014b). Functional analysis of *OsPGIP1* in rice sheath blight resistance. *Plant Mol. Biol.* 87, 181–191. doi: 10.1007/s11103-014-0269-7
135. Wang, X., Zhu, X., Tooley, P., and Zhang, X. (2013). Cloning and functional analysis of three genes encoding polygalacturonase-inhibiting proteins from *Capsicum annuum* and transgenic CaPGIP1 in tobacco in relation to increased resistance to two fungal pathogens. *Plant Mol. Biol.* 81, 379–400. doi: 10.1007/s11103-013-0007-6

Chapter 8

UNRAVELING INCOMPATIBILITY BETWEEN WHEAT AND THE FUNGAL PATHOGEN ZYMOSEPTORIA TRITICI THROUGH APOPLASTIC PROTEOMICS

Fen Yang[1], Wanshun Li[2], Mark Derbyshire[3], Martin R Larsen[4], Jason J Rudd[3] and Giuseppe Palmisano[4,5]

[1]Department of Plant and Environmental Sciences, University of Copenhagen, 1871 Frederiksberg C, Denmark

[2] BGI-tech, BGI, 518083 Shenzhen, China

[3] Department of Plant Biology and Crop Science, Rothamsted Research, Harpenden, Hertfordshire AL5 2JQ, United Kingdom

[4] Department of Biochemistry and Molecular Biology, University of Southern Denmark, 5230 Odense M, Denmark

[5] Present address: Institute of Biomedical Science, Dep

ABSTRACT

Background

Hemibiotrophic fungal pathogen *Zymoseptoria tritici* causes severe foliar disease in wheat. However, current knowledge of molecular mechanisms involved in plant resistance to *Z. tritici* and *Z. tritici* virulence factors is far from being complete. The present work investigated the proteome of leaf apoplastic fluid with emphasis on both host wheat and *Z. tritici* during the compatible and incompatible interactions.

Results

The proteomics analysis revealed rapid host responses to the biotrophic growth, including enhanced carbohydrate metabolism, apoplastic defenses and stress, and cell wall reinforcement, might contribute to resistance. Compatibility between the host and the pathogen was associated with inactivated plant apoplastic responses as well as fungal defenses to oxidative stress and perturbation of plant cell wall during the initial biotrophic stage, followed by the strong induction of plant defenses during the necrotrophic stage. To study the role of anti-oxidative stress in *Z. tritici* pathogenicity in depth, a YAP1 transcription factor regulating antioxidant expression was deleted and showed the contribution to anti-oxidative stress in *Z. tritici*, but was not required for pathogenicity. This result suggests the functional redundancy of antioxidants in the fungus.

Conclusions

The data demonstrate that incompatibility is probably resulted from the proteome-level activation of host apoplastic defenses as well as fungal incapability to adapt to stress and interfere with host cell at the biotrophic stage of the interaction.

BACKGROUND

The ascomycete fungus *Zymoseptoria tritici* causes Septoria tritici blotch (STB) in wheat, a foliar disease that poses a significant threat to global food production. Leaf penetration occurs by means of fungal hyphae emerging from geminating, surface-attached spores that enter via stomata. The fungus has a slow intercellular biotrophic symptomless growth, for typically up to 10 days, as hyphae extend in close contact with mesophyll cells, probably utilizing lipid and fatty acid stores for growth [1,2]. Subsequently, the fungus suddenly switches to the necrotrophic growth associated with leakage of nutrients from dying plant cells into the apoplastic spaces, an increase in fungal biomass, enhanced signaling, metabolism and defense responses in host, the appearance of lesions on the leaf surface, and the collapse of the plant tissue [3,4]. Disease transition and appearance of symptoms have been suggested to be triggered by fungal small protein effectors secreted into apoplast [2-4]. Differing from many other phytopathogenic fungi, *Z. tritici* does not form any specialized penetration or feeding structures and remains strictly apoplastic throughout the entire infection cycle.

The plant apoplast is potentially important as a bridge that perceives and transduces signals from the environment to the symplast. Under stress conditions, complex mechanisms, including accumulation of reactive oxygen species (ROS) and changes in the synthesis of extracellular proteins, are activated in the apoplast as a first line of defenses. The secreted plant apoplastic proteins predominantly represent functional categories associated with carbohydrate metabolism, cell wall metabolism, defense, and programmed cell death [5]. As an apoplast-inhabiting fungus, *Z. tritici* need to acquire apoplastic nutrients, shape the plant cell structures, and overcome the activated apoplastic defenses to survive, possibly via secretion of effectors involved in detoxification of defense-related molecules as well as protection against recognition by the plant. This highly dynamic compartment serves as the molecular battlefield that contributes to the success of infection or plant resistance.

Given the crucial role of leaf apoplast in wheat-*Z. tritici* interaction, it is of particular interest to investigate the molecular basis underlying plant apoplastic immunity and the counter defenses that *Z. tritici* evolves at different growth stages. Although there have been advances in understanding mechanisms of wheat responses to *Z. tritici*, involving programming cell death, ROS accumulation, activation of signal transduction, transport and energy metabolism, expression of a broad spectrum of pathogenesis-related (PR) proteins, antioxidants and jasmonic acid biosynthesis genes, and production of small signaling and defense compounds [1,2,4,6-10], well-characterized systematic apoplastic responses to *Z. tritici* in wheat, which are essential for determining the plant fate, are currently lacking.

Considerable studies have been performed to understand *Z. tritici* gene functions, mainly focusing on the necrotrophic growth due to low fungal biomass hardly detectable at the biotrophic stage. Until recently, the emerging high throughput 'omics' and sequencing technologies partly address the issue and enable the discovery of several fungal genes and proteins expressed at different growth stages [2,4,10]. Compared to uncovering the expression of *Z. tritici* genes including the genes encoding cell-wall-degrading enzymes (CWDEs), ROS-scavenging enzymes and putative effector proteins such as LysM as well as the production of the secondary metabolites during the compatible interaction [1-4,11], *Z. tritici in planta* proteins, particularly secreted protein effectors, have not been fully explored at a systematic level. The only proteomic report identified thirty-one proteins and five phosphoproteins of *Z. tritici* mainly involved in basic cellular machinery and signaling at the biotrophic stage of the compatible and incompatible interactions [10]. This study revealed a similarity in fungal protein profiles between two interactions, possibly due to the fact that analysis of whole inoculated leaves resulted in the dominance of most abundant plant and fungal proteins, which largely diluted the information about low abundant fungal proteins likely essential for pathogenicity. A deeper insight into *Z. tritici*-specific strategies of colonization can be achieved by a detailed analysis of host apoplast for the enrichment of fungal identifications, in particular secreted molecules.

The expression of genes encoding ROS-scavenging enzymes is one of the multiple defense systems evolved by *Z. tritici* to thrive in the oxidative apoplastic environment. The YAP1 transcription factor is one of the most important determinants of oxidative stress responses, responsible for transcriptional activation of oxidative stress-associated genes in fungi [12]. YAP1 undergoes a conformational change due to the formation of disulfide bonds upon exposure to oxidative stress and is thus transported from the cytoplasm into the nucleaus to activate gene transcription. YAP1-mediated detoxification of ROS is essential in the virulence of many pathogenic fungi including biotroph *Ustilago maydis*, necrotroph *Alternaria alternate* and human pathogen *Candida albcans* [13]. However, the role of YAP1 in oxidative stress responses during *Z. tritici* colonization is poorly defined.

In order to gain further insights into host resistance and *Z. tritici* pathogenicity, we explored the proteome from apoplastic washing fluid (AWF) with emphasis on the host and the pathogen during the compatible and incompatible interactions. This is a perspective focusing on the direct battle ground and differing from the previous genome-wide studies on plant-*Z. tritici*interaction. The analysis uncovers apoplastic regulatory networks that shape the aspects of the plant physiology in response to the fungus and the principles of fungal modulation of apoplastic immunity. The combat between the host and*Z. tritici* through the activation of their defense mechanisms at the biotrophic stage likely determines their fates. Our data have also expanded on the current models of fungal apoplastic proteome.

RESULTS

Wheat Proteins from AWF in Response to *Z. tritici*

It was difficult to avoid the leakage of symplastic compounds in leaf apoplast during the extraction, particularly from the dying leaf cells in the necrotrophic phase. In order to obtain the important apoplastic components at the necrotrophic stage but avoid severe symplastic contamination, we sampled the leaves at the early time point of the necrotrophic growth (14 days after inoculation, 14 dai) when immature asexual sporulation structure started to form. Malate dehydrogenase (MDH) was used as a cytosolic enzyme marker to evaluate the level of contamination. MDH activities in AWF contained max. 1% of those from the whole leaf extracts in all samples (Additional file 1). The ratio below 1% is considered as little damage to the cells and is acceptable for plant apoplast studies [14]. Approx. 1% MDH activity related to the total leaf extracts was also observed in the apoplastic tissue of ryegrass leaves during senescence [15], and of *V. longisporum*-infected oilseed rape leaves [16] and *A. thaliana* [17]. This recurrent and unavoidable cytoplasmic contamination has already been highlighted by several previous studies of plant apoplast [18-22]. Therefore, we considered apoplastic proteins were enriched in the AWF in the present study.

We applied a mass spectrometry (MS)-based shotgun quantitative proteomics using isobaric tags for relative and absolute quantitation (iTRAQ) labeling. This approach has been developed to analyze the proteome and phosphoproteome of *Z. tritici*-inoculated wheat leaves [10]. Missing reporter ions of the peptide from some of the samples in the MS/MS spectrum can occur in the iTRAQ-based proteomics due to the incomplete labeling and inefficient fragmentation of the tags, which results in no quantitative ratio. Therefore, the criterion defined for the reliable quantified proteins was that proteins had to be identified in at least two biological replicates with quantitative ratio. This resulted in the identification of 2122 and 2071 wheat proteins and quantification of 607 and 575 proteins in Stakado (resistant cultivar) and Sevin (susceptible cultivar), respectively (Additional file 2). Not surprisingly, cytosolic proteins such as Rubisco and ribosomal proteins were identified in all the samples by this highly sensitive MS approach, which were eliminated for the further analysis.

Based on the selection criteria, 45 proteins in Stakado and 100 proteins in Sevin were found to change in abundance in response to the fungus (Additional file 3). Eleven were detected in both cultivars. Of the regulated proteins, 42% and 60% from Stakado and Sevin were predicted with signal peptides using SignalP program, respectively. Different percentages of proteins with predicted signal sequences were found in the secretome of soybean (65%), grapevine (66%), *Arabidopsis*(47%) and rice (37%) [18]. There were 43 and 96 differentially expressed proteins identified from Stakado and Sevin whose corresponding transcripts were identified from the previous transcriptome dataset, respectively [4].

Expression cluster analysis revealed that a substantial number of proteins were up-regulated only at the necrotrophic stage (14 dai) in contrast to minor changes at the biotrophic stage (5 dai) in Sevin, whereas proteins could be up- or down- regulated at both time points in Stakado (Figure 1A). The major proteins in each cluster were annotated in carbohydrate metabolism, stress, defense, cell wall metabolism, protein process, and other metabolic processes (Figure 1B). In Stakado, at either time point, the fungus caused a significantly increased expression of several carbohydrate metabolic proteins involved in glycolysis (*i.e.*, 2,3-bisphosphoglycerate-independent phosphoglycerate mutase and phosphoglycerate kinase), pentose phosphate pathway (*i.e.*, 6-phosphogluconate dehydrogenase) and TCA (*i.e.*, aconitate hydratase), as well as stress/defense-related proteins including PR-1, PR-2, PR-3, peroxidase 3, heat shock protein and glycolate oxidase (clusters 1, 3 and 4), in contrast to a decreased expression of cell-wall metabolic proteins including α-L-arabinofuranosidase, β-D-glucan exohydrolase,

1,3(4)-β-glucanase, and β-D-xylosidase (cluster 2 and 5). The homologs of aconitate hydratase, 2,3-bisphosphoglycerate-independent phosphoglycerate mutase and phosphoglycerate kinase were also identified in the study of poplar apoplastic proteome [5]. Additionally, a 1-deoxy-D-xylulose 5-phosphate reductoisomerase responsible for terpenoid biosynthesis displayed a strong accumulation (11-fold change) at 14 dai, indicating the possible involvement of terpenoid in wheat resistance to *Z. tritici*. The pathogen-induced expression of terpenoid synthase gene and production of diterpenoid phytoalexins have been seen in maize and rice [23,24]. Moreover, integration of proteome and previous transcriptome datasets revealed that the majority of regulated proteins during the incompatible interaction did not change at transcription level in the compatible interaction. This result further emphasizes an important role of these proteins in plant resistance to *Z. tritici*. In Sevin, up-regulated proteins at 14 dai consisted of proteins involved in protein process, cell wall metabolism, stress, and defense, including a cold acclimation induced protein, peroxidases and numerous PR-proteins belonging to PR families 1, 2, 3, 4, 5, and 17 (clusters 1, 3, 4 and 5). A good correlation between protein and transcript regulation at the necrotrophic stage was observed, indicating the great amplitude of activation of host defense responses. On the other hand, most proteins that did not change in abundance at the biotrophic stage of the compatible interaction were transcriptionally suppressed, suggesting a complexity of molecular mechanisms in host plant triggered by *Z. triticibiotrophic* growth.

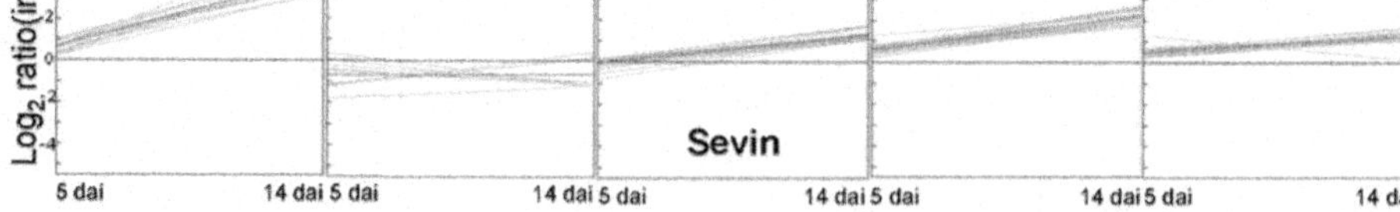

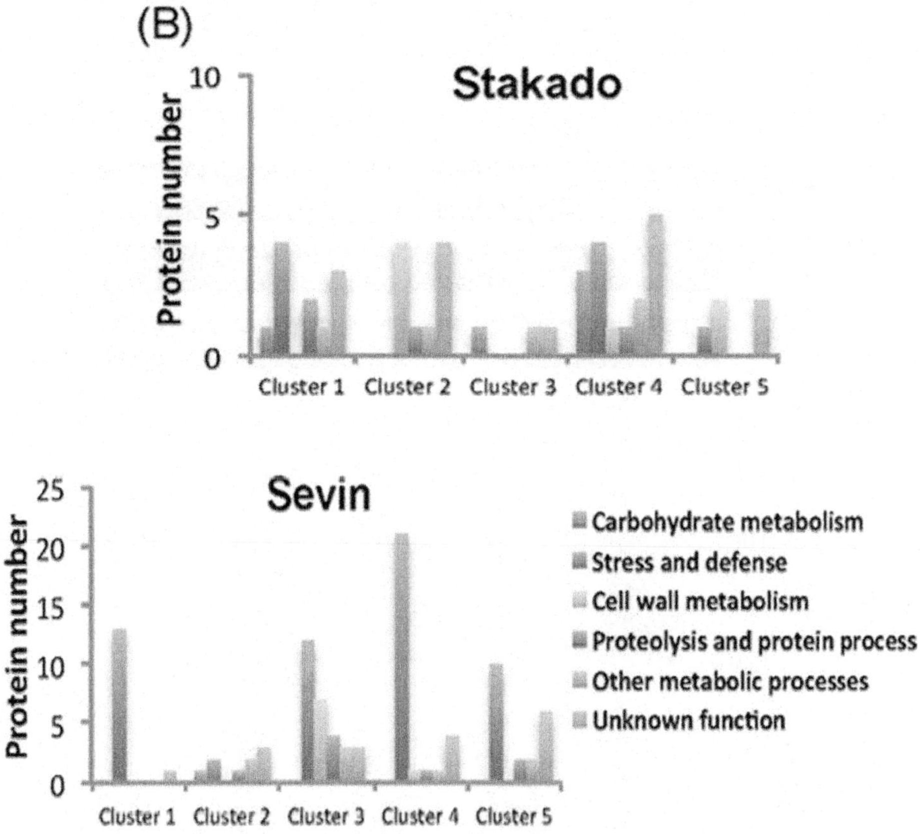

Figure 1: Differentially expressed plant proteins identified from AWF. AWF was isolated from the leaves of wheat cvs. Stakado (resistant) and Sevin (susceptible) inoculated with *Z. tritici* or water (control) at 5 and 14 dai. Three biological samples were prepared. The proteins from AWF were labeled with iTRAQ and subjected to LC-MS/MS analysis. Differentially expressed plant proteins were analyzed by expression clustering **(A)** and functional classification **(B)**.

In order to validate our proteomics data, western blot analysis of PR-1, PR-2, and PR-3 proteins was conducted with three biological replicates of AWF samples. In agreement with the proteomics analysis, the abundance of PR-1, −2, and −3 increased slightly at the early biotrophic stage of the incompatible interaction and markedly at the late stage of the compatible interaction (Figure 2).

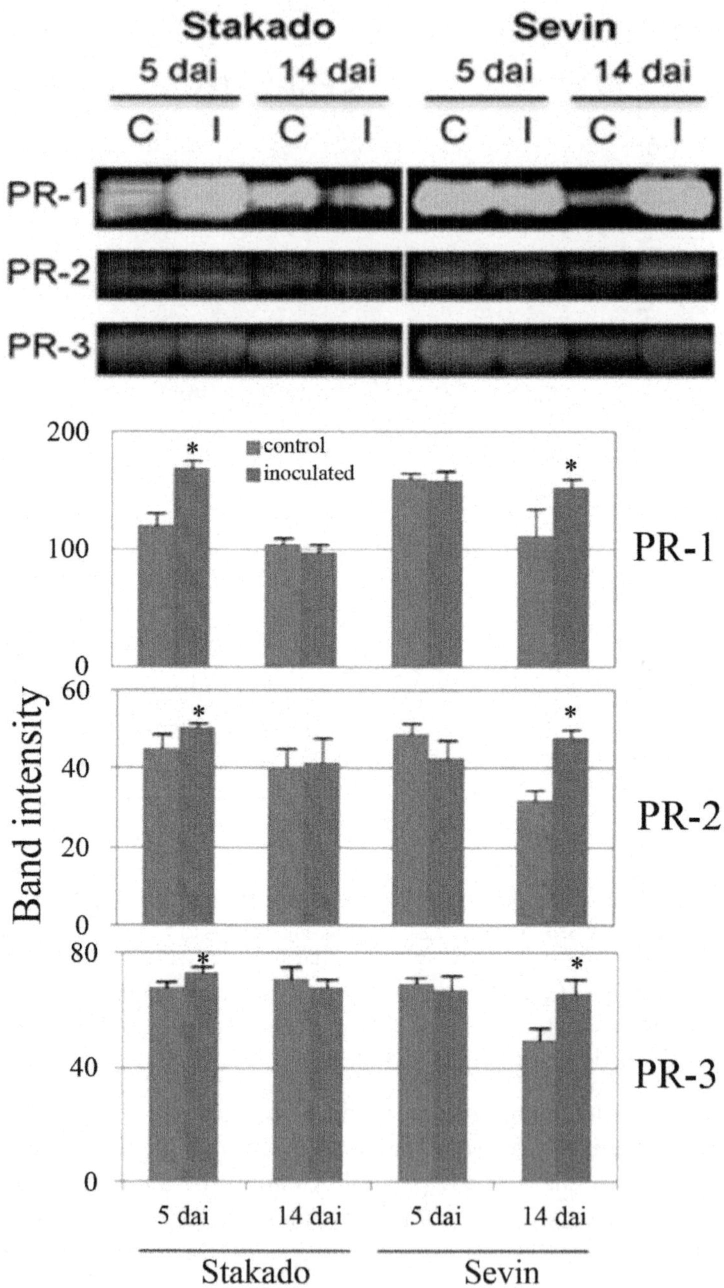

Figure 2: Western blotting validation of proteome data. Protein was extracted from three biological replicates of control (C) and *Z. tritici*-inoculated (I) wheat cvs. Stakado (resistant) and Sevin (susceptible) at 5 and 14 dai and used in western blot analysis.

The representative membranes of PR-1, PR-2 and PR-3 protein expression are shown. Quantification of signals on the membranes was performed by using ImageJ program based on three biological samples. The asterisks indicate significant differences in signal intensity (*P* 0.05) between control and inoculated samples.

Fungal Proteins Identified from Inoculated Leaf AWF

It is difficult to distinguish whether the protein is produced by the plant, the fungus or both in a plant-fungus interaction system, when the peptide matches to the protein sequences from both host and pathogen databases. Therefore, we have showed the proteins with peptides confidently identified from fungal database only, since the proteins identified from both databases are more likely of plant origin due to high biomass ratio of plant to the fungus. Twenty-four and thirty-one fungal proteins were identified from the inoculated Stakado and Sevin, respectively (Table 1). There were eighteen proteins whose corresponding transcripts were identified from the previous RNA-seq-based transcriptome dataset (Table 1, Additional file 4). Twelve proteins had an N-terminal signal peptide sequence predicted by SignalP, indicating they could be secreted proteins. The identified fungal proteins with known functions could be assigned to biological processes, encompassing metabolism, degradation of host cell wall, signaling, stress, defense, transport, cell mobility, cell component organization, and cell wall degradation and remodeling. Despite similar functional categories of fungal proteins and nine proteins identified from both inoculated cultivars, the overviews of fungal protein profiles were distinct between the compatible and incompatible interactions (Table 1). Apart from the categories of primary carbohydrate, amino acid and protein metabolisms regarded as no significant changes between two interactions, the fungus seemed to preferentially express the proteins involved in signal transduction, transport, and cell wall remodeling during the incompatible interaction, whereas the proteins involved in nucleic acid metabolism, transcription, perturbation of plant cell wall, anti-oxidative stress, and cellular component organization could only be identified in the compatible interaction.

Table 1: *Z. tritici* proteins identified from *Z. tritici* -inoculated wheat leaves

		Ratio (14 d/ 5 d)			
ID	**Annotation**	**S**	**SK**	**SV**	**Biological process**
91239	Myosin			1.36	Cell mobility
105948	Actin		1.15	1.11	Cell mobility

110409	Myosin	Y	1.17	1.33	Cell mobility
45905	Chitinase		0.52		Cell wall remodeling
86354	SWI/SNF chromatin re-modeling complex protein			N/A	Cellular component organization
87313	Rrp15p domain-containing protein			1.04	Cellular component organization
68922	Glycoside hydrolase family 62	Y		4.18	Degradation of plant cell wall
70396	α-L-arabinofuranosidase B	Y		5.67	Degradation of plant cell wall
99970	β-glucosidase	Y		2.70	Degradation of plant cell wall
44922	N6 adenine-specific DNA methylase			N/A	Metabolism
46697	ATPase		0.81		Metabolism
52059	Formate-tetrahydrofolate ligase		N/A	1.05	Metabolism
55968	Triosephosphate isomerase		1.63	1.38	Metabolism
64923	Glucose/ribitol dehydro-genase			7.74	Metabolism
68338	Isopenicillin N synthase	Y	1.23		Metabolism
69333	Glycine dehydrogenase		0.67		Metabolism
73114	Hydantoinase/oxoprolinase			2.31	Metabolism
74730	ATP-citrate lyase/succinyl-CoA ligase		1.81		Metabolism
76530	D-isomer specific 2-hy-droxyacid dehydrogenase	Y		N/A	Metabolism
77172	Aldo-keto reductase		1.37	1.70	Metabolism
96092	Trypsin	Y	1.80	0.78	Metabolism
102177	20S proteasome		1.58	0.99	Metabolism
106055	Aconitate hydratase		1.05		Metabolism
107787	Mediator16 domain-con-taining protein			1.86	Metabolism
110230	α -isopropylmalate/homoci-trate synthase			1.35	Metabolism
65824	Serine/threonine-specific protein phosphatase			1.07	Signaling

98343	Ras GTPase		1.64	1.22	Signaling
99493	Ras GTPase		2.33		Signaling
99564	Calmodulin		1.02		Signaling
110139	14-3-3 protein		N/A		Signaling
67250	Catalase/peroxidase	Y		36.9	Stress and defense
75170	Chaperonin 60		0.81		Stress and defense
99959	Heat shock protein 90	Y		1.04	Stress and defense
103593	Copper/Zinc superoxide dismutase			4.33	Stress and defense
105409	Catalase/peroxidase			9.33	Stress and defense
105895	Heat shock protein 70			N/A	Stress and defense
77089	Adaptin			1.05	Transport
83550	Golgi transport complex		1.07		Transport
94552	Ca^{2+}-modulated nonselective polycystin	Y	1.30		Transport
102764	Ankyrin		0.63		Transport
67764	Protein of unknown function DUF185			0.42	--
69789	Hypothetical protein	Y	0.64		--
90006	Hypothetical protein		0.94	1.33	--
92804	Hypothetical protein			0.75	--
103686	Hypothetical protein		0.45		--
109652	Membrane protein containing DUF221	Y		0.45	--

Fungal proteins were identified from wheat leaves of resistant cultivar Stakado (SK) and susceptible cultivar Sevin (SV) inoculated with *Z. tritici* at 5 and 14 days by LC-MS/MS. ID is protein accession number from Joint Genome Institute gene index for *Zymoseptoria tritici.* Proteins whose corresponding transcripts were identified from RNA-seq-based transcriptome dataset [4] are indicated in bold. The expression of these transcripts marked in bold is shown in Additional file 4. Y indicates protein identification containing a signal peptide examined by SignalP (S). The present ratios of protein abundance between 14 days and 5 days were calculated from at least two biological replicates by iTRAQ-117/iTRAQ-115 in labeling-based quantitative proteomics analysis.

The ratios≥2 or≤0.5 were defined as significant change. N/A indicates the identified protein without quantitative data. No ratio value shown indicates the protein was not identified in the cultivar.

Furthermore, the changes in fungal protein expression at two time points were measured. A signaling-related protein Ras GTPase was found to increase the expression during the incompatible interaction. The predominant fungal proteins identified from the inoculated Stakado, did not significantly change in abundance, indicating the restraint fungal growth during the incompatible interaction. A hydantoinase, a glucose/ribitol dehydrogenase, three CWDEs and three ROS-scavenging enzymes strongly accumulated during the compatible interaction, suggesting an increase in carbohydrate metabolism, interfering with host cell and oxidative stress. By comparing the changes in fungal proteins and the corresponding transcripts from the previous study during the compatible interaction, it clearly illustrates a strong multilevel induction of the mechanisms of anti-oxidative stress and host cell wall perturbation in *Z. tritici*, the greater depth and sensitivity of detection of molecules by RNA-seq approach, and the tendency for proteomics methodologies to preferentially detect proteins of higher abundance (Additional file 4).

ZtYAP1 Contributes to Resistance to Oxidative Stress but is not Required for Fungal Pathogenicity

Since the fungus distinctively secreted antioxidants during the compatible interaction, a gene encoding transcription factor YAP1 activating the expression of genes encoding antioxidants was targeted for deletion to examine the role of anti-oxidative stress mechanisms in fungal pathogenicity. The *ZtYAP1* mutant slightly reduced growth on regular Potato Dextrose Agar (PDA) compared to wild type (Figure 3A). Growth of the mutant was strongly inhibited in the presence of H_2O_2, oxidizing compound cumene hydroperoxide, Rose Bengal diacetate generating singlet oxygen or SDS (Figure 3A). Additionally, the mutant retained normal formation and germination of conidia at a rate and magnitude similar to the wild type *in vitro* (data not shown).

To determine whether *ZtYAP1* gene is required for fungal pathogenicity and lesion development, conidial suspensions prepared from the wild type and the mutant were spray-inoculated onto detached susceptible wheat leaves. Disease symptoms on the leaves inoculated with the wild type or the mutant appeared at 10 dai and developed rapidly afterwards (Figure 3B). No clear difference in leaf lesion caused by the wild type and the mutant was observed throughout the entire infection process. The spore production of the mutant in 21-d-infected wheat leaves was approximately three times as high as that of the wild type.

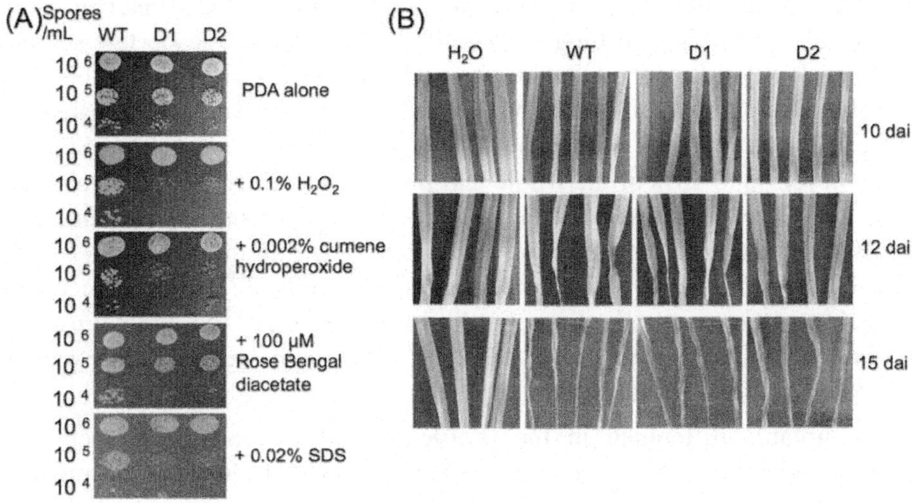

Figure 3: Characterization of *ΔZtYAP1* mutant. **(A)** *ΔZtYAP1* mutant is hypersensitive to oxidants *in vitro*. Sensitivity of *Z. tritici* wild type (WT) and *ΔZtYAP1* deletion strains D1 and D2 was determined by radial growth on PDA supplemented with oxidants or compounds as indicated. Results from one representative of two technical replicates are shown. **(B)** *ΔZtYAP1* mutant is as virulent as WT *in planta*. Fungal pathogenicity was assayed on wheat leaves spray-inoculated with spore suspension (1 x 10^6 spores/mL) prepared from WT, D1, and D2 strains. Photos were taken at 10, 12 and 15 dai.

Discussion

Because of the economic importance as a major fungal pathogen of wheat, *Z. tritici* has emerged as model for studying plant-pathogen interaction. Understanding the molecular mechanisms underlying plant defenses and fungal pathogenicity is a prerequisite for understanding the host-pathogen interaction and can contribute to the development of new strategies of crop protection against pathogen. Thus, we have presented a detailed analysis of proteome from the battle ground apoplast between host wheat and *Z. tritici* during the compatible and incompatible interactions. A comparison of the apoplastic protein profiles of both host and the pathogen between two interactions allows us to gain further insights into such mechanisms.

The analysis of cellular location of protein identifications in the present study shows that approx. half of the differentially expressed plant proteins and fungal proteins from the AWF do not contain signal peptides. This could a consequence of cell lysis leading to the release of cytosolic proteins in the apoplast. Alternatively, evidences are emerging to support that proteins identified in the AWF without symplastic contamination and lacking signal peptides could be secreted by a non-classical secretory mechanism as described in bacteria and fungi [5,18,25]. They are named leaderless secreted proteins (LSPs). Some predicted intracellular proteins such as carbohydrate metabolic proteins found in the AWF can be in fact actively translocated into the extracellular space. Inventories of plant secretome reveal that LSPs may account for up to 50% of the whole leaf proteins identified in the extracellular fluid [25]. The secretome of plants submitted to stresses usually contains more LSPs than unstressed plants, which has been shown in the pathogen-infected maize [26] and salicylic acid-treated *Arabidopsis* cells [27]. With respect to the fungal secretome, investigation of culture supernatant of the fungus*Fusarium graminearum* has indicated that 30% of the protein identifications may be secreted but do not contain signal peptides [28].

A notable difference in the profiles of differentially expressed plant proteins identified from the AWF between compatible and incompatible interactions is the occurrence of regulation of the proteins implicated in carbohydrate metabolism, defense, and stress. It was very likely that the carbohydrate metabolic proteins identified in the AWF were translocated from intracellular space into the apoplast [25]. Plant defense against pathogens is known to be costly in terms of energy, assimilates, reducing equivalents, and carbon skeleton components that are provided by the primary metabolism [29]. Rapid mobilization and metabolism of the carbohydrates are important factors determining the outcome of plant-pathogen interactions. In the incompatible interaction between wheat and *Z. tritici*, it has been shown that accelerated sugar production associated with signal transduction cascades and expression of defense responses, occurs rapidly and intensively, compared to the compatible interaction [10]. Consistent with this, enhanced carbohydrate metabolism occurring in the intracellular space was accompanied by the activation of defense and anti-oxidative stress responses in the apoplast during the biotrophic stage of the incompatible interaction observed in the present study. By contrast, in the compatible interaction, apoplastic defenses were impeded or not activated at protein level and suppressed at transcription level during the establishment of biotrophic growth, but strongly triggered at both levels in the necrotrophic phase [2,4]. These results collectively suggest a contribution of rapid apoplastic defenses and energy metabolism to host resistance to*Z. tritici*.

Host plant cell wall metabolism and remodeling were also distinct in response to the fungus between two interactions. The plant cell wall represents a first line of defense against microbial pathogens since it is a preformed, passive physical barrier limiting access of pathogens and is actively remodeled and reinforced specifically at discrete sites of interaction with the pathogens [30]. During the incompatible interaction, the abundance of the plant proteins that hydrolyze the carbohydrate moieties of arabinogalactan, arabinoxylan, and other polysaccharides diminished, possibly preventing the degradation of cell wall polysaccharides, and thus stabilizing the plant cell wall. Similar results were seen in a study of rice root apoplastic proteome in response to salt stress [31]. Conversely, these proteins largely increased the abundance at the necrotrophic stage of the compatible interaction. On one hand, this phenomenon can result in the release of plant oligosaccharides triggering a set of basal defenses. On the other hand, in agreement with the previous transcriptome analysis [4], the necrotrophic growth of *Z. tritici* induces the high metabolic activity of host cell wall, giving rise to the loss of control of host cell permeability and host necrosis, which allows the fungus thrives on the dead plant.

Moreover, a large group of responsive proteins manifesting increased abundance in both interactions were peroxidases. In addition to catalyzing formation and the consumption of ROS, peroxidases have been implicated in the modification of plant cell wall structures [17]. Via oxidative cross-linking of monolignols, ploysacchraides, and cell wall proteins, peroxidases can reinforce the cell wall to restrict pathogen invasion. Accumulation of certain classes of peroxidases is involved in plant cell wall lignification that occurs as stress response to prevent the spread of fungal pathogens including*Verticillium* and *Melampsora* species [5,17]. Despite the up-regulation of diverse peroxidases as well as other defense-related protein like PR proteins in the necrotrophic phase of the compatible interaction, the velocity and magnitude of activated apoplastic defenses were apparently insufficient to prevent the necrotrophic growth of *Z. tritici*.

An important aspect in studying AWF of the plant challenged with the apoplast-inhabiting pathogens is to obtain the knowledge concerning the molecular compounds produced or secreted by the pathogens, which can provide the candidates essential for the pathogenicity. Here, proteomics analysis showed that approx. half of the fungal proteins whose corresponding transcripts were not identified in the previous RNA-seq-based transcriptome dataset [4]. Similar results were observed in the global analysis of transcript and protein levels across the *Plasmodium falciparum* lifecycle [32]. The little correlation between mRNA and protein abundance has been well demonstrated due to the regulation of transcription and translation, mRNA processing, mRNA stability and degradation, and protein modifications and turnover [4]. Furthermore, the bias in methodologies of RNA-seq and proteomics can result in the discordance between detected transcripts and proteins. Combined with the poor correlation between the regulation of host apoplastic proteins and corresponding transcripts at the biotrophic stage of the compatible interaction, the results strongly suggest that the integrated 'omics' studies are required to comprehensively understand the molecular processes during *Z. tritici* colonization of host wheat.

Based on the fungal protein identifications, there is little evidence for fungal nutrient acquisition from the plant throughout symptomless biotrophic colonization by *Z. tritici* in the incompatible and compatible interactions, which further supports the previous findings from transcriptome and proteome datasets [2,10]. Different protein profiles of *Z. tritici* between two interactions were observed, suggesting specific strategies of *Z. tritici* for a successful colonization. Most notably, signaling, transport and cell wall remodeling were particularly identified in the incompatible interaction, whereas genes and proteins involved in anti-oxidative stress and modification of plant cell wall were identified and induced during the compatible interaction [4]. Fungal pathogens often mount effective responses to counter the defenses of the host plants. A well-established colonization requires circumventing and overcoming host defense responses such as generation of ROS, basically via hiding, detoxification and inhibition [33]. ROS accumulation is considered as one of the primary responses to *Z. tritici,* starting in the apoplast during the biotrophic stage and spreading to the entire leaf tissue at the late necrotrophic stage. Therefore, *Z. tritici* expresses a number of ROS-scavenging proteins and CWDEs that are known to aid invasion into plant cells by hydrolyzing the plant cell wall polymers. The increased expression of CWDEs may also play a role, although not essential, in nutrient acquisition in *Z. tritici* during the necrotrophic growth [4]. The ability to adapt to the oxidative stress, interfere with host cell structure, and overcome other host defense responses such as expression of apoplastic PR proteins probably results in compatibility with the host wheat. On the other hand, it seems that the resistant plant deploys a highly stressful apoplastic environment to put pressure on the pathogen, as evidenced by the active signaling, transport, and cell wall remodeling in *Z. tritici* during the incompatible interaction. The fungal pathogens are known to sense the stress by rapid signal transduction contributing to the regulation of stress functions including glycerol accumulation and cell wall remodeling [34].

The hypothesis of significance of anti-oxidative stress mechanisms in *Z. tritici* colonization was tested by deletion of a gene encoding YAP1 transcription factor regulating expression of antioxidants. The ability to defend against oxidative stress *in vitro* was partially compromised in *YAP1*-dirupted mutant, which, however, showed normal virulence as the wild type strain, suggesting a non-essential role of *YAP1* in *Z. tritici* pathogenicity and functional redundancy of antioxidants. Similarly, disruption of a *YAP1* homolog did not affect the virulence of the phytopathogenic fungus *Cochliobolus heterostrophus* [35] or *Fusarium graminearum* [36]. These findings highlight the sophisticated mechanisms of the involvement of YAP-mediated ROS detoxification in pathogenicity in fungi kingdom. Further investigations on anti-oxidative stress are highly required to elucidate its clear role in *Z. tritici* pathogenicity.

CONCLUSIONS

We have reported an extensive survey of leaf apoplastic proteome in resistant and susceptible wheat in response to *Z. tritici*. The results collectively demonstrate that plant resistance to *Z. tritici* is correlated with rapid activation of responses at proteome level, including enhanced carbohydrate metabolism, cell wall reinforcement and remodeling, production of PR proteins in the apoplast, and generation of a stressful apoplastic environment. On the other hand, the fungus has to interfere with host cell wall and overcome host defenses and stress at the biotrophic stage, for example, by detoxifying ROS and producing CWDEs, to achieve a successful colonization. Taken together, our work provides the valuable insights into STB resistance and *Z. tritici in planta* proteome, which form a fundamental and prerequisite step for the further research of plant apoplastic immunity and *Z. tritici* pathogenicity.

METHODS

Plant Growth, Fungal Inoculation and Extraction of AWF

Growth of wheat cultivars Sevin (susceptible) and Stakado (resistant), preparation of the inoculum of *Z. tritici* isolate IPO323, and inoculation were preformed as described [8]. Separate control plants were mock inoculated with water. Approximately 60 leaves were collected from four separate pots, serving as one biological replicate. Three biological replications were harvested separately at 5 and 14 dai prior to the extraction of AWF by water as described [37]. The remaining leaf samples were ground into fine powder in liquid nitrogen. The collected AWF and leaf powder were stored at −80°C until use.

Malate Dehydrogenase Assay

Total soluble protein was extracted from the leaf powder in 50 mM phosphate buffer, pH 7.5. MDH activity assay was performed to assess the contamination of AWF by intracellular proteins. Five μL of AWF or total leaf protein extract was added to 200 μL of the reaction buffer containing 0.17 mM oxalacetic acid, 0.094 mM β-NADH disodium salt and 0.1 M phosphate buffer, pH 7.5

[14]. The change of absorbance at 340 nm was monitored for 5 min in a spectrophotometer.

Shotgun Proteomics Analysis of AWF

Protein concentration in AWF was determined by Bio-Rad Protein Assay (Bio-Rad) with bovine serum albumin as standard. Fifty micrograms protein was precipitated by 5 volumes of 10% TCA (w/v) in acetone at −20°C overnight. The protein pellet was washed three times in cold 100% acetone followed by solubilization in the buffer containing 6 M urea and 2 M thiourea. Protein samples were treated with 10 mM DTT for 30 min at room temperature and 40 mM iodoacetamide for 30 min in the dark prior to digestion with trypsin (2%, w/w) at 37°C overnight. The resulting peptides were purified on Poros Oligo R3 microcolumn and vacuum-dried prior to amino acid analysis using a Biochrom 30+ Amino Acid Analyzer (Biochrom, UK). Six micrograms peptides from each biological replicate of inoculated and control samples were labeled with iTRAQ® 4-plex (Applied Biosystems) according to the manufacturer's protocol (114 for the control at 5 dai, 115 for the inoculated sample at 5 dai, 116 for the control at 14 dai, 117 for the inoculated sample at 14 dai). Labeled peptides were combined and desalted on Poros Oligo R3 microcolumn.

The labeled peptides were fractionated using hydrophilic interaction liquid chromatography (HILIC) fractionation and analyzed by liquid chromatography (LC)-MS/MS as described [10] with modifications. Briefly, isobaric-labeled peptides obtained from biological replicate 1 were fractionated on a TSKGel Amide 80 HILIC-HPLC column by using the Agilent 1200 microHPLC instrument. Samples were suspended in solvent B (90% acetonitrile and 0.1% trifluoroacetic acid), and peptides were eluted at 6 μL/min by decreasing the solvent B concentration (100−60%) over 26 min. Fractions were collected and lyophilized.

Peptides from biological replicate 1 were resuspended in 0.1% formic acid and separated by reversed-phase liquid chromatography on a Reprosil-Pur C18 (3 μC; Dr. Maisch GmbH - Ammerbuch, Germany) column (22 cm x 100 μm inner diameter, in-house packed). The chromatographic gradient was 0−34% solvent B (90% acetonitrile and 0.1% formic acid) for 90 min at a flow rate of 300 nL/min. A LTQ-Orbitrap XL mass spectrometer (Thermo Fisher Scientific) was operated in a data-dependent mode automatically switching between MS and MS/MS. A survey MS scan (400–1800 m/z) was acquired in the Orbitrap analyzer with a resolution of 30000 at 400 m/z. The top three most intense ions with a threshold of 5000 were selected for low-resolution CID at normalized collision energy of 35 and high-resolution HCD (normalized collision energy=55; 7500 resolution at 400 m/z). Peptides from biological replicates 2 and 3 without HILIC fractionation were performed using the same LC configuration with a 120 min gradient but coupled to a LTQ-Orbitrap Velos mass spectrometer. Following a survey MS scan at a resolution of 30000 at 400 m/z, the top seven most intense ions were selected for high-resolution HCD–MS/MS (normalized collision energy=48; 7500 resolution at 400 m/z). These two biological replicates were run twice using this setup as technical replicates. Raw data were viewed in Xcalibur v2.0.7 (Thermo Fisher Scientific, USA).

Raw MS/MS spectra were processed and quantified using Proteome Discoverer (Version 1.2, Thermo Fisher) software with quantification setup as ratios of 115/114 (inoculated/control at 5 dai) and 117/116 (inoculated/control at 14 dai). Peptide identification was performed with MASCOT (v2.2, Matrix Science Ltd. - London, UK) and Sequest algorithms, searching against a target and decoy TaGI wheat gene index Release 12.0 (released on 18th April, 2010; TC sequences, 93508; ESTs, 128166; ETs, 251; http://compbio.dfci.harvard.edu/tgi/) and DOE Joint Genome Institute gene index for *Z. tritici* (released on 10th September, 2008; 10933 genes; http://genome.jgi-psf.org/Mycgr3) databases. The following parameters were set for searching: 2 missed cleavages, S-carbamidomethyl-cysteine as a fixed modification, oxidation (M), deamidation (N and Q), iTRAQ® reagents (protein N-terminus and Lys side-chain), peptide mass tolerance 10 ppm, and fragment ion mass tolerance 0.5 Da for CID and 0.05 Da for HCD. False discovery rates were obtained using Percolator selecting identifications with a q-value$\leq$0.01. Only high-confidence peptide sequences with a Mascot ion score$\geq$23, Sequest Xcorr value>2.2 and rank 1 were considered for the further analysis. The identified plant and fungal proteins must contain at least two confidentially identified peptides. Protein quantification data were normalized using log2-transformed median. The statistics analysis was performed on the data merged from all biological and technical replicates by using R program. The plant proteins identified in at least two biological replicates and with average ratios$\geq$2 or$\leq$0.5 at either time point were defined as regulated proteins. The fungal proteins identified in at least two biological replicates with quantitative data were shown in Table 1. SignalP (http://www.cbs.dtu.dk/services/SignalP) was performed to examine signal peptides of fungal and wheat regulated proteins. The differential expression profiles of host proteins at two time points were clustered by k-means algorithm using Euclidean distance.

Western Blot Analysis

Western blotting was performed with three biological replicates as described

[38]. Five micrograms proteins from AWF were separated on CriterionTM XT Precast Gels (12% Bis-Tris, Bio-Rad) followed by blotting to nitrocellulose membranes (GE Healthcare). Membranes were blocked followed by incubation with rabbit antibodies against barley PR-1, PR-2 (β-1,3-glucanase), and PR-3 (chitinase). After extensive washes, the membranes were incubated with anti-rabbit secondary antibodies conjugated to horseradish peroxidase (1:2000, Dako, Denmark) and detected using the Immun-Star HRP Substrate Kit (Bio-Rad). Quantification of the signals on the membranes was carried out by using ImageJ program.

Agrobacterium-Mediated Targeted Gene Deletion of *ZtYAP1*

Approx. 700-bp and 1400-bp of flanking fungal genomic DNA were amplified upstream and downstream of *YAP1* (Zt35076) opening reading frame by PCR, respectively, and cloned into vector pCHYG containing the hph cassette conferring resistance to hygromycin [39]. The final plasmid pCHYG-YAP1 was sequenced to confirm the correct insert. The plasmid was transformed to Agrobacterium strain Agl-1 via the freeze-thaw method. Agrobacterium transformation of *Z. tritici* was performed as described [39]. Fungal genomic DNA was isolated and targeted insertion of the T-DNA was initially confirmed by PCR on genomic DNA directed against Hph and *ZtYAP1*. All the primers used for study of fungal mutant are listed in Additional file 5.

Characterization of *ZtYAP1* Mutant

Preparation of spores of *ZtYAP1* mutant and inoculation of 14-day-old wheat cv. Sevin seedling were performed as described above to evaluate fungal pathogenicity. Three independent pots of plant (ten leaves per pot) were prepared for inoculation with the wild type, the mutant or water, serving as three biological replicates. Leaves were photographed at 10, 12 and 15 days. Infected leaves were harvested at 21 dai. The spores were washed out from leaves and counted under microscope.

Sensitivity assay to oxidative stress was conducted by applying a 5 μL droplet of fungal spore suspension onto PDA containing compounds at appropriate concentrations and incubating under constant fluorescent light. Photographs of relative colony densities were taken after 6 days.

ACKNOWLEDGEMENTS

This work was funded by a postdoctoral grant from the Danish Research Council for Technology and Production (11–105997) to Dr. Fen Yang.

AUTHORS' CONTRIBUTIONS

FY designed the experiments, performed plant inoculation, protein extraction, cloning, and fungal mutant characterization, coordinated the study, and drafted and finalized the manuscript. WL performed bioinformatics analysis of the proteome data. MD and JJR performed agrobacterium-mediated transformation and fungal mutant generation. MRL and GP performed LC-MS/MS analysis. All authors have read and approved the manuscript.

REFERENCES

1. Keon J, Antoniw J, Carzaniga R, Deller S, Ward JL, Baker JM, et al. Transcriptional adaptation of *Mycosphaerella graminicola* to programmed cell death (PCD) of its susceptible wheat host. Mol Plant Microbe Interact. 2007;20:178–93.
2. Rudd JJ, Kanyuka K, Hassani-Pak K, Derbyshire M, Andongabo A, Devonshire J, et al. Transcriptome and metabolite profiling the infection cycle of *Zymoseptoria tritici* on wheat (*Triticum aestivum*) reveals a biphasic interaction with plant immunity involving differential pathogen chromosomal contributions, and a variation on the hemibiotrophic lifestyle definition. Plant Physiol. 2015;167:1158–85.
3. Marshall R, Kombrink A, Motteram J, Loza-Reyes E, Lucas J, Hammond-Kosack KE, et al. Analysis of two in *planta* expressed LysM effector homologs from the fungus *Mycosphaerella graminicola* reveals novel functional properties and varying contributions to virulence on wheat. Plant Physiol. 2011;156:756–69.
4. Yang F, Li WS, Jørgensen HJ. Transcriptional reprogramming of wheat and the hemibiotrophic pathogen *Septoria tritici* during two phases of the compatible interaction. PLoS One. 2013;8, e81606.
5. Pechanova O, Hsu CY, Adams JP, Pechan T, Vandervelde L, Drnevich J, et al. Apoplast proteome reveals that extracellular matrix contributes to multistress response in poplar. BMC Genomics. 2010;11:674.
6. Lee WS, Rudd JJ, Hammond-Kosack KE, Kanyuka K. *Mycosphaerella graminicola* LysM effector-mediated stealth pathogenesis subverts recognition through both CERK1 and CEBiP homologues in wheat. Mol Plant Microbe Interact. 2014;27:236–43.
7. Rudd JJ, Keon J, Hammond-Kosack KE. The Wheat mitogen-activated protein kinases TaMPK3 and TaMPK6 are differentially regulated at multiple levels during compatible disease interactions with *Mycosphaerella graminicola*. Plant Physiol. 2008;147:802–15.

8. Shetty NP, Kristensen BK, Newman MA, Møller K, Gregersen PL, Jørgensen HJ. Association of hydrogen peroxide with restriction of *Septoria tritici* in resistant wheat. Physiol Mol Plant Pathol. 2003;62:333–46.
9. Shetty NP, Mehrabi R, Lütken H, Haldrup A, Kema GH, Collinge DB, et al. Role of hydrogen peroxide during the interaction between the hemibiotrophic fungal pathogen *Septoria tritici* and wheat. New Phytol. 2007;174:637–47.
10. Yang F, Melo-Braga MN, Larsen MR, Jørgensen HJ, Palmisano G. Battle through signaling between wheat and the fungal pathogen *Septoria tritici* revealed by proteomics and phosphoproteomics. Mol Cell Proteomics. 2013;12:2497–508.
11. Kema GH, van der Lee TA, Mendes O, Verstappen EC, Lankhorst RK, Sandbrink H, et al. Large-scale gene discovery in the septoria tritici blotch fungus *Mycosphaerella graminicola* with a focus on *in planta* expression. Mol Plant Microbe Interact. 2008;21:1249–60.
12. Toone WM, Morgan BA, Jones N. Redox control of AP-1- like factors in yeast and beyond. Oncogene. 2001;20:2336–46.
13. Lin CH, Yang SL, Chung KR. The YAP1 homolog-mediated oxidative stress tolerance is crucial for pathogenicity of the necrotrophic fungus *Alternaria alternata* in citrus. Mol Plant Microbe Interact. 2009;22:942–52.
14. Husted S, Schjoerring JK. Apoplastic pH and ammonium concentration in leaves of *Brassica napus* L. Plant Physiol. 1995;109:1453–60.
15. Mattsson M, Schjørring JK. Senescence-induced changes in apoplastic and bulk tissue ammonia concentrations of ryegrass leaves. New Phytol. 2003;160:489–99.
16. Floerl S, Druebert C, Majcherczyk A, Karlovsky P, Kües U, Polle A. Defence reactions in the apoplastic proteome of oilseed rape (*Brassica napus* var. *napus*) attenuate *Verticillium longisporum* growth but not disease symptoms. BMC Plant Biol. 2008;8:129.
17. Floerl S, Majcherczyk A, Possienke M, Feussner K, Tappe H, Gatz C, et al. *Verticillium longisporum* infection affects the leaf apoplastic proteome, metabolome, and cell wall properties in *Arabidopsis thaliana*. PLoS One. 2012;7, e31435.
18. Delaunois B, Colby T, Belloy N, Conreux A, Harzen A, Baillieul F, et al. Large-scale proteomic analysis of the grapevine leaf apoplastic fluid reveals mainly stress-related proteins and cell wall modifying enzymes. BMC Plant Biol. 2013;13:24.

19. Witzel K, Shahzad M, Matros A, Mock HP, Muhling KH. Comparative evaluation of extraction methods for apoplastic proteins from maize leaves. Plant Methods. 2011;7:48.

20. Djordjevic MA, Oakes M, Li DX, Hwang CH, Hocart CH, Gresshoff PM. The glycine max xylem sap and apoplast proteome. J Proteome Res. 2007;6:3771–9.

21. Soares NC, Francisco R, Ricardo CP, Jackson PA. Proteomics of ionically bound and soluble extracellular proteins in *Medicago truncatula* leaves. Proteomics. 2007;7:2070–82.

22. Casasoli M, Spadoni S, Lilley KS, Cervone F, De Lorenzo G, Mattei B. Identification by 2-D DIGE of apoplastic proteins regulated by oligogalacturonides in *Arabidopsis thaliana*. Proteomics. 2008;8:1042–54.

23. Köllner TG, Schnee C, Li S, Svatoš A, Schneider B, Gershenzon J, et al. Protonation of a neutral (S)-β-bisabolene intermediate is involved in (S)-β-macrocarpene formation by the maize sesquiterpene synthases TPS6 and TPS11. J Biol Chem. 2008;283:20779–88.

24. Schmelz EA, Kaplan F, Huffaker A, Dafoe NJ, Vaughan MM, Ni X, et al. Identity, regulation, and activity of inducible diterpenoid phytoalexins in maize. Proc Natl Acad Sci U S A. 2011;108:5455–60.

25. Agrawal GK, Jwa NS, Lebrun MH, Job D, Rakwal R. Plant secretome: unlocking secrets of the secreted proteins. Proteomics. 2010;10:799–827.

26. Chivasa S, Simon W, Yu XL, Yalpani N, Slabas A. Pathogen elicitor-induced changes in the maize extracellular matrix proteome. Proteomics. 2005;5:4894–904.

27. Cheng FY, Blackburn K, Lin YM, Goshe MB, Williamson JD. Absolute protein quantification by LC/MSE for global analysis of salicylic acid-induced plant protein secretion responses. J Proteome Res. 2009;8:82–93.

28. Yang F, Jensen JD, Svensson B, Jørgensen HJ, Collinge DB, Finnie C. Secretomics identifies *Fusarium graminearum* proteins involved in the interaction with barley and wheat. Mol Plant Pathol. 2012;13:445–53.

29. Bolton MD. Primary metabolism and plant defense-fuel for the fire. Mol Plant Microbe Interact. 2009;22:487–97.

30. Underwood W. The plant cell wall: a dynamic barrier against pathogen invasion. Frontier Plant Sci. 2012;3:85.

31. Song Y, Zhang C, Ge W, Zhang Y, Burlingame AL, Guo Y. Identification of NaCl stress-responsive apoplastic proteins in rice shoot stems by 2D-DIGE. J Proteomics. 2011;74:1045–67.

32. Le Roch KG, Johnson JR, Florens L, Zhou Y, Santrosyan A, Grainger M, et al. Global analysis of transcript and protein levels across the*Plasmodium falciparum* life cycle. Genome Res. 2004;14:2308–18.

33. Doehlemann G, Hemetsberger C. Apoplastic immunity and its suppression by filamentous plant pathogens. New Phytol. 2013;198:1001–16.

34. Brown AJ, Budge S, Kaloriti D, Tillmann A, Jacobsen MD, Yin Z, et al. Stress adaptation in a pathogenic fungus. J Exp Biol. 2014;217:144–55.

35. Lev S, Hadar R, Amedeo P, Baker S, Yoder OC, Horwitz BA. Activation of an AP-1-like transcription factor of the maize pathogen*Cochliobolus heterostrophus* in response to oxidative stress and plant signals. Eukaryot Cell. 2005;4:443–54.

36. Montibus M, Ducos C, Bonnin-Verdal M-N, Bormann J, Ponts N, Richard- Forget F, et al. The bZIP transcription factor Fgap1 mediates oxidative stress response and trichothecene biosynthesis but not virulence in *Fusarium graminearum*. PLoS One. 2013;8:e83377.

37. Shetty NP, Jensen JD, Knudsen A, Finnie C, Geshi N, Blennow A, et al. Effects of β-1,3-glucan from *Septoria tritici* on structural defence responses in wheat. J Exp Bot. 2009;60:4287–300.

38. Yang F, Jensen JD, Svensson B, Jørgensen HJ, Collinge DB, Finnie C. Analysis of early events in the interaction between *Fusarium graminearum* and the susceptible barley *(Hordeum vulgare)* cultivar Scarlett. Proteomics. 2010;10:3748–55.

39. Motteram J, Küfner I, Deller S, Brunner F, Hammond-Kosack KE, Nürnberger T, et al. Molecular characterization and functional analysis of MgNLP, the sole NPP1 domain-containing protein, from the fungal wheat leaf pathogen *Mycosphaerella graminicola*. Mol Plant Microbe Interact. 2009;22:790–9.

CITATION

CHAPTER 1

Chung et al., Resistance loci affecting distinct stages of fungal pathogenesis: use of introgression lines for QTL mapping and characterization in the maize - Setosphaeria turcica pathosystem BMC Plant Biology 2010, 10:103.

CHAPTER 2

Raquel González-Fernández, Elena Prats, and Jesús V. Jorrín-Novo, "Proteomics of Plant Pathogenic Fungi,"Journal of Biomedicine and Biotechnology, vol. 2010, Article ID 932527, 36 pages, 2010. doi:10.1155/2010/932527.

CHAPTER 3

Richard J O'Connell, Michael R Thon, Stéphane Hacquard, Stefan G Amyotte, Jochen Kleemann, Maria F Torres, Ulrike Damm, Ester A Buiate, Lynn Epstein, Noam Alkan, Janine Altmüller, Lucia Alvarado-Balderrama, Christopher A Bauser, Christian Becker, Bruce W Birren, Zehua Chen, Jaeyoung Choi, Jo Anne Crouch, Jonathan P Duvick, Mark A Farman, Pamela Gan, David Heiman, Bernard Henrissat, Richard J Howard, Mehdi Kabbage, "Lifestyle transitions in plant pathogenic Colletotrichum fungi deciphered by genome and transcriptome analyses," Nature Genetics 44, 1060–1065 (2012) doi:10.1038/ng.2372.

CHAPTER 4

Walters DR, Havis ND, Paterson L, Taylor J, Walsh DJ and Sablou C (2014) Control of foliar pathogens of spring barley using a combination of resistance elicitors. Front. Plant Sci. 5:241. doi: 10.3389/fpls.2014.00241.

CHAPTER 5

Asghar Heydari and Mohammad Pessarakli, 2010. A Review on Biological Control of Fungal Plant Pathogens Using Microbial Antagonists. Journal of Biological Sciences, 10: 273-290.

CHAPTER 6

Shaowu Meng, Trudy Torto-Alalibo, Marcus C Chibucos, Brett M Tyler and Ralph A Dean, "Common processes in pathogenesis by fungal and oomycete plant pathogens, described with Gene Ontology terms," BMC Microbiology20099(Suppl 1):S7, DOI: 10.1186/1471-2180-9-S1-S7.

CHAPTER 7

Kalunke RM, Tundo S, Benedetti M, Cervone F, De Lorenzo G and D'Ovidio R (2015) An update on polygalacturonase-inhibiting protein (PGIP), a leucine-rich repeat protein that protects crop plants against pathogens. Front. Plant Sci. 6:146. doi: 10.3389/fpls.2015.00146.

CHAPTER 8

Fen Yang, Wanshun Li, Mark Derbyshire, Martin R Larsen, Jason J Rudd and Giuseppe Palmisano, "Unraveling incompatibility between wheat and the fungal pathogen Zymoseptoria tritici through apoplastic proteomics," BMC Genomics201516:362, DOI: 10.1186/s12864-015-1549-6.

INDEX

J

L

M

N

O

P

Q

R

S

T

U

Y

Encyclopaedia of GENETIC RESOURCES, CHROMOSOME ENGINEERING, AND CROP IMPROVEMENT

Encyclopaedia of GENETIC RESOURCES, CHROMOSOME ENGINEERING, AND CROP IMPROVEMENT

Volume III: Managing Global Genetic Resources: Agricultural Crop Issues and Policies

Contributors

Teresa Cuellar,Chia-Lin Chung,and Luciano Lourenço Nass et al.

AURIS Reference

www.aurisreference.com

Encyclopaedia of Genetic Resources, Chromosome Engineering, and Crop Improvement

Volume III: Managing Global Genetic Resources: Agricultural Crop Issues and Policies

Contributors: Teresa Cuellar,Chia-Lin Chung,and Luciano Lourenço Nass et al.

Published by Auris Reference Limited

www.aurisreference.com

United Kingdom

Printed in 2017 for Sale in the Indian Subcontinent

Managing Global Genetic Resources: Agricultural Crop Issues and Policies

ISBN: 978-1-78154-957-5 (Set of 3 Volumes)

British Library Cataloguing in Publication Data
A CIP record for this book is available from the British Library

Printed in the United Kingdom
Exclusively distributed by CBS Publishers & Distributors Pvt. Ltd.

Sales & Distribution Rights only for India, Pakistan, Bangladesh, Sri Lanka, Nepal and Bhutan.This book is not to be sold outside these territories.

Contents

List of Abbreviations

APSIM	Agricultural Production Systems Simulator
AIC	Akaike information criterion
ASC	Alternative-specific constants
BIC	Bayesian information criterion
CT	Canopy temperature
CE	choice experiment
CSA	Climate-smart agriculture
CGIAR	Consultative Group on International Agricultural Research
CWR	Crop wild relatives
FGB	field gene banks
CGN	Genetic Resources of the Netherlands
GMO	Genetically Modified Organisms
IAC	Instituto Agronômico de Campinas
IFPRI	International Food Policy Research Institute's
IMPACT	International Model for Policy Analysis of Agricultural Commodities and Trade
KSEP	Key Science Engineering Program
LN	liquid nitrogen
LCFS	Low carbon fuel standard
MOA	Ministry of Agriculture
MOST	Ministry of Science and Technology
MS	Murashige and Skoog
NAGS	National Active Germplasm Sites
NBPGR	National Bureau of Plant Genetic Resources
NGOs	nongovernmental organizations
NDVI	Normalized Difference Vegetation Index
PGR	plant genetic resources
PGRFA	Plant Genetic Resources for Food and Agriculture
PCR	polymerase chain reaction
PEs	Primary somatic embryos
SFTPRC	Special Foundation of Transgenic Plants Research and Commercialization
SDPC	State Development and Planning Commission
SSTC	State Science and Technology Commission
SGSV	Svalbard Global Seed Vault
SGRP	System-Wide Genetic Resources Programme
TCD	Trinity College Dublin
VPD	vapour pressure deficit
WTP	willingness to pay

List of Contributors

Luciano Lourenço Nass
Empresa Brasileira de Pesquisa Agropecuária (Embrapa), Secretaria de Relações Internacionais, 70.770-901, Brasília, DF, Brazil

Mário Sérgio Sigrist
DuPont Pioneer - Pesquisa Soja, Rodovia DF 250, km 20, 73.301-970, Brasília, DF, Brazil

Cláudia Silva da Costa Ribeiro
Embrapa Hortaliças, Rodovia Brasília/Anápolis BR 060, km 09, 70.359-970, Gama, DF, Brazil

Francisco José Becker ReifschneiderI
Embrapa Hortaliças, Rodovia Brasília/Anápolis BR 060, km 09, 70.359-970, Gama, DF, Brazil

Rukhsar Ahmad Dar
KVK, Neyoma, Ladakh, India

Mushtaq Ahmad
Division of Genetics and Pant Breeding, Sher-e-Kashmir University of Agricultural Sciences and Technology, Kashmir Shalimar campus, Srinagar -191 121, India

Sanjay Kumar
KVK, Neyoma, Ladakh, India

Monica Reshi
KVK, Neyoma, Ladakh, India

Kerri L Steenwerth
Crops Pathology and Genetics Research Unit, c/o Department of Viticulture and Enology, United States Department of Agriculture (ARS/USDA), Agricultural Research Service

Amanda K Hodson
Department of Land, Air and Water Resources, University of California at Davis

Arnold J Bloom
Department of Plant Sciences, University of California at Davis

Michael R Carter
Department of Agricultural and Resource Economics, University of California at Davis

Andrea Cattaneo
Climate Smart Agriculture Project, Food and Agriculture Organization of the U.N., Viale delle Terme di Caracalla

Colin J Chartres
eWater, University of Canberra Innovation Centre, University Drive South

Jerry L Hatfield
National Laboratory for Agriculture and the Environment, ARS/USDA

Kevin Henry
Where the Rain Falls, CARE France
School of Global Envirfonmental Sustainability, Colorado State University

Jan W Hopmans
Department of Land, Air and Water Resources, University of California at Davis

William R Horwath
Department of Land, Air and Water Resources, University of California at Davis

Bryan M Jenkins
Department of Biological and Agricultural Engineering, University of California at Davis

Ermias Kebreab
Department of Animal Science, University of California at Davis

Rik Leemans
Environmental Sciences, Wageningen University

Leslie Lipper
Agricultural and Development Economic Analysis Division, Food and Agriculture Organization of the U.N., Viale delle Terme di Caracalla

Mark N Lubell
Department of Environmental Science and Policy, Center for Environmental Policy and Behavior, University of California at Davis

Siwa Msangi
Environment and Production Technology Division, International Food Policy Research Institute (IFPRI)

Ravi Prabhu
World Agroforestry Center (ICRAF)

Matthew P Reynolds
Plant, International Maize and Wheat Improvement Center, Consultative Group on International Agricultural Research (CGIAR) Apdo, Postal

Samuel Sandoval Solis
Department of Land, Air and Water Resources, University of California at Davis

William M Sischo
Food- and Water-borne Disease Research Program, College of Veterinary Medicine, Washington State University

Michael Springborn
Department of Environmental Science and Policy, University of California at Davis

Pablo Tittonell
Plant Sciences, Wageningen University

Stephen M Wheeler
Department of Landscape Architecture, University of California at Davis

Sonja J Vermeulen
Department of Plant and Environmental Sciences, Climate Change, Agriculture and Food Security, Consultative Group on International Agricultural Research (CGIAR), University of Copenhagen

Eva K Wollenberg
Gund Institute for Ecological Economics and Rubenstein School of Environment and Natural Resources, University of Vermont

Louise E Jackson
Department of Land, Air and Water Resources, University of California at Davis

Mohammed Kasso
Department of Zoological Sciences, Addis Ababa University, Addis Ababa, Ethiopia

Mundanthra Balakrishnan
Department of Zoological Sciences, Addis Ababa University, Addis Ababa, Ethiopia

Eija Pouta
MTT Agrifood Research Finland, Helsinki, Finland

Annika Tienhaara
MTT Agrifood Research Finland, Helsinki, Finland

Heini Ahtiainen
MTT Agrifood Research Finland, Helsinki, Finland

Jikun Huang
Chinese Academy of Sciences

Qinfang Wang
Chinese Academy of Sciences

K. Nirmal Babu
Indian Institute of Spices Research, Kerala

D. Minoo
Providence Women's College, Kerala India

G. Yamuna
Indian Institute of Spices Research, Kerala

K. Praveen
Indian Institute of Spices Research, Kerala

P.N. Ravindran
Indian Institute of Spices Research, Kerala

K.V. Peter
Indian Institute of Spices Research, Kerala

Maia Akhalkatsi
Ilia State University Georgia

Jana Ekhvaia
Ilia State University Georgia

Zezva Asanidze
Ilia State University Georgia

Ola T. Westengen
Nordic Genetic Resource Center (NordGen), Alnarp, Sweden
Centre for Development and the Environment, University of Oslo, Oslo, Norway

Simon Jeppson
Nordic Genetic Resource Center (NordGen), Alnarp, Sweden

Luigi Guarino
The Global Crop Diversity Trust, Bonn, Germany

Preface

Encyclopaedia of Genetic Resources, Chromosome Engineering, and Crop Improvement summarizes landmark research and describes medicinal plants as nature's pharmacy. Genetic resources (GRs) refer to genetic material of actual or potential value. Genetic material is any material of plant, animal, microbial or other origin containing functional units of heredity. The text Managing Global Genetic Resources: Agricultural Crop Issues and Policies examines the structure that underlies efforts to preserve genetic material, including the worldwide network of genetic collections, the role of biotechnology, and a host of issues that surround management and use. First chapter emphasizes the importance of plant genetic resources and discusses about aspects of costs involved in conservation and suggest recommendations for strengthening the area in Brazil. Second chapter focuses on agriculture germplasm resources. Third chapter deals with climate-smart agriculture global research agenda. Fourth chapter discusses ex situ conservation of biodiversity with particular emphasis to Ethiopia. The aim of fifth chapter is to emphasize the importance of in situ conservation of native cattle breeds and plant varieties in developing conservation policies. The objectives of sixth chapter are to review the status of China's agricultural biotechnology research and commercialization, and to gain a better understanding of China's policies governing both agricultural biotechnology research and its applications (or commercialization). Cryopreservation of spices genetic resources has been presented in seventh chapter. Diversity and genetic erosion of ancient crops and wild relatives of agricultural cultivars for food have been proposed in last chapter.

Chapter 1

GENETIC RESOURCES: THE BASIS FOR SUSTAINABLE AND COMPETITIVE PLANT BREEDING

Luciano Lourenço Nass[1], Mário Sérgio Sigrist[2]; Cláudia Silva da Costa Ribeiro[3]; Francisco José Becker Reifschneider[1,3]

[1]Empresa Brasileira de Pesquisa Agropecuária (Embrapa), Secretaria de Relações Internacionais, 70.770-901, Brasília, DF, Brazil

[2]DuPont Pioneer - Pesquisa Soja, Rodovia DF 250, km 20, 73.301-970, Brasília, DF, Brazil

[3]Embrapa Hortaliças, Rodovia Brasília/Anápolis BR 060, km 09, 70.359-970, Gama, DF, Brazil

ABSTRACT

Plant genetic resources are the fuel for breeding, which in the search for higher yield and adapted genotypes, manipulates genes in order to meet the needs of farmers, and especially, of the current market. However, the use of accessions available in germplasm banks is low. Topics discussed in this paper emphasize the importance of plant genetic resources, and warn about problems related to genetic vulnerability; also, they discuss about aspects of costs involved in conservation and suggest recommendations for strengthening the area in Brazil.

INTRODUCTION

The Importance of Agrobiodiversity, an Unexplored Treasure

Agriculture and biodiversity are intimately linked to each other. Biodiversity in agriculture is essential for human development. Agricultural biodiversity, or agrobiodiversity, is a generic term which basically includes all components of biological diversity (plants, animals, microorganisms) that are important for food and agriculture itself, as well as all components of ecological biodiversity that make up agroecosystems. The importance of agricultural biodiversity is evident, since it supplies food, wood, fiber, oil, medicine, and fuel. Furthermore, agrobiodiversity contributes to ecosystem services, being the conservation of water and soil the most known, as well as pollination.

The beginning of agriculture 10,000 years ago is undoubtedly one of the major events in the history of human life on Earth. Approximately 300,000 species of plants have been described, of which 3,000 have been used by humans for food. Currently, around 300 species are used, and out of these, only 15 are responsible for 90% of all human food. The 15 most used species are rice, wheat, corn, sorghum, barley, sugarcane, beet, potato, sweet potato, cassava, common bean, soybean, peanut, coconut, and banana (Goodman 1990). Together, rice, potato, corn and wheat, represent 60% of this total and are the staples of human consumption.

Brazil, as a megadiverse country, has extensive economic opportunity for the development of new food, fiber, drugs products; these unique opportunities also come accompanied by greater responsibility (Silva et al. 2011) to national and international society. The importance and the economic potential of this heritage for current and future Brazilian generations are monumental. For a significant portion of the population, agricultural biodiversity is the primary income source. In 2010, Brazil's agricultural gross domestic product (agricultural GDP) reached US$ 821 billion, a total GDP of US$3.7 trillion (http://www.cepea.esalq.usp.br/pib/); and millions of Brazilians rely on jobs related to agriculture. Historically, Brazil's agriculture has been one of the primary engines of economic growth. In 2010, the three main export items were soybean, iron ore and oil. Besides, coffee and meats were also among the most important agricultural export products.

Agriculture, when properly carried out, produces a range of environmental services and contributes significantly to the conservation and use of biodiversity. At the same time, agriculture can generate negative impacts to this same biodiversity, by the expansion of the agricultural frontier, the misuse of available technologies, or simply by greed, which is not controlled by the political system in which we live. This loss of diversity is a concern, since it puts at risk the agriculture and ecosystem services provided by it.

Brazil is the leading country among the megadiverse countries, holding in its territory an estimate of 13% of all species on the planet. Over 20% of the world flora, in a total of approximately 55,000 described species, are found in Brazil. Despite having the largest biodiversity on the planet, with a huge range of native species,Brazilian agriculture is highly dependent on exotic species[1], both from the Americas and from other continents.

Brazil's agriculture is based, among others, on: sugar cane, from New Guinea; coffee, from Ethiopia; rice, from the Asian continent; soybeans and oranges, from China; corn, from Mexico; wheat, from Minor Asia. Several native species are important for human consumption with regional and local importance, such as cassava, pineapple, peanuts, cocoa, cashew, cupuaçu, passion fruit, guaraná, among others. Native forage species also contribute to the support of a good part of the livestock sector in Brazil (MMA 2011). Livestock depends on cattle from India and forage grasses brought originally from Africa. Fish farming depends on tilapia, from Eastern Africa, and carp, brought from China; moreover, apiculture and pollination of major crops are based on Africanized bees. However, this is not a Brazilian weakness, since the dependence of exotic genetic resources is a global phenomenon.

Brazilian agriculture would never have reached today´s stage without a systematic and growing import of these genetic resources for food, oil, fiber and energy. There are many good examples of this rich history (Reifschneider et al. 2010): it is noteworthy the introduction of cattle in Brazil, and the germplasm introduction and adaptation gardens of the 16th century and beyond, exemplified by the Quinta do Tanque, demonstrating that the concern with agrobiodiversity in Brazil is historic!

The «Quinta do Tanque» The Quinta do Tanque, located in Salvador, Bahia, is considered one of the most important civilian monuments of Brazil. Its history begins in 1555, when the Jesuits built a cottage for the college of Bahia. A Quinta, measuring about two acres, had large gardens and a dam, or a tank. There, the Jesuits cultivated fruits and vegetables in an orchard irrigated by springs and precious water reservoirs, which is the origin of the name Quinta do Tanque. At the Quinta, plants from Europe, Asia and America were also cultivated. Later, the Quinta also served as experimental garden, in which species from all over the world were tested and selected. The cultivation of cinnamon in Brazil started at the Quinta, and cocoa started to be grown in Bahia (which was previously grown in Maranhão) by intervention of the Jesuits (Reifschneider et al. 2010).

The genetic vulnerability of agriculture nowadays and the successful use of native and exotic germplasm

The use of 15 species that represents 90% of all human food makes clear the narrow basis on which we depend. A limited number of species that we use for our livelihood presents a huge concern about the genetic vulnerability of agriculture; a concern that once belonged to researchers, but which today permeates society. Despite the conservation of a vast amount of genetic variability in germplasm collections or banks, the economically important crops continue becoming more uniform. Thus, despite all the efforts made in the establishment of germplasm banks in the international arena, this was not enough to make the world agriculture less vulnerable to diseases and pests. The existence of this vulnerability is due to the use of uniform genotypes in extensive farming areas, and there are many examples well-studied examples of problems arising from such genetic uniformity. The disaster involving the use of potato clones susceptible to the fungus *Phytophthora infestans* in Ireland, the ruin of the grapevines caused by an insect parasite that feed on root of grapevines (*Phylloxera*) observed in France, and the use of corn hybrids with only one source of male-sterility susceptible to *Helminthosporium maydis* race T fungus are classic examples of genetic vulnerability.

Nevertheless, there are many examples of successful use of native and exotic germplasm to support the development of national agriculture. The research in plant breeding and genetic resources is one of the most relevant innovation activities for the country, having produced results that have contributed significantly to the main qualitative and quantitative gains achieved by Brazilian agriculture - and Brazilian development - over the past decades. Plant breeding in Brazil is among the best in the world, with significant contributions, highlighting the well-trained human resources and the development of a large diversity of plants adapted to tropical conditions (Queiroz and Lopes 2007). Thus, it is important to mention the tropicalization of soybean, which allowed its cultivation in low latitude regions; the development of numerous varieties of cassava and beans with distinctive traits, including resistance to several pathogens; and the new cultivars of *Capsicum* peppers and peanuts (*Arachis*) for forage.

All this work aiming at the sustainable use of germplasm is based on the existence of collections or germplasm banks, used by the present generation and maintained for the future generations, and properly characterized so they can be effectively used in plant breeding. The availability of germplasm is of fundamental importance for the improvement of any species. Thus, every breeding program depends, ultimately, on genetic resources from germplasm banks or from those that are in use by farmers. Nass (2001) discussed in details the use of plant genetic resources for breeding.

The Importance of Plant Genetic Resources

Plant genetic resources (PGR) are the basis of food security and global energy. It is essential that these resources are properly preserved and characterized for the current and future demand, since they serve as raw material for plant breeding. Given their importance, it is expected that issues related to PGR are discussed frequently in various media in order to encourage debate in society. The degree of importance that society gives to certain subjects can be estimated, for example, through field surveys using structured questionnaires. However, a research of this kind has a high cost to obtain acceptable margins of error. Alternatively, the degree of exposure of a particular subject in the media can be used as a proxy for its significance to society. This latter approach was used here to illustrate the attention given to PGR by scientific journals. We evaluated papers published by five scientific journals of wide circulation nationally and internationally: PAB (Pesquisa Agropecuária Brasileira), CBAB (Crop Breeding and Applied Biotechnology), Horticultura Brasileira, Crop Science and HortScience. All articles published from January 2008 to the most recent volume were evaluated, regardless of the section in which the articles were allocated, making up a total of 4,777 papers analyzed. For each volume, the total number of published papers and the number of papers related to PGR were counted and classified by scanning the titles of articles, seeking for keywords such as *germplasm, accession* and *pre-breeding*. Abstracts were read in case of doubt. Articles related to PGR were then divided into three categories in order to identify the main objective of the study: *conservation methods*, *evaluation/characterization* and *use in breeding*. Manuscripts involving tissue culture, collection techniques, or improved cryopreservation protocols used in germplasm banks were classified as*conservation methods*. In turn, articles included in the category *evaluation/characterization* presented as objectives the characterization and evaluation of collections or sets of materials using molecular markers or agromorphological descriptors, and also the evaluation of genotypes regarding biotic and abiotic stresses. Finally, examples of manuscripts in the category use in breeding involved the use of native or exotic germplasm to develop populations and lines from backcrossing programs.

In general, there was a variation in the quantity of articles associated to PGR in relation to the journals. However, when the same journal was analyzed over the years, the number of articles related to genetic resources remained almost stable (Figure 1). These results suggest that, given the time horizon considered, there were no major changes related to the importance given to the subject PGR. Nevertheless, the results reflect the differences of the proposed objectives for each journal, which may be more focused on genetics and conservation or on the management and cultivation of crops. In absolute terms, the international journals evaluated showed a greater amount of published articles on genetic resources. In part, this happens due to the fact that these journals have specific sections for PGR in each volume, which indicates that the importance of this subject is consolidated in such journals. However, the evaluation of the importance given to the topic brings biased results in favor of international journals, as they also showed higher total number of papers. Thus, it is necessary to evaluate the importance given in relative terms (Figure 2). In this case, CBAB had the highest proportion of articles related to the topic, around 30%, followed by international journals, with approximately 20% each.

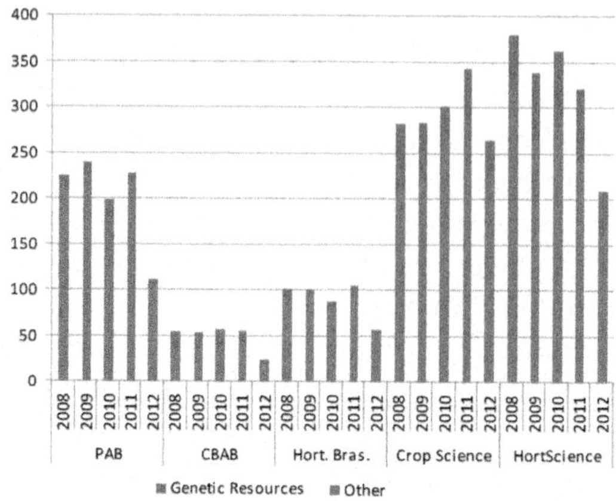

Figure 1: Amount of articles regarding plant genetic resources in relation to the total number of papers published according to the journal and year.

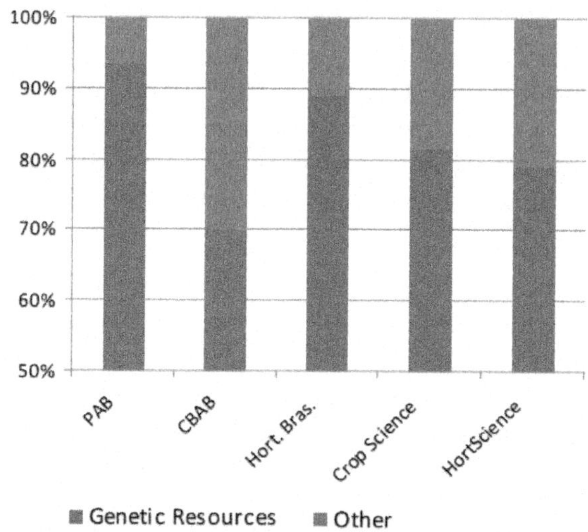

Figure 2: Percentages of articles related to plant genetic resources by journal, grouping all the years evaluated.

Considering only those papers which discussed genetic resources, it was verified that the majority aimed at the evaluation or characterization of the genetic diversity of germplasm, followed by the use of the PGR in breeding, and finally, the development of methods and techniques of conservation (Figure 3). In general, this ranking of objectives remained consistent, both over the five years of evaluation as well as in relative terms within each journal (Figure 4).

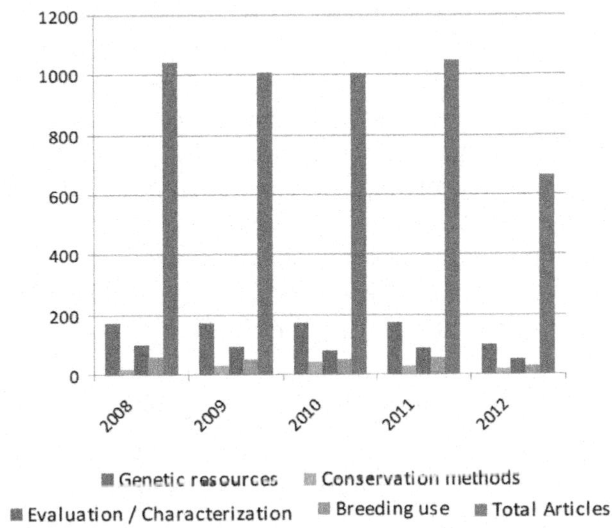

Figure 3: Annual quantity of articles related to plant genetic resources according to the proposed objectives, grouping all journals.

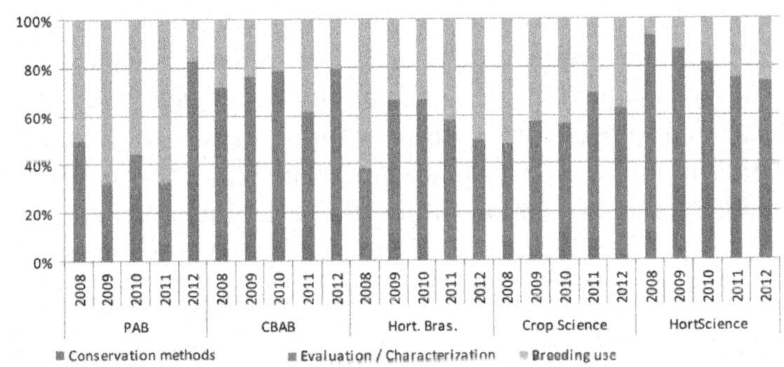

Figure 4: Percentage of articles' objective related to plant genetic resources according to the year and the journal.

Based on the results, it can be considered that, although the journals evaluated dedicate a space for studies related to PGR, the number of manuscripts focused on this topic could definitely increase. In a way, this should be the likely trend for the coming years, mainly due to the increasing importance of PGR in relation to intellectual property and as an alternative to the search for genes resistant to biotic and abiotic factors to face the expected consequences of t climate change. Considering the papers which were focused on PGR, it is clear the need for a greater number of studies aimed at improvement and innovation in conservation methods and techniques, which is important for reducing operating costs in germplasm collections and banks. The major weight of studies aimed at the characterization of germplasm was expected due to the amount of unknown accessions and reduced costs of using molecular markers. However, it is necessary to question the real need to characterize all genotypes stored in banks, although the characterization is one of the conditions to promote the use of PGR. Based on the amount of accessions currently stored and on the costs involved in phenotypic and molecular characterization, it is necessary to set priorities for a more effective allocation of the limited financial resources of the collections (Koo and Wright 2008).

Brazilian Germplasm Collections - a Snapshot

Until the 1940s, the centers of origin of cultivated plants were considered unlimited sources of genetic variability. The expansion of the agricultural frontier without the concern for preserving the environment and natural resources endangered wild relatives of many domesticated species. The possibility of exhaustion of genetic variability of many species and related wild species led the scientific community, in the late 1960s, to promote the conservation of the hereditary material of many plant species, especially those of agricultural importance, under controlled environmental conditions (Ribeiro 2000). The routine activities of germplasm banks like collection, characterization, evaluation, documentation and conservation of accessions require qualified professionals from various fields of knowledge; these activities present high cost and the return is almost always in long term (Nass 2007). Besides the conservation of genetic variability for future use, another objective is that accessions available are used by their customers.

The first Brazilian germplasm collections were organized in research institutes and universities aiming primarily to support several existing breeding programs. Among them, the contribution of the Instituto Agronômico de Campinas (IAC) in the development of agriculture in São Paulo and in Brazil is indisputable and is due to the research on genetic resources and the creation of active germplasm banks (AGBs) of various plant species, such as coffee, sugarcane, rubber, cassava, rice, soybean, common bean, various fruits, vegetables, and ornamental plants. IAC›s coffee AGB was established in 1932 at Fazenda Santa Elisa, in Campinas, SP, aiming the research in different areas, including genetic improvement, besides conservation of germplasm.

IAC's coffee AGB (exotic species - example of successful use of the collection in breeding program, with the release of several cultivars important for the development of Brazilian coffee) IAC›s program for genetic improvement started about 80 years ago, and it has released several cultivars, which have been recommended for planting in different regions of the country. It is estimated that 90% of the Arabica type coffee in Brazil come from cultivars developed at the Institute. The increase of genetic diversity of the collection is a concern of researchers from IAC, and it has had partnership with various institutions in order to establish exchange of coffee germplasm. The short stature cultivars «Catuaí Amarelo» and «Catuaí Vermelho» have modified systems of coffee production in the country and allowed the introduction of cultivation in large areas of Cerrado in São Paulo, Minas Gerais and Goiás.

Embrapa Vegetables' Capsicum AGB (some autochthonous species - an example of successful use of CapsicumAGB in the breeding program of Embrapa Vegetables)

The Capsicum breeding program led by Embrapa Vegetables started about 30 years ago; it has a collection of more than 4,000 germplasm accessions and large genetic variability. The program involves researchers from different fields and various units of Embrapa, research and educational institutions, cooperatives, producers and processing companies. In the last 20 years lines with resistance to diseases have been made available; also, several cultivars of different types of peppers important for Brazilian agriculture have been released. One of the released cultivars is today responsible for over 50% of all the pepper sauce produced in the country.

Although the private sector has participated significantly in the development and availability of cultivars and hybrids of corn in the Brazilian market, in the public sector, IAC was one of the pioneer institutions to develop double hybrids in the late 1930›s to the 60›s (Queiroz and Lopes 2007). The corn breeding program of the Department of Genetics of the Escola Superior de Agricultura «Luiz de Queiroz» (ESALQ/USP) also made a great contribution, allowing the creation of one of the first corn germplasm collections in the country, in 1952. Research at ESALQ involved both the collection of corn landraces and Brazilian indigenous varieties and the development of cultivars. In 1975, the database was transferred to Embrapa Maize and Sorghum, in Sete Lagoas, Minas Gerais (Andrade et al. 1994). The Department of Genetics of ESALQ also pioneered the establishment of germplasm collections of vegetables, which served as the basis for genetic improvement of various vegetables, with emphasis on onion and brassica programs (Queiroz and Lopes 2007). The Universidade Federal de Viçosa (UFV) was one of the first national institutions to work with soybean breeding. The program started in 1963 and over 20 cultivars have been developed. IAC and Embrapa Soybean began studies with adaptation of soybean cultivars in the 1970s (Borém 2001, Queiroz and Lopes 2007). UFV also keeps some germplasm collections, for example, beans, soy and various vegetables.

The Empresa Pernambucana de Pesquisa Agropecuária (IPA) stood out in the national scenario for developing breeding programs oriented to the Northeast, such as cowpea (Vigna unguiculata), onion and tomato for the industry. Several bean and tomato cultivars were developed and occupied significant acreages due to different traits such as fruit setting at high temperatures, fruit firmness and resistance to several diseases (Queiroz and Lopes 2007), attributes essential for a successful crop in the Northeast. IPA maintains germplasm collections of cowpea (Assunção et al. 2005), sweet potato (Ritschel and Huaman 2002), several tropical fruits regionally important like Brazilian cherry (Eugenia uniflora), Brazilian guava (Psidium araça), yellow mombin (Spondias mombin), star fruit (Averrhoa carambola), jackfruit (Artocarpus heterophyllus), soursop (Annona muricata) , sugar apple (Annona squamosa), pomegranate (Punica granatum), «umbu» (Spondias tuberosa), guava (Pisidium guajava), sapodilla (Manilkara sapota), among others (Ferreira et al. 2005).

Since 1972, conservation of genetic resources and genetic improvement of cotton, rice, coffee, winter cereals, beans, temperate fruits, cassava, corn and sorghum have been the main lines of research at the Instituto Agronômico do Paraná (IAPAR).

According Valls (2007), the creation of the National Research Center for Genetic Resources (Cenargen, now Embrapa Genetic Resources and Biotechnology), in 1974, by the Empresa Brasileira de Pesquisa Agropecuária (Embrapa), enabled the interaction of this center, with other State and Federal institutions, in order to consolidate the philosophical basis on enrichment and long-term conservation of genetic variability. Such interaction has generated mutual benefits, being it the possibility of establishing a vast network of active germplasm banks (Table 1), or by creating safe conditions for storage of accessions in the long term, under appropriate infrastructure built in Brasília-DF. Currently, the long term collection of Embrapa (Colbase) maintains approximately 120,000 accessions, representing 222 genera and approximately 700 species. It is estimated that, in Brazil, the total number of accessions available in the active germplasm banks is around 300,000.

Using the Treasure

The development of more competitive, better adapted cultivars is a continuous process which requires adaptation to new and often unexpected challenges faced by modern agriculture. Despite the great genetic variability in germplasm collections, underutilization of these resources in genetic breeding programs have been registered in Brazil and other countries, and may have several reasons: practical problems in processing and shipping of samples; legal aspects and quarantine-related issues; lack of information and description of accessions in the banks; restricted adaptability of accessions; insufficient number of breeders to meet the large amount of crops/plant species in the country (even having an available germplasm collection), low seed quality and/or insufficient number of seeds kept in collections; difficulty of crossing exotic germplasm and improved genotypes (Nass 2001, 2011, Valls 2007).

Many plant breeders do not make direct use of the genetic resources since they consider their working collections significant, or because most accessions, despite showing some characteristic of agronomic interest, bring together many undesirable traits that are difficult to manage (linkage drag). However, pre-breeding programs have been shown to be efficient in the use of genetic resources available in AGBs for enlarging the genetic breeding programs of different species (Nass and Paterniani 2000, Nass 2001, 2011, Nass et al. 2007, Nass and Sigrist 2012). The second Brazilian report on the use of genetic resources in Brazil emphasizes several examples of the use of Brazil's collection in pre-breeding programs (Mariante et al. 2009).

Economics of conserving plant genetic resources

The demand for proper valuation of PGR has increased due to new applications offered by biotechnology. Biodiversity held in germplasm banks is the essential raw material for agribusiness and for society in general. When assigning value, one should keep in mind the expected benefit for society. Besides the use related to the production of food, fiber, medicines and bioenergy, the total value of PGR should also consider the cultural, aesthetic and even psychological aspects of the awareness of conserving something that can be extinguished (Brown 1991). But the determination of this value is not trivial, as is determining the cost/benefit of preservation.

Although there is little information about these issues, some studies have sought to reconcile the knowledge of biodiversity with economic principles to enhance and optimize the use and conservation of genetic resources (Evenson and Gollin 2003). For example, given that the budget of germplasm bank is often limited, it is necessary to decide what to conserve. If all accessions in a bank present equal value, accessions with higher maintenance costs would be undesirable. In case all of them have the same cost of conservation, materials which are most likely to be used in the present would be favored. Additionally, similar accessions have less value when compared with rare accessions (Brown 1991, Smale and Koo 2003).

Benefits: Valuation of Plant Genetic Resources

The total value resulting from the conservation of a particular accession can be categorized into non-use value and use value. Non-use, or existence value, reflects the satisfaction of individuals or societies in simply knowing that something exists and is maintained regardless of use. In turn, use value can be divided into direct and indirect. The indirect use value reflects the contribution of PGR for habitats or ecosystems, while the direct use value is the value generated by the development of food, fiber, biofuels and drugs. Both values have dimensions in the present and in the future. A third value is known by option value, which considers the value in having flexibility of use in relation to a future unexpected demand. Based on these definitions, it is noteworthy the complexity in defining, with confidence, the total value of a single accession or whole germplasm banks. Although the theory for estimating these categories have already been proposed (Brock and Xepapadeas 2003), ultimately, many model parameters may be subjective, especially in relation to non-use, indirect use and option values. This presents a difficulty, since one of the main functions of germplasm banks is to meet future challenges. Therefore, these important components of the total value of a germplasm collection are usually underestimated.

Based on various assumptions, it can be estimated the current value of future benefits of germplasm in breeding by combining the probability of finding a useful accession with its expected benefit (e.g. yield increase). However, the time necessary to explore and incorporate useful genes into agronomically elite materials affects the magnitude of the expected benefits due to the time value of money. Of all the values associated with PGR, the most tangible in practical terms is its direct use, usually measured based on financial gain generated by increased production, improved quality, resistance to biotic and abiotic stresses or other characteristics incorporated into elite materials. These benefits are originated from the use of wild relatives and landraces to introduce desirable traits by breeders, using accessions and information generated by curators of germplasm banks. Estimating the benefits of plant breeding by separating the contribution of breeders to the direct use value of germplasm is a complex task, so most studies aim at estimating the value of genetic improvement as a whole, and not only the use value of genetic resources *per se* (Rubenstein et al. 2005). In general, the direct increase in productivity achieved with the new variety, for example, can be estimated by multiplying the productivity excess, in relation to a control variety, at market price and quantity sold in a given period. This value is then compared with the total cost of the breeding program for obtaining the variety, resulting in a rate of return. Although these studies present an overview of genetic gain for production, they do not always define correctly the use value and benefits derived from genetic resources. Considering a characteristic controlled by a single gene, the estimate is easily obtained, but there is greater complexity as an increased amount of genes is introduced. Additionally, many of the estimates are based on experimental data, under controlled environmental and management conditions different from those faced by farmers (Alston et al. 1995). Finally, the supply chain and demand in agribusiness is variable, so that a single estimate of gain for the entire agribusiness system is unlikely.

In any case, the results considering the contribution of genetic resources and genetic improvement as a whole emphasize the economic importance of the activity. In most cases, the rate of return obtained from plant breeding has been extremely positive, adding value to the entire production chain. Byerlee and Traxler (1995) estimated a rate of return of 52% in an international cooperation for wheat breeding in developing countries. Pardey et al. (1996) evaluated data from two centers of CGIAR, reporting benefit and cost ratios of 48:1 for rice and 190:1 for wheat. Brennan et al. (1997) suggested that 64% of the gains from rice breeding in Australia were due to the acquisition of international germplasm, generating benefits of US$ 848 million. Evenson and Gollin (1997) estimated that without the efforts of an international network for evaluation of rice germplasm, 20 new commercial varieties would not have been developed. In this case, the present value of production loss for a period of 20 years would be US$ 1.9 billion.

Aiming at identifying the intrinsic use value of genetic resources, several empirical approaches have been proposed (see Evenson et al. 1998). However, such methods are poorly used in practice, based on real data. Evenson and Gollin (1997) examined the genealogies of rice varieties produced by IRRI between 1965 and 1990, concluding that the addition of 1,000 accessions to the bank was associated with the obtainment of 5.8 varieties. Assuming a period of 10 years after acquisition and a discount rate of 10%, gains were estimated at US$ 325 million. Using similar methodology, Lamarié and Evenson (1998) estimated that rice production in India was 5.6% higher due to the incorporation of landraces in the collection throughout the 1980s.

As seen, the theoretical basis for estimating the total value of genetic resources is still being developed. Since PGRs are considered public goods that are not directly marketed, the biggest limitation is to define a total value for the accessions or genes preserved in a germplasm collection, that considers both use and non-use values.

Costs: The Management of Collections

Since quantifying the benefits (products) generated from PGR has shown to be a challenging task, a more effective approach to increase the value of these resources is to correctly manage the costs involved in *ex-situ*conservation. Cost analysis is derived from microeconomic theory of production, in which a genebank can be treated similarly to any other company (Pardey et al. 2001). Germplasm banks are organized to produce outputs, represented by viable accessions and their information. Based on this simple statement, it is concluded that genebank managers often have to decide about the financial resources needed to produce a certain amount of outputs. As part of the decision process, managers should seek to minimize operational costs using the same infrastructure and maximize production using the same budget. In both cases, the final result will be gain in efficiency.

The total costs of operating a germplasm bank are generally classified into human capital, variables (production labor and operating expenses) and fixed (Koo et al. 2003). The costs involved in the operations of a germplasm bank depend on a number of factors, such as the biological characteristics of the species preserved, preservation methods used, social and economic conditions of the country and local edaphoclimatic factors. Consequently, cost surveys should be performed for each genebank, since conditions and conservation objectives are rarely similar (Koo et al. 2004). For example, the cost of long term conservation of species with vegetative propagation or recalcitrant seeds, which demands special techniques of tissue culture or *in vivo* conservation, is usually greater than the orthodox seed storage (Smale and Horna 2010).

Despite the difficulties, all germplasm banks share a set of basic activities, which can be summarized to better estimate the components of total operating cost on an annualized basis (Table 2). Variable costs are easier to estimate, since the manager is aware of costs of supplies purchase and employees payment. In turn, the fixed costs can be estimated based on the purchase price of the good, which is depreciated by a determined interest rate (Koo and Smale 2003).

Based on cost information in Table 2, the average cost can be calculated by dividing the total cost by the total number of accessions. For comparison purposes, the mean cost for each activity can also be calculated, or the mean costs of different years of the same activity could be compared to measure the efficiency. For example, the average annual cost of storage is calculated as the total cost of storage in a given year divided by the total number of accessions stored in the collection. The average annual cost can be divided into average variable cost and average fixed cost depending on the components of the total cost considered (Smale and Koo 2003). The mean fixed cost tends to decrease when increasing the number of accessions stored, unless it is necessary to purchase new equipment or expand facilities. Moreover, the average variable cost tends to decrease to a certain quantity of stored material due to efficiency gain (economies of scale). However this cost increases again due to excessive use of variable resources under the same fixed production factors. At this point, managers must make a decision of expanding the facilities or reducing the number of stored accessions. Further clarification on the costs associated with *ex-situ* conservation is available in a study by Horna and Smale (2010), where costs of conservation of PGR in CGIAR centers were estimated.

Cost analysis has proven to be an essential tool for better management of activities and financial resources spent on the conservation of genetic resources. An inefficient financial management of collections incurs not only in excessive cost, but also in the incapability of the bank in disseminating feasible accessions containing all the information necessary to meet the demand. Aiming at better management of collections worldwide the System-Wide Genetic Resources Programme (SGRP) of the Consultative Group on International Agricultural Research (CGIAR) has recently developed a database and tools (including a decision-support tool) to assist PGR managers, the Crop GeneBank Knowledge Database (http://cropgenebank.sgrp.cgiar.org/).

A Vision on the Future - Recommendations for Strengthening the Area

Agrobiodiversity has been, since its discovery, one of the pillars of national development. In recent years, the development of national agriculture, both for corporate and family farming, had a great improvement through the adoption of appropriate public policies, higher and more easily accessible credit, continuous and exemplary entrepreneurship of the Brazilian producer, increasing external and internal demand for products traditionally produced in the country, availability of appropriate technologies to tropical agriculture and many other factors. However, the basis of this agriculture has used just a little of our agrobiodiversity and it depends, substantively, on a limited number of species.

An optimistic view of the future:

- Predicts increasing use of agrobiodiversity in favor of Brazilian society and the world's population;
- Predicts the launch of a national program of recovery of germplasm collections of Federal and State public institutions, renewing the seeds and vegetative materials, strengthening the characterization of accessions, and modernizing and integrating the databases so they can be easily accessible to public and private users;
- Predicts the release of a call for proposals, by Federal and State agencies that support research, in a coordinated way, in order to strengthen the

capacity of young professionals in national and foreign institutions that have expertise in the subject;

- Predicts the simplification of legislation, which due to its complexity has tremendously hindered the use of plant genetic resources for the benefit of Brazilian society;
- Continues the strengthening program of Active Germplasm Banks of Embrapa (the largest AGB collection of Brazil), which started with the Agroverde program, in 2010, which supported the AGBs of cassava, cowpea, grape, rice and soybeans;
- Estimates the inclusion of a growing number of plants and their derivatives in the Brazilian diet, which are now forgotten, unknown, unfamiliar, or restricted to specific regions, and makes use of genetic variability to increase the quality of food by the use of traditional techniques and new biotechnology;
- Projects an increase in national and global agriculture production and productivity, based on the rational use of agrobiodiversity and technology, meeting the demand for food, oil, fiber, energy and nutrients of the world's population, and particularly for the less favored ones; and finally,
- Predicts the establishment of ConservaBrasil©, an audacious and futuristic vision for conservation and use of agrobiodiversity. ConservaBrasil©'s mission is to preserve the long-term national biodiversity, in particular the collections of species of agricultural interest, both native and exotic, for future use for the benefit of Brazilian society; in its vision, ConservaBrasil©will have under its tutelage, in 2020, the largest collection of Brazilian germplasm for future use by national agriculture. This initiative will be responsible for keeping thousands of plant, animal and microbial accessions used in Brazil. The germplasm collection maintained by ConservaBrasil©will be one of the largest and best preserved in the world. The collection will have captured significant fraction of the genetic diversity of species relevant for Brazilian agriculture. The results of research on long-term conservation of germplasm will enable the Brazilian society to ensure that the varieties, lines, clones, hybrids, semen, ova and microorganisms of interest are being preserved for future use with the highest technical accuracy. Ongoing operations in 2020 will ensure, in the future, the supply of accessions fundamental for maintaining the competitiveness of domestic agriculture. In this process, Brazilian society, with support from Embrapa, will be participant in the efforts of conservation and use of germplasm that will ensure the future of Brazilian agriculture.

REFERENCES

1. Alston JM, Norton GW and Pardey PG (1995) Science under scarcity: principles and practice for agricultural research evaluation and priority setting. Cornell University Press, Ithaca, 618p.
2. Andrade RV, Azevedo JT, Borba CS and Andreoli C (1994) Banco ativo de germoplasma. In Embrapa (ed.)Relatório técnico anual do Centro Nacional de Pesquisa de Milho e Sorgo 1992-1993. Embrapa, Sete Lagoas, p. 276-277
3. Assunção IP, M-Filho LR, Resende LV, Barros MCS, Lima GSA, Coelho RSB and Lima JAA (2005) Genes diferentes podem conferir resistência ao Cowpea severe mosaic vírus em caupi Fitopatologia Brasileira 30:274-278.
4. Borém A (2001) Melhoramento de plantas. 3rd ed., UFV, Viçosa, 500p.
5. Brennan J, Singh IP and Lewin LG (1997) Identifying international rice research spillovers in New South Wales.Agricultural Economics 17:35-44
6. Brock WA and Xepapadeas A (2003) Valuing biodiversity from an economic perspective: a unified economic, ecological and genetic distance approach. American Economic Review 93:1597-1614.
7. Brown GM (1991) Valuation of genetic resources. In Orians GH, Brown Jr. GM, Kunin WE and Swierbinski JE (eds.)The preservation and valuation of biological resources. University of Washington Press, Seattle, p. 203-245.
8. Byerlee D and Traxler G (1995) National and international wheat improvement research in the post green revolution period: evolution and impacts. American Journal of Agricultural Economics 77:268-278.
9. Evenson R and Gollin V (1997) Genetic resources, international organizations and improvement in rice varieties.Economic Development and Cultural Change 45:471-500.
10. Evenson R and Lemarié S (1998) Optimal collection and search for crop genetic resources. In Smale M (ed.)Farmers, gene banks and crop breeding: economic analyses of diversity in wheat, maize and rice. Kluwer, London, p. 79-82.
11. Evenson R, Gollin D and Santaniello V (1998) Agricultural values of plant genetic resources. CABI, Wallingford, 285p.
12. Ferreira MAJF, Wetzel MMVS and Valois ACC (2005) El estado del arte de los recursos genéticos en las Américas: conservación, caracterización y utilización. Embrapa Recursos Genéticos e Biotecnología e Programa

Cooperativo de Investigación y Transferencia de Tecnología para los Trópicos Suramericanos (PROCITROPICOS), Brasília, 100p

13. Gollin D and Evenson R (2003) Valuing animal genetic resources: lessons from plant genetic resources. Ecological Economics 45:353-363.
14. Goodman MM (1990) Genetic and germplasm stocks worth conserving. Journal of Heredity 81:11-16.
15. Horna D and Smale M (2010) Evaluating cost-effectiveness of collection management: a methodological framework. Available at <http://cropgenebank.sgrp.cgiar.org/images/file/management/DST/framework_dst.pdf> Assessed on Oct 14, 2012.
16. Koo B and Smale M (2003) Economic costs of genebank operation. In Engels JMM and Visser L (eds.) A guide to effective management of germplasm collections. IPGRI, Rome, p. 93-106 (IPGRI handbooks for genebanks, 6).
17. Koo B and Wright BD (2008) The optimal timing of evaluation genebank accessions and the effects of biotechnology. American Journal of Agricultural Economics 82:797-811.
18. Koo B, Pardey PG, Wright BD, Bramel P, Debouck D, van Dusen ME, Jackson MT, Rao NK, Skovmand B, Taba S and Valkoun J (2004) Saving seeds: the economics of conserving crop genetic resources ex-situ in the future harvest centres of the CGIAR. CABI Publishing, Oxfordshire, p. 7-20.
19. Koo B, Pardey PG and Wright BD (2003) The economic costs of conserving genetic resources at the CGIAR centres.Agricultural Economics 29:287-297.
20. Mariante AS, Sampaio MJA and Inglis MCV (2009) The state of Brazil´s plant genetic resources. Embrapa Technological Information, Brasília, 163p.
21. MMA(2011) Quarto relatório nacional para a convenção sobre diversidade biológica: Brasil/Ministério do Meio Ambiente. MMA, Brasília, 248p.
22. Nass LL (2001) Utilização de recursos genéticos vegetais no melhoramento. In Nass LL, Valois ACC, Melo IS and Valadares-Inglis MC (eds.) Recursos genéticos e melhoramento: plantas. Fundação MT, Rondonópolis, p. 30-55.
23. Nass LL (2007) Recursos genéticos vegetais. Embrapa Recursos Genéticos e Biotecnologia, Brasília, 858p.
24. Nass LL (2011) Pré-melhoramento vegetal. In Lopes MA, Fávero AP, Ferreira MAJF, Faleiro FG, Folle SM and Guimarães EP (eds.) Pré-

melhoramento de plantas; estado da arte e experiências de sucesso. Embrapa Informação Tecnológica, Brasília, p. 23-38.

25. Nass LL and Paterniani E (2000) Breeding: a link between genetic resources and maize breeding. Scientia Agricola 57:581-587

26. Nass LL and Sigrist MS (2012) Wild species: potential use in pre-breeding. In Borén A, Lopes MTG, Clement CR and Noda H (eds.) Domestication and breeding: amazon species. Suprema, Viçosa, p. 101-115.

27. Nass LL, Nishikawa MAN, Fávero AP and Lopes MA (2007) Pré-melhoramento de germoplasma vegetal. In: Nass LL (ed.) Recursos genéticos vegetais. Embrapa Recursos Genéticos e Biotecnologia, Brasília, p. 683-744

28. Pardey PG, Alston JM, Christian JE and Fan S (1996) Hidden harvest: U.S. benefits from International Research Aid. International Food Policy Research Institute, Washington, 17p.

29. Pardey PG, Bonwoo Koo, Wright BD, Van Dusen ME, Skovmand B and Taba S (2001) Costing the conservation of genetic resources: CIMMYT›s ex situ maize and wheat collection. Crop Science 41:1286-1299

30. Queiroz MA and Lopes MA (2007) Importância dos recursos genéticos vegetais para o agronegócio. In Nass LL (ed.) Recursos genéticos vegetais. Embrapa Recursos Genéticos e Biotecnologia, Brasília, p. 61-119.

31. Reifschneider FJB, Henz GP, Ragassi CF, Anjos UG and Ferraz RM (2010) Novos ângulos da história da agricultura no Brasil. Embrapa Informação Tecnológica, Brasília, 112p.

32. Ribeiro CSC (2000) Criando novas variedades. In Reifschneider FJB (org.) Capsicum: pimentas e pimentões no Brasil. Embrapa Transferência de Tecnologia/Embrapa Hortaliças, Brasília, p. 68-80

33. Ritschel PS and Huamán Z (2002) Variabilidade morfológica da coleção de germoplasma de batata-doce da Embrapa-Centro Nacional de Pesquisa de Hortaliças. Pesquisa Agropecuária Brasileira 37:485-492

34. Rubenstein KD, Heisey P, Shoemaker R, Sullivan J and Frisvold G (2005) Crop genetic resources: an economic appraisal. USDA, Washington, 41p. (Economic Information Bulletin, 2).

35. Silva JAA, Nobre AD, Manzatto CV, Joly CA, Rodrigues RR, Skorupa LA, Nobre C, Ahrens S, May PH, Sá TDA, Cunha MC and Rech Filho EI (2011) O código florestal e a ciência: contribuições para o diálogo. Sociedade Brasileira para o Progresso da Ciência, São Paulo, 124p

36. Smale M and Koo B (2003) Biotechnology and genetic resource policies: what is a genebank worth?International Food Policy Research Institute (IFPRI), Brief 7:1-5.

37. Sousa GS (1879) Tratado descriptivo do Brazil em 1587. 2nded., João Ignácio da Silva, Rio de Janeiro, 382p.

38. Valls JFM (2007) Caracterização de recursos genéticos vegetais. In Nass LL (ed.) Recursos genéticos vegetais. Embrapa Recursos Genéticos e Biotecnologia, Brasília, p. 281-342.

Chapter 2

AGRICULTURE GERMPLASM RESOURCES: A TOOL OF CONSERVING DIVERSITY

Rukhsar Ahmad Dar[1],Mushtaq Ahmad[2],Sanjay Kumar[1],Monica Reshi[1]

[1]KVK, Neyoma, Ladakh, India

[2]Division of Genetics and Pant Breeding, Sher-e-Kashmir University of Agricultural Sciences and Technology, Kashmir Shalimar campus, Srinagar -191 121, India

ABSTRACT

Three major physical resources in the world comprise land, water and the biological diversity. Agricultural biodiversity is an important component of biodiversity, which has a more direct link to the well being and livelihood of mankind than other forms of bio¬diversity. In fact, it is one of our most fundamental and essential resources, one that has enabled farming systems to evolve since the birth of agriculture about 10,000 years ago. Food plant and animal species have been collected, used, domesticated and improved through traditional sys¬tems of selection over many generations. The resulting diversity of genetic resources developed by early farmers now forms the basis on which modern high yielding and disease resistant varieties have been produced to feed the growing human population, expected to reach 9.1 billion by 2050. According to the Convention on Biological Diversity (CBD), "agricultural biodiversity in¬cludes all components of biological diversity of relevance to food and agriculture, and all components of biological biodiversity that constitute agro-ecosystems: the variety and variability of animals, plants and micro-organisms, at the genetic, species and ecosystem levels, which are neces¬sary to sustain key functions of the agricultural ecosystem, its structure and processes". The effective conservation and use of agricultural biodiversity is very important in ensuring sustainable increases in the productivity and produc¬tion of healthy food by and for mankind as well as contrib¬uting to increased resilience of agricultural ecosystems.

INTRODUCTION

There are many threats or drivers of changes on biodiversity that have been recognized and intensified in recent years (Millennium Ecosystem Assessment, 2005). With regard to agriculture the most important ones include changes in land use, replacement of traditional varieties by modern cultivars, agricultural intensification, increased population, poverty, land degradation and environmental change (including climate change) (FAO, 2010). It is predicted that climate change will have a significant impact on agriculture with temperatures rising on average by 2 to 4°C over the next 50 years, causing significant changes in regional and seasonal patterns of precipitation (Burke et al., 2014). Climate change will also impact agricultural biodiversity in a major way. Model projections carried out by Lane and Jarvis (2007) based on global distribution of suitable cultivated areas of 43 crops, highlight that more than 50% may decrease in extent. Evidence based on bioclimatic modelling suggests that climate change could cause a marked contraction in the distribution ranges of CWR. In the case of wild populations of peanut (Arachis spp.), potato (Solanum spp.) and cowpea (Vigna spp.), studies suggest that 16 to 22% of these species may go extinct by 2055, with most species possibly losing 50% of their range size (Jarvis et al., 2008). These threats or drivers of change are leading to large scale degradation and loss of agricultural biodiversity and consequently its genetic variability (Millennium Ecosystem Assessment, 2005; van de Wouw et al., 2009). Information regarding the threat and rate of genetic erosion among various components of agricultural biodiversity is important, yet very little work has been carried out to quantify the magnitude of any trends. The availability of large gene pools, including CWR, is becoming even more important as farmers will need to adapt to changing conditions that result from these pressures. It is likely that many of the genetic traits which will be necessary to adapt our crops to changing climate will be found in CWR. Hence, it is widely urged that such strategies be adopted which may be used to get maximum crop stand and economic returns from adverse environments. Major strategies which may be used to overcome the adverse effects of such stressful environments may include screening and selection of well adopted existing germplasm of potential crops (Ahmad et al., 2014).

There are two main strategies for conserving agricultural biodiversity, namely ex situ and in situ conservation, both of which are equally important and should be regarded as complementary (Thormann et al., 2006; Engelmann and Engels, 2002). Ex situconservation is the conservation of components of biodiversity outside their natural habitats. It is generally used to safeguard populations that are at present or potentially under threat and need to be collected and conserved in genebanks in the form of seeds, live plants, tissues, cells and/or DNA materials. Article 2 of the CBD defines in situ conservation as "the conservation of ecosystems and natural habitats and the maintenance and recovery of viable populations of species in their natural surroundings and, in the case of domesticated or cultivated species, in the surroundings where they have developed their distinctive properties" (UNCED, 1992). It thus refers to the maintenance of a species in its natural habitat. This can be either on farm, requiring the maintenance of the agro-ecosystem along with the cultivation and selection processes on local varieties and landraces, or in the wild, which involves the maintenance of the ecological functions that allow species to evolve under natural conditions.

STATUS AND TRENDS OF AGRICULTURAL BIODIVERSITY

Little is known about the global status of agricultural biodiversity. Although the CBD recognize genetic diversity as one of the fundamental levels of biodiversity, actions to protect genetic diversity are lacking (Laikre et al., 2010). Policy makers and scientists require a better understanding of how the intraspecific diversity is changing over time and space in order to make informed decisions for their conservation. However, there is no routine global scale monitoring of genetic diversity over time (Frankham, 2010; Laikre et al., 2010), except for a few target species at national level (Laikre et al., 2008). A major challenge remains to develop simple inexpensive means to monitor genetic diversity at a global scale (Frankham, 2010). Several efforts under the 2010 Biodiversity Indicators Partnership (http://www.twentyten.net) have been made to identify indicators useful to detect changes in species and ecosystem diversity, but there are only two initiatives that are explicitly working on developing indicators that deals with genetic variation for agricultural biodiversity (Laikre, 2010; Walpole et al., 2009).

The only authoritative account of agricultural biodiversity status at the global level is represented by the First and Second reports on the State of the World's Plant Genetic Resources for Food and Agriculture published by the Food and Agriculture Organization of the United Nations (FAO, 1998, 2010). The Second Report mention that there are about 7.4 million accessions conserved in over 1750 gene banks around the world in either seed banks, field collections, and in vitro and cryopreservation conditions (FAO, 2010). This represents an increase of more than 1.4 million accessions added to ex situ collection since publication of the first report on theState of the World's Plant Genetic Resources for Food and Agriculture. Although reportedly over-represented, a large part of the genetic diversity of major food crops is stored in ex situ collections. The exact proportion is still uncertain, but estimates suggest that more than 70% of the genetic diversity of some 200 to 300 crops is already conserved in gene banks (SBSTTA, 2010). In addition there are over 2,500 botanic gardens maintaining samples of some 80,000 plant species (FAO, 2010). However, regeneration of gene bank accessions remains a major problem, threatening collections (FAO, 1998). In the past decade there have been significant advances made in regenerating collections at risk, in part due to efforts made by the Global Crop Diversity Trust (CGDT) in supporting regeneration programmes of globally important priority gene bank collections for 22 priority crops for which crop specific regeneration guidelines have recently been produced (Dulloo et al., 2013). Another major achievement has been the creation of the Svalbard Global Seed Vault (SGSV) in 2008, established to serve as the ultimate safety net for seeds samples from the world's most important collections (GCDT, 2010).

Great efforts for the conservation of many CWRs and wild species have been made by the Millennium Seedbank (MSB) at Wakehurst Place, Royal Botanic Gardens, Kew, UK which aims to house up to 10% of the world's seed-bearing flora, principally from arid zones by 2010. Genetic erosion has also been prevented by the significant amount of crop genetic diversity in the form of traditional varieties and neglected and underutilized species (NUS) that continues to be maintained on-farm. Yet, in spite of these advances, important reservoirs of adaptive variation such as CWR, landraces and NUS, which are increasingly recognized by the global scientific community as key resources for the maintenance of agro biodiversity, remain under-represented (FAO, 2010). CWR in particular, which have avoided the genetic bottleneck of domestication, contain greater genetic variation than their cultivated relatives and represent an important reservoir of genetic resources for breeders (Maxted and Kell, 2009). Yet to retain the genetic characteristics that make them so valuable for crop improvement, it is now widely recognized that populations of CWR are best conserved in situ, in their wild habitats, where they can continue to adapt and evolve along with their natural surroundings, thus ensuring new variation is generated in the gene pool and the continued supply of the novel genetic material critical for future crop improvement. The underpinning of the conservation strategy of most countries is a protected areas system and this is reflected in the CBD, where the main thrust of biodiversity conservation is in situ, through the development of such protected systems. Populations of many CWR occur in existing protected areas, but this alone does not in many cases represent effective in situconservation without some degree of management or intervention targeted at the populations of the particular target species, particularly if the species is threatened. Despite protected areas being in existence for many years we still have not been able to undertake significant actions to conserve the CWR they contain, except a few cases.

Despite this, the in situ conservation of CWR has gained increasing attention in many countries, as demonstrated by their inclusion in the many national reports drafted for the Second report on the State of the World's Plant Genetic Resources for Food and Agriculture (FAO, 2010). Unfortunately, little quantitative data were provided by countries on the changing status of CWR, but several reports indicated that specific measures had been taken to promote their conservation. The Second report also mentions that surveys and inventories of CWR were carried out in at least 28 countries and many new priority sites for conserving CWR in situhave been identified over the last decade. There is also evidence that public awareness of the importance of CWR, and neglected and under-utilized species such as traditional vegetables and fruits, is growing both in developing and developed countries (FAO, 2010). This has been furthered by a number of global initiatives aimed at conserving CWR, such as the proposed establishment of a global network for the in situ conservation of CWR (Maxted and Kell, 2009), and more concretely by the creation of web-based international platforms for the exchange of CWR information and data. These include the European platform "An Integrated EuropeanIn Situ Management Work Plan: Implementing Genetic Reserves and On Farm Concepts" (AEGRO) and the CWR Global Portal, developed as part of the UNEP/GEF Crop Wild Relative Project, that provides access to CWR information and data at the global level (Thormann et al., 2012). The significant increase in number of scientific articles published on CWR and on specific actions targeting their conservation is also a testimony to the renewed interest in CWR, however, to the best of our knowledge few of the recommendations have been implemented, largely due to a lack of funds and capacity.

Over the last decade, the number and coverage of protected areas has increased by approximately 30% (United Nations, 2010), yet limited efforts have been made to target CWR, whose conservation remains unplanned and largely an indirect effect of protecting flagship species or threatened habitats. For example, despite the increase in isolated activities targeting CWR conservation, the formal recognition and/or the adoption of appropriate management regimes to protect CWR are largely lacking. Furthermore, considering that national parks and other conservation areas cover only 12 to 13% of the earth's surface, it is clear that these areas alone will not be able to ensure the continued existence of CWR species, the majority of which occur in marginal lands outside protected areas, where no form of legal protection is offered. If protected areas are to ensure the long-term survival of CWR they will need to become more flexible in size and scale and a connected network of habitats will need to be established to allow species to migrate and adjust their ranges in response to global change and anthropogenic disturbances, along with the development of effective management strategies targeting their conservation (that is, off-reserve management). The success of this strategy will depend largely on promoting more biodiversity-sensitive management of ecosystems outside protected areas, and successfully engaging private landowners and local communities living around protected areas in the conservation process. Finally, more effective policies, legislation and regulations that take into account the impacts of global changes on future species distribution and that govern the in situ conservation of CWR, both inside and outside of protected areas, are needed, along with closer collaboration and coordination between the agriculture and environment sectors.

Formidability of Genetic Resources

Wild plants have often played an important role in many diets due to their higher nutritional value than cultivated species. These are, at the same time, hardy and resilient. Crop varieties are improved by the suitable recombination of genes from the wild, made more productive, resistance to biotic stresses, tolerant to abiotic stresses, and better nutritional and keeping quality. Such characteristics, needed to improve crop varieties, may be found in a range of cultivated as well as wild plants. This broad variability provides essential link in the food chain, which, in turn, provides the basic for world food and nutritional security. Plant genetic resources essentially constitute the prime components of the food chain ever since the dawn of agriculture. In the history of some 12,000 years, nearly 30,000 edible plant species have been utilized as a source of food. However, merely a hundred odd plant species out of these have been propagated to provide about 90% of the world food and, further, only three species among these, namely, rice, wheat and maize produced the two-third. An assessment of the contribution of different plant sources towards the dietary energy supply at the global level shows predominance of only two crops, that is, rice (26%) and wheat (23%) (FAO, 1996). The search for new diversity is, therefore, important.

In the developing and the economically weaker parts of the world, the discovery of wild species for food may have coincided with the hunger season, such as, those preceding the crop harvest particularly when drought or flood situations occurred. Mother Nature provided food for people at such junctures when they badly needed it and the resultant discovery of plant species or their diversity became the automatic human choices for further propagation. Even today, though agriculture has advanced so much, humans still gather many wild and semi-wild plants or plant parts like fruit, leaf root, seed, nut, wood etc. for use. About 80,000 species of plants have been used to meet the routine needs by the human beings. Of these, 30,000 species so far have been identified as edible and about 7,000 species have been cultivated and/or collected for food at one time or the other. Presently only 20 to 30 crops, such as cereals (wheat, rice, maize, millets, sorghum), root/tuber crops (potato, sweet potato, cassava), legumes (pea, beans, peanut, soybean) and sugarcane, sugar beet, coconut and banana are mainly used to feed the world (NAS, 1975).

Synoptic View of Plant Diversity

There are 425,000 species in living plants in the plant kingdom from unicellular algae to the highly evolved flowering plants. The flowering plants are a diverse group which are seed producing plants that have evolved in synchronization with the evolution of insects which help the plants in cross-pollination assuring heterozygous population. Hence, this group of flowering plants (about 250,000 species) have developed great plasticity for adaptation for different climatic regimes. They consist of a variety of life forms from the minute Wolffia (1 mm long) to the largest Eucalyptus regnant growing to height of 100 m. This spectrum of flowering plants includes humble herbaceous species, beautiful orchids, parasitic Rafflesia arnoldii having largest flowers (1 m across), plants of medicinal value, trees of timber importance, food plants, fodder species, gums and resin producing plants. The 250,000 flowering plant species are packed in about 17,000 genera and 300 to 400 families. Some of the economically important families which hold life supporting food sustenance species are Gramineae, Legumonosae, Criciferae, Cucurbitaceae, Rosaceae, Brassicaceae and Rutaceae. The drug yielding families cover a spectrum of alkaloids producing crops such as, Apocynaceae, Papavaraceae, Asteraceae, Cannabinaceae, Piperaceae, Zingiberaceae and Rubiaceae. Gums and resins occur in several families as Eurphorbiceae, Dipterocarpaceae, Mimosaceae and Sapotaceae. The families vary in size from monotypic ones to large families having 25,000 to 35,000 species. The family Orchidaceae has about 25,000 to 35,000 species, Compositae has about 20,000 species, Legumionceae has about 14,00 species and Gramineae has about 8,000 species, while there are about 35 families which has only one species.

Plant Diversity in India

Indian subcontinent has a rich and varied heritage of biodiversity, encompassing a wide spectrum of habitats from tropical rainforests to alpine vegetation and from temperate forests to coastal wetlands. It is one of the eight centres of origin (Vavilov, 1951) and is one of the 12 mega gene centres of the world; possess 11.9% of world's flora. About 33% of the country's recorded flora are endemic to the region and are concentrated mainly in the North-East, Western Ghats, North-West Himalays and Andaman and Nicobar islands, nurish one third of the human population on this earth (Damania, 2002; Mayres et al.,

2000) have brought out an updated list of 25 global hotspots of diversity out of which 8 hotspots are in figured in India. Indian sub-continent is a centre of domestication and diversification of several economically useful wild plant species comprising about 3,000 plants of edible value, 4,000 species having known reputed medicinal value, 700 plants of traditional and social significance, 500 fibre yielding species, 400 fodder species, 40 species having insectivorous uses, 300 gum and dye yielding plants, 100 aromatic and essential oil yielding species (MoEF, 1994; Chowdary and Murti, 2000). Indian diversity comprises of 49,219 higher plant species, out of which, 5,725 are endemic and belonging to 141 genera under 47 families. Of these endemic species, 3,500 are found in Himalayas and adjoining regions and 1,600 in the Western Ghats alone.

India is a homeland of 167 cultivated species and 329 wild relatives of crop plants (Arora, 1991). It has around 30,000 to 50,000 landraces of rice, wheat, Pigeonpea, mango, turmeric, ginger, sugarcane, etc. and ranks 7th in terms of contribution to world agriculture. Further, around 1,000 wild edible plant species are exploited by native tribes. These include 145 species of roots and tubers, 521 of leafy vegetables/greens, 101 of buds and flowers, 647 of fruits and 118 of seeds and nuts (Arora and Pandey, 1996). In addition, nearly 9,500 plant species of ethano-botanical uses are reported from the country of which around 7,500 are of ethano-medical importance and 3,900 are multipurpose edible species.

The endemic plant wealth of the country has also been supplemented with the species/forms that had been introduced from abroad. These species got naturalized over time and have undergone the process of domestication on being isolated climatically and spatially. Prominent among these aer apple, pear, peach, apricot, grape, almond, date palm, maize, potato, sweet potato, tomato, bean, onion, garlic, chilli, lentil, clove, coriander, cumin, fennel, coffee, cocoa, cashew nut, litchi, cinchona, strawberry, blueberry, tea, rubber and pineapple.

BIODIVERSITY IN JAMMU AND KASHMIR

The State of Jammu and Kashmir has been regarded as heaven on earth, and is also called the bio-mass state of India. This area, located in the far north of the Indian republic, is a mountainous zone in the north-west Himalayas that shares international boundaries with Pakistan in the west, Chinese autonomous region of Xinjiang in the north and Tibet in the north-east. The North-western-Himalayan region being the rich repository of biological heritage, particularly in respect of agri-horticultural crops and was recognized that collection and maintenance of germplasm is essential to provide genetic diversity within a crop and to reduce chances of genetic vulnerability. Exploration and collection of native biodiversity, particularly in agri-horticultural crops of the region, including the wild relatives, rare/endangered plants together with the documentation of related ethano-botanic information for exercising to concomitant with regeneration and preliminary evaluation of the collected genetic resources to ensure their long term conservation as well as use in crop breeding programmes recognizing the fact that improvement and sustenance of cultivated crop species requires variability.

A total of 1911 germplasm accessions comprising local cultivars that were in cultivation before introduction of improved cultivars , old varieties, land races, wild crop relatives and under-utilized crops of agri-horticultural significance were collected in respect of various field, vegetable , and horticultural crops as well as medicinal and aromatic plants. The collected biodiversity included 742 accessions in cereals, 38 in pseudo cereals, 28 in millets,71 in oilseeds, 358 in pulses, 377 in vegetable crops, 21 in spices and condiments, 13 in fodder crops, 204 in medicinal and aromatic plants, 55 in fruits crops and 4 in others. The collection of agro-biodiversity in different crops has not only helped in ensuring their conservation on a long term basis but their use may also increase productivity, food security and economic returns. The valuable biological resources will make the farming systems more stable and sustainable. By establishing suitable linkages with user scientists in the university and sister institutions in the region a total of 382 accessions in cereals, 135 pulses, 78 vegetables, 26 horticultural crops, 110 in medicinal and aromatic plants were made available for use in respective crop improvement programmes. Their eventual use in the development of varieties with high yield potential and improved quality characteristics may diversify production and income opportunities for the end user.

Conservation of Germplasm

Global concern about loss of valuable genetic resources prompted international action. Programs for conservation of plant genetic resources for food and agriculture were thus initiated and gene banks established in many countries. The main objective was to collect and maintain the genetic diversity in order to ensure its continued availability to meet the needs of different users. The concept of germplasm conservation demands that collection methods initially capture maximum variation and subsequently, conservation and regeneration techniques minimize losses through time. To this effect, plant genetic resources (PGR) conservation activities comprise collecting, conservation and management, identification of potentially valuable material by characterization, and evaluation for subsequent use.

There are two approaches for conservation of plant genetic resources, namely in situ and ex situ. In situ conservation involves maintaining genetic resources in the natural habitats where they occur, whether as wild and uncultivated plant communities or crop cultivars in farmers' fields as components of the traditional agricultural systems. Ex situ conservation on the other hand, involves conservation outside the native habitat and is generally used to safeguard populations in danger of destruction, replacement or deterioration. Approaches to ex situ conservation include methods like seed storage, field gene banks and botanical gardens. DNA and pollen storage also contribute indirectly to ex situ conservation of PGR.

Ex situ Conservation Approach

Ex situ conservation refers to the conservation of germplasm away from its natural habitat. This complementary approach for conservation had begun on a wide scale about three decades ago and is now practised, to some extent, in almost all countries as a means to conserve crop species diversity for posterity. This strategy is particularly important for crop gene pools, and can be achieved by propagating/ maintaining the plants in genetic resource centre, botanical gardens, tissue culture repositories or in seed gene banks. The Second Report mention that there are about 7.4 million accessions conserved in over 1750 genebanks around the world in either seed banks, field collections, and in vitro and cryopreservation conditions, (FAO, 2010).

Various approaches are employed for the ex situ conservation depending upon the mode of reproduction and nature of plants to be conserved. Ex situ conservation approach generally comprises the following methods: seed storage, field gene banks, in vitro storage, pollen storage, DNA storage and botanical gardens.

Seed Storage

In the past, many collections were maintained without the help of storage facilities which would affect the viability of seeds. Due to this, the conserved accessions had to be regenerated very frequently leading to loss of genetic diversity in gene banks (Frankel and Hawkes, 1975). In maintaining genetic purity of the conserved accessions, problems arise due to differential survival in storage, selection during regeneration, out-crossing with other entries and genetic drift (Allard, 1970). Good storage conditions coupled with proper grow-outs are expected to reduce the effects of such problems (Rao, 1980).

Storing orthodox seeds at low moisture content and at subzero temperature is the most convenient and widely used method of genetic conservation.

Orthodox seeds are the seeds which can withstand dehydration without damage. This type of seeds can be stored in the dry state on long term basis (indefinite period) which can be prolonged by decreasing their moisture content and storage temperature (at sub zero temperature).

The number of seed storage facilities has increased dramatically over the last two decades. Today, according to the WIEW – World Information and Early Warning System on Plant Genetic Resources for Food and Agriculture – databases of the FAO, there are 1320 national, regional and international germplasm collections in the seed form, 397 of which are maintained under long- or medium-term storage conditions. Over 6.1 million accessions have been conserved as seeds.

As opposed to common orthodox seeds, there are a number of species whose seeds are unable to withstand desiccation, that is, cannot be dried to low levels for optimum storage. Such seeds are referred to as 'recalcitrant' seeds (Roberts and King, 1986). Mainly these seeds originate from the plants grown in tropical and sub-tropical regions. These seeds can be stored for short duration (up to several months) by imbibed storage (at higher levels of seed moisture/hydrated state) and relatively warm conditions (well above zero temperatures) because they are often chilling sensitive. e.g., rubber, cocoa, coconut. Seeds such as oil palm and coffee, showing intermediate storage behaviour (Ellis et al., 1990, 1991) were grouped as recalcitrant until recently. Careful adjustment of the desiccation level and storage conditions allowed their storage for increased period (1 to 2 years).

Very low temperature storage using liquid nitrogen, called cryo-preservation, appears to be promising, with a more extended life span, described as long-term storage (-20°C). Another area in which considerable work is required is on storage of ultra dry seeds (dried to seed moisture content of 2 to 5%) at room temperature conditions and in hermetically sealed containers (Zhou et al., 1995). However, more research will be necessary before ultra dry seed technology can be adopted (Zheng et al., 1998). Prior to embarking on any seed conservation programme, a decision is to be made on how long it will be necessary to maintain the germination capacity of the seed lots, because longer storage requires more exacting storage conditions. This shall be determined by the objective of the conservation which could be research, introduction, breeding, etc.

Field Gene Bank Conservation

Many important varieties of field, horticultural and forestry species are either difficult or impossible to conserve as seeds (that is, no seeds are formed or if formed, the seeds are recalcitrant) or reproduce vegetatively. Hence they are conserved in field gene banks (FGB). FGBs provide easy and ready access to conserved material for research as well as for use. It is one of the options of a complementary strategy for the conservation of germplasm of many plant species. The conservation of germplasm in field gene bank involves the collecting of materials and planting in the orchard or field in another location. Field gene bank has traditionally been used for perennial plants, including:

•Species producing recalcitrant seeds;

•Species producing little or no seeds or sterile seeds;

•Species that are preferably stored as clonal material;

•Species that have a long life cycle to generate breeding and/or planting material.

Field gene banks are commonly used for such species as cocoa, rubber, coconut, coffee, sugarcane, banana, cassava, sweet potato, yam, tropical and temperate fruits, vegetatively propagated crops (e.g. wild onion and garlic) and forage grasses (e.g. sterile hybrids or shy seed producers). This is the traditional method of conservation to keep the germplasm in plantations as mature trees. It provides mature material for vegetative propagation, hybridization and characterization. The site for a field gene bank should have a suitable climate and soil for the species and should have an adequate water supply. The site should be chosen in a location with little or no threat of pests, diseases, bush fire and vandalism.To avoid loss of vigour as well as to prevent the incidences of attack by pests the plants have to be replanted routinely, and this adds to the cost further.

Botanical Gardens

There are about 2500 botanic gardens and arboreta worldwide. It is estimated that these gardens maintaining samples of some 80,000 plant of threatened species in botanical gardens and arboreta. The objectives of most of the gardens include:

•Maintaining essential ecological processes and life support systems,

•Preserving genetic diversity, and

•Ensuring sustainable utilization of species and ecosystem.

However, the botanical gardens may play a limited role in the context of conservation and propagation and probably a greater role in public awareness and education. Botanical gardens may mainly be used to display a great number of different and exotic species. As the number that can be maintained in this manner is limited, it cannot reflect or conserve genetic diversity. There is a possibility that a few well-managed gardens can emphasise on conservation of certain groups of species as living collections (that is, field gene banks). Often botanical gardens focus their conservation efforts on wild, ornamental, rare and endangered species. Indeed botanical garden conservation could be considered as field gene bank and/or seed gene bank, depending on the conservation method being used. The living plant collections in botanic gardens and arboreta may be considered as field collections, but the original purpose of the gardens and arboreta is not for germplasm conservation. Most of the germplasm conserved in botanical gardens do not belong to the PGRFA.

In vitro Storage

Research on finding solutions to better conserve these difficult-to-store seeds has focused on the use of biotechnology (Engelmann and Engels, 2002). In vitro slow-growth conservation methods, involving culturing different parts of the plant (meristem, tissues, cells) into pathogen-free sterile culture in a synthetic medium with growth retardants have been cited as good ways of complementing and providing backup to field collections. It has long been known that in vitro slow growth method suffers high risks of somaclonal variation (Withers, 1993) and also from the need to develop individual maintenance protocols for the majority of species (Thormann et al., 2006).

Slow growth

Slow growth procedures allow clonal plant material to be held for 1 to 15 years under tissue culture conditions with periodic sub-culturing, depending on species. There are several methods by which slow growth can be maintained. In most cases, a low temperature often in combination with low light intensity or even darkness is used to limit growth. Temperatures in the range of 0 to 5°C are employed with cold tolerant species, but for tropical species which are generally sensitive to cold, temperatures between 15° and 20°C are used. It is also possible to limit growth by modifying the culture medium, mainly by reducing the sugar and/or mineral elements concentration and reduction of oxygen level available to cultures by covering explants with a layer of liquid medium or mineral oil (Withers and Engelmann, 1993). Although slow growth procedures have been developed for a wide range of species, they are routinely used for conservation of genetic resources of only a few species including Musaspp., potato, sweet potato, cassava, yam, Allium spp. and temperate tree species. About 37,600 accessions are reportedly conserved by in vitro techniques in gene banks, worldwide (FAO, 1996).

Cryopreservation

Cryopreservation, the process in which living tissues are conserved at very low temperatures (-196°C) in liquid nitrogen (LN) or in vapour phase (-150°C) to arrest mitotic and metabolic activities, provides a more promising option (Thormann et al., 2006). Significant progress has been made in cryopreservation research over the past twenty years and much of that research has been focusing on understanding the desiccation sensitivity of recalcitrant seeds and on the underlying mechanism of desiccation tolerance (Engelmann and Panis, 2009). The techniques for cryopreservation currently in use are quite varied and include the older classical techniques based on freeze-induced dehydration of cells as well as newer techniques based on vitrification (Engelmann, 2000). In classical techniques, tissues are cooled slowly at a controlled rate (usually 0.1-4°C/min) down to about -40°C, followed by rapid immersion of samples in liquid nitrogen. Slow freezing is carried out using a programmable freezing apparatus. Cryoprotectants are added to the freezing mixtures to maintain membrane integrity and increase osmotic potential of the external medium. Classical cryopreservation procedures have been successfully applied to undifferentiated culture systems such as cell suspensions and calluses (Kartha and Engelmann, 1994). However, in case of differentiated structures, they have been employed for storage of apices or embryonic axes of only cold-tolerant species (Reed and Chang, 1997), and their utilization for tropical species has been limited (Escobar et al., 1997). Vitrification-based procedures involve removal of most or all free able water by physical or osmotic dehydration of explants, followed by ultra-rapid freezing which results in vitrification of intracellular solutes, that is, formation of an amorphous glassy structure without occurrence of ice crystals which are detrimental to cellular structural integrity. These techniques are more appropriate for complex organs like embryos and shoot apices; they are also less complex and do not require a programmable freezer, hence are suited for use in any laboratory with basic facilities for tissue culture.

DNA Storage

With the rapid development in the field of molecular genetics and genomics, DNA material is becoming more and more in demand for molecular studies and is one of the most requested materials from gene banks (Andersson, 2006). The establishing of a DNA storage facility as a complementary "back-up" to traditional ex situ collections has been suggested (Dulloo et al., 2013), but little effort has been made to collect and conserve DNA as a genetic resource. Some efforts have been made to establish DNA banks for endangered animals (Ryder et al., 2000) and a few plant DNA banks including Missouri Botanic

Garden, Royal Botanic Gardens - Kew, Australian Plant DNA Bank and Trinity College Dublin (TCD) (Rice et al., 2006; Hodkinson et al., 2007). The Global Biodiversity Information Facility (GBIF) in Germany has establish a DNA Bank Network in 2007, last accessed 22 September 2010, and offers a worldwide central web portal, providing DNA samples of complementary collections (microorganisms, protists, plants, algae, fungi and animals). The GBIF Germany DNA network would provide a good mechanism to link both to the scientific community conserving genotypes in genebanks and to breeders and molecular biologists that use the resources for genetic improvement.

Pollen Storage

Pollen storage was mainly developed as a tool for controlled pollination of asynchronous flowering genotypes, especially in fruit tree. Even if it may not be considered to be a viable method for meaningful genetic conservation of genotypes, cryopreservation is likely to be more successful than other storage techniques routinely employed for pollen. Pollen can be easily collected and cryopreserved in large quantities in a relatively small space. In addition, exchange of germplasm through pollen poses fewer quarantine problems compared with seed or other propagules.

The pollen longevity of different species varies between minutes and years depending on the taxonomic status of the plant and on abiotic environmental conditions. For some crops, the storage of pollen grains is possible in appropriate conditions, allowing their subsequent use for crossing with living plant material. It is also possible to regenerate haploid plants from pollen culture for some crops. By controlling the storage temperature and relative humidity (0 to 10°C, 10 to 30% RH, depending on species), pollens of Citrus spp., Cocos nucifera, Fragaria sp., Olea europea, Pinus silvestris, Pistachio altantica, Pyrus malus and Vitis vinifera could maintain their viability for more than 1 year.

For long-term conservation, cryopreservation seems to be the most efficient method. For example, maize pollen could be dried to 50% of its original water content in an air current for 1 h and then stored at -196°C in liquid nitrogen. Deep-frozen maize pollen can be used for fertilization after 10 years storage. Successful cryopreservation of pollen from various 24 crops has been reported (Barnabas and Kovacs, 1997).

In situ Conservation

In situ conservation refers to conservation of genetic resources within their ecosystem and natural habitats. These techniques involve maintenance of genetic variation at location where it is encountered, either in wild or traditional farming systems.

- **Genetic reserves:** in this type of conservation location, management, and monitoring of genetic diversity is carried in natural wild populations within defined areas designated for active, long-term conservation.
- **On-farm conservation:** This refers to the sustainable management of genetic diversity of locally developed traditional crop varieties with associated wild and weedy species or forms by farmers within traditional agricultural, horticultural or agrisilvicultural cultivation systems.
- **Home gardens:** This type of conservation is similar to on-farm conservation, involves smaller scale but more species-diverse genetic conservation in home, kitchen, backyard or door-yard gardens.
- Complementary conservation

For ex situ conservation of PGR in a crop or crop group, a gene pool approach has to be followed for safe and effective conservation. Following this approach, it is very likely that a range of ex situ conservation methods would be applicable to satisfy the needs of a gene pool. For example, the rice gene pool consists of self-pollinated cultigens and a range of wild Oryza spp. habitat to range of climatic conditions with breeding ranging from obligate vegetative to facultative and obligate self-pollination. In a situation, it is quite logical to have an approach, which is appropriate and has balanced application of both in situ and ex situ conservation methods. This will lead to the adoption of a more "holistic" approach to conservation. Even withex situ, a balance has to be struck as per the need. For example, in case of wild Oryza species, it has to be assessed, whether they would be best conserved in a field gene bank or in vitro as cell, tissue, organ, pollen or perhaps as DNA or in combination thereof. Therefore, a network of complementary and comprehensive strategy is needed to ensure effective conservation and sustainable use of PGR for food and agriculture by present and future generations.

Svalbard Global Seed Vault

A major achievement for the conservation of the germplasm have been the creation of the of the Svalbard Global Seed Vault (SGSV) in 2008, established to serve as the ultimate safety net for seed samples from the world's most important collections (GCDT, 2010). The Svalbard Global Seed Vault is a secure seedbank located on the Norwegian Island of Spitsbergen near the town of Longyearbyen in the remote Arctic Svalbard archipelago, about 1,300 km (810 miles) from the North Pole. The facility preserves a wide variety of plant seeds in an underground cavern. The seeds are duplicate samples, or «spare» copies, of seeds held in genebanks worldwide. The seed vault will provide insurance against the loss of seeds in genebanks, as well as a refuge for seeds in the case of large scale regional or global crises. The seed vault is managed under terms spelled out in a tripartite agreement between the Norwegian government, the Global Crop Diversity Trust (GCDT) and the Nordic Genetic Resource Center (also known as NordGen and previously named the Nordic Gene Bank, a cooperative effort of the Nordic countries under the Nordic Council of Ministers). The Prime Ministers of Norway, Sweden, Finland, Denmark, and Iceland participated in a ceremonial «laying of the first stone» on 19 June, 2006. The Svalbard Global Seed Vault opened officially on February 26, 2008. The first seeds arrived in January 2008. Five percent of the seeds in the Vault, about 18,000 samples with 500 seeds each, come from the Centre for Genetic Resources of the Netherlands (CGN), part of Wageningen University, Netherlands.

Construction of SGSV

The seedbank is constructed 120 m (390 ft) inside a sandstone mountain at Svalbard on Spitsbergen Island. The bank employs a number of robust security systems. Seeds are packaged in special four-ply packets and heat sealed to exclude moisture. The facility is managed by the Nordic Genetic Resource Center, though there is no permanent staff on-site.

Spitsbergen was considered ideal due to its lack of tectonic activity and its permafrost, which will aid preservation. The location 130 m (430 ft) above sea level will ensure that the site remains dry even if the icecaps melt. Locally mined coal provides power for refrigeration units that further cool the seeds to the internationally recommended standard -18°C (0°F). Even if the equipment fails, at least several weeks will elapse before the temperature rises to the -3°C (27°F) of the surrounding sandstone bedrock. Prior to construction, a feasibility study determined that the vault could preserve seeds from most major food crops for hundreds of years. Some seeds, including those of important grains, could survive far longer, possibly thousands of years.

Mission and Seed Storage

The Svalbard Global Seed Vault's mission is to provide a safety net against accidental loss of diversity in traditional genebanks. While the popular press has emphasized its possible utility in the event of a major regional or global catastrophe, it will certainly be more frequently accessed when gene banks lose samples due to mismanagement, accident, equipment failures, funding cuts and natural disasters. Such events occur with some regularity. In recent years, some national genebanks have also been destroyed by war and civil strife. There are some 1,400 crop diversity *collections* around the world, but many are in politically unstable or environmentally threatened nations. The seeds are stored in four-ply sealed envelopes, then placed into plastic tote containers on metal shelving racks. The storage rooms are kept at -18°C (-0°F). The low temperature and limited access to oxygen will ensure low metabolic activity and delay seed aging. The permafrost surrounding the facility will help maintain the low temperature of the seeds if the electricity supply should fail. Approximately 1.5 million distinct seed samples of agricultural crops are thought to exist. The variety and volume of seeds stored will depend on the number of countries participating – the facility has a capacity to conserve 4.5 million.

Gene Bank Standards

Research on seed storage has indicated that the potential of seeds to store, that is, retaining genetic integrity and seed viability, is influenced by storage seed moisture content and temperature. Germplasm is generally conserved as a base collection or an active collection. Base collections are those that are being conserved on a long-term basis for posterity. These are unique accessions that are closest to the original samples and are not to be disturbed except for regeneration of active collections. Active or working collections are those that are immediately available for multiplication and distribution for use in research and crop improvement. To minimise the alteration in genetic structure and loss of viability in germplasm accessions during storage, the seed genebanks (that are part of the national network) preferably follow the genebank standards as recommended by FAO/IPGRI (Anonymous, 2001) in relation to various factors important to the good maintenance of active and base collections. The base collections are being stored in modules maintained at -20°C. Such a low temperature minimises metabolic activities and is expected to enable the seed to retain viability for 50 to 100 years without any change in genetic structure. Active collections are stored in modules maintained at 4°C and 35 to 40% relative humidity, under which seeds are expected to remain viable for 15 to 50 years without substantial change in viability and genetic integrity. For both types of collections, seed is processed after validating physical and genetic purity of seed, assessment of seed viability and seed moisture content. In most crops, seed samples with more than 85% seed viability are conserved. However, recognising inherent problems, such as indeterminate nature, which limits the harvest of physiologically mature seed of the same age in certain crops like cotton, several forages and vegetable crop species, the initial viability standards have been lowered down to between 50 to 75%. For long-term storage, the seed moisture content is brought down to 3 to 7%, while for medium-term storage the seed moisture content is brought down to 8 to 10%. For base collections to be put under long-term conservation, the preferable size of accession is 2,000 seeds in the case of self-pollinated and 4,000 in the case of cross-pollinated crops. However, in many cases, such as groundnut and castor, because of large seed size and low multiplication rates, the sample-size of the accessions has been reduced to between 1,000 to 1,500 seeds. The base and active collections are regularly monitored for seed viability, seed quantity, seed health, etc., at recommended intervals of 10 and 5 years, respectively. However, the monitoring of accessions at the National Seed Genebank (NSGB) in the Germplasm Conservation Division, NBPGR has generated valuable information on storability in a number of crop species, such as wheat, minor millets, cotton, grain legumes etc. (Anonymous, 2001). These results suggest a revision of the exact period of monitoring intervals. This information will be useful in revising the seed genebank standards in relation to other components and make seed conservation more cost effective. Seed storage problems are more common in India, because a large part of the country has a predominantly hot and humid, tropical and sub-tropical climate with great variation in temperature, rainfall and relative humidity across the year.

NATIONAL NETWORK ON CONSERVATION OF PGR

Efficient conservation of PGR in a country of the size and dimension of India, one of the 12 mega-centres of plant biodiversity and where 384 crop plants are reported to be cultivated (of which 168 species were earlier reported under the Hindustani centre, one of the eight Vavilovian centres of origin and diversity (Paroda et al., 1999), essentially requires a network approach. Network facilitates short-, medium-, and long-term conservation requirements, the division of responsibilities, application of complementary conservation strategies, and access for the use of these genetic resources in crop improvement programmes. The national network consists of the NSGB at NBPGR headquarters, New Delhi, 11 NBPGR Regional Stations situated in different agro-climatic zones of the country, and 40 crop-based National Active Germplasm Sites (NAGS), located generally at various ICAR institutes. The network is linked with the All India Co-ordinated Crop Improvement Projects, various research institutes (crop-based institutes, project directorates and national research centres; multi-crop based institutes) in the ICAR, SAUs, etc. All network components operate in close collaboration to ensure the efficient conservation and sustainable use of germplasm in crop improvement, in which the National Seed Genebank plays a pivotal role in conservation.

The National Seed Genebank

The NSGB is responsible for conservation of seeds of unique accessions on a long-term basis, as base collections for posterity. In addition, it provides technical support to the network in the planning, development and operation of medium-term genebank facilities, in human resource development, and in provision of accessions for the regeneration of active collections. The Indian NSGB has 12 modules with a capacity to hold around 1 million accessions.

NBPGR Regional Stations

The NBPGR has 11 regional stations/base centres/satellite stations located in different agroecological and phytogeographical zones of the country. They are responsible for the collection, characterisation, evaluation and/or conservation of germplasm in the region. The regional stations also coordinate various PGR activities in the region with other partners. Seven of the regional stations have medium-term seed storage modules for the conservation of active collections to meet the requirement of the region for various crops. The regional stations hold around 98,498 active collections. In addition, plant quarantine is looked after at the NBPGR headquarters, New Delhi and at the NBPGR regional station, Hyderabad.

National Active Germplasm Sites (NAGS)

The NAGS are based at ICAR institutes, at All India Co-ordinated Crop Improvement Projects and at SAUs. They are entrusted with the responsibility of crop specific collection, multiplication, evaluation, maintenance and conservation of active collections and their distribution to users at a national level. Large multiplications of active collections are preferred to reduce the number of regeneration cycles that can cause possible genetic changes and to meet the demand of seed distribution. The NAGS have a multidisciplinary team of scientists to study all the aspects of crop improvement, production and management. Therefore, the NAGS, in addition to their conservation role, are well equipped for the evaluation of germplasm and the generation of information on the potential value of accessions. This information forms the basis for use of accessions in research and crop improvement. Eleven of the NAGS have been provided with medium-term seed storage modules, to facilitate the use of active collections in research and breeding programmes.

Safety Duplicates of Crop Species

There is a built-in duplicity of accessions in the system, wherein the accessions conserved at NAGS and the crop-based institutes as active collection are conserved as base collection in the National Genebank. The active collections are used in research and crop improvement and the National Genebank helps in restoration of lost accessions to the active sites. This also serves as safety mechanism. There exists medium to high capability for research and use of improved methodologies for ex situ conservation. Nevertheless, strengthening of technical and infrastructure capabilities is required in some cases.

The capacity building in genebank management and information systems has been carried out satisfactorily, though there is a need for extension of medium-term facilities to more crop based institutes to cover larger number of crops. In last ten years 196,745 accessions were collected under 166 projects involving 599 professional and of these 104 084 accessions have been conserved. The maximum number of accessions conserved in ex situ is in the category of traditional cultivars and landraces. A significant number of collections belonging to wild and weedy relatives and advanced and improved cultivars developed using various genetic resources is also being conserved.

Major Achievements Through Germplasm Conservation

Plant Genetic Resources for Food and Agriculture (PGRFA) are vital to the development and welfare of human society. They contribute enormously towards achieving the global objectives of food security and poverty alleviation, environment protection and sustainable development. The local communities and farmers in India have sustained and enriched the diversity of these resources which they domesticated, used, conserved and made available to meet the ever increasing needs of the present and future generations. Characterization and evaluation of germplasm is required to know its worth or usefulness and availability of information on characterization and evaluation of conserved genetic resources is the key to utilization. Plant breeding provides many examples of the use of genetic resources for the improvement of the varieties of crop plants. There are examples that range from highly specific improvement to one major factor such as susceptibility to a pest or disease to all round improvement in yield, agronomical traits, disease resistance and to changes in the form and structure of the plant type.

FUTURE THRUST

- Endangered germplasm from the threatened areas of diversity to be salvaged and conserved for future use.
- Morphological and molecular characterization of germplasm to enhance their utilization in crop improvement.
- Conservation, management and protection of bio-resources especially

plant resources, through the participation of the people.

- Conservation and use of diversity needed to be addressed in a holistic manner and to meet the demands of the users of germplasm.
- Research on core and mini-core collections and identification of new diverse sources.
- Public awareness of the importance of CWR and neglected and underutilized species.
- Need to maximize synergy through appropriate collaboration between various national, sub-regional and international levels.

REFERENCES

1. Ahmad M, Zaffar G., Razvi SM, Mir SD, Bukhari SA, Dar ZA, Habib M (2014). Resilience of cereal crops to abiotic stress: A review. Afr. J. Biotechnol. 13(29):2908-2921.
2. Allard RW (1970). Populations structure and sampling methods. in Genetic Resources in Plants (O.H. Frankel and E. Bennett, eds.) 97–107. Blackwell, Oxford. Council, National Research. 1993. Managing Global Genetic Resources, Agricultural Crop Issues and Policies. National Academy Press, Washington D.C.
3. Andersson MS, Fuquen EM, de Vicente MC (2006). State of the art of DNA storage: results of a worldwide survey, 6-10. In: M.C. de Vicente Eds. DNA Banks- providing novel options for genebanks. Topical reviews in agricultural biodiversity. International Plant Genetic Resources Institute, Rome, Italy.
4. Anonymous (2001). Annual report 2000. National Bureau of Plant Genetic Resources, New Delhi, India.
5. Arora RK (1991). Plant Diversity in Indian Gene Centre. In: Paroda, R.S. and R.K. Arora (eds) Plant Genetic Resources - Conservation and Management. IPGRI, Reginoal Office for South Asia, New Delhi, India, pp. 25-54.
6. Arora RK, Pandey A (1996). Wild Edible Plants of India: Diversity, Conservation and Use. NBPGR, New Delhi, India.
7. Burke MB, Lobell DB Guarino L (2014). Shifts in African crop climates by 2050 and the implications for crop improvement and genetic resources conservation. Global Environmental Change. 19(3):317-325.
8. Damania AB (2002).the Hindustan centre of origin of important plants. Asian Agri-history 6:333-341.
9. Dulloo ME, Thormann I, Jorge AM, Hanson J (2013) Crop specific

regeneration guidelines [CD-ROM]. CGIAR System-wide Genetic Resource Programme (SGRP), Rome, Italy. (CD-ROM)

10. Ellis RH, Hong TD. Roberts EH (1990). An intermediate category of seed storage behaviour. J. Exp. Bot. 41:1167-1174.
11. Ellis RH, Hong TD, Roberts EH, Soetisna U (1991). Seed storage ehaviour in Elaeis guineensis. Seed Sci. Res. 1:99-104'.
12. Engelmann F (2000). Importance of cryopreservation for the conservation of plant genetic resources. In Engelmann F, Takagi H (eds) Cryopreservation of tropical plant germplasm. Current research progress and application. IPGRI, Rome, Italy, pp. 8-20.
13. Engelmann F, Engels JMM (2002). Technologies and strategies for ex situ Conservation, 89-103 In: Engels, J.M.M., V. Rao Ramanatha, A.H.D. Brown, M.T. Jackson, Eds. Managing Plant Genetic Diversity. CABI Publishing, UK.
14. Engelmann F, Panis B (2009). Strategy of Bioversity International cryopreservation research - a discussion paper. Bioversity international, Rome, Italy. Unpublished. 8.
15. Escobar R, Mafla G, Roca WM (1997). Methodology for recovering cassava plants from shoot tips maintained in liquid nitrogen. Plant Cell Rep. 16:474-478.
16. FAO (1996). Report on the state of the World's Plant Genetic Resources, Rome Italy.
17. FAO (1998). The state of the world's plant genetic resources for food and agriculture. Food and Agriculture Organization of the United Nations, Rome, Italy.
18. FAO (2010). The second report on the state of the world's plant genetic resources for food and agriculture. Food and Agriculture Organization of the United Nations, Rome, Italy.
19. Frankel OH, Hawkes JG (1975). Crop Genetic Resources for Today and Tomorrow. Cambridge University Press, Cambridge. P. 492.
20. Frankham R (2010). Challenges and opportunities of genetic approaches to biological conservation. Biological conservation 143:1919-1927.
21. GCDT (2010) http://www.croptrust.org/main/arcticseedvault.php?itemid=211, (last accessed 20 August 2010)
22. Hodkinson TR, Waldren S, Parnell JAN, Kelleher CT, Salamin K, Salamin N (2007). DNA banking for plant breeding, biotechnology and biodiversity evaluation. J. Plant Res. 120:17-29.
23. Jarvis DI, Hodgkin T (2008). The maintenance of crop genetic diversity

24. Kartha KK, Engelmann F (1994). Cryopreservation and germplasm storage. In: (Vasil IK, Thorpe TA (eds) Plant Cell and Tissue Culture. Kluwer Academic Publishres, Netherlands, pp. 195-230.

25. Laikre L, Allendorf FD, Aroner LC, Baker CS, Gregovich DP, Hansen MH, Jackson J, Kendal AKC, Mckelvey K, Neel MC, Olivieri I, Ryman NM, Schwartz K, Bull R S, Jeffrey B Stetz JB, Tallmon DA, Taylor BL, Vojta CD, Waller DM, Waples RS (2010). Neglect of genetic diversity in implementation of the convention of biological diversity. Conservation Biology 24(1):86-88.

26. Laikre L, Larsson LC, Palme A, Charlier J, Josefsson M, Ryman N (2008). Potentials for monitoring gene level biodiversity: using Sweden as an example. Biodivers. Conserv. 17:893-910.

27. Lane A, Jarvis A (2007). Changes in climate will modify the geography of crop suitability: agricultural biodiversity can help with adaptation. J. SAT Agric. Res. 4(1):1-12.

28. Maxted N, Kell SP (2009). Establishment of a Global Network for the In Situ Conservation of Crop Wild Relatives: Status and Needs. FAO Commission on Genetic Resources for Food and Agriculture, Rome, Italy.

29. Millennium Ecosystem Assessment (2005). Ecosystems and Human Well-being: Biodiversity Synthesis. World Resources Institute, Washington, DC.

30. MoEF (1994). Ethnobotany in India – A status report. All India Coordinated Research Project on Ethnobotany. Ministry of Environment and Forest, Government of India, New Delhi, India.

31. NAS (1975). Under-exploited Tropical Plants with Promising Economic Value. National Academy of Sciences, Washington DS, USA.

32. Paroda RS, Rai M, Gautam PL, Kochher S, Singh AK (1999). National action plan on agrobiodiversity in India. National Academy of Agricultural Sciences, Indian

33. Rao VR (1980). Groundnut genetic resources at ICRISAT. in Proceedings of the International Workshop on Groundnuts, 13–17 October, 1980. 47–57. ICRISAT, Patancheru, 502324, A.P., India.

34. Rice N, Henry R, Rossetto M (2006). DNA Banks: a primary resources for conservation research, 41-48. In: M.C. de Vicente Eds. DNA Banks-providing novel options for genebanks. Topical reviews in agricultural biodiversity. International Plant Genetic Resources Institute, Rome, Italy.

35. Roberts EH, King MW (1986). Storage of recalcitrant seeds. in Crop

Genetic Resources – The Conservation of Difficult Material (L.A. Withers and J.T. Williams, eds.) 39–48. IUBS/IGF/IBPGR, Paris.

36. Ryder OA, McLaren A, Brenner S, Zhang Y, Benirschke K (2000). DNA Banks for Endangered Animal Species. Science 288 (5464):275-277.
37. SBSTTA (2010). Proposals for a consolidated update of the global strategy for plant conservation. UNEP/CBD/SBSTTA/14/9,-http://www.cbd.int/doc/?meeting=SBSTTA-14.
38. Thormann I, Hunter D, Danielian A, Djataev S, Ramelison J, Wijesekara A, Zapata Ferrufino B, Borelli T (2010). Meeting the information challenge for crop wild relatives in situ conservation: A global portal hosting national and international data on CWR. Presented at the IAALD XIIIth World Congress - 26-29 April 2010, Montpellier, France.
39. Thormann I, Dulloo ME, Engels JMM (2006). Techniques of ex situ plant conservation, 7-36. In: Robert Henry (ed.) "Plant Conservation Genetics". Centre for Plant Conservation Genetics, Southern Cross University, Lismore, Australia. The Haworth Press inc.
40. UNCED (1992). Convention on Biological Diversity. United Nations Conference on Environment and Development, Geneva.
41. United Nations (2010). The Millennium Development Goals Report 2010. New York. USA.
42. Van de Wouw M, Kik C, Hintum T, Van Treuren R, Visser B, (2009).
43. Plant genetic resources; Characterization and utilization 2009:1-15.
44. Vavilov NI (1951). Phyto-geographical basis of plant breeding. In: (selected Writings of N.I. Vavilov and translated by K.S. Chester) The Origin, Variation, Immunity and Breeding of Cultivated Plants, Chronica Botanica 13:364.Waltham Mass., USA.
45. Walpole Matt, Rosamunde EAA, Besançon C, Butchart SHM, Campbell-Lendrum D, Carr GM, Collen B, Collette L, Davidson NC, Dulloo E, Fazel AM; Galloway JN, Gill M, Goverse T, Hockings M, Leaman DJ, Morgan DHW, Revenga C, Rickwood C J, Schutyser F, Simons S, Stattersfield AJ, Tyrrell TD, Vié JC and Zimsky M (2009). Tracking progress towards 2010 biodiversity target and beyond. Science 135:1503-1504.
46. Withers LA, Engelmann L (1993). In vitro conservation of plant genetic Resources. In biotechnology in Agriculture, (A. Altman, ed.), Marcel Dekker Inc., New York, in press.
47. Zheng GH, Jing XM, Tao KL (1998). Ultradry seed storage cuts costs of gene bank. Nature 393:223-224.
48. Zhou M, Engels JMM, Ramanatha RaoV 1(995). Effect of low seed

moisture content on seed storage. 14 (ab) in International Symposium on Research and Utilization of Crop Germplasm Resources, 1–3 June, Beijing, China.

Chapter 3

CLIMATE-SMART AGRICULTURE GLOBAL RESEARCH AGENDA: SCIENTIFIC BASIS FOR ACTION

Kerri L Steenwerth[1], Amanda K Hodson[2], Arnold J Bloom[3], Michael R Carter[4], Andrea Cattaneo[5], Colin J Chartres[6], Jerry L Hatfield[7], Kevin Henry[8,9], Jan W Hopmans[2], William R Horwath[2] Bryan M Jenkins[10], Ermias Kebreab[11], Rik Leemans[12], Leslie Lipper[13], Mark N Lubell[14], Siwa Msangi[15], Ravi Prabhu[16], Matthew P Reynolds[17], Samuel Sandoval Solis[2], William M Sischo[18], Michael Springborn[19], Pablo Tittonell[20], Stephen M Wheeler[21], Sonja J Vermeulen[22], Eva K Wollenberg[23], and Louise E Jackson[2]

[1]Crops Pathology and Genetics Research Unit, c/o Department of Viticulture and Enology, United States Department of Agriculture (ARS/USDA), Agricultural Research Service

[2]Department of Land, Air and Water Resources, University of California at Davis

[3]Department of Plant Sciences, University of California at Davis

[4]Department of Agricultural and Resource Economics, University of California at Davis

[5]Climate Smart Agriculture Project, Food and Agriculture Organization of the U.N., Viale delle Terme di Caracalla

[6]eWater, University of Canberra Innovation Centre, University Drive South

[7]National Laboratory for Agriculture and the Environment, ARS/USDA

[8]Where the Rain Falls, CARE France

[9]School of Global Envirfonmental Sustainability, Colorado State University

[10]Department of Biological and Agricultural Engineering, University of California at Davis

[11]Department of Animal Science, University of California at Davis

[12]Environmental Sciences, Wageningen University

[13]Agricultural and Development Economic Analysis Division, Food and Agriculture Organization of the U.N., Viale delle Terme di Caracalla

[14]Department of Environmental Science and Policy, Center for Environmental Policy and Behavior, University of California at Davis

[15]Environment and Production Technology Division, International Food Policy Research Institute (IFPRI)

[16]World Agroforestry Center (ICRAF)

[17]Plant, International Maize and Wheat Improvement Center, Consultative Group on International Agricultural Research (CGIAR) Apdo, Postal

[18]Food- and Water-borne Disease Research Program, College of Veterinary Medicine, Washington State University

[19]Department of Environmental Science and Policy, University of California at Davis

[20]Plant Sciences, Wageningen University

[21]Department of Landscape Architecture, University of California at Davis

[22]Department of Plant and Environmental Sciences, Climate Change, Agriculture and Food Security, Consultative Group on International Agricultural Research (CGIAR), University of Copenhagen

[23]Gund Institute for Ecological Economics and Rubenstein School of Environment and Natural Resources, University of Vermont

ABSTRACT

Background

Climate-smart agriculture (CSA) addresses the challenge of meeting the growing demand for food, fibre and fuel, despite the changing climate and fewer opportunities for agricultural expansion on additional lands. CSA focuses on contributing to economic development, poverty reduction and food security; maintaining and enhancing the productivity and resilience of natural and agricultural ecosystem functions, thus building natural capital; and reducing trade-offs involved in meeting these goals. Current gaps in knowledge, work within CSA, and agendas for interdisciplinary research and science-based actions identified at the 2013 Global Science Conference on Climate-Smart Agriculture (Davis, CA, USA) are described here within three themes: (1) farm and food systems, (2) landscape and regional issues and (3) institutional and policy aspects. The first two themes comprise crop physiology and genetics, mitigation and adaptation for livestock and agriculture, barriers to adoption of CSA practices, climate risk management and energy and biofuels (theme 1); and modelling adaptation and uncertainty, achieving multifunctionality, food and fishery systems, forest biodiversity and ecosystem services, rural migration from climate change and metrics (theme 2). Theme 3 comprises designing research that bridges disciplines, integrating stakeholder input to directly link science, action and governance.

OUTCOMES

In addition to interdisciplinary research among these themes, imperatives include developing (1) models that include adaptation and transformation at either the farm or landscape level; (2) capacity approaches to examine multifunctional solutions for agronomic, ecological and socioeconomic challenges; (3) scenarios that are validated by direct evidence and metrics to support behaviours that foster resilience and natural capital; (4) reductions in the risk that can present formidable barriers for farmers during adoption of new technology and practices; and (5) an understanding of how climate affects the rural labour force, land tenure and cultural integrity, and thus the stability of food production. Effective work in CSA will involve stakeholders, address governance issues, examine uncertainties, incorporate social benefits with technological change, and establish climate finance within a green development framework. Here, the socioecological approach is intended to reduce development controversies associated with CSA and to identify technologies, policies and approaches leading to sustainable food production and consumption patterns in a changing climate.

Introduction

Globally, agricultural and forestry systems are expected to change significantly in response to future climate change, manifesting as major transitions in livelihoods and landscapes [1–4]. During the few past decades, crop yields have been reduced because of warming [5], and the results of modelling studies suggest that climate change will reduce food crop yield potential, particularly in many tropical and midlatitude countries [6–9]. Rising atmospheric CO_2 concentrations will decrease food and forage quality [10]. Price and yield volatility likely will continue to rise as extreme weather continues, further harming livelihoods and putting food security at risk [11]. Global demand for agricultural products, be they food, fibre or fuel, continues to increase because of population growth, changes in diet related to increases in per capita income and the need for alternative energy sources while there is less and less additional land available for agricultural expansion. Agriculture thus needs to produce more on the same amount of land while adapting to a changing climate and must become more resilient to risk derived from extreme weather events such as droughts and floods.

The term *climate-smart agriculture* (CSA) has developed to represent a set of strategies that can help to meet these challenges by increasing resilience to weather extremes, adapting to climate change and decreasing agriculture's greenhouse gas (GHG) emissions that contribute to global warming (Figures 1 and 2). CSA also aims to support sustainable and equitable transitions for agricultural systems and livelihoods across scales, ranging from smallholders to transnational coalitions. Forming a core part of the broader green development agenda for agriculture [12–14], CSA focuses on meeting the needs of people for food, fuel, timber and fibre through science-based actions; contributing to economic development, poverty reduction and food security; maintaining and enhancing the productivity and resilience of both natural and agricultural ecosystem functions, thus building natural capital; and reducing the trade-offs involved in meeting these goals. It invokes a continuous, iterative process for stakeholders, researchers and policymakers to meet the challenges presented by climate change and collectively transform agricultural and food systems towards sustainability goals [15]. Increased awareness and adaptive management are essential components of the CSA strategy. Yet, CSA is controversial. Such a broad agenda can be appropriated to support conflicting agendas or promote specific ecosystem services [16]. GHG emission mitigation by resource-poor farmers raises equity as an issue in developing countries because it may bring farmers little benefit unless it directly provides them with adaptive capacity.

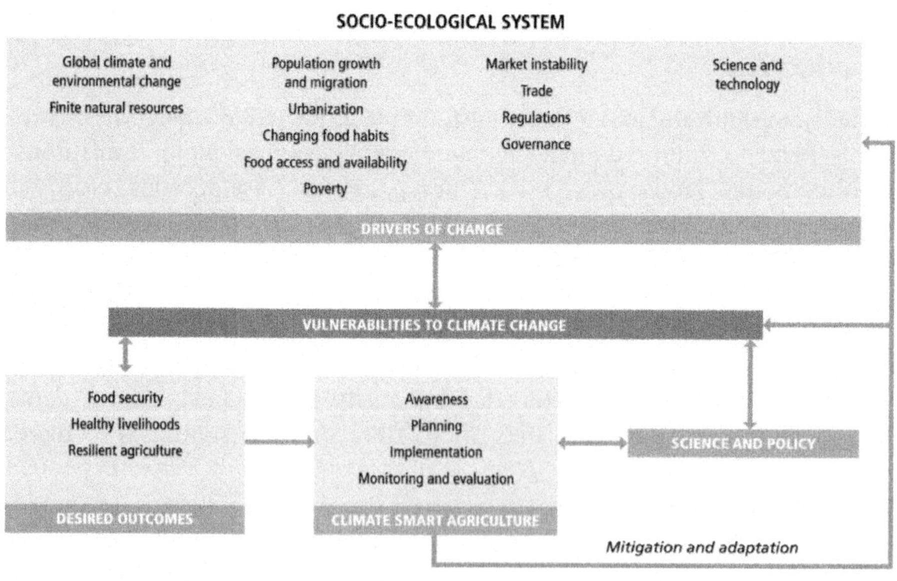

Figure 1: Diagram illustrating how climate-smart agriculture can be utilised as an agent for developing resilience, mitigation and adaptation within the socioecological system. Although not exclusively within the purview of climate-smart agriculture (CSA), 'adaptation and mitigation' in this figure are implied to be derived from an

iterative CSA strategy. Adaptation and mitigation affect 'drivers of change' to diminish existing 'vulnerabilities to climate change' in the socioecological system, leading to the long-term goals of CSA in 'desired outcomes'. The arrow between CSA and 'science and policy' indicates the vital role of novel science–policymaking partnerships and science-based actions in CSA.

Setting CSA in the context of a safe operating space for humanity with socioecological systems that support adaptive management and governance will require scientific metrics and science–policymaking dialogues [16] that depend on strong engagement of the scientific community.

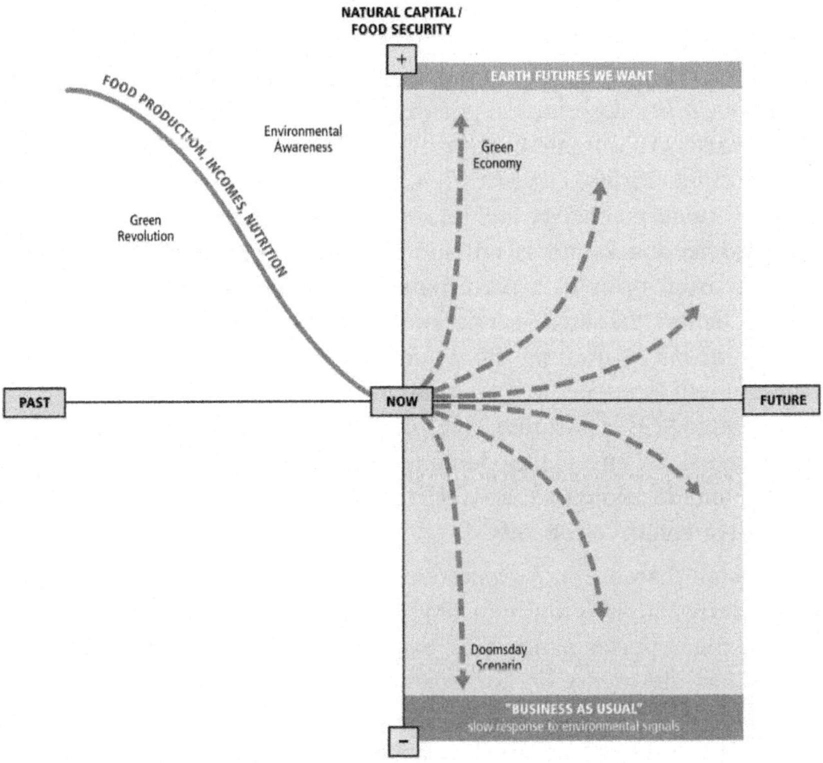

Figure 2: Schematic of the historical trajectory of food production, incomes and nutrition within the socioecological system with respect to food security and natural capital over time juxtaposed with potential future outcomes. Placement of 'environmental awareness' indicates that this factor gained prominence in the socioecological system as food security and natural capital diminished with respect to the trajectory of food production, income and nutrition. Positive outcomes are shaded in green, and the 'business as usual' and 'doomsday' outcomes are shaded in orange. Climate-smart agriculture plays a role in building positive outcomes such as the 'green economy'. Adapted with permission from M van Noordwijk (personal communication, 2014).

At the 2013 Global Science Conference on Climate-Smart Agriculture (Davis, CA, USA), participants examined the state of global science and best practices concerning climate and agriculture worldwide. Participants built on the consensus achieved at the 2011 Global Science Conference on Climate-Smart Agriculture conference (Wageningen, the Netherlands), agreeing on a broad strategy for science and policymaking to strengthen food security, mitigation and adaptation [17]. Participants further examined current gaps in knowledge, identified existing and promising work within CSA and formulated agendas for interdisciplinary research and science-based actions to support CSA.

The relationship between vulnerability, resilience and adaptation was an overarching theme echoed across the conference and is crucial to CSA. *Vulnerability* describes exposure, sensitivity and capacity to respond to negative impacts of climate change, and *adaptation* is the means by which to reduce the vulnerability. Here *resilience* is regarded as the capacity to tolerate disturbance, undergo change and retain the same essential functions, structure, identity and feedback and is not indicative solely of returning to the same state that existed prior to a perturbation or disturbance [18–20]. Resilience focuses on factors that enable functioning despite adverse conditions [21, 22], provides a means of framing the dynamic relationships between humans and the environment (socioecological systems) and considers society's capacity to manage change [23]. Thus, the principle of resilience can guide transformative change needed to meet the demands of food security, natural resource protection, and development, as well as to diminish vulnerability and promote adaptation (or adaptive capacity).

The recent increase in extreme weather events (climate shocks) threatens disruptive impacts on agriculture [24, 25]. Projected adaptive actions include improving plant performance (for example, nutrition, yields, food quality) in response to elevated CO_2 and rising temperatures [26–28]; avoiding pest damage and food waste [28, 29]; developing forecasting, management and insurance options to decrease the risk due to unexpected rainfall patterns, higher temperatures and shifting length in growth seasons [14, 28, 30]; and managing natural resources at the landscape and regional levels to assure the environmental quality and ecosystem services upon which agriculture depends [31–33]. Solutions involve trade-offs. For instance, planning now for higher temperatures and declining precipitation in arid zones may reduce water deficits for agriculture, but it will require institutional investment to support both the intensified demand for ground and surface waters [34, 35] and the necessary improvements in irrigation efficiencies [36]. Along with these adaptive actions, CSA seeks to contribute to the mitigation and reduction of GHG, mainly nitrous oxide (N_2O) and methane (CH_4) emissions, and to balance trade-offs with food security and livelihoods [7, 37, 38]. For example, combining agroforestry, afforestation and conservation efforts with agriculture to meet global food demand will help to mitigate GHG emissions, support biodiversity and concomitantly preserve ecosystem services [39, 40]. Other trade-offs that occur when abrupt environmental changes stress agricultural systems include changes in rural and urban human migration patterns, as well as loss of cultural resources, which reduces the ability to manage land use effectively [41–43]. Without doubt, the development status of a country or region will influence the approach to mitigating and adapting to climate uncertainty and will affect the implementation and focus of the CSA strategy. For example, industrialized nations focus more strongly on mitigation of climate change through reduction of agriculture's environmental impacts, whereas developing countries' approaches to climate uncertainty emphasize stabilizing and boosting food production, improving incomes and building adaptive capacity [7, 15, 44]. Gender can also influence decisions and capacity for mitigation and adaptation. Women in some regions in Africa have experienced greater exposure and vulnerability, especially to extreme events, than men, but they also have demonstrated greater collective action in farming decisions linked to social networking [45, 46].

Crucial science questions and challenges for food systems in the face of climate change and uncertainty require comprehensive, collaborative investments and science-based actions. In the past few years, policies and programmes have included landscape-scale research on food security and natural resources, policy and governance to achieve agricultural resilience to climate change and capacity building [47]. Under CSA, transformative changes to achieve food security, poverty relief, mitigation and adaptation target novel types of science–policymaking partnerships and involve stakeholders and decision-makers in the public and private sectors to gain long-term commitment and investment to carry the new actions to fruition. CSA emphasizes the involvement of scientists with farmers, land managers, agroforesters, livestock keepers, fishers, resource managers and policymakers (stakeholders) to empower them in the formation of palatable choices to enact adaptive capacity and resilience 'on the ground' and within broader policies [14, 15]. Farmer-led innovative approaches and social learning are crucial parts of this process, where social learning represents a 'change in understanding that goes beyond the individual to become situated within wide social units or communities of practice through social interactions between actors within social networks' [48, 49]. In this article, we summarize and synthesize the discussions and ideas presented at the 2013 CSA conference by an international community of scientists, growers, policymakers, research scientists, government officials, nonprofit entities and students who are working to achieve food security, poverty reduction, mitigation and adaptation within the CSA context. The three sections of this article reflect the scientific themes presented at the conference: (1) farm and food systems, (2) landscape and regional issues and (3) institutional and policy-related aspects. Within the first and second themes, parallel sessions at the conference charged participants to identify knowledge gaps, research initiatives and transformative actions required to address these specific issues. We provide a summary of the 12 sessions and highlights of the oral presentations by subject experts, and we conclude with recommendations offered during discussions as well as a consensus agenda for future actions [50]. Finally, broad outcomes and messages are presented, largely adhering to the actual proceedings to reflect the spirit and outcomes of this conference. Thus, the emphasis is on structuring disciplinary and interdisciplinary science in a CSA context rather than mechanisms for implementing science in action. This article is intended to serve as a benchmark and guide for future CSA research activities.

THEME 1

Farm and Food System Issues: Sustainable Intensification, Agro-ecosystem Management and Food Systems

Considerable research on climate change and agriculture exists at the farm and food system levels, including topics such as farming practices for mitigation of agricultural GHG emissions, choice and adaptation of crops and livestock to new climate regimes, decision-making by farmers and life-cycle assessments [51–55]. The tendency has been to apply disciplinary science that informs particular problems and solutions for agriculture, as demonstrated by the topics of the six sessions in theme 1. Sustainable intensification, focused initially on increased agricultural production and food security, has now moved to a broader set of goals with multiple social, ethical and environmental dimensions [56, 57]. The integrative challenge for CSA is to better understand the trade-offs and choices farmers must make for greater multifunctionality and resilience to climate change. Because planning for climate change can be highly farm-, commodity- and context-specific, especially in response to extreme events, CSA is committed to new ways of engaging in participatory research and partnerships with producers [14].

Crop Physiology and Genetics under Climate Change

Responding to effects of climate change (for example, changes in nutrient availability and plant nutrient acquisition, higher CO_2 concentrations and temperatures, water deficits and flooding) that influence the closure of the yield gap between potential and actual production will require continuation of existing 'best management practices' coupled with improvements in agronomic management practices and crop-breeding [58, 59]. Uncertain is the degree to which advances in crop physiology and genetics will continue to support higher agricultural production in response to more frequent climate shocks. Whereas successful crop adaptation to new production locations may be a good predictor of future outcomes, much higher CO_2 concentrations and temperatures are conditions beyond our current set of experiences [21,60]. Molecular approaches and genetic engineering will foster better understanding and manipulation of physiological mechanisms responsible for crop growth and development, as well as the breeding of stress-adapted genotypes [61–63], but there are social controversies surrounding the use of some of these technologies. High-throughput phenotyping platforms and comprehensive crop models will lead to more rapid exploration of genetic resources, enabling both gene discovery and better physiological understanding of how crop improvement can increase tolerance to environmental stress [64–68]. Development of new crop genotypes to meet the need to thrive under future management and climate conditions, the expected increases in the frequency of climate shocks and the uncertainty of rates of climate change presents a challenge. The specific examples set forth in the following paragraphs demonstrate how greater understanding of biochemical pathways, plant traits and phenotypes and germplasm evaluation could help overcome bottlenecks in both yield and development of physiological resilience to environmental stresses.

Molecular approaches provide opportunities to establish linkages between biochemical pathways and physiological responses. In cereals such as rice, grain yield is highly dependent on the carbohydrate source (top leaves) and sink (florets) relationship, which is strongly influenced by the plant hormone cytokinin [69]. Cytokinin production also affects drought tolerance and senescence, and isopentenyl transferase (IPT) expression controls upregulation of pathways for cytokinin degradation. Therefore, it follows that tolerance of abiotic stress by delaying stress-induced senescence through manipulation of IPT expression in transgenic lines could maintain optimal levels of cytokinin, resulting in greater fitness and more seed and grain production [62]. When exposed to varying drought intensities pre- and postflowering, transgenic rice with higher IPT expression maintained consistently higher grain yields and concentrations of sucrose and starch compared to the wild-type genotype. The delayed onset of drought-related symptoms in the transgenic lines caused positive source–sink relationships for a relatively longer period with higher photosynthetic rates than the wild type.

Combinations of multiple plant traits to survive stress, however, may produce more resilient crop production in the face of climate change [64]. Survival strategies employed by plants include early flowering to escape drought periods, stomatal control to prevent water loss, enhanced root growth in deeper soil layers to access water [70] and reduced leaf growth to minimize the transpiring surface [71]. These adaptations come at a cost, where reductions in the growth cycle, light interception and carbon (C) assimilation by photosynthesis are often accompanied by a higher C requirement to build additional plant roots, especially under nutrient stress [72]. Thus, the trade-offs of introducing new plant traits must be considered for specific types of environmental stress [65].

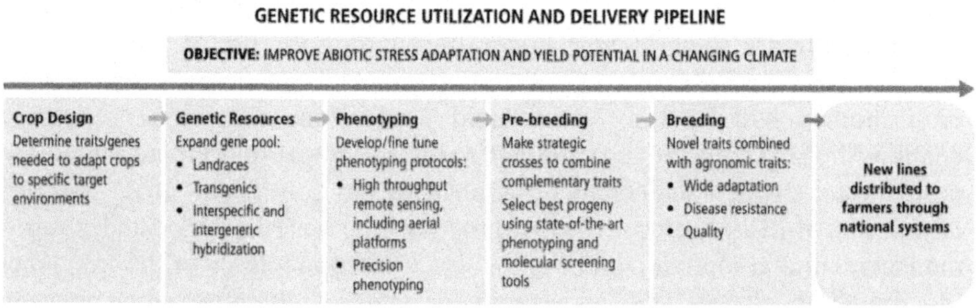

Figure 3: The genetic resource and utilization pipeline reflects the combination of physiological, molecular and traditional breeding approaches. Adapted with permission from M Reynolds (personal communication, 2014).

By examining the genetic basis of physiological mechanisms and environmentally induced stress responses, crops such as maize, wheat and other cereals can be bred to produce better yields and tolerances through targeted accumulation of alleles that confer robust responses to environmental stressors such as drought [73] (Figure 3). This approach is used by the International Maize and Wheat Improvement Center for the discovery and accumulation of drought-adaptive traits in wheat and maize germplasm from wild-type crop relatives and cultivars from a wide range of climates and growing conditions [65, 67, 74]. Screening for physiological traits can be highly effective in selecting such lines for a breeding programme. Canopy temperature (CT) is an example of a widely used, high-throughput germplasm screening tool. CT is linked to stomatal conductance, an indirect indicator of water uptake by roots, especially under drought and heat stress [75, 76]. In one study, researchers found that 60% of variation in yield from recombinant inbred lines grown under drought conditions was explained by CT [77]. Screening for physiological traits in candidate genotypes as an initial step may thus accelerate the search for novel genes [75] and genotypes that will be needed to deal with rapid changes in climate, such as the greater intensity and frequency of drought. Trait-based breeding programmes will be most effective when approaches are developed to simultaneously screen a broad array of genotypes for phenotypic responses to environmental stresses quickly (for example see, [78]).

Complementary approaches are necessary for solving complex physiological plant responses to climate and management. Changes in temperature, precipitation, water delivery, salinity and CO_2 concentrations will occur simultaneously. Direct experimentation, high-throughput screening platforms using molecular-based techniques and predictive modelling are a set of tools for achieving multiple goals [79–81], which include exploration of genetic resources for broader use and dissemination, gene pool expansion and yield stability in the face of interannual weather variation. In addition, these tools can help with other crop selection criteria, including quality of food and feed, source–sink relationships, pest and disease resistance, plant–microbe interactions that reduce CH_4 and N_2O emissions, and postharvest storage [60, 81]. Regional networks that examine environmental and physiological tolerances and yield potentials, as well as their coalescence into global crop improvement networks [82], will provide large-scale screening approaches to assessing both germplasm and phenotypic responses of crop plants. These networks already exist in representative target environments, such as the Network for the Genetic Improvement of Cowpea for Africa, Sorghum and Millet Networks, International Wheat Improvement Network, International Maize Improvement Network; and other regional networks linked to CGIAR that focus on grains and legumes in Africa, Latin America, the Caribbean and Asia. They also include networks for research and extension supported by Association for Strengthening Agricultural Research in Eastern and Central Africa. Participatory breeding by farmers and other stakeholders will eventually be an essential way to advance this agenda [83, 84].

Livestock Management and Animal Health

Livestock production not only contributes to climate change via GHG emissions (see [85]) but also suffers due to extreme weather events and disease related to climate change. Direct and indirect challenges in both mitigation and adaptation include fluctuating feed prices, habitat changes, expansion of vector-borne diseases in warm climates, impaired reproduction, pasture quality and availability and physiological heat stress [86, 87]. Opportunities for mitigating emissions include dietary manipulation, genetic improvement and mortality reduction to enhance overall production potential; manure management; and reduction of deforestation and pasture burning through payments for ecosystem services [88,89]. Adaptation strategies include income and livelihood diversification by mixing crop and livestock production; sustainable intensification through pasture regeneration or destocking; diversifying livestock feeds; manipulation of rumen microbial composition; matching animal breeds to local environments and moving animals to other sites; and better risk management and transformative change (for example, exit from or entry into animal agriculture) [88, 90–92]. These strategies rely heavily on sustainable intensification, as in the improvement of productivity and efficiency that exists in conjunction with incentives and investments that allow systems to intensify and in the development of regulations and limits on intensifying systems, among other aspects [93]. Access to credit or savings, land and resource inputs, and livelihood diversification are other potential pathways towards adaptation and food security [94, 95]. Technology, supporting policies and investments will require varied mixtures of strategies, as shown by the examples described in the following paragraphs.

Flexibility in livelihood options for pastoralist, agropastoralist and ranching communities can increase a household's capacity to manage risk and adapt in the face of burgeoning external stress [96]. Adaptation options depend on household objectives and attitudes; local access to natural resources, inputs and output markets; and sustainable intensification. Nutrient management is fundamental to maintaining a livelihood in livestock production. In Madagascar, external nitrogen (N) inputs are not commonly used to replenish the N losses that occur through erosion, leaching, GHG emissions and harvest. Hence, Alvarez *et al.* [97] examined N flows through crop-livestock systems to determine management scenarios leading to improvement in their N use efficiency, productivity and economic viability. They evaluated four intensification scenarios for system productivity, food self-sufficiency and gross margins: (1) using supplementary feed (N inputs) to increase dairy production; (2) applying mineral N fertilizer to increase crop production; (3) improving conservation of manure N during storage and soil application; and (4) combining scenarios 1 and 3. They found that gross margin increased in response to improved retention of manure N and that increased N supply through supplementary feeding (scenario 4) across farm types led to increases in whole-farm N use efficiencies from 2% to 50%, in N cycling from 9% to 68% and in food self-sufficiency from 12% to 37%. An example of adaptation to manage risk in East Africa is pastoralists who have shifted from cows to camels, which are better-adapted to survive periods of water scarcity and able to consistently provide more milk [98]. Risk adaptation by farmers may also involve changing from cultivated crops to livestock, as crops may be more environmentally and spatially constrained in the pastoralists' home regions [99].

Mitigation options at farm to regional scales form a large part of Brazil's multifaceted approach to managing direct and indirect GHG emissions from livestock. Brazil's commercial cattle industry is the largest in the world (more than 170 million head in 2006), and emissions from raising cattle are responsible for about half of the country's total emissions [100]. The principal targets for mitigating GHG emissions associated with cattle production in Brazil are reduction in deforestation and enteric fermentation, regeneration of secondary forest, recuperation of degraded pasture and soils and elimination of fire in pasture management. Maintenance of grazing productivity and high stocking rates through pasture reclamation and adoption of integrated crop-livestock systems, such as rotational grazing and introduction of legumes in pastures, buffers pressure on deforestation. Such pasture regeneration creates a potential for increasing soil C storage, with increases of up to 0.72 Mg of $C \cdot ha^{-1} \cdot yr^{-1}$ reported under improved management [101]. However, other pasture maintenance practices increase emissions. For example, burning accounted for 1.69 CO_2eq (Mt from total biome) in the Cerrado ecosystem from 2003 to 2008. Key mitigation efforts include reduction in enteric CH_4 emissions by genetic stock improvement and dietary manipulation [91]. This dietary manipulation through grain supplementation increases forage digestibility and reduces enteric fermentation, but it leads to greater emissions of N_2O through the use of fertilizers to grow the grain [100]. Several other promising technologies include grass and legume species with lower GHG emission potential, additives (for example, ionophores and secondary plant compounds such as tannins) and use of propionate precursors in feed to reduce methanogenesis [102]. To complement farm-based efforts, uniform and fair economic procurement and incentivized policies must be in place and enforced across the supply chain in order to establish supply and trade chains with low C footprints. Regional and national policies must contain mechanisms that balance market pressure to convert from low-impact land uses (for example, forests) to relatively more intensive uses (for example, ranching). The Norwegian Agency for Development Cooperation (NORAD) and the Brazilian organization Aliança da Terra, which includes farmers, researchers and agribusiness entrepreneurs, are partnering to increase contributions by private landowners throughout the Brazilian Amazon and Cerrado to the goals of the Reducing Emissions from Deforestation and Forest Degradation (REDD+) programme by combining sustainable rangeland economic development with forest and water protection [103].

Effectively managing emerging zoonotic diseases and outbreaks due to climate change is a strong component of maintaining agropastoralist livelihoods. Increased temperatures due to climate change will affect the survival of pests in the winter and thus distribution of pests and diseases (for example, zoonotic, endemic, emerging, foodborne and noninfectious diseases), though some regions may find relief from these existing pressures in a changing climate. The Emerging Pandemic Threats Program (PREDICT) addresses the broad geographic issues in disease emergence from farm to national to global scales related to a shifting climate [104, 105]. The programme, which is run within the US Agency for International Development, leverages existing networks within national and local governments as well as networks of scientists and specialists involved in outbreak reporting, microbial characterization and pathogen discovery [104]. The programme includes 20 developing countries where hotspots of emerging infectious diseases exist and is focused on surveillance of human–animal interfaces where transmission is more likely. PREDICT is focused on the prevention of pandemics by addressing underlying ecological, economic and social drivers of change, such as shifts in land use from forested systems to livestock. It is used to develop and deliver new technologies (such as information management and communication tools) to improve control efforts close to the pathogen's source. This type of interdisciplinary effort that moves from farm-level to broad spatial scales is considered necessary for creating comprehensive strategies for the control and prevention of emerging zoonotic diseases in a changing climate [104].

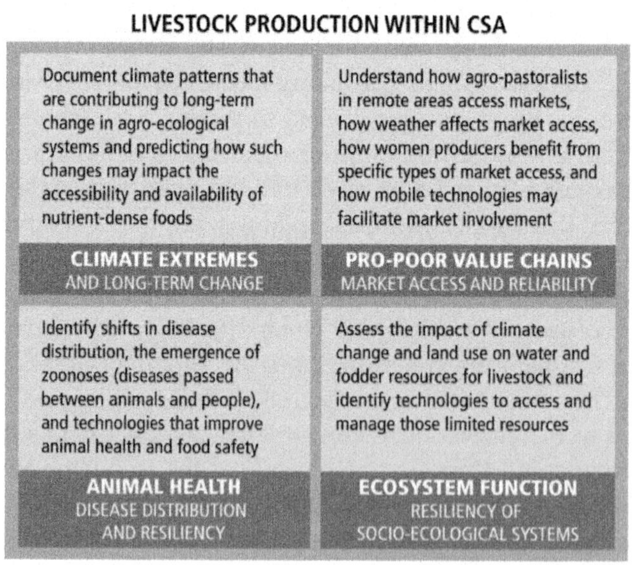

Figure 4: Four aspects of livestock management in climate-smart agriculture. Adapted from [111] with permission from R Bowen (personal communication, 2014). CSA, Climate-smart agriculture.

In support of agropastoral farming systems, models must integrate mitigation options, alternative intensification pathways, zoonotic disease and vector ecology (for example, genetic shifts, patterns of emergence); mechanisms of effecting behavioural change; and adaptation to future climate change scenarios. Some existing models for predicting regional GHG emissions from livestock production include BEEFGEM (Ireland; [106]), IFSM (USA; [107]) and $SIMS_{DAIRY}$ (UK; [108]). Reisinger*et al.* [109] recently evaluated different metrics on the integrated assessment model, MESSAGE, and the land-use model, the Global Biosphere Management Model (GLOBIOM), to examine the global costs of abatement strategies used to reduce the magnitude of climate change and subsequent effects on regional food production and supply prices for livestock products and other agricultural commodities. Other transformative approaches to livestock production include identifying the value of a blend of market-orientated smallholders vs. large-scale farms, evaluating ecosystem services payments as a means of income diversification, forming institutional and market mechanisms for reaching smallholders to foster technological change, finding the best locations for both livestock production and marginal land rehabilitation, and creating new capacity of the livestock sector for mitigation and adaptation in the face of climate change [90, 110] (Figure 4). The social and economic impacts (for example, income stability, human nutrition, product value chains, transaction and opportunity costs of other alternatives) of land-sparing and reducing livestock consumption, two recently suggested mitigation options, merit further investigation, especially with respect to gender, region and income differentiation [92].

Nitrogen Management: Agricultural Production, Greenhouse Gas Mitigation, and Adaptation

Future food security will continue to rely on N fertilizer inputs, but cropping systems must achieve yield potential (that is, close the yield gap) while minimizing trade-offs in air, water and soil quality [58, 59]. The long-term ramifications of N-related GHG emissions; off-site movement of N on eutrophication, acidification and pollution of aquatic and terrestrial ecosystems; and human health problems have led to a recommendation that anthropogenic inputs of reactive N to terrestrial ecosystems be reduced by up to one-fourth of present quantities, or about 35 million tonnes of N per year [112]. Even if this reactive, anthropogenic N entering agroecosystems is emitted as N_2 rather than N_2O, the energy associated with the Haber-Bosch process and transport of fertilizers will still contribute to GHG emissions [113]. Cropping system diversification, careful selection of crop rotations to reduce nutrient loss, and improved soil organic matter content are means by which to promote sustainable intensification. Yet, this often involves a set of complex trade-offs for producers and their livelihoods [114], emphasizing the need for a CSA strategy that involves stakeholders from the beginning to develop viable scenarios that include both mitigation and adaptation to climate change. The examples presented here demonstrate how strategies for N fertilization practices provide both mitigation and adaptation benefits by decreasing GHG emissions, reducing reliance on synthetic mineral fertilizer and enhancing food security.

Enhanced-efficiency fertilizers (EEFs), such as slow-release fertilizers or those containing nitrification inhibitors and urease inhibitors, hold potential to mitigate GHG emissions. According to the Intergovernmental Panel on Climate Change (IPCC) Fourth Assessment Report [114], the mean mitigation potential of N_2O by nutrient management using nitrification inhibitors and slow-release fertilizers has been estimated to be 0.07 t CO_2-eq·ha^{-1}·yr^{-1} (as a reference, agriculture accounted for an estimated 5.1 to 6.1 GtCO_2eq·yr^{-1} in 2005, which amounts to 10% to 12% of total global anthropogenic emissions of GHGs). In practice, N_2O emissions decreased by 54% from a no-till corn–dry bean rotation receiving urea, urease and nitrification inhibitors in comparison to a urea-only application in Colorado (USA) [115]. According to a recent global meta-analysis of enhanced-efficiency fertilizers, nitrification inhibitors can reduce N_2O emissions by 38% and polymer-coated fertilizers by 35%, on average, compared to conventional fertilizer, but urease inhibitors alone are not as effective in reducing N_2O emissions [116]. Nitrification inhibitors are compatible with both chemical and organic fertilizers, making them a seemingly attractive mitigation option, but their efficacy varies with edaphic factors. For example, EEF materials were applied to rainfed corn in the central Corn Belt (Midwest region, USA), a more humid region than Colorado [117]. Although all EEF treatments had lower cumulative emissions than the treatment that did not include EEFs, episodic N_2O emissions from EEF treatments corresponded to rainfall patterns, and the relative effectiveness among EEF materials was similar. Together, these findings suggest that the impact of EEF materials may be diminished in rainfed agriculture systems compared to irrigated systems with regulated water availability. Although yield responses to EEF materials may also vary with respect to crops and location, consistent yield increases in corn (central Corn Belt) grown with EEF materials were reported to occur as a result of the increased duration of photosynthetic leaf area during grain-filling [118]. In microirrigation systems, the results are less impressive, likely due to the increased efficiency of EEF where fertilization by fertigation matches crop needs more precisely, leaves less residual fertilizer and avoids its loss [119]. With this emerging evidence that use of EEF materials could have a positive impact on crop production and limit N_2O emissions, the results of research on understanding the conditions for which these materials are useful could underpin both the development of risk assessment tools and the feasibility of grower adoption of these technologies.

Mitigating GHG emissions through C sequestration depends on the stability of soil C pools. Declining productivity in the rice–wheat cropping systems of India's Indo-Gangetic plains has been attributed to reductions in soil C [120]. Mandal *et al.* [121] found that, to combat this, addition of NPK fertilizer during double-rotations of rice led to increases in soil organic C stocks compared to adding just N or NP alone [121]. When compost was applied during rice production, as much as 29% of compost-derived C was stabilized [121]. This was attributed to high lignin and polyphenol content in crop residue and compost and also to the diminished soil C decomposition stemming from anaerobic conditions due to soil submergence under rice cultivation. Crop residue management improves poor soil fertility through soil organic matter accumulation, leading to reductions in soil N loss by leaching and gaseous emissions; in many situations in developing countries, however, crop residues are used to feed animals, to provide fuel for cooking or are turned into biochar [122, 123]. Developers of mitigation strategies for increasing soil C and decreasing N_2O emissions have to take into account the dynamics of crop residue, tillage and nutrient management, along with climate, in order to evaluate the efficacy of different practices across locations [124].

Legumes, a form of ecological intensification, offer both mitigation and adaptation options, especially to smallholder farms susceptible to deficits in soil fertility, climatic uncertainty and reduced economic access to agricultural inputs such as mineral fertilizer. The biologically fixed N from legumes is often tightly synchronized with plant N demand and has a much lower C footprint than industrially produced synthetic N fertilizers [125]. For instance, intercropping with N-fixing trees in Sub-Saharan Africa were found not only to reduce reliance on fertilizers but also to enhance soil C sequestration and reduced N_2O emissions [126]. In this intercropping system, 10.9 Mg $C{\cdot}ha^{-1}{\cdot}yr^{-1}$ were sequestered in the soil. The potential for N_2O mitigation was only 0.12 to 1.97 kg $N_2O{\cdot}ha^{-1}{\cdot}yr^{-1}$[126]. However, the authors of a review of 71 site-years of pasture, cropping and agroforestry systems indicated that providing N additions via legumes can increase accumulation of soil C at rates greater than can be achieved with other crops, such as cereals or grasses, even when they are supplied with N fertilizer [125]. Furthermore, intercropped mixtures of peas and barley (*Hordeum vulgare* L.), compared to the respective sole crops, were found to lead to effective weed suppression in weed communities across sites in Western Europe [127]. Adaptation options that include legumes to reduce dependence on fossil-fuel derived fertilizers include integration of intercropped or rotational legumes into management regimes, development and facilitation of access to new legume cultivars with broader stress tolerance and removal of barriers to legume use and consumption in the food system (for example, competing uses, seed availability, labour).

The design of more efficient N management strategies will only be conducive to climate change solutions if based on knowledge systems and participatory research with stakeholders to ensure viable action and adaptive management. Although decision-making support tools and metrics are being developed to aid producers in tempering N inputs for the desired outcomes of higher crop production (for example, quantity and nutritional quality) and lower environmental impacts [128], adoption is a major obstacle. When extension agents are involved in troubleshooting with and training of participants, the new knowledge systems that are created begin to delineate clear pathways that benefit farmers' livelihoods. In regions dominated by smallholder farmers who are already experiencing climate impacts such as increased drought, flooding or heat waves, the priority is on adaptive measures for reliable N availability to support food security and minimize vulnerability. Combining low inputs of synthetic N fertilizers with practices that increase soil quality through organic matter management and acquisition of N from biological N fixation allows adaptation measures to contribute to GHG mitigation. However, synthetic N sources are fraught with constraints such as high cost, price fluctuations and availability, whereas biological N sources are affected by constraints of labour, time and physiological tolerance. Future food security also will depend on a substantial rate of yield gains for major cereal crops. Maintaining these yield increases above a 1% annual growth rate will require constant improvement in crop yields, stress avoidance and agronomic management to achieve physiological yield potential [129]. However, maintaining a compounding rate of yield increases is not consistent with historical trends and likely is not achievable without great effort [130]. Therefore, the limits of current crop productivity need to be estimated using potential yield and water-limited yield levels as benchmarks. Determining and closing the yield gap, especially in developing countries, is fundamental to achieving food security because variety improvement through breeding and genetic modification might be insufficient [129, 131, 132].

Farmer Decision-Making and Barriers to the Adoption of Climate-Smart Agriculture Practices

Climate change challenges farmers' decisions by altering risks and uncertainty and incorporating new information into their traditional knowledge-processing systems. The unfolding of the decision-making process and its translation into action depends on the socioecological context in which farmers are embedded. How well innovation models apply to all climate-related behaviours is a major question, especially given that governance regimes at the national and international levels strongly influence farmers' actions [133]. The massive literature on innovation systems has established the basic hypothesis that farmers evaluate the costs and benefits of different practices in light of information accessed through social networks and other communication channels. The diffusion of innovation model can provide critical insights into adoption decisions. In this model, adoption of innovations follows a sequence of stages: knowledge, persuasion, decision, implementation and confirmation [134]. Innovations generated by agricultural research are communicated by extension agents to farmers. This approach may place too much emphasis on traditional socioeconomic variables and ignore how other social factors (for example, networks, gender, social norms, values, climate-change attitudes), and uncertainty may be implicated by practices that are ostensibly consistent with CSA priorities (for example, adoption of new crops and cultivars or changes in N fertilization) [135–137]. Effective outreach strategies will manifest with greater understanding of farmers' beliefs about climate change and their readiness to respond to climate change through mitigation and adaptation. Little is known about farmers' and their advisors' willingness to use outreach tools, their information needs with respect to climate change or their ability to incorporate this knowledge into existing decision-making processes. A survey of almost 5,000 farmers in 22 top corn-producing watersheds across the United States showed that farmers' climate change beliefs correlated with both their perceptions of climate risk and their willingness to respond and adapt to changing conditions [138]. Farmers who believed that climate change is occurring, and is due in large part to human activity, were significantly more likely to support both mitigation and adaptation actions and also more likely to support government- and farm-level GHG reduction efforts. Most farmers supported adaptive strategies, with two-thirds agreeing that they should take efforts to protect land from increased weather variability. Many (59%) expressed lower levels of support, however, for mitigation through GHG reduction. These farmers obtained much of their information through social networks that included professional advisors. A survey of corn grower advisors, including government, nonprofit, for-profit and agricultural extension personnel, found that advisors are more influenced by current weather conditions and 1- to 7-day forecasts than by longer-term climate outlooks [139]. The advice given to farmers has been based predominately on historical weather trends and focused on short-term operational decisions rather than on long-term strategies. For climatic data to be useful to such populations, designing outreach strategies that target extension agents and other professional advisors will increase the potential to influence beliefs and practices of farmers. Furthermore, though mitigation policies alone might not resonate with farmers, those that combine mitigation with adaptation could be effective. In general, adoption of best management practices can be promoted by focusing on implementation among farmers most likely to adopt them, followed by leveraging social networks to inform other farmers about the benefits of adoption [140].

The constraints that farmers face when making decisions, such as whether to use conservation agricultural techniques, may create barriers to practices that could improve resilience to climate change. Conservation agriculture includes practices such as minimum mechanical soil disturbance, permanent organic soil cover and crop rotation, all of which typically increase soil C storage, especially when applied in concert [141, 142]. Cited benefits of conservation agriculture in Sub-Saharan Africa include increased yields, reduced labour, improved soil fertility, reduced erosion and land-saving [141–143]. Reports of conservation agriculture's widespread adoption may be overrated, though, because many farmers seem to adopt technologies only while incentives are offered and the project is actively supported, and then they quickly return to their former crop management practices once project support ceases [144]. Constraints to adoption include strong competition for mulched crop residues for livestock feeding, increased labour demand for weeding (which often changes cultural gender divisions of agricultural work) and lack of access to and/or use of herbicides and other inputs [143,144]. Although there are some recognized factors that influence adoption (for example, larger farm size and more education), no universal variables seem to explain adoption [145], leading some to suggest that conservation farming may be successful only under certain agroecological conditions [144, 146].

Recent work in Zambia may help to explain regional variation in farmer adoption and rejection of conservation agriculture practices. Analysis of surveys of rural incomes and livelihoods revealed that rates of rejection in Zambia were high (approximately 95%), and practice dropped from 13% to 5% of farmers between 2004 and 2008 [145]. Rainfall data reveal that, during the past 10 years, the onset of the first rains needed for planting have been progressively delayed. Although adoption decisions are not strongly or explicitly based on labour constraints, farmer age or education level, farmers in districts that experience more rainfall variability are more likely to adopt conservation agriculture practices and to implement those practices with greater intensity [147]. Because conservation agriculture allows planting to occur as soon as the rains begin, it offers an adaptive response to changing rainfall regimes [148].

Fundamentally, an existing lack of food security and farmers' concerns about poor health will counteract incentives to their adoption of new farming technology [149, 150]. Although many farmers believe climate risk is real, they are less likely to believe it is caused by human behaviour. They have paid the most attention to climate variables that have traditionally constrained their operations and have relied on an existing suite of adaptive behaviours [53]. Thus, knowledge networks are especially critical to their understanding of trade-offs between the short-term costs and longer-term benefits of adopting new farming technology and practices that will help them mitigate and adapt to the effects of climate change as well as to increases in climate variability. Adaptation to climate change and the idea of climate change itself define and change human cultures. Indeed, cultural factors (for example, place attachment, value systems, individual and collective identities) shape how people support and respond to adaptation interventions and must be woven into climate change policies and programmes [151]. Key to this effort is linking science, technology and decision-making to the context of socioecological systems to better achieve balance between economic, cultural and social needs [152]. Systems that effectively leverage science and technology in support of sustainability efforts create salience, credibility and legitimacy across boundaries where boundaries exist between science and policy, disciplines, public and private sectors, and/or organizational hierarchies. Actions employed within these systems include convening (bringing all stakeholders in the CSA context together to foster communication and build trust), translation (defining a shared ontology and language), collaboration (actors working together to produce applied knowledge and specific outcomes, with specific mechanisms in place to facilitate interactions across multiple boundaries) and mediation. Specifically, mediation is 'a process by which different interests are represented and evaluated so that mutual gains can be crafted and value created in a way that leads to perceptions of fairness and procedural justice by multiple parties' [152], p. 470. These components, as well as broad stakeholder engagement from the initiation of a project, are keys for linking science with action, developing knowledge networks and forming critical capacity to reach desired outcomes also see [135–137, 152]. Other approaches for forming new knowledge networks and adaptive capacity in the socioecological system combine both back-casting and explorative scenarios [137]. Interactions between climate change and culture, as well as ideas regarding the ethics and morality involved with climate change and the role of these constructs in stakeholders' and the larger society's adoption of actions related to mitigation and adaptation, are outside the scope of this article, but they are discussed by Hayward [153] and Markowitz and Shariff [154].

Climate Risk Management: Financial Mechanisms, Insurance and Climate Services for Farmers

An alternative to emergency aid in the face of climate shocks is reliable programmes developed to minimize farmer risk, which could prove to be more effective by preventing the slide into poverty traps [155]. The uncertainty of climate change, especially extreme events, makes it difficult for individual farmers to incorporate risk into their decision-making [156, 157]. Vulnerabilities to climate effects on production, pests, disease and price volatility depend on farmers' assets and natural resource base [158]. Appropriate risk management tools, such as improved forecasts and extension support, and appropriately designed safety nets or insurance instruments must revolve around the vulnerabilities in specific farming situations. Rural households in developing countries, limited in both resources and access to information, could be disproportionately affected unless appropriate measures are introduced to manage the additional risk and uncertainty related to climate change [159–161]. Innovative management of risk and uncertainty employs financial mechanisms (for example risk transfer or insurance contracts) that use several types of methods to understand investment decisions, technology choices, and risk perceptions. These methods include remote-sensing technology, micro-level household data, analysis of diversification, and farm surveys. Implementation of such insurance instruments requires appropriate technical innovation, building awareness and trust, ensuring viable market demand, and enhancing local capacity building among local financial institutions [162, 163].

Index insurance is one such instrument that effectively reduces farmers' risk under a changing climate and generally has many advantages. With index insurance, indemnity payments are decoupled from actual crop losses, instead of linking payments to changes in attributes that impact or reflect crop growth or survival over a given spatial extent. This then reduces transaction costs associated with verifying ownership and losses, removes the opportunity for individuals to change their risk behaviours to increase the likelihood of receiving a payout, and allays the problem of adverse selection, in which high risk individuals are disproportionately represented in the insured pool. Most vitally, the rural poor are no longer widely excluded from insurance by the need to demonstrate assets as a prerequisite to purchasing a policy [161]. For example, the Index Based Livestock Insurance (IBLI) programme recently launched by The Index Insurance Innovation Initiative seeks to accurately represent the insured's loss experience through the use of landscape-level data derived from measures such as the Normalized Difference Vegetation Index (NDVI) (Figure 5). NDVI is a satellite-derived indicator of photosynthetic activity or a proxy for plant production to feed livestock, which is available in real time every 10 days [165]. Livestock in Northern Kenya's arid and semiarid lands account for more than two-thirds of average income, with most livestock mortality associated with severe drought [164]. Herd losses that push a household below a certain threshold tend to result in long-term consequences, including destitution, which can trap the household in poverty. The data derived from the developed index showed that the NDVI performed well when tested against other herd mortality data from the same region, and, when compared to drought experiences over the past 27 years, removed 25% to 40% of total livestock mortality risk in simulations. The IBLI programme has been implemented, with initial payouts issued to households in October 2011 [166]. Actions needed to facilitate establishment of the IBLI include identification of systematic criteria for end users to evaluate whether they need to purchase this insurance product [167] and development of programmes for client recruitment, low-cost marketing, and claim settlements.

To provide long-term farm and community security in support of CSA, bundling agronomic breeding programmes for drought tolerance and financial programmes with index-based drought insurance will maximize farms' resilience to financial shocks due to drought, especially as the drought tolerance of crops diminishes with more severe drought stress.

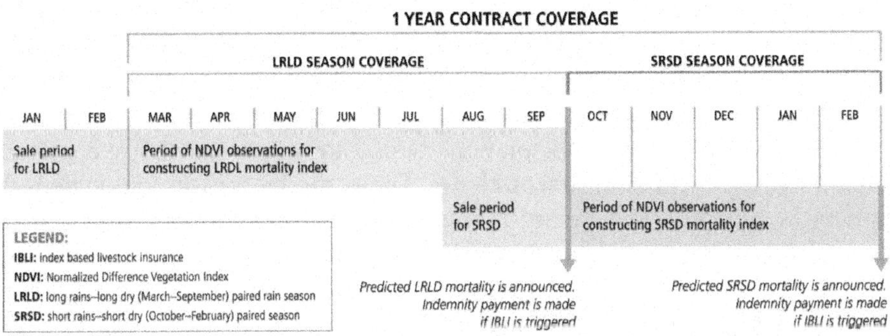

Figure 5: Depiction of a 1-year contract for index-based livestock insurance and its implementation. Adapted from Chantarat *et al.* [164] with permission from John Wiley & Sons.

Developing and planting crops with drought tolerance is primarily a more cost-effective risk management tool than index insurance in the face of less extreme climatic events; however, index insurance could complement both private and public crop improvement programmes by providing assistance when even drought-tolerant varieties fail during extreme climatic events. Demand for bundled strategies seems likely to be high [168, 169], thus creating a sustainable market for both drought-tolerant varieties and index insurance. To assess how bundled strategies affect household welfare and operate in practice in a drought-prone region of Ecuador, Carter and Lybbert [170] estimated the underlying probability structure for traditional maize yields from yield data collected annually by the Ecuadorian government from random samples of producers in different regions of the country. The certainty equivalent of the drought-tolerant technology was 6% higher than that of traditional technology. Incomes were most stable under drought pressure when drought-tolerant and insurance index technologies were combined, but interactions of such bundled strategies with other risk management and safety net programmes remain to be determined.

Uncertainty influences individual farmers' expectations of yield and dramatically impacts their adaptation behaviour. For example, government policies to protect farmers against climate-change risks, such as insurance programmes and direct*ex post facto* payments after extreme climate shocks, may reduce farmers' incentives to diversify farm production away from more climate-sensitive crops. Antón *et al.* [30, 171] examined farmers' responses to agricultural risk management policies under conditions of climate change using a stochastic microeconomic simulation model calibrated with data derived from farming in Australia, Canada and Spain. They distinguished between farming risk and uncertainty with regard to climate and farmers' beliefs. They examined the impacts of *ex post facto* disaster payments and three types of crop insurance (individual yields, area-based yield and weather index) utilizing a combination of climate-change scenarios (no change, marginal change, change with an increase in extreme events) and farmers' behavioural response options (lack of adaptation due to misalignment of expectations, diversification, structural adaptation). Their model results indicated that farmers in Australia and Spain, in the absence of government policy, would respond by increasing diversification, assuming they correctly anticipated climate change. The introduction of risk management policies in these two countries tended to crowd out diversification, and this effect increased with climate change. The relative cost-effectiveness of policies depended strongly on the extent of extreme events and farmers' misperceptions of climate (that is, misalignment), which can greatly inflate a policy's budget. Reducing the uncertainty that farmers face, with regard to how climate change will affect them, by developing information strategies will aid in the design of robust risk management policies and will limit the excessive financial costs brought on by misperceptions [30].

The goal in using the risk management instruments described here is to promote resilience of rural households to weather shocks and climatic variability, a key premise of CSA. Although not addressed here, other index insurance products promote the integration of rural households into market production and often are used in concert with programmes aimed at promoting agricultural value chains and supply chain risk management [162]. These kinds of programmes consolidate and facilitate the participation in the agricultural value chain by specific populations in discrete regions, and they are intended to help increase farmers' access to credit and to encourage investment in appropriate technology to increase productivity.

Energy and Biofuels: Development of Production Methods and Technologies to Cut Emissions without Interfering With Food Production

Bioenergy is the native energy resource embedded within agriculture, but, more fundamentally, agriculture is itself an energy conversion process with the capacity to develop a rich portfolio of products for diverse markets, including markets for food and energy. The role of biofuels in achieving reduction goals (that is, mitigation) for GHG emissions and meeting future energy needs (that is, adaptation), as well as their impact on food commodity prices, remains a principally global issue [172, 173]. Estimates of increases in food and commodity prices suggest that between 3% and 70% of retail food price increases can be attributed to biofuels; however, this wide range stems from differences in time periods, data sets using different price series (export, import, wholesale, retail) and different food products [174–176]. Global models used to predict mid- to long-term effects of biofuel production growth on prospective prices, production of feedstocks (for example, maize, sugar cane, oilseeds), mitigation and adaptation measures, and land-use change are general or partial equilibrium (PE) models. General equilibrium models encompass supply, demand and prices in the entire economy and take into account multiple markets and associated inputs; PE models are focused on equilibrium conditions in an individual market or sector of a national economy, in which prices, quantities under demand and product supply remain constant. Along with models used to assess land-use change in response to bioenergy production [177] are models such as the Asia-Pacific integrated model, which is used for analysis of global and national CO_2 emissions, mitigation costs and C taxes [178]; the Modular Applied General Equilibrium Tool which is used to examines links between agricultural markets, the general economy and agricultural policy issues [179]; the Global Change Assessment Model, which is an integrated assessment model of energy, agriculture and climate used extensively by IPCC and others [180, 181]; GLOBIOM, which is used in analysis of mid- to long-term land-use change scenarios in agriculture, forestry and bioenergy [182]; and the Model of Agricultural Production and its Impact on the Environment which is utilized in evaluating spatially explicit patterns of production, land-use change and water use in different global regions and linking economic development with food and energy demand [183, 184]. These models can provide information regarding uncertainties, costs and trade-offs crucial to CSA for (1) climate policymaking, GHG mitigation and sustainable energy futures and (2) projections regarding agriculture, agricultural markets and the future of the world's food and feed supplies. The case studies described here are used to assess costs and trade-offs of biofuel expansion at the farm and global scales as well as the impacts of enacted policies in the European Union (EU) and the State of California in the United States.

Increased future demands for food, fibre and fuels from biomass can only be met if the available land and water resources on a global scale are used and managed much more efficiently than they are now. Therefore, developers of an integrated bioenergy framework must incorporate not only bioenergy's mitigation potential but also its costs and trade-offs with food, water security and land use. To assess the cost-effectiveness of bioenergy for climate change mitigation, Popp *et al.* [184] coupled global models of vegetation and hydrology [185, 186], land-use optimization (MAgPIE) and the energy–economy–climate interface [187]. If all suitable land for agricultural production was made available, bioenergy from specialized grassy and woody bioenergy crops, such as *Miscanthus* (poplar), could produce 100 EJ globally by 2055 and up to 300 EJ by 2095. However, bioenergy cropland would grow from 1.52 billion ha to 1.83 billion ha, thereby increasing CO_2 emissions due to deforestation. Meeting bioenergy needs while preserving intact and frontier forests would require higher rates of technological change in agriculture (by 0.9% per year until 2095), thus leading to additional costs. The potential trade-offs of conserving forests and cultivating bioenergy crops on a large scale include conflicts with respect to food supply, food prices (especially in the tropics) and water resource management [188].

In the EU, market demand for biofuels and biomass will likely increase as the region becomes less reliant on fossil fuels and the EU implements targets for renewable energy, such as the Renewable Energy Directive and the ensuing national renewable energy action plans. This demand was first met with imported biomass sources from residue streams, such as palm kernel shells and wood pellets, and industrially produced biomass, such as palm oil and ethanol [189]. In an analysis conducted for the International Energy Agency, Hoefnagels *et al.* estimated future intra and inter European trade of solid bioenergy biomass by combining geographic information system models of transport routes with models of supply and demand for energy crops, forestry products and/or residues and agricultural residues [189]. They estimated that intra European biomass trade could increase to 6,560 kilotonnes of oil equivalent (ktoe) by 2020 in the low-import scenario and to 5,640 ktoe in the high-import scenario. Transportation costs could contribute substantially to these totals (for example, up to 75% (9 €/GJ) of the total cost (12 €/GJ) in the case of forestry residues). However, they determined that the lower transportation costs of pelletized biomass would not make up for its high production costs. In both scenarios, the chief future exporting regions for inter European biomass trade included Poland, Estonia, Hungary and Slovakia and the major importing regions included Germany, Italy, the United Kingdom and the Netherlands. Within the CSA strategy, these modelled outcomes can help in the identification of the issues and stakeholders that should be involved in the development of future energy use and policy.

Newly enacted low carbon fuel standard (LCFS) policies in California and the EU offer promising approaches to reducing the C footprint of transportation fuels. The LCFS applies to itself a direct life-cycle C intensity analysis that captures all GHGs emitted per unit of fuel energy during extraction, cultivation, land-use conversion, processing, transport and fuel use [190]. Both California's LCFS and the European Parliament's revised fuel-quality directive require a 10% reduction in GHG emissions by 2020, and both allow credit-trading. These standards differ from previous policies aimed at reducing petroleum fuels, which comprised volumetric mandates and only indirectly required reductions in GHG emissions. As a case in point, the US renewable fuels standard requires annual sales of 36 billion gallons of biofuels by 2022, 21 billion gallons of which must derive from advanced biofuels and achieve a 50% reduction from baseline life-cycle GHG emissions. The other 15 billion gallons must come from corn ethanol [190]. With this focus on total GHG emissions rather than on volume, biofuels under LCFS will not be forced into a small number of categories, and transformative innovation, a key part of the CSA strategy, will be promoted. The flexibility and performance-based nature of the LCFS allows industry, rather than government, to pick the likely biofuel winners [190]. If implemented on a global scale, such changes in biofuel policies will heavily influence agricultural markets and environmental outcomes. Tokgoz *et al.* [191] simulated a reduction in maize ethanol production of the magnitude suggested by the LCFS analysis by utilizing a modified version of the International Food Policy Research Institute's (IFPRI) PE model, or the International Model for Policy Analysis of Agricultural Commodities and Trade (IMPACT). IMPACT was developed to project future global food supply, demand and security in 115 country regions. Holding biodiesel production levels constant at 2010 levels in this model dramatically decreased rapeseed and soybean oil prices and increased the availability of food calories. Building future international policies upon the LCFS policies implemented in Europe and California will further the demonstrated benefits of reducing fuel C intensity rather than promoting policies that benefit biofuel producers who pursue ongoing profit-driven growth.

Policymakers and financial institutions have been hesitant to invest in bioenergy, owing to negative press and the resultant uncertainty about its long-term sustainability. In response, the scientific community must present a balanced perspective of how bioenergy can (or cannot) be managed as part of CSA (for example, see [172, 173]). Models comprising the global impacts of bioenergy, along with agricultural productivity at local, regional and country scales, can be utilized to effectively assess the realization of environmental and economic objectives via policy and technology [192]. Separate consideration of bioenergy in the agricultural context will lead to suboptimization of the system with the likelihood of realizing lower environmental and economic benefits [193]. The viability of biofuels will be achieved when their cost is competitive with those of fossil fuels when it includes both the cost of the feedstock seed and the value of coproducts derived from the biofuel by-products, which can provide additional revenue. In some cases, large subsidies are required to make biofuels competitive with fossil fuels (for example, *Jatropha*-based oil in Senegal) and/or feedstock seeds must be imported to satisfy demand, suggesting that alternative feedstocks should be adopted [193]. A stable supply of feedstock, determination whether other industries strongly compete for the same feedstock and access to a well-functioning value chain for the product are all crucial to facilitating vertical integration of production, conversion and processing, as observed in Brazil's biofuel sector. Msangi and Evans [194] suggested that growing a biofuel feedstock that can serve as a food product with coproducts will create greater stability for the farmer and that solving problems of food security in developing countries will lead to a flourishing biofuel sector. Furthermore, increases in food crop production and efficiency underpin the success of increased reliance on bioenergy and the conservation of forested lands in lieu of expansion of agricultural lands [188]. Lignocellulosic biofuels also can be a strong component of GHG mitigation with small impacts on global food prices, especially if sufficient land for feedstock production exists and does not compete with land devoted to food production, as indicated by modelled outcomes [173]. It is imperative to engage producers and affiliated industries in research to better understand how markets for new development of bioenergy and nontraditional biological products can become an integral part of energy-efficient agriculture.

THEME 2

Landscape and Regional Issues: Land Use, Ecosystem Services and Regional Resilience

Recently, extensive research on climate impacts on landscape and regional scales has been stimulated in part by policies that require institutional action to mitigate and adapt to climate change [14, 195]. Such research includes use of remote sensing to analyse land-use mosaics, inventory approaches to assessing C stocks and water resources, and models to examine the potential of land-use change in different climate scenarios [196–198]. These techniques are being combined with farm- and field-scale data on crop performance, soil biogeochemistry and irrigation use to analyse if and how mitigation and/or adaptation strategies build food security and ecosystem services [34, 199–201]. Interdisciplinary science underpins an integrated landscape approach, along with involvement of stakeholders who hold key information for developing climate-change scenarios and innovation pathways [202, 203]. Landscape approaches that expand beyond agriculture itself are needed to understand how extreme events trigger rural outmigration and create new types of rural–urban connections. The development of metrics and indicators to track responses of climate change and ecosystem services is accelerating with broader recognition of the need for greater accessibility of data, formation of more types of socioecological assessments [203–205] and charting of the progress of climate-change policies.

Climate Change and Food Security: Modelling Adaptation and Uncertainty

Determining the adaptive capacity of mitigation and adaptation scenarios that will evolve with CSA's participatory processes rely, in part, on biophysical models. Models that will be used to examine the limits to crop adaptation as well as the impacts of climate change on biodiversity, land use and ecosystem services are now available [2, 206]. They still contain much uncertainty due to (1) the ability of process models to accurately simulate the growth and development of crops when exposed to very high temperatures and elevated CO_2 levels, (2) the rate and degree to which agricultural productivity and development can progress in concert with reductions in GHG emissions and (3) the ramifications of successful agricultural adaptation to climate change for land-use change and associated ecosystem services [207–209]. Despite these uncertainties, the use of models and scenario-building has led to the exploration of potential synergies and obstacles to coping strategies in agricultural that would not have been possible with empirical data alone [210, 211]. Here we present modelling approaches to evaluating adaptation scenarios across the EU, the Mediterranean region and the United States.

Modelling can be used to identify climate-change impacts and sensitivities as well as possible adaptation strategies. Rather than being focused solely on climate-change constructs, such vulnerability assessments also include changes in CO_2concentrations, GHG emission management, N deposition, land use, and socioeconomic trends to manage vulnerability. The Advanced Terrestrial Ecosystem Analysis and Modelling (ATEAM) program produced a new set of climate scenarios for Europe in multiple global change scenarios and ecosystem models [212]. A dialogue among relevant stakeholders from the private sector, governmental and nongovernmental organizations and policymakers was conducted. Unlike global trends, European trends included moderate or no population increase, little urbanization, increased forest area and decreasing demand for agricultural land. The modelled outcomes allowed for changes in land management that could decrease vulnerability, such as C sequestration due to reforestation. Modelled outcomes indicated that the Mediterranean region could face increased risks of forest fires, water shortages, changes in tree species distribution and losses of agricultural potential. Under the different scenarios, which ranged from business as usual to greatly reduced GHG emissions, 20% to 38% of the population in the Mediterranean would live in watersheds under stress and experience water scarcity exacerbated by increased tourism and demand for irrigation. Mountain regions would be especially vulnerable because of less snow cover and subsequent changes in river runoff. These modelled outcomes provide opportunities for back-casting and identification of sensitivities where mitigation and adaptation efforts should be focused, as well as how subsequent research could inform policies around such efforts.

The participants in the EU SmartSOIL project [213] employ a CSA-like strategy that includes stakeholder involvement and is used to examine the implications of findings for economics and policy implementation. As of 2012, consultation with policymakers and advisors had begun in six case study regions [214]. The creators of SmartSOIL developed a framework of C flows and stocks informed by new data and meta-analysis of long-term European experiments that are relevant to short- and long-term CSA management decisions. This framework will be used to improve existing soil and crop simulation models out of which a simplified model will be derived to predict scenarios for future management systems to improve productivity and enhance C sequestration. As an example of modelling for C sequestration, Lugato *et al.* [197] used the CENTURY model to inform proposed European policies on the mitigation potential of agricultural soils through C sequestration and to assist in the evaluation of the agricultural sector's deployment of 'greening' measures for agriculture that benefit climate and environment as required in the EU's post-2013 Common Agricultural Policy. Nearly 16 soil–climate–land combinations in the EU and neighbouring countries (Serbia, Bosnia and Herzegovina, Croatia, Montenegro, Albania, former Yugoslav Republic of Macedonia and Norway) were used in calculations, including the main arable crops, orchards and pastures as well as management practices (for example, irrigation, mineral and organic fertilization, tillage) (nearly 164,000 scenarios). Testing modelled results against soil inventories collected using comprehensive and standardized approaches (the European Environment and Observation Network and the Land Use/Cover Statistical Area Frame Survey) strengthened the examination of the uncertainty of modelled outcomes. Consideration of a broad spatial extent (pan-EU scale) allowed for better evaluation of C sequestration, in which an estimated current stock of 17.63 Gt C is predicted to increase through 2100. Within the pan-EU region, stocks will diminish in the southern and eastern parts because of higher soil respiration, whereas these losses will be offset by increases in the central and northern regions due to increased CO_2atmospheric concentration and favourable crop-growing conditions. Such survey and monitoring programmes support the need for further spatiotemporal analysis of climate trends and stakeholder dialogue in modelling efforts so that proposed adaptation strategies are relevant to economic and socioecological contexts such as local, national and EU-wide policies and regulations.

Many climate modelling studies are focused on yield variations in response to changes in mean climate conditions [215]; yet, this approach overlooks several key factors, such as the occurrence of extreme events in which variance is changing [216]. Empirical approaches that capture the effects of extreme temperatures can be used to more efficiently assess climate impacts and adaptation. For example, in Mediterranean sunflower and wheat, an increase in both mean temperatures and climate extremes modelled under A2 and B2 scenarios (year 2100 business-as-usual and reduced GHG emissions scenarios, respectively) would cause severe yield reductions by shortening growing seasons and intensifying heat stress [217]. In the United States, yield patterns of rainfed maize have been explained by accounting for extreme events using the process-based Agricultural Production Systems Simulator (APSIM). With APSIM, observed negative yield responses to extreme heat shocks (measured as accumulated extreme degree days) were best explained by increased vapour pressure deficit (VPD). VPD contributed to water stress by increasing plant demand for soil water and reducing future water supply as a consequence of higher plant transpiration rates [218]. The ratio of water supply to demand, as modelled with APSIM, was three times more responsive to a 2°C mean warming than to a 20% reduction in rain. The results of these studies direct researchers, policymakers and extension agents to take science-based actions that rely on climate scenarios and predicted outcomes that are not based solely on the change in climatic means but include climate extremes. Despite incongruences between actual biological patterns and model simulations, model outputs provide an evolving information base for planning strategies and new research directions.

Quantitative assessments of adaptation to consider the effects of extreme events on agriculture can inform policymaking by providing a much wider set of outcomes than is possible with perceptions or projected impacts. Modelled outcomes evaluated in a socioecological context allow investigation into the limits of adaptation and related consequences for agricultural productivity, other economic sectors and land use (for example, an indicator-based, spatially explicit and scenario-driven adaptive capacity model [211]). Coordinated cycles of model improvement and projection across multiple spatial scales (global, regional, local) will facilitate model validation and calibration as well as effective use of studies with different geographical domains [219]. Challinor *et al.* [219] recommended that different model intercomparisons and improvement programmes (MIPs) form separate but linked strategies, that detailed modelling studies of response mechanisms (for example biophysical processes, crop yields) and robust experimental data (for example, see [208]) underpin the models and that systematic comparisons of impact studies and their outcomes be used to address sources of models' uncertainties. Involvement of stakeholders at the outset of model development also aids in development of relevant scenarios and tools [152]. Modelled outcomes form a key part of the climate policy and governance process necessary to attain the Copenhagen 2°C target, which requires 70% to 90% worldwide emission reduction targets and which has been questioned as being too weak [220]. A wider set of options for targets will be facilitated by examining options for climate governance (that is, institutional mechanisms to guide and direct societal policymaking) and their societal implications, as well as assessing the potential for success in achieving the target within existing political structures (for example, democracy, autocracy) [221, 222]. For example, the 'mitigate for 2°C but adapt for 4°C' option implies that society will take steps to adapt to existing in a warmer world, but will maintain the goal of reaching the current 2°C target. This approach will diminish conflicts and trade-offs associated with the water–food–energy nexus. Yet, if it is perceived that the 2°C target is unattainable and investment is strongly supportive of adaptation, then acceptance of even higher target values and a greater need for adaptation-related burden-sharing will be a consequence [221].

Soil Carbon and Achieving Multifunctionality through Mitigation and Adaptation

Soil resource degradation has led to loss of functions and ecosystem services, such as water availability, water-holding capacity, C storage, mitigation of GHG emissions and sustained agricultural productivity [223, 224]. Soil degradation limits resilience to climate change and extreme events, such as drought, and therefore impacts food security and augments susceptibility to poverty, especially in vulnerable regions such as Sub-Saharan Africa. Better understanding of the biophysical capacity of agricultural landscapes to act as C sinks through capture and storage of atmospheric CO_2 in soils and perennial vegetation leads to strategic design and operational management for both mitigation and adaptation actions [122, 225]. Improving biophysical capacity for desired functions such as GHG mitigation, food production and maintenance of soil and ecosystem biodiversity is a form of ecological intensification and is enhanced within a multifunctional landscape. Ecological intensification builds resilience by leveraging ecological processes to increase outputs from agricultural lands to promote (provisioning supporting, and regulatory ecosystem services) and decrease dependence on external inputs [93]. Balancing trade-offs between the different types of services can be facilitated by assessing indicators such as soil organic C (SOC). Trade-off analysis can employ simulation methods and modelling tools (for example, the Agricultural Model Intercomparison and Improvement Project, known as AgMIP; see [9]) to evaluate existing and alternative agricultural systems, changes in market conditions affecting supply and demand, and related policies in relation to climate change. The negative trade-offs can be minimized when landscapes are managed to achieve multifunctionality objectives, such as by a diverse set of land-use types, each providing a different combination of services [31]. The case studies below are focused on tools for accounting for GHG emissions and soil C storage, processes to enhance soil C storage and use of a paired economic-biophysical model to assess impacts of mitigation efforts within multifunctional landscapes.

Climate-change mitigation and adaptation within multifunctional landscapes depends on the multiple roles of SOC, which include a reservoir for plant nutrients (N and P) to support crop production and reduce external inputs, a substrate for soil organisms affecting their activity and diversity, and a promoter of soil physical structure leading to enhanced water quality and reduced erosion [223]. To maximize mitigation efforts, accurate GHG calculation can engage stakeholders and other end users to form a database with which to understand the C budget of their practices, such as SOC sequestration and storage and CO_2 emissions from fossil fuel combustion. Many models used to calculate GHG emissions and SOC are designed for specific geographical areas to meet distinct needs. Colomb *et al.* [226] provided information on the features of 18 available calculators and created a framework for choosing the most suitable GHG and C calculators for a given situation. They found that major sources of GHG emissions were usually well-identified, but that the calculators used failed to account for landscape effects due to land-use change. Few calculators accounted well for emissions from the loss of previous biomass, which is especially crucial in cases of deforestation–reforestation or rehabilitating and restoring grazed and ungrazed grasslands. To illustrate this point, Colomb *et al.* [226] used seven calculators to assess the GHG balance of replacing grassland by wheat, a case where the average emissions due to land-use change were greater than those that occurred during the production of wheat itself. Owing to differences in reporting units, measurement of emissions and scope, the results obtained with different calculators could not be directly compared and uncertainty levels were very high. Minimizing uncertainty in C and GHG accounting methods will provide reliable data to aid global markets and agencies for use in developing GHG- and C-footprinting and life-cycle assessment criteria. Greater standardization of metrics will also help in the enumeration of trade-offs in balancing between crop management and land use.

The design of multifunctional, ecologically intensive landscapes when providing ecosystem services of local and global interest is informed by analysing synergies between agricultural practices and landscape attributes [58]. For example, an analysis of carbon stocks and flows in smallholder farms in Kenya revealed positive synergies between agricultural production, on-farm biodiversity and above-ground C storage [227]. Dominant land-use types considered included home gardens, food-crop plots, cash-crop plots, pasture plots and woodlots. Close to the homestead, home gardens received the most organic nutrients in the form of compost, kitchen waste and manure, and downslope and farthest away maize, vegetables and eucalyptus woodlots were planted. Tree species diversity was highest in home gardens and near crop fields. Although such trees contributed up to 39% of total aboveground C storage, the greatest contribution came from monospecific woodlots dominated by *Eucalyptus saligna* (which contributed up to 81% of total aboveground farm C). In a landscape survey of 250 farms across 6 regions in Kenya, SOC, available P and exchangeable K^+ varied widely but generally varied by management practice and reflected diminished soil fertility with greater distance from the homestead [58]. Thus, a combination of land-use practices contributed to C storage below and above ground as well as to multiple functions on the farm (Figure 6). Including the diverse agricultural landscapes in such studies leads to understanding of how management practices support ecological processes for C storage, and farmer participation supports identification of economically viable options for smallholder farmers [58].

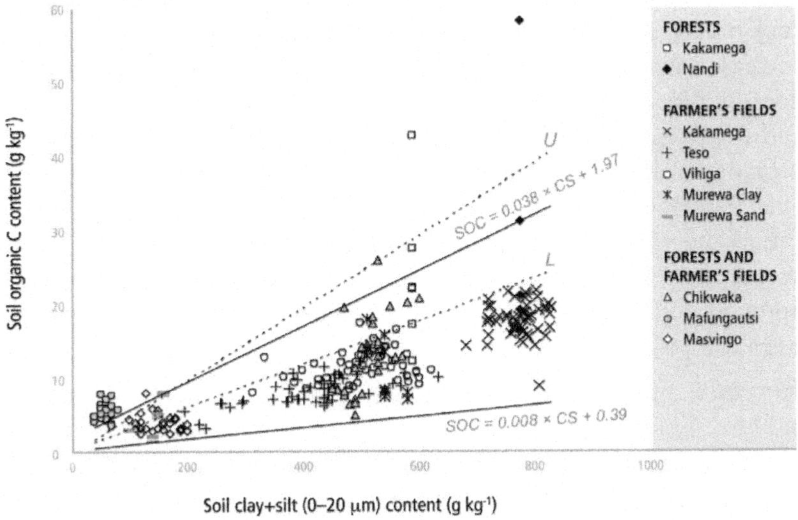

Figure 6: Soil organic C from the upper 0 to 20 cm as a function of clay plus silt. The tropical forests (Forests) and cultivated fields converted from the forests (Farmer's Fields) represented here were in Zimbabwe (Chikwaka, Mafungautsi, Masvingo, Murewa) and Kenya (Kakemega, Nandi, Teso, Vihiga). The Zimbabwean forests were the Miombo woodlands with unimodal rainfall of 800 mm. The Kenyan forests were rain forests with bimodal rainfall of 1,800 mm. Upper (U) and lower (L) boundary lines were fitted to the 95th and 5th quantiles, respectively. Samples collected at Chikwaka, Mafungautsi and Masvinga (Forests and Farmer's Fields) were collected along a temporal gradient from 0 to 60 years after conversion from tropical forests, but were not differentiated by land use in this figure. Adapted from Tittonell [228] with permission from the author.

Trade-offs between mitigation and adaptation occur often in agricultural systems, notably in the allocation of scarce resources between competing activities. The Trade-off Analysis model for Multi-Dimensional Impact Assessment is used to evaluate climate-change impacts and the viability of adaptation strategies by combining survey, experimental and modelling data [229]. Its next step is calculation of future land use, output, output price, cost of production and farm and household sizes for different climate-change and socioeconomic scenarios. The authors applied the model to the Vihiga and Machakos districts in Kenya to simulate changes in crop and livestock productivity and the effects of climate change to 2030. Climate change was projected to have a negative economic impact for 62% of farmers in Machakos and 76% in Vihiga, but these modelled effects could be partially offset by specific adaptation strategies. The most viable adaptation strategies included introduction of an improved maize variety or low-yielding, dual-purpose sweet potatoes in Machakos and improved feed quantity and quality combined with livestock breeds adapted to increased drought and high temperatures in Vihiga. In some cases, mitigation activities result in negative trade-offs, such as organic practices that increase SOC offset net GHG emissions, leading to competition for feed for livestock or fuel, or even to decreases in average yields, thereby exacerbating forest conversion to agricultural land [122]. Agroforestry, however, contributes to multifunctional landscapes that support mitigation and adaptation and can lead to improvements in livelihoods, whereby provision of fuel wood, timber, fruits and/or fodder is often associated with the cobenefits of improved soil fertility, water infiltration and below- and above-ground C sequestration [40, 150].

Currently, agricultural decision-makers and policymakers rarely consider SOC to be a major factor in agricultural management or land-use change, and the concept of multifunctional landscapes is an emerging idea in the science-based policymaking realm. Yet, the study of SOC formation, its functions, its physical and chemical protection and identification of those fractions most susceptible to degradation is an area of active research. Through various international conventions, this scientific knowledge is slowly becoming part of the science–policymaking interface relevant to climate-change mitigation and adaptation (for example, the United Nations Framework Convention on Climate Change, the United Nations Convention to Combat Desertification, and the Global Soil Partnership). However, the complex trade-offs in land-use decision-making regarding provision of multiple ecosystem services in a given landscape are usually local, so new interdisciplinary and socioecological research approaches are needed in order to downscale information and options regarding how to best manage soil C in relation to other ecosystem services and farmers' livelihoods [40, 122, 150,230–232].

Water Management for Food and Fishery Systems

The effects of climate change on hydrology are far more uncertain than temperature change, and yet, global irrigation water demand will likely increase by approximately 10% by midcentury [233]. IFPRI models indicate that calorie availability in developing countries could potentially reach almost 85% of that in developed countries by 2050, but in more pessimistic scenarios, calorie availability will decline in all regions, due in part to lesser water availability [36]. IPCC models for irrigated areas within this same time frame indicate that the gap between potential evapotranspiration and effective rainfall will be about 17% by 2050 under a high-emission scenario, placing extra stress on demand for irrigation water [234]. Taylor *et al.* [235] asserted that land-use change may have even more noticeable impacts on the hydrological cycle than climate change itself, but that, given the strong focus of mitigation and adaptation planning on land-use change, the two will remain intimately linked. For example, following conversion of forests and grasslands to agriculture in the West African Sahel [236], Southeastern Australia [237], New Zealand [238] and Southwest USA [239], runoff and/or groundwater recharge increased up to two orders of magnitude. Such increases are not always sustained, owing to a range of vegetation cover and hydrological response factors [240]. Forests and woodland cover can also support water quality and, in some cases, can assist in reducing dryland salinization and water-quality decline in semiarid environments [241–243]. Massive abstraction of groundwater and redistribution to agricultural land (nearly 70% of global freshwater withdrawal and 90% of consumptive water use for irrigation) has led to groundwater depletion in regions with primarily groundwater-fed irrigation (for example, regions of China and in the Ogallala Aquifer region in the United States). With projected increases in drought incidence and severity, changes in rainfall patterns and intensification, and decreases in snowpack, agricultural areas that are currently irrigated with surface water will become heavily reliant on groundwater. In Mediterranean-type climate regions in California's Central Valley and in southern Europe, groundwater recharge will be highly dependent on uncertain changes in precipitation patterns [235]. Aquifer salinization is also predicted to increase, at least in California's Central Valley [244]. Sea-level rise also threatens groundwater and surface water with saltwater inundation [245]. The case studies here depict adaptation measures that have been employed to meet the challenge of water management in the face of climate change across a range of spatial scales.

In the Central Valley of Chile, multidisciplinary teams have enacted a CSA-like strategy to address climate-related changes in water [245]. In Chile, farmers' permanent water rights are determined by estimates of minimum stream flow. In a high-emissions scenario, the Central Valley may experience temperature increases of 4°C by the end of this century [4], which would lead to decreases in water supply and thus challenge the existing system of determining water rights and their allocation. In the Maipo basin of Chile, snowmelt from the mountains will be reduced, affecting both river discharge and water demand. In a moderate climate-change scenario (B2), modelled reference evapotranspiration, an indicator metric of irrigation demand, was discovered to potentially increase by 10% to 15%, whereas under the high-emissions scenario (A2), increases ranged from 14% to almost 20% [31]. Permanent water rights vulnerability under the two scenarios, on the basis of data for monthly mean river flow and an agricultural census, indicated that water demands would be inadequately met in 40% to 50% of years under the more severe climate-change scenario. In response, farmers could change crops and/or cultivars, increase irrigation or sell their land and water rights. Even under current climatic conditions, farmers' existing water rights have been questioned because of increasing demand by urban users [245]. To address this issue of failing water rights and limited availability in future climate scenarios, a 'science-policy' strategy has been employed that involves civil society, scientists and policymakers in an iterative dialogue to identify the challenge and its solutions (Figure 7). Since 2008, annual meetings have been conducted with researchers and stakeholders from the national water service, irrigation commission, and environment ministry in Chile). The result has been increased inclusivity and quality of overall participation in topics such as climate-change impact assessment, water-allocation system reliability and water-sector adaptation evaluation, leading to improvements in decision-makers' support of studies on uncertainty in evaluating irrigation projects and future reservoir operations. The science-policy approach supports dissemination of information and projects to strengthen vulnerability assessment tools and coping strategies for irrigated agriculture.

In the Mekong River Delta in Vietnam, more than 700,000 ha of coastal habitats used for aquaculture are threatened by rising sea levels due to climate change. Kam *et al.* [246] analysed the farm-level economic costs and benefits of several alternatives: (1) autonomous adaptation, that is, spontaneous adoption or response, to climate change; (2) no climate change; and (3) planned, or policy-driven, adaptive strategies in which costs are distributed more equitably across the supply chain or are borne by government and other entities. Here 'autonomous adaptation' includes farmers' responses to changes in land and water availability, commodity prices, market incentives, and climate variability. Such responses incur incremental capital costs and include using different levels and combinations of inputs, altering species and production systems, adjusting the height of pond dikes, and increasing water volumes pumped into ponds. Shrimp farmers will be better able to bear the cost of autonomous adaptation than catfish farmers because they sustain relatively higher profit margins and require lower capital investments than catfish farmers. However, without government intervention to prevent flooding and salinity intrusion, the shrimp industry in aggregate will likely experience higher adaptation costs, as it covers more area. Planned adaptive strategies include genetic improvement of breeding stock and pathogen control. Although constructing dikes would reduce river and coastal flooding and salinity intrusion in support of fish production (a provisioning service), opportunities for expansion in both brackish-water and mangrove aquaculture systems that are key to coastal preservation (supporting service) will be lost. In general, evaluating adaptive planning with many types of metrics, including those for ecosystem services through restoration of coastal and intertidal vegetation, were found to provide more data to inform the final choices made by stakeholders [247].

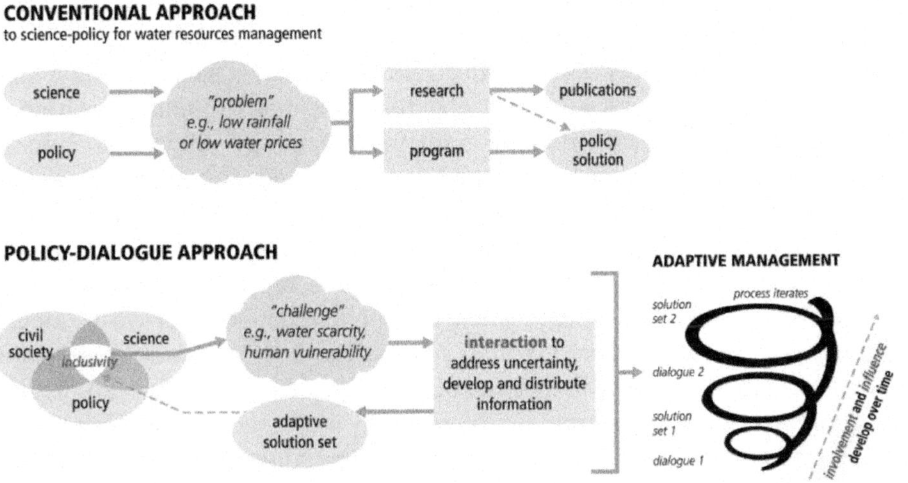

Figure 7: A comparison of the conventional approach and the policy-dialogue approach. The policy-dialogue approach led to the development of greater adaptive capacity and stakeholder engagement described by Scott *et al.* [245] and is also being employed in CSA. From Scott *et al.* [245]. Reproduced with permission from Taylor & Francis.

Recently, the concept of rainbow water, or terrestrial and oceanic evaporation as a source of atmospheric moisture and subsequent precipitation, has emerged. This conceptualization frames how to harmonize the interests of all users of the hydrologic cycle [248]. Available blue water sources—water used for irrigation, industrial or domestic use—and grey water sources cannot support the rate of agricultural intensification, so interest in green water—rainfall used by forests and other vegetation—has grown. Although controversial, passage of air over vegetation with a specific leaf index of 1 in the 10 days preceding rainfall was observed to lead to increased precipitation in Africa [249]. It follows that assessments of climate must take into consideration whether, where and how landscape changes alter large-scale atmospheric circulation patterns of water far from where the land use and cover changes occur to avoid misalignment of investment in climate mitigation and adaptation [248].

Given that climate change is likely to reduce water availability across many agricultural regions, it is critical that water policy and management practices focus on efficient and equitable water rights and allocation policies; increasing water productivity via more and better irrigation storage, conveyance and delivery systems that reduce evaporative losses; in-field water-use efficiency improvements; and technologies that reduce seawater intrusion in coastal environments. These challenges are equally important in the quest to increase agricultural productivity to feed a growing global population, irrespective of the degree of climate-change impact. Responses to the spatial and temporal shifts in water quantity and quality due to climate change involve many scales and stakeholders, and the need for coordinated planning at regional and national scales will increase with growth in the urban and industrial sectors. Approaches to increasing the efficiency of water used for food supply must employ drought-tolerant crops and irrigation technology (for example, water-conserving irrigation systems, crop coefficients and surface renewal [250, 251]). They also need to address both consumptive behaviour (that is, overconsumption and resource-intensive food selection) and waste incurred during postharvest and along the supply chain (for example, threshing, transport, storage) [252]. Other adaptive strategies include the involvement of communities and government agencies in increasing storage capacity via small-scale reservoir projects, rainwater harvesting, groundwater banking through artificial and/or natural aquifer recharge and flood harvesting (that is, directed capture of floods in floodplains) and restoration of coastal vegetation to promote opportunities for aquaculture [242, 244,252]. Additional adaptation options include reduction in end-user demand, deengineering and reoperation of water systems to create adequate supply and distribution, improved wastewater treatment plants to facilitate wastewater reuse, desalination plants and targeted water-conservation projects [253].

Managing Forest Biodiversity to Increase Ecosystem Services and Resilience

Forest loss and degradation cause GHG emissions and loss of C stocks, biodiversity and ecosystem services. Trees and forests buffer microclimates, regulate water quality and flows, store C and provide habitat for plants and animals in protected areas and corridors [248, 254, 255]. When landscapes are managed to contain a mosaic of forestry and agroforestry ecosystems, the diversification of food, feed and timber production, income sources, and markets promotes greater resilience to environmental uncertainty [149, 256]. REDD+programmes to pay developing countries for conservation and sustainable use of forests have evolved over the past decade toward greater attention on (1) increased interactions between institutional networks and (2) achieving reduced GHG emissions along with improvement of livelihoods of local communities and biodiversity conservation [257]. A systems approach involving biophysical and social sciences, as well as indigenous knowledge, is fundamental to demonstrating that REDD+projects are performance-based, fair and equitable [33]. Although afforestation and reforestation are often considered in REDD+projects, trees on farms are usually not included, owing to strict 'forest' definitions. Yet, agroforestry systems offer many REDD+-related benefits. Intentional integration of trees on farms and in agricultural landscapes increases C sequestration, along with greater food security and resilience [40, 229] (for example, see Figure 8). Assessing such multifaceted trade-offs across an agricultural landscape is relevant to the CSA strategy, but will require greater coordination on local, regional and international levels to be incorporated into REDD+.

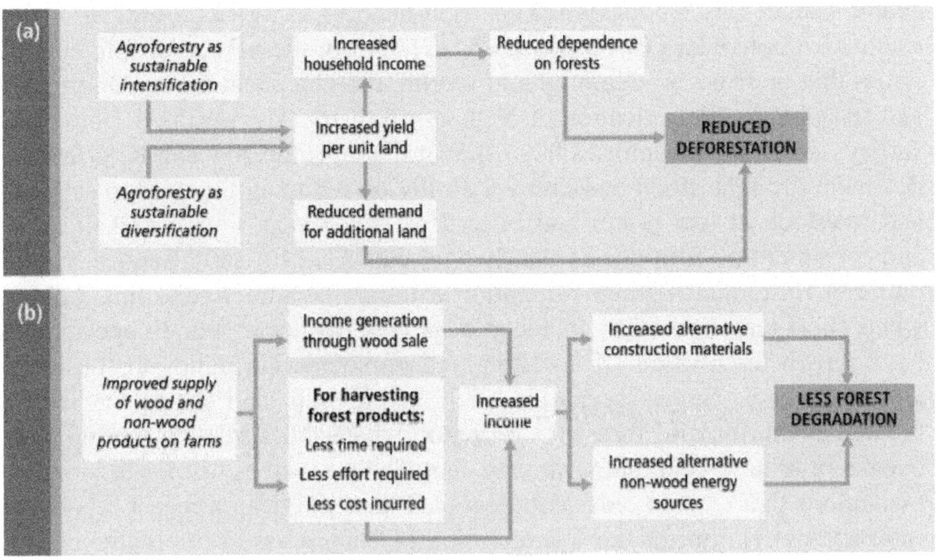

Figure 8: Contribution of agroforestry to the REDD+mechanism. (a) Sustainable intensification and diversification pathway. (b) Source of wood and nontimber forest products. Deforestation is avoided through sustainable intensification and diversification. Reproduced from Minang *et al.* [33] with permission from P Minang (personal communication, 2014).

Examples of agroforestry types in agricultural landscapes include remnant forest or savanna, agroforests, tree crops, home gardens and boundary plantings [258]. Tree species and densities for each type are selected by desired ecological processes, farmers' criteria and land-use policies. An integrated landscape approach allows valuation of the ecosystem services derived from these management options and can be used to determine potential trajectories of tree-cover transitions [31, 149]. It permits the nesting and spanning of spatial scales of different agroforestry types, the confrontation of biases for C benefits versus livelihood choices, and the optimization of tree-diversity exploration. It also opens opportunities to identify synergies and trade-offs and helps sidestep definitional challenges that result in negotiation platforms for proactive actions that reduce vulnerability and increase benefits (for example, see [259]). The landscape perspective is useful for scenario-building, such as comparing financial incentives that emphasize economic efficiency for agricultural and timber purposes versus socially 'green' and 'rights-based' approaches that support resilient livelihoods and broader sustainable development goals. The current scientific literature does not adequately detail these socioecological and community-based processes or how they underpin decision-making.

Examining trade-offs in REDD+can provide scientific information to enable science-based policies and decision-making, as well as coordination and standardization of REDD+practices. Many of the trade-offs involve livelihood issues that increase productivity and wealth, thereby encouraging land tenure and sustainable intensification through agroforestry. The results of household surveys and farm inventories have shown that agroforestry can help farmers deal with drought, flood and rain variability by reducing the need to sell land and livestock at low prices and instead sell seedlings, timber and firewood and consume tree fruit during the 'hunger gap' [33, 40, 260]. Sequestering C on farms for climate-change mitigation will only be attractive to smallholders when short-term increases in income or welfare occur. Landscape models have shown the impacts of investing and implementing policy in 'business-as-usual' versus 'green' scenarios, such as allowing land swaps for permits granted within natural forest for oil palm expansion, so that plantations can expand only onto land that is already degraded, as well as tax concessions for plantations that expand only onto degraded land [261]. In a recent report, the International Union for the Conservation of Nature assessed climate-change mitigation activities across many regions of the world where REDD+policies likely would be implemented [262]. Examination of the social, economic and environmental trade-offs and potential synergies revealed that clear tenure and property rights, including rights of access, use and ownership, are essential for effective REDD+implementation To benefit local communities, including the most vulnerable, REDD+policies must enhance the ecosystem services upon which the rural poor are most dependent and leverage new financial resources to reward local communities for management. These opportunities can easily be lost if the vulnerable are explicitly excluded as beneficiaries (for example, because of unclear tenure) or high barriers to entry (for example, forest certification) [263].

Participatory, transparent, accountable governance can help achieve benefits of implementing REDD+policy by creating synergy between parties at multiple scales. A governance approach that facilitates harmonized goals and policies between civil society and engaged stakeholders focuses on the relationships among organizations rather than on new organizational structures and financing mechanisms. Public–private partnerships can improve the effectiveness of the biodiversity governance system and complement regional and multinational efforts [263]. In Cameroon, for example, nongovernmental organizations are implementing REDD+pilot projects and acting as bridges between the public and the state, both to create awareness among local communities and to voice concerns about social safeguards [264]. Such partnerships have helped government institutions organize international biodiversity governance around an ecosystem approach, largely by changing the scale and nature of the dialogue through a community of practice with institutions outside the immediate REDD+network [257].

Although REDD+will benefit from institutional interactions that build trust and reach eventual consensus on forming, coordinating and integrating policies that support livelihoods and resilience while sequestering C in forests, the definition of appropriate ecosystems for payments still is a major issue. As pointed out by Visseren-Hamakers and Verkooijen [257], it remains to be seen whether CSA, with its integrated planning of land, agriculture, forests, fisheries and water, will be included in policymaking steps towards broadening of the REDD+agenda.

Rural Migration Due To Climate Change

A worldwide transition toward urbanization is occurring, partly in response to climate change, although rural outmigration due to climate shocks, such as hurricanes, is better documented than gradual changes, such as lower rainfall in arid areas [43]. Migration within countries is complex, having both positive and negative impacts on adaptation and household resilience. Climate shocks and disasters can propel people living under vulnerable conditions into poverty traps that force migration out of rural areas [265], where men most often migrate, leaving the women and children with increased household and farming burdens [45]. Migration can be a beneficial strategy that spreads risks through resource diversification, such as remittances that bring money back to the household [266]. Livelihood and food security, as well as culture, affect who migrates, when, for what reasons and to which destinations [267] (Figure 9). Despite the material benefits that can result from mobility and migration, displacement of people from places that they value reduces culturally based activities, such as preplanning for specific climate-change events [42]. Migration can lead to inhabiting vulnerable urban locations, such as flood-prone areas [268], and increase inequities due to poverty and lack of social networks. Opportunities exist to improve structural and institutional frameworks to reduce migration from rural areas, including greater diversification of rural livelihood systems [149, 269]; opportunities for public health, social equity and environmental welfare [270]; and connection of urban populations with local or regional food sources to support rural incomes [3, 11, 28].

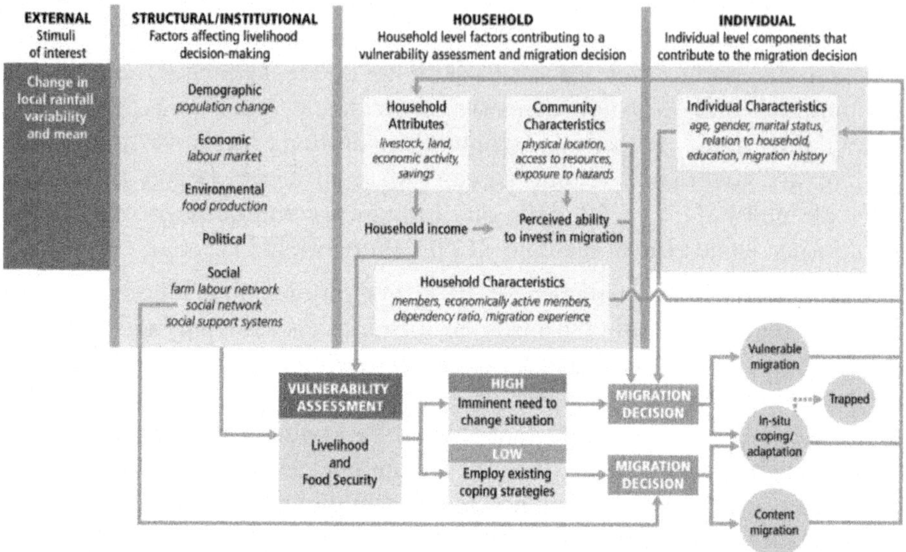

Figure 9: Decision pathway for rural migration in response to external stimuli, often related to climate change. Factors that affect decisions occur at institutional, household and individual levels. Adapted from Warner *et al.* [267] with permission from K Henry.

Land scarcity and degradation are conducive to outmigration. In Guatemala, people from households affected by flooding or soil degradation were found to be more likely to leave settled rural areas for the forest frontier to engage in clearing of forests for agriculture [271]. Surprisingly, on the basis of employing a remote-sensing approach across Central and South America over a 10-year period, rural–urban migration was not observed to strongly affect the recovery of forest vegetation [272]. The researchers in that study found that a significant increase in woody vegetation occurred in only about half of the municipalities that lost population. Thus, depopulation does not necessarily imply land-use change. In their analysis of annual satellite land cover maps, they found that 180,000 km^2 of forest was lost between 2001 and 2010, with the majority of deforestation occurring in South America (92%), particularly in Argentina, Brazil, Bolivia and Paraguay. Much of this land is in soybean production and cattle-grazing to meet the increasing global demand for meat. DeFries *et al.*[273] recently demonstrated that increases in rates of deforestation are closely linked to increases in urban populations and their demand for agricultural products rather than changes in rural populations. In Central America, temporary international migration of members of smallholder households has been indirectly associated with a lack of reforestation; remittances are spent on owning more land, and less household labour favours a transition to cattle production. This is relatively safe and risk-averse compared to row crop production, but it increases forest loss and land degradation and thus decreases the mitigation and adaptation potential [274].

Rural–rural migration offers a livelihood adaptation strategy for rural people facing stresses and shocks due to climate change, but it can also increase migrants' vulnerability. In Vietnam, migrants to the fertile Central Highlands aim to increase their economic livelihoods by producing coffee destined for international markets. Instead of settling permanently, many circulate between their new and origin communities because their social networks that remain at home allow them to avoid some of the risks of permanent relocation [275]. For example, family members in the community of origin may look after the migrants' children, take care of land and assets and provide access to loans. The lack of formal credit institutions at the new destination means that the community of origin may provide continual financial support instead of successful migrants' sending remittances home. Such social networks expose remaining household members to risk if ventures fail because of economic, social and environmental conditions. Both the migrants and origin households may then require loans to take further livelihood risks. In these cases, migration may drive both households into further poverty. Reforming Vietnam's household registration system to allow migrants access to banking, lending and other public services at their new locations could reduce the risks of such outcomes [275].

In the project 'Where the Rain Falls: Climate Change, Food and Livelihood Security, and Migration', researchers have examined rainfall, food security and human migration in eight countries in Asia, Africa and Latin America [267], mainly in agricultural areas. Four distinct household migration profiles were identified, varying along a spectrum from resilience, where migration is one of a variety of adaptation measures that progressively reduce climate sensitivity, to vulnerability, where migration either is difficult or exacerbates sensitivity to climatic stressors. Although national and regional contexts affect migration, household characteristics were discovered to be most important for migration related decisions and outcomes. For example, migration was generally erosive for the poor and those with small land holdings. Household size and composition, land ownership, asset base, degree of livelihood diversity and education levels were associated with migration strategies that increased resilience, such as nonagricultural jobs or diversified livelihoods [267]. One of the 'Where the Rain Falls' project case studies is the Mantaro Basin of Peru, where pressures to migrate stem from lower precipitation that reduced farmer and herder incomes [276]. Two livelihood and migration profiles in the Mantaro Basin were identified in response to climatic vulnerability. Lowland farmers who often commuted on a daily basis for casual urban employment used their proximity to the city to diversify their livelihoods. In contrast, herders farther from the city were forced to migrate for longer periods or permanently, in the absence of other options, and therefore were generally more vulnerable.

The act of migration has a risk dimension, whether it is a positive form of adaptation or part of erosive coping strategies. Understanding the cultural dimensions of risk-taking under climate uncertainty is crucial for determining migration decisions, especially as the necessity for climate-driven planned resettlement becomes more urgent [42]. Although outmigrants are mainly men, the outcomes of climate-change–induced migration are likely to be highly gendered because women are disproportionately affected. Women tend to be poorer and less educated and to have lower health status and limited direct access to, or ownership of, natural resources [45]. It will become more feasible to identify risk-prone agricultural areas and circumstances if models of biophysical aspects of climate change and land use also take into consideration factors that influence migration decisions, such as landlessness, land tenure and distribution issues, as well as the role of social networks that facilitate resilience and adaptation in rural areas as well as escape from poverty traps [167].

Climate-induced outmigration from rural areas involves mitigation and adaptation issues related to urban and periurban outcomes, such as increased GHG emissions due to urban sprawl on land that once supported food production [11]. Interdisciplinary work is needed to understand effective strategies for developing and preserving smallholder agriculture near cities, expanding urban and periurban agriculture, managing urban growth for farmland preservation, connecting agricultural producers with local urban markets, ensuring availability of agricultural labour and enabling diversified rural livelihood systems. Such strategies will have combined benefits for climate change mitigation and adaptation.

Metrics for Vulnerability Assessment, Food Security and Ecosystem Services in Agricultural Landscapes

Science-based actions within CSA require integrated data sets and sound metrics for testing hypotheses about feedback regarding climate, weather data products and agricultural productivity, such as the nonlinearity of temperature effects on crop yield [277], and the assessment of trade-offs and synergies that arise from different agricultural intensification strategies. Approaches range from the development of broad indicators for identifying differences in climate vulnerability over large spatial scales down to the use of finely disaggregated spatial metrics [278]. New and innovative research and policy designs, as well as cooperative arrangements among and between government agencies, research institutions and civil society, have the potential to implement monitoring and assessment systems for decision-making. Examples presented here demonstrate how biometeorological, economic and sociological indicators can be used in vulnerability assessments and show nuances that must be addressed with respect to scale.

Novel outcomes, such as nonlinear effects of climate change on agricultural productivity (for example, US maize), are emerging based on the use of large-scale data sets, indicating that environmental change may drive agricultural productivity in unexpected ways [277]. For example, Lobell *et al.* [5] examined harvest and daily weather data derived from more than 20,000 historical maize trials conducted by the International Maize and Wheat Improvement Center and private seed companies in Sub-Saharan Africa from 1999 to 2007. 'Optimal management' and 'drought stress' were the two most common scenarios under which maize was grown. Final yield was reduced to the following different extents due to warmer temperatures: by 1% under optimal rain-fed conditions and 1.7% under drought conditions for every degree day spent above 30°C. Lobell *et al.* [5] suggested that a 1°C warming would lead to negative yield where maize is presently grown under optimal management (roughly 65% of the area) in Sub-Saharan Africa, whereas all areas in this region would show decreased yield of as much as 20% under drought stress. Similarly, in the United States, which generates 40% of global maize production, predicted increases in interannual weather variability (temperature and precipitation) could result in an 18% decrease in maize yields by 2030 to 2050 in comparison to the period from 1980 to 2000, along with increasing volatility in annual yields [279]. Expansion of cropland in other regions and retention of speculative inventories (that is, holding volumes for higher price earnings) may offset the volatility. Here metrics of climate and indicators of crop productivity and other agronomic factors predicted crop response to climate warming and drought over a widespread region, setting the stage for more research on how adaptation measures, such as improving soil moisture and breeding for drought and heat tolerance, could be used to reduce vulnerability in the future [5].

Metrics that incorporate human ecology are integral to enabling the CSA strategy. Vital Signs [280] is a monitoring programme for changes in human well□being, agriculture and ecosystem services and is designed to provide metrics in rapidly expanding and intensifying agricultural landscapes in Africa, leading to integrated approaches that support food security (Figure 10). A primary goal of Vital Signs is characterizing the uncertainty and quantifying the sampling intensity needed to achieve different levels of accuracy and statistical power to detect change. Information gathered in the initial phase will be further evaluated for its overall utility and delivery cost. Measurements collected by Vital Signs participants are based on hierarchical spatial scales to provide integrated information that can inform structural relationships and counterfactuals involved in decision-making from the global to household scale. The global perspective facilitates comparisons between different regions (250,000 $km^2 \cdot region^{-1}$), whereas regional measurements deliver information at the scale on which agricultural investments are made. Information collected at the landscape scale (10 to 20 units per region) measures the relationships between agricultural intensification, water availability, soil health and other ecosystem services, together with human well-being. Plot-level (1 ha) data reflect agricultural production, including seed selection, fertilizer type and application rate, as well as crop yield response. At the household level, surveys are employed to collect information on health, nutritional status, income and assets. Stakeholder planning meetings and participatory research established both at the onset and throughout the project are integral to garnering active engagement in Vital Signs.

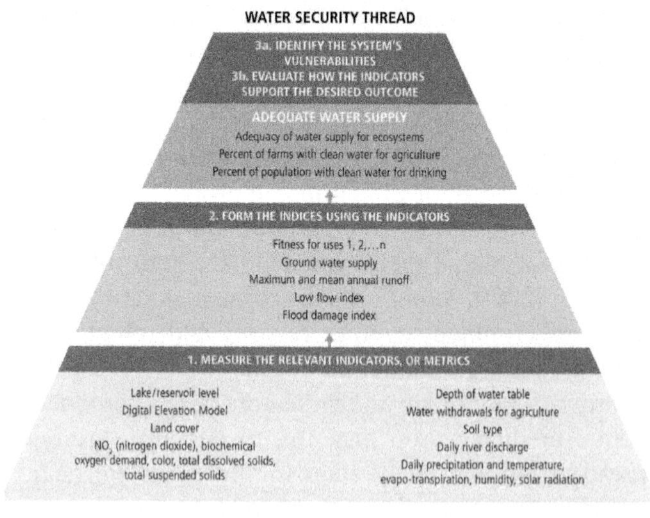

Figure 10: Water security thread from Vital Signs. The pyramid of the water security thread depicts the integration of metrics (*1*) that build the desired indices (*2*) with the outcomes of interest (*3a* and *3b*). Adapted with permission from the Vital Signs programme (S Barbour, personal communication, 2014).

Prioritizing allocation of resources and focusing policies on vulnerable regions requires metrics to assess susceptibility to a lack of food security due to climate change [281]. Biophysical climate indicators derived from global climate-change models and food insecurity indicators (that is, availability, access and utilization) can serve as such metrics. As an example of this approach, Ericksen *et al.* evaluated hotspots of vulnerability using the overlap among indicators of global climate (for example, rainfall variability, number of reliable growing degree days, and change in mean annual temperature) and food security indicators across the global tropics [281]. The latter were composed of availability (for example, crop yield and mean food production indices), access (for example, GDP per capita, transport time to markets, and monthly staple food prices) and utilization (for example, malnutrition prevalence and proportion of the population using unimproved water source). Future vulnerability was depicted by existing resource pressure (for example, annual population growth and agricultural area per capita). The resulting index of vulnerability reflected three central components: exposure of populations to the impacts of climate change, sensitivity of food systems to these impacts and coping capacity of populations to address these impacts. With this vulnerability index, it was possible to rank the most highly exposed regions, leading to the emergence of southern Africa as a highly exposed region, as well as areas within Brazil, Mexico, Pakistan, India and Afghanistan. This approach is limited by the following factors: The data represent only current food security levels; data are gathered only at the national level, which masks variability within regions and among households; and other data are needed on climate-change exposure and on food security variables other than crop yields and utilization, such as food distribution and equity.

Systems delivering real-time indicators and metrics that are tied closely to management decisions and current conditions allow science and policymaking entities to progress from using lagging indicators to finding leading indicators that can be used to identify when and where thresholds of climate-change responses will occur [112]. Indicators and metrics are often used to support public goods and services, so better standards and codified practices that support shared vocabulary and ontology will reduce the costs and streamline efforts for curating and disseminating such information. Research designed to develop metrics that inform global to local social networks for data collection, sharing and integration can also be leveraged for extension efforts. The identification of efficient and location- and situation-specific sets of indicators will complement efforts to construct human capital, social awareness and consensus regarding specific issues, leading to action strategies and policy guidelines across various temporal and spatial scales.

THEME 3

Integrative and Transformative Institutional and Policy Issues: Bridging Across Scales

Figures 11 and 12 provide an overview of some of the main points covered in each session of the 2013 Global Science Conference on Climate-Smart Agriculture. The relative emphasis on mitigation of GHG emissions versus adaptive capacity to climate change (or both) varied depending on the session topic. CSA strives for food security, adaptation, mitigation and resilience, but not all of these are achieved in the same context. The session topics often invoked multiple scientific disciplines to inform further action and problem-solving strategies in support of CSA goals in the context of the session topic, but further integration across these topics and disciplines is necessary. Scientific uncertainties are inherent in climate science, given the difficulty of forecasting climate and its interactions with other aspects of human-induced environmental change. The examples that are mentioned here require intensified scientific activity, formation of knowledge networks, and involvement of many relevant stakeholders to obtain better information to support decision-making (see also [135–137,152]). Also, there are clear social controversies challenging CSA, often derived from assumptions and questions of equity and legitimacy, such as who will implement a response to climate change and how this will occur. To obtain buy-in from vulnerable populations and countries, such issues must gain the forefront in discussions of CSA science and policy among the diverse set of stakeholders described in the Introduction section above. Many of the stakeholder-driven programmes mentioned in the conference sessions exist at regional and global levels, as climate science is often funded for large-scale initiatives. As stated previously, this article and the conference presentations do not emphasize the local knowledge-to-action processes that are essential for transformations towards climate preparedness. Nonetheless, some of the possible pathways towards such socioecological approaches to fostering greater participation and advancement of CSA objectives are shown for each of the session topics. Clearly, science must play an active and central role in developing the information base that will support food security, adaptation and mitigation in CSA and new types of inclusive, participatory decision-making as well as knowledge exchange processes [135, 152].

SESSION CONTENT FROM THEME 1 (2.0):
FARM AND FOOD SYSTEM ISSUES: SUSTAINABLE INTENSIFICATION, AGRO-ECOSYSTEM MANAGEMENT, AND FOOD SYSTEMS

Mitigation and/or adaptation	Interdisciplinary science	Scientific uncertainties	Social controversies	Stakeholder programs mentioned	Social-ecological pathways
2.1 CROP PHYSIOLOGY & GENETICS					
Mainly adaptation	Molecular biology, ecophysiology, agronomy	Plant response to high [CO2] + high temperatures	Genetic engineering	Regional crop breeding networks	Participatory plant breeding
2.2 LIVESTOCK MANAGEMENT & ANIMAL HEALTH					
Both	Animal science, range science, veterinary medicine, economics	Feasibility of scaling up of combined mitigation and adaptation strategies	Emphasis on low GHG emissions under poverty	Global programs for REDD+; for managing diseases	Management decisions based on livelihood criteria
2.3 NITROGEN MANAGEMENT					
Both	Agronomy, plant nutrition biogeochemistry	Decision-support tools effective across agroecosystems	Emphasis on low GHG emissions under poverty	Agricultural extension programs	Knowledge systems to enhance adoption of practices
2.4 FARMER DECISION MAKING					
Both	Social & political sciences, agronomy	Overcoming barriers to adoption of new practices	Dealing with perceptions of low climate risk	Social networks in farming communities	Short- & long-term planning for uncertainty
2.5 CLIMATE RISK MANAGEMENT					
Adaptation	Economics, remote-sensing, agronomy	Design of safety nets for uncertain climate shocks	Insurance reduces incentives for adaptation	Index-based livestock insurance programs	Bundled strategies for household welfare
2.6 ENERGY & BIOFUELS					
Mainly mitigation	Biogeochemistry, life-cycle analysis, agronomy, economics, geography	Projections of costs & benefits from biofuels	Food price increase attributed to biofuels	Energy directives and action plans	Biofuel feedstock value chains that support food security

Figure 11: Conference session 2.0 content within theme 1. Farm and food system issues: sustainable intensification, agroecosystem management and food systems[a]. [a]GHG, Greenhouse gas; REDD+, Reducing Emissions from Deforestation and Forest Degradation.

Inter- and transdisciplinary scientific approaches are principal both to our understanding of how socioecological systems support the adaptive management and governance that are essential to long-term human provisioning of food and to the establishment of science–policymaking dialogues to plan for the future [47, 282–284]. These actions are keys to assessing trade-offs of mitigation in context-specific situations, such that resource-poor farmers are supported rather than undermined by CSA. To realize the CSA objectives of increased food security, resilience, mitigation and adaptation, scientific research supports awareness, analytical capacity and the evidence base to understand the impacts of climate change on agricultural growth strategies and food security, and identify climate smart options suitable to the local context [15].

SESSION CONTENT FROM THEME 2 (3.0):
LANDSCAPE AND REGIONAL ISSUES: LAND USE, ECOSYSTEM SERVICES AND REGIONAL RESILIENCE

Mitigation and/or adaptation	Interdisciplinary science	Scientific uncertainties	Social controversies	Stakeholder programs mentioned	Social-ecological pathways
3.1 CLIMATE CHANGE & FOOD SECURITY					
Adaptation	Agroecology, soil science, land use & geography, political science	Ability of models to simulate multiple types of processes	'Mitigate for 2 °C but adapt for 4 °C'	Policy makers	Strategize with spatial/temporal modeling & stakeholder dialog
3.2 SOIL CARBON					
Both	Soil science, soil microbiology, biogeochemistry, agroecology	Accurate models of C, GHG emissions & tradeoffs	Support to create multifunctional landscapes	Science-policy interface of UN conventions	Participatory research across landscapes
3.3 WATER MANAGEMENT					
Mainly adaptation	Hydrology, land use & geography, atmospheric science, economics	Forecasting future water availability and demand	Governance, e.g. failing water rights, access & storage	Regional civil society & policy makers	Multi-stakeholder dialog for vulnerability and adaptation scenarios
3.4 MANAGING FOREST BIODIVERSITY					
Both	Forestry, agroecology, land use & geography, economics, political science	Appropriate methods & ecosystems for payments & rewards	Excluding the vulnerable as beneficiaries of REDD+	Multi-scale & multi-institutional planning for REDD+	Public-private partnerships for socially 'green' & 'rights-based' approaches
3.5 RURAL MIGRATION					
Adaptation	Social and political sciences, land use & geography, economics	Quantifying impact of climate shocks on out-migration	Disproportionate adverse effects on women and children	Stakeholder participation not clearly identified	Social networks that provide income, ties to the land & resilience
3.6 METRICS FOR VULNERABILITY					
Both	Ecology, social sciences, remote-sensing, land use & geography, economics	Biophysical & social data at scales that detect climate effects	Choice of indicators to direct policy decisions	Government and policy makers	Stakeholder involvement in data collection & for use of metrics

Figure 12: Conference session 3.0 content from theme 2. Landscape and regional issues: land use, ecosystem services and regional resilience[a]. [a]GHG, Greenhouse gas; REDD+, Reducing Emissions from Deforestation and Forest Degradation.

How does research better inform the institutional, financial and knowledge-sharing arrangements to create a sense of possibility for transformative processes that reduce vulnerability and increases climate preparedness? Truly transformative solutions tap into a sense of possibility for positive action, and, as in business value propositions, there is a promise of goods and services to be delivered and experienced [47]. Yet, a 'doomsday' attitude has permeated much of the agricultural science regarding climate change, emphasizing harsh potential impacts under business-as-usual scenarios (Figure 2). Although it is effective in stimulating awareness and action in some sectors, CSA research is potentially more conducive to achieving food security, adaptation, mitigation and resilience. Examples include models that go beyond impacts to include adaptation and transformation at either the farm or landscape scale (for example, see [211]), capacity approaches to examine multifunctional solutions within the socioecological system and direct evidence for situations, options and scenarios which increase human behaviours that build natural capital and resilience. Action-oriented research can also show how public-private partnerships can be used successfully to develop technologies, policies and approaches that may lead to sustainable food production and consumption patterns in a changing climate.Uncertainty is one of the most difficult obstacles to determining priorities for CSA research. Not only is future climate uncertain, but so also is the existence and operation of the institutions that are and will be involved in adaptation, mitigation and resilience. Uncertainty can breed scepticism about the urgency to plan for climate change, especially in agricultural communities and industries that already deal with large annual variability in production and prices. Thus, uncertainty is a barrier to mitigation and adaptation among some of the stakeholders whose investment, engagement and broad agricultural knowledge are critical for designing better research on coping strategies. CSA recognizes that the unfolding of decision-making processes, their translation into action and the formation of adaptive capacity depend on the socioecological contexts in which farmers are embedded (for example, the vital role of social networks in rural communities) (Figure 1). The ways of addressing uncertainty are likely to differ greatly among communities and socio-ecological systems, and research is needed to understand how to approach uncertainty in different contexts.

Although poverty can sometimes drive collective action, such as for improved food security in Kenya and Uganda through risk-sharing and pooling of labour and other limited assets [16], the least food-secure may be less likely to adopt new CSA practices because innovation implies additional costs before benefits can be realized [285]. Research on adoption of new farming technology and practices is needed to understand how upfront costs, lost income, worries about personal health and additional risks assumed during the conversion period can present formidable barriers to farmers [149], even if the new practices leverage ecological processes to improve sustainability and production [14]. For instance, diffusion of new germplasm with specialized traits (for example, drought tolerance) to targeted end users may suffer slow adoption even though new regional and local cultivars will likely be adapted to the range of conditions and management practices employed during climate change. To illustrate this point, modelled diffusion of a drought-tolerant variety among vulnerable (highly risk averse) farmers took four times longer than it did among those less vulnerable (less risk averse), underscoring the need for consideration of how seed prices affect the access of vulnerable farmers to new crop varieties [286]. Synthesis of information on how CSA practices have been facilitated by specific policy interventions, leading to broad community support, also is needed. A better understanding of how social benefits such as access to food and healthcare, rights to land and water, markets, and financing situations facilitate adoption of new farming practices or technologies will inform governance decisions [14].

Collective action for climate preparedness and problem-solving has already been effective in some situations. Safety nets for the poorest and most vulnerable households usually occur in the form of humanitarian relief and food aid, cash payments, agricultural inputs and public works [14], often after a critical event has occurred. Instead, communities can collectively plan safety-net strategies and resource transfers that are predictable and flexible enough to be scaled up and then scaled down when the crisis subsides [13, 14]. CSA research on learning, knowledge-sharing and social network analysis can help build awareness, early-warning indicators and criteria for benefit transfers for disaster responses and also effectively combine local collective action with national and/or international aid. Enhancing human and social capital, such as for childhood nutrition, entrepreneurship by women, and synergies between fuel use and C sequestration in trees, also rehabilitates household and community assets. Proactive planning will be more effective than reactive responses to a disastrous climate event, and research can help increase understanding of how adaptation policies must be designed accordingly [204].

Furthermore, collective action at the institutional scale is essential to avoiding conflicts that result from climate change. For example, institutional transboundary water agreements are associated with lower risk of conflict during water scarcity, but even one weak link in the communication, coordination and cooperation between coriparian nations will reduce their adaptive capacity to respond to new changes in hydrology, thus increasing the potential for risk and disputes [206]. So far, climate change has rarely been incorporated into such agreements. Collective action at the institutional scale could also address changing migration patterns of rural–urban connections that are likely due to extreme climate events and climate change and which will have potentially large ramifications on food production and food security, land tenure and cultural integrity [42] (Figure 10). At this point, research is needed to better understand how climate affects the dynamics of the rural labour force and thus on the stability of local food production for rural communities and nearby cities [267, 287].

To realize CSA, research on targeted financing is essential, especially in support of the most vulnerable. Upfront investment to plan and start implementation strategies is required, as is research to develop monitoring systems designed to track climate-related human responses by utilizing consistent metrics that demonstrate private benefits along with public goods (for example, GHG mitigation). Already existent funds, such as the Adaptation Fund established under the Kyoto Protocol [288], and the International Fund for Agricultural Development's Adaptation for Smallholder Agriculture Program [289], can improve smallholders' access to climate-smart assessments, technologies and institutions related to sustainable management of forests, providing up to 16 million additional jobs globally and increasing household income in rural areas as a result of restoring degraded forest [290]. Larger-scale investments, such as financing infrastructure for water resources and carbon capture, can potentially be provided by the Green Climate Fund [291], and private finance may also play a role. As climate finance develops, research shares a role in prioritizing investments and effective financing solutions and in monitoring outcomes.

Investment in research on food systems that are resilient to climate shocks may be more likely to occur if CSA expands beyond the agricultural sector. As examples, CSA research could more explicitly involve issues related to: (1) local, national and regional food trade, including governance and regulations, food safety, roads and infrastructure, and value chain coordination; (2) flexibility in financial arrangements, insurance and planning to cope with, and be responsive to, variability in climate and markets; and (3) integration of the interdisciplinary research to form a more holistic and service-oriented approach based on science to inform policy. For research to be utilized most effectively in policies related to CSA, pathways for communication of the latest scientific progress and research results must be established within relevant time frames. Communication must span sectors and scales in which policymakers and other stakeholders operate, crossing boundaries between scientists and local, regional and global actors such as nongovernmental organizations, governmental agencies, corporations and broad social and media networks [290].

CSA strategies support the realization of a broader green economy concept that acknowledges 'the sum total of all ecosystem services and how they collectively provide the complete life support system we need' [292], p. 9. In practice, market prices, costs, and benefits for the ecosystem services related to carbon sequestration, clean water production, flood protection and grass forage have been quantified. In Cameroon, for instance, the value (in US$\cdot ha^{-1}\cdot yr^{-1}$) attributed to the forest's contribution to climate and flood control is 1.3- to 2.6-fold greater than that of the timber, fuel wood and nontimber products. Coordinated action resulting from CSA and green economy research not only realizes the improvement of livelihoods and food security through mitigation and adaptation to climate change but also creates cobenefits for ecosystem services and sustainable use of natural capital and enables evaluation of a broader set of trade-offs associated with a certain course of action.

Conclusions

Disciplinary, interdisciplinary and transdisciplinary scientific approaches play a fundamental and profound role in developing understanding of the processes underlying CSA and serve as partners in enumerating priorities for CSA. They form a crucial element in the knowledge base needed to implement CSA actions and manifest future transformative changes in agriculture in a changing climate. Global science conferences on CSA have already been influential in assembling scientists and other stakeholders to share knowledge [17, 49]. A third conference in Montpellier, France, is planned for 2015 with the following agenda items: discussion key scenarios in agriculture and food systems, identifying priorities for early action and designing a roadmap for moving forward with an action plan. These objectives set the stage for a much stronger emphasis on knowledge-to-action frameworks, capacity-building and the changes in human behaviour and social infrastructure that are necessary for adaptation and resilience [133, 152, 293]. The momentum that has already built among the science community for CSA forms the foundation for critical engagement by more researchers in fundamental and applied studies. To this end, establishing a more formal governance mechanism to embed science in the information base for the CSA Alliance, would be a vital step in developing priorities, scientific engagement and funding to support the knowledge needed for policymaking decisions.

ACKNOWLEDGEMENTS

We acknowledge Jill E Walker for graphic design, Ria D'Aversa for maintaining the bibliographic content and Kayla Burns for copyediting. We also acknowledge all conference participants who shared their knowledge and ideas at the Global Science Conference for Climate-Smart Agriculture held in Davis, CA, USA, in March 2013. We thank the session leaders, who also are authors on this manuscript, for composing summaries of the information presented in individual sessions at the conference and synthesizing discussions by participants. These summaries formed the basis of this paper. This article was funded by a Programmatic Initiative from the College of Agriculture and Environmental Science at the University of California, Davis (to LEJ).

AUTHORS' CONTRIBUTIONS

KLS, AKH and LEJ cowrote the manuscript. AJB, MRC, AC, CJC, JLH, KH, JWH, WRH, LSJ, BMJ, EK, RL, LL, MNL, SM, RP, MPR, SSS, WMS, MS, PT, SJV, SMW and EW contributed content. All authors read and approved the final manuscript.

REFERENCES

1. Morton JF: The impact of climate change on smallholder and subsistence agriculture. Proc Natl Acad Sci U S A. 2007, 104: 19680-19685.
2. Reidsma P, Ewert F, Lansink AO, Leemans R: Adaptation to climate change and climate variability in European agriculture: the importance of farm level responses. Eur J Agron. 2010, 32: 91-102.
3. Vermeulen SJ, Campbell BM, Ingram JSI: Climate change and food systems. Annu Rev Environ Resour. 2012, 35: 195-222.
4. Solomon S, Qin D, Manning M, Chen Z, Marquis M, Averyt KB Tignor MMB, Miller HL: Climate Change 2007: The Physical Science Basis (Contribution of Working Group I to the Fourth Assessment Report of the Intergovernmental Panel on Climate Change (IPCC)). 2007, (accessed 25 July 2014), Cambridge, UK: Cambridge University Press, accessed 25 July 2014)
5. Lobell DB, Schlenker W, Costa-Roberts J: Climate trends and global crop production since 1980. Science. 2011, 333: 616-620.
6. Cline WR: Global Warming and Agriculture: Impact Estimates by Country. 2007, Washington DC: Center for Global Development
7. Jarvis A, Lau C, Cook S, Wollenberg E, Hansen J, Bonilla O, Challinor A: An integrated adaptation and mitigation framework for developing agricultural research: synergies and trade-offs. Exp Agric. 2011, 47: 185-203.
8. Knox J, Hess T, Daccache A, Wheeler T: Climate change impacts on crop productivity in Africa and South Asia. Environ Res Lett. 2012, 7: 034032-
9. Rosenzweig C, Elliott J, Deryng D, Ruane AC, Müller C, Arneth A, Boote KJ, Folberth C, Glotter M, Khabarov N, Neumann K, Piontek F, Pugh TAM, Schmid E, Stehfest E, Yang H, Jones JW: Assessing agricultural risks of climate change in the 21st century in a global gridded crop model intercomparison. Proc Natl Acad Sci U S A. 2014, 111: 3268-3273.
10. Myers SS, Zanobetti A, Kloog I, Bloom AJ, Carlisle EA, Dietterich LH, Fitzgerald G, Hasegawa T, Holbrook NM, Huybers P, Leakey

ADB, Nelson R, Ottman MJ, Raboy V, Sakai H, Sartor KA, Schwartz J, Seneweera S, Tausz M, Usui Y: Rising CO_2 threatens food quality. Nature. 2014, 510: 139-142.

11. Wheeler T, von Braun J: Climate change impacts on global food security. Science. 2013, 341: 508-513.
12. World Bank: Climate-Smart Agriculture: Increased Productivity and Food Security, Enhancing Resilience and Reduced Carbon Emissions for Sustainable Development, Opportunities and Challenges for a Converging Agenda: Country Examples. 2011, Washington, DC: World Bank
13. World Bank: World Development Report 2014: Risk and Opportunity: Managing Risk for Development. 2013, (accessed 25 July 2014), Washington DC: World Bank, (accessed 25 July 2014)
14. Food and Agriculture Organization of the United Nations (FAO): Climate-Smart Agriculture Sourcebook. 2013, Rome: FAO
15. Climate Smart Agriculture: The 2nd Global Science Conference on Climate-Smart Agriculture. 2014, (accessed 5 August 2014), (accessed 5 August 2014)
16. Neufeldt H, Jahn M, Campbell BM, Beddington JR, DeClerck F, De Pinto A, Gulledge J, Hellin J, Herrero M, Jarvis A, LeZaks D, Meinke H, Rosenstock T, Scholes M, Scholes R, Vermeulen S, Wollenberg E, Zougmoré R: Beyond climate-smart agriculture: toward safe operating spaces for global food systems. Agric Food Secur. 2013, 2: 12-
17. Wageningen Statement: Climate-Smart Agriculture – Science for Action. The Global Science Conference on Climate-Smart Agriculture (GSCSA). 2011, (accessed 25 July 2014), (accessed 25 July 2014)
18. Carpenter S, Walker B, Anderies JM, Abel N: From metaphor to measurement: resilience of what to what?. Ecosystems. 2001, 4: 765-781.
19. Holling CS: Understanding the complexity of economic, ecological, and social systems. Ecosystems. 2001, 4: 390-405.
20. Walker B, Holling CS, Carpenter SR, Kinzig A: Resilience, adaptability, and transformability in social-ecological systems. Ecol Soc. 2004, 9: 5-
21. Obrist B: Multi-layered social resilience: a new approach in mitigation research. Progr Dev Stud. 2010, 10: 283-293.
22. Cumming GS: Spatial Resilience in Social-Ecological Systems. 2011, New York: Springer Science
23. Cabell JF, Oelofse M: An indicator framework for assessing agroecosystem resilience. Ecol Soc. 2012, 17: 18-

24. Easterling DR: Climate extremes: observations, modeling, and impacts. Science. 2000, 289: 2068-2074.
25. Battisti DS, Naylor RL: Historical warnings of future food insecurity with unprecedented seasonal heat. Science. 2009, 323: 240-244.
26. Bloom AJ, Burger M, Rubio Asensio JS, Cousins AB: Carbon dioxide enrichment inhibits nitrate assimilation in wheat and *Arabidopsis*. Science. 2010, 328: 899-903.
27. Yin X: Improving ecophysiological simulation models to predict the impact of elevated atmospheric CO_2 concentration on crop productivity. Ann Bot. 2013, 112: 465-75.
28. Vermeulen SJ, Aggarwal PK, Ainslie A, Angelone C, Campbell BM, Challinor AJ, Hansen JW, Ingram JSI, Jarvis A, Kristjanson P, Lau C, Nelson GC, Thornton PK, Wollenberg E: Options for support to agriculture and food security under climate change. Environ Sci Policy. 2012, 15: 136-144.
29. Gutierrez AP, Ponti L, d'Oultremont T, Ellis CK: Climate change effects on poikilotherm tritrophic interactions. Clim Change. 2007, 87 (1 Suppl): S167-S192.
30. Antón J, Cattaneo A, Kimura S, Lankoski J: Agricultural risk management policies under climate uncertainty. Glob Environ Change. 2013, 23: 1726-1736.
31. Jackson LE, Pulleman MM, Brussaard L, Bawa KS, Brown GG, Cardoso IM, de Ruiter PC, García-Barrios L, Hollander AD, Lavelle P, Ouédraogo E, Pascual U, Setty S, Smukler SM, Tscharntke T, van Noordwijk M: Social-ecological and regional adaptation of agrobiodiversity management across a global set of research regions. Glob Environ Change. 2012, 22: 623-639.
32. Stringer LC, Dougill AJ, Thomas AD, Spracklen DV, Chesterman S, Speranza CI, Rueff H, Riddell M, Williams M, Beedy T, Abson DJ, Klintenberg P, Syampungani S, Powell P, Palmer AR, Seely MK, Mkwambisi DD, Falcao M, Sitoe A, Ross S, Kopolo G: Challenges and opportunities in linking carbon sequestration, livelihoods and ecosystem service provision in drylands. Environ Sci Policy. 2012, 19–20: 121-135.
33. Minang PA, Duguma LA, Bernard F, Mertz O, van Noordwijk M: Prospects for agroforestry in REDD+landscapes in Africa. Curr Opin Environ Sustain. 2014, 6: 78-82.
34. Elliott J, Deryng D, Müller C, Frieler K, Konzmann M, Gerten D, Glotter M, Flörke M, Wada Y, Best N, Eisner S, Fekete BM, Folberth C, Foster I, Gosling SN, Haddeland I, Khabarov N, Ludwig F, Masaki Y, Olin S,

Rosenzweig C, Ruane AC, Satoh Y, Schmid E, Stacke T, Tang Q, Wisser D: Constraints and potentials of future irrigation water availability on agricultural production under climate change. Proc Natl Acad Sci U S A. 2013, 111: 3239-3244.

35. Meza FJ, Wilks DS, Gurovich L, Bambach N: Impacts of climate change on irrigated agriculture in the Maipo Basin, Chile: reliability of water rights and changes in the demand for irrigation. J Water Resour Plan Manag. 2012, 138: 421-430.

36. Nelson GC, Rosegrant MW, Palazzo A, Gray I, Ingersoll C, Robertson R, Tokgoz S, Zhu T, Sulser TB, Ringler C, Msangi S, You L: Food Security, Farming, And Climate Change to 2050: Scenarios, Results, Policy Options. 2010, (accessed 25 July 2014), Washington DC: International Food Policy Research Institute, (accessed 25 July 2014)

37. Thornton PK, Gerber PJ: Climate change and the growth of the livestock sector in developing countries. Mitig Adapt Strateg Glob Change. 2010, 15: 169-184.

38. Valin BH, Havlík P, Mosnier A, Herrero M, Schmid E, Obersteiner M: Agricultural productivity and greenhouse gas emissions: trade-offs or synergies between mitigation and food security?. Environ Res Lett. 2013, 8: 035019-

39. Kraxner F, Nordström EM, Havlík P, Gusti M, Mosnier A, Frank S, Valin H, Fritz S, Fuss S, Kindermann G, McCallum I, Khabarov N, Böttcher H, See L, Aoki K, Schmid E, Máthé L, Obersteiner M: Global bioenergy scenarios–future forest development, land-use implications, and trade-offs. Biomass Bioenergy. 2013, 57: 86-96.

40. Mbow C, Van Noordwijk M, Luedeling E, Neufeldt H, Minang PA, Kowero G: Agroforestry solutions to address food security and climate change challenges in Africa. Curr Opin Environ Sustain. 2014, 6: 61-67.

41. Adger WN: Social and ecological resilience: are they related?. Prog Hum Geogr. 2000, 24: 347-364.

42. Adger WN, Barnett J, Brown K, Marshall N, O'Brien K: Cultural dimensions of climate change impacts and adaptation. Nat Clim Chang. 2012, 3: 112-117.

43. Gray CL, Mueller V: Natural disasters and population mobility in Bangladesh. Proc Natl Acad Sci U S A. 2012, 109: 6000-6005.

44. Ifejika Speranza C: Drought coping and adaptation strategies: Understanding adaptations to climate change in agro-pastoral livestock production in Makueni District. Kenya. Eur J Dev Res. 2010, 22: 623-642.

45. Chindarkar N: Gender and climate change-induced migration: proposing a framework for analysis. Environ Res Lett. 2012, 7: 025601-

46. Villamor GB, van Noordwijk M, Djanibekov U, Chiong-Javier ME, Catacutan D: Gender differences in land-use decisions: shaping multifunctional landscapes?. Curr Opin Environ Sustain. 2014, 6: 128-133.

47. Vermeulen SJ, Challinor AJ, Thornton PK, Campbell BM, Eriyagama N, Vervoort JM, Kinyangi J, Jarvis A, Läderach P, Ramirez-Villegas J, Nicklin KJ, Hawkins E, Smith DR: Addressing uncertainty in adaptation planning for agriculture. Proc Natl Acad Sci U S A. 2013, 110: 8357-8362.

48. Reed MS, Evely AC, Cundill G, Fazey I, Glass J, Laing A, Newig J, Parrish B, Prell C, Raymond C, Stringer LC: What is social learning?. Ecol Soc. 2010, 15: r1-

49. Speelman EN, Groot JCJ, García-Barrios LE, Kok K, van Keulen H, Tittonell P: From coping to adaptation to economic and institutional change–trajectories of change in land-use management and social organization in a Biosphere Reserve community, Mexico. Land Use Policy. 2014, 41: 31-44.

50. The Davis Statement–Climate-Smart Agriculture Global Research Agenda: Science for Action. Climate-Smart Agriculture: Global Science Conference, 19–23 March 2013, University of California, Davis, USA. (accessed 26 July 2014), Climate-Smart Agriculture: Global Science Conference, 19–23 March 2013, University of California, Davis, USA. (accessed 26 July 2014)

51. de Graaf MA, van Groenigen KJ, Six J, Hungate B, van Kessel C: Interactions between plant growth and soil nutrient cycling under elevated CO_2: a meta-analysis. Glob Chang Biol. 2006, 12: 2077-2091.

52. Thornton PK, Jones PG, Alagarswamy G, Andresen J, Herrero M: Adapting to climate change: agricultural system and household impacts in East Africa. Agric Syst. 2010, 103: 73-82.

53. Haden VR, Niles MT, Lubell M, Perlman J, Jackson LE: Global and local concerns: what attitudes and beliefs motivate farmers to mitigate and adapt to climate change?. PLoS One. 2012, 7: e52882-

54. Stavi I, Lal R: Agroforestry and biochar to offset climate change: a review. Agron Sustain Dev. 2012, 33: 81-96.

55. Weber CL, Matthews HS: Food-miles and the relative climate impacts of food choices in the United States. Environ Sci Technol. 2008, 42: 3508-3513.

56. Garnett T, Appleby MC, Balmford A, Bateman IJ, Benton TG, Bloomer P, Burlingame B, Dawkins M, Dolan L, Fraser D, Herrero M, Hoffmann I, Smith P, Thornton PK, Toulmin C, Vermeulen SJ, Godfray HCJ: Sustainable intensification in agriculture: premises and policies. Science. 2013, 341: 33-34.

57. Godfray HC, Garnett T: Food security and sustainable intensification. Philos Trans R Soc Lond B Biol Sci. 2014, 369: 20120273-

58. Tittonell P, Giller KE: When yield gaps are poverty traps: the paradigm of ecological intensification in African smallholder agriculture. Field Crops Res. 2013, 143: 76-90.

59. George T: Why crop yields in developing countries have not kept pace with advances in agronomy. Glob Food Sec. 2014, 3: 49-58.

60. Newton AC, Johnson SN, Gregory PJ: Implications of climate change for diseases, crop yields and food security. Euphytica. 2011, 179: 3-18.

61. Mittler R, Blumwald E: Genetic engineering for modern agriculture: challenges and perspectives. Annu Rev Plant Biol. 2010, 61: 443-462.

62. Peleg Z, Reguera M, Tumimbang E, Walia H, Blumwald E: Cytokinin-mediated source/sink modifications improve drought tolerance and increase grain yield in rice under water-stress. Plant Biotechnol J. 2011, 9: 747-758.

63. Reguera M, Peleg Z, Blumwald E: Targeting metabolic pathways for genetic engineering abiotic stress-tolerance in crops. Biochim Biophys Acta. 2012, 1819: 186-194.

64. Cossani CM, Reynolds MP: Physiological traits for improving heat tolerance in wheat. Plant Physiol. 2012, 160: 1710-1718.

65. Reynolds M, Foulkes J, Furbank R, Griffiths S, King J, Murchie E, Parry M, Slafer G: Achieving yield gains in wheat. Plant Cell Environ. 2012, 35: 1799-1823.

66. Pask AJD, Reynolds MP: Breeding for yield potential has increased deep soil water extraction capacity in irrigated wheat. Crop Sci. 2013, 53: 2090-2104.

67. Pask A, Pietragalla J, Mullan D, Reynolds M: Physiological Breeding II: A Field Guide To Wheat Phenotyping. 2011, (accessed 26 July 2014), Mexico City: International Maize and Wheat Improvement Center (CIMMYT), (accessed 26 July 2014)

68. Araus JL, Cairns JE: Field high-throughput phenotyping: the new crop breeding frontier. Trends Plant Sci. 2014, 19: 52-61.

69. Peleg Z, Blumwald E: Hormone balance and abiotic stress tolerance in

crop plants. Curr Opin Plant Biol. 2011, 14: 290-295.

70. Reynolds M, Dreccer F, Trethowan R: Drought-adaptive traits derived from wheat wild relatives and landraces. J Exp Bot. 2007, 58: 177-186.

71. Sadok W, Naudin P, Boussuge B, Muller B, Welcker C, Tardieu F: Leaf growth rate per unit thermal time follows QTL-dependent daily patterns in hundreds of maize lines under naturally fluctuating conditions. Plant Cell Environ. 2007, 30: 135-146.

72. Wang X, Taub DR: Interactive effects of elevated carbon dioxide and environmental stresses on root mass fraction in plants: a meta-analytical synthesis using pairwise techniques. Oecologia. 2010, 163: 1-11.

73. Roy SJ, Tucker EJ, Tester M: Genetic analysis of abiotic stress tolerance in crops. Curr Opin Plant Biol. 2011, 14: 232-239.

74. Reynolds M, Manes Y, Izanloo A, Langridge P: Phenotyping approaches for physiological breeding and gene discovery in wheat. Ann Appl Biol. 2009, 155: 309-320.

75. Pinto RS, Reynolds MP, Mathews KL, McIntyre CL, Olivares-Villegas JJ, Chapman SC: Heat and drought adaptive QTL in a wheat population designed to minimize confounding agronomic effects. Theor Appl Genet. 2010, 121: 1001-1021.

76. Rebetzke GJ, Chenu K, Biddulph B, Moeller C, Deery DM, Rattey AR, Bennett D, Barrett-Leonard G, Mayer JE: A multisite managed environmental facility for targeted trait and germplasm phenotyping. Funct Plant Biol. 2013, 40: 1-13.

77. Olivares-Villegas JJ, Reynolds MP, McDonald GK: Drought-adaptive attributes in the Seri/Babax hexaploid wheat population. Funct Plant Biol. 2007, 34: 189-203.

78. Bouteillé M, Rolland G, Balsera C, Loudet O, Muller B: Disentangling the intertwined genetic bases of root and shoot growth in Arabidopsis. PLoS One. 2012, 7: e32319-

79. Tester M, Langridge P: Breeding technologies to increase crop production in a changing world. Science. 2010, 327: 818-822.

80. Langridge P, Fleury D: Making the most of "omics" for crop breeding. Trends Biotechnol. 2011, 29: 33-40.

81. Heslot N, Akdemir D, Sorrells ME, Jannink JL: Integrating environmental covariates and crop modeling into the genomic selection framework to predict genotype by environment interactions. Theor Appl Genet. 2014, 127: 463-480.

82. Philippot L, Hallin S: Towards food, feed and energy crops mitigating

climate change. Trends Plant Sci. 2011, 16: 476-80.

83. Reynolds MP, Hellin J, Govaerts B, Kosina P, Sonder K, Hobbs P, Braun B: Global crop improvement networks to bridge technology gaps. J Exp Bot. 2012, 63: 1-12.

84. Ceccarelli S, Galie A, Grando S: Participatory breeding for climate change-related traits. Genomics and Breeding for Climate-Resilient Crops: Concepts and Strategies, Volume 1. Edited by: Kole C. 2013, 331-376. Berlin: Springer Science & Business, 8

85. Opio C, Gerber P, Mottet A, Falcucci A, Tempio G, MacLeod M, Vellinga T, Henderson B, Steinfeld H: Greenhouse Gas Emissions From Ruminant Supply Chains: A Global Life Cycle Assessment. 2013, (accessed 26 July 2014), Rome: Food and Agriculture Organization of the United Nations (FAO), (accessed 26 July 2014)

86. Thornton PK, van de Steeg J, Notenbaert A, Herrero M: The impacts of climate change on livestock and livestock systems in developing countries: a review of what we know and what we need to know. Agric Syst. 2009, 101: 113-127.

87. Sutherst RW: Implications of global change and climate variability for vector-borne diseases: generic approaches to impact assessments. Int J Parasitol. 1998, 28: 935-945.

88. Hristov AN, Oh J, Lee C, Meinen R, Montes F, Ott T, Firkins J, Rotz A, Dell C, Adesogan A, Yang W, Tricarico J, Kebreab E, Waghorn G, Dijkstra J, Oosting S: FAO Animal Production and Health Paper No 177. Mitigation of Greenhouse Gas Emissions in Livestock Production: A Review of Technical Options for Non-CO2 Emissions. Edited by: Gerber PJ, Henderson B, Makkar HPS. 2013, (accessed 26 July 2014), Rome: Food and Agriculture Organization of the United Nations, (accessed 26 July 2014)

89. Niles MT: Achieving social sustainability in animal agriculture: challenges and opportunities to reconcile multiple sustainability goals. Sustainable Animal Agriculture. Edited by: Kebreab E. 2013, 193-211. Wallingford, UK: CABI

90. Herrero M, Thornton PK, Bernués A, Baltenweck I, Vervoort J, van de Steeg J, Makokha S, van Wijk MT, Karanja S, Rufino MC, Staal SJ: Exploring future changes in smallholder farming systems by linking socio-economic scenarios with regional and household models. Glob Environ Change. 2014, 24: 165-182.

91. Thornton PK, Herrero M: Potential for reduced methane and carbon dioxide emissions from livestock and pasture management in the tropics.

Proc Natl Acad Sci U S A. 2010, 107: 19667-19672.

92. Herrero M, Thornton PK: Livestock and global change: emerging issues for sustainable food systems. Proc Natl Acad Sci U S A. 2013, 110: 20878-20881.

93. Doré T, Makowski D, Malézieux E, Munier-Jolain N, Tchamitchian M, Tittonell P: Facing up to the paradigm of ecological intensification in agronomy: revisiting methods, concepts and knowledge. Eur J Agron. 2011, 34: 197-210.

94. Silvestri S, Osano P, de Leeuw J, Herrero M, Ericksen P, Hariuki J, Njuki J, Notenbaert A, Bedelian C: ILRI Position Paper. Greening Livestock: Assessing the Potential of Payment for Environmental Services in Livestock-Inclusive Agricultural Production Systems in Developing Countries. 2012, Nairobi: International Livestock Research Institute (ILRI)

95. Rufino MC, Thornton PK, Ng'ang'a SK, Mutie I, Jones PG, van Wijk MT, Herrero M: Transitions in agro-pastoralist systems of East Africa: impacts on food security and poverty. Agric Ecosyst Environ. 2013, 179: 215-230.

96. Thornton PK, Boone RB, Galvin KA, Burnsilver SB, Waithaka MM, Kuyiah J, Karanja S, González-Estrada E, Boone B: Coping strategies in livestock-dependent households in East and Southern Africa: a synthesis of four case studies. Hum Ecol. 2007, 35: 461-476.

97. Alvarez S, Rufino MC, Vayssières J, Salgado P, Tittonell P, Tillard E, Bocquier F: Whole-farm nitrogen cycling and intensification of crop-livestock systems in the highlands of Madagascar: an application of network analysis. Agric Syst. 2014, 126: 25-37.

98. Bengoumi M, Vias G, Faye B: Camel milk production and transformation in Sub-Saharan Africa. Desertification Combat and Food Safety: the Added Value of Camel Producers. Edited by: Faye B, Esenov P. 2005, 200-208. Amsterdam: IOS Press

99. Jones PG, Thornton PK: Croppers to livestock keepers: livelihood transitions to 2050 in Africa due to climate change. Environ Sci Policy. 2009, 12: 427-437.

100. Bustamante MMC, Nobre CA, Smeraldi R, Aguiar APD, Barioni LG, Ferreira LG, Longo K, May P, Pinto AS, Ometto JPHB: Estimating greenhouse gas emissions from cattle raising in Brazil. Clim Change. 2012, 115: 559-577.

101. Maia SMF, Ogle SM, Cerri CEP, Cerri CC: Effect of grassland management on soil carbon sequestration in Rondonia and Mato Grosso

states, Brazil. Geoderma. 2009, 149: 84-91.

102. Newbold CJ, López S, Nelson N, Ouda JO, Wallace RJ, Moss AR: Propionate precursors and other metabolic intermediates as possible alternative electron acceptors to methanogenesis in ruminal fermentation *in vitro*. Br J Nutr. 2007, 94: 27-35.

103. Norad: the Norwegian Agency for Development Cooperation. (accessed 26 July 2014), (accessed 26 July 2014)

104. Morse SS, Mazet JAK, Woolhouse M, Parrish CR, Carroll D, Karesh WB, Zambrana-Torrelio C, Lipkin WI, Daszak P: Prediction and prevention of the next pandemic zoonosis. Lancet. 2012, 380: 1956-1965.

105. US Agency for International Development (USAID): Pandemic Influenza and Other Emerging Threats. 2013 (accessed 26 July 2014), Washington, DC: USAID, (accessed 26 July 2014)

106. Foley P, Crosson P, Lovett DK, Boland TM, O'Mara FP, Kenny D: Whole-farm systems modelling of greenhouse gas emissions from pastoral suckler beef cow production systems. Agric Ecosyst Environ. 2011, 142: 222-230.

107. Rotz C, Montes F, Chianese DS: The carbon footprint of dairy production systems through partial life cycle assessment. J Dairy Sci. 2010, 93: 1266-1282.

108. Del Prado A, Misselbrook T, Chadwick D, Hopkins A, Dewhurst RJ, Davison P, Butler A, Schröder J, Scholefield D: $SIMS_{DAIRY}$: a modelling framework to identify sustainable dairy farms in the UK framework description and test for organic systems and N fertiliser optimisation. Sci Total Environ. 2011, 409: 3993-4009.

109. Reisinger A, Havlik P, Riahi K, Vliet O, Obersteiner M, Herrero M: Implications of alternative metrics for global mitigation costs and greenhouse gas emissions from agriculture. Clim Change. 2012, 117: 677-690.

110. Havlík P, Valin H, Herrero M, Obersteiner M, Schmid E, Rufino MC, Mosnier A, Thornton PK, Böttcher H, Conant RT, Frank S, Fritz S, Fuss S, Kraxner F, Notenbaert A: Climate change mitigation through livestock system transitions. Proc Natl Acad Sci U S A. 2014, 111: 3709-3714.

111. Collaborative Research on Adapting Livestock Systems to Climate Change. Feed the Future Innovation Lab Progress Report: Alignment with Feed the Future. 2013, (accessed 26 July 2014), Fort Collins, CO: Colorado State University, US Agency for International Development, (accessed 26 July 2014)

112. Rockström J, Steffen W, Noone K, Persson A, Chapin FS, Lambin EF, Lenton TM, Scheffer M, Folke C, Schellnhuber HJ, Nykyist B, de Wit CA, Hughes T, van der Leeuw S, Rodhe H, Sorlin S, Snyder PK, Costanza R, Svedin U, Falkenmark M, Karlberg L, Corell RW, Fabry VJ, Hansen J, Walker B, Liverman D, Richardson K, Crutzen P, Foley JA: A safe operating space for humanity. Nature. 2009, 461: 472-475.

113. Sutton MA, Oenema O, Erisman JW, Leip A, van Grinsven H, Winiwarter W: Too much of a good thing. Nature. 2011, 472: 159-161.

114. Smith P, Martino D, Cai Z, Gwary D, Janzen H, Kumar P, McCarl B, Ogle S, O'Mara F: Climate Change 2007: Mitigation. Edited by: Metz B, Davidson O, Bosch P, Dave R, Meyer L. 2007, 497-540. (26 July 2014), Agriculture, Cambridge, UK: Cambridge University Press,Contribution of Working Group III to the Fourth Assessment Report of the Intergovernmental Panel on Climate Change, (26 July 2014)

115. Halvorson AD, Del Grosso SJ, Alluvione F: Tillage and inorganic nitrogen source effects on nitrous oxide emissions from irrigated cropping systems. Soil Sci Soc Am J. 2010, 74: 436-445.

116. Akiyama H, Yan X, Yagi K: Evaluation of effectiveness of enhanced-efficiency fertilizers as mitigation options for N_2O and NO emissions from agricultural soils: meta-analysis. Glob Chang Biol. 2009, 16: 1837-1846.

117. Parkin TB, Hatfield JL: Enhanced efficiency fertilizers: effect on nitrous oxide emissions in Iowa. Agron J. 2014, 106: 694-702.

118. Hatfield JL, Parkin TB: Enhanced efficiency fertilizers: effect on agronomic performance of corn in Iowa. Agron J. 2014, 106: 771-780.

119. Kallenbach CM, Rolston DE, Horwath WR: Cover cropping affects soil N_2O and CO_2 emissions differently depending on type of irrigation. Agric Ecosyst Environ. 2010, 137: 251-260.

120. Singh A, Kaur J: Impact of conservation tillage on soil properties in rice-wheat cropping system. Agric Sci Res J. 2012, 2: 30-41.

121. Mandal B, Majumder B, Adhya TK, Bandyopadhyay PK, Gangopadhyay A, Sarkar D, Kundu MC, Choudhury SG, Hazra GC, Kundu S, Samantaray RN, Misra AK: Potential of double-cropped rice ecology to conserve organic carbon under subtropical climate. Glob Chang Biol. 2008, 14: 2139-2151.

122. Smith P, Haberl H, Popp A, Erb KH, Lauk C, Harper R, Tubiello FN, de Siqueira Pinto A, Jafari M, Sohi S, Masera O, Böttcher H, Berndes G, Bustamante M, Ahammad H, Clark H, Dong H, Elsiddig EA, Mbow C, Ravindranath NH, Rice CW, Robledo Abad C, Romanovskaya A, Sperling

F, Herrero M, House JI, Rose S: How much land-based greenhouse gas mitigation can be achieved without compromising food security and environmental goals?. Glob Chang Biol. 2013, 19: 2285-302.

123. Joseph S, Khoi DD, Hien NV, Anh ML, Nguyen HH, Hung TM, Yen NT, Thomsen MF, Lehmann J, Chia CH: North Vietnam Villages Lead the Way in the Use of Biochar: Building on an Indigenous Knowledge Base. (accessed 26 July 2014), International Biochar Initiative, (accessed 26 July 2014)

124. Hatfield JL, Parkin TB, Sauer TJ, Prueger JH: Mitigation opportunities from land management practices in a warming world: increasing potential sinks. Managing Agricultural Greenhouse Gases: Coordinated Agricultural Research through GRACEnet to Address Our Changing Climate. Edited by: Liebig MA, Franzluebbers AT, Follett RF. 2012, 487-504. Waltham, MA: Academic Press

125. Jensen ES, Peoples MB, Boddey RM, Gresshoff PM, Hauggaard-Nielsen H, Alves B, Morrison MJ: Legumes for mitigation of climate change and the provision of feedstock for biofuels and biorefineries: a review. Agron Sustain Dev. 2012, 32: 329-364.

126. Kim DG: Estimation of net gain of soil carbon in a nitrogen-fixing tree and crop intercropping system in Sub-Saharan Africa: results from re-examining a study. Agrofor Syst. 2012, 86: 175-184.

127. Corre-Hellou G, Dibet A, Hauggaard-Nielsen H, Crozat Y, Gooding M, Ambus P, Dahlmann C, von Fragstein P, Pristeri A, Monti M, Jensen ES: The competitive ability of pea-barley intercrops against weeds and the interactions with crop productivity and soil N availability. Field Crops Res. 2011, 122: 264-272.

128. Delgado JA, Kowalski K, Tebbe C: The first Nitrogen Index app for mobile devices: using portable technology for smart agricultural management. Comput Electron Agric. 2013, 91: 121-123.

129. Fischer RA, Edmeades GO: Breeding and cereal yield progress. Crop Sci. 2010, 50 (Suppl 1): S85-S98.

130. Grassini P, Eskridge KM, Cassman KG: Distinguishing between yield advances and yield plateaus in historical crop production trends. Nat Commun. 2013, 4: 2918-

131. van Ittersuma MK, Cassman KG, Grassini P, Wolfa J, Tittonell P, Hochman Z: Yield gap analysis with local to global relevance—a review. Field Crops Res. 2013, 143: 4-17.

132. van Warta JK, Kersebaum C, Peng S, Milner M, Cassman KG: Estimating crop yield potential at regional to national scales. Field Crops Res. 2012,

143: 34-43.

133. Reed MS, Podesta G, Fazey I, Geeson N, Hessel R, Hubacek K, Letson D, Nainggolan D, Prell C, Rickenbach MG, Ritsema C, Schwilch G, Stringer LC, Thomas AD: Combining analytical frameworks to assess livelihood vulnerability to climate change and analyse adaptation options. Ecol Econ. 2013, 94: 66-77.

134. Rogers EM: Diffusion of Innovations. 2003, New York: Free Press, 5

135. Kristjanson P, Reid RS, Dickson N, Clark WC, Romney D, Puskur R, MacMillan S, Grace D: Linking international agricultural research knowledge with action for sustainable development. Proc Natl Acad Sci U S A. 2009, 106: 5047-5052.

136. Spielman DJ, Ekboir J, Davis K: The art and science of innovation systems inquiry: applications to Sub-Saharan African agriculture. Technol Soc. 2009, 31: 399-405.

137. Vervoort JM, Thornton PK, Kristjanson P, Forch W, Ericksen PJ, Kok K, Ingram JSI, Herrero M, Palazzo A, Helfgott AES, Wilkinson A, Havlík P, Mason-D'Croz D, Jost C: Challenges to scenario-guided adaptive action on food security under climate change. Glob Environ Change. in press. doi:10.1016/j.gloenvcha.2014.03.001

138. Arbuckle JG, Prokopy LS, Haigh T, Hobbs J, Knoot T, Knutson C, Loy A, Mase AS, McGuire J, Morton LW, Tyndall J, Widhalm M: Climate change beliefs, concerns, and attitudes toward adaptation and mitigation among farmers in the Midwestern United States. Clim Change. 2013, 117: 943-950.

139. Prokopy LS, Haigh T, Mase AS, Angel J, Hart C, Knutson C, Lemos MC, Lo YJ, McGuire J, Morton LW, Perron J, Todey D, Widhalm M: Agricultural advisors: a receptive audience for weather and climate information?. Weather Clim Soc. 2013, 5: 162-167.

140. Baumgart-Getz A, Prokopy LS, Floress K: Why farmers adopt best management practice in the United States: a meta-analysis of the adoption literature. J Environ Manage. 2012, 96: 17-25.

141. Food and Agriculture Organization of the United Nations (FAO): Investing in Sustainable Agricultural Intensification: The Role of Conservation Agriculture. A Framework for Action. 2008, (accessed 26 July 2014), Rome: FAO, (accessed 26 July 2014)

142. Shaxson F, Kassam A, Friedrich T, Adekunle A: Conservation Agriculture: Looking beneath the Surface. 2008, (accessed 4 August 2014), Nakuru, Kenya: Food and Agriculture Organization of the United Nations (FAO). Conservation Agriculture Workshop, (accessed 4 August 2014)

143. Ndlovu PV, Mazvimavi K, An H, Murendo C: Productivity and efficiency analysis of maize under conservation agriculture in Zimbabwe. Agric Syst. 2014, 124: 21-31.

144. Giller KE, Witter E, Corbeels M, Tittonell P: Conservation agriculture and smallholder farming in Africa: the heretics' view. Field Crops Res. 2009, 114: 23-34.

145. Knowler D, Bradshaw B: Farmers' adoption of conservation agriculture: a review and synthesis of recent research. Food Policy. 2007, 32: 25-48.

146. Giller KE, Tittonell P, Rufino MC, van Wijk MT, Zingore S, Mapfumo P, Adjei-Nsiah S, Herrero M, Chikowo R, Corbeels M, Rowe EC, Baijukya F, Mwijage A, Smith J, Yeboah E, van der Burg WJ, Sanogo OM, Misiko M, de Ridder N, Karanja S, Kaizzi C, K'ungu J, Mwale M, Nwaga D, Pacini C, Vanlauwe B: Communicating complexity: integrated assessment of trade-offs concerning soil fertility management within African farming systems to support innovation and development. Agric Syst. 2011, 104: 191-203.

147. Arslan A, McCarthy N, Lipper L, Asfaw S, Cattaneo A: Adoption and Intensity of Adoption of Conservation Farming Practices in Zambia. ESA Working Paper No. 13-01. 2013, (accessed 26 July 2014), Rome: Food and Agriculture Organization of the United Nations, (accessed 26 July 2014)

148. Harvey CA, Chacón M, Donatti CI, Garen E, Hannah L, Andrade A, Bede L, Brown D, Calle A, Chará J, Clement C, Gray E, Hoang MH, Minang P, Rodríguez AM, Seeberg-Elverfeldt C, Semroc B, Shames S, Smukler S, Somarriba E, Torquebiau E, van Etten J, Wollenberg E: Climate-smart landscapes: opportunities and challenges for integrating adaptation and mitigation in tropical agriculture. Conserv Lett. 2013, 7: 77-90.

149. Jerneck A, Olsson L: Food first! Theorising assets and actors in agroforestry: risk evaders, opportunity seekers and "the food imperative" in Sub-Saharan Africa. Int J Agric Sustain. 2013, 12: 1-22.

150. Jerneck A, Olsson L: More than trees! Understanding the agroforestry adoption gap in subsistence agriculture: Insights from narrative walks in Kenya. J Rural Stud. 2013, 32: 114-125.

151. Adger N, Barnett J, Dabelko G: Climate and war: a call for more research. Nature. 2013, 498: 171-

152. Cash DW, Borck JC, Patt AG: Countering the loading-dock approach to linking science and decision making: comparative analysis of El Niño/ Southern Oscillation (ENSO) forecasting systems. Sci Technol Human Values. 2006, 31: 465-494.

153. Hayward T: Climate change and ethics. Nat Clim Chang. 2012, 2: 843-848.

154. Markowitz EM, Shariff AF: Climate change and moral judgment. Nat Clim Chang. 2012, 2: 243-247.

155. Barnett BJ, Barrett CB, Skees JR: Poverty traps and index-based risk transfer products. World Dev. 2008, 36: 1766-1785.

156. Lybbert TJ, McPeak J: Risk and intertemporal substitution: livestock portfolios and off-take among Kenyan pastoralists. J Dev Econ. 2012, 97: 415-426.

157. Carter MR: Designed for development impact: next generation index insurance for smallholder farmers. Protecting the Poor: A Microinsurance Compendium, Volume II: Part IV. General Insurance. Edited by: Churchill C, Matul M. 2012, 238-257. ISBN 978-92-2-125744-8 (accessed 26 July 2014), Geneva: International Labour Office, ISBN 978-92-2-125744-8 (accessed 26 July 2014)

158. McDowell JZ, Hess JJ: Accessing adaptation: multiple stressors on livelihoods in the Bolivian highlands under a changing climate. Glob Environ Change. 2012, 22: 342-352.

159. Adger WN: Vulnerability. Glob Environ Change. 2006, 16: 268-281.

160. Valdivia C, Seth A, Gilles JL, García M, Jiménez E, Cusicanqui J, Navia F, Yucra E: Adapting to climate change in Andean ecosystems: landscapes, capitals, and perceptions shaping rural livelihood strategies and linking knowledge systems. Ann Assoc Am Geogr. 2010, 100: 818-834.

161. Johnson L: Index insurance and the articulation of risk-bearing subjects. Environ Plan A. 2013, 45: 2663-2681.

162. Patt A, Suraez P, Hess U: How do small-holders understand insurance, and how much do they want it? Evidence from Africa. Glob Environ Change. 2010, 20: 153-161.

163. Traerup S: Informal networks and resilience to climate change impacts: a collective approach to index insurance. Glob Environ Change. 2012, 22: 255-267.

164. Chantarat S, Mude AG, Barrett CB, Carter MR: Designing index-based livestock insurance for managing asset risk in Northern Kenya. J Risk Insur. 2013, 80: 205-237.

165. BASIS, United States Agency for International Development, Food and Agriculture Organization of the United Nations, Micro-Insurance Innovation Facility of the International Labour Organization, Oxfam America: Feed the Future BASIS Assets and Market Innovation Lab:

Index Insurance: Feed the Future BASIS Assets and Market Innovation Lab: Index Insurance Innovation Initiative (I4). (accessed 26 July 2014), Davis, CA: BASIS Assets and Market Access CRSP, (accessed 26 July 2014)

166. International Livestock Research Institute (ILRI): Index Based Livestock Insurance.

167. apital, economic mobility and poverty traps. J Econ Inequal. 2011, 10: 299-342.

168. Karlan D, Osei RD, Osei-Akoto I, Udry C: Agricultural Decisions after Relaxing Credit and Risk Constraints (NBER Working Paper No 18463). 2012, 1-65. Cambridge, MA: National Bureau of Economic Research

169. McIntosh C, Sarris A, Papadopoulos F: Productivity, credit, risk, and the demand for weather index insurance in smallholder agriculture in Ethiopia. Agric Econ. 2013, 44: 399-417.

170. Carter MR, Lybbert TJ: Consumption versus asset smoothing: testing the implications of poverty trap theory in Burkina Faso. J Dev Econ. 2012, 99: 255-264.

171. Antón J, Kimura S, Lankoski J, Cattaneo A: A comparative study of risk management in agriculture under climate change. OECD Food, Agriculture and Fisheries Papers. 2012, 58: 1-89.

172. Klein D, Luderer G, Kriegler E, Strefler J, Bauer N, Leimbach M, Popp A, Dietrich JP, Humpenöder F, Lotze-Campen H, Edenhofer O: The value of bioenergy in low stabilization scenarios: an assessment using REMIND-MAgPIE. Clim Change. 2014, 123: 705-

173. Lotze-Campen H, von Lampe M, Kyle P, Fujimori S, Havlik P, van Meijl H, Hasegawa T, Popp A, Schmitz C, Tabeau A, Valin H, Willenbockel D, Wise M: Impacts of increased bioenergy demand on global food markets: an AgMIP economic model intercomparison. Agric Econ. 2014, 45: 103-116.

174. US Department of Agriculture (USDA): World Agricultural Supply and Demand Estimates (WASDE-459). 2008, (accessed 26 July 2014), Washington, DC: USDA, (accessed 26 July 2014)

175. Mitchell D: A Note on Rising Food Prices (Policy Research Working Paper 4682). 2008, (accessed 26 July 2014), Washington, DC: World Bank Development Prospects Group, (accessed 26 July 2014)

176. Zhang W, Yu EA, Rozelle S, Yang J, Msangi S: The impact of biofuel growth on agriculture: Why is the range of estimates so wide?. Food Policy. 2013, 38: 227-239.

177. Schmitz C, van Meijl H, Kyle P, Nelson GC, Fujimori S, Gurgel A, Havlik P, Heyhoe E, D'Croz DM, Popp A, Sands R, Tabeau A, van der Mensbrugghe D, von Lampe M, Wise M, Blanc E, Hasegawa T, Kavallari A, Valin H: Land-use change trajectories up to 2050: insights from a global agro-economic model comparison. Agric Econ. 2014, 45: 69-84.

178. Fujimori S, Matsuoka Y: Development of method for estimation of world industrial energy consumption and its application. Energy Econ. 2011, 33: 461-473.

179. van Meijl H, van Rheenen T, Tabeau A, Eickhout B: The impact of different policy environments on agricultural land use in Europe. Agric Ecosyst Environ. 2006, 114: 21-38.

180. Clarke L, Edmonds J, Krey V, Richels R, Rose S, Tavoni M: International climate policy architectures: overview of the EMF 22 International Scenarios. Energy Econ. 2009, 31 (Suppl 2): S64-S81.

181. Edmonds JA, Reilly JM: Global Energy: Assessing the Future. 1985, New York: Oxford University Press

182. Havlik P, Valin H, Mosnier A, Obersteiner M, Baker JS, Herrero M, Rufino MC, Schmid E: Crop productivity and the global livestock sector: implications for land use change and greenhouse gas emissions. Am J Agric Econ. 2012, 95: 442-448.

183. Lotze-Campen H: Improved data for integrated modeling of global environmental change. Environ Res Lett. 2011, 6: 041002-

184. Popp A, Dietrich JP, Lotze-Campen H, Klein D, Bauer N, Krause M, Beringer T, Gerten D, Edenhofer O: The economic potential of bioenergy for climate change mitigation with special attention given to implications for the land system. Environ Res Lett. 2011, 6: 034017-

185. Bondeau A, Smith P, Zaehle SON, Schaphoff S, Lucht W, Cramer W, Gerten D, Lotze-Campen H, Muller C, Reichstein M: Modelling the role of agriculture for the 20th century global terrestrial carbon balance. Glob Chang Biol. 2007, 13: 679-706.

186. Rost S, Gerten D, Bondeau A, Lucht W, Rohwer J, Schaphoff S: Agricultural green and blue water consumption and its influence on the global water system. Water Resour Res. 2008, 44: W09405

187. Leimbach M, Bauer N, Baumstark L, Edenhofer O: Mitigation costs in a globalized world: climate policy analysis with REMIND-R. Environ Model Assess. 2009, 15: 155-173.

188. Popp A, Krause M, Dietrich JP, Lotze-Campen H, Leimbach M, Beringer T, Bauer N: Additional CO_2 emissions from land use change—forest

conservation as a precondition for sustainable production of second generation bioenergy. Ecol Econ. 2012, 74: 64-70.

189. Hoefnagels R, Junginger M, Resch G, Matzenberger J, Panzer C: Report for IEA Bioenergy Task 40. Development of a Tool to Model European Biomass Trade. 2011, (accessed 26 July 2014), Utrecht, the Netherlands: Science Technology and Society, Copernicus Institute, University of Utrecht, (accessed 26 July 2014)

190. Sperling D, Yeh S: Toward a global low carbon fuel standard. Transp Policy. 2010, 17: 47-49.

191. Tokgoz S, Zhang W, Msangi S, Bhandary P: Biofuels and the future of food: competition and complementarities. Agriculture. 2012, 2: 414-435.

192. Lotze-Campen H, Popp A, Beringer T, Müller C, Bondeau A, Rost S, Lucht W: Scenarios of global bioenergy production: the trade-offs between agricultural expansion, intensification and trade. Ecol Modell. 2010, 221: 2188-2196.

193. van Dam J, Junginger M, Faaij APC: From the global efforts on certification of bioenergy towards an integrated approach based on sustainable land use planning. Renew Sustain Energy Rev. 2010, 14: 2445-2472.

194. Msangi S, Evans M: Biofuels and developing economies: is the timing right?. Agric Econ. 2013, 44: 501-510.

195. Beddington JR, Asaduzzaman M, Clark M: The role for scientists in tackling food insecurity and climate change. Agric Food Secur. 2012, 1: 10-

196. Verburg PH, Ellis EC, Letourneau A: A global assessment of market accessibility and market influence for environmental change studies. Environ Res Lett. 2011, 6: 034019-

197. Lugato E, Panagos P, Bampa F, Jones A, Montanarella L: A new baseline of organic carbon stock in European agricultural soils using a modelling approach. Glob Chang Biol. 2014, 20: 313-26.

198. Smith P, Davies CA, Ogle S, Zanchi G, Bellarby J, Bird N, Boddey RM, McNamara NP, Powlson D, Cowie A, van Noordwijk M, Davis SC, Richter DDB, Kryzanowski L, Wijk MT, Stuart J, Kirton A, Eggar D, Newton-Cross G, Adhya TK, Braimoh AK: Towards an integrated global framework to assess the impacts of land use and management change on soil carbon: current capability and future vision. Glob Chang Biol. 2012, 18: 2089-2101.

199. Ewert F, van Ittersum MK, Heckelei T, Therond O, Bezlepkina I, Andersen E: Scale changes and model linking methods for integrated assessment

of agri-environmental systems. Agric Ecosyst Environ. 2011, 142: 6-17.

200. Therond HB, Oomen R, Russell G, Ewert F: Using a cropping system model at regional scale: low-data approaches for crop management information and model calibration. Agric Ecosyst Environ. 2011, 142: 85-94.

201. Grace PR, Antle J, Aggarwal PK, Ogle S, Paustian K, Basso B: Soil carbon sequestration and associated economic costs for farming systems of the Indo-Gangetic Plain: a meta-analysis. Agric Ecosyst Environ. 2012, 146: 137-146.

202. de Groot RS, Alkemade R, Braat L, Hein L, Willemen L: Challenges in integrating the concept of ecosystem services and values in landscape planning, management and decision making. Ecol Complex. 2010, 7: 260-272.

203. Olsson L, Jerneck A: Farmers fighting climate change-from victims to agents in subsistence livelihoods. Wiley Interdiscip Rev Clim Chang. 2010, 1: 363-373.

204. You L, Wood S, Wood-Sichra U: Generating plausible crop distribution maps for Sub-Saharan Africa using a spatially disaggregated data fusion and optimization approach. Agric Syst. 2009, 99: 126-140.

205. Milman A, Bunclark L, Conway D, Adger WN: Assessment of institutional capacity to adapt to climate change in transboundary river basins. Clim Chang. 2013, 121: 755-770.

206. Rivington M, Matthews KB, Buchan K, Miller DG, Bellocchi G, Russell G: Climate change impacts and adaptation scope for agriculture indicated by agro-meteorological metrics. Agric Syst. 2013, 114: 15-31.

207. Trnka M, Brázdil R, Olesen JE, Eitzinger J, Zahradníček P, Kocmánková E, Dobrovolný P, Štěpánek P, Možný M, Bartošová L, Hlavinka P, Semerádová D, Valášek H, Havlíček M, Horáková V, Fischer M, Žalud Z: Could the changes in regional crop yields be a pointer of climatic change?. Agric For Meteorol. 2012, 166–167: 62-71.

208. Del Prado A, Crosson P, Olesen JE, Rotz CA: Whole-farm models to quantify greenhouse gas emissions and their potential use for linking climate change mitigation and adaptation in temperate grassland ruminant-based farming systems. Animal. 2013, 7: 373-385.

209. Gourdji SM, Sibley AM, Lobell DB: Global crop exposure to critical high temperatures in the reproductive period: historical trends and future projections. Environ Res Lett. 2013, 8: 024041-

210. Smith P, Olesen JE: Synergies between the mitigation of, and adaptation

to, climate change in agriculture. J Agric Sci. 2010, 148: 543-552.

211. Acosta L, Klein RJT, Reidsma P, Metzger MJ, Rounsevell MDA, Leemans R, Schröter D: A spatially explicit scenario-driven model of adaptive capacity to global change in Europe. Global Environ Change. 2013, 23: 1211-1224.

212. Schröter D, Cramer W, Leemans R, Prentice IC, Araújo MB, Arnell NW, Bondeau A, Bugmann H, Carter TR, Gracia CA, de la Vega-Leinert AC, Erhard M, Ewert F, Glendining M, House JI, Kankaanpää S, Klein RJT, Lavorel S, Lindner M, Metzger MJ, Meyer J, Mitchell TD, Reginster I, Rounsevell M, Sabaté S, Sitch S, Smith B, Smith J, Smith P, Sykes MT: Ecosystem service supply and vulnerability to global change in Europe. Science. 2005, 310: 1333-1337.

213. Sustainable farm Management Aimed at Reducing Threats to SOILs underclimate change: SmartSOIL. (accessed 26 July 2014), (accessed 26 July 2014)

214. Ingram J, Mills J, Frelih-Larsen A, Davis M: Uptake of Soil Management Practices and Experiences with Decisions Support Tools: Analysis of the Consultation with the Farming Community (Project No 289684). 2012 (accessed 26 July 2014), SmartSOIL, (accessed 26 July 2014)

215. White JW, Hoogenboom G, Kimball BA, Wall GW: Methodologies for simulating impacts of climate change on crop production. Field Crops Res. 2011, 124: 357-368.

216. Hansen J, Sato M, Ruedy R: Perception of climate change. Proc Natl Acad Sci U S A. 2012, 109: 415-423.

217. Moriondo M, Giannakopoulos C, Bindi M: Climate change impact assessment: the role of climate extremes in crop yield simulation. Clim Change. 2010, 104: 679-701.

218. Lobell DB: Errors in climate datasets and their effects on statistical crop models. Agr Forest Meteorol. 2013, 170: 58-66.

219. Challinor A, Martre P, Asseng S, Thornton P, Ewert F: Making the most of climate impacts ensembles. Nat Clim Change. 2014, 4: 77-80.

220. Peters GP, Andrew RM, Boden T, Canadell JG, Ciais P, Le Quéré C, Marland G, Raupach MR, Wilson C: The challenge to keep global warming below 2°C. Nat Clim Chang. 2012, 3: 4-6.

221. Jordan A, Rayner T, Schroeder H, Adger N, Anderson K, Bows A, Le Quéré C, Joshi M, Mander S, Vaughan N, Whitmarsh L: Going beyond two degrees? The risks and opportunities of alternative options. Clim Policy. 2013, 13: 751-769.

222. Petherick A: Seeking a fair and sustainable future. Nat Clim Change. 2014, 4: 81-83.

223. Lal R, Lorenz K, Hüttl RF, Schneider BU, von Braun J: Ecosystem Services and Carbon Sequestration in the Biosphere. 2013, New York: Springer Science

224. Qadir M, Noble AD, Chartres C: Adapting to climate change by improving water productivity of soils in dry areas. Land Degrad Develop. 2013, 21: 12-21.

225. Cochard R: Natural hazards mitigation services of carbon-rich ecosystems. Ecosystem Services and Carbon Sequestration in the Biosphere. Edited by: Lal R, Lorenz K, Hüttl RF, Schneider BU, von Braun J. 2013, 221-293. New York: Springer Science

226. Colomb V, Touchemoulin O, Bockel L, Chotte JL, Martin S, Tinlot M, Bernoux M: Selection of appropriate calculators for landscape-scale greenhouse gas assessment for agriculture and forestry. Environ Res Lett. 2013, 8: 015029-

227. Henry M, Tittonell P, Manlay RJ, Bernoux M, Albrecht A, Vanlauwe B: Agriculture, ecosystems and environment biodiversity, carbon stocks and sequestration potential in aboveground biomass in smallholder farming systems of western Kenya. Agric Ecosyst Environ. 2009, 129: 238-252.

228. Tittonell PA: 2007, Msimu wa Kupanda, targeting resources within diverse, heterogeneous and dynamic farming systems of East Africa, (accessed 26 July 2014)

229. Claessens L, Antle JM, Stoorvogel JJ, Valdivia RO, Thornton PK, Herrero M: A method for evaluating climate change adaptation strategies for small-scale farmers using survey, experimental and modeled data. Agric Syst. 2012, 111: 85-95.

230. Maskell LC, Crowe A, Dunbar MJ, Emmett B, Henrys P, Keith AM, Norton LR, Scholefield P, Clark DB, Simpson IC, Smart SM: Exploring the ecological constraints to multiple ecosystem service delivery and biodiversity. J Appl Ecol. 2013, 50: 561-571.

231. Setälä H, Bardgett RD, Birkhofer K, Brady M, Byrne L, de Ruiter PC, de Vries FT, Gardi C, Hedlund K, Hemerik L, Hotes S, Liiri M, Mortimer SR, Pavao-Zuckerman M, Pouyat R, Tsiafouli M, Putten WH: Urban and agricultural soils: conflicts and trade-offs in the optimization of ecosystem services. Urban Ecosyst. 2013, 17: 239-253.

232. Williams A, Hedlund K: Indicators and trade-offs of ecosystem services in agricultural soils along a landscape heterogeneity gradient. Appl Soil Ecol. 2014, 77: 1-8.

233. Wada Y, Wisser D, Eisner S, Flörke M, Gerten D, Haddeland I, Hanasaki N, Masaki Y, Portmann FT, Stacke T, Tessler Z, Schewe J: Multimodel projections and uncertainties of irrigation water demand under climate change. Geophys Res Lett. 2013, 40: 4626-4632.

234. Sood A, Muthuwatta L, McCartney M: A SWAT evaluation of the effect of climate change on the hydrology of the Volta River basin. Water Int. 2013, 38: 297-311.

235. Taylor RG, Scanlon B, Doll P, Rodell M, van Beek R, Wada Y, Longuevergne L, Leblanc M, Farniglietii JS, Edmunds M, Konikow L, Green TR, Chen J, Taniguchi M, Bierkens MFP, MacDonald A, Fan Y, Maxwell RM, Yechieli Y, Gurdak JL, Allen DM, Shamsudduha M, Hiscock K, Yeh PJF, Holman I, Treidel H: Ground water and climate change. Nat Clim Chan. 2013, 3: 321-330.

236. Dye PJ: Climate, forest, and streamflow relationships in South African afforested catchments. Commonw Forest Rev. 1996, 75: 31-38.

237. Khan S, Hanjra MA: Footprints of water and energy inputs in food production–global perspectives. Food Policy. 2009, 34: 130-140.

238. Duncan MJ: Hydrological impacts of converting pasture and gorse to pine plantation, and forest harvesting, Nelson, New Zealand. J Hydrol. 1995, 34: 15-41.

239. Bosch JM, Hewlett JD: A review of catchment experiments to determine the effect of vegetation changes on water yield and evapotranspiration. J Hydrol. 1982, 55: 3-23.

240. Scanlon BR, Jolly I, Sophocleous M, Zhang L: Global impacts of conversions from natural to agricultural ecosystems on water resources: quantity versus quality. Water Resour Res. 2007, 43: W034037

241. Jobbágy EG, Jackson RB: Patterns and mechanisms of soil acidification in the conversion of grasslands to forests. Biogeochemistry. 2003, 64: 205-229.

242. Jobbágy EG, Jackson RB: Groundwater use and salinization with grassland afforestation. Glob Chan Biol. 2004, 10: 1299-1312.

243. Farley KA, Kelly EF, Hofstede RGM: Soil organic carbon and water retention following conversion of grasslands to pine plantations in the Ecuadorian Andes. Ecosystems. 2004, 7: 729-739.

244. Schoups G, Hopmans JW, Young CA, Vrugt JA, Wallender WW, Tanji KK, Panday S: Sustainability of irrigated agriculture in the San Joaquin Valley, California. Proc Natl Acad Sci U S A. 2005, 102: 15352-15356.

245. Scott CA, Varady RG, Meza F, Montaña E, De Raga B, Luckman B,

Martius C: Science-policy dialogues for water security. Sci Policy Sustain Dev. 2012, 54: 37-41.

246. Kam SP, Badjeck M, Teh L, Teh L, Tran N: Working Paper 2012-24. Autonomous Adaptation to Climate Change by Shrimp and Catfish Farmers in Vietnam's Mekong River Delta. (accessed 26 July 2014), Penang, Malaysia: Worldfish

247. Duarte CM, Losada IJ, Hendriks IE, Mazarrasa I, Marbà N: The role of coastal plant communities for climate change mitigation and adaptation. Nat Clim Chang. 2013, 3: 961-968.

248. van Noordwijk M, Namirembe S, Catacutan D, Williamson D, Gebrekirstos A: Pricing rainbow, green, blue and grey water: tree cover and geopolitics of climatic teleconnections. Curr Opin Environ Sustain. 2014, 6: 41-47.

249. Spracklen DV, Arnold SR, Taylor CM: Observations of increased tropical rainfall preceded by air passage over forests. Nature. 2012, 489: 282-285.

250. McElrone AJ, Shapland TM, Calderon A, Fitzmaurice L, Paw UKT, Snyder RL: Surface renewal: an advanced micrometeorological method for measuring and processing field-scale energy flux density data. J Vis Exp. 2013, 82: e50666-

251. Shapland TM, Snyder RL, Paw UKT, McElrone AJ: Thermocouple frequency response compensation leads to convergence of the surface renewal alpha calibration. Agric For Meteorol. 2014, 189–190: 36-47.

252. Clausen J, Jägerskog A: SIWI Report No 31. Feeding a Thirsty World: Challenges and Opportunities for a Water and Food Secure Future. 2012, (accessed 26 July 2014), Stockholm: Stockholm International Water Institute (SIWI), (accessed 26 July 2014)

253. Mukherji A, Facon T, de Fraiture C, Molden D, Chartres C: Growing more food with less water: how can revitalizing Asia's irrigation help?. Water Policy. 2012, 14: 430-446.

254. Mbow C, Smith P, Skole D, Duguma L, Bustamante M: Achieving mitigation and adaptation to climate change through sustainable agroforestry practices in Africa. Curr Opin Environ Sustain. 2014, 6: 8-14.

255. Mbow C, van Noordwijk M, Prabhu R, Simons T: Knowledge gaps and research needs concerning agroforestry's contribution to sustainable development goals in Africa. Curr Opin Environ Sustain. 2014, 6: 162-170.

256. van Noordwijk M, Hoang MH, Neufeldt H, Öborn I, Yatich T: How Trees

and People Can Co-adapt to Climate Change: Reducing Vulnerability in Multifunctional Agroforestry Landscapes. 2011, (accessed 26 July 2014), Nairobi: World Agroforestry Centre, (accessed 26 July 2014)

257. Visseren-Hamakers I, Verkooijen P: The practice of interaction management: enhancing synergies among multilateral REDD+ institutions. Forest and Nature Governance: A Practice Based Approach (World Forests Series Vol 14). Edited by: Arts B, Behagel J, van Bommel S, de Koning J, Turnhout E. 2013, 133-149. New York: Springer Science

258. Ordonez JC, Luedeling E, Kindt R, Tata HL, Harja D, Jamnadass R, van Noordwijk M: Constraints and opportunities for tree diversity management along the forest transition curve to achieve multifunctional agriculture. Curr Opin Environ Sustain. 2014, 6: 54-60.

259. Geldenhuys CJ, Ham C, Ham H: Sustainable Forest Management in Africa: Some Solutions to Natural Forest Management Problems in Africa. Proceedings of the Sustainable Forest Management in Africa Symposium. 3–7 November 2008, Stellenbosch, South Africa. Edited by: Geldenhuys CJ, Ham C, Ham H. 2011, 1-538. (26 July 2014), Matieland, South Africa: Department of Forest and Wood Science, Stellenbosch University, (26 July 2014)

260. Thorlakson T, Neufeldt H: Reducing subsistence farmers' vulnerability to climate change: evaluating the potential contributions of agroforestry in western Kenya. Agric Food Secur. 2012, 1: 15-

261. van Paddenburg A, Bassi AM, Buter E, Cosslett CE, Dean A: Heart of Borneo: Investing in Nature for a Green Economy. 2012, (accessed 26 July 2014), WWF Heart of Borneo (HoB) Global Initiative, (accessed 26 July 2014)

262. Parrotta JA, Wildburger C, Mansourian S: A Global Assessment Report Prepared by the Global Forest Expert Panel on Biodiversity, Forest Management, and REDD+ (IUFRO World Series Vol 31). Understanding Relationships between Biodiversity, Carbon, Forests and People: The Key to Achieving REDD+ Objectives. 2012, (accessed 26 July 2014), Vienna: International Union of Forestry Research Organizations (IUFRO), (accessed 26 July 2014)

263. Visseren-Hamakers IJ, Gupta A, Herold M, Peña-Claros M, Vijge MJ: Will REDD+work? The need for interdisciplinary research to address key challenges. Curr Opin Environ Sustain. 2012, 4: 590-596.

264. Somorin OA, Visseren-Hamakers IJ, Arts B, Sonwa DJ, Tiani A-M: REDD+policy strategy in Cameroon: actors, institutions and governance. Environ Sci Policy. 2014, 35: 87-97.

265. Wrathall DJ: Migration amidst social-ecological regime shift: the search for stability in Garífuna villages of Northern Honduras. Hum Ecol. 2012, 40: 583-596.

266. Gibson MA, Gurmu E: Rural to urban migration is an unforeseen impact of development intervention in Ethiopia. PLoS One. 2012, 7: e48708-

267. Warner K, Afifi T, Henry K, Rawe T, Smith C, de Sherbinin A: Global Policy Report of Where the Rain Falls Project. Where the Rain Falls: Climate Change, Food and Livelihood Security, and Migration. 2012, (accessed 26 July 2014), Bonn: CARE France and United Nations University Institute for Environment and Human Security (UNU-EHS), (accessed 26 July 2014)

268. Jha AK, Bloch R, Lamond J: Cities and Flooding: A Guide to Integrated Urban Flood Risk Management for the 21st Century. 2012, (accessed 26 July 2014), Washington, DC: World Bank, (accessed 26 July 2014)

269. Laube W, Schraven B, Awo M: Smallholder adaptation to climate change: dynamics and limits in Northern Ghana. Clim Change. 2011, 111: 753-774.

270. Jankowska MM, Lopez-Carr D, Funk C, Husak GJ, Chafe ZA: Climate change and human health: Spatial modeling of water availability, malnutrition, and livelihoods in Mali, Africa. Appl Geogr. 2012, 33: 4-15.

271. López-Carr D: Agro-ecological drivers of rural out-migration to the Maya Biosphere Reserve, Guatemala. Environ Res Lett. 2012, 7: 045603-

272. Aide TM, Clark ML, Grau R, López-Carr D, Levy MA, Redo D, Bonilla-Moheno M, Riner G, Andrade-Núñez MJ, Muñiz M: Deforestation and reforestation of Latin America and the Caribbean (2001–2010). Biotropica. 2013, 45: 262-271.

273. DeFries RS, Rudel T, Uriarte M, Hansen M: Deforestation driven by urban population growth and agricultural trade in the twenty-first century. Nat Geosci. 2010, 3: 178-181.

274. Davis J, Lopez-Carr D: Migration, remittances and smallholder decision-making: implications for land use and livelihood change in Central America. Land Use Policy. 2014, 38: 319-329.

275. Winkels A: Migration, social networks and risk: the case of rural-to-rural migration in Vietnam. J Vietnam Stud. 2013, 7: 92-121.

276. Ho R, Milan A: "Where the Rain Falls" Project. Results from Huancayo Province, Junín Region (Report no 5). Rainfall, Food Security and Human Mobility. Case Study: Peru. 2012, (accessed 26 July 2014), Bonn: United Nations University Institute for Environment and Human

Security (UNU-EHS), (accessed 26 July 2014)

277. Schlenker W, Roberts MJ: Nonlinear temperature effects indicate severe damages to U.S. crop yields under climate change. Proc Natl Acad Sci U S A. 2009, 106: 15594-15598.
278. Auffhammer M, Hsiang SM, Schlenker W, Sobel A: Using weather data and climate model output in economic analyses of climate change. Rev Environ Econ Policy. 2013, 7: 181-198.
279. Urban D, Roberts MJ, Schlenker W, Lobell DB: Projected temperature changes indicate significant increase in interannual variability of U.S. maize yields. Clim Change. 2012, 112: 525-533.
280. Vital Signs. (accessed 26 July 2014), (accessed 26 July 2014)
281. Ericksen P, Thornton P, Notenbaert A, Cramer L, Jones P, Herrero M: Mapping Hotspots of Climate Change and Food Insecurity in the Global Tropics. (CCAFS Report no 5). 2011, (accessed 26 July 2014), Copenhagen: CGIAR Research Program on Climate Change, Agriculture and Food Security (CCAFS), (accessed 26 July 2014)
282. Holm P, Goodsite ME, Cloetingh S, Agnoletti M, Moldan B, Lang DJ, Leemans R, Moeller JR, Buendía MP, Pohl W, Scholz RW, Sors A, Vanheusden B, Yusoff K, Zondervan R: Collaboration between the natural, social and human sciences in global change research. Environ Sci Policy. 2013, 28: 25-35.
283. Leemans R, Solecki W: Redefining environmental sustainability. Curr Opin Environ Sustain. 2013, 5: 3-4.
284. Andersson E, Gabrielsson S: 'Because of poverty we had to come together': collective action for improved food security in rural Kenya and Uganda. Int J Agr Sustain. 2012, 10: 245-262.
285. Kristjanson P, Neufeldt H, Gassner A, Mango J, Kyazze FB, Desta S, Sayula G, Thiede B, Förch W, Thornton PK, Coe R: Are food insecure smallholder households making changes in their farming practices? Evidence from East Africa. Food Secur. 2012, 4: 381-397.
286. Lybbert TJ, Bell A: Why drought tolerance is not the new *Bt*. Nat Biotechnol. 2010, 28: 553-554.
287. Shah AK, Mullainthan S, Shafir E: Some consequences of having too little. Science. 2012, 338: 682-685.
288. Adaptation Fund. (accessed 26 July 2014), (accessed 26 July 2014)
289. International Fund for Agricultural Development (IFAD): Adaptation for Smallholder Agriculture Programme. (accessed 26 July 2014), Rome, (accessed 26 July 2014)

290. UNEP International Resource Panel Working Group on Reducing Emissions from Deforestation and Forest Degradation (REDD+) and a Green Economy: Building natural capital: how REDD+ can support a green economy. 2014, (accessed 26 July 2014), Nairobi: United Nations Environment Programme (UNEP), (accessed 26 July 2014)

291. Green Climate Fund. (accessed 26 July 2014), (accessed 26 July 2014)

292. United Nations Environment Programme (UNEP): UNEP Policy Series: Ecosystems Management Policy Brief 2-2010. The Role of Ecosystems in Developing a Sustainable 'Green Economy'. 2010, (26 July 2014), Nairobi: UNEP, (26 July 2014)

293. Bernard F, van Noordwijk M, Luedeling E, Villamor GB, Sileshi GW, Namirembe S: Social actors and unsustainability of agriculture. Curr Opin Environ Sustain. 2014, 6: 155-161.

Chapter 4

EX SITU CONSERVATION OF BIODIVERSITY WITH PARTICULAR EMPHASIS TO ETHIOPIA

Mohammed Kasso and Mundanthra Balakrishnan

Department of Zoological Sciences, Addis Ababa University, Addis Ababa, Ethiopia

ABSTRACT

Biodiversity encompasses variety and variability of all forms of life on earth that play a great role in human existence. Its conservation embraces maintenance, sustainable utilization, and restoration, of the lost and degraded biodiversity through two basic and complementary strategies called in situ and ex situ. Ex situconservation is the technique of conservation of all levels of biological diversity outside their natural habitats through different techniques like zoo, captive breeding, aquarium, botanical garden, and gene bank. It plays key roles in communicating the issues, raising awareness, and gaining widespread public and political support for conservation actions and for breeding endangered species in captivity for reintroduction. Limitations of ex situ conservation include maintenance of organisms in artificial habitats, deterioration of genetic diversity, inbreeding depression, adaptations to captivity, and accumulation of deleterious alleles. It has many constraints in terms of personnel, costs, and reliance on electric power sources. Ethiopia is considered to be one of the richest centers of genetic resources in the world. Currently, a number of stakeholders/actors are actively working on biodiversity conservation through ex situ conservation strategies by establishing gene banks, botanical garden, and zoo.

INTRODUCTION

According to the Convention on Biological Diversity, biodiversity refers to the variability among living organisms (animals, plants, and microorganisms) including inter alia, terrestrial, marine, and other aquatic ecosystems with their ecological complexes. In another expression, biodiversity encompasses the variety and variability of all forms of life on earth that play a great role in human existence [1, 2]. It also includes the ethnical value of biodiversity such as tradition and traditional knowledge of the indigenous and local communities [2] and the diversity within species (genetics), between species and of ecosystems [3].

Genetic diversity refers to the variation within species of any plant, animal or microbes in the functional units of heredity. Species diversity refers to the variety of species within a geographical area, which become central in the evaluation of diversity, and used as a point of reference in biodiversity conservation. Finally, ecosystem diversity refers to the variety of life forms in a given territory or area with all its functional ecological processes, which is often evaluated based on the diversity of all of its components [1].

Biodiversity is important for the maintenance of a healthy environment and used for direct human benefits like food, medicine, and energy. It is also used for recycling of different essential elements, for mitigation of pollution, for protection of watersheds, to mitigate soil erosion and to control excessive variations in climate and catastrophic events. For example, biodiversity provides different services free of charge worth of billions dollar every year for crucial well-being of the society. Some of these services are providing clean water and air, soil formation and protection, pollination, pest control, food, fuel, fibers, medicine, and construction and industry raw materials [4]. Agricultural biodiversity is another important component of biodiversity, which has a more direct link to the well-being and livelihood of mankind than other forms of biodiversity. Food plant and animal species have been collected, used, domesticated, and improved through traditional systems of selection over many generations [5].

However, today much of the lines of evidences are increasingly pointing out a significant global decline in biodiversity by numerous, varied, and interacting drivers [6]. More than half of the habitable surface of the earth has already been significantly altered by human activities. As a consequence, biodiversity of our planet is on the verge of decline and extinction despite our limited and incomplete knowledge on them [7]. Biodiversity loss and extinction processes can occur in two phases. The first phase is known as deterministic and often resulted from human threats such as habitat loss, fragmentation and degradation, direct exploitation of the species, competition from exotic and domestic species, and persecution and killing due to human animal conflicts. The second phase is known as deterministic that resulted from failures in mitigating threats that eventually result in very small, fragmented, and isolated remnant populations. Then these small remnant populations become vulnerable to a number of other, nonhuman caused threats mainly stochastic, genetic (genetic drift and inbreeding), and demographic events [8]. Thus, small, fragmented, isolated populations can find themselves being dragged into an extinction vortex whereby genetic and demographic stochastic events can cause the species to go extinct. During this second phase of the extinction process, very intensive management of populations and individuals is often necessary to prevent extinction [9].

Several human induced impacts are leading to a mass extinction process affecting global biodiversity. The major reasons for rapid diminishing of biodiversity are attributed to conversion of land for agriculture, wild fires, poor management of available land, over-exploitation for food, fuel-wood, medicine, construction, overgrazing by cattle, displacement and loss of landraces, lower yielding varieties, pests and diseases, global climate change, pollution (e.g., acid rain), and gap of scientific knowledge on some of the biological resources [1, 7, 10]. Human beings are destroying biodiversity, particularly during livelihood activities with or without knowledge of the consequences of their actions [6]. Agriculture is one of the most important land-use that results in detrimental environmental consequences from increased use of fertilizers and biocides, land draining, irrigation, and the loss of many biodiversity-rich landscape features [6]. There are many threats to biodiversity as a result of agricultural practice through changes in land-use, replacement of traditional varieties by modern cultivars, agricultural intensification, increased population, poverty, land degradation, and environmental changes (including climate change) [5]. Recent estimates indicate that humans use more than 40% of the terrestrial components and significantly modified global biodiversity [7]. As a consequence, many species of living organisms are classified as threatened today and this has become a central concern for conservation [11].

Conserving biodiversity has economic, social, and cultural values. Conservation of biodiversity is integral to the biological and cultural inheritance of many people and the critical components of healthy ecosystems that are used to support economic and social developments. Moreover, it is used to maintain the earth's genetic library from which society has derived the basis of its agriculture and medicine [5, 12]. The twenty-first century is predicted to be an era of bioeconomy driven by advances of bioscience and biotechnology. Bio-economy may become the fourth economy form after agricultural, industrial, and information technology economies, having far-reaching impacts on sustainable development in agriculture, forestry, environmental protection, industry, food supply, health care and other micro-economy aspects. Thus, a strategic vision for conservation and sustainable use of biodiversity in the 21st century is of far-reaching significance for sustainable development economy and society [13].

Biodiversity conservation refers to the management of human use of biodiversity in order to get the greatest sustainable benefit to present and future generations. Thus, conservation of biodiversity embraces the protection, maintenance, sustainable utilization, restoration, and enhancement of biodiversity [1]. Biodiversity conservation mainly focuses on genetic conservation with its diverse life-support systems (ecosystems) for the connotation of human well-being [3].

Conservation techniques can be grouped into two basic, complementary strategies: in situ and ex situ [14]. As also outlined in the articles 8 and 9 of the Convention of Biological Diversity (CBD), biodiversity is conserved by two major methods called in situ and ex situ. The conservation efforts, either in situ or ex situ, involve the establishment and management of protected areas and relevant research institutes or academic institutions, which establish and manage arboreta, botanical or zoological gardens, tissue culture, and gene banks [1]. The concept of ex situ conservation is fundamentally different from that of in situ conservation; however, both are important complementary methods for conservation of biodiversity. The principal difference (and hence the reason for the complementarities) between the two lies in the fact that ex situ conservation implies the maintenance of genetic materials outside of the "normal" environment where the species has evolved and aims to maintain the genetic integrity of the material at the time of collection, whereas in situ conservation (maintenance of viable populations in their natural surroundings) is a dynamic system, which allows the biological resources to evolve and change over time through natural or human-driven selection processes [5].

In situ conservation is defined as conservation of ecosystems and natural habitats, the maintenance of viable populations of the species in their natural surroundings and, in the case of the cultivated species, in the surroundings where they have developed their distinctive properties. In situ conservation can be done in farmers' fields, in pasture lands, and in protected areas [15]. For cultivated species, in situ conservation concerns the maintenance of the local intra- and inter-population diversity available in various ecological and geographical sites [1, 16]. Thus, it allows ongoing host-parasite coevolution, which is likely to provide material resistance to pests and diseases, and CBD recognized it as a primary approach to conserve biodiversity [4]. However, in situ conservation has certain limitations like more difficult access to breeders requiring the application of its complimentary technique. For example, some of the natural habitats or wild habitats are very risky when compared to relatively safe captive environment [9]. The second biodiversity conservation technique receiving the most attention to conserve biodiversity is ex situ. Ex situ conservation techniques are mostly used to be applied to species with one or some of the following characteristics: endangered species, species with a past, present or future local importance, species of ethno-botanical interest, species of interest for the restoration of local ecosystems, symbolic local species, taxonomically isolated species, and monotypic or oligotypic genera [17]. Intensive conservation and management of populations and individuals can come in many different forms, like translocation, breeding in a fenced wild habitat, supplementary feeding, captive hand rearing of young of wild parents to become pregnant sooner, and captive breeding [9].

EX SITU CONSERVATION

Ex situ conservation is a technique of conservation of biological diversity outside its natural habitats, targeting all levels of biodiversity such as genetic, species, and ecosystems [1, 2, 16]. Its concept was developed earlier before its official adoption under the Convention on Biological Diversity signed in 1992 in Rio de Janeiro [2]. In general, ex situ conservation is applied as an additional measure to supplement in situconservation, which refers to conservation of biological diversity in its natural habitats [16]. In some cases, ex situ management will be central to a conservation strategy and in others it will be of secondary importance [18]. Broadly, ex situ conservation includes a variety of activities, from managing captive populations, education and raising awareness, supporting research initiatives and collaborating with in situ efforts [19]. It is used as valuable tools in studying and conserving biological resources (plants, animals, and microorganisms) for different purposes [2] through different techniques such as zoos, captive breeding, aquarium, botanical gardens, and gene banks [16].

Types of Ex Situ Conservation

Zoos

Zoos or zoological gardens or zoological parks in which animals are confined within enclosures or semi-natural and open areas, displayed to the public, and in which they may also breed. They are considered by universal thinkers and environmentalists as important means of conserving biodiversity [19–21]. Zoos attract as many as 450 million visitors each year and so are uniquely placed to have very large educational and economic values [22]. Zoos not only act as places of entertainment and observing animal behavior, but are also as institutions, museums, research laboratories, and information banks of rare animals [23]. Although some people dislike zoos, many people enjoy them. Over the last several decades, zoos have made significant progress in its cooperative management of ex situ populations of a variety of biodiversity [9]. Zoos breed many endangered species to increase their numbers. Such captive breeding in zoos has helped to save several species from extinction [19]. Management of animals in zoos includes animal identification, housing, husbandry, health, nutrition as well as addressing and ways of interaction with the public [20]. There are various processes and mechanisms used to determine whether a species or taxon is included within a zoo's collection plan. The frequently used criteria include how the species is valued, according to its uniqueness, contribution to research or education, and conservation status [19]. Zoos help the animal to secure food, shelter, social contact and mates, and to be motivated by desire (appetitive behavior), which is reinforced by pleasure (consummative behavior) [21].

In the past, some zoos paid little attention to the welfare of the animals, and some zoos today have poor environments for animals [24]. They were also once reliant on harvest from the wild to populate their exhibits and reliance on continued wild collection to breeding closed populations [23]. Many zoo animals also became endangered or extinct due to visitor disturbances, unfavorable climate and due to insufficient space [20]. From this aspect, many scholars state on the negative features of keeping animals in zoo as it causes pain, stress, distress, sufferings and evolutionary impacts [21]. Animal welfare, education, conservation, research, and entertainment are major goals of modern zoos, but these can be in conflict. For example, visitors enjoy learning about and observing behavior in captive animals, but visitors often want to observe and interact with the animals in close proximity. Unfortunately, proximity to and interactions with humans induce stress for many species [25]. The same is true for Addis Ababa Lion Zoo Park.

However, progressive zoos are engaged in education, research, and conservation, with the aim of maintaining healthy animals, which behave as if they are in their natural habitats [24]. The current paradigm for managing essential populations is to minimize the rate of genetic decay, slow adaptation to the captive environment, and retain typical behaviors [23, 26]. It is widely accepted that the more generations a population spends in captive breeding, the less suitable it is for attempted restorations in the wild. Hence, population management is designed not to deplete too quickly the resource obtained from the founders [23]. Thus, for true sustainability of the species for the purpose of conservation, display, education, and research, constant refreshing of populations is required [9, 23]. Majority of the current breeding programs base on the genetic management of populations by the analysis of individual pedigrees in order to minimize kinship [9].

Captive Breeding

Captive breeding is an integral part of the overall conservation action plan for a species that helps to prevent extinction of species, subspecies, or population. It is an intensive management practice for threatened individuals, populations, and species by anthropogenic and natural factors [9]. In small and fragmented populations, even if the human caused threats could be magically reversed, the species would still have a high probability of extinction by random demographic and genetic events, environmental variations, and catastrophes. Thus, under sufficient knowledge on the biology and husbandry of the species, captive breeding helps individuals in the relative safety of captivity, under expert care and sound management by providing an insurance against extinction [9]. Stock for reintroduction or reinforcement efforts, opportunities for education, raising of awareness, scientific and husbandry research, and other contributions to conservation are also possible through captive breeding [9, 27].

Environmental enrichment strategies are used to improve both physiological and psychological welfares of captive animals, which can be achieved by increasing the expression of natural behavior and decreasing abnormal behaviors. Successful environmental enrichment includes the improvement of enclosure design and the provision of feeding devices, novel objects, appropriate social groupings, and other sensory stimuli [27]. The minimum requirement for successful ex situ management, particularly in the captive populations, is the inclusion of as much of the genetic diversity present in wild populations. Genetic sustainability (retention of 90% of the gentic diversity of the wild population for 100 years) in captive breeding is maintained if consideration is given on number of founders, population growth rate, effective population size, and duration of the captive program [19]. However, even if at least 30 founders in captive breeding are recommended to ensure the representation of large enough proportion of the genetic diversity of the wild population, for critically endangered species, actively removing individuals from the wild population to serve as founders may compromise the survival of the wild population [9]. For example, the Arabian Oryx captive breeding program was based on fewer founders and grew to a couple of thousand individuals through breeding management, which helped to reduce risks.

However, there are several challenges (biological and environmental) that are limiting factors to the attainment of the goal of captive breeding for many species [19]. One of the major challenges is a circular consequence of small-population management that has inherent genetic and demographic problems due to genetic diversity loss and demographic stochasticity [19]. In addition, individuals that are well adapted to the circumstances in captivity may also be less well adapted to the circumstances in the wild and may show lower fitness upon reintroduction [28]. Most notably, within the captive environment, housing and husbandry will also have significant impacts on birth and death rates [19].

Aquarium

An aquarium is an artificial habitat for water-dwelling animals. It can also be used to house amphibians or large marine mammals and plant species for tourist attractions. It is usually found in zoos or marine parks with different size. The 15,750 described species of freshwater fish comprise around 25% of living vertebrate species diversity and a key for global economic and nutritional resources of which more than 11% is threatened (60-extinct, 8-extinct in the wild and 1679-threatened) [22]. Fresh waters (0.3%) of available global surface water support 47–53% of all extant fish species that are threatened by overfishing, pollution, habitat loss, damming, alien invasive species, and climate change. This requires world's zoos and aquariums to identify the potential targets (species or areas) for in situ and ex situ conservation program [18, 22]. Aquarium is used to admire at home by hobbyists, to portray as public exhibits, to provide large quantities of human food and animal fodder [18].

Fishes are often overlooked within the development of conservation priorities. This leads to the low focus on meaningful conservation efforts rather than giving more attention for their importance to food supply and livelihoods. For example, it provides job opportunity for over 60 million people, as source of food for over 200 million people in Africa, for US$1.5 billion income from trade of 4000 species global ornamental fish industry, and many are displayed in the world's public zoos and aquaria to a global audience of as many as 450 million people per year [22, 29]. However, despite the clear value of freshwater fish diversity, wetland habitats and their associated freshwater-fish species continue to be lost or degraded at an alarming rate [30]. One recommendation is for aquariums to set up sustainable breeding program that prioritizes threatened species (VU, EN, and CR) and those classified as EW to support species conservation in situ and aid the recovery of species via collaborative reintroduction or translocation efforts when appropriate [22].

Botanical Gardens

Botanical gardens consist of living plants, grown out of doors or under glass in greenhouses and conservatories. They are used to grow and display plants primarily for scientific and educational purposes. They also include herbarium, lecture rooms, laboratories, libraries, museum and experimental or research plantings. It can be taxonomic collection of a particular family, genus or group of cultivars, native plants, wild relatives, medicinal, aromatic, or textile plants [4]. There are over 2,000 botanic gardens, holding 80,000 plant species in their living collections and receiving hundreds of millions of visitors per annum [14, 31]. Furthermore, they have valuable and distinctive mix of officials dedicated to plant research, systematics, conservation education, and public awareness [31]. They are now extremely well networked both among themselves and with other professionals, conservation organizations, and nongovernmental organizations (NGOs) [31]. They provide different services for sectors that utilize and conserve plant diversity like agriculture, forestry, pharmaceutical and biofuel industries, protected area management, and ecotourism. They have a unique opportunity as visitor attraction places and scientific institutions for documentation and conservation of plant diversity by shaping and mobilizing citizens to the current environmental challenges [31]. They also play a great role in attaining target of the Global Strategy for Plant Conservation for 2020 to cultivate 75% of the world's threatened plant species in ex situ [14]. Botanical gardens give opportunity for arable plants to be grown under relatively modified environmental conditions (intense cultivation, relatively high fertility, and high levels of disturbance) [14].

However, most of the cultivated taxa are held in a small number of collections and mostly only in small populations. Lack of genetic exchange and stochastic processes in small populations make them susceptible to detrimental genetic effects [14]. The low number of ex situ populations in most botanical gardens poses a fundamental problem for conservation. The total ex situ breeding collection is therefore very small with respect to the stated aim of conserving regional gene pools [14]. The striking lack of information on source populations casts doubt on the value of using such ex situ populations for potential reintroductions. They also require testing for fitness and similarity to wild populations before they are brought to the field [14]. Thus, conservation actions of botanic gardens such as training and capacity building, needs to be better understood and better coordinated [31].

Gene Banks

Genome resource banking is another management technique used for biodiversity conservation. Different types of gene banks have been established for the storage of biodiversity, depending on the type of materials conserved. These include seed banks (for seeds), field gene banks (for live plants), in vitro gene banks (for plant tissues and cells), pollen, chromosome, and deoxyribonucleic acid (DNA) banks for animals (living sperm, eggs, embryos, tissues, chromosomes, and DNA) that are held in short term or long term laboratory storage; usually cryopreserved or freeze-dried [32]. Currently, there are about 7.4 million PGRFA accessions conserved in over 1750 gene banks

[5]. The Genome 10K (G10K) project aims to sequence the genomes of 10,000 vertebrates, of which 4,000 will be fish. To date, 60 fish genomes have been sequenced in laboratories worldwide and added to the database, and a further 100 are targeted. The Frozen Ark database holds details of 28,060 frozen DNA samples. Among these 6,997 are from species listed in the IUCN Red List [18]. The principal aim of gene bank conservation is to maintain genetic diversity alive as long as possible and to reduce the frequency of regeneration that may cause the loss of genetic diversity [5].

With the rapid development in the field of molecular genetics and genomics, DNA is becoming more and more in demand for molecular studies and is one of the most requested materials from gene banks. Establishing DNA storage facility as a complementary "backup" to traditional ex situ collections has been suggested [5]. Some efforts have been made to establish DNA banks for endangered animals [33], and a few plant DNA banks in different parts of the world such as the Missouri Botanic Garden, Kew Royal Botanic Garden and Australian Plant DNA Bank. Many research groups are already developing their own archives of extracted genomic DNA. Recently, the Global Biodiversity Information Facility in Germany has established a DNA bank network, which provides DNA samples of complementary collections (microorganisms, protists, plants, algae, fungi, and animals) [5].

Seeds are usually the most convenient and easiest material to collect and to maintain in a viable state for long periods of time and that makes it preferred for conservation in gene banks [14, 34]. Seed banking techniques rely on the storage of dried seeds of threatened or other plants at low temperatures as the most important factors influencing seed longevity are temperature, seed moisture content, and relative humidity [14, 35, 36]. Seeds are typically conserved at moisture content between 3 and 7 percent and stored at 4 degrees Celsius for short-term conservation, and between −18 and −20 degrees Celsius for long-term conservation [5, 34]. Current research is showing that there exists variability in seed longevity for different species being conserved under similar conditions. In addition, it has been found that the type of seed (endospermic or non-endospermic) and intraspecific variation may also affect accessions longevity [5]. In addition, high initial quality seeds are a major prerequisite for ensuring seed longevity in seed banks [5].

Plants that cannot be conserved as seeds because of their recalcitrant nature (i.e. seeds that are desiccation and/or cold sensitive) or are clonally propagated are traditionally conserved as live plants in ex situ field gene banks. But, field gene banks present real logistical challenges; they require large areas and are costly, they are vulnerable to pests and diseases, natural disasters, political unrest, extreme weather, fire, vandalism, and theft, and they often are at risk due to policy changes on land use [5]. In vitro conservation refers to one type of gene bank known as slow-growth conservation method. It involves culturing of different parts of the plant (meristem, tissues, and cells) into pathogen-free sterile culture in a synthetic medium with growth retardants, which has been cited as a good way of complementing and providing backup to field collections [5]. The other genome conservation technique is cryopreservation, in which living tissues are conserved at very low temperatures (−196°C) in liquid nitrogen to arrest mitotic and metabolic activities [5]. It is now realized that cryopreservation method can offer greater security for long-term, cost effective conservation of plant genetic resources, including orthodox seeds [5]. The storage in liquid nitrogen clearly prolonged shelf life of lettuce seeds with half-lives projected as 500 and 3400 years for fresh lettuce seeds stored in the vapor and liquid phases of liquid nitrogen, respectively [5].

Advantages of Ex Situ Conservation

It is generally preferred to conserve threatened species in situ, because evolutionary processes are more likely to remain dynamic in natural habitats [14, 19]. However, considering the rate of habitat loss worldwide, ex situ cultivation is becoming increasingly important [14]. Further more, as many of the taxa are located outside natural parks or reserves, in situ measures are not enough to assure their conservation. Translocation, introduction, reintroduction, and assisted migrations are species conservation strategies that are attracting increasing attention, especially in the face of climate change [37].

As approximately 450 million people per year visit zoos and aquaria globally, their education and marketing services play a key role in communicating the issues, raising awareness, changing behavior, and gaining widespread public and political support for conservation actions. Zoos support conservation by educating the public, raising money for conservation programs, developing technology that can be used to track wild populations, conducting scientific research, advancing veterinary medicine, and developing animal handling techniques [22]. By studying animals in captivity and applying that knowledge to their husbandry, zoos can provide valuable and practical information that may be difficult or impossible to gather from the wild [24]. Zoos and aquaria have significant roles to play in improveing public awareness of the issue facing species and their habitats; for example, through presentation of maps and photographs of species recently extinct as a result of anthropogenic impacts. A similar display of threatened species, even if not currently in the collections of the zoo, would help convey to the public the magnitude of the threat facing the species [18, 22]. It also reaches a wide cross-section of the society, because zoo audiences are not limited to those who are already passionately interested in wildlife and because many zoo visitors are children. Some of these children may become committed conservationists. Some may grow up to be oil company tycoons, politicians, or movie stars, with great potential influence. Some may even live next door to a poacher or wildlife dealer. Thus, instilling an interest in conservation of wildlife in people from all walks of life while they are young is one vital role zoos can play [38]. It is often claimed that zoos perform valuable conservation work by breeding endangered species and returning them to the wild. Zoos can also be used for businesses that make money. This means that animals are often bred for commercial purposes because the public like to see new-born animals. Such breeding leads to a surplus of animals, and in order to keep numbers down sold to private collectors, circuses, or even research laboratories. A zoo with good and attractive entertainments encourages initial visits and subsequent returns to the zoo, which is used to get more revenue for conservation efforts, research, and general animal care and welfare and also to develop more positive perceptions of animals in zoos and become more supportive of conservation efforts [25, 39].

Disadvantages of Ex Situ Conservation

Some ex situ conserved collections showed lower resistance levels, although still others showed higher resistance levels than their in situ conserved counterparts mainly due to the high evolutionary drive and complex nature of evolutionary scenario [40].

The behavior of animals in the zoo may be affected by the frequent arrival of large number of people, who are unfamiliar to the animals [41]. Animals housed in artificial habitats are confronted by a wide range of potentially offensive environmental challenges such as artificial lighting, exposure to loud or aversive sound, arousing odors, and uncomfortable temperatures or substrates. In addition, confinement-specific stressors such as restricted movement, reduced retreat space, forced proximity to humans, reduced feeding opportunities, maintenance in abnormal social groups, and other restrictions of behavioral opportunity are considered [42]. However, over the course of the twentieth century, as knowledge of wildlife biology improved, zoo animals began to be kept in more natural surroundings and social groupings, and diets and veterinary care began to improve. Thus, survival and breeding rates of captive populations improved [38]. Evidence mainly from studies of rodents and primates strongly indicates that prenatal stress can impair stress-coping ability and is able to cause a disruption of behavior in aversive or conflict-inducing situations. Prenatally stressed animals show retarded motor development, reduced exploratory and play behavior, and impairments of learning ability, social behavior, and sexual and maternal behavior. Prenatal stress may also affect the sex ratio at birth and the reproductive success [43].

Although populations of some species managed in ex situ may have the best hope for their long-term survival, they might be challenged if not properly managed during translocation and reintroduction with the effects of climate change [22]. Some species may lose their biological integrity particularly on morphology. For example, an experimental study on black-footed ferrets (Mustela nigripes) in ex situ indicated a decrease of 5–10% body size than pre-captive, in situ animals [44]. In other words, the small cage size and environmental homogeneity inhibit mechanical stimuli necessary for long bone development. Thus, in the absence of such an environment, “unnatural” morphologies can result that may contribute to poor fitness or perhaps even for domestication and reintroduction and relocation [44]. It would be very difficult to reintroduce some zoo-reared animals to their natural habitats because, after generations of captivity, many have lost the necessary skills to survive in their original habitats [22]. For naturally out-breeding species, the high levels of inbreeding in captivity often have negative effects on life history traits related to reproduction and survival [11]. It makes the population in captivity deteriorate due to loss of genetic diversity, inbreeding depression, genetic adaptations to captivity, and accumulation of deleterious alleles [17]. For plants, ecological shifts, small population size, genetic drift, inbreeding, and gardener-induced selection may negatively affect population structure after several generations of ex situcultivation [16, 45, 46]. These factors could seriously put at risk the success of ex situ conservation [17].

Captive breeding of threatened species has used increasingly sophisticated technologies and protocols in recent years [47]. Although, this has blurred the dichotomy between in situ and ex situ species management, the value of captive breeding as a conservation tool remains controversial [48]. It is recognized that ex situconservation has many constraints in terms of personnel, costs, and reliance on electric power sources (especially in many developing countries where electricity power can be unreliable) for gene banks. It requires high facilities and financial investments. It cannot also conserve all of the thousands of plant and animal species that make up complex ecosystems such as tropical rainforests [49]. Capture of individuals from the wild for captive breeding or translocation some times can have detrimental effects on the survival prospects of the species as a whole through disease infection [50].

Even though the management of irreplaceable animal populations in zoos and aquarium has focused primarily on minimizing genetic decay with the use of advanced technologies, recent analyses have shown that as most zoo programs are not projected to meet the stated goals due to lack of achieving “sustainability” of the populations [23]. Thus, managing zoo populations as comprehensive conservation strategies for the species requires research on determinants of various kinds of genetic, physiological, behavioral, and morphological variations, and their roles in population viability, development of an array of management techniques, tools, and training of managers [23].

Challenges to Ex Situ Conservation

Ex situ conservation requires different kinds and levels of intensity of management, and a multistakeholder approach like the input from experts on aquarium and zoo husbandry, ex situ breeding, gene-banking, reintroduction, and habitat restoration [51]. Other expert input may include taxonomy, ecology and conservation, ethnography, and sociology. For outreach program, there is a need to liaise with local communities and national government fisheries and wildlife departments; with international (nongovernmental and intergovernmental) conservation bodies [18].

The most important challenges of applying ex situ conservation (captive breading) are the difficulty in recognizing the right time, identifying the precise role of the conservation efforts within the overall conservation action plan, and setting realistic targets in terms of required time span, population size, founder numbers, resources, insurance of sound management and cooperation, and the development of much needed new technical methods and tools [9]. In captive breeding to achieve the retention of 90% of the wild gentic diversity, it is necessary to incorporate sufficient number of founders, careful pair combinations and management [9]. Evidence also exists, which demonstrates that manipulation of housing and husbandry variables can also have significant positive influence on animal reproduction in captivity [19].

In many cases, ex situ populations are founded from only a few individuals, which cause genetic bottlenecks. Small populations are exposed to threats such as stochastic demographic events as well as genetic effects, including loss of genetic diversity, inbreeding depression or accumulation of new, potentially deleterious mutations [11]. More specific problems in garden populations include poorly documented or even unknown sources of material, accidental hybridization of material from various localities, and or unintended selection for traits more suited to garden conditions [14]. In every region, most of the cooperatively managed breeding programs have too few animals, too few animals in appropriate situations for breeding, too few successful breeders, too few founders, and too many animals with undocumented ancestries and/or too little cooperation with scientifically designated breeding recommendations. These deficiencies are resulting in declining populations or declining gentic diversity or both [23, 52]. Problems associated with small founder populations such as inbreeding depression, removal of natural selection, and rapid adaptation to captivity pose considerable challenges for managers of captive populations of threatened species [48]. Equally, reintroduction of captive-bred stock to the wild may require implementation of rigorous protocols that embrace acclimation, pre- and post-release training, health screening, genetic management, long-term monitoring, and involvement of local stakeholders [53, 54]. Shortfalls in implementing such protocols may jeopardize the likelihood of achieving success [47].

Inbreeding due to the mating between two related individuals is unavoidable in small, fragmented, or isolated populations typical of many threatened species, and it can lead to a significant reduction in fitness. The deleterious effects of inbreeding on individual fitness can be large and may be an important factor contributing to population extinction. Inbreeding depression has potential significance for the management and conservation of endangered species [55]. As populations get smaller, the probability increases for all offspring in a given generation are of the same sex [19].

Evaluating the long-term efficiency of ex situ conservation is important, but is complicated because of the difficulty of finding more than one sample of a documented (origin and cultivation) ex situ population and its corresponding still-existing in situ source population [46].

Animal translocations are usually risky and expensive, and a number of biological and nonbiological factors can influence success. Biological considerations include knowledge of genetics, demography, behavior, disease, and habitat requirements. It also includes legal framework, fiscal and intellectual resources, monitoring capacity, goal of the translocation, logistic challenges, and organizational structure of decision making [56].

The regeneration process is one of the most critical steps and a major challenge in gene bank management, during which there is the highest probability for genetic erosion [57]. It is equally important to understand how different conservation methods (seed, field, and cryopreservation) and their management can affect or change the genetic make up, thereby reducing the effective population size (Ne). This will also contribute to decision-making process for determining which methods to use for conservation of the wide diversity [5].

If people are discouraged or prevented from interacting with the resident animals, fewer visitors attend, decreasing public financial support. The visitors' noise and crowding become a source of stress for many species that affects both their welfare and the enjoyment of the visitor [25].

EX SITU CONSERVATION PRACTICE IN ETHIOPIA

Ethiopia is considered to be one of the richest centers of genetic resources in the world. It is believed that indigenous crops such as teff (Eragrostis tef), Noug (Guizotia abyssinica), and Enset (Ensete ventricosum) were first domesticated in Ethiopia. Numerous major crop species including durum wheat (Triticum durum), barley (Hordeum vulgare), sorghum (Sorghum bicolor), sesame (Sesamum indicum), castor (Ricinus communis), and coffee (Coffea arabica) are also known to show significant diversity in the Ethiopian region [58]. Almost 85% of the populations of Ethiopia live in rural areas and most of this population depends directly or indirectly on biodiversity. Biodiversity also plays a crucial role in the different sectors like energy, agriculture, forestry, fisheries, wildlife, industry, health, tourism, commerce, irrigation, and power [15].

The records on biodiversity conservation efforts in Ethiopia date back to the days of Emperor Zera-Yakob (1434–1468 E.C.). The Emperor brought juniper seedlings from Wof Washa of North Shewa and planted in Managesha-Suba area. Modern conservation intervention began by Emperor Menilik in 1908 E.C. and eventually evolved to the establishment of protected areas in the 1960s [4]. Currently, a number of stakeholders are actively working on biodiversity related issues at the federal government level. These include Institute of Biodiversity Conservation (IBC), Ethiopian Institute of Agricultural Research (EIAR), Ethiopian Wildlife Conservation Authority (EWCA), Ministry of Agriculture and Rural Development (MoARD), Ministry of Science and Technology (MoST), Higher Learning Institutions (HLIs), particularly Addis Ababa University (AAU), and offices in various regional states of Ethiopia [59]. Ex situ conservation as complementary to the rehabilitation and restoration of degraded ecosystems and the recovery of threatened species was started 1976 with the establishment of IBC [4].

Ex situ conservation activities mostly focus on high socioeconomic value and internationally important crop types that are considered to be facing immediate danger of genetic erosion [15]. The collections held at IBC are mostly of indigenous landraces some of which are not seen today in farmlands. The collections of root crops, medicinal plants, weedy species, and wild relatives of cultivated species are still relatively inadequate within the existing ex situ collections [60]. However, appropriate emphasis is being placed on conservation and sustainable use of all forms of plant biological resources [60]. Since the establishment of IBC, systematic crop germplasm exploration and collection operations have been undertaken in the different administrative regions of the country, covering a wide range of agroecological zones. Collection priorities were set based on factors like economic importance, degree of genetic erosion and diversity, researchers' needs, the rate of diffusion of improved varieties, clearing of natural vegetation, agricultural policies, natural disasters, and resettlement program [4].

Currently about 68,014 seed accessions of 200 plant species, 6,704 accessions of 205 species of forestry, medicinal, forage and pasture plants (in field gene bank), 290 species of microbial genetic resources, and three semen of threatened breed of domestic animals are conserved by IBC. About 90% of the total germplasm holdings in the gene bank consist of field crops. The total collection is composed of cereal seeds, pulses, oil crops, spices, and species of medicinal and industrial value. Aside from the crop collections, the gene bank also holds 650 collections of micro-organisms. Over 9,000 accessions of horticultural crops, medicinal plants, and herbs are maintained in field gene banks. The type and nature of collection missions and number and lists of plant species and landraces collected have been documented in manuals and reports. Regular monitoring activities are performed for seed viability [15].

For plant species with recalcitrant and intermediate storage behavior, there are ten field gene banks under IBC control and small sized fields in the various research stations of the Ethiopian Institute of Agricultural Research (EIAR) and at universities. The plan for the immediate future is to increase the number of field gene banks in different agroecological zones. Community gardens, backyards, and holy places are being considered for inclusion in the future plan. Spices, vegetables and medicinal plants require management on a large scale and with the full involvement of the local communities [15]. The initiative at national level is still in its infancy and there is currently no well-established national botanical garden in Ethiopia including the Gulele Botanical Garden Center [4, 59]. The Gulele Botanic Garden Center was established through the Proclamation No. 18/2005 E.C. in October 30/2002 E.C. in a 705 hectare land at Gulele and Kolfe-Kernayo subcities. It was established with a vision to see the center to be developed as an exemplary garden in terms of education, ecotourism attraction and center for originality of the Ethiopian plant species, and to be a place of research and nurturing of plant species. The center also has a mission to provide persistent ecotourism services to tourists by taking care of plant species and carrying out educational and research works [59].

Although there is no well-established zoo or zoological garden in Ethiopia, the Addis Ababa Lion Zoo Park can be dominantly cited [59]. The Addis Ababa Lion Zoo Park was established in 1948 with five founder lions presented to Emperor Haile Selassie as gifts. The park accommodates lions with cubs, tortoises, baboons, monkeys, apes, and ducks. A team of international researchers has provided the first comprehensive DNA evidence from 15 (eight males and seven females) samples of Addis Ababa lion indicating the genetically unique samples that requires immediate conservation action. Both microsatellite and mitochondrial DNA data suggest that the zoo lions are genetically distinct from all existing lion populations for which comparative data exist.

Desiccation-intolerant seeds and species that do not readily produce seeds are conserved ex situ in field gene banks. For example, accessions of coffee (Coffea arabica), root crops such as yam (Dioscorea bulbifera) and "Oromo dinich" (Coleus edulis), and spices like ginger (Zingiber officinale) and Ethiopian cardamom (Aframomum corrorima) are conserved in agro-ecological zones in field gene banks [60].

The need for action for global biodiversity conservation is now well understood, and government agencies, nongovernmental organizations, and botanic gardens have all been working in various ways to promote environmental sustainability and reduce species and habitat loss [31].

Seed banking is the major ex situ conservation method employed in Ethiopia. There are three major seed banks operating in Ethiopia. The National Tree Seed Project processes seeds from a narrow range of tree species and uses short-term storage facilities. It aims to cater for the annual seed demand from commercial and small-scale forestry enterprises. Of the 70 species regularly collected and processed, 20 are indigenous. The Forage Genetic Resources Centre maintained by the Consultative Group on International Agricultural Research at the International Livestock Research Institute maintains long-term conservation of a wide range of native and exotic forage species. The Institute for Biodiversity Conservation and Research holds active collections of seeds mainly for research and distribution and as a base collection for long-term conservation [60].

For security reasons, the collected and stored germplasm need to be conserved in duplicate gene banks. However, except for the limited samples of the Ethiopian germplasm held by the Consultative Group on International Agricultural Research, United States Agency for International Development, and the Nordic Gene Bank, majority of the Ethiopian collections are still kept in a single copy at the National Gene Bank. Greater efforts need to be made to store duplicate collections to avoid future genetic erosion [15]. For the continuing power supply, the Ethiopian Gene Bank has independent power supply in the form of a stand by generator to overcome power cuts [15].

Lack of adequate knowledge with respect to collection, handling, and treatment of seeds often impedes the planting of indigenous trees and shrubs. Inadequate work has been done on establishing the seed storage behavior of native species resulting in only limited availability of ex situ conservation seed collections especially with respect to native forest species [60] and lack of alternative storage facilities for the existing conventional cold rooms (e.g., in vitro and cryo-preservation methods) [4]. The current holdings of the IBC gene bank reach over 60,000 accessions of plant species. Some collections are in the medium-term storage mainly due to insufficient seed samples [15].

CONCLUSION

Biodiversity plays a great role in human existence and in healthy function of natural systems although it is on the way of depletion dominantly due to anthropogenic activities. This requires conservation of biodiversity either in in situ or ex situ or both methods in combination based on the conservation situation and its objective. Although in situ conservation is more encouraged to be used for biodiversity conservation, ex situconservation is recommended as it complements through different techniques like zoo, captive breeding, aquarium, botanical garden, and gene bank. Ex situ conservation has its own advantages, disadvantages, and challenges making decision on its application by evaluating advantages, disadvantages and challenges. Although, Ethiopia is rich in biodiversity resources, more people depend on it for their livelihood directly or indirectly causing a great loss. Even if the conservation of biodiversity in Ethiopia has long-time history, its progress, coverage, and enforcement of the rule for conservation seem to be weak. Despite of good progress made in gene bank conservation, it is yet to be developed. In the same way, attention should be given for developing a National Zoological Park and a Botanical Garden.

ACKNOWLEDGMENTS

The authors are extremely grateful to Dr. Habte Jebessa for his valuable comments, suggestions and corrections on the draft of this paper. Our appreciation also goes to Professor. Afework Bekele for his remarkable encouragements and critical comments on the draft of this paper.

REFERENCES

1. T. I. Borokini, A. U. Okere, A. O. Giwa, B. O. Daramola, and W. T. Odofin, "Biodiversity and conservation of plant genetic resources in Field Genebank of the National Centre for Genetic Resources and Biotechnology, Ibadan, Nigeria," The International Journal of Biodiversity and Conservation, vol. 2, pp. 37–50, 2010.
2. M. Antofie, "Current political commitments' challenges for ex situ conservation of plant genetic resources for food and

agriculture," Analele Universității din Oradea—Fascicula Biologie, vol. 18, pp. 157–163, 2011.

3. C. A. Tisdell, "Core issues in the economics of biodiversity conservation," Annals of the New York Academy of Sciences, vol. 1219, no. 1, pp. 99–112, 2011. ·
4. Institute of Biodiversity Conservation, National Biodiversity Strategy and Action Plan, Institute of Biodiversity Conservation, Federal Democratic Republic of Ethiopia, Addis Ababa, Ethiopia, 2005.
5. M. E. Dulloo, D. Hunter, and T. Borelli, "Ex situ and in situ conservation of agricultural biodiversity: major advances and research needs," Notulae Botanicae Horti Agrobotanici Cluj-Napoca, vol. 38, no. 2, pp. 123–135, 2010.
6. J. Young, C. Richards, A. Fischer et al., "Conflicts between biodiversity conservation and human activities in the central and eastern European countries," Ambio, vol. 36, no. 7, pp. 545–550, 2007.
7. H. Debela, "Human influence and threat to biodiversity and sustainable living," Ethiopian Journal of Education and Sciences, vol. 3, no. 1, pp. 85–95, 2007.
8. R. C. Lacy, "Considering threats to the viability of small populations using individual-based models,"Ecological Bulletin, vol. 48, pp. 39–51, 2000.
9. K. Leus, "Captive breeding and conservation," Zoology in the Middle East, vol. 54, supplement 3, pp. 151–158, 2011.
10. K. A. Wilson, J. Carwardine, and H. P. Possingham, "Setting conservation priorities," Annals of the New York Academy of Sciences, vol. 1162, pp. 237–264, 2009. ·
11. R. Frankham, J. D. Ballou, and D. A. Briscoe, Introduction to Conservation Genetics, Cambridge University Press, Cambridge, UK, 2002.
12. T. Getachew, "Biodiversity hotspots: pitfalls and prospects," IBC News Letter, vol. 1, pp. 14–16, 2012.
13. H. Huang, "Plant diversity and conservation in China: planning a strategic bioresource for a sustainable future," Botanical Journal of the Linnean Society, vol. 166, no. 3, pp. 282–300, 2011.
14. C. Brutting, I. Hensen, and K. Wesche, "Ex situ cultivation affects genetic structure and diversity in arable plants," Plant Biology, vol. 15, pp. 505–513, 2013.
15. Institute of Biodiversity Conservation, Ethiopia: Second Country Report on the State of PGRFA to FAO, FAO, Rome, Italy, 2007.

16. E. D. Kjaer, L. Graudal, and I. Nathan, Ex Situ Conservation of Commercial Tropical Trees: Strategies, Options and Constraints, Danida Forest Seed Centre, Humlebaek, Denmark, 2001.

17. J. Hakansson, Genetic Aspects of Ex Situ Conservation: Introductory Paper, Department of Biology, Linköping University, 2004.

18. G. M. Reid, T. C. Macbeath, and K. Csatadi, "Global challenges in freshwater-fish conservation related to public aquariums and the aquarium industry," International Zoo Yearbook, vol. 47, pp. 6–45, 2013.

19. V. A. Melfi, "Ex situ gibbon conservation: status, management and birth sex ratios," International Zoo Yearbook, vol. 46, no. 1, pp. 241–251, 2012.

20. C. Ratledge, "Towards conceptual framework for wildlife Tourism," Tourism Management, vol. 22, pp. 31–40, 2001.

21. J. Balcombe, "Animal pleasure and its moral significance," Applied Animal Behaviour Science, vol. 118, no. 3-4, pp. 208–216, 2009.

22. S. F. Carrizo, K. G. Smith, and W. R. T. Darwall, "Progress towards a global assessment of the status of freshwater fishes (Pisces) for the IUCN Red List: application to conservation programmes in zoos and aquariums," International Zoo Yearbook, vol. 47, pp. 46–64, 2013.

23. R. C. Lacy, "Achieving true sustainability of zoo populations," Zoo Biology, vol. 32, pp. 19–26, 2013.

24. R. L. Eaton, "An overview of zoo goals and exhibition principles," International Journal for the Study of Animal Problems, vol. 2, pp. 295–299, 1981.

25. E. J. Fernandez, M. A. Tamborski, S. R. Pickens, and W. Timberlake, "Animal-visitor interactions in the modern zoo: conflicts and interventions," Applied Animal Behaviour Science, vol. 120, no. 1-2, pp. 1–8, 2009.

26. L. A. Dickie, "The sustainable zoo: an introduction," International Zoo Yearbook, vol. 43, no. 1, pp. 1–5, 2009.

27. M. Claxton, "The potential of the human-animal relationship as an environmental enrichment for the welfare of zoo-housed animals," Applied Animal Behaviour Science, vol. 133, no. 1-2, pp. 1–10, 2011.

28. S. E. Williams and E. A. Hoffman, "Minimizing genetic adaptation in captive breeding programs: a review," Biological Conservation, vol. 142, no. 11, pp. 2388–2400, 2009.

29. R. J. Whittington and R. Chong, "Global trade in ornamental fish from an Australian perspective: the case for revised import risk analysis and

management strategies," Preventive Veterinary Medicine, vol. 81, no. 1–3, pp. 92–116, 2007.

30. D. Dudgeon, A. H. Arthington, M. O. Gessner et al., "Freshwater biodiversity: importance, threats, status and conservation challenges," Biological Reviews of the Cambridge Philosophical Society, vol. 81, no. 2, pp. 163–182, 2006.
31. S. Blackmore, M. Gibby, and D. Rae, "Strengthening the scientific contribution of botanic gardens to the second phase of the Global Strategy for Plant Conservation," Botanical Journal of the Linnean Society, vol. 166, no. 3, pp. 267–281, 2011.
32. G. Clarke, "The Frozen Ark Project: the role of zoos and aquariums in preserving the genetic material of threatened animals," International Zoo Yearbook, vol. 43, no. 1, pp. 222–230, 2009.
33. O. A. Ryder, A. McLaren, S. Brenner, Y.-P. Zhang, and K. Benirschke, "DNA banks for endangered animal species," Science, vol. 288, no. 5464, pp. 275–277, 2000.
34. C. W. Vertuccj, "Predicting the optimum storage conditions for seeds using thermodynamic principles," Journal of Seed Technology, vol. 17, pp. 41–52, 1993.
35. R. H. Ellis and E. H. Roberts, "Improved equations for the prediction of seed longevity," Annals of Botany, vol. 45, no. 1, pp. 13–30, 1980.
36. J. B. Dickie, R. H. Ellis, H. L. Kraak, K. Ryder, and P. B. Tompsett, "Temperature and seed storage longevity," Annals of Botany, vol. 65, no. 2, pp. 197–204, 1990.
37. M. L. Moir, P. A. Vesk, K. E. C. Brennan et al., "Considering extinction of dependent species during translocation, ex situ conservation, and assisted migration of threatened hosts," Conservation Biology, vol. 26, no. 2, pp. 199–207, 2012.
38. S. Christie, "Why keep tigers in zoos?" in Tigers of the World: The Science, Politics and Conservation of Panthera Tigris, R. Tilson and P. Nyhus, Eds., pp. 205–214, Elsevier Inc., Amsterdam, The Netherlands, 1998.
39. U. S. Anderson, A. S. Kelling, R. Pressley-Keough, M. A. Bloomsmith, and T. L. Maple, "Enhancing the zoo visitor›s experience by public animal training and oral interpretation at an otter exhibit,"Environment and Behavior, vol. 35, no. 6, pp. 826–841, 2003.
40. H. R. Jensen, A. Dreiseitl, M. Sadiki, and D. J. Schoen, "The Red Queen and the seed bank: pathogen resistance of ex situ and in situ conserved

barley," Evolutionary Applications, vol. 5, no. 4, pp. 353–367, 2012.

41. G. R. Hosey, "How does the zoo environment affect the behaviour of captive primates?" Applied Animal Behaviour Science, vol. 90, no. 2, pp. 107–129, 2005.·

42. K. N. Morgan and C. T. Tromborg, "Sources of stress in captivity," Applied Animal Behaviour Science, vol. 102, no. 3-4, pp. 262–302, 2007.

43. B. O. Braastad, "Effects of prenatal stress on behaviour of offspring of laboratory and farmed mammals," Applied Animal Behaviour Science, vol. 61, no. 2, pp. 159–180, 1998.·

44. S. M. Wisely, R. M. Santymire, T. M. Livieri et al., "Environment influences morphology and development for in situ and ex situ populations of the black-footed ferret (Mustela nigripes)," Animal Conservation, vol. 8, no. 3, pp. 321–328, 2005.

45. K. Helenurm and L. S. Parsons, "Genetic variation and the reintroduction of Cordylanthus maritimus ssp. maritimus to Sweetwater Marsh, California," Restoration Ecology, vol. 5, no. 3, pp. 236–244, 1997. ·

46. D. Lauterbach, M. Burkart, and B. Gemeinholzer, "Rapid genetic differentiation between ex situ and their in situ source populations: an example of the endangered Silene otites (Caryophyllaceae),"Botanical Journal of the Linnean Society, vol. 168, no. 1, pp. 64–75, 2012.

47. Balmford, G. M. Mace, and N. Leader-Williams, "Designing the ark: setting priorities for captive breeding," Conservation Biology, vol. 10, no. 3, pp. 719–727, 1996.

48. R. A. Griffiths and L. Pavajeau, "Captive breeding, reintroduction, and the conservation of amphibians," Conservation Biology, vol. 22, no. 4, pp. 852–861, 2008.

49. R. J. Probert, M. I. Daws, and F. R. Hay, "Ecological correlates of ex situ seed longevity: a comparative study on 195 species," Annals of Botany, vol. 104, no. 1, pp. 57–69, 2009.

50. L. M. Clayton, E. J. Milner-Gulland, D. W. Sinaga, and A. H. Mustari, "Effects of a proposed ex situconservation program on in situ conservation of the babirusa, an endangered suid," Conservation Biology, vol. 14, no. 2, pp. 382–385, 2000.

51. W. G. Conway, "Buying time for wild animals with zoos," Zoo Biology, vol. 30, no. 1, pp. 1–8, 2011. ·

52. N. C. Ellstrand and D. R. Elam, "Population genetic conseqences of small population size: implications for plant conservation," Annual Review of Ecology and Systematics, vol. 24, pp. 217–242, 1993.

53. A. Cunningham, "Disease risks of wildlife translocations," Conservation Biology, vol. 10, no. 2, pp. 349–353, 1996.
54. R. P. Reading, T. W. Clark, and B. Griffith, "The influence of valuational and organizational considerations on the success of rare species translocations," Biological Conservation, vol. 79, no. 2-3, pp. 217–225, 1997.
55. L. I. Wright, T. Tregenza, and D. J. Hosken, "Inbreeding, inbreeding depression and extinction,"Conservation Genetics, vol. 9, no. 4, pp. 833–843, 2008. ·
56. B. Miller, K. Ralls, R. P. Reading, J. M. Scott, and J. Estes, "Biological and technical considerations of carnivore translocation: a review," Animal Conservation, vol. 2, no. 1, pp. 59–68, 1999.
57. L. Laikre, L. C. Larsson, A. Palmé, J. Charlier, M. Josefsson, and N. Ryman, "Potentials for monitoring gene level biodiversity: using Sweden as an example," Biodiversity and Conservation, vol. 17, no. 4, pp. 893–910, 2008.
58. N. I. Vavilov, "The origin, variation, immunity and breeding of cultivated plants," Chronica Botanica, vol. 13, pp. 1–366, 1951.
59. S. Demissew, "How has government policy post-global strategy for plant conservation impacted on science? The Ethiopian perspective," Botanical Journal of the Linnean Society, vol. 166, no. 3, pp. 310–325, 2011.
60. B. Girma, T. Pearce, and D. Abebe, "Biological diversity and current ex situ conservation practices in Ethiopia," in Seed Conservation Turning Science into Practice, R. D. Smith, J. B. Dickie, S. H. Linington, H. W. Pritchard, and R. J. Probert, Eds., pp. 849–856, Kew Publishing, Kew, UK, 2003.

Chapter 5

CITIZENS' PREFERENCES FOR THE CONSERVATION OF AGRICULTURAL GENETIC RESOURCES

Eija Pouta, Annika Tienhaara and Heini Ahtiainen

MTT Agrifood Research Finland, Helsinki, Finland

ABSTRACT

Evaluation of conservation policies for agricultural genetic resources (AgGR) requires information on the use and non-use values of plant varieties and animal breeds, as well as on the preferences for in situ and ex situ conservation. We conducted a choice experiment to estimate citizens' willingness to pay (WTP) for AgGR conservation programmes in Finland, and used a latent class model to identify heterogeneity in preferences among respondent groups. The findings indicate that citizens have a high interest in the conservation of native breeds and varieties, but also reveal the presence of preference heterogeneity. Five respondent groups could be identified based on latent class modeling: one implying lexicographic preferences, two with reasoned choices, one indicating uncertain support and one with a preference for the current status of conservation. The results emphasize the importance of in situ conservation of native cattle breeds and plant varieties in developing conservation policies.

INTRODUCTION

The intensification of agriculture has led to marked changes in the utilization of agricultural genetic resources (AgGR), and many previously common cultivated plant varieties as well as native animal breeds that are of interest in terms of food and agricultural production have become rare or even endangered (Drucker et al., 2001; FAO, 2007, 2010). In Finland, several native breeds, such as the Eastern and Northern Finncattle, the Kainuu Gray Sheep and the Åland Sheep, are endangered according to the FAO classification (FAO, 2007), and the majority of old Finnish crop varieties as well as the Finnish landrace pig are already extinct.

Decisions on the focus and extent of genetic resource conservation should consider both the costs and benefits of conservation. The full benefits of conserving AgGR are not revealed by markets, as the resources are either not traded in the markets or the price of agricultural products does not completely capture their value (Oldfield, 1989; Brown, 1990; Drucker et al., 2001). These market failures result in an inefficient allocation of resources, i.e., the level of conservation is too low as the full benefits are not considered. Although the importance of economic analyses has been recognized, the literature on the monetary value of genetic resources in agriculture is still relatively limited (e.g., Evenson et al., 1998; Rege and Gibson, 2003; Ahtiainen and Pouta, 2011).

Conservation policies for AgGR in Finland, as in many other European countries, are currently based on international agreements such as the Convention on Biological Diversity (1992) and the Global Plan of Action for Animal Genetic Resources (FAO, 2007). National genetic resource programmes were initiated for plants in the year 2003 and for farm animals in 2005 to strengthen the conservation of genetic resources in Finland. Although there has been some progress in putting the programmes into action, they have not been fully implemented. This may reflect, for example, the lack of political interest in the conservation.

To evaluate conservation policies, there is a need for monetary benefit estimates that encompass both use and non-use values associated with genetic resources. Use values refer to the benefits obtained from current and future use of genetic resources in production and breeding, while non-use values are generated from the knowledge that genetic resources, e.g., certain breeds, exist and are saved for future generations. Stated preference methods, such as the discrete choice experiment (CE) method, are capable of estimating both use and non-use values in monetary terms. A choice experiment is a survey-based method whereby respondents are asked to choose between two or more discrete alternatives that are described with attributes. By varying attribute levels and including a price variable as one of the attributes, respondents› willingness to pay (WTP) for a policy alternative or attribute level is indirectly revealed based on the choices they make (e.g.,Hanley et al., 2001). The CE method has been found suitable for valuing genetic resources due to its flexibility and ability to value the different traits that breeds or varieties may have. The CE method can also be used to evaluate the means of conservation in situ (live animals and plants) and ex situ(as seeds, cryopreserved embryos and other genetic material), and both plant genetic resources (PGR) and animal genetic resources (AnGR). Previous choice experiments have focused on valuing breeds or varieties and their attributes, especially related to their use in agriculture (Birol et al., 2006; Ouma et al., 2007), and applications focusing on consumer or citizen values for AgGR are rare. Valuation studies on biodiversity have found heterogeneity in consumer preferences, and even identified lexicographic preferences toward conservation (Hanley et al., 1995; Sælensminde, 2006). Lexicographic preferences imply that people are unwilling to accept any trade-offs for changes in environmental goods, such as biodiversity, and may arise when an individual believes that the environment should be protected without regard to the costs. In the context of AgGR, preference heterogeneity has mainly been studied among farmers (e.g., Ouma et al., 2007; Omondi et al., 2008; Roessler et al., 2008), and there have been only few empirical studies of heterogeneity of citizen preferences (Zander et al., 2013) or lexicographic preferences.

In this paper, we present the results of a choice experiment conducted to estimate the benefits of genetic resource conservation programmes in Finland. We tested the effect of in situ and ex situconservation on citizens› choices between programmes. We also analyzed whether plant varieties and animal breeds are perceived as equally valuable by citizens. As heterogeneity in the preferences for the conservation of AgGR is likely, we tested for the existence of citizen segments that place different values on the conservation of genetic resources.

We expected that AgGR would be rather unfamiliar to some of the respondents of the valuation survey. However, in valuation surveys, respondents are assumed to make "informed" choices when responding to value elicitation questions (e.g., Blomquist and Whitehead, 1998). To obtain informed choices that produce valid estimates of WTP, surveys need to provide a sufficient amount of neutral information on the environmental good while avoiding information overload. Providing more information on the quality (characteristics and services) of an environmental good can increase the stated WTP, have no effect, or in some cases reduce WTP (Blomquist and Whitehead, 1998).

There is a substantial body of literature on the effects of information and respondent effort in contingent valuation studies (e.g., Cameron and Englin, 1997; Blomquist and Whitehead, 1998;Berrens et al., 2004), and some choice experiment studies have also examined the issue, mainly focusing on respondent effort (Hu et al., 2009; Vista et al., 2009). Hu et al. (2009) used data from a choice experiment concerning genetically modified food to simultaneously model voluntary information access and product choices. They found that information was accessed rather infrequently, and that those who held critical views on GM food accessed information more often. There were interlinkages between information access and choices, but they were complex and varied between individuals. Vista et al. (2009) examined the effect of time spent on attribute information, choice questions or completing the survey, finding no significant effects on parameter estimates. Here, we were particularly interested in examining how the use of information differs between respondent segments. In the survey, respondents had the opportunity to obtain additional information on genetic resources by accessing a hyperlink to a web page. The Internet survey allowed us to measure whether the respondents accessed the additional information and how much time they used to read it. Offering the opportunity for voluntary access to information instead of using different information treatments for split samples has the advantage of not assuming that respondents read all the information that is provided (Hu et al., 2009). Furthermore, we tested the effects of response certainty and self-perceived carefulness in filling the survey as sources of preference heterogeneity. The rest of the paper is organized as follows. Section Materials and Methods introduces the data and statistical models used in the analysis. Results are presented in section Results, and section Discussion and Conclusions provides discussion and conclusions.

MATERIALS AND METHODS

Data Collection

The survey data were collected using an Internet survey during the summer of 2011. The sample was drawn from the Internet panel of a private survey company, Taloustutkimus, which comprises 30,000 respondents who have been recruited to the panel using random sampling to represent the population (Taloustutkimus, 2013). After a pilot survey of 138 people, a random sample of 6200 respondents was selected, of which 2426 completed parts of the survey and 1495 completed the survey entirely. These numbers correspond to response rates of 39 and 24%, respectively. Based on the socio-demographic variables, the data represented the population rather well (Table 1).

Table 1: Descriptive statistics (n = **1608**)

	In the data	In the population[a]
Proportion of females, %	48	51
Mean age, years	52	47
Proportion of people with a higher educational level, %	24	23
Proportion of people living in households with a gross income under €40,000, %	43	53
Proportion of people with children (<18 years) in the family, %	35	40
Proportion of people living in South Finland, %	40	41

[a] *Statistics Finland 2010, www.stat.fi.*

Survey Design

In the first section, the survey introduced the most common Finnish native animal breeds and plant varieties by explaining what landraces are and giving examples. After asking the respondents about their familiarity with PGR and AnGR, all respondents were offered a short piece of information on the conservation of these breeds and varieties. Next, the respondents were given the opportunity to obtain further information by clicking on two hyperlinks, one for PGR and the other for AnGR. Providing voluntary access to additional information made it possible to identify those respondents who accessed the information, and the time spent on the information page was also recorded (Hu et al., 2009). The additional information provided in our survey included motives for conservation, descriptions of the in situ and ex situ conservation methods and facts about the sustainable use of genetic resources. After several questions concerning perceptions of genetic resources, the survey proceeded to the choice experiment.

The choice experiment was framed by telling respondents that the conservation of native plant varieties and animal breeds is not yet comprehensive in Finland. The survey presented a programme that would conserve the majority of the varieties and breeds on farms and in gene banks. The operation of gene banks would be extended to missing plants and varieties, and conservation on farms would be enhanced by developing the support provided to farmers for conservation activities. Furthermore, those who are using native varieties in gardens were stated to be supported monetarily and by providing guidance.

The survey explained that the conservation programme would be financed with an increase in income tax between the years 2012 and 2021, and that depending on the extent of the programme, the cost to taxpayers would vary, but all taxpayers would participate in financing the programme. The conservation measures (attributes) of the alternative programmes were illustrated to the respondents using a table.

Table 2 presents the attributes together with their descriptions and levels. The first attribute level is always the level specified in the status quo alternative (current state). The attributes included conservation measures of both plant varieties and animal breeds in gene banks and farms. Instead of having a separate attribute for each native breed, only one attribute for breeds in gene banks and one on farms was included to have the same number of attributes for varieties and breeds, and to ease the cognitive burden of the respondents. The native breeds in gene banks attribute had eight levels and native breeds on farms nine levels, including the status quo attribute level.

Table 2: Attributes of conservation programmes and their levels

Attribute	Description	Current state	Levels (unit)
Native food plant varieties in gene banks	Native food plants are stored in a gene bank, either as seeds or plant parts.	The gene bank contains seeds from about 300 landrace varieties. Plants that are added vegetatively (e.g., berry and apple varieties) are missing.	300, 400, 500 (number of plants)
Farms growing native food plants	Farmers and hobby gardeners cultivate native food plants on farms or in gardens.	Seven farms grow seeds of native food plants with agri-environmental support. Other activities than growing seeds are not supported.	7, 500, 1000 (number of farms)
Native ornamental plant varieties mapped and in gene banks	Scientists identify and register native ornamental plants. Varieties are preserved in a gene bank, either as seeds or plant parts.	Only a small proportion of the native ornamental plants are known. Storage in the official gene bank is not provided.	small proportion, about half, majority (proportion of plants)
Native breeds in gene banks	Landrace breeds are kept in a gene bank as gametes and embryos.	The gene bank contains Western, Eastern and Northern Finncattle, as well as Finn-, Åland and Kainuu sheep. Native chicken, goat and horse breeds are missing from the gene bank.	3 cattle breeds and 3 sheep breeds (status quo level), + all combinations of goat, horse and chicken breeds
Native breeds on farms	Native breeds are kept on farms in their natural environment. A breed is considered to be endangered if the number of females is less than 1000.	Farms secure goat, horse and chicken breeds, Finnish sheep and Western Finncattle. Eastern and Northern Finncattle, as well as Åland and Kainuu sheep, are endangered.	1 cattle breed, 1 sheep breed, goat, horse and chicken (status quo level), + all combinations of additional 1-2 cattle and sheep breeds
Cost	Cost for taxpayers, €/year during 2012–2021.	No additional costs.	0, 5, 20, 40, 80, 100, 150, 300 (€)

After introducing the attributes, the respondents were presented with six choice tasks. Each choice task included three alternatives: the status quo alternative, described as maintaining the current situation, and two policy alternatives describing an improved level of conservation compared to the current level. Each alternative was described with five conservation attributes, their levels and the cost attribute. The status quo alternative was uniform across choice tasks. An example of a choice task is shown in Table 3.

Table 3: Example of a choice set

		Current state	Conservation programme A	Conservation programme B
Native food plant varieties in gene banks		Approximately 300	400	400
Farms growing native food plants		7 farms	2000 farms	1000 farms
Native ornamental plant varieties mapped and in gene banks		Some	Majority	About half
Native breeds in gene banks		3 cattle breeds, 3 sheep breeds	3 cattle breeds, 3 sheep breeds, chicken, goat, horse	3 cattle breeds, 3 sheep breeds, goat
Native breeds on farms		Goat, horse, chicken, 1 cattle breed, 1 sheep breed	Goat, horse, chicken, 3 cattle breeds, 1 sheep breed	Goat, horse, chicken, 2 cattle breeds, 3 sheep breeds
Cost for taxpayers, €/year during 2012–2021	€	€0/year	€80/year	€200/year
I support the alternative		( )	( )	( )

We employed an efficient experimental design to allocate the attribute levels to the choice tasks in the choice experiment survey. Efficient designs aim to generate parameter estimates with standard errors that are as low as possible, and thus produce the maximum information from each choice situation (e.g., Rose and Bliemer, 2009). The generation of efficient designs requires the specification of priors for the parameter estimates. In the pilot survey, we employed zero priors in the design, and used the parameter estimates obtained in the pilot study to construct the final experimental design. In the final study, we employed a Bayesian D-efficient design using Ngene (v. 1.0.2), taking 500 Halton draws for the prior parameter distributions. Bayesian designs take into account the uncertainty related to the parameter priors. Instead of fixed priors, they make use of random priors by specifying a mean and standard deviation for the prior.

In the design phase, animal breeds in gene banks and on farms were treated as separate attributes, but were later combined to the "Native breeds in gene banks" and the "Native breeds on farms" attributes in the choice tasks presented to the respondents. Bayesian priors were employed for the chicken attribute and the number of cattle breeds on the farm attribute, and fixed priors for all other attributes. We generated 180 choice tasks, blocking them into 30 subsets, which resulted in six choice situations presented for each respondent. The final design had a D-error of 0.002.

Statistical Models

The choices between environmental programmes were originally modeled with a conditional logit model (also called a multinomial logit model) (McFadden, 1974). The conditional logit, however, assumes a similar preference structure for all respondents, which implies that they have similar tastes for the attributes of conservation. In this study, we were particularly interested in defining heterogeneous citizen segments, which have a similar preference structure within each segment. One approach that allows this heterogeneity is the latent class model (Boxall and Adamowicz, 2002), which has frequently also been applied in choice experiment models of environmental conservation programmes (e.g., Garrod et al., 2012; Grammatikopoulou et al., 2012). In the latent class model, preferences are assumed to be homogeneous in each segment, but to vary between the segments.

In the modeling, price was treated as a continuous variable and the other attributes were effects-coded, implying that the parameters will sum to zero over the categories of the nominal variable concerned. The status quo attribute levels were thus included in the model, and could obtain either negative or positive coefficients depending on their effect on respondent's utility. Alternative-specific constants (ASC) were included for all alternatives in order to allow systematic choice tendencies not explained with the parameters describing the attributes.

Heterogeneity was statistically included in the latent class model by simultaneously dividing individuals into behavioral groups or latent segments, and estimating a choice model for each of these classes. The estimation was carried out by assuming first one class, then two classes, three classes and so forth. In each step, the explanatory power of the model was assessed to decide on the optimal number of classes. For this purpose, we used the Bayesian information criterion (BIC) and Akaike information criterion (AIC), which are log-likelihood scores with correction factors for the number of observations and the number of parameters. The latent class model also enables the calculation of the WTP for the attributes for each citizen segment.

The relationship between the individual characteristics and the latent classes was examined a posteriori of the actual estimation of the latent class model in order to describe the heterogeneous citizen segments. Thus, the segments were formed solely based on the conservation program choices. The membership in the most probable segment was regressed using a logistic regression to characterize each class compared to the rest of the respondents. The explanatory variables for the class memberships included respondents' socioeconomic characteristics, perceived values and responsibilities, use of provided information, response certainty and self-reported perception of the carefulness of completing the survey. The independent variables in the logistic regression models and their descriptive statistics are presented in Table 4.

Table 4: Variables in the logistic regression models

Characteristic	Description	Mean	Standard deviation	Min	Max
Female	1 if the respondent is female, if male	0.49	0.50	0	1
Year of birth	Respondents year of birth, continuous	1960	15	1931	1992
High income	1 if household income is over €50,000 per year, 0 otherwise	0.45	0.49	0	1
High education	1 if respondents education level is university education, 0 otherwise	0.24	0.46	0	1
Eastern Finnish	1 if respondents lives in Eastern Finland, 0 otherwise	0.11	0.32	0	1
Childhood in city	1 if respondent spent his/her childhood in a city, 0 otherwise	0.41	0.49	0	1
Certainty	Mean of respondent's certainty in the conservation programme choices, on a scale of 10 completely certain—1 not at all certain.	6.85	2.23	1	10
Agri-environmental attitude	Importance of environmental issues in agriculture, mean of nine measures on scales from 1 to 4	3.26	0.44	1	4
Relative importance of preserving AgGR	The importance of preserving native breeds and varieties relative to other environmental protection measures, 1 if both equally important, >1 if preserving native breeds and varieties more important, <1 if other environmental protection measures more important	0.94	0.16	0.36	1.66
Existence value	Factor score based on 8 measures of the importance of existence values, continuous*	0.00	1.00	−4.38	2.39
Use values	Factor score based on 8 measures of the importance of use values, continuous*	0.00	1.00	−3.78	2.62
Citizen responsibility	Factor score based on 9 measures of stakeholder responsibilities in conservation*	0.00	1.00	−3.38	2.30
Consumer responsibility	Factor score based on 9 measures of stakeholder responsibilities in conservation*	0.00	1.00	−5.27	2.01
Farmer responsibility	Factor score based on 9 measures of stakeholder responsibilities in conservation*	0.00	1.00	−3.12	2.88
Familiarity of products	Familiarity of AgGR products, mean of 10 measures on scales from 1 to 3	2.03	0.42	1	3
Info use (animals) > 0.5 min	1 if respondent used more than 30 s for additional information about breeds, 0 otherwise	0.33	0.47	0	1
Info use (plants) > 0.5 min	1 if respondent used more than 30 s for additional information about varieties, 0 otherwise	0.35	0.48	0	1
Hasty response	1 if respondent evaluated his/her response as hasty, 0 if careful	0.05	0.22	0	1

*Detailed description of these variables can be found in Tienhaara et al. (2014).

Results

In 24% of the choice sets, the respondents chose the status quo option, i.e., the current state without any additional program to conserve AgGR. The probability of choosing one of the two alternative conservation programs varied between 46% for the lowest cost level of €5 and 28% for the highest cost level of €300.

Table 5 presents the conditional logit model results for the choice of the conservation programme. As expected, an increase in the programme cost negatively affected the probability of choosing it. Turning to consider the genetic resource attributes, the number of food plants in the gene bank was not statistically significant. All other attributes were significant in determining respondents› choices. A higher number of farms growing native plant varieties increased the choice probability. The larger the number of ornamental plants to be mapped and conserved in gene banks, the more probable it was that the respondent would choose the programme. Conserving native breeds of Finnish goats, horses and chickens in the gene bank all increased the support for the programme. The effect was highest for horse, followed by chicken and goat. The guaranteed existence of cattle breeds on farms had a positive and significant effect on choice. As expected, the effect was greater if the number of conserved cattle breeds was three instead of two. This was also the case with sheep breeds, although the conservation of two breeds did not have a positive effect on choice compared with the status quo of one conserved breed.

Table 5: Conditional logit (CL) model results

Variable	Coefficient	Wald *p*-value
ASC1 (SQ)	−0.263***	0.000
ASC2	0.291***	
ASC3	−0.028	
Cost	−0.005***	0.000
300 plants in bank (SQ)	0.002	1.000
400 plants in bank	−0.002	
500 plants in bank	0.000	
7 plants on farms (SQ)	−0.199***	0.000
500 plants on farms	0.075***	
1000 plants on farms	0.124***	
Ornamental plants in bank (SQ)	−0.057**	0.008
Ornamental plants in bank L2	−0.004	
Ornamental plants in bank L3	0.061***	
Goats (SQ)	−0.039***	0.005
Goats in bank	0.039***	
Horses (SQ)	−0.075***	0.000
Horses in bank	0.075***	
Chickens (SQ)	−0.047***	0.001
Chickens in bank	0.047***	
1 cattle breed on farms (SQ)	−0.114***	0.000
2 cattle breeds on farms	0.025	
3 cattle breeds on farms	0.089***	
1 sheep breed on farms (SQ)	0.020	0.027
2 sheep breeds on farms	−0.052***	
3 sheep breeds on farms	0.032	
No. of respondents	1608	
No. of observations	9484	
Correct predictions %	48	
R^2	0.04	

*z-test: *** 99% significance level; ** 95% significance level.*
SQ, attribute level in the status quo alternative.

The alternative specific constants (ASC) capture the tendency to choose one of the alternatives which is not explained by the attributes. The negative ASC1

(SQ) coefficient showed the reluctance to choose the status quo alternative regardless of the attribute levels in the policy alternatives. Furthermore, the ASC2 and ASC3 coefficients differed unexpectedly in sign and significance. The positive coefficient for ASC2 and negative for ASC3 indicated that the conservation programme that was presented first received more support. This was surprising, as the programmes were not presented in a specific order in the survey. The model predicted 48% of the choices right, clearly exceeding the probability of correct random choices of 33%, leading to a relatively weak goodness of fit.

The homogeneity of preferences was tested in the estimation of the latent class models. Based on the AIC and BIC, the estimation process showed that a model of five citizen clusters provided the best fit of the data. Table 6 presents the latent class model results with the cluster names, and the logit model for the membership of each cluster is presented in Table 7.

Table 6: Latent class models for conservation programme choice

	Class 1	Class 2	Class 3	Class 4	Class 5	Overall	
Pseudo R^2	0.131	0.288	0.019	0.015	0.472	0.559	
Class size	0.27	0.26	0.17	0.17	0.13		
Class names	**Conserva-tionists**	**Bid-sensitive animal conservers**	**Uncertain supporters**	**Status quo preferers**	**Bid sensitives**	**Wald *p*-value**	**Wald (=) *p*-value**
Attributes			**Coefficients and significance levels**				
ASC 1 (SQ)	−0.000***	2.097***	0.841***	1.668***	−0.554**	0.000	0.000
ASC 2	0.332***	1.499***	1.757***	−0.414**	0.478***		
ASC 3	0.658***	1.438***	−0.916***	−1.254***	0.076		
Cost	0.000	−0.018***	−0.003*	−0.001	−0.041***	0.000	0.000
300 plants in bank (SQ)	−0.162***	0.138**	0.018	0.412**	−0.322***	0.003	0.001
400 plants in bank	0.025	−0.007	0.078	−0.166	0.225*		
500 plants in bank	0.137**	−0.131*	−0.096	−0.245	0.097		
7 plants on farms (SQ)	−0.621***	−0.120*	−0.261**	−0.006	−0.169	0.000	0.000
500 plants on farms	0.125**	0.208***	0.237*	0.003	0.104		
1000 plants on farms	0.496***	−0.088	0.024	0.003	0.066		
Ornamental plants in bank (SQ)	−0.462***	0.015	0.116	−0.004	−0.332**	0.000	0.000
Ornamental plants in bank L2	0.158***	0.002	0.023	−0.053	0.16		
Ornamental plants in bank L3	0.304***	−0.017	−0.139	0.067	0.172		
Goats (SQ)	−0.063***	−0.063***	−0.063***	−0.063***	−0.063***	0.001	C.i.
Goats in bank	0.063***	0.063***	0.063***	0.063***	0.063***		
Horses (SQ)	−0.152***	−0.128***	−0.075	0.447***	−0.256***	0.000	0.000
Horses in bank	0.152***	0.128***	0.075	−0.447***	0.256***		
Chickens (SQ)	−0.062***	−0.062***	−0.062***	−0.062***	−0.062***	0.001	C.i.
Chickens in bank	0.062***	0.062***	0.062***	0.062***	0.062***		
1 cattle breed on farms (SQ)	−0.144***	−0.144***	−0.144***	−0.144***	−0.144***	0.000	C.i.
2 cattle breeds on farms	0.034	0.034	0.034	0.034	0.034		
3 cattle breeds on farms	0.110***	0.110***	0.110***	0.110***	0.110***		
1 Sheep breed on farms (SQ)	−0.213***	0.046	−0.036	0.581***	−0.245**	0.000	0.001
2 Sheep breeds on farms	0.056	−0.04	−0.156	−0.282	0.116		
3 Sheep breeds on farms	0.167***	−0.007	0.192	−0.300	0.128		
No. of respondents	1608						
No. of observations	9484						
Correct predictions %	85						

*z-test: *** 99% significance level; ** 95% significance level; * 90% significance level.*

SQ, attribute level in the status quo alternative.

C.i., class independent.

Table 7: Logistic regression models profiling consumer classes

Class	Class 1	Class 2	Class 3	Class 4	Class 5
Variable	**Coefficients and significance levels**				
Constant	−2.76***	−43.31***	48.77***	39.90**	−29.46**
Female	−0.46***				
Year of birth		0.02***	−0.02***	−0.02**	0.02*
High income			−0.39**		
High education				−0.72***	
Eastern Finnish			0.40*		
Childhood in city				−0.68**	
Certainty	0.12***		−0.09**		−0.08**
Agri-environmental attitude	0.37*	0.43**			
Relative importance of AgGR		−1.482***	1.412**	−1.82**	
Existence values	0.32***			−0.50***	
Use values	0.38***			−0.39***	
Citizen responsibility		0.29***	0.21**	−1.06***	−0.43***
Consumer responsibility		0.17**		−0.31**	−0.38***
Farmer responsibility	−0.16**			0.27**	
Familiarity of products					−0.48**
Info use (animals) > 0.5 min				−0.39*	
Info use (plants) > 0.5 min		0.54***	−0.47***		
Hasty response			0.70*		−1.08**
N	1088	1201	1098	1077	1199
Nagelkerke R^2	0.103	0.083	0.071	0.397	0.104
Chi-squared	81.99	71.44	46.48	252.37	68.25
p-Value	0.000	0.000	0.000	0.000	0.000
Correctly classified (cut 0.5)	69.6	71.6	83.9	90.4	86.8

*Variables are significant at the *** 99% level, ** 95% level, * 90% level.*

The latent class model showed that although preferences for some attributes, such as conserving goat and chicken breeds in gene banks and cattle breeds on farms, did not differ significantly between clusters, there was significant heterogeneity in preferences for most of the attributes. The first class, named as "conservationists," comprised 27% of the respondents. They did not take the personal cost of the conservation programme into account in their decision process, as the coefficient of the cost variable was not significant. Instead, almost all the conservation attributes had significant and positive signs. Contrary to other clusters, most plant-related attributes were significant for conservationists. They also valued the conservation of ornamental plants. Table 5also shows that this cluster perceived higher use and existence values from genetic resource conservation than respondents in other segments, and also higher than average certainty in their responses to the choice tasks. This class contained more men than women and considered the conservation not to be a responsibility of farmers. For this cluster we also tested the effect gardening as a hobby, but it did not turn out to be significant. Thus, it seems that these respondents did not support the program because of the possible private good aspect of measures to support native varieties in gardens.

The second cluster, covering 26% of the respondents, was named as "bid-sensitive animal conservers." This group had a higher tendency to choose the improvement programmes compared to the status quo. The coefficient of the bid was significant and the second smallest of all clusters. In this cluster, the emphasis of preferences was on the conservation of animal breeds. The conservation of plant varieties in gene banks was even valued negatively. These respondents perceived more often than average that citizens and consumers should be responsible for the conservation of genetic resources. They also had positive agri-environmental attitudes. Furthermore, the respondents in this cluster used more than the average time to familiarize themselves with the information available in the survey concerning PGR, and they were slightly younger than the average respondent.

A confusing aspect in the third cluster was the large difference between the ASC for the two conservation programmes. This group, comprising 17% of the respondents, had a considerably greater tendency to choose conservation programme A rather than B or the status quo, although this could not be explained by the experimental design and attribute levels. The bid variable followed expectations, but for the other attributes, only plants on farms and the class-independent variables were significant. The logistic regression revealed that members of this cluster were older and had a lower income, and they emphasized the responsibility of citizens in conservation. Geographically, this cluster had more members who lived in Eastern Finland. The respondents in this group were relatively uncertain of their preferences, used the additional information less, and responded, according to their self-evaluation, less carefully than other respondents. As there were random tendencies in their support for a programme (ASC), but they still preferred an increase in several conservation attributes, they were named as "uncertain supporters."

The fourth class, with 17% of respondents, clearly preferred the status quo option, as the ASC for the programme options were negative. The coefficient of the bid variable was not significant. Among these "status quo preferers," the choice was consistent with their negative attitudes, as the relative importance of AgGR was low, as well as the perceived existence and use values. Citizens and consumers were less frequently seen as those responsible for conservation; instead, it was perceived as a responsibility of the farmers. This class was characterized by an older age, lower educational level and growing up on a farm. The fifth class of respondents (13%), named as "bid sensitives," were the most sensitive to the cost of the programme of all groups. Nevertheless, the ASC revealed that they were interested in conservation, and almost all conservation attributes had significant coefficients. Among these respondents, particularly the ex situ conservation of Finnhorse positively affected their choices. In this class, the conservation of genetic resources was not seen as a responsibility of citizens or consumers. The logit model for this group showed that they evaluated themselves as careful respondents but felt somewhat uncertain of their choices. They were younger than average and less familiar with products from traditional breeds and varieties.

Table 8: Annual willingness to pay (in 2009 €) for attributes

	Conditional logit model	Latent class model, Class 2	Latent class model, Class 3	Latent class model, Class 5
Plants in bank (400)	–	–	–	13
Plants in bank (500)	–	-15	–	–
Plants on farms (500)	60	19	7	–
Plants on farms (1000)	70	–	–	–
Ornamental plants (majority) inventoried and in bank	14	–	–	–
Goats in bank	17	7	105	3
Horses in bank	33	15	–	12
Chickens in bank	20	7	104	3
3 cattle breeds on farms	44	14	211	6
2 sheep breeds on farms	−15	–	–	–

–, Indicates that the estimate is missing due to the non-significance of the cost coefficient.

WTP for different attributes was calculated based on the conditional logit model and the latent class model for those classes for which the cost coefficient was significant (Table 8). WTPs based on the conditional logit model indicated that plants on farms, cattle breeds and horses were most highly valued. In general, there was substantial variation in WTPs between the classes. In class 3, WTPs were higher due to the low importance of the cost attribute.

Discussion and Conclusions

The results of a choice experiment concerning agricultural genetic resource policies showed that citizens are interested in the conservation of native breeds and varieties in agriculture. However, there was considerable variation in preferences between citizen segments. Of the five identified groups, two groups covering over half of the respondents had a high interest in the conservation of native breeds and varieties. Respondents in one of the segments clearly preferred the current state of conservation to additional conservation efforts, while one group had a favorable attitude toward conservation if the expenses were on a low level, and respondents in one segment were supportive but wavering in their preferences. The respondent groups were identified based on their preferences for conservation, and they also differed with respect to the use of additional information, their response carefulness and the certainty of the stated WTP. Similar to previous studies of consumer preferences on biodiversity (e.g., Hanley et al., 1995), we also found lexicographic preferences for conserving AgGR. Those were expressed by the largest group of respondents (27%), as their interest in conservation was high regardless of the costs. Lexicographic choices can occur as a result of simplification if the respondent finds the choice task too difficult to handle or as a result of actual lexicographic preferences (Sælensminde, 2006). In our case, it is difficult to determine whether respondents exhibited lexicographic preferences because they wanted to simplify the choice tasks or because the differences in the attribute levels were large. Respondents in the group which exhibited lexicographic preferences were more certain about their preferences, which supports the phenomenon of actual lexicographic preferences as the reason for their choices. In addition, their positive perceptions concerning the existence and use values of genetic resources support the observation of actual lexicographic preferences.

Due to the preference structures, WTP estimates were only obtained for three respondent groups and some of the attributes. In those groups where the cost variable was significant and meaningful WTP estimates could thus be estimated, the marginal WTPs were considerably lower than the WTPs of the whole sample based on the conditional logit model. This implies that in the whole sample, the results were influenced by the groups that were insensitive to the costs of conservation. Our results can be compared with those obtained by Zander et al. (2013), who assessed the economic value of conservation programs for two Italian cattle breeds using a choice experiment directed to citizens. Zander et al. (2013) also found preference heterogeneity for most of the attributes of the conservation programs, as well as differences in the sensitivity to the cost attribute. According to their findings, 85% of the respondents supported increased conservation, and the mean WTP was 90€ for conserving each breed. The present results can also be linked to previous results of heterogeneity among farmers using native breeds and varieties. Soini et al. (2012)identified a segment of production-oriented farmers among European cattle breeders that would benefit from increased subsidies for keeping native breeds on farms. If the subsidies were increased to correspond to citizens› WTP, it would help particularly this subsidy-dependent group of farmers.

As the survey was Internet-based, we were able to obtain information on the time used for obtaining additional information about plant and AnGR. These variables, combined with certainty, could partly explain the membership in the latent classes. However, similarly to Hu et al. (2009)and Vista et al. (2009), there were no clear tendencies for the use of information to be associated with a lower or higher WTP. Further research is, however, needed to clarify the associations between preferences, uncertainty and information acquisition in the case of genetic resources.

The results provide implications concerning how to direct the conservation policies for AgGR in Finland. The WTP estimates for the attributes of the conservation programmes indicated that the participants valued particularly in situ conservation in the case of PGR, which would also imply the existence of native plant varieties in the landscape. However, a moderate level of this in situconservation would be sufficient, as the highest level increased the WTP only slightly. For the conservation of animal breeds, the results emphasize the importance of in situ conservation of cattle breeds. The weak support for the conservation of sheep breeds compared to cattle breeds was understandable, as Finnsheep breeds are less familiar to the public. However, the low, even negative, WTP for the conservation of sheep breeds is in contradiction with the importance of Finnsheep in breeding (e.g., Thomas, 2010). Ex situ conservation of those animal breeds that are at present insufficiently protected in gene banks was perceived as important, particularly the conservation of the genetic material of the Finnhorse.

Although the cost-effectiveness of AgGR conservation is case-dependent, some previous studies have recommended ex situ conservation in gene banks as a less expensive, less vulnerable and less policy-sensitive method of conservation (Dulloo et al., 2010; Silversides et al., 2012). These cost-effectiveness considerations do not, however, take into account the additional benefits that may be associated with in situ conservation, such as the visibility of local breeds and varieties in the landscape or the opportunity to use local breed products. Thus, taking into account citizens' preferences for in situ and ex situ conservation and using cost-benefit analysis in policy evaluation may shift the priorities of agricultural genetic resource conservation policies.

In this study, the conservation policies were based on equal participation of all citizens, as the policy was financed with taxes. An alternative approach would be to apply market-based incentives, e.g., payments for environmental services (PES) for the conservation of genetic resources (McNeely, 2006; Wunder, 2007; Narloch et al., 2011). PES would imply that actors who are major users of the resources are involved in making and adapting rules for conservation markets. For future experiments of PES, our results of the citizen groups that are most interested provide information for identifying the interested parties for the markets of AgGR.

Conflict of Interest Statement

The Guest Associate Juha Kantanen declares that, despite being affiliated to the same institution as the authors, the review process was handled objectively and no conflict of interest exists. The authors declare that the research was conducted in the absence of any commercial or financial relationships that could be construed as a potential conflict of interest.

REFERENCES

1. Ahtiainen, H., and Pouta, E. (2011). The value of genetic resources in agriculture: a meta-analysis assessing existing knowledge and future research needs. Int. J. Biodivers. Sci. Ecosyst. Serv. Manage. 7, 27–38. doi: 10.1080/21513732.2011.593557
2. Berrens, R. P., Bohara, A. K., Jenkins-Smith, H. C., Silva, C. L., and Weimer, D. L. (2004). Information and effort in contingent valuation surveys: application to global climate change using national internet samples. J. Environ. Econ. Manage. 47, 331–363. doi: 10.1016/S0095-0696(03)00094-9
3. Birol, E., Smale, M., and Gyovai, Á. (2006). Using a choice experiment to estimate farmers' valuation of agrobiodiversity on hungarian small farms. Environ. Resour. Econ. 34, 439–469. doi: 10.1007/s10640-006-0009-9
4. Blomquist, G., and Whitehead, J. (1998). Resource quality information and validity of willingness to pay in contingent valuation.Resour. Energy

Econ. 20, 179–196. doi: 10.1016/S0928-7655(97)00035-3

5. Boxall, C. P., and Adamowicz, L. W. (2002). Understanding heterogeneous preferences in random utility models: a latent class approach. Environ. Resour. Econ. 23, 421–446. doi: 10.1023/A:1021351721619
6. Brown, G. M. (1990). "Valuation of genetic resources," in The Preservation and Valuation of Biological Resources, eds G. H. Orians, G. M. Jr. Brown, W. E. Kunin, and J. E. Swierzbinski (Seattle, WA: University of Washington Press), 203–228.
7. Cameron, T. A., and Englin, J. (1997). Respondent experience and contingent valuation of environmental goods. J. Environ. Econ. Manage. 33, 296–313. doi: 10.1006/jeem.1997.0995
8. Drucker, A. G., Gomez, V., and Anderson, S. (2001). The economic valuation of farm animal genetic resources: a survey of available methods. Ecol. Econ. 36, 1–18. doi: 10.1016/S0921-8009(00)00242-1
9. Dulloo, M. E., Hunter, D., and Borelli, T. (2010). Ex situ and in situ conservation of agricultural biodiversity: major advances and research needs. Notulae Botanicae Horti Agrobotanici Cluj-Napoca 38, 123–135.
10. Evenson, R. E., Gollin, D., and Santaniello, V. (1998). "Introduction and overview: agricultural values of plant genetic resources," in Agricultural Values of Plant Genetic Resources, eds R. E. Evenson, D. Gollin, and V. Santaniello (Wallingford: CABI Publishing), 1–25.
11. FAO, (2007). The State of the World›s Animal Genetic Resources for Food and Agriculture. Rome: Commission on Genetic Resources for Food and Agriculture, FAO.
12. FAO, (2010). The Second Report on the State of the World›s Plant Genetic Resources for Food and Agriculture. Available online at: http://www.fao.org/docrep/013/i1500e/i1500e.pdf [Accessed 11 October 2011].
13. Garrod, G., Ruto, E., Willis, K., and Powe, N. (2012). Heterogeneity of preferences for the benefits of environmental stewardship: a latent-class approach. Ecol. Econ. 76, 104–111. doi: 10.1016/j.ecolecon.2012.02.011
14. Grammatikopoulou, I., Pouta, E., Salmiovirta, M., and Soini, K. (2012). Heterogeneous preferences for agricultural landscape improvements in southern Finland. Landsc. Urban Plan. 107, 181–191. doi: 10.1016/j.landurbplan.2012.06.001
15. Hanley, N., Mourato, S., and Wright, R. E. (2001). Choice modelling approaches: a superior alternative for environmental valuation? J. Econ. Surv. 15, 435–462. doi: 10.1111/1467-6419.00145

16. Hanley, N., Spash, C., and Walker, L. (1995). Problems in valuing the benefits of biodiversity protection. Environ. Res. Econ. 5, 249–272. doi: 10.1007/BF00691519

17. Hu, W., Adamovicz, W. L., and Veerman, M. M. (2009). Consumers› preferences for GM food and voluntary information access: a simultaneous choice analysis. Can. J. Agri. Econ. 57, 241–267. doi: 10.1111/j.1744-7976.2009.01150.x

18. McFadden, D. (1974). "Conditional logit analysis of qualitative choice behaviour," in Frontiers in Econometrics, ed P. E. Zarembka (New York, NY: Academic Press), 105–142.

19. McNeely, J. A. (2006). Using economic instruments to overcome obstacles to in situ conservation of biodiversity. Integr. Zool. 1, 25–31. doi: 10.1111/j.1749-4877.2006.00009.x

20. Narloch, U., Drucker, A., and Pascual, U. (2011). Payments for agrobiodiversity conservation services for sustained on-farm utilization of plant and animal genetic resources. Ecol. Econ. 70, 1837–1845. doi: 10.1016/j.ecolecon.2011.05.018

21. Oldfield, M. L. (1989). The Value of Conserving Genetic Resources. Sunderland, MA: Sinauer Associates Inc.

22. Omondi, I., Baltenweck, I., Drucker, A. G., Obare, G., and Zander, K. K. (2008). Valuing goat genetic resources: a pro-poor growth strategy in the Kenyan semi-arid tropics. Trop. Anim. Health Prod. 40, 583–589. doi: 10.1007/s11250-008-9137-2

23. Ouma, E., Abdulai, A., and Drucker, A. (2007). Measuring heterogeneous preferences for cattle traits among cattle-keeping households in East Africa. Am. J. Agric. Econ. 89, 1005–1019. doi: 10.1111/j.1467-8276.2007.01022.x

24. Rege, J. E. O., and Gibson, J. P. (2003). Animal genetic resources and economic development: issues in relation to economic valuation. Ecol. Econ. 45, 319–330. doi: 10.1016/S0921-8009(03)00087-9

25. Roessler, R., Drucker, A. G., Scarpa, R., Markemann, A., Lemke, U., Thuy, L. T., et al. (2008). Using choice experiments to assess smallholder farmers› preferences for pig breeding traits in different production systems in North-West Vietnam. Ecol. Econ. 66, 184–192. doi: 10.1016/j.ecolecon.2007.08.023

26. Rose, J. M., and Bliemer, M. C. J. (2009). Constructing efficient stated choice experimental designs. Trans. Rev. 29, 597–617. doi: 10.1080/01441640902827623

27. Sælensminde, K. (2006). Causes and consequences of lexicographic choices in stated choice studies. Ecol. Econ. 59, 331–340. doi: 10.1016/j.ecolecon.2005.11.001

28. Silversides, F. G., Purdy, P. H., and Blackburn, H. D. (2012). Comparative costs of programmes to conserve chicken genetic variation based on maintaining living populations or storinf cryopreserved material. Br. Poult. Sci. 53, 599–607. doi: 10.1080/00071668.2012.727383

29. Soini, K., Diaz, C., Gandini, G., de Haas, Y., Lilja, T., Martin-Collado, D., et al. (2012). Developing a typology for local cattle breed farmers in Europe. J. Anim. Breed. Genet. 129, 436–447. doi: 10.1111/j.1439-0388.2012.01009.x

30. Taloustutkimus. (2013). Internet Panel. Available online at: http://www.taloustutkimus.fi/in-english/products_services/internet_panel/ [Accessed 28 August 2013].

31. Thomas, D. L. (2010). Performance and utilization of Northern European short-tailed breeds of sheep and their crosses in North America: a review. Animal 4, 1283–1296. doi: 10.1017/S1751731110000856

32. Tienhaara, A., Ahtiainen, H, and Pouta, E. (2014). "The consumer and citizen roles and motives in valuation of agricultural genetic resources in Finland," in Paper Presented in the 20th Annual Conference of the European Association of Environmental and Resource Economists (Toulouse). Available online at: http://www.webmeets.com/EAERE/2013/m/viewpaper.asp?pid=870

33. Vista, A. B., Rosenberger, R. S., and Collins, A. R. (2009). If you provide, will they read it? Response time effects in a choice experiment. Can. J. Agri. Econ. 57, 365–377. doi: 10.1111/j.1744-7976.2009.01156.x

34. Wunder, S. (2007). The efficiency of payments for environmental services in tropical conservation. Conserv. Biol. 21, 48–58. doi: 10.1111/j.1523-1739.2006.00559.x

35. Zander, K., Signorello, G., De Salvo, M., Gandini, G., and Drucker, A. (2013). Assessing the total economic value of threatened livestock breeds in Italy: Implications for conservation policy. Ecol. Econ. 93, 219–229. doi: 10.1016/j.ecolecon.2013.06.002

Chapter 6

AGRICULTURAL BIOTECHNOLOGY DEVELOPMENT AND POLICY IN CHINA

Jikun Huang and Qinfang Wang

Chinese Academy of Sciences

ABSTRACT

This article provides an overview of China's agricultural biotechnology development policies. Research goals, strategies, priorities, commercialization, and China's organizational framework for agricultural biotechnology development are presented. Included is a description of the evolution of China's biosafety regulations as well as China's research capacity building and public investment—one of the largest public research efforts on agricultural biotechnology in the world.

INTRODUCTION

In the past three years, the growth rate has slowed for acres planted with genetically modified (GM) crops globally, in contrast to its rapid increase in the late 1990s (James, 2002). This slowdown may be due to worldwide conflicting views on biotechnology that not only affect global investment in the biotechnology industry, but also impact farmers' adoption of this technology. Some, but not all, of these issues are present in China; however, China's development of its biotechnology industry has been unique, catalyzed by the active involvement of the public sector.

A survey of China's plant biotechnologists by the authors and their collaborators in 2000 shows that China is developing the largest plant biotechnology capacity outside of North America (Huang, Rozelle, Pray, & Wang, 2002). In 1997, when the National Genetically Modified Organisms (GMO) Biosafety Committee was established, this committee immediately approved 46 cases for field trials, environmental release, and commercialization, which covered 12 GM crops. Among them three cases of cotton, tomato, and petunia were approved for commercialization in certain locations (Huang, Wang, & Keeley, 2001).[1] A number of earlier studies concluded that China adopted a promotional policy to embrace the benefits of biotechnology (Chen, 2000; Huang, Wang, Zhang, & Zepeda, 2001; Paarlberg, 2000). China became one of the world›s leading countries in biotechnology development. China also received criticism from biotechnology opponents for not paying enough attention to biosafety, the environment, consumer and food safety, and the potential impacts of biotechnology on China›s future agricultural trade position.

However, the above perceptions regarding China›s position on agricultural biotechnology lasted for only a few years. In May 2001, China›s State Council decreed a new rule—Regulation on Safety Administration of Agricultural GMOs. And in early 2002, the Ministry of Agriculture (MOA) issued three detailed regulations on the biosafety management, trade and labeling of GM farm products.[2] After these events, China received more criticism than support from both proponents and critics of biotechnology. For example, biotech scientists and biotech industry representatives criticized China's new regulations as too restrictive to provide a favorable environment for the development of biotechnology. They called the period following 1999 as the "winter of biotechnology." Alternatively, Greenpeace and environmental agencies continuously warned China of the potential risks associated with GMOs.

International trade impacts occurred for both imports and exports. New regulations required importers of GM agricultural products to apply for official safety verification approval from China›s Ministry of Agriculture. This led the US government to accuse Beijing of using these new rules to hinder imports and protect Chinese soybean farmers.[3]Pressure was also raised on the export side. China was frequently asked to certify that its agricultural exports to Japan and EU markets were free of GMOs. In addition, there has been growing criticism of China's financial and institutional ability to label its GM farm products.

Additionally, the media has claimed that China had reversed its former enthusiastic embrace of biotechnology by imposing extra restrictions on both domestic and imported varieties of genetically modified crops. These claims stated that China made a decisive shift away from its intentions to become the developing world's leader in biotechnology.[4]After 15 years of nationwide promotion of agricultural biotechnology in China, the current policy debate appears confusing to many observers. The industry wonders whether China will continue to advance its biotechnology, and some scientists question how to proceed in the near future. Given the above background, the objectives of this article are to review the status of China's agricultural biotechnology research and commercialization, and to gain a better understanding of China's policies governing both agricultural biotechnology research and its applications (or commercialization).[5] In order to achieve these objectives, this article is organized as follows. The next section provides an overview of China's agricultural biotechnology development policies. We argue that despite the slight adjustment of GM strategies for commercialization policy in the short run, the overall goal of China's biotechnology development has not been altered. The growth of China's public investment in agricultural biotechnology has not slowed, but instead accelerated. These arguments are further discussed and supported by information provided in the third section of this article, focusing on agricultural biotechnology research capacity building and public investment. The fourth section examines specific cases in China's agricultural biotechnology development, research priorities, and commercialization. The final section provides concluding remarks.

An Overview of China›s Agricultural Biotechnology Development Strategies and Policies

China's leaders have paid great attention to agricultural technology. Among various agricultural technologies, agricultural biotechnology is one of the priority areas that have received the greatest attention. For example, in response to *Science* Editor Ellis Rubenstein's question about concerns in the West regarding GMOs and criticisms of biotechnology, China's President Jiang Zemin stated, «We are also very much concerned about these.... I think it is important to uphold the principle of freedom of science. But advances in science must serve, not harm humankind. The Chinese government is now mulling over new rules and regulations to guide, promote, regulate, and guarantee a healthy development of science. I believe biotechnology—especially gene research—will bring good to humanity...» (Rubenstein, 2000).[6] This statement reflects China's position on biotechnology development: promoting the technology but showing appropriate precaution for biosafety, the environment, food safety, and the commercialization of biotechnology.

Goals and Strategies

Beginning in the early 1980s when China prepared to initiate its national biotechnology program, its biotechnology developmental goals were multifaceted. The government defined its goals of in terms improving the nation's food security, promoting sustainable agricultural development, increasing farmers' income, improving the environment and human health, and raising its competitive position in international agricultural markets along with other public agricultural development programs. And from the point of view of the technology itself, the most frequently stated goal was to create a modern, market-responsive, and internationally competitive biotechnology research and development system in China (Ministry of Science and Technology [MOST], 1990, 2000; State Science and Technology Commission [SSTC], 1990).

To meet these goals, the government's plan to modernize its agricultural biotechnology system was composed of several key measures. These included measures to establish a comprehensive public financed research system, investment to enhance the innovative capacity (both human and physical capacity) of the national biotechnology research program, and creation of institutions and regulations to ensure healthy development of the technology that contributes to human welfare (MOST, 2000).

National Agricultural Biotechnology Research Institutions

The earliest plan to promote biotechnology research was initiated in the beginning of the "Seventh Five-year Plan" (1986-1990) when the first comprehensive National Biotechnology Development Policy Outline was issued (SSTC, 1990). This outline was prepared by more than 200 scientists and officials under the leadership of the Ministry of Science and Technology (MOST), the State Development and Planning Commission (SDPC), and the State Economic Commission in 1985 and further revised in 1986 (Table 1). The outline defined research priorities (see later part of this section), the development plan (e.g., the "863 Plan"), and measures to achieve targets or goals.

Under this outline, a number of high-profile technology programs were launched after the middle 1980s. Some of the most significant programs included the "863 High-tech Plan," the "973 Plan," Natural Science Foundation of China, the Initiative of National Key Laboratories on Biotechnology, the Special Foundation for Transgenic Plants Research and Commercialization, the Key Science Engineering Program, the Special Foundation for High-tech Industrialization (or Commercialization), the Bridge Plan, and others (Table 1).

Table 1: Major policy measures related to biotechnology in China since the early 1980s

Key Breakthrough Science & Technology Projects	Started in 1982 by SDPC. Updated every five years. One of major components of these projects is biotechnology R&D.

Patent system	Patent law promulgated 1985. A total of 1,599 applications on genetic engineering for invention patents were filed between 1985 and 1999.
National Biotechnology Development Policy Outline	Prepared by scientists and officials led by MOST, SDPC, and others in 1985. Formally issued by the State Council in 1988. The Outline defined the research priorities, development plan and measures to achieve targets.
National Key Laboratories (NKLs) on Biotechnology	Started in 1985 under MOST. Thirty National Key Laboratories in biotechnology (15 on agriculture or agriculture related) have been established. NKLs are open laboratories, inviting both domestic and international visiting fellows.
The Climbing Program	A National Program for Key Basic Research Projects, including biotechnology program, initiated in the early 1980s.
High Technology Research and Development Plan (863 Plan)	Approved in March 1986 with 10 billion RMB for 15 years to promote high-technology R&D in China. Biotechnology is one of seven supporting areas, with a total budget of about 1.5 billion RMB from 1986-2000.
Natural Science Foundation of China	Established in 1986 to support basic science research. Life science and agronomy are two support areas related to agrobiotechnology.
Biosafety regulations	MOST issued the Biosafety Regulations on Genetic Engineering in July of 1993, which include the biosafety grading and safety assessment, application and approval procedure, safety control measures, and legal regulations.
Agricultural biosafety regulations	MOA issued the Safety Administration, Implementation, and Regulations on Agricultural Biological Genetic Engineering in July 1996.

973 Plan	Initiated in March 1997 to support basic science and technology research. Life science is one of the key supporting areas.
Agricultural GMO Biosafety Committee	Ministry-level Agri GMO Biosafety Committee was set up in MOA in 1997. The Committee was updated in 2002 to national level with its office in MOA.
Special Foundation for Transgenic Plant Research and Commercialization	A five-year program launched in 1999 by MOST to promote the research and commercialization of transgenic plants in China. The total budget of this program in the first five years is 500 million RMB.
Key Science Engineering Program	Started in the late 1990s under MOST and SDPC to promote basic research, including biotechnology program. The first project on biotech (crop germplasm and quality improvement) was funded in 2000 with 120 million RMB.
Foundation for high-tech commercialization	A special program supported by the SDPC to promote the application and commercialization of technologies, started from 1998.
Seed Regulation and Law	Regulation on the Protection of New Varieties of Plants was issued in 1999. The first Seed Law was issued in 2000.
Updated and amended agricultural biosafety regulations	1996 MOA's biosafety regulation was amended and issued by the State Council in May 2001. Three regulations on the biosafety management, trade, and labeling of GM farm products were issued by MOA to take effect after March 20, 2002.

Foreign investment in GMOs	In April 2002, the SDPC, State Economic and Trade Commission, and MOTEC jointly issued a Guideline List of Foreign Investment, which puts GMO as a prohibited area for foreign investment

The 863 Plan, also called National High-Tech Research and Development Plan, was approved in March 1986. The 863 Plan supports a large number of applied as well as basic research projects with a 10 billion RMB yuan budget (equivalent to US$ 3 billion, based on the official exchange rate of 3.4 in 1985, or US$ 1.2 billion, based on the official exchange rate of 8.27 in 2000) over 15 years to promote high technology research and development (R&D) in China. Biotechnology is one of seven supporting areas, with a budget of 1.3 billion RMB yuan in 1986-2000, with 50% of this budget focused on agricultural biotechnology.

The National Basic Sciences Initiative, also called the 973 Plan, with a total budget of 2.5 billion yuan (US$ 302 million, converted at the1997-2002 average exchange rate) in the period of 1997-2002, was another high-tech research plan initiated in March 1997. This plan is complementary to the 863 and many other national initiatives on high-tech development, as it exclusively supports basic research. Life science, with biotechnology as a priority, constitutes one of the key programs under this plan.

In contrast to the perception that China's biotechnology development is shifting towards a "go slow" approach, our review of recent biotechnology research programs indicates that China instead has accelerated its biotechnology development since the late 1990s. The view suggesting that progress in biotechnology research has slowed is unfounded. For example, a new program aimed at strengthening the national research and industrialization of China's agricultural biotechnology, the Special Foundation of Transgenic Plants Research and Commercialization (SFTPRC), was initiated in 1999 by the Ministry of Science and Technology. This new program is a unique foundation to promote both research and commercialization of transgenic plants. Only those projects that are jointly submitted by research institutes and companies are eligible to receive funding from about half of the programs under SFTPRC. The foundation also requires a significant financial commitment from companies to commercialize technology generated by a project, a reflection of China's aim to accelerate the diffusion of biotechnology. The total budget of SFTPRC during its first five years (1999-2003) was 500 million RMB yuan (about US$ 60 million). Concurrently, the Ministry of Science and Technology and the State Development and Planning Commission jointly sponsored the Key Science

Engineering Program (KSEP), a national program to promote the fundamental construction for research in the late 1990s. As an example, one extremely large biotechnology project on crop germplasm and quality improvement through biotechnology received 140 million RMB yuan (US$ 17 million) from KSEP in 2000. Moreover, the State Council passed a new Agricultural Science and Technology (S&T) Development Compendium in 2001. The compendium reemphasizes the importance of agricultural biotechnology in improving the nation's agricultural productivity, food security, and farmers' income, and has led to a new decision to further increase the research budget for the development of biotechnology. The proposed biotechnology development budget for the Tenth Five-year Plan (2001-2005) is far more than all prior budgets over the past 15 years (see the next section for more detail).

With the above efforts, by 2001 there were about 150 laboratories at national and local levels located in more than 50 research institutes and universities across China working on agricultural (plant and animal) biotechnology. Over the last two decades, China established 30 National Key Laboratories (NKL). Among these NKLs, 12 are exclusively working on, and three have major activities in agricultural biotechnology (Huang, Wang, Zhang, & Zepeda, 2001). Besides NKLs, there are numerous Key Biotechnology Laboratories and programs within ministries and local provinces.

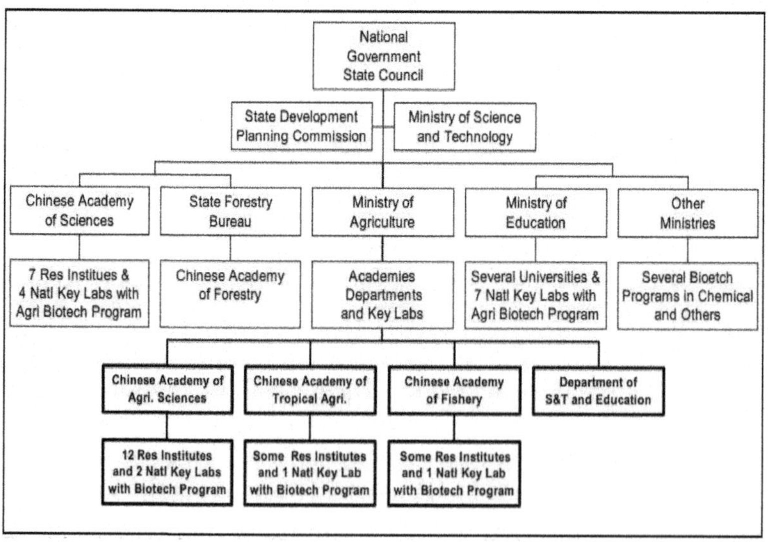

Figure 1: Organization chart for agricultural biotechnology research at national level.

At the national level, the Ministry of Agriculture (MOA), the Chinese Academy of Sciences (CAS), the State Forestry Bureau (SFB), and the Ministry of Education (MOE) are the major authorities responsible for agricultural biotechnology research (Figure 1). Under the Ministry of Agriculture, there are three large academies—the Chinese Academy of Agricultural Sciences (CAAS, which employs about 8,000 research and support staff), the Chinese Academy of Tropical Agriculture (CATA), and the Chinese Academy of Fisheries (CAFi). Among the 37 institutes in CAAS, there are 12 institutes, two National Key Laboratories and five Key Ministerial Laboratories conducting biotechnology research programs. The CAFi and the CATA also have several biotechnology laboratories or programs, and each has one NKL for biotechnology.

Agricultural biotechnology research is also conducted by national institutes external to the Ministry of Agriculture's research system. For example, under the Chinese Academy of Sciences there are at least seven research institutes and four NKLs that focus on agricultural biotechnology. Research institutes within the Chinese Academy of Forestry (CAFo) under the State Forest Bureau and numerous universities (i.e., Beijing University, Fudan University, Nanjing University, Central China Agricultural University, and China's Agricultural University) under the Ministry of Education are examples of other institutions conducting agricultural biotechnology research. There are seven NKLs located in seven leading universities conducting agricultural biotechnology or agriculturally related basic biotechnology research. Other public biotechnology research efforts on agriculturally related topics include agrochemical (e.g., fertilizer) research by institutes in the State Petro-Chemical Industrial Bureau.

Although the programs at the national level presented in Table 1 and Figure 1 and discussed above constitute China's mainstream agricultural biotechnology research, research at the provincial level also contributes to the development of China's agricultural biotechnology. They follow a similar institutional framework to that at the national level (Figure 1). Each province has its own provincial academy of agricultural sciences and at least one agricultural university. Each academy or university at the provincial level normally has one or two institutes or laboratories focused on agricultural biotechnology. Provincial biotechnology research is funded by both local governments (core funding and research projects) and the central government (research projects only).

Finally, it is worth noting that the numbers of both national and provincial biotechnology programs and institutes continue to increase. China is even considering establishing a new national agricultural biotechnology research center—a megaresearch center over the current 150 agricultural biotech laboratories. Based on these developments, if there were shifts in China›s biotechnology developmental plan, it is towards probiotechnology research.

Biosafety Management Institutions and Regulations

Institutional Setting. Although the Ministry of Science and Technology is mainly responsible for biotechnology research, the Ministry of Agriculture is the primary institution in charge of the formulation and implementation of biosafety regulations on agricultural GMOs and their commercialization, particularly after 2000. In order to incorporate representation of stakeholders from different ministries, the State Council established an Allied Ministerial Meeting comprised of leaders from the MOA, the SDPC, the MOST, the Ministry of Public Health, the Ministry of Foreign Economy and Trade (MOFET), the Inspection and Quarantine Agency, and the State Environmental Protection Authority (SEPA). This Allied Ministerial Meeting coordinates key issues related to biosafety of agricultural GMOs, examines and approves the applications for GMO commercialization, determines the list of GMOs for labeling, and establishes import or export policies for agricultural GMOs and their products.

However, routine work and daily operations are handled by the Office of Agricultural Genetic Engineering Biosafety Administration (OGEBA). The National Agricultural GMO Biosafety Committee (BC) is the major player in the process of biosafety management.[7]Currently, the Committee comprises of 56 members.[8] The committee meets twice each year to evaluate all biosafety assessment applications related to experimental research, field trials, environmental release, and commercialization of agricultural GMOs. It provides approval or disapproval of recommendations to OGEBA based on the results of its biosafety assessments. OGEBA is responsible for the final approval of decisions.

The Ministry of Public Health (MPH) is responsible for food safety management of biotechnology products. The Appraisal Committee, consisting of food health, nutrition, and toxicology experts nominated by MPH, is responsible for reviewing and assessing GM foods as they have been designated a Noval Food. The State Environmental Protection Authority participates in GMO biosafety management through the Allied Ministerial Meeting and through their members on the National Agricultural GMO Biosafety Committee. Although SEPA has taken the responsibility of international biosafety protocol, its focus on biotechnology in China is limited to biodiversity.

Concerning the institutional setting of agricultural GMO biosafety management, China has several unique elements compared to the US and the EU. The Ministry of Agriculture in China appears to have more power than its counterparts in the US and the EU. The leaders in the State Council of the previous government believed that the MOA is more familiar with, and has more expertise in agriculture and agricultural GMOs than any other ministry. Moreover, because the MOA is also in charge of pesticide use and its environmental assessment in agricultural production, national leaders consider the MOA a major player in China›s agricultural biosafety management.

Critics (i.e., SEPA) of this system argue that this institutional setting might result in less attention paid to environmental risks of GMOs and may have a potential conflict of interest, as the MOA is primarily responsible for agricultural production, and many biotechnologies are developed under the MOA›s own research system. The debate on whether SEPA or MOA is a more appropriate institution to take a lead role on biosafety has continued since the biosafety management system was set up in 1997. However, under the current national administrative system, it is unlikely for SEPA to take a significant role in biotechnology, unless there is significant reform of the government›s structure by China›s new leaders, who have been in office since early 2003.

The other unique aspect is that China›s National Agricultural GMO Biosafety Committee plays a critical role in the biosafety decision-making process. As most of its 56 current members (29 for GM plants, nine for recombined microorganisms for plant, 12 for transgenic animals and recombined microorganisms for animals, and six for GM aquatic organisms) are experts from various research institutes within the public sector, its GMO biosafety assessment provides key information for decision makers on whether OGEBA should approve or disapprove GMO application cases. However, the weakness of this approach is the time constraint from BC members who often are leading scientists in various disciplines. Other concerns include the heavy service burden of a few key individual scientists and too many biotechnologists on the Biosafety Committee.

Biosafety Regulations. Before 2002, the principle governing China›s agricultural GMO biosafety was to adopt a product-based GMO management system. However, China has attempted to impose labeling regulations on GMOs and GM products since March 2002. By imposing a compulsory labeling policy on GMOs, China›s biosafety management partially shifts towards a process-based GMO management system. This adjustment has led to wide debate within China and between China and many other countries, as we described above in the introduction. Before we discuss this new labeling policy, it is worth reviewing briefly the evolution of China›s agricultural GMO biosafety regulations and policies in the past.

Evolution of China›s Biosafety Regulations. In response to the emerging progress in China›s agricultural biotechnology, the first biosafety regulation, «Safety Administration and Regulation on Genetic Engineering,» was issued by the Ministry of Science and Technology in 1993. This regulation consisted of general principles, safety categories, risk evaluation, application and approval, safety control measures, and legal responsibilities. After the above regulation was decreed, MOST required relevant ministries to draft and issue corresponding biosafety regulations on biological engineering (i.e., the Ministry of Agriculture for agriculture and the Ministry of Public Health for food safety). Following MOST›s guidelines, the MOA issued the Implementation Regulations on Agricultural Biological Engineering in 1996. This regulation is similar in many aspects to the US GMO biosafety regulations. Labeling was not part of this regulation, nor was any restriction imposed on imports or exports of GMO products. The regulation also did not regulate processed food products that use GMOs as inputs.

Under the 1996 GMO biosafety regulation policy, OGEBA received 433 applications for field trials, environmental release, or commercialization in 1997-2000 (Table 2). Among them, 322 cases had been approved, covering more than 60 crops and several animals, as well as numerous microorganisms. It is interesting to note that both the number of cases applied or submitted and cases approved increased persistently over time. Imposing more GMO restrictions did not reduce the number of applications. However, if we decompose this data into different stages of GMO development and by crop, they do show that the numbers of cases approved for commercialization declined in 2000 and that no new GMO crops have been approved since 1999, excluding cotton, tomatoes, sweet peppers, and petunias.

China›s Stance on Biotechnology Development—For or Against? Using solely the above approval case numbers may lead to erroneous conclusions regarding China's stance on GMO development. Indeed, the small number of approvals for commercialization of GM crops in 2000 (seven cases, Table 2) was due to many factors. First, in the prior year, 1999, a large number of cases were approved for commercialization (27 cases, Table 2), almost all for Bt cotton. As expected, there were fewer applications for Bt cotton commercialization in 2000, as most of the Bt cotton varieties (nearly 20 varieties from both CAAS and Monsanto) had been earlier approved for commercialization. Second, as argued by OGEBA and the Biosafety Committee, the existing food-related GM crops were not ready for commercialization due to unclear issues over their food safety. For example, food safety testing managed by Ministry of Public Health has not come to a conclusion on whether the current GM rice is not substantially different from non-GM rice. Research on GM rice›s food safety is still ongoing. Third, the testing on environmental safety and biodiversity has been limited to a very small scale and in few locations. Lastly, Bt cotton had been tested and adopted widely in other countries before China approved its commercialization, and given that cotton is a nonfood crop, the context surrounding GM rice differs than that for cotton. Rice is the most important food crop in both China and the rest of Asia. Additionally, GM rice has never been commercialized anywhere in the world.

Table 2: The number of cases in agricultural (plant, microorganism, and animal) biotechnology submitted and approved for field trials, environmental release, and commercialization in 1997-2000

	1997	1998	1999	2000	To-tal
Submitted					
Field trial	14	41	28		
Environmental release	37	18	63		
Commercialization	6	9	35		
Total	57	68	126	182	433
Approved					
Field trial	12	40	22	115	189
Environmental release	30	10	34	19	93
Commercialization	4	2	27	7	40
Total	46	52	83	141	322

More recently, our communication with OGEBA's officials and members of the Biosafety Committee reveal that China is badly in need of institutional and capacity building for GMO biosafety management. During the 7th International Symposium on the Biosafety of Genetically Modified organisms held in Beijing in October 2002, an official from OGEBA concluded his speech with five major challenges that OGEBA currently faces: "an appropriate regulatory approach *(to improve current practices),*[9] a science-based safety assessment, capacity building, transparency, communication and information exchange (Cheng & Peng, 2002).

Given the above discussion, it is no surprise that OGEBA declined three applications for GM rice commercialization in 1999-2000. We believe that China›s current adjustment in biosafety management is just one effort to establish a more comprehensive GMO biosafety management system that provides a firm base for future sustainability. Current adjustment is also partly in response to the growing worldwide debate on GMOs and their potential risks, as well as China›s agricultural trade.[10]

Chinese policymakers are concerned about environmental and food safety in response to the debate on the potential risks of GMOs recently raised by the Chinese media. The debate in China has involved scientists, government officials, and newspaper reporters; responses and reactions vary among stakeholders and change over time as more information becomes available on biotechnology (Huang, Wang, & Keeley, 2001). A consensus seems to be growing in China that the most important task a scientist or biotechnologist can do is to reduce the potential negative effects and demonstrate the safety of GMOs.

As a consequence of this consensus, research budgets allocated to biosafety management and the study of biosafety have increased. Since 1999-2000, nearly all biotechnology research programs have expanded their scope into biosafety issues, particularly for the following programs: 863, 973, and the Special Foundation for Transgenic Plants Research and Commercialization. A number of national institutes under the Ministry of Agriculture, the Ministry of Public Health and the State Environmental Protection Agency have launched various biosafety programs, including capacity building for biosafety management and risk assessment, research studies on environmental safety and food safety, detection technology for GMOs and GMO products, and monitoring of international practices.

However, arguing for a more comprehensive and science-based safety assessment as reasons for the recent adjustment of China's GMO commercialization does not imply that there is no concern over the impacts of GMO development on agricultural trade. Issues such as labeling of GM products and possible trade barriers resulting from biotechnology concerns in countries that follow precautionary and preventive policies do have impacts on the current (short run) pace of GMO commercialization in China. Agricultural trade had been an important contributor to the aggregate Chinese economy and trade.

It appears that international trade concerns may have been one of the important factors, but not the dominant factor, in recent agricultural biotechnology policy processes. The critical event here appears to have been the EU's decision to ban Chinese soy sauce imports produced with GM soybeans imported from the United States. Additionally, the recent decision by Thailand (the world's leading rice exporter) to halt further development of GM rice may also have been significant. It is unclear whether public attitudes toward GMOs in Europe are now softening or whether policies may soon change; hence, a short-run "wait and see" tactic is probable in China.

New Biosafety Regulations. In response to the above concerns, in May 2001 the State Council decreed a new and general rule of Regulation on Safety Administration of Agricultural GMOs to replace an early regulation issued by the Ministry of Sciences and Technologies in 1993 (Safety Administration Regulation on Genetic Engineering, Table 1). The Ministry of Agriculture then announced three new implementation regulations on biosafety management, trade, and labeling of GM farm products that were planned to take effect after March 20, 2002.[11] There were several important changes to existing procedures included in these guidelines, as well as details of regulatory responsibilities after commercialization. These included the addition of an extra preproduction trial stage prior to commercial approval, new processing regulations for GM products, labeling requirements for marketing, new export and import regulations for GMOs and GMO products, and local- and provincial-level GMO monitoring guidelines. In the meantime, the Ministry of Public Health also promulgated its first regulation on GMO food hygiene in April 2002, to take effect after July 2002.

By late 2002, the system of biosafety regulation in China had clearly become progressively more elaborate and sophisticated. Many provinces have established provincial biosafety management offices under provincial agricultural bureaus. These biosafety management offices collect local statistics on and monitor the performance of research and commercialization of agricultural biotechnology in their provinces and assess and approve (or disapprove) all applications of GM related research, field trials, and commercialization in their provinces. Only those cases that are approved by provincial biosafety management offices are submitted to the National Biosafety Committee for further assessment. However, China still has a long way to go before all decreed regulations could be fully implemented. Our three years of Bt cotton farm surveys across five provinces during 1999-2001 found that about half of Bt cotton varieties had been adopted by farmers, but they did not apply to the National Biosafety Committee for commercialization. Seeds are distributed to farmers mainly by local seed companies, the extension system, research institutes, and small traders. The institutions, human capacity, and financial support for implementation of GMO regulations are far away from the necessary requirements. In addition to this, collaboration and coordination between ministries on research, commercialization, and biosafety management needs to be further strengthened.

Agricultural Biotechnology Capacity Building and Public Investment

Creation of a modern and internationally competitive biotechnology research and development system requires substantial investments in human and financial capacities. Since the early 1980s, China's public investment in, and the number of research staff working on biotechnology has increased significantly, in contrast to stagnating trends for general agricultural research expenditures in the late 1980s and early 1990s (Huang, Hu, & Rozelle, 2002). For example, based on our 2000 survey of 29 research institutes in plant biotechnology[12] and on extensive interviews with ministries and research institutes in 2002, we estimate that the number of plant biotechnology researchers tripled in the past 15 years (Table 3). More than 2,100 researchers are now working on plant biotechnology alone. If we include biotechnology from the animal sector, the number of agricultural biotechnology researchers may reach 3,000, and may be one of the largest biotechnology research efforts in the world.

Table 3: Estimated number of research staff and expenditures on plant biotechnology research in China, 1986-2000

Year	Number of staff	Research expenditure		
		Million RMB at current price	Million RMB at 2000 price	Million US$
1986	740	14	38	4.2
1990	1067	40	68	8.3
1995	1447	88	87	10.5
2000	2128	322	322	38.9

Note. Expenditures include both project grants and costs related to equipment and buildings. Both staff and research expenditures are estimated by the authors based on our earlier studies (Huang, Wang, Zhang, & Zepeda, 2001) and recent interviews in China. The results from our recent interviews show that the data in Table 2 are higher than our earlier estimates. Official exchange rate in the corresponding year is used to convert the domestic currency to US dollars.

Similar to other agricultural research programs in China, agricultural biotechnology research is primarily built upon research institutes. Among the 29 institutes surveyed, the number of agricultural biotechnology researchers in universities accounted for only 10% of total research staff.[13] Among total researchers, nearly 60% are professionals, and the share of the professional staff has been increasing over time (Huang, Wang, Zhang, & Zepeda, 2001), again indicating growing human capacity in biotechnology research.

The quality of human capacity to conduct biotechnology research has improved over time. Among professional staff, the share of researchers with Ph.D. degrees increased from only 2% in 1986 to more than 20% in 2000. This share is expected to continue to increase in the future. Although the share of researchers with biotechnology Ph.D. degrees is still low by international standards, it is interesting to note that this share is much higher than those in the general agricultural research system. In China›s national agricultural research system, Ph.D. researchers accounted for only 1.1% of the total professional staff in 1999 (Huang, Hu, & Rozelle, 2001).

Even more dramatic growth has occurred in China›s biotechnology research investment (Table 3). China›s biotechnology research investment was trivial in the early 1980s (MOST, 1990). Although there are no statistics available from official sources, our estimates show that biotechnology investment has grown substantially. For example, the estimated investment in plant biotechnology research was only US$ 4.2 million in 1986 when China formally started its 863 Plan (Table 3). By 1990, China›s investment grew to US$ 8.3 million. During this period, the research project budget nearly tripled, and equipment expenses nearly doubled (Huang, Wang, Zhang, & Zepeda, 2001). Although the growth rate of biotechnology research investment slowed between 1990 and 1995 (this is expected as the large investment in biotechnology equipment was nearly complete in the early 1990s), the annual growth rate in the research project budget in real terms remained as high as 10% during this period.

China›s biotechnology research investment increased considerably from US$ 10.5 million in 1995 to US$ 38.9 million in 2000, representing an annual growth rate of about 30%. This investment in China›s biotechnology is mainly due to government sources. According to our survey of 29 biotech research institutes, public investment accounted for 94% of the total plant biotechnology budget in 1999, and this share has been increasing over our study period, from 1986 to 1999 (Huang, Wang, Zhang, & Zepeda, 2001). Budgets from competitive grants for research projects accounted for two thirds of the total budget and this share also has shown an increase over time, reflecting China›s biotechnology development moving from a capacity-building stage to a research stage. Our recent interviews with officials and research administrators from the Ministry of Science and Technology confirm that the Ministry is accelerating its investment in national biotechnology program: the Tenth Five-year Plan (2001-2005) for biotechnology development. Under this plan, the total investment in agricultural biotechnology is targeted to be four times as much as the total amount spent on agricultural biotechnology in the past 15 years (1985-2000). If this goal is realized, China will account for more than one fourth of the world›s current public spending on agricultural biotechnology.

Agricultural Biotechnology Development, Research Priorities and Commercialization

An Overview

The focus of China's biotechnology development in its early stages (in the early 1970s) was on cell engineering, tissue culture, and cell fusion and emphasized crops such as rice, wheat, maize, cotton, and vegetables (Key Laboratory of Crop Molecular and Cell Biology, 1996). However, the most significant progress in agricultural biotechnology was made following the development of transgenic techniques after 1983. The pace of biotechnology research accelerated significantly after China initiated the 863 Plan in 1986 (Table 1). Bt cotton is a most successful story of agricultural biotechnology in China. In response to rising pesticide use and the emergence of a pesticide resistant bollworm population in the late 1980s, China's scientists began research on GM cotton. Starting with a synthesized gene originally from the bacterium *Bacillus thuringiensis* (Bt), China›s scientists transferred this modified Bt gene into major cotton cultivars by the so-called pollen tube pathway transformation. Greenhouse testing began in the early 1990s. The first commercial use of GM cotton was approved in 1997. During the same year, Bt cotton varieties from publicly funded research institutes and from a joint venture with Monsanto became available to farmers. The release of Bt cotton began China›s first large-scale commercial experience with a product of the nation›s biotechnology research program.

In addition, other transgenic plants with resistance to insects, disease or herbicides, stress tolerance, or plants with improved quality have been approved for field release, and some are nearly ready for commercialization. These include transgenic cotton lines resistant to fungal disease, rice resistant to rice stem borer or bacteria blight, diseases, herbicide, and salt tolerance, wheat resistant to barley yellow dwarf virus (Cheng, He, & Chen, 1997), maize resistant to insects and with improved quality (Zhang, Liu, & Zhao, 1999), poplar trees resistant to gypsy moths, soybeans resistant to herbicides, transgenic potato resistant to bacterial disease or Colorado beetles, among others (Ministry of Agriculture [MOA], 1999; National Center of Biological Engineering Development [NCBED], 2000; Li, 2000).

Progress in plant biotechnology has also been made in recombinant microorganisms such as soybean nodule bacteria (nitrogen-fixing bacteria for rice and corn) and phytase from recombinant yeasts for feed additives (Huang, 2002). Genetically modified nitrogen-fixing bacteria and phytase have been commercialized since 1999. In animals, transgenic pigs and carp have been produced since 1997 (NCBED, 2000). Recently, Chinese researchers also announced the successful sequencing of the rice genome (Yu et al., 2002). They have produced a draft sequence of the rice genome for the most widely cultivated subspecies in China, *Oryza sativa L.* ssp. *indica*, by whole-genome shotgun sequencing.

According to a nationwide survey conducted by the MOA in 1996, Chinese scientists have tried to use more than 190 genes transferring to more than

100 organisms (103 genes used in 47 plants, 32 genes used in 22 animals, 56 genes used in 31 species of microorganisms). These figures have been further expanded after 1996 (Cheng & Peng, 2002). By 2001, there were more than 60 plants under research and 121 genes used for transformation (Peng, 2002). The list of GM crops in trials is also impressive and differs from those being worked on in other countries.

Research Priorities and Products in the Research Continuum

Huang, Wang, Zhang, and Zepeda (2001) summarized research priorities for plant biotechnology identified in various Biotechnology Development Outlines over the past 15 years in China (Table 4). Since the mid-1980s, cotton, rice, wheat, maize, soybean, potato, and rapeseed have been consistently listed as priority crops for biotechnology research funding. Functional genomics for major plants and animals, crop genetic breeding through the application of gene transformation, chromosome hybridization and marker assisted selection, cultivation of crops with improved resistance or quality and genetic breeding of animals, use of animal and plant cells as bioreactors in producing secondary products, special proteins and vaccines, recombined microorganisms to produce biofertilizers and biopesticides, and others have been identified as priority technologies for public funding.

Among crops, cotton is listed as a priority crop not only because of its importance by sown area and its contributions to the textile industry and trade, but also because of the serious problems with the associated rapid increase in pesticide applications to control insects (i.e., bollworm and aphids). Per-hectare pesticide expenditures for cotton production in China increased considerably over recent decades, reaching 834 RMB yuan (approximately US$ 100) in 1995. This amount is much higher than comparable expenditures for grain crop production but lower than horticultural production. Cotton production alone consumed about US$ 500 million annually in pesticides in recent years.

Rice, wheat, and maize are the three most important crops in China. Each accounts for about 20% of the total area planted. Production and market stability of these three crops are a primary concern of the Chinese government, as they are central to China's food security. National food security, particularly related to grains, has been a central goal of China's agricultural and food policy and has been incorporated into biotechnology research priority setting.

Among all traits, pest resistance traits have top priority (Table 4). Recently, quality improvement traits have been included as priority traits in response to increased market demand for quality foods. Quality improvements have been targeted particularly for rice and wheat, as consumer income rises in China. In addition, stress tolerance traits—particularly resistance to drought—are gaining attention, particularly with the growing concern over water shortages in Northern China. In addition, Northern China is a major wheat and soybean production region with significant implications for China's future food security and trade.

Table 4: Research focus of plant biotechnology programs in China

	Prioritized areas
Crops	Cotton, rice, wheat, maize, soybean, potato, rape-seed, cabbage, tomato
Traits	
Insect resistance	Cotton bollworm, boll weevil, and aphids Rice stem borer Wheat aphids Maize stem borer Soybean moth Potato beetle Poplar gypsy moth
Disease resistance	Rice bacteria blight and blast Cotton fungal disease Cotton yellow dwarf Wheat yellow dwarf and rust Soybean cyst nematode Potato bacteria wilt Rapeseed sclerosis CMV and TMV
Stress tolerance	Drought, salinity, cold
Quality improvement	Cotton fiber quality Rice cooking quality Wheat quality Maize quality Corn with phytase or high lysine
Herbicide resistance	Rice, soybean
Functional genomics	Rice, rapeseed, and arabidopsis

Newer research focuses on the isolation and cloning of new disease and insect resistance genes, including the new genes conferring resistance to cotton bollworm (Bt, CpTI and others), rice stem borer (Bt), rice bacterial blight (Xa22 and Xa24), rice plant hopper, wheat powdery mildew (Pm20), wheat yellow mosaic virus, and potato bacterial wilt (cecropin B) (MOA, 1999; NCBED, 2000). These genes have been applied in plant genetic engineering since the late 1990s. Significant progress has also been made in the functional genomics of arabidopsis and in plant bioreactors, especially in utilizing transgenic plants to produce oral vaccines (Biotechnology Research Institute, 2000).

By the end of 2001, GM plants from 13 plant species and more than 50 genes were approved for field trial, environmental release, and commercialization. Thirty-six recombined microorganism species and 51 strains have been involved in research with 89 genes for insect and disease resistance or nitrogen fixation.

Commercialization of Agricultural Biotechnology

By 2002, 18 transgenic cotton varieties generated by Chinese institutions and five varieties from Monsanto with resistance to bollworm have been approved for commercialization in China. Although several GM varieties of tomato, sweet pepper, chili pepper, and petunia have also been approved for commercialization since 1997, the area planted with these four crops remains small. Personal communications with several member of the agricultural Biosafety Committee show that the economic benefits of adopting the current three GM crops are minimal or nonexistent; no private companies have been attracted to invest in their commercialization.

Table 5 presents our most updated estimates of Bt cotton areas sown in China in 1997-2001. After the Bt cotton variety was approved for commercialization in 1997, the total area planted using Bt cotton increased to 0.65 million hectares in 1999. In 2001, the area reached more than 2 million hectares and accounted for 45% of China›s cotton area. China›s GM crop area follows that of the US, Argentina, and Canada. Although less than 4% of the total global area of GM crops was grown in China in 2001, we estimate that nearly 5 million Chinese farmers planted Bt cotton, as the average farm size is only about 0.5 hectares and includes several crops.

Table 5: Bt cotton adoption in China, 1997-2001

	Cotton area (000 hectare)		
Year	**Total**	**Bt cotton**	**Bt cotton share (%)**
1997	4491	34	1
1998	4459	261	6
1999	3726	654	18
2000	4041	1216	30
2001	4810	2174	45

Concluding Remarks

Chinese policymakers consider agricultural biotechnology as a strategically significant tool for improving national food security, raising agricultural productivity, and creating a competitive position in international agricultural markets. Consistent with these aims, China also intends to be one of world leaders in biotechnology research and major domestic supplier of biotechnologies. This objective is closely linked to the perception by Chinese policymakers that there are risks associated with reliance on imported technologies to guarantee national food security. Despite the growing debate worldwide on GM crops, China has developed agricultural biotechnology decisively since the mid-1980s. By 2001, China had the fourth largest sown area of GM crops in the world. Research and development has continued apace, and China now has several genetically modified plants that are in the pipeline for commercialization.

The institutional framework for supporting agricultural biotechnology research program is complex both at national and local levels. The growth of government investment in agricultural biotechnology research has been remarkable. However, coordination among institutions and consolidation of agricultural biotechnology programs will be essential for China to create an even stronger and more effective biotechnology research program in the future.

Examination of the research foci of agricultural biotechnology research reveals that food security objectives and farmers' current demands for specific traits and crops have been incorporated into priority setting. Moreover, the current priority setting for investments in agricultural biotechnology research has been directed at commodities for which China does not have a relative comparative advantage in international markets (such as grain, cotton, and oil crops). This implies that China is targeting its GMO products at the domestic market. The emphasis on developing drought-resistant and other stress-tolerant GM crops also suggests that biotechnological products are not only being geared to high-potential areas, as critics argue, but also at the needs of poorer farmers.

Many competing factors are exerting pressure on Chinese policymakers to continue with research and commercialization of transgenic crops. The demand of producers (for productivity-enhancing technology) and consumers (for cost savings), the current size and rate of increase of research investments, and past success in developing technologies suggest that products from China's plant biotechnology industry are likely to become widespread in China in the near future. Although China is still struggling with issues of environmental and consumer safety, and the system of biosafety regulation has become progressively more elaborate and sophisticated, the system might not work well and might eventually hurt its national biotechnology application in the future if biosafety management capacity is not improved as much as research capacity. Investment in China's biotechnology R&D is essential for the nation to promote its biotechnology industry; investment in biosafety management capacity and policy implementation are also critical factors for health and sustainable development of this industry.

ENDNOTES

[1] In 1998, GM sweet peppers were approved for commercialization.

[2] Also see Marchant, Fang, and Song (2002), in this issue, for information on the evolution of China›s agricultural biotechnology policies.

[3] In 2001, China imported about 14 million metric tons of soybeans from the US, Argentina, and Brazil. Most of these imports were Roundup Ready soybeans. After two months of intensive negotiations between China and the US, an interim agreement was reached in early 2002. China in effect temporarily waived its import and export regulations of GMOs until December 2002, and this was further postponed to September 2003. Concurrently, China has agreed to recognize US assurances that its soybeans are safe for human consumption.

[4] See the recent report in the *Washington Post* (Goodman, 2002), the *New York Times*(Kahn, 2002), and a front-page article in *China Daily* (Zhigang, 2002).

[5] Issues related to impacts of biotechnology are not discussed in this paper. They can be found in a series of papers written by the authors with their collaborators, including Pray, Ma, Huang, and Qiao (2001); Huang, Rozelle, Pray, and Wang (2002); Huang, Hu, Rozelle, Qiao, and Pray (2002); Huang, Hu, Pray, Qiao, and Rozelle (in press); and Pray, Huang, and Rozelle (2002).

[6] In his opening speech at the International Rice Conference held in Beijing on September 15, 2002, President Jiang Zemin restated the importance of agricultural biotechnology in boosting agricultural productivity growth and food security.

[7] The Biosafety Committee was established in 1997 under the Ministry of Agriculture; it was a ministry-level institution. Since June 2002, the Committee was upgraded to a national-level institution.

[8] Biosafety Committee members work part-time for the BC and are scientists from different disciplines including agronomy, biotechnology, plant protection, animal science, microbiology, environmental protection, and toxicology. A few members are also agricultural administrations. All BC members are nominated by the Ministry of Agriculture.

[9] Because biosafety management is a new activity for OGEBA, it is

understandable that they are seeking a more appropriate approach even after years of commercialization of nonfood crops (Bt cotton).

[10] So far, Chinese consumers have not created many problems for GMO development in China.

[11] These three new regulations replaced the Safety Administration, Implementation, and Regulation on Agricultural Biological Genetic Engineering issued by the Ministry of Agriculture in July 1996.

[12] The survey was conduced by the Center for Chinese Agricultural Policy and the International Service for National Agricultural Research; detailed results are reported in Huang, Wang, Zhang, and Zepeda (2001).

[13] In terms of the overall agricultural research system in China, researchers in universities account for about 8% of the nation's total agricultural researchers.

REFERENCES

1. Biotechnology Research Institute. (2000). *Research achievements of biotechnology*(working paper). Beijing: Chinese Academy of Agricultural Sciences.
2. Chen, Z. (2000, July-August). *Review of R&D on plant genetic engineering in China*. Paper presented at the China-Asean workshop on transgenic plants, Beijing, China.
3. Cheng, Z, He, X., & Chen, C. (1997). Transgenic wheat plants resistant to barley yellow dwarf virus obtained by pollen tube pathway-mediated transformation. In *Chinese Agricultural Science for the Compliments to the 40th Anniversary of the Chinese Academy of Agricultural Science* (pp. 98-108). Beijing: China Agricultural SciTech Press.
4. Cheng, J., & Peng, Y. (2002, October). *Biosafety regulation in China*. Paper presented at the 7th International Symposium on the Biosafety of Genetically Modified Organisms, Beijing, China.
5. Goodman, P. (2002, September 25). China's new economy begins on the farm: Growers bear burden of being first as trade brings opportunity, risk. *Washington Post*, p. A01.
6. Huang, D. (2002, October). *Research and development of recombinant microbial agents and biosafety consideration in China*. Paper presented at the 7th International Symposium on the Biosafety of Genetically Modified Organisms, Beijing, China.
7. Huang, J., Hu, R., & Rozelle, S. (2002). *China's agricultural research investment: Challenges and prospects*. Beijing: China Financial and Economic Press.

8. Huang, J., Hu, R., Pray, C., Qiao, F., & Rozelle, S. (in press). Biotechnology as an alternative to chemical pesticides: A case study of Bt cotton in China. *Agricultural Economics*.

9. Huang, J., Hu, R., Rozelle, S., Qiao, F., & Pray, C.E. (2002). Transgenic varieties and productivity of smallholder cotton farmers in China. *Australian Journal of Agricultural and Resource Economics*, *46*(3), 367-388.

10. Huang, J., Rozelle, S., Pray, C., & Wang, Q. (2002). Plant biotechnology in China. *Science*, *295*, 674-677.

11. Huang, J., Wang, Q., & Keeley, J. (2001). *Agricultural biotechnology policy processes in China* (working paper). Beijing: Chinese Academy of Sciences Center for Chinese Agricultural Policy.

12. Huang, J., Wang, Q., Zhang, Y., & Zepeda, J. (2001). *Agricultural biotechnology development and research capacity in China* (working paper). Beijing: Chinese Academy of Sciences Center for Chinese Agricultural Policy.

13. James, C. (2002). *Global review of commercialized transgenic crops: 2001*. International Service for the Acquisition of Agro-Biotech Applications.

14. Kahn, J. (2002, October 22). The science and politics of super rice. *The New York Times*.

15. Key Laboratory of Crop Molccular and Ccll Biology. (1996). *Thc research and prospects of crop genetic engineering*. Beijing: Chinese Agricultural Science and Technology Press.

16. Li, N. (2000, July-August). *Review on safety administration implementation regulation on agricultural biological genetic engineering in China*. Paper Presented at the China-ASEAN Workshop on Transgenic Plants, Beijing, China.

17. Marchant, M.A., Fang, C., & Song, B. (2002). Issues on adoption, import regulations, and policies for biotech commodities in China with a focus on soybeans. *AgBioForum*, *5*(4), 167-174.

18. Ministry of Agriculture [China]. (1990). *The guideline for the development of science and technology in middle and long terms: 1990-2000*. Beijing: MOA.

19. Ministry of Agriculture [China]. (1999). The application and approval on agricultural biological genetic modified organisms and its products safety. *Administrative Office on Agricultural Biological Genetic Engineering*, *4*, 35-37.

20. Ministry of Science and Technology [China]. (1990). *Biotechnology development policy*. Beijing: China S&T Press.

21. Ministry of Science and Technology [China]. (2000). *Biotechnology development outline*. Beijing: MOST.

22. National Center of Biological Engineering Development. (2000). The research progress in biotechnology. *Biological Engineering Progress, 20*.

23. Paarlberg, R.L. (2000). *Governing the GM crop revolution: Policy choices for developing countries* (2020 Discussion Paper #33). Washington DC: International Food and Policy Research Institute.

24. Peng, Y. (2002, October). *Strategic approaches to biosafety studies in China*. Paper presented at the 7th International Symposium on the Biosafety of Genetically Modified Organisms, Beijing, China.

25. Pray, C., Ma, D., Huang, J., & Qiao, F. (2001). Impact of Bt cotton in China. *World Development, 29*(5), 813-825.

26. Pray, C., Huang, J., & Rozelle, S. (2002). Five years of Bt cotton production in China: The benefits continue. *The Plant Journal, 31*(4), 423-430.

27. Rubenstein, E. (2000). China's leader commits to basic research, global science. *Science,288*, 1950-1953.

28. State Development and Planning Commission [China]. (2003). *Development of biotechnology industry in China—2002*. Beijing: Chemical Industry Press.

29. State Science and Technology Commission [China]. (1990). *Development policy of biotechnology.* Beijing: The Press of Science and Technology.

30. Yu, J., Hu, S., Wang, J., Wong, G.K.S., Li, S., Liu, B., et al. (2002). A draft sequence of the rice genome (Oryza sativa L. ssp. Indica). *Science, 296*, 79-108.

31. Zhang, X., Liu, J., & Zhao, Q. (1999). Transfer of high lysine-rich gene into maize by microprojectile bombardment and detection of transgenic plants. *Journal of Agricultural Biotechnology*, *7*(4), 363-367.

Chapter 7

CRYOPRESERVATION OF SPICES GENETIC RESOURCES

K. Nirmal Babu[1], G. Yamuna[1], K. Praveen[1], D. Minoo[2], P.N. Ravindran[1] and K.V. Peter[1]

[1] Indian Institute of Spices Research, Kerala

[2]Providence Women's College, Kerala India

INTRODUCTION

Plant genetic resources - constituting genotypes or populations of cultivars (landraces, advance/improved cultivars), genetic stocks, wild and weedy species, which are maintained in the form of plants, seeds, tissues, etc. - hold key to food security and sustainable agricultural development (Iwananga, 1994). They are non-renewable and are among the most essential of the world's natural resources. Due to deforestation, spread of superior varieties and selection pressure, genetic variability is gradually getting eroded. This demands priority action to conserve germplasm be it at species, genepool or ecosystem level, for posterity (Frankel, 1975). Whilst ecologists focused on in situ conservation might argue that ex situ conserved germplasm cannot offer the advantages afforded by selection and adaptation as a result of environmental pressures, there is no denying that if species are under threat—or worse, near extinction—then ex situ conservation of even limited germplasm is preferable to extinction. The opportunities offered by conservation biotechnology should not be missed or restricted by lack of interconnectivity between traditional and contemporary conservation practitioners.

SPICES AND GERMPLASM CONSERVATION

Spices and herbs are aromatic plants–fresh or dried plant parts like foliage, young shoots, roots, bark, buds, seeds, berries and other fruits of which are mainly used to flavour our culinary preparations, confectionary. They are also major ingredients in indigenous medicine and perfumery. Spices and herbs are grown throughout the world–different plant species in different regions.

Peninsular India is a rich repository of spices and over 100 species of spices and herbs are grown. The other major spice growing countries are Brazil, China, Guatemala, Indonesia, Madagascar, Nigeria, West Indies, Malaysia, Sri Lanka, Spain, Turkey, Mediterranean region and the Central America. Black pepper, cardamom, ginger, turmeric, vanilla, capsicum, cinnamon, clove, nutmeg, tamarind, coriander, cumin, fennel, fenugreek, dill, caraway, anise and herbs like saffron, lavender, thyme, oregano, celery, anise, sage and basil are important as spices. India being the native home of many spices, their conservation and characterization are one of the priority programmes. Deforestation, habitat degradation and overexploitation caused considerable loss of diversity in spices.

In many spices, conventional seed storage can satisfy most of the conservation requirements. But in crops with recalcitrant seeds and those having conservation needs cannot be satisfied by seed storage, have to be stored in vitro. Most field gene banks are prone to high labour cost, vulnerable to hazards like natural disasters, pests and pathogens attack (especially viruses and systemic pathogens), to which they are continuously exposed and require large areas of space. This supports in vitro and cryo conservation. In addition, other resources like continuous supply of standard stock cultures for experiments to examine physiological and biochemical processes, cell and callus lines developed for in vitro synthesis of valuable secondary products, flavours and other important compounds will benefit strongly from in vitro cultures. Most of the spice crops are either vegetatively propagated or have recalcitrant seeds. The spices germplasm is mostly conserved in field gene banks. Most of the spices are plagued by destructive and epidemic diseases caused by viruses, bacteria and fungi. This makes germplasm conservation in field gene bank risky. Thus in vitro and cryo storage system becomes important in the overall strategy of conserving genepool. Each technology should be chosen on the basis of utility, security and complementarily to other components of the strategy. A balance needs to be struck between seed, field gene bank, in vitro and cryo conservation of propagules, tissues, pollen, cell lines and DNA storage for overall objective of conserving gene pool.

METHODOLOGIES

Micropropagation

Plant regeneration and successful cloning of genetically stable plantlets in tissue culture is an important pre-requisite in any conservation effort of recalcitrant species. These techniques form the base for establishing tissue cultures and developing in vitro and cryo conservation technology for conservation. Simultaneously these tissue-cultured plants should be evaluated for their morphological and genetic stability in culture. The in vitro storage experiments, as much as possible, use growth regulators free media to reduce the rate of multiplication which in turn will reduce the extent of variation. Micropropagation (culture initiation, multiplication, plant regeneration and in vitro rooting) form the cycle of events that form the backbone of cryopreservation studies. For initial culture establishment earlier protocols developed by Nirmal Babu et al., 1997 can be used. Murashige and Skoog (1962), Woody Plant (McCown and Amos, 1979) and Schenk and Hildebrandt (1972) media can be used depending upon the crop for micropropagation Table 1. The miniaturized in vitro grown shoots can be used for cryopreservation. Micropropagation protocols for stable cloning of elite genotypes of spice crops were standerdised. Protocols were available for black pepper and its related species cardamom, ginger, turmeric and related genera, large cardamom, kasturi turmeric, mango ginger, Kaempferia galanga, K. rotunda, Alpinia spp, large. Cardamom, vanilla and related species, cinnamon, camphor, cassia seed and herbal spices like lavender, celery, thyme, mint, anise, savory, spearmint and oregano (Nirmal Babu et al., 1997, 2005, Minoo 2002). These techniques form the base for establishing tissue cultures and developing in vitro technology for conservation.

Table 1: Composition of MS*, WPM* and SH* basal media

Composition	Molecular formula	Concentration (mgl^{-1}) MS	Concentration (mgl^{-1}) WPM	Concentration (mgl^{-1}) SH
Macronutrients				
Ammonium nitrate	NH_4NO_3	1650.00	400.00	-
Ammonium phosphate	$NH_4H_2PO_4$	-	-	300.00
Potassium nitrate	KNO_3	1900.00	-	2500.00
Calcium chloride	$CaCl_2.2H_2O$	440.00	-	-
Calcium chloride	$CaCl_2$	-	72.50	151.00
Calcium nitrate	$Ca(NO_3)_2.4H_2O$	-	386.00	-
Potassium di hydrogen orthophosphate	KH_2PO_4	170.00	170.00	-
Potassium sulfate	K_2SO_4	-	990.00	-
Magnesium sulphate	$MgSO_4.7H_2O$	370.00	180.70	195.40
Micronutrients				
Sodium EDTA	Na_2EDTA	37.30	37.30	20.00
Ferrous sulphate	$FeSO_4.7H_2O$	27.80	27.800	15.00
Boric acid	H_3BO_3	6.20	6.20	5.00
Manganese sulphate	$MnSO_4.4H_2O$	22.30	22.30	10.00
Potassium iodide	KI	0.83	-	1.00
Zinc sulphate	$ZnSO_4.7H_2O$	8.60	8.60	1.00
Sodium molybdate	$Na_2MoO_4.2H_2O$	0.25	0.25	0.10
Copper sulphate	$CuSO_4.5H_2O$	0.025	0.25	0.20
Cobalt chloride	$CoCl_2.6H_2O$	0.025	-	0.10
Vitamins				
Myo-inositol	$C_6H_{12}O_6$	100.00	100.00	1000
Thiamine HCl	$C_{12}H_{17}ClN_4OS.HCl$	0.10	0.50	100
Nicotinic acid	$C_6H_5NO_2$	0.50	0.025	1.00
Pyridoxine HCl	$C_6H_{11}NO_3.HCl$	0.50	0.025	1.00
Amino acid				
Glycine	$C_2H_5NO_2$	2.00	1	-

*Murashige and Skoog, 1962, McCown and Amos, 1979, Schenk and Hildebrandt 1972

The basal media used are MS (Murshige and Skoog, 1962) for crops like cardamom, ginger, turmeric, kasturi turmeric, mango ginger, large cardamom, Kaempferia, Vanilla spp. seed and herbal spices and WPM-Woody Plant Medium (Mc Cown and Amos, 1979) for black pepper and its related species, cinnamon, camphor and cassia. Simultaneously these tissue-cultured plants are being evaluated for their morphological and genetic stability in culture (Luckose et al, 1993, Chandrappa et al, 1997, Nirmal babu et al 2003, Madhusoodanan et al 2005). Though micropropagation protocols were standardized using growth regulators, all the in vitro storage experiments were carried out using growth regulators free media to reduce the rate of multiplication which in turn will reduce the extent of variation.

Protocols are available for micropropagation and multiplication of many endangerd species like Piper hapnium, P. silent vallyensis, P.schmidtii, P. wightii, P. barberi , Vaniilla aphylla, V. pilifera, V. walkyrie, V. wightiana, K. rotunda and Alpinia galanga are available (Peter et al 2002, Minoo 2002, Nirmal Babu et al 1999, 2005). Bertaccini et al (2004), Du et al (2004) reported micropropagation and establishment of mitebrone virus-free garlic.

Callus and Cell Culture Systems

Quatrano (1968) and Nag and Street (1973) reported the first successful experiments on cryopreservation of plant cells. Since then a large number of cell suspension and calli cultures have been successfully cryopreserved (Engelmann et al 1994). In general, callus cultures are more difficult to cryopreserve than cell suspensions, because of the relative volume of the callus, its slow growth rate and the cellular heterogeneity (Withers 1987). One successful cryopreservation procedure that is applicable to all different cell suspensions or calli cultures has not been developed yet. Research focuses on optimizing the factors on which successful cryopreservation of plant organs cells suspensions and calli depends, such as: (i) starting material, (ii) pretreatment, (iii) cryopreservation procedure, and (iv) postthaw treatment. Plant cells cultured in vitro produce wide range of primary and secondary metabolites of economic value. Production of phytochemicals from plant cell cultures has been presently used for pharmaceutical products. Production of flavour components and secondary metabolites in vitro using immobilised cells is an ideal system for spices crops. Production of saffron and capsaicin was reported using such system (Ravishankar et al., 1988; 1993, Johnson et al., 1996; Venkataraman and Ravishankar 1997). Johnson et al (1996) reported biotransformation of ferulic acid vanillamine to capsacin and vanillin in immobilised cell cultures of Capsicum frutescens. Reports on the in vitro synthesis of crocin, picrocrocin and safranel from saffron stigma (Himeno and

Sano, 1995) and colour components from cells derived from pistils (Hori et al, 1988) are available for further scaling up. Callus and cell cultures were established in nutmeg, clove, camphor, ginger, lavender, mint, thyme, celery etc. Cell immobilization techniques have been standardized in ginger, sage, anise and lavender (Ilahi and Jabeen, 1992; Ravindran et al, 1996; Sajina et al, 1997). Studies on conservation of cell lines is yet o become popular in spices. Suspensions of embryogenic cell lines of fennel, conserved at 4 0C for up to 12 weeks produced normal plants upon transfer to normal laboratory conditions (Umetsu et al, 1995).

Somatic Embryogenesis and Plant Regeneration

In black pepper primary embryogenic cultures can be established as per the method described by Nair and Dutta Gupta (2003). Culture the surface sterilized seeds on agar gelled full-strength, PGR-free SH (Schenk and Hildebrandt, 1972) medium containing 3.0% (W/V) sucrose under darkness. Primary somatic embryos (PEs) derived from micropylar tissues of germinating seeds after 90 days could be utilized for inducing secondary somatic embryogenic cultures. Primary somatic embryo clumps having pre-globular to torpedo shaped embryos (5–6 visible embryos per seed) were carefully detached and inoculated on half strength PGR-free SH medium containing 1.5 % sucrose and gelled with 0.8% agar (Bacteriological grade, Himedia). The pH of the medium was adjusted to 5.9 prior to autoclaving. Cultures were maintained at darkness at a temperature of 25±2oC. The culture conditions remained the same for all further experiments unless otherwise specified. While inoculating, the PEs were uniformly spread on the surface of the medium. Secondary embryogenic cultures were further maintained by subculturing on SH medium containing 1.5% sucrose at intervals of 20 d. The proliferating SEs were spread periodically on the surface of the medium, to facilitate proliferation.

Somatic Embryogenesis and Plant Regeneration

In black pepper primary embryogenic cultures can be established as per the method described by Nair and Dutta Gupta (2003). Culture the surface sterilized seeds on agar gelled full-strength, PGR-free SH (Schenk and Hildebrandt, 1972) medium containing 3.0% (W/V) sucrose under darkness. Primary somatic embryos (PEs) derived from micropylar tissues of germinating seeds after 90 days could be utilized for inducing secondary somatic embryogenic cultures. Primary somatic embryo clumps having pre-globular to torpedo shaped embryos (5–6 visible embryos per seed) were carefully detached and inoculated on half strength PGR-free SH medium containing 1.5 % sucrose and gelled with 0.8% agar (Bacteriological grade, Himedia). The pH of the medium was adjusted to 5.9 prior to autoclaving. Cultures were maintained at darkness at a temperature of 25±2oC. The culture conditions remained the same for all further experiments unless otherwise specified. While inoculating, the PEs were uniformly spread on the surface of the medium. Secondary embryogenic cultures were further maintained by subculturing on SH medium containing 1.5% sucrose at intervals of 20 d. The proliferating SEs were spread periodically on the surface of the medium, to facilitate proliferation.

Pollen Storage

Pollen storage can be considerable value supplementing the germplasm conservation strategy by facilitating hybridisation between plants with different time of flowering and to transport pollen across the globe for various crop improvement programmes in addition to developing haploid or homozygous lines. No significant work was done in India, except a few initial reports. The technique of pollen storage is comparable with that of seed storage, since pollen can be dried (less than 5% moisture content on a dry weight basis) and stored below 0oC. There are limited reports on the survival and fertilizing capacity of cryopreserved pollen more than five years old. Pollen might represent an interesting alternative for the long-term conservation of problematic species (IPGRI, 1996). However, pollen has a relatively short life compared with seeds (although this varies significantly among species) and viability testing can be time-consuming and uneconomical. Other disadvantages of pollen storage are the small amount produced by many species, the lack of transmission of organelle genomes via pollen, the loss of sex-linked genes in dioecious species and the general inability to regenerate into plants. Pollen, therefore, has been used to a limited extent in germplasm conservation (Hoekstra ,1995). An advantage is that pests and diseases are rarely transferred by pollen (excepting some virus diseases). This allows safe movement and exchange of germplasm as pollen.

Cryo Preservation

For long-term conservation of the problem species, cryopreservation is the only method currently available. Dramatic progress has been made in recent years in the development of new cryopreservation techniques and cryopreservation protocols have been established for over 100 different plant species. Cryopreservation is an attractive option for long-term storage. Liquid nitrogen (–196°C) is routinely used for cryogenic storage, since it is relatively cheap and safe, requires little maintenance and is widely available. Below –120°C the rate of chemical or biophysical reactions is too slow to cause biological deterioration (Kartha 1985). Only in the long term might there be a small risk of ionising radiation causing genetic changes in materials stored at cryogenic temperatures (Grout 1995). An array of plant material could be considered for cryopreservation as dictated by the actual needs vis-a-vis preservation. These include meristems, cell, callus and protoplast cultures, somatic and zygotic embryos, anthers, pollen or microspores and whole seeds (Withers, 1985; Kartha, 1985). Plant germplasm stored in liquid nitrogen (-196°C) does not undergo cellular divisions. In addition, metabolic and most physical processes are stopped at this temperature. As such, plants can be stored for very long time periods and both the problem of genetic instability and the risk of loosing accessions due to contamination or human error during subculturing are overcome. Most cryopreservation endeavours deal with recalcitrant seeds, in vitro tissues from vegetatively propagated crops, species with a particular gene combination (elite genotypes) and dedifferentiated plant cell cultures. Care must be taken to avoid ice crystallisation during the freezing process, which otherwise would cause physical damage to the tissues. The existing cryogenic strategies rely on air-drying, freeze dehydration, osmotic dehydration, addition of penetrating cryoprotective substances and adaptive metabolism (hardening), encapsulation, vitification or combinations of these processes. Cryopreservation methods have been developed for more than 80 different plant species in various forms like cell suspensions, calluses, apices, somatic and zygotic embryos (Kartha and Engelmann, 1994; Engelmann, 1997, 2000, Engelmann et al 1994, 1995). However, their routine utilisation is still restricted almost exclusively to the conservation of cell lines in research laboratories For small volumes, long-term storage is practicable through storage of cultures in cryopreservation at ultra-low temperature, usually by using liquid nitrogen (-196oC). At this temperature all cellular divisions and metabolic processes are virtually halted and consequently, plant material can be indefinitely stored without alteration or modification. The normal approach of tissue culture is to find a medium and set of conditions that favour the most rapid rate of growth with a subculture interval of 20 – 30 days. For cryopreservation storage biological materials are stored in liquid nitrogen for long term with out subculturing. Cryopreservation, i.e., the storage of biological material at ultra low temperature usually that of liquid nitrogen (-1960C) can be achieved by different techniques like direct freezing, encapsulation- dehydration, encapsulation- vitrification and vitrification.

Encapsulation – Dehydration

A simplified methodology for vitrification is given below (Yamuna 2007). The in vitro plants already established were used as mother plants for source of explants. This in turn facilitates the reduction in size of the plantlets and smaller somatic embryos which made them suitable for cryopreservation.

- Suspend in vitro grown shoots/ somatic embryos in MS basal medium supplemented with 4% (w/v) Na alginate, 2M Glycerol and 0.4 M sucrose.
- Drop the mixturecontining microshoots, with a sterile pipette into 0.1M CaCl2 solution containing 2M Glycerol and 0.4M sucrose and left for 20 min to form beads about 4 mm in diameter, each bead containing at least one shoot.
- Preculture the encapsulated shoots – stepwise - on MS medium enriched with different concentration of 0.3, 0.5, 0.75 and 1.0M for four days with one day on each.
- Place the precultured beads on sterile fitter paper in Petridishes (diameter 90mm) and dehydrated by air drying on a flow bench (at room temperature and humidity) for periods of 0-10 h to determine the optimal dehydration time.
- Measure the water content of the beads was by weighing them prior and after drying in an oven at 800C for 48h.
- Transfer the dehydrated beads into a 2 ml cryovial (ten beads per tube) and directly immerse in liquid nitrogen for 24h.

Vitrification

A simplified methodology for vitrification is given below (Yamuna 2007).

- Shoots (1-2mm)/ somatic embryos were excised and cultivated on MS medium supplemented with 0.3 M sucrose for 24h at 250C.
- The treated explants were then cultured on MS medium supplemented with sucrose at 0.75 M for 1 day in the same conditions.
- After pretreatments explants were transferred to a cryovial with 1.8 ml of loading solution (2 M Glycerol + 0.4 M sucrose) and kept for 15 min.
- Different incubation periods in PVS2 (40-100 minutes) were tested for osmoprotected explants
- Cryovials containing 8-10 explants were directly immersed in liquid nitrogen and kept for 24 h.

Encapsulation – Vitrification

A simplified methodology for encapsulation - vitrification is given below (Yamuna 2007).

- Suspend pre-cultured shoots (1-2mm)/ somatic embryos with 2-3 apical domes on 0.3M sucrose for 16h in MS basal medium supplemented with 4% sodium alginate and 0.3 M sucrose.

- Dispense the mixture including shoots, were with a sterile pipette into MS medium supplemented with 0.1M CaCl2 and 0.4 to 1.0M sucrose, with or without 2M Glycerol gently shaken (20 rpm) on a rotary shaker for 1h at 250C.
- The encapsulated and osmo-protected shoots were dehydrated with 20 ml PVS2 in a 100 ml Erlenmeyer flask at 250C and plunged into LN and held for at least 24 h at - 1960C.

Thawing and Recovery of Conserved Materials

After LN storage, cryovials warm rapidly in a 40 0C water bath for 2-3 minutes. The solution was drained from the cryovials and replace twice at 10 min intervals with 1 ml 1.2 M sucrose solution in the case of encapsulation-vitrification and vitrification methods. The composition of recovery medium was MS/WPM/SH basal medium supplemented with 2.22 – 4.44 μM and BA, 2.69- 5.37 μM NAA. In the Encapsulation - dehydration, Encapsulation - vitification and vitrification procedures, surviving shoots can be identified by greening of explants following 2 weeks of post culture. Regrowth can be defined as the shoots that regenerated to shoots in 6 weeks of postculture. Elongated shoots can be used for micropropagation and rooting and subculture was done every 4 weeks. For rooting well grown shoots can be transferred to solid MS medium used for multiplication.

Genetic Stability of Conserved Materials

An important prerequisite for any conservation technique is that the regenerants produced from the conserved material should be true-to-type. There are ample evidences to indicate that under certain culture conditions the materials undergo genetic changes (somaclonal variations) and as a consequence lose their integrity and uniformity. This would be highly undesirable in spices varieties where the purpose is not only to conserve a genotype but also retain its specific quality traits. Thus testing for the genetic stability of in vitro conserved materials is of utmost importance. Besides morphology, cytology and isozyme profiling sophisticated biochemical and DNA-based techniques have enabled more critical analysis of the genetic stability of in vitro materials.

RAPD, ISSR and SSR analysis can be done to evaluate genetic fidelity of the cryopreserved lines of Spices. DNA isolation can be done as per CTAB method (Ausubel et al., 1995 or Sambrook et al. 1989). RAPD and ISSR, SSR profiles were developed as per the method suggested by Williams et al., (1990),

Nirmal babu et al., (2003, 2007) and Ravindran et al., (2004). Morphological characters coupled with RAPD profiles using 24 operon primers have indicated genetic fidelity among randomly selected micropropagated plants of Subhakara and Aimpiriyan, indicating that micropropagation protocol can be used for commercial cloning of black pepper (Nirmal Babu et al., 2003). Genetic uniformity of micropropagated Piper longum using RAPD profiling was reported by Ajith (1997) and Parani et al. (1997) for conservation. Peter et al (2001) and Ravindran et al (2004) reported that the conserved materials of all the species conserved by them showed normal rate of multiplication when transferred to multiplication medium after storage. The normal sized plantlets when transferred to soil established with over 80% success. They developed into normal plants without any deformities and were morphologically similar to mother plants. RAPD profiling of these conserved plants also showed their genetic uniformity. Ravindran et al (2004), Yamuna et al (2007) and Yamuna (2007) reported genetic uniformity was observed in cryo preserved and recovered plants of cardamom, ginger, black pepper and endangered species of Piper, P. barberi based on RAPD and ISSR profiling.

Status of Cryo Conservation in Spices

Reports on cryopreservation of spices are meager and limited. The present status of cryo preservation in major spices is given Table 2. The number of accessions conserved in cryo genebank at the National Bureau of Plant Genetic Resources (NBPGR), New Delhi are given in Table 3.

Table 2: Present status of information on cryo conservation of spices

Application	Technique	Reference
Black pepper (*Piper nigrum*) and related species	Meristem culture	Philip *et al.*, 1992
Disease eradication propagation	Shoot Culture Leaf/root	Broome and Zimmerman, 1978
Cryopreservation	Seeds	Chaudhury and Chandel ,1994
Cryopreservation	synseeds	Ravindran *et al.*, 2004; Nirmal Babu *et al.*, 2007; Yamuna 2007
Cryopreservation	Seed	Decruse and Seeni, 2003
Slow growth storage and cryopreservation	Plantlets and shoot tips	Ravindran *et al.*, 2004; Nirmal Babu *et al.*, 2007; Yamuna 2007
***Allium* Spp**		
Disease eradication	Meristem culture and themotherapy	Conci and Nome, 1991
Cryopreservation	Shoot culture,	Keller 1991

Application	Technique	Reference
	microbullbets	
Cryopreservation	Shoot tips	Niwata 1995
Cardamom (*Elettaria cardamomum* Maton)		
Disease eradication	Meristem culture	Nadagauda *et al.* 1983
Cryopreservation	Seeds	Chaudhury and Chandel, 1995
Slow growth storage and cryopreservation	Plantlets and shoot tips	Ravindran *et al.*, 2004; Nirmal Babu *et al.*, 2007; Yamuna 2007
***Zingiber* spp.**		
Disease eradication	Shoot cultures, shoot buds	Balachandran *et al* .,1990
Propagation	Somatic embryo regeneration	Hosoki and Sagawa ,1977, Nirmal Babu, 1997
Cryopreservation	Synseeds	Sharma *et al.*, 1994
Slow growth storage and cryopreservation	Plantlets and shoot tips	Ravindran *et al.*, 2004; Nirmal Babu *et al.*, 2007; Yamuna *et al* 2007 ; Yamuna 2007
***Curcuma* spp**		
Slow growth storage and cryopreservation	Plantlets and shoot tips	Ravindran *et al.*, 2004 ; Nirmal Babu *et al.*, 2007
***Vanilla* spp.**		
Disease eradication	Apical meristem	Cereveta and Madrigal, 1981
Cryo preservation	Synthetic seeds	Ravindran *et al.*, 2004
Pollen Cryo preservation	Pollen	Minoo, 2002; Minoo *et al* 2011
Slow growth storage and cryopreservation	Plantlets and shoot tips	Ravindran *et al.*, 2004; Nirmal Babu *et al.*, 2007, Minoo and Babu 2009
Herbal spices		
Slow growth storage	*In vitro* plantlets	Nirmal Babu *et al.*1996
Capsicum		
Cryopreservation	Seed	Peter *et al* 2002 ; Ravindran *et al* 2004
Cryopreservation	Pollen	Alexander *et al.*, 1991
Cryopreservation	Pollen	Rajasekharan and Ganeshan, 2003
Fennel (*Foeniculum vulgare*)		
Cold storage	Embryogenic	Umetsu *et al.*, 1995
	suspension cells	
Coriander (*Coriandrum sativum*)		
Cryopreservation	somatic embryos	Elena *et al.*, (2010)
Mint (*Mentha* spp.)		
Cryopreservation	Somatic embryos	Leigh and Remi 2003
***Ocimum* spp**		
Slow growth	Encapsulated beads	Mandal *et al* (2000)
Syzygium francissi		
	Shoot tips	Shatnawi *et al* (2004).
Armoracia rusticana		
Cryopreservation	Hairy root cultures	Phunchindawan *et al*
***Crocus* spp.**		
	Encapsulated calluses	Chand *et al* (2000); Baghdadi et al., (2010)

*Ashmore, 1997, 2002 and Nirmal Babu *et al*, 1999, 2007, Yamuna 2007; Yamuna et al 2007

Table 3: Present status of Spices in in vitro and Cryo genebank at NBPGR

Species	No.of accessions
Maintained as in vitro cultures	
Spices and industrial crops	380 accessions (7 genera, 27 species)
Medicinal and Aromatic plants	169 accessions (21 genera, 28 species)
Maintained in cryo bank	
Spices and Condiments	148 accessions
Medicinal and Aromatic plants	5 accessions
Total	702

Source: Annual Report NBPGR 2010-11

Black Pepper and Related Species

Cryopreservation of black pepper (Piper nigrum L.)seeds in liquid nitrogen (LN2) was reported by Choudhary and Chandel, (1994), and Choudhury and Malik (2004). Pepper seeds are recalcitrant and the seed viability decreases with reduction in moisture content. Seeds desiccated to 12% & 6%moisture contents were successfully cryopreserved in liquid nitrogen at –1960C, with a survival rate of 45% & 10.5% respectively (Chaudhury and Chandel 1994).

Figure. 1: Cryopreservation of black pepper somatic embryos by encapsulation de-

hydration. a) Somatic embryos used for cryopreservation, b) Somatic embryos encapsulated in Naalginate, c) Encapsulated and dehydrated somatic embryos, d) Viable somatic embryo stained in red colour after cryopreservation, e), f), g), h) & i) Various stages of development of somatic embryos to plantlet after cryopreservation, j) Fully developed plantlet from a somatic embryo cryopreserved by encapsulation dehydration, k) A cluster of somatic embryos at different stages of development, originated from an embryogenic line after cryopreservation

laminar air flow for 6 h which resulted in 21 % moisture content. In the vitrification procedure, the somatic embryos were precultured for 3 days on SH basal medium containing 0.3 M sucrose and subjected to vitrification treatment for 60 minutes at 250C resulted in 71 % survival after cryopreservation. The study concluded that the embryogenic lines of Piper nigrum cultivar karimunda can be successfully cyopreserved following an encapsulation dehydration/ desiccation procedure (62 % success). This success rate can be enhanced to 71 % using a vitrification/one step freezing in liquid nitrogen (Fig. 1).This was mainly because of the nature of somatic embryos which is more suitable to cryopreservation compared to shoot buds. The genetic stability of the conserved somatic embryos was proved by RAPD and ISSR profiling. Cryopreservation of encapsulated shoot buds of endangered Piper barberi was reported by Peter et al (2001) and Ravindran et al (2004). Encapsulated shoot tips of Piper barberi were cryopreserved with 60% success using vitrification technique. In encapsulation vitrification the encapsulated shoot tips were precultured on MS medium, supplemented with 0.3 M, 0.5 M and 0.7 M sucrose (pH 5.8) for three days followed by dehydration with PVS2 solution (100%) at 00 C for 3 hours. After dehydration the beads (10 encapsulated shoot tips in 0.8 ml PVS2 solution per 1.5 ml cryotube) were frozen rapidly by direct immersion in to liquid nitrogen (- 196 0C) and kept for one hour (Peter et al 2001 and Ravindran et al 2004). Yamuna 2007 also reported that studies on cryopreservation of endangered P.barberi shoot tips revealed that, the encapsulation- vitrification procedure produced higher survival (70 %) of cryopreserved shoot tips (Fig. 2) compared to encapsulation - dehydration which gave 40 % survival. Genetic fidelity studies showed that the regenerated plants were similar to the controls. Thus encapsulation - vitrification as a simple and efficient method for long term preservation of P.barberi propagules.

Cardamom and Related Species

Choudhary and Chandel (1995) attempted cryo-conservation of cardamom (Elettaria cardamomum Maton.) seed. They tried to conserve seeds at ultra-low temperature by suspending seeds in cryovials in vapor phase of liquid nitrogen (-150oC) by slow freezing and also by direct immersion in liquid nitrogen

(-196oC) by fast freezing. The result showed that seeds possessing 7.7-14.3% moisture content could be successfully cryo-preserved with 80% germination when tested after one-year storage in vapor phase of liquid nitrogen (at-150oC). Shoot tips(1.0-2.0mm) from in vitro grown plantlets of cardamom were subjected to progressive increase of sucrose concentrations (0.1, 0.3, 0.5, 0.7, 0.9, and 1.0) for two days each under the same cultural conditions as the parent plantlets. These shoot tips were transferred to 1.8ml cryotube containing ice cold PVS2 solution (30%(v/v) glycerol + 15% (v/v) ethylene glycol + 15% (v/v) DMSO in culture medium with 0.4 M sucrose, pH (5.8)) at 00C for 3 hours. After 3 hours equilibration at 00C, the shoot tips were directly immersed into liquid nitrogen for 1 hour. Vials were thawed in 400C water for 1 minute. The cryoprotectant was removed and the shoot tips were washed 2-3 times in 1.2M sucrose solution. About 70%Shoot tips were recovered on MS medium supplemented with BAP and NAA. But the encapsulation vitrification method gave only 60% success (Ravindran et al 2004).

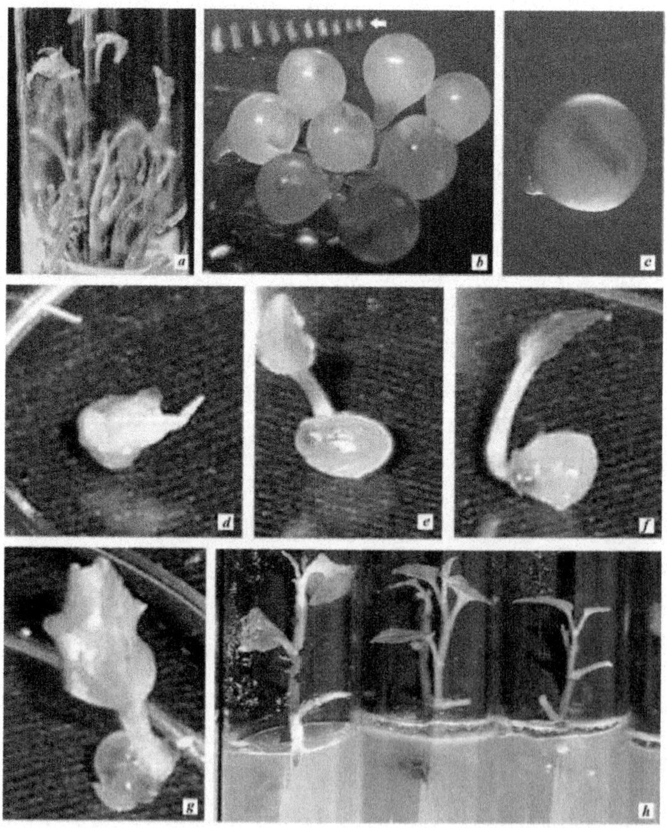

Figure. 2: Cryopreservation of Piper barberi by encapsulation vitrification. a) In vitro

culture of P. barberi, b) & c) Shoot tips encapsulated in Na-alginate, arrow indicates shoot tip used as explants, d), e), f) & g) Various stages of development of cryopreserved shoot tips after post culturing, h) Regenerated plantlets after 3 months of post culturing

Yamuna (2007) tested the effect of encapsulation – dehydration, encapsulation vitrification and vitrification methods on cryopreservation of cardamom. In the vitrification treatment, to enhance tolerance to vitrification solution (PVS2), a two step sucrose preculture with 0.3 M and 0.75 M sucrose for one day each and an osmo protection step with a loading solution (LS) of 2 M glycerol and 0.4 M sucrose were performed prior to PVS2 treatment. The shoots

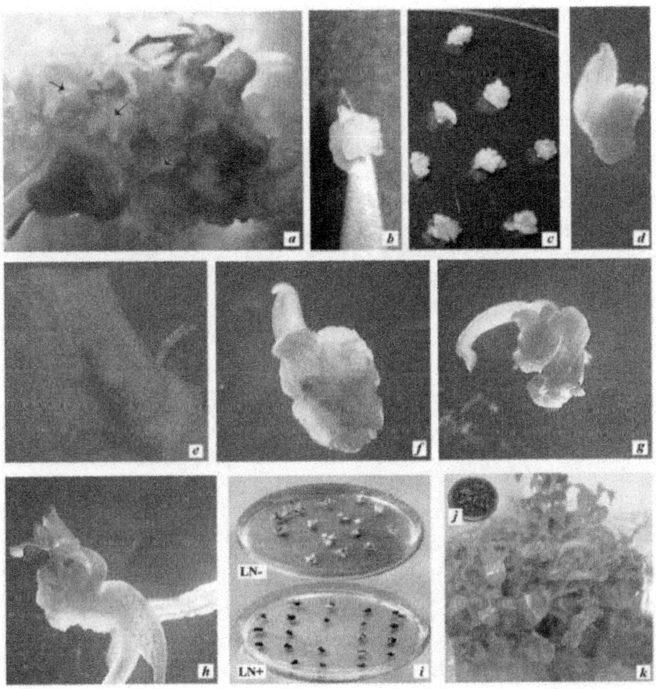

Figure. 3: Plant regeneration from cryopreserved miniature shoots of cardamom by vitrification. a) Cardamom culture with miniature shoots, b) & c) Excised meristematic clumps used for cryopreservation, d) Explant turned brown after cryopreservation, e) Viable tissues stained in TTC after cryopreservation, f), g), h), & i) Shoot development after 10, 14 and 25 days of post culturing , j) regenerating shoot buds in a petridish, k) Development of multiple shoots after 4 months of post culturing

dehydrated with PVS2 for 60 min retained a high level of shoot formation (70 %). The vitrification procedure resulted in higher regrowth (70 %) (Fig.3) when compared to encapsulation vitrification (62 %) and encapsulation dehydration

(60 %). In all the three cryopreservation procedures tested, shoots grew after cryopreservation without intermediary callus formation. The genetic stability of cryopreserved cardamom shoots were confirmed using ISSR and RAPD profiling.

Figure. 4: Plant regeneration from cryopreserved shoot buds of ginger by encapsulation vitrification. a) In vitro culture, b) A typically excised shoot bud used for cryopreservation, c) & d) Shoot buds encapsulated in Na-alginate, e) & f) Shoot buds turned brown after thawing, g) Viable apical dome stained in red colour after liquid nitrogen storage (TTC staining), h) Regenerating shoot bud 20 days after post culturing, i) & j) Elongated shoot with no intermediary callus formation, k) & l) Regenerating shoot buds in petriplates, m) Plantlets regenerating from cryopreserved shoot bud

Ginger, Turmeric and Related Species

Cryopreservation of Ginger (Zingiber officinale Rosc) and turmeric (Curcuma longa L.) shoot tips was successfully done with 80% of recovery using vitrification method. But the rate of recovery was only 40% when encapsulated shoot tips were dehydrated in progressive increase of sucrose concentration together with 4- 8 hrs. of desiccation (Peter et al 2001 and Ravindran et al 2004). Efficient cryopreservation techniques were developed for in vitro grown shoots of ginger based on encapsulation dehydration, encapsulation vitrification and vitrification procedures (Yamuna et al 2007 and Yamuna 2007. The vitrification procedure resulted in higher regrowth (80 %) when compared to encapsulation vitrification (66 %) and encapsulation dehydration (41 %). The genetically stability of shoot apices was confirmed by molecular profiling. The RAPD and ISSR assays performed suggested that no genetic aberrations originated in ginger plants during culture and cryopreservation (Fig. 4).

Vanilla and Related Species

Technology for cryopreservation of vanilla germplasm - using encapsulation and vitrification methods – were available. Encapsulated in vitro grown shoot tips of vanilla could be cryo preserved with 70% success when pretreated with progressive increase of sucrose concentration (0.1M-1.0M) for one day each and dehydrated for 8 hrs (Peter et al 2001; Minoo 2002 and Ravindran et al 2004) (Fig. 5).

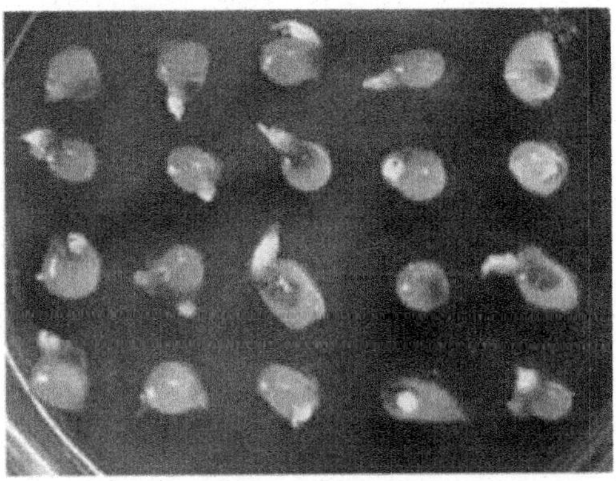

Figure. 5: Germination of cryopreserved encapsulated shoot tips protocorms of vanilla

Ginzalez-Arnao, et al., (2009) attempted to cryo-preserve V. planifolia Andr. using in vitro fragmented explants (IFEs) and the apices derived from them. Cryopreservation of apices from in vitro grown plants was achieved using the droplet vitrification protocol. Maximum survival (30%) and further regeneration (10%) of new shoots were obtained for apices derived from clusters of in vitro plantlets produced from microcuttings through a three-step droplet vitrification protocol: 1-d preculture of apices on solid MS medium with 0.3 M sucrose; loading with a 0.4 M sucrose + 2 M glycerol solution for 20–30 min; and exposure to plant vitrification solution PVS3 for 30 min at room temperature. Minoo (2002) reported cryopreservation of vanilla pollen for conservation (Fig. 6) of haploid genome as well as assisted pollination between species that flower at different seasons and successful fertilisation using cryopreserved pollen (Minoo, 2002, Minoo et al 2011). Pollen from two asynchronously flowering species of Vanilla viz., cultivated V. planifolia and its wild relative V. aphylla, were cryopreserved after desiccation to 12 % moisture content, pretreated with cryoprotectant Dimethyl sulphoxide (5%) and cryopreserved -196°C in Liquid Nitrogen. This cryopreserved pollen was latter thawed and tested for their viability both in vitro and in vivo. A germination percentage of 82.1% and 75.4% in V. planifolia and V.aphylla pollen respectively were observed indicating their viability(Fig.6). This cryopreserved pollen of V. planifolia was used successfully to pollinate V.aphylla flowers resulting in fruit set (Fig.7). The seeds thus obtaines were sussfully cultured to develop hybrid plantlets. This system is of great importance and can be used for conserving the haploid gene pool of Vanilla in cryobanks and their subsequent utility in crop improvement (Fig. 6 and 7)

Figure. 6: Germination of cryopreserved Vanilla pollen

Figure. 7: Fruit set after pollination with cryopreserved pollen

Capsicum

Plants could be successfully regenerated (Fig 8) from cryopreserved seeds of capsicum (Peter et al 2001 and Ravindran et al 2004). Alexander et al (1991) and

Rajasekharan and Ganeshan. (2003) reported freeze preservation of capsicum pollen (Capsicum annuum) in liquid nitrogen (–1960C) for 42 months

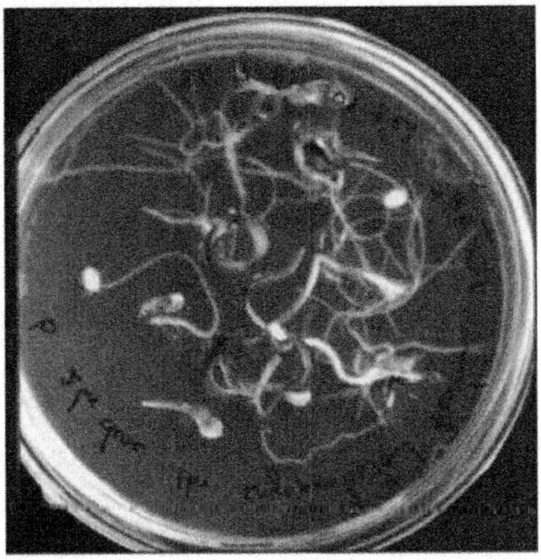

Figure. 8: Successful germination of cryopreserved seeds of capsicum

Seed Herbal and Other Spices

Elena et al., (2010) successfully cryopreserved coriander (Coriandrum sativum L.) somatic embryos using sucrose pre-culture and air desiccation procedure utilized embryo clumps (ECs). The regrowth after cryopreservation and average number of new embryos developed from cryopreserved ECs were retained at the level of the untreated control (98% and 13 embryos per clump, respectively). Both normal and abnormal plants were produced from control and cryopreserved cultures, indicating that appearance of abnormalities was not related to cryopreservation. The regenerants with normal phenotype showed the same peaks of relative DNA content regardless of cryopreservation. The results suggest that simple desiccation method is effective for cryopreservation of coriander somatic embryos with subsequent regeneration. Plants could be regenerated from cryopreserved seeds of Anise.(Peter et al 2001). Successful Cryopreservation of seeds, meristems, somatic or zygotic embryos were reported in Allium Spp (Niwata, 1995, Hyung et al 2003, Haeng et al 2003, 2004.2005, Jung et al 2005, Gayle et al 2004). Preliminary success was reported in cryo preservation of Mint (Leigh and Remi 2003). Most of the reports are confined to a few genotypes and hence the techniques standardized needs to be extended to more genotypes before adopting them for routine conservation. Reports of cryoconservation of spices like Ocimum, Lavendula, Salvia are available from National Bureau of Plant Genetic Resources (NBPGR), New Delhi.

Mandal et al (2000) reported propagation and conservation of four pharmaceutically important herbs, Ocimum americanum L. syn. O. canum Sims. (hoary basil); O basilicum L. (swett basil); O. gratissimum L. (shrubby basil); and O. sanctum L. (sacred basil) using synthetic seed technology. Synthetic seeds were produced by encapsulating axillary vegetative buds harvested from garden-grown plants of these four Ocimum species in calcium alginate gel. The gel contained Murashige and Skoog (MS) nutrients and 1.1-4.4 μM benzyladenine (BA). Shoots emerged from the encapsulated buds on all six planting media tested. However, the highest frequency shoot emergence and maximum number of shoots per bud were recorded on media containing BA. Of the six planting media tested, both shoot and root emergence from the encapsulated buds in a single step was recorded on growth regulator-free MS medium as well as on vermi-compost moistened with halfstrength MS medium. Rooted shoots were retrieved from the encapsulated buds of O. americanum, O. basilicum, and O. sanctum on these two media, whereas shoots of O. gratissimum failed to root. The encapsulated buds could be stored for 60 d at 4°C. Plants retrieved from the encapsulated buds were hardened off and established in soil. An efficient procedure for the in vitro propagation and cryogenic conservation of Syzygium francissi was developed by Shatnawi et al (2004). Shoot tips excised from in vitro-grown plants were successfully cryostoraged at −196°C by the encapsulation-dehydration method. A preculture of formed beads on MS medium containing 0.75 M sucrose for 1 d, followed by 6 h dehydration (20% moisture content) led to the highest survival rate after cryostorage for 1h. This method is a promising technique for in vitro propagation and cryopreservation of shoot tips from in vitro-grown plantlets of S. francissi germplasm. Hairy root cultures of Armoracia rusticana Gaertn. Mey. et Scherb. (horseradish) were successfully cryopreserved by two cryogenic procedures (Phunchindawan et al., 1997). Encapsulated shoot primordia were precultured on solidified Murashige-Skoog medium supplemented with 0.5M sucrose for 1 day and then dehydrated with a highly concentrated vitrification solution (PVS2) for 4 h at 0°C prior to a plunge into liquid nitrogen. The survival rate of encapsulated vitrified primordia amounted to 69%. In a revised encapsulation-dehydration technique, the encapsulated shoot primordia were precultured with a mixture of 0.5M sucrose and 1M or 1.5M glycerol for 1 day to induce dehydration tolerance and then subjected to air-drying prior to a plunge into liquid nitrogen. The survival rate of encapsulated dried primordia was more than 90%, and the revived primordia produced shoots within 2 weeks after plating. A long-term preservation of shoot primordia was also achieved by the technique. Thus, this revised encapsulation-dehydration technique appears promising as a routine method for the cryopreservation of shoot primordia of hairy roots The effect of sucrose concentration and dehydration period on survival and regrowth of encapsulated calluses were also studied in 2 species of Crocus (Chand et al 2000). Highest survival (83.3; 88.9%) and regrowth (77.6; 83.3%) rates were obtained when encapsulated unfrozen calluses of Crocus hyemalis and C. moabiticus precultured with 0.1 M sucrose for two days without further air dehydration. After cryopreservation, the highest survival (55.6; 61.1%) and regrowth (16.7; 27.8%) rates were achieved when calluses of C. hyemalis and C. moabiticus were pretreated with 0.5 M sucrose for two days after two hours of dehydration. Viability of crocus decreased with increased sucrose concentration and dehydration period. Dehydration of encapsulated calluses of C. hyemalis and C. moabiticus with silica gel for one hour prior to freezing resulted in maximum rates of survival (77.8; 83.3%) and re-growth (33.3; 72.1%). However, further studies should be initiated to improve regrowth of surviving embryogenic calluses and to study genetic stability after cryopreservation.

DNA BANK

Concurrent with the advancements in gene cloning and transfer has been the development of technology for the removal and analysis of DNA. DNAs from the nucleus, mitochondrion and chloroplast are now routinely extracted and immobilized onto nitrocellulose sheets where the DNA can be probed with numerous cloned genes. In addition, the rapid development of polymerase chain reaction (PCR) now means that one can routinely amplify specific oligonucleotides or genes from the entire mixture of genomic DNA.These advances, coupled with the prospect of the loss of significant plant genetic resources throughout the world,have led to the establishment of DNA bank for the storage of genomic DNA. The conserved DNA will have numerous uses viz, molecular phylogenetics and systematics of extinct taxa, production of previously characterized secondary compounds in transgenic cell cultures, production of transgenic plants using genes from gene families, in vitro expression and study of enzyme structure and function and genomic probes for research laboratories. The vast resources of dried specimens in the world's herbaria may hold considerable DNA that would be suitable for PCR. It seems likely that the integrity of DNA would decrease with the age of specimens. Because there are many types of herbarium storage environments, preservation and collections, there is a need for systematic investigations of the effect of modes of preparation, collection and storage on the integrity of DNA in the world's major holdings. The advantage of storing DNA is that it is efficient and simple and overcomes many physical limitations and constraints that characterize other forms of storage (Adams 1988, 1990, 1997, Adams and Adams 1991, Adams et al 1994). The disadvantage lies in problems with subsequent gene isolation, cloning and transfer but, most importantly, it does not allow the regeneration of live organisms (Maxted et al., 1997). DNA banking is yet to catch up in spices. DNA samples of over 600 genotypes of spices is stored in the DNA bank of Indian Institute of Spices Research (IISR), Calicut.

FUTURE FOCUS

In contrast to the prevailing attitude among conservation biologists, globally there is considerable interest among cryobiologists in the use of in vitro, cold and ultra-cold technology for germplasm conservation. The procedures for plant material are given indepth coverage by Reed et al. (2004) who stress equally the ecological and plant/germplasm health aspects preceeding and following storage. Panis and Lambardi (2006) discussed the evolution of technologies for plant material, covering cell suspensions and callus cultures of herbaceous species, pollen, shoot meristems, woody species, as well as seed and embryonic axes. The ex situ gene bank at Gatersleben in Germany houses 986 potato accessions are cryopreserved and trials on other species are performed (Börner 2006). The National Bureau of Plant Genetic Resources (NBPGR), New Delhi has over 702 accesion of various spices, medicinal and aromatic crops in its cryo gene bank (Table. 3). Keller et al. (2008) make the point that cryopreservation affords the best of conditions for the long-term maintenance of plant material, particularly for vegetatively propagated species. Cryopreservation is the only viable method available for long-term preservation of the both plant and animal origin species. As an ultimate aim of cryoconservation is the reintroduction of preserved material into the field, it is appropriate at this point to consider the concept of restoration a little more closely. In terms of ultimate ecosystem restoration, the possibilities raised by in vitro conservation, including cryoconservation, do not mean that species selection should merely take random advantage of what germplasm has or can be conserved as there are many genetic, physiological and phenotypic considerations to be taken into account (Kramer and Havens 2009). The establishment and maintenance of biological resource centers (BRCs) or germplasm conservatories requires careful attention to implementation of reliable preservation technologies and appropriate quality control to ensure that recovered cultures and other biological materials perform in the same way as the originally isolated culture or material. There are many types of BRC that vary both in the kinds of material they hold and in the purposes for which the materials are provided. All BRCs are expected to provide materials and information of an appropriate quality for their application and work to standards relevant to those applications. There are important industrial, biomedical, and conservation issues that can only be addressed through effective and efficient operation of BRCs in the long term. This requires a high degree of expertise in the maintenance and management of collections of biological materials at ultra-low temperatures, or as freeze dried material, to secure their long-term integrity and relevance for future research, development, and conservation. The application of cryogenic preservation in biotechnology and medicine has recently been a topic of interest. The use of cryogenic preservation in this area has given new horizon to this field of applications.

REFERENCES

1. Adams, R.P. and Adams, J.E., 1991. Conservation of Plant Genes: DNA Banking and In Vitro Biotechnology. Academic Press, New York.
2. Adams, R.P., 1988. The preservation of genomic DNA: DNA Bank Net. Amer. J. Bot., 75: 156.
3. Adams, R.P., 1990. The preservation of Chihuahuan plant genomes through in vitro biotechnology:DNA Bank-Net, a genetic insurance policy.
4. In: Third Symposium on Resources of the Chihuahuan Desert Region, (Eds.) Powell, A.M., Hollander, R.R., Barlow, J.C., McGillivray, W.B. and Schmidly, D.J. Printech Press, Lubbock, TX, pp.1–9.
5. Adams, R.P., 1997.Conservation of DNA: DNA banking. In: Biotechnology and Plant Genetic Resources: Conservation and Use, (Eds.) Callow, J.A. Ford-Loyd, B.V. and Newbury, H.J. Biotechnology in Agriculture Series, No. 19. CAB International, pp. 163–174.
6. Adams, R.P., Miller, J.S., Golenberg, E.M. and Adams, J.E., 1994. Conservation of Plant Genes 11: Utilization of Ancient and Modern DNA. Missouri Botanical Garden Press. St.Louis, MI, 276 pp. 227.
7. Ajith, A., 1997. Micropropagation and genetic fidelity studies in Piper longum L. In: Biotechnology of Spices, Medicinal and Aromatic Plants, (Eds.) Edison, S., Ramana,K.V., Sasikumar, B., Nirmal Babu K. and Santhosh J. Eapen. Indian Society for Spices, Calicut, India, p. 94–97.
8. Alexander, M.P., Ganeshan, S. and Rajasekharan, P.E., 1991. Freeze preservation of capsicum pollen (Capsicum annuum) in liquid nitrogen (–196°C) for 42 months:
9. Effect on viability and fertility. PlantCell Incompatibility Newsletter, 23: 1–4. NBPGR2011 Annual Report, National Bureau of Plant Genetic Resources, 2010-11
10. Ashmore, S.E., 1997. Status report on the development and application

of in vitro techniques for conservation and use of plant genetic resources. International Plant Genetic Resources Institute, Rome, Italy. 67 p.

11. Ausubel F M, Brent R, Kingston R E, Moore D D, Seidman, J G, Smith J A & Struhl K 1995 Short Protocols in Molecular Biology, John Wiley & Sons, New York, Ch.2.4.
12. Balachandran, S.M., Bhat, S.R. and Chandel, K.P.S., 1990. In vitro clonal multiplication of turmeric (Curcuma longa) and ginger (Zingiber officinale Rosc.). Plant Cell Reports, 3: 521–524.
13. Bertaccini, A., Botti, S., Tabanelli, D., Dradi, G., Fogher, C., Previati, A., da Re, F., Nicola, S., Nowak, J. and Vavrina, C.S., 2004. Micropropagation and establishment of miteborne virus-free garlic Allium sativum), Acta Horticulturae, 631: 201–206.
14. Börner A. Preservation of plant genetic resources in the biotechnology era. Biotech J 1: 1393– 1404; 2006.
15. Broome, O.C. and Zimmerman, R.N., 1978. In-vitro propagation of black pepper. Hort. Sci., 43: 151–153.
16. Chand, P.K., Mandal, J. and Pattnaik, S., 2000. Alginate encapsulation of axillary buds of Ocimum americanum L. (hoary basil), O.basilicum L. (sweet basil), O. gratissimum L.
17. (shrubby basil), and O. sanctum L. (sacred basil). In vitro Cellular and Developmental Biology (Plant), 36(4): 287–292.
18. Chandrappa, H.M., Shadakshari, Y.G., Sudharshan, M.R. and Raju, B. ,1997. Preliminary yield trial of tissue cultured cardamom selections. In: Biotechnology of Spices, Medicinal and Aromatic Plants,(Eds.) Edison, S., Ramana, K.V., Sasikumar, B., Nirmal Babu K. and Santhosh J. Eapen. Indian Society for Spices, Calicut, India,p. 102–105.
19. Chaudhury, R. and Chandel, K.P.S., 1994. Germination studies and cryopreservation of seeds of black pepper (Piper nigrum L.): A recalcitrant species. Cryoletters, 15: 145–150.
20. Choudhary, R. and Chandel, K.P.S., 1995. Studies on germination and cryopreservation of cardamom (Elletaria cardamomum Maton.) seeds. Seed Science and Biotechnology,23(1): 235–240.
21. Choudhury, R. and Malik, S.K., 2004. Genetic conservation of plantation crops and spices using cryopreservation. Ind. J. Biotechnology, 3: 348–358.
22. Conci, V.C. and Nome, S.F., 1991. Virus free garlic (Allium sativum L.) plants obtained by thermotherapy and meristem tip culture. J. Phytopathol., 132: 186–192.

23. Decruse, S W and Seeni, S 2003. Seed cryopreservation is a suitable storage procedure for arange of Piper species. Seed Sci. Technol. 31, 213-217.

24. Du, Y.Q., Zhu, J.Z. and Shen, W.P., 2004. Study on virus elimination technology of garlic in Jiading by shoot tip culture. Acta Agriculturae Shanghai, 20(1): 9–12.

25. Elena Popova, Haeng-Hoon Kim and Kee-Yoeup Paek 2010. Cryopreservation of coriander (Coriandrum sativum L.) somatic embryos using sucrose pre-culture and air desiccation Scientia Horticulturae 124 (4), 522-528.

26. Engelmann F. Use of biotechnologies for conserving plant biodiversity. Acta Hort 812: 63–82; 2009.

27. Engelmann, F., 1997. In vitro conservation methods. In: Biotechnology and Plant Genetic Resources:Conservation and Use, (Eds.) Callow, J.A., Ford-Loyd, B.V. and Newbury, H.J. Biotechnology in Agriculture Series, No. 19. CAB International, UK, pp. 119–161.

28. Engelmann, F., 2000. Importance of cryopreservation for the conservation of plant genetic resources. In: Cryopreservation of Tropical Plant Germplasm, (Eds.) Engelmann, F. and Takagi, H. Current Research Progress and Application. Japan International Research Centre for Agricultural Sciences, Japan/International Plant Genetic Resources Institute, Rome, Italy, pp. 8–20.

29. Engelmann, F., Benson, E.E., Chabrillange, N., Gonzalez-Arnao, M.T., Mari, S., MichauxFerriere, N., Paulet, Glazmann, J.C. and Charrier, A., 1994. Cryopreservation of several tropical plant species using encapsulation/dehydration of apices. In. Proc.VIIIth IAPTC Meeting, Firenze, Italy.

30. Engelmann, F., Dumet, D., Chabrillange, N., Abdelnour-Esquivel, A., Assy-Bah, B., Dereuddre, J. and Duval, Y., 1995. Cryopreservation of zygotic and somatic embryos from recalcitrant and intermediate-seed species. Plant Genet. Resources Newsletter, 103: 27–31.

31. Frankel, O.H., 1975. Genetic resources survey as a basis for exploration. In: Crop Genetic Resources for Today and Tomorrow, (Eds.) O.H. Frankel and J.H. Hawkes. Cambridge University Press, Cambridge. pp. 99.

32. Gayle, M., Volk, Nicholas Maness and Kate Rotindo, 2004. Cryopreservation of Garlic (Allium sativum L) using Plant Vitrification Solution 2. CryoLetters, 25: 219–226.

33. Gonzalez-Arnao M.T, Claudia Esther Lazaro-Vallejo, Florent Engelmann, Roberto GamezPastrana, Yolanda Maria Martinez-Ocampo, Miriam

Cristina Pastelin-Solano and

34. Carlos Diaz-Ramos 2009. Multiplication and cryopreservation of vanilla (Vanilla planifolia 'Andrews') In Vitro Cellular & Developmental Biology - Plant 45 (5): 574-582.
35. Grout B W W 1995 Introduction to the in vitro preservation of plant cells, tissues and organs. In: Grout, B.W.W. (Ed.), Genetic Preservation of Plant Cells in Vitro.Springer–Verlag, Berlin. pp : 1–20.
36. Haeng-Hoon Kim, Eun-Gi Cho, Hyung-Jin Baek, Chang-Yung Kim, ER Joachim Keller and Florent Engelmann, 2004. Cryopreservation of garlic shoot tips by vitrification:
37. Effects of dehydration, Rewarming, unloading and regrowth conditions CryoLetters, 25: 59–70.
38. Haeng-Hoon Kim, Jung-Bong Kim, Hyung-Jin Baek, Eun-Gi Cho, Young-Am Chaand Florent Engelmann, 2004. Evalution of DMSO Concentration in Garlic Shoot tips during a Vitrification Procedure. CryoLetters, 25: 91–100.
39. Haeng-Hoon Kim, Ju-Won Yoon, Jung-Bong Kim, Florent Engelmann and Eun-Gi Cho, 2005. Thermal analysis of Garlic Shoot tips during a Vitrification Procedure.CryoLetters, 26(1): 33–44.
40. Himeno, H. and Sano, K., 1995. Synthesis of crocin, picrocrocin and safranal by saffron stigma like structures proliferated in vitro. Agricultural Biology and Chemistry, 51 (9):2395–2400.
41. Hoekstra, F.A., 1995. Collecting pollen for genetic resources conservation. In: Collecting Plant Genetic Diversity: Technical Guidelines, (Eds.) Guarino, L., Rao, V.R. and Reid, R.CAB International, Wallingford, UK, pp. 527–550.
42. Hori, H., Enomoto, K. and Nakaya, H., 1988. Induction of callus from pistils of Crocus sativus L. and production of colour components in the callus. Plant Tissue Culture Letters, 5:72 –77.
43. Hosoki, T. and Sagawa, Y., 1977. Clonal propagation of ginger (Zingiber officinale Rosc.) through tissue culture. Horticulture Science, 12: 451-452.
44. Hyung-Jin Baek, Haeng-Hoon Kim, Eun-Gi Cho, Young-Am Chae and Florent Engelmann, 2003. Importance of explant size and origin and of preconditioning treatments for cryopreservation of garlic shoot apices by vitrification. CryoLetters,24: 381–388.
45. Ilahi, I. and Jabeen, M., 1992. Tissue culture studies for micropropagation and extraction of essential oils from Zingiber officinale Rosc. Pakistan

Journal of Botany, 24(1): 54–59.

46. IPGRI, 1996 Pragamme activities, germplsm maintenance and use. In: Annual Report, IPGRI. Rome. pp. 56–65.

47. Iwananga, M., 1994. Role of International organisations in global genetic resource management. In:Proc. 27th International Symposium on Tropical Agriculture Research, Japan International Research Centre for Tropical Agricultural Sciences, Ministry of Agriculture, Forestry and Fisheries, Tsukuba, Japan, August, 25–26,1993, p.1–6.

48. Johnson, T.S., Ravishanker, G.A and Venkataraman, L.V., 1996. Biotransformation of ferulic acid vanillamine to capsaicin and vanillin in immobilised cell cultures of Capsicum frutescens PlantCell, Tissue and Organ Culture, 44(2): 117–123

49. Jung-Bong Kim, Haeng-Hoon Kim, Hyung-Jin Baek, Eun-Gi Cho, Yong-Hwan Kim and Florent Engelmann, 2005. Changes in sucrose and glycerol content in garlic shoot tips during freezing using pvs3 solution. CryoLetters, 26(2): 103–112.

50. Kartha K K & Engelmann F 1994 Cryopreservation and germplasm storage. In Plant Cell and Tissue Culture. Vasil, I. K. and Thorpe, T. A. (eds.), Kluwer Academic Publishers, Dordrecht/Boston/London. pp : 195230.

51. Kartha, K.K., 1985. In: Cryopreservation of Plant Cells and Organs, (Ed.) Kartha, K.K. CRC Press, Boca Raton, Florida, pp. 243–267.

52. Keller E. R. J.; Kaczmarczyk A.; Senula A. 2008. Cryopreservation for plant genebanks—a matter between high expectations and cautious reservation. CryoLetts 29: 53–62.

53. Keller, J., 1991. In vitro conservation of haploid and diploid germplasm in Allium cepa L. Acta Hortic.,289: 231–232.

54. Leigh, E. Towill and Remi Bonnart, 2003. Cracking in a vitrification solution during cooling or warming does not effect growth of cryopreserved mint shoot tips. CryoLetters, 24: 341–346.

55. Lukose, R., Saji, K.V., Venugopal, M.N. and Korikanthimath, V.S., 1993. Comparative field performance of micropropagated plants of cardamom (Elettaria cardamomum).Indian Journal of Agricultural Sciences, 63(7): 417–418.

56. Madhusoodanan, K.J., Kuruvilla, K.M., Vadiraj, B.A., Radhakrishnan, V.V. and Thomas, J., 2005. On farm evaluation of tissue culture vanilla plants vis-à-vis vegetative cuttings. Proceedings of ICAR National Symposium on Biotechnological Interventions for Improvement of

Horticultural Crops: Issues and Strategies, Kerala Agricultural University, Trissur, Kerala, India, p. 89–90.

57. Mandal, J., Pattnaik, S.and Chand, P.K. 2000. Alginate encapsulation of axillary buds of Ocimum americanum L. (hoary basil), O. Basilicum L. (sweet basil), O. Gratissimum L.

58. (shrubby basil), and O. Sanctum. L. (sacred basil) In Vitro Cellular & Developmental Biology - Plant 36(4): 287-292.

59. Maxted N, Brian, F and Hawkes, J G. 1997. Plant Genetic Conservation : the in situ approach. Springer. 446 pp

60. Mc Cown, B.H. and Amos, R., 1979. Initial trials of commercial micropropagation with birch. Proc.Inter. Plant Pro. Soc., 29: 387–393.

61. Minoo Divakaran and Nirmal Babu, K. 2009.Micropropagation and In Vitro Conservation of Vanilla (Vanilla planifolia Andrews) pp 129-138. In SM Jain and PK Saxena (eds)

62. Springer Protocols, Methods in Molecular Biology 547, Protocols for In Vitro Cultures and Secondary Metabolite Analysis of Aromatic and Medicinal Plants, The Humana Press, (Springer), USA.

63. Minoo Divakaran, Nirmal Babu K. and Pete,r K. V. 2011. Cryopreservation of pollen and inter specific hybridization in important orchid species V. planifolia and V. aphylla.

64. National Consultation for Production and Utilisation of Orchids 19-21 February, 2011 National Research Center for Orchids, Pakyong- 737106, Sikkim. Abstract.pp.98-99.

65. Minoo Divakaran, Nirmal Babu K and Michel Grisoni 2010. Biotechnological applications In Vanilla, pp. 51-73 In Eric Odoux and Michel Grisoni (eds) Vanilla, CRC Press,Boca Raton, USA.

66. Minoo, D., 2002. Seedling and somaclonal variation and their characterization in Vanilla. Ph.D. Thesis, Calicut University, Kerala, India.

67. Murashige, T. and Skoog, F., 1962. A revised medium for rapid growth and bioassays with tobacco tissue cultures. Physiol Plant, 15: 473–493.

68. Nadagauda, R S., Mascarenhas, A F & Madhusoodanan K J. 1983. Clonal multiplication of Elettaria cardamomum Maton. by tissue culture. J. Plantn Crops 11 : 60-64.

69. Nag, K.K. and Street, H.E., 1975. Freeze preservation of cultured plant cells, 11. The freezing and thawing phases. Physiol. Plantarum, 340: 254–260

70. Nair RR and Dutta Gupta 2003 Somatic embryogenesis and plant

regeneration in black pepper (Piper nigrum L.): Direct somatic embryogenesis from tissues of germinating seeds and ontogeny of somatic embryos. J. Hortic. Sci. Biotechnol, 78 : 416-421

71. Nirmal Babu, K., Geetha, S.P., Minoo D., Yamuna, G., Praveen, K., Ravindran, P.N. and Peter, K.V. 2007. Conservation of Spices Genetic Resources through In Vitro Conservation and Cryo- Preservation. pp. 210-233. In KV Peter and Z Abraham (ed). Biodiversity in Horticultural Crops Vol 1, Daya Publishing house, New Delhi Nirmal Babu, K., Geetha, S.P., Minoo, D., Ravindran, P.N. and Peter, K.V., 1999. In vitro conservation of germplasm. In: Biotechnology and its Application in Horticulture, (Ed.) S.P. Ghosh. Narosa Publishing House, New Delhi, pp. 106–129.
72. Nirmal Babu, K., Minoo, D., Geetha, S.P., Ravindran, P.N. and Peter, K.V., 2005. Advances in Biotechnology of Spices and Herbs. Ind. J. Bot. Res., 1(2): 155–214.
73. Nirmal Babu, K., Ravindran, P.N. and Peter, K.V., 1997. Protocols for Micropropagation of Spices and Aromatic Crops. Indian Institute of Spices Research, Calicut, India, p. 35.
74. Nirmal Babu, K., Ravindran, P.N. and Sasikumar, B., 2003. Field evaluation of tissue cultured plants of spices and assessment of their genetic stability using molecular markers. Final Report submitted to Department of Biotechnology, Government of India, pp. 94.
75. Nirmal Babu, K., Rema, J., Sree Ranjini, D.P., Samsudeen, K. and Ravindran, P.N., 1996. Micropropagation of an endangered species of Piper, P. barberi Gamble and its conservation. Journal of Plant Genetic Resources, 9(1): 179–182.
76. Niwata, E., 1995. Cryopreservation of apical meristems of garlic (Allium sativum L.) and high subsequent plant regeneration. CryoLetters, 16: 102–107.
77. Panis B.; Lambardi M. Status of cryopreservation technologies in plants (crops and forest trees). In: Ruane J.; Sonnino A. (eds) 2006.The role of biotechnology in exploring and protecting agricultural genetic resources. FAO, Rome, pp 61–78;
78. Parani, M., Anand, A. and Parida, A., 1997. Application of RAPD finger printing in selection of micropropagated plants of Piper longum for conservation. Current Science, 73(1): 81–83.
79. Peter, K.V., Ravindran, P.N., Nirmal Babu, K., Sasikumar, B., Minoo, D., Geetha, S.P. and Rajalakshmi, K., 2002. Establishing In vitro Conservatory of Spices Germplasm. ICAR Project report. Indian Institute

of Spices Research, Calicut, Kerala, India, pp. 131.

80. Philip, V.J, Joseph, D., Triggs, G.S. and Dickinson, N.M., 1992. Micropropagation of black pepper (Piper nigrum L.) through shoot tip cultures. Plant Cell Report, 12: 41-44.

81. Phunchindawan, M., Hirata, K., Sakai, A. and Miyamot K. Cryopreservation of encapsulated shoot primordia induced in horseradish (Armoracia rusticana) hairy root cultures Plant Cell Reports 16(7): 469-473. DOI: 10.1007/BF01092768

82. Quatrano, R.S., 1968. Freeze preservation of cultured flax cells using DMSO. Plant, 43: 2057– 2061.

83. Rajasekharan, P.E. and Ganeshan, S., 2003. Pollen cryopreservation in Capsicum species: A feasibility study. Capsicum and Eggplant Newsletter, 22: 87–90.

84. Ravindran, P.N., Nirmal Babu, K., Saji, K.V., Geetha, S.P., Praveen, K. and Yamuna, G., 2004. Conservation of Spices genetic resources in in vitro gene banks. ICAR Project report. Indian Institute of Spices Research, Calicut, Kerala, India pp. 81.

85. Ravishankar, G A., Sarma K S., Venkataraman L V. and Kalyan A K. 1988. Effect of nutritional stress on capsaicin production in immobilized cell cultures of Capsicum annuum. Curr. Sci. 57 : 381-383.

86. Ravishankar, G.A., Sudhakar, J.T. and Venkataraman, L.V., 1993. Biotechnological approach of in vitro production of capsaicin. In: Proceedings of the National Seminar on Post Harvest Technology of Spices, Trivandrum, p. 75–82.

87. Reed B. M.; Engelmann F.; Dulloo M. E.; Engels J. M. M. 2004. Technical guidelines for the management of field and in vitro germplasm collections. IPGRI Handbooks for Genebanks No. 7. IPGRI, Rome, Italy.

88. Rekha Chaudhury and Malik, S.K., 2004. Genetic conservation of plantation crops and spices using cryopreservation. Indian Journal of Biotechnology, 3(3): 348–358.

89. Sajina, A., Minoo, D., Geetha, S.P., Samsudeen, K., Rema, J., Nirmal Babu, K., Ravindran, P.N. and Peter, K.V., 1997. Production of synthetic seeds in a few spice crops. In: Biotechnology of Spices, Medicinal and Aromatic Plants, (Eds.) Edison, S., Ramana, K.V., Sasikumar, B., Nirmal Babu, K. and Santhosh, J.E. Indian Society for Spices, Calicut, India, p. 65–69.

90. Sambrook J, Fritsch E F & Maniatis T 1989 Molecular Cloning – a laboratory manual. Vol.3.Cold Spring Harbor Laboratory Press, New

York.

91. Schenk R U & Hildebrandt A C 1972 Medium and techniques for induction and growth of monocotyledonous and dicotyledonous plant cell cultures. Can. J. Botany, 50:199-204.

92. Sharma, T.R., Singh, B.M. and Chauhan, R.S., 1994. Production of encapsulated buds of Zingiber officinale Rosc. Plant Cell Reports, 13: 300–302.

93. Shatnawi, M. A., Johnson K.A. and Torpy F.R. (2004) In vitro propagation and cryostorage of Syzygium francissi (Myrtaceae) by the encapsulation-dehydration method, In Vitro Cellular & Developmental Biology - Plant 40 (4): 403-407.

94. Baghdadi, S.H, Shibli, R.A, Syouf M.Q., Shatnawi MA., Arabiat, A. and Makhadmeh I.M (2010) Cryopreservation by encapsulation-vitrification of embryogenic callus of wild crocus (Crocus hyemalis and C. moabiticus) Jordan J. Agricultural Sciences. 6 (3) 436-442.

95. Umetsu, H., Wake, H., Saitoh, M., Yamaguchi, H. and Shimomura, K., 1995. Characteristics of cold preserved embryogenic suspension cells in fennel Foeniculum vulgare Miller.Journal of Plant Physiology, 146(3): 337–342.

96. Venkataraman, L.V., Ravishanker, G.A., Sarma, K.S. and Rajasekaran, T., 1989. In vitro metabolite production from saffron and capsicum by plant tissue and cell cultures.

97. In: Tissue Culture and Biotechnology of Medicinal and Aromatic plants, (Eds.) Kukreja et al. CIMAP, Lucknow, India, p. 147–151.

98. Williams J G K, Kubelik A R, Livak K J, Rafalski J A & Tingey S V 1990 DNA polymorphisms amplified by arbitrary primers are useful as genetic markers.Nucleic acid research 18:6531-6535

99. Withers, L.A., 1985. Cryopreservation of cultured cells and meristems. In: Cell Culture andSomatic Cell Genetics of Plants, Vol. 2: Cell Growth, Nutrition, Cyto-differentiation and Cryopreservation, (Ed.) I.K. Vasil. Academic Press, Orlando, Florida, pp. 253–316.

100. Withers, L.A., 1987. Long-term preservation of plant cells, tissues and organs. Oxford Surveys of Plant Mol. and Cell Biology, 4: 221–272.

101. Withers, L.A., 1991. Biotechnology and plant genetic resources Conservation. In: Plant Genetic Resources Conservation and Management: Concepts and Approaches, (Eds.) R.S. Paroda and R.K. Arora. IBPGR, New Delhi, pp. 273–297.

102. Yamuna, G. 2007. Studies on Cryopreservation of Spices Genetic

Resources Ph. D Thesis, University of Calicut, Kerala, India.

103. Yamuna, G, Sumath,i V., Geetha, S. P., Praveen, K., Swapna, N. and Nirmal Bab, K. 2007. Cryopreservation of In Vitro grown shoot of Ginger (Zingiber officinale Rosc). CryoLetters.28(4):241-252

Chapter 8

DIVERSITY AND GENETIC EROSION OF ANCIENT CROPS AND WILD RELATIVES OF AGRICULTURAL CULTIVARS FOR FOOD: IMPLICATIONS FOR NATURE CONSERVATION IN GEORGIA (CAUCASUS)

Maia Akhalkatsi, Jana Ekhvaia and Zezva Asanidze

Ilia State University Georgia

INTRODUCTION

The interpretation of a healthy diet is one of the dilemmas for our modern civilization. Advances in agriculture are mainly directed at increasing food production to solve problems of a growing human population. However, food security remains a problem to ensure healthy food and to prevent human disease. These two tendencies often do not coincide. At present, the selective breeding programs of crops are mainly oriented toward the production of high-yielding varieties of genetically enhanced cultivars of cereals that have increased growth rates, increasing the percentage of usable plant parts and resistance against crop diseases. This initiative is linked to what began in the 1960s and was named by William Gaud (of USAID) a "Green Revolution" (Davies, 2003). It was a product of globalization as evidenced in the creation of international agricultural research centres to introduce new crop varieties around the world. This process caused a significant increase in total cereal production and daily calorie supply in developing countries between the 1960s and 1990s (Davies, 2003). However, this process has caused the gradual replacement of traditional crop varieties, and as a result has had a dramatic effect on agrobiodiversity in many countries. Particularly impacted have been the traditional landraces used by local peoples for thousands of years and this has affected the health of these communities. Georgia, located in the South Caucasus, owns one of the oldest agricultural traditions. The name of the country is "Sakartvelo" in the Georgian language but its common name "Georgia" is semantically linked to Greek (γεωργία, transliterated geōrgía) and Latin (georgicus) roots meaning "agriculture" (Javakhishvili, 1987). Many Georgian endemic species and local varieties of wheat, barley, legumes, grapevine and fruits are known (Ketskhoveli, 1957). The traditional use of local cultivars is considered to be a reason for human longevity in the Caucasus region (Fox, 2004). Over five percent of the male Georgian centenarians were reputed to have been over age 120 in 1959 (Garson, 1991). The percentage of males over age 70 was 0.9% in 1959 and 1.07% of women were over 70. However, these values had diminished by 1970 to 0.66% and 0.86%, respectively. At present, no exact data are available, but longevity has obviously diminished (Fox, 2004).

Archaeological data clearly show that the Caucasus region (and Georgia in particular) was settled from prehistoric time and agriculture was developed during the early Neolithic era (Javakhishvili, 1987). The information about the wide chronological intervals in the archaeological materials connected with the history of mankind in the Caucasus starts from the Early Pleistocene. The 1.7-Myr-old specimens of small-brained hominids are found in the Caucasus at Dmanisi, located in Southern Georgia (Fig.1), which is the earliest known hominid site outside of Africa (Gabunia & Vekua, 1995; Finlayson, 2005). This speciman has been classified as Homo erectus senso lato, which is a very early type of H. ergaster and/or a new taxon, H. georgicus (Gabunia et al., 2002). The next chronological interval in the archaeological materials is connected with the period of Late Middle Palaeolithic and Early Upper Palaeolithic periods demonstrating patterns of mobility, land-use, and hunting of Neanderthal and modern human competition within the South Caucasus (Adler & Bar-Oz, 2009). Neanderthals invaded the Caucasus region at an unknown time and modern humans may have occupied the region alongside them from ~40 Ka before the present (BP). According to the archaeological material from different caves in Georgia (Tushabramishvili, 2011) and the northern Caucasus (Ovchinnikov et al., 2000), the final replacement of the Neanderthals by modern humans might be occurred here ~28 Ka BP. The Upper Palaeolithic archaeological findings at Dzudzuana Cave (Fig.1), Imereti region, Georgia, revealed remnants of wool (Capra caucasica) and dyed fibers of wild flax (Linum usitatissimum) dated to ~36–34 Ka BP (Adler & Bar-Oz, 2009). The Dzudzuana Cave flax fibers have clearly been modified, cut, twisted and dyed black, gray, turquoise and pink, most likely with locally available natural plant-derived pigments (Kvavadze et al., 2009). E. Kvavadze and colleagues (2009) surmise that this represents the production of colourful textiles for some purpose, perhaps clothing. In general, it is supposed that the microscopic flax fibres are the remains of linen and thread, which would have been used in clothing for warmth, for shoes, to sew together pieces of leather or to tie together packs. The archaeological findings from

Neolithic and Early Bronze periods are rich with plant fossils and seeds of both wild species and local landraces. The ancient findings from Neolithic period of cereal grains in Georgia were discovered from Trialeti Range, Kvemo Kartli region (Arukhlo excavations, Bolnisi district; Fig.1) and Samegrelo region (Dikha-Gudzuba, Zugdidi district; Fig.1) from 6th up to 2nd millennium BC (Melikishvili, 1970). Seven species of cultivated wheat - Triticum aestivum, T. spelta, T. carthlicum, T. macha, T. monococcum, T. dicoccum, T. compactum and one wild relative Aegilops cylindrica have been discovered in Arukhlo, Kvemo Kartli region. Other cereals: millet - Panicum milleaceaum, barley - Hordeum vulgare, Italian millet - Setaria italica, oats - Avena sativa, wild lentil - Lens ervoides and pea -Pisum sativum have been found in the same site. The wheat fields in Arukhlo were irrigated. Very recent studies on einkorn wheat domestication using amplified fragment length polymorphism (AFLP) show that T. boeoticum was domesticated in southeast Turkey in the Karacadag Mountains close to Diyarbakir (Heun et al., 1997). Old Georgian kingdom Diauehi (Diaokhi) is adjacent region to this place. Therefore, it might be considered to be an area where cultivation of cereals occurred in very early historical time. The earliest archaeological finding of cultivar grapevine pips are found in Shulaveri (Fig.1), located near Dmanisi in southern Georgia and dated to ~8.000 years BP (Ramishvili, 1988). A wide range of carbonised seeds, including wild and domesticated grape (Vitis vinifera, V. vinifera subsp. sylvestris), wheat (Triticum sp.), pea (Pisum sativum), rowan (Sorbus sp.) and walnut (Juglans regia) are found in soil samples in Nokalakevi (Fig.1), Western Georgia, dated to the Hellenistic period (Grant et al., 2009).

Figure. 1: Map of Georgia. The administrative regions: 1. Abkhazia; 2. Samegrelo-Upper Svaneti; 3. Guria; 4. Adjara; 5. Racha-Lechkhumi; 6. Imereti; 7. Meskheti- Javakheti; 8. Shida Kartli; 9. Kvemo Kartli; 10. Mtskheta-Mtianeti; 11. Kakheti. The places of archaeological excavations are indicated: Dikha-Gudzuba, Nokalakevi, Dzudzuana cave, Arukhlo, Dmanisi and Shulaveri.

According to N. I. Vavilov (1992), the origin of ancient crop varieties and landraces in Georgia coincides with the period of their primary domestication. Georgia is often considered part of the Near East where many field crops were domesticated. N. I. Vavilov (1992) determined 8 centres of crop origin and diversity. Among them was the fourth centre, which included the South Caucasus, Asia Minor, Iran and Turkmenistan. The main crops domesticated in this centre (which includes Georgia) are wheat, rye, oats, seed and forage legumes, herbs, fruits, and grapes for winemaking; 83 species all tolled. The problem is that there are no concrete data to assess either the current status of local varieties or information about the domestication process in Georgia. The fundamental work on domestication and origin of wheat and barley in this region was done by the famous Georgian botanist V. Menabde (1938, 1948). The agricultural evidence was reported by several other Georgian authors (Ketskhoveli, 1957; Khomizurashvili, 1973; Akhalkatsi et al., 2010). We have studied domestication of wild grapevine (Vitis vinifera subsp. sylvestris) and wild pear (Pyrus caucasica) using morphometric and systematic molecular methods (Ekhvaia & Akhalkatsi, 2010; Ekhvaia et al., 2010; Asanidze et al., 2011) confirming genetic relationships between wild populations and local cultivars of grape and pear. However, complete evaluation of diversity of Georgian local cultivars and crop wild relatives (CWR) has not yet been complete. There are many threats to these oldest of crops in the modern period. In our opinion, the main threat to agrobiodiversity in Georgia is the loss of landraces and ancient crop varieties. Protection measures in the country are still not being implemented at an appropriate rate. National policies and comprehensive measures are urgently needed to address the problem of conserving the genetic resources of ancient crops in Georgia. Thus, we suggest that it is necessary to establish a general overview of the types of crops that are current landraces and primitive forms occurring in Georgia and to publish lists of indigenous landraces and CWRs of cereals, legumes, vegetables and fruits representing direct ancestors, and endemic, rare or endangered species, in order to evaluate the sustainability of their traditional use in terms of nature conservation.

LANDRACE ASSESSMENT

Agriculture in Georgia is characterized by a great diversity of local landraces, varieties and even endemic species of crops. These varieties reveal a high level of adaptation to local climatic conditions and often have high resistance to diseases. Georgians have used these crops for a very long time and their healthy life, reflected by the longevity of individuals in the population, was considered to be connected to their good food. However, there are many threats to these oldest of crops in the modern period, particularly since the

1950s. The loss of local and ancient crop varieties should be considered to be the main threat to agrobiodiversity in Georgia. The known diversity and distribution of local landraces is based on data obtained from archaeological reports, historical manuscripts, ethnography, and botanical field expeditions in different regions of Georgia since 1920s. The oldest known text about Georgian cultivars is a XVII century by the work of Vakhushti Batonishvili "Geographic Description of Georgian Kingdom" (Batonishvili, 1991). Active investigation of Georgian crops began in the 1920s (Ketskhoveli, 1928, 1957; Menabde, 1938, 1948; Dekaprelevich, 1947). These investigations revealed that ancient cultivars of grapevine, fruits, wheat, barley, rye, oats, common millet, Italian millet, legumes, flax, and a number of herbal and spice plants, were still being cultivated in Georgia. The rapid loss of local cultivars of cereals, legumes and flax began in the 1950s and reached an extreme in the 1990s (Akhalkatsi et al., 2010). At present, almost all of Georgia's ancient crops are maintained in conservatory collections and seed banks, but none are present in peasant house gardens in lowland areas. Only the mountain areas contain depositories of the ancient crops of Georgia, where some number of landraces still exist. The process of genetic erosion of ancient crop varieties, however, has begun even in these regions since the 1990s and this presents great concern about the loss of aboriginal crops adapted to high mountain areas (Pistrick et al., 2009). Monitoring of crop diversity is now conducted by international nature conservation institutions and Georgian scientific and nongovernmental organizations to preserve the genetic resources of local cultivars. One of the problems is the deficit of information about the current state of ancient crops and recommendations for their conservation are inadequate. Therefore, it is necessary to assess research needs and implications for conservation and to formulate recommendations for the conservation and on-farm maintenance of Georgian landraces.

Diversity of Ancient Crop Varieties

Reports of the diversity of local landraces in Georgia has to present been published primarily in Georgian- and Russian-language scientific publications (Ketskhoveli, 1928, 1957; Menabde, 1938, 1948; Dekaprelevich, 1947; Kobakhidze, 1974). Databases and international periodicals lack descriptions of this diversity, taxonomy and discussions of the conservation value of landraces. In our opinion it is important to spread information about diversity and conservation needs of local cultivars of Georgia worldwide to support the evaluation of their roles in healthy life of human. Some crops, such as grapevine, wheat, barley and fruit trees are characterized by the highest diversity of landraces in Georgia. Grapevine - Vitis vinifera L. (Vitaceae) shows greatest genetic and morphological variability. About 500 names of autochthonous grapevine varieties known from Georgia are characterized by a wide range of colour gamma and shapes of berries and pips (Javakhishvili, 1987; Ketskhoveli et al., 1960), which points to an evolutionary centre in this region (Vavilov, 1992). These cultivars showed great ampelometric variability and broad adaptability to different climate and soil conditions (Ketskhoveli et al., 1960). Each province of Georgia possesses its own grapevine cultivars adapted to local climate. The varieties are of three forms: 1) Babilo is an old grapevine with stem more than 20 cm in diameter clambering on trees (Fig.2A). 2) Maghlari represents varieties that climb tree trunks (alder, persimmon, mulberry, cherry, beech, chestnuts, etc.) distributed mainly in peasant orchards in western and southern Georgia (Fig.2B). 2) Dablari is used to create typical vineyards (Fig.2C) found in commercial areas. The total area of vineyards in Georgia was 40.000 ha in the 1980s. It has diminished to ca. 25.000 ha today (Bedoshvili, 2010). Forty-four percent of this territory is located in Kakheti region, 26% in Imereti, 15% in Kartli and 15% in almost all regions of Georgia except in the high mountain regions of Khevi, Khavsureti and Tusheti. Forty-one cultivars of grapevine are used as commercial varieties in Georgia. Among them, 27 are technical varieties used for winemaking and 14 are table grapes (Bedoshvili, 2010). Ninetyseven percent of total annual yield is used for winemaking and only 3% as table grapes. Wine is made from landraces: 'Rkatsiteli' (55%); 'Tsolikauri' (10.2%); 'Chinuri' (7%); 'Saperavi' (4%); 'Kakhuri Mtsvane' (3.3%); and, several local and introduced cultivars (20.5%).

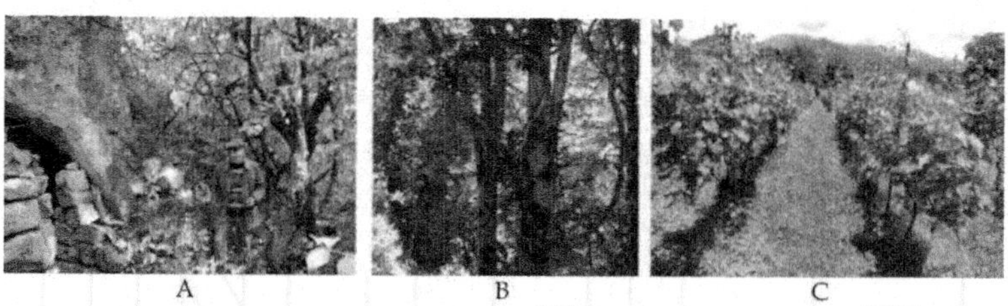

Figure. 2: Different types of vineyeards: A - Babilo, an old grapevine 'Meskhuri Shavi' (diameter 32 cm) belonged to the family of Gogi Natenadze in village Chachkari, Meskheti; B - Maghlari, clambering introduced cultivar Vitis labrusca 'Odessa' on beech tree in village Naghvarevi, Adjara. C - Dablari, typical vineyeard of landraces 'Rkatsiteli' and 'Saperavi' in village Shilda, Kakheti. Photos by Maia Akhalkatsi.

The first known threat to the grapevine in Georgia occurred during the occupation of Georgian territory by the Muslims during the medieval period. They destroyed vineyards and/or led to the destruction of human settlements, where until now are presented local wilding grape cultivars clambering on trees. The vineyards grow on the terraces of Mediterranean type in the historic province of Tao-Klarjeti located now in southern Georgia and in the province of Artvin, Turkey. Since 15th century, the Seljuk Turks occupied this territory and vine terraces disappeared and were covered with trees or grasses. However, we have found peasants in some villages of Meskheti province searching for old cultivars in abandoned settlements. They are replanting naturalized grape cultivars to house gardens. We have found ancient grapevine varieties 'Meskhuri Shavi' and 'Meskhuri Mtsvane' to be frost resistant and growing in high mountain areas in villages Zemo Vardzia (1322 m a.s.l.) and Chachkari (1264 m a.s.l.), Aspindza district; village Zazalo, (1486 m a.s.l.), Adigeni district; and, Karzameti castle near boundary to Turkey, 1450 m a.s.l. One of the oldest Georgian grape cultivar 'Krikina', which is morphologically nearly identical to wild grapevine (V. vinifera subsp. sylvestris [C. C. Gmel.] Hegi), was found in village Rveli, Borjomi district, in house garden of Gaioz Tabatadze who replanted this cultivar from the ruins of the historic village Baniskhevi where his ancestors lived. The grape variety 'Shonuri' ('Lushnu' in Svanetian) adapted to high mountain areas grows in Upper Svaneti province from 1045 to 1400 m a.s.l. Landraces adapted to high elevations are rare and are usually replaced with the introduced grape cultivar V. labrusca 'Odessa'. It is widespread in mountainous villages in all regions of Georgia. Some other rare grapevine cultivars - 'Kachichi', 'Saperavi', 'Sebeli', 'Jvarisa', have been found on lower elevations in village Gvimbrala, Lentekhi district, Lower Svaneti province. Rare grape cultivars - 'Aladasturi', 'Tsulukidzis Tetra' and 'Tskhvedanis Tetra' have been found in village Tabori and 'Usakhelouri' in village Zubi, Tsageri district, Lechkhumi province. Rare landrace 'Chkhaveri' was found in village of Merisi, at 474 m a.s.l. Adjara province. In 1860, the V. vinifera was virtually wiped out in the places of its origin, when an aphid, Phylloxera vastatrix was accidentally introduced into France, and within a few years had ravaged all vineyards in Europe and in Georgia as well. Currently, almost all Georgian grape varieties are grafted on rootstocks of American grapevines - V. riparia, V. rupestris and V. berlandieri and their hybrids are resistant to Phylloxera. This disaster made it necessary to undertake urgent steps for ex situ conservation of old, endangered and autochthonous grapevine varieties by establishing living collections in Georgia; this was begun in the 1930s. The collections of plant genetic resources were established in research institutes, which have been under reforms since 1990s and operating with diminishing funding to maintain the collections. In 2003, 929 varieties were protected in the living collections. Among them, 701 were cultivars obtained from selective breeding and only 248 of the 524 autochthonous Georgian varieties remain. These collections of the State Agricultural University were located in Dighomi (573 cultivars) and Mukhrani (155 cultivars), and, the collections in Telavi (226 cultivars) and Skra (75 cultivars) belonged to the Georgian research Institute of Horticulture, Viticulture and Winemaking (Maghradze, et al., 2010). Recently, these collections have been closed. Nevertheless, some effort has been made to establish new collections in Telavi (573 accessions), Skra (440) and Vachebi (312) in 2008. Three other new collections were set up by Saguramo "Centre for Grapevine and Fruit Tree Planting Material Propagation" (ca 400 accessions), "Kindzmarauli" and "Shumi" wineries (as a total 149 accessions). Two new collections were established in Italy by the University of Milan (Maghradze, et al., 2010). Some Georgian cultivars are in living collections abroad in Russia, Moldova and Germany as well.

Table 1: List of wheat species distributed in Georgia by V. Menabde (1948, 1961). The accepted names are added from web-page: http://www.ars-grin.gov/cgibin/npgs/html/splist.pl?28515. The status of species is based on phylogenetic studies of V. Menabde (1948, 1961): EG - endemic of Georgia; W- wild; PS - primary species; SP - secondary species; IS - introduced species. Ploidy levels are indicated

N	*Taxon names by Menabde, 1948, 1961*	*Taxon accepted name by ARS-GRIN, USDA 2011*	*Status*	*Ploidy levels*
	T. aegilopoides Balansa ex Körn. (=*T. boeticum* Boiss.)	*T. monococcum* subsp. *aegilopoides* (Link) Thell.	W	2n
	T. monococcum L.	*T. monococcum* L.	PS	2n
	T. timopheevii Zhuk.	*T. timopheevii* (Zhuk.) Zhuk. subsp. *timopheevii*	EG, PS	4n
	T. chaldicum Menabde	*T. timopheevii* (Zhuk.) Zhuk. subsp. *armeniacum* (Jakubz.) Slageren	W	4n
	T. dicoccoides Körn.	*T. turgidum* L. subsp. *dicoccoides* (Körn. ex Asch. & Graebn.) Thell.	W	4n
	T. palaeo-colchicum Menabde	*T. turgidum* L. subsp. *palaeocolchicum* Á. Löve & D. Löve	EG, SP	4n
	T. dicoccum Schuebl.	*T. turgidum* L. subsp. *dicoccon* (Schrank) Thell.	SP	4n
	T. durum Desf.	*T. turgidum* L. subsp. *durum* (Desf.) Husn.	SP	4n
	T. turgidum L.	*T. turgidum* L. subsp. *turgidum*	SP	4n
	T. carthlicum Nevski (=*T. ibericum* Menabde; *T. persicum* Vavilov ex Zhuk.)	*T. turgidum* L. subsp. *carthlicum* (Nevski) Á. Löve & D. Löve	EG, SP	4n
	T. polonicum L.	*T. turgidum* L. subsp. *polonicum* (L.) Thell.	SP	4n
	T. turanicum Jakubz.	*T. turgidum* L. subsp. *turanicum* (Jakubz.) Á. Löve & D. Löve	IS	4n
	T. abyssinicum Vavilov	*T. dicoccon* subsp. *abyssinicum* Vavilov	IS	4n
	T. vulgare Villars	*T. aestivum* L.	SP	6n
	T. macha Dekapr. & Menabde	*T. aestivum* L. subsp. *macha* (Dekapr. & V.L. Menabde) Mackey	EG, PS	6n
	T. spelta L.	*T. aestivum* subsp. *spelta* (L.) Thell.	IS	6n
	T. sphaerococcum Percival	*T. aestivum* subsp. *sphaerococcum* (Percival) Mackey	IS	6n
	T. compactum Host	*T. aestivum* subsp. *compactum* (Host) Mackey	SP	6n
	T. zhukovskyi Menabde & Ericzjan	*T. zhukovskyi* V.L. Menabde & Eritzjan	EG, SP	6n

A small living grapevine collection exists in the G. Eliava National Museum in Martvili district, Samegrelo province, founded in 1972 and containing 24 old Colchic grapevine varieties (Eliava, 1992). Seven cultivars of Meskheti region were collected in the research station of Biological Farming Association

Elkana in village Tsnisi, Akhaltsikhe district. Many grape landraces are extinct and do not exist even in living collections. Wheat - Triticum L. (Poaceae) also shows high diversity in Georgia. Nineteen species of wheat from the 26 known species of the genus Triticum have been historically distributed in Georgia (Tab.1). Some of them are endemic species: T. timopheevii, T. zhukovskyi, T. macha, T. carthlicum and T. palaeo-colchicum. Sixteen species, 144 varieties, and 150 forms of wheat were registered in Georgia in the 1940s (Menabde, 1948).

According to V. Menabde (1948), three species from the list are wild – T. boeticum (2n=14), T. dicoccoides (2n=28), T. timopheevii subsp. armeniacum (2n=28); they were mixed with cultivars in the wheat fields and did not exist in natural habitats in Georgia. Sites of T. boeoticum are concentrated in south-eastern Turkey, where this species was probably domesticated (Heun et al., 1997). The current distribution indicates that its weedy races have spread with cultivated cereals far to the west and east. There is evidence that it was found in fields with T. monococcum in Georgia (Menabde, 1948). Since the 1930s their number has diminished and all of these species had disappeared after the 1960s, when non-aboriginal cultivars were introduced in kolkhozis—agricultural farming corporations in Soviet times, changing the species composition in wheat fields. At present, none of these species occur in agricultural fields of Georgia. Three species from the list (Tab.1) are considered by V. Menabde (1948) as primary species (close to the first domesticated species): T. monococcum (2n=14), T. timopheevii (2n=28) and T. macha (2n=42). First two species, T. monococcum - 'Gvatsa Zanduri' and T. timopheevii - 'Cheltha Zanduri', in Georgian, are old autochthonous wheat species distributed mainly in western Georgia - Racha-Lechkhumi, Imereti and Samegrelo. T. timopheevii was growing in a small area in western Georgia together with its hexaploid derivative - T. zhukovskyi, and cultivated einkorn - T. monococcum (Menabde, 1961). These three species represent polyploid series of wheat Zanduri, which was possible to find in peasant farms till 1990s. T. macha is archaeological findings in Dikha-Gudzuba and Shulaveri excavations dated by Neolithic period and was cultivated in Racha-Lechkhumi, Imereti and Samegrelo up to 1950s (Dekaprelevich, 1947). Nine native species of wheat - T. palaeo-colchicum, T. dicoccum, T. durum, T. turgidum, T. carthlicum, T. polonicum, T. aestivum, T. zhukovskyi and T. compactum, are considered by V. Menabde (1948, 1961) as secondary species originated by hybridization with wild and primary species of Triticum, Aegilops spp., Thinopyrum intermedium (Host) Barkworth & D. R. Dewey subsp. intermedium (=Agropyron glaucum [Desf. ex DC.] Roem. & Schult.), and Thinopyrum elongatum (Host) D. R. Dewey (=Agropyron elongatum [Host] P. Beauv.). T. aestivum, T. carthlicum and T. durum have many varieties and cultivars. The four species in the list (Tab.1)

- T. abyssinicum, T. spelta, T. sphaerococcum and T. turanicum represent geographical races introduced from different regions in the historically different times. The traditional wheat fields in all regions of Georgia usually contain several species and varieties (Eritzjan, 1956; Zhizhilashvili & Berishvili, 1980). Bread wheat fields contain: T. aestivum var. erythrospermum 'Tetri dolis puri', T. aestivum var. ferrugineum 'Tsiteli dolis puri', T. aestivum var. lutescens 'Upkho tetri dolis puri', T. aestivum var. milturum 'Upkho tsiteli dolis puri', T. compactum 'Kondara khorbali'. Usually, this combination of wheat taxa is associated with wild weed Makhobeli - Cephalaria syriaca (L.) Schrad. ex Roem. & Schult. (Dipsacaceae) occurring most often in such wheat fields. The seeds of this species are of the same size as wheat and after threshing remain in the harvest. Seeds are ground into a powder and used with wheat to make bread, cakes, etc. It adds a nice flavour but quickly goes rancid. Another combination of varieties was dominated by T. durum 'Shavpkha' composed by T. durum var. apulicum, T. durum var. leucurum, T. durum var. murciense, T. aestivum var. erythrospermum, T. aestivum var. pseudo-barbarossa, T. aestivum var. lutescens, T. compactum var. erinaceum. This population is adapted to dry climate in the lowland areas and in the high elevations up to 1800 m a.s.l. in Javakheti Plateau, where it is sown in early spring. The same character of adaptation to high elevation is typical for the wheat species, T. carthlicum 'Dika', sown on high mountain areas in spring. The combination of varieties dominated by 'Dika' is as follows: T. carthlicum var. rubiginosum, T. carthlicum var. stramineum, T. aestivum var. erythrospermum, T. aestivum var. ferrugineum, T. compactum var. erinaceum.

A B C

Figure. 3: A- Wheat field of the Georgian endemic Triticum carthlicum 'Dika' in research station of the Biological Farming Association Elkana, village Tsnisi, Meskheti; B - Traditional wheat bread in Meskheti; C- Six row barley field with mixture of wild rye Svila (Secale segetale) in village Shilda, Kakheti. Photos by Maia Akhalkatsi.

Wheat fields were planted throughout Georgia at elevations from 300 m to 2160 m a.s.l. We have found this highest location of soft wheat field in the Eastern Greater Caucasus, village Chero in Tusheti (Akhalkatsi et al., 2010). At present, almost none of these traditional wheat varieties and species occur in the territory of Georgia. Only aboriginal varieties of bread wheat still exist in several high mountain regions like Tusheti, Meskheti, Javakheti and Svaneti (Pistrick et al., 2009). Living collections and gene banks preserve the local varieties. The living collection of the Biological Farming Association Elkana has many landraces in village Tsnisi, Akhaltsikhe district (Fig.3A). In 2010, they sowed a 10 ha wheat field. The harvest from this field contained local cultivar T. aestivum var. ferrugineum 'Akhaltsikhis tsiteli dolis puri' and weed Makhobeli. The flour was baked as bread in Tbilisi and as traditional bread in Meskheti (Fig.3B). Barley – Hordeum vulgare L. (Poaceae) is an ancient agricultural crop in Georgia. It was the second most important cereal in Georgia after wheat and main crop in high mountain regions used for bread, forage and production of beer, as well as an attribute of religious rituals and in the folk medicine (Javakhishvili, 1987). Two different names were used for barley in Georgian language - Krtili and Keri. Krtili denotes six-row winter barley (H. vulgare subsp. hexastichon [L.] Čelak.) that was sowed in autumn; Keri refers to two-row summer barley (H. vulgare subsp. distichon [L.] Körn.) sowed in spring (Menabde, 1938). Six-row barley was sown in lowland areas but was cultivated up to 2130 m a.s.l. in Svaneti. Two-row barley was cultivated mainly in high mountain regions. The cultivars of two-row barley H. vulgare var. nutans ‹Akhaltesli› and H. vulgare var. nigrum Willd. ‹Dzveltesli shavpkha› are distributed up to 2100 m a.s.l. in all high mountain areas. H. vulgare var. nutans is mixed in the field with wheat - T. carthlicum ‹Dika›, and the flour is produced from mixed wheat and barley seeds. H. vulgare var. nudum Spenn. ‹Kershveli› was cultivated in Meskheti and Svaneti. Four-row barley (H. vulgare subsp. tetrastichon [Stokes] Čelak.) is rare and the cultivar - H. vulgare var. pallidum Ser. ‹Tetri Keri› occurs only in the high mountain region of Meskheti, Tusheti and Svaneti up to 2100 m a.s.l. These cultivars persist today only in high mountain regions. However, their distribution has been seriously diminished. At present, introduced varieties of barley are widely cultivated in the lowlands and their names are unknown to the local population. Rye – Secale cereale L. (Poaceae) is only a local cultivar of high mountain regions of Georgia (1800-2200 m a.s.l.). Fields of S. cereale (2n=14) are now found only in Upper and Lower Svaneti and Meskheti. Rye was used for making alcohol and as forage. The wild species, S. segetale (Zhuk.) Roshev. (2n=42), called Svila is widespread in wheat and barley fields and is harvested together with them (Fig.3C). The bread of wheat with Svila is considered to be very nutricious and has good taste. An endemic species of rye is S. vavilovii Grossh. (2n=14). It is also called Caucasian rye. This species was found in wheat field in Georgia (Bockelman et al., 2002). We have monitored the place in village Beghleti, Khashuri district in 2008, where Georgian botanists had noted the presence of this species in the wheat fields, but cultivated plots no longer exist in that area. The village has lost of most of its residents and no agriculture is undertaken there. Introduced cultivars and commercial varieties of rye are not used in Georgia. Oats – Avena sativa L. (Poaceae) is a traditionally cultivated plant distributed from 400 to 1400 m a.s.l. It is used only as forage for horses and poultry. Two varieties of oats have been described for Upper Svaneti (Ketskhoveli, 1957) - A. sativa var. aurea Körn. and A. sativa var. krausei Körn. In lowlands, usually, the origin of the seeds is unknown to local farmers. It is purchased in the market and farmers receive no information about their origin. Millet – Panicum miliaceum L. (Poaceae) is very old agricultural plant cultivated in all regions of Georgia. It was used as a supplementary feed (for animals and poultry) and for making alcoholic drinks. At present, it is cultivated only in high mountain regions (1000-1800 m a.s.l.). Several varieties are described in Upper and Lower Svaneti: P. miliaceum var. aureum V.M. Arnold & Shibaiev. - grain yellow or cream; P. miliaceum var. subaereum Körn. - grain grey; P. miliaceum var. griseum Körn. - grain brown; P. miliaceum var. atrocastaneum Batalin ex V.M. Arnold & Shibaiev. - grain black; P. miliaceum var. badium Körn. - grain white (Zhizhilashvili & Berishvili, 1980). The acreage of millet fields declined after introduction of maize in Georgia in 17th century. Italian millet - Setaria italica (L.) P. Beauv. (Poaceae) was cultivated in Colchis, Samegrelo since ancient times. The cultivar - S. italica subsp. colchica (Dekapr. & Kaspar.) Maisaya & Gorgidze was represented with 32 landraces (Maisaia et al., 2005). It can currently be found in the Samegrelo region of western Georgia. Another subspecies - S. italica subsp. moharia (Alef.) H. Scholz., is called Kvrima in Georgian. It was cultivated for a long time but was replaced by maize.

Legumes - peas, lentils, chickpeas, faba beans, common vetch, bitter vetch, chickling vetch, alfalfa, sainfoin and blue fenugreek are traditional crops in Georgia (Tab.2). Green Pea (Pisum sativum) is originated in the South Caucasus. It is grown in house gardens in small amounts for food today. Two species of pea are cultivated in Georgia - P. sativum with white flowers, round white or yellow seeds, and P. arvense with purple flowers, ridged dark coloured seeds. The third wild species P. elatius Steven ex M. Bieb. with dark purple flowers is often found in locations of old settlements, ruins of monasteries and churches and inside castle walls. The local cultivar of green pea, P. sativum subsp. transcaucasicum Govorov, has 14 varieties (Kobakhidze, 1974). Local varieties of Chickpea (Cicer arietinum) are rarely cultivated today. Three subspecies and 24 varieties were available in western Georgia - Racha-Lechkhumi, Svaneti and Imereti up to 1920s (Dekaprelevich & Menabde, 1929). In the 1970s, the same three subspecies - C. arietinum subsp. mediterraneum G. Pop., C. arietinum subsp. eurasiaticum G. Pop. and C. arietinum subsp. orientalis G. Pop., remained, but included only 6 of 24 varieties - C. arietinum subsp. mediterraneum var. ochroleucum A.

Kob., C. arietinum subsp. mediterraneum var. rozeum G. Pop., C. arietinum subsp. eurasiaticum var. aurantiacum G. Pop., C. arietinum subsp. orientalis var. fulvum G. Pop., C. arietinum subsp. orientalis var. rufescens G. Pop., and C. arietinum subsp. orientalis var. rufescens brunneopunctatus A. Kob. (Kobakhidze, 1974). Chickpeas were traditionally available in Svaneti, but by the 1970s only one farmer was sowing it in Kala community village Khe (Zhizhizlashvili & Berishvili, 1983). The Biological Farming Association Elkana is producing local cultivars of chickpea and selling them in market. Lentil (Lens culinaris) was represented in Georgia by two subspecies - L. culinaris subsp. macrosperma N.F. Mattos and L. culinaris subsp. microsperma N.F. Mattos; and 15 varieties (Kobakhidze, 1974). The last subspecies with small seeds was sown in high mountain areas in Javakheti. It was available in Meskheti till 1970s. Lentils were cultivated in Upper Svaneti from prehistoric times, but, at present, it is nearly extinct. In 1980s, three cultivars were described in Svaneti - 1. L. culinaris var. persica Bar. - reddish-brown seeds; 2. L. culinaris var. ochroleucus nigro-punctulata A. Kob. - light brown seeds with black dots; 3. L. culinaris var. nigro-marmorata A. Kob. - seeds have reddish-yellow background with black marbling (Zhizhizlashvili & Berishvili, 1983). The Biological Farming Association Elkana is producing local cultivars of lentil for the market. Faba bean (Vicia faba) is one of the oldest cultivated plants. Faba bean is ancient agricultural plant in western and southern Georgia. Three varieties and 31 subvarieties were described in Georgia with small (V. faba var. minor Beck.), medium (V. faba var. equina Pers.) and large (V. faba var. major Harz.) seeds (Kobakhidze, 1974). At present, the large seed Faba bean is widely distributed in Upper and Lower Svaneti. Two varieties are found in the Lower Svaneti: 1. V. faba var. minor subvar. straminea A. Kob. - compressed on sides, tip obtuse, colour light cream. 2. V. faba var. equina subvar. ochroleucus A. Kob. - slightly compressed on sides, tip rounded, colour yellowish (Zhizhizlashvili & Berishvili, 1983). Chickling vetch (Lathyrus sativus) is used as human food in a soup to called shechamandi. It is also a green forage, used as silage and fed as seed flour to pigs and poultry. It is now available only at the research station of the Biological Farming Association Elkana. Bitter vetch – Vicia ervilia is distributed in Meskheti and Javakheti. There are cultivated and wild forms of this species. It is used as a forage and for soil enrichment with nitrogen. Common vetch (Vicia sativa) is used as forage and for hay, especially in Upper and Lower Svaneti and Javakheti. It is a valuable forage crop, rich in proteins. More often it appears as a weed in the fields of high mountain areas among grain crops – millet, barley, rye. Sainfoin (Onobrychis spp.), alfalfa (Medicago sativa) and clover (Trifolium spp.) are forage legumes. A local variety of Onobrychis transcaucasica Grossh. ‹Akhalkalakuri›, is widely used. Blue fenugreek (Trigonella caerulea) is traditional spice plant used in almost all of the foods of Georgian cuisine. It is available in all regions of Georgia.

Flax – Linum usitatissimum L. (Linaceae), was one of the oldest and important field crops in Georgia. Since prehistoric times, it was used to produce excellent linens (Kvavadze et al., 2009) and to make oil from its seeds. Big millstones were used to extract the oil from the flax seeds and they remain in many historical ruins. Until recently, flax was cultivated only in Javakheti, where flax seeds were used to produce pharmacologically pure oil for medicines. According to the eighteenth century Georgian scientist and geographer Vakhushti Batonoshvili (1991), several volatile oil-bearing plants were cultivated in Georgia - roses, camphor, lavender and basil. Kenaf - Cannabis sativa L. (Cannabaceae), was used to produce fiber for cord and thread for sacks. The seeds were used to produce oil. A traditional use of kenaf seeds was to mix them with wheat flour and making breads that had antidepressant effects. Traditional vegetables (Tab.2) are represented by sugar beets, spinach, carrots, radishes, turnips, onions, Welsh onion, leeks and garlic. Beet - Beta vulgaris, is an ancient cultivated plant whose tubers and young leaves were used in Georgian cooking. Leaves primarily came from the variety B. vulgaris subsp. cicla (L.) W.D.J. Koch ‹Tsiteli Mkhali› that was grown in lower elevations up to 1400 m a.s.l. Another beet variety - B. vulgaris L. subsp. vulgaris ‹Sasufre Charkhali› is rare. Carrot - Daucus carota, was edible as a wild species in Georgia since prehistoric times. The cultivated carrot is widespread in peasant›s house gardens in lowland areas. Onion - Allium cepa and garlic - A. sativum, are ancient cultivated plants available in all regions of Georgia. Red onions are very popular in Georgian people. A. sativum is called ‹Georgian garlic›. Another variety is ‹Russian garlic› representing A. ampeloprasum L. Leek - A. porrum, is typical in western Georgia. Welsh onion - A. fistulosum is currently grown in several high mountain areas. Until the 1970s, it was widespread in Imereti, but at present, Chinese shallot - A. cepa var. aggregatum G. Don has completely supplanted Welsh onion. Radish - Raphanus sativus, is grown in lower elevations in gardens and is cultivated by farmers for the market.

Table 2: Seed and forage legumes and traditional vegetables of Georgia

N	Latin name	Family	English common name	Georgian common name
	Allium cepa L.	Liliaceae	*Onion*	*Khakhvi*
	Allium fistulosum L.	Liliaceae	*Welsh Onion*	*Chlakvi*
	Allium porrum L.	Liliaceae	*Leek*	*Prasi*
	Allium sativum L.	Liliaceae	*Garlic*	*Niori*
	Beta vulgaris L. subsp. *vulgaris*	Chenopodiaceae	*Beet*	*Charkhali*
	Brassica rapa L. subsp. *rapifera* Metzger	Brassicaceae	*Turnip*	*Talgami*
	Cannabis sativa L.	Cannabaceae	*Kenaf*	*Kanafi*
	Cicer arietinum L.	Fabaceae	*Chickpea*	*Mukhudo*
	Daucus carota L.	Apiaceae	*Carrot*	*Stafilo*
	Lathyrus sativus L.	Fabaceae	*Chickling vetch*	*Tsulispira*
	Lens culinaris Medik.	Fabaceae	*Lentil*	*Ospi*
	Linum usitatissimum L.	Linaceae	*Flax*	*Seli*
	Medicago sativa L.	Fabaceae	*Alfalfa*	*Ionja*
	Onobrychis transcaucasica Grossh.	Fabaceae	*Sainfoin*	*Espartseti*
	Pisum arvense L.	Fabaceae	*Pea*	*Barda*
	Pisum sativum L.	Fabaceae	*Pea*	*Barda*
	Raphanus sativus L.	Brassicaceae	*Radish*	*Boloki*
	Spinacia oleracea L.	Chenopodiaceae	*Spinach*	*Ispanakhi*
	Vicia ervilia (L.) Willd.	Fabaceae	*Bitter vetch*	*Ugrekheli*
	Vicia faba L.	Fabaceae	*Faba bean*	*Tsertsvi*
	Vicia sativa L.	Fabaceae	*Common vetch*	*Tsertsvela*

Herbs are represented by numerous species (Tab.3) - parsley, coriander, tarragon, sweet basil, savory, gardencress pepperweed, dill, fennel, celery, garden lettuce, peppermint. Herbs are cultivated in small sections of house gardens even in urban settlements. Sometimes, people have herbs indoors in pots.

Table 3: List of traditionally cultivated herbs in Georgia

N	Latin name	Family	English common name	Georgian common name
	Anethum graveolens L.	Apiaceae	*Dill*	*Kama*
	Apium graveolens L.	Apiaceae	*Celery*	*Niakhuri*
	Artemisia dracunculus L.	Asteraceae	*Tarragon*	*Tarkhuna*
	Coriandrum sativum L.	Apiaceae	*Coriander*	*Kindzi*
	Foeniculum vulgare Mill.	Apiaceae	*Fennel*	*Didi Kama*
	Lactuca sativa L.	Asteraceae	*Garden lettuce*	*Salati*
	Lepidium sativum L.	Brassicaceae	*Gardencress pepperweed*	*Tsitsmati*
	Mentha piperata L.	Lamiaceae	*Peppermint*	*Pitna*
	Ocimum basilicum L.	Lamiaceae	*Sweet basil*	*Rehani*
	Petroselinum crispum (Mill.) A.W. Hill	Apiaceae	*Parsley*	*Okhrakhushi*
	Satureja hortensis L.	Lamiaceae	*Savory*	*Kondari*
	Trigonella caerulea (L.) Ser.	Fabaceae	*Blue fenugreek*	*Utskho Suneli*

Fruits are valuable cultivars in Georgia. Wild and cultivated fruit crops reveal high species and genetic diversity in Georgia and represent rich material for future breeding activities. Many fruits have wild relatives representing the same species and direct ancestors of local cultivars (Tab.4,5).

Introduced Cultivated Plants

Georgia is located at the crossroads of Europe and Asia. Many cultivated plants have been introduced since ancient times to Georgia from other regions of the world (Javakhishvili 1987). Some introduced crops have become very popular and widespread. They are introduced from different countries. Such crops as cucumber (Cucumis sativus), found in Georgia since medieval times, eggplant (Solanum melongena), marigold (Tagetes patula), used in almost all traditional meals; and black pepper (Piper nigrum) were introduced from India. Watermelon (Citrullus lanatus) from South Africa was cultivated in the Caucasus since medieval times. Maize (Zea mays), sunflower (Helianthus annuus), tomato (Solanum lycopersicum), bean (Phaseolus vulgaris), pepper (Capsicum annuum), and potato (Solanum tuberosum) were introduced to Georgia from the Americas at about the same time as in Europe (Javakhishvili, 1987). Tea (Camellia sinensis) and citrus fruits (Citrus limon, Citrus reticulata, Citrus sinensis) came from China in the 1830s (Bakhtadze, 1947). Nicotiana rustica, (tutuni in Georgian) has been cultivated for a long time and is found in the most regions, including high mountain areas, of Georgia. N. tabacum, was introduced during the Soviet period and was cultivated in kolkhozis for commercial use. Georgia has become a secondary centre of diversity for most of these crops. Landraces of bean, maize, potato, tomato, and cucumber that do not exist in their countries of origin can be found in Georgia. Bean (Phaseolus spp.; Vigna spp.) was introduced via Turkey to Guria and Samegrelo during the second half of the XVI century (Javakhishvili, 1987). At present, 61 varieties and 406 forms of common bean had originated in Georgia due to widespread distribution and hybridization of different species of bean: Phaseolus vulgaris L., P. lunatus L., P. coccineus L. (=P. multiflorus Lam.), P. acutifolius A. Gray, Vigna radiata (L.) R. Wilczek var. radiata (=P. aureus Roxb.), V. angularis (Willd.) Ohwi & H. Ohashi var. angularis (=P. angularis [Willd.] W. Wight) and V. umbellata (Thunb.) Ohwi & H. Ohashi (=P. calcaratus Roxb.) (Kobakhidze, 1965). Beans are cultivated in gardens in large amounts providing sufficient food for families and representing a cash crop for additional income. Diversity of beans remains high. Maize (Zea mays L.) was introduced to Georgia in 1633-1650 (Javakhishvili, 1987). The Georgian name Simindi originated from the old name for flour Samindo as flour was introduced earlier to Georgia than the initial cultivation of maize. Besides landraces such as 'Kazha simindi' from Svaneti there are some cultivars that originated in Georgia: 'Ajametis tetri', 'Abashis kviteli', 'Kartuli kruki', 'Gegutis kviteli', 'Imeruli hibridi' and 'Lomtagora'. Many cultivars were introduced from Russia, Hungary, Yugoslavia, etc., during Soviet time. The last introduction occurred in 2011, when the high-yield US corn hybrid 'Pioneer' was sown in Georgia. Corn had replaced common and Italian millet and is used as an everyday food, especially in western Georgia. Potato (Solanum tuberosum L.) was introduced to Georgia during the second half of the XIX century. Several landraces of high quality are grown in high-mountain regions: Svaneti, Racha-Lechkhumi, Khevsureti, Khevi and Adjara. Breeder's cultivars were introduced into lowland areas during Soviet time. Recently, genetically modified potato cultivars have been introduced in Georgia by international seed-distribution organizations. These modern cultivars have almost completely supplanted local landraces even in high mountain regions. Tea and citrus had high commercial value in Georgia, but in the 1990s these crops were abandoned and tea was not produced in Georgia until recently. Citrus (lemons, oranges and mandarins) were sold only in the local Georgian market. At present, this business is restored. Information about introduced varieties has been published annually during the XX century. The latest official edition of the Catalogue of the Georgian Released Varieties of 1997 (published in 1996) listed 195 varieties of field and vegetable crops and 195 varieties of fruits. These varieties were part of the collections that existed at the end of the 1980s and beginning of the 1990s. At present, only a few of these varieties exist. The data about modern breeder's varieties introduced into Georgia during last decade are usually absent and a number of varieties have been cultivated in Georgia without going through the official procedures for release. Therefore, it is difficult to evaluate the diversity of recently introduced cultivars.

CROP WILD RELATIVE ASSESSMENT

The CWR are taxa related to species of direct socio-economic importance, which includes the progenitors of crops. According to modern concept of wild relatives, under CWR we should understand all species related to any cultivated plants, as well as to wild species of ornamental, food, fodder and forage, medicinal plants, condiments, forestry species and plants used for industrial purposes, such as oils and fibers i.e. to all plants of economic importance (Laguna, 2004). Although, "classical" definition of CWR is restricted only to species related to cultivated crops, including such important field crops as wheat, barley, rye, oats, sorghum, common and Italian millet, grain and forage legumes (such as Phaseolus, Vicia, Vigna, Lens, Lathyrus, Cicer) and some vegetables and industrial crops.

The flora of the Caucasus harbours a remarkable concentration of economically important plants, particularly CWRs such as cereals, legumes, fruits, vegetables, herbs and technical plants like flax. The list of CWRs in Georgia was published in Plant Genetic Resources (PGR) Forum - CWR Catalogue of Europe and the Mediterranean (Maxted et al., 2008). This catalogue listed 1784 species of vascular plants, representing 43% of the 4130 vascular plant species found in Georgia. These are mainly wild species that also have considerable economic importance providing food, fuel, timber, forage, hay and habitats for animal life. A large number of taxa used in folk and scientific medicine are also included among economically valuable plants. However, this list is not detailed enough to assess the economic value of CWRs representing the same species or direct ancestors of crop plants. There is no information on the status of endangered and endemic species. Thus, we developed a general description of vegetation types and separated CWR endemic species and species genetically closely related to crops.

Flora and Vegetation

Georgia is located between 41°02' and 43°34' latitude north and between 40° and 46°43' longitude east. It borders the Russian Federation to the north, Turkey and Armenia to the south, Azerbaijan to the east, and has approximately 310 km of coastline along the Black Sea to the west. Georgian territory (69.700 km2) covers two separate mountain systems: the Greater Caucasus Range which trends north-west to east-southeast between the Black Sea and Caspian Sea; and the Lesser Caucasus Mountains, which run parallel to the greater range at a distance to the south that averages about 100 kilometres. Two thirds of the country is mountainous with an average height of 1.200 m.a.s.l., with the highest peaks of Mount Shkhara (5.184 m.a.s.l.) in the western Greater Caucasus and Mount Didi Abuli (3.301 m.a.s.l.) in the Lesser Caucasus. Colchis, Kartli and Alazeni valleys and Iori plateau represent intermontane lowlands located between these two mountain systems. Geologically, the Caucasus consists of Meso- to Cenozoic deposits. Ancient Precambrian and Paleozoic formations are rare (Neidze et al., 2008). The Likhi Range divides the country into eastern and western halves that differ in climate and landscapes. Western Georgia has a humid subtropical climate with annual precipitation ranging from 1000–4000 mm. Temperatures fluctuate between the winter averages of 2.8° to 6.7° C and the summer averages of 22.7° to 23.8° C. Eastern Georgia has a more continental climate, due to the barrier of the Likhi Range, which bars the warm Black Sea winds from this area. The temperatures vary from the January averages of 0-2.2° C to the July mean of 27.8° C. Annual precipitation is considerably less in eastern Georgia and ranges from 400– 1600 mm. Southern Georgia has a continental climate. The local winters are cold. The frosts are - 25° C and in July temperatures rise to 40° C. Annual rainfalls are usually less than 600- 1000 mm (Neidze et al., 2008). Soil types vary in Georgia. The most widespread types in the lowlands of western Georgia are bog, podzolic, red, yellow and hilly piedmonts, which are mainly acidic. Mountainforest and mountain-meadow soils occur in higher elevations. Chestnut and chernozem soils are widespread in the lowlands of eastern Georgia and are characterized by neutral pH. Brown humid-sulphates, saline soils of steppes and semi-deserts, as well as intermediate forest-steppe and mountain-meadow soils occur in semi-desert areas. Alluvial soils are found along the rivers throughout Georgia. Brown soils are typical for the Georgian forest zone in the range of 800-2000 m a.s.l. (Sabashvili, 1970). Western Georgia's landscape ranges from see-level swamps and lowland temperate rainforests to eternal subnival zone and glaciers, while the eastern part of the country contains temperate forests and semi-arid plains in lower elevations and alpine and subnival zones. Main rivers are R. Mtkvari, R. Rioni, R. Enguri and R. Alazani. There are 70 natural lakes and 11 artificial reservoirs. The lower section of the Rioni River is located in the Colchis valley and was naturally occupied by marshes and lagoons, but in 1960s this area was the site of a large reclamation-drainage project and it was converted to agricultural land. The majority of the forests that covered the Colchis plain are now virtually gone; the exceptions are those included in national parks and reserves. The Mtkvari River basin, which includes the major parts of southern and eastern Georgia, is drier. It is covered with semi-arid vegetation and temperate forests. Forests, in total, amount to 40% of Georgia's territory while the alpine zone accounts for roughly 10% of the land. Much of the natural habitat in the lowland areas of western Georgia has disappeared over the last 100 years because of agricultural development and urbanization. The vegetation of Georgia belongs to three floristic provinces – Euxine, Caucasian and Armeno-Iranian (Takhtajian, 1986). The Euxine and Caucasian provinces belong to the Circumboreal Region, the Boreal Subkingdom and the Holarctic Kingdom and the ArmenoIranian Province belongs to the Irano-Turanian Region, the Tethyan (Mediterranean) Subkingdom and the Holarctic Kingdom. There are following vegetation zones: 1. Colline zone (0-400 m a.s.l.), which includes coastal and halophytic habitats in western Georgia and dry open woodlands and semi-deserts in eastern Georgia; 2. Lower montane zone (400-800 m a.s.l.) is used as arable land. The natural vegetation in western Georgia is represented by small remnant areas of Colchic broad-leaved mixed forest. Oak-hornbeam forests and dry scrublands occur in eastern Georgia; 3. Middle montane zone (800-1500 m a.s.l.) is primarily used for agriculture. Broad-leaved mixed forests, mountain xerophytes scrublands, and mountain steppes are represented; 4. Upper montane zone (1200-2050 m a.s.l.) is covered by beech and broadleaf-coniferous mixed forests; 5. Subalpine zone (1900-2400[2500] m a.s.l.) is a treeline ecotone, with tall herbaceous vegetation, shrublands and polydominant subalpine grass and herb meadows used as pastures or arable land; 6. Alpine zone (2500–2900 m a.s.l.) has alpine meadows and snowbed communities. Vegetation is mostly used for grazing and is of considerably lower quality than the subalpine vegetation, both by biomass volume and typological diversity; 7. Subnival zone (2900-3300 m a.s.l.) is patchy highest limits of vegetation. 8. Nival zone (3300-5184 m a.s.l.) covered by glaciers. 9. Azonal vegetation type is represented by fragments of wetlands rich in boreal type flora, halophytic desert vegetation and rocky areas (Nakhutsrishvili, 1999). Flora of Georgia is represented by 4,130 species of vascular plants. Among them are 79 pteridophytes, 17 gymnosperms and 4034 angiosperms (Nakhutsrishvili, 1999). The 10 leading families are Asteraceae (538 species), Poaceae (332 species), Fabaceae (317 species), Rosaceae (238), Brassicaceae (183), Scrophulariaceae (179), Apiaceae (177), Lamiaceae (149), Caryophyllaceae (135) and Liliaceae (129). High endemism is characteristic of the Caucasus and represents one of the world's hot spots of biodiversity. Out of all, 1304 (32.3%) species are endemics of the Caucasus ecoregion and 261 (6.6%) are endemics of Georgia (Schatz et al., 2009). There are 17 endemic genera in the flora of the Caucasus. Most of them are represented by one species: Agasyllis latifolia (M. Bieb.) Boiss., Alboviodoxa elegans (Albov) Woronow, Charesia akinfievii (Schmalh.) E. Busch, Cladochaeta candissima (M. Bieb.) DC., Gadellia lactiflora (M. Bieb.) Schulkina, Mandenovia komarovii (Manden.) Alava, Paederotella pontica (Rupr. ex Boiss.) Kem.-Nath., Petrocoma hoefftiana (Fisch.) Rupr., Pseudobetckea caucasica (Boiss.) Lincz., Pseudovesicaria digitata (C. A. Mey.) Rupr., Sredinskya grandis (Trautv.) Fed., Symphyoloma graveolens C. A. Mey., Trigonocaryum involucratum (Steven) Kusn., Woronowia speciosa (Albov) Juz. Two genera contain two species of each: Chymsydia agasylloides (Albov) Albov, C. colchica (Albov) Woronow, Grossheimia macrocephala (Muss.- Puschk. ex Willd.) Sosn. & Takht., G. polyphylla (Ledeb.) Holub. One endemic genus is represented by 5 species: Kemulariella caucasica (Willd.) Tamamsch., K. rosea (Steven ex M. Bieb.) Tamamsch., K. abchasica (Kem.-Nath.) Tamamsch., K. tugana (Albov) Tamamsch., and K. colchica (Albov) Tamamsch. For conservation action to be effective, it is important to understand not just the needs of individual species, but also the context in which conservation efforts will need to take place. Therefore, it is important to evaluate the conservation values of the species that contribute most to human health and to develop conservation measures to avoid their extinction.

Diversity of Crop Wild Relatives in Georgia

Flora of the Caucasus region is rich as there are high concentrations of economically important and edible plants, particularly wild crop relatives such as grapevine, wheat, barley, rye, oats, seed and forage legumes, fruits and vegetables. The Caucasus is considered to be the centre of evolution for many unique life forms and is a natural museum for rich genetic resources (Vavilov, 1992). We identified the number of species of the genera that are traditional crops in Georgia (Tab.4). A total of 20 plant families, 76 genera and 479 species were identified as wild relatives of ancient crops in Georgia. Most of these plant species are closely related genetically to landraces and might be their progenitor species.

Table 4: List of wild relatives of ancient crops in Georgia

Families	*Number of genera*	*Number of species*	*Genera with number of species*
Apiaceae	8	17	*Anethum* (1), *Apium* (2), *Carum* (5), *Coriandrum* (1), *Daucus* (1), *Foeniculum* (1), *Pastinaca* (5), *Petroselinum* (1)
Asparagaceae	1	3	*Asparagus* (3)
Asteraceae	3	16	*Cichorium* (1), *Lactuca* (7), *Scorzonera* (8)
Betulaceae	1	6	*Corylus* (6)
Brassicaceae	5	20	*Brassica* (4), *Lepidium* (8), *Raphanus* (2), *Rorippa* (4), *Sinapis* (2),
Cannabaceae	2	3	*Cannabis* (2), *Humulus* (1)
Chenopodiaceae	2	3	*Beta* (2), *Spinacia* (1)
Cornaceae	1	1	*Cornus* (1)
Fabaceae	10	154	*Cicer* (1), *Lathyrus* (20), *Lens* (3), *Lotus* (5), *Medicago* (21), *Onobrychis* (19), *Pisum* (1), *Trifolium* (40), *Trigonella* (10), *Vicia* (34)
Grossulariaceae	2	4	*Grossularia* (1), *Ribes* (3)
Juglandaceae	1	1	*Juglans* (1)
Lamiaceae	4	19	*Mentha* (4), *Origanum* (1), *Satureja* (3), *Thymus* (11)
Liliaceae	2	39	*Allium* (36)
Linaceae	1	12	*Linum* (12)
Moraceae	2	3	*Ficus* (1), *Morus* (2)
Poaceae	16	64	*Aegilops* (7), *Agropyron* (2), *Avena* (8), *Brachypodium* (3), *Cynosorus* (2), *Elymus* (4), *Elytrigia* (9), *Echinochloa* (3), *Hordeum* (5), *Hordelymus* (1), *Panicum* (5), *Psathyrostachis* (1), *Secale* (5), *Setaria* (6), *Sorghum* (1), *Tueniatherum* (2)
Punicaceae	1	1	*Punica* (1)
Rosaceae	12	110	*Amygdalus* (1), *Cerasus* (4), *Crataegus* (8), *Cydonia* (1), *Fragaria* (3), *Malus* (1), *Mespilus* (1), *Prunus* (2), *Pyrus* (11), *Rosa* (30), *Rubus* (36), *Sorbus* (12)
Staphyleaceae	1	2	*Staphylea* (2)
Vitaceae	1	1	*Vitis* (1)
Total: 20	76	479	

CWR are commonly defined in terms of wild species related to agricultural and horticultural crops. As such a broad definition of a CWR would be any wild taxon belonging to the same genus as a crop. A working definition of a CWR was provided by N. Maxted and colleagues (Maxted et al., 2006): "A crop wild relative is a wild plant taxon that has an indirect use derived from its relatively close genetic relationship to a crop; this relationship is defined in terms of the CWR belonging to gene pools 1 or 2, or taxon groups 1 to 4 of the crop". According to gene pool concept three gene pools are distinguished as follows: (1) Primary Gene Pool (GP-1) within which GP-1A are the cultivated forms and GP-1B are the wild or weedy forms of the crop; (2) Secondary Gene Pool (GP-2) which includes the coenospecies (less closely related species) from which gene transfer to the crop is possible but difficult using conventional breeding techniques; (3) Tertiary Gene Pool (GP-3) which includes the species from which gene transfer to the crop is impossible, or if possible requires sophisticated techniques, such as embryo rescue, somatic fusion or genetic engineering. The taxon group concept is used to establish the degree of CWR relatedness of a taxon. Application of the taxon group concept assumes that taxonomic distance is positively related to genetic distance. CWR rank of taxon groups is defined as follows: (1) Taxon Group 1A – crop; (2) Taxon Group 1B – same species as crop; (3) Taxon Group 2 – same series or section as crop; (4) Taxon Group 3 – same subgenus as crop; (5) Taxon Group 4 – same genus; (6) Taxon Group 5 – same tribe but different genus to crop (Maxted et al., 2006). Diversity and Genetic Erosion of Ancient Crops and Wild Relatives of Agricultural Cultivars for Food: Implications for Nature Conservation in Georgia (Caucasus)

Thus, the combined use of the gene pool and taxon group concept proposed above provide the most pragmatic means available to determine whether a species is a CWR and how closely related a CWR is to its crop. We have determined 66 species of CWR belonging to 43 genera and 17 families, which can be assigned as Primary (GP-1) and Secondary Gene Pool (GP-2) and Taxon Group 1 and 2 (Tab.5). Seventeen (25.75%) are wild species but used as crops by collecting in the natural habitats and they were identified as GP1A. The same CWR species as crop were 19 (28.8%). Different species were 30 (45.45%), but representing direct progenitors whose genome is involved in the evolution of cultivars. Almost the same numbers were obtained during taxon group classification: TG1A - 17 species (25.75%), TG1B - 16 (24.25%), TG2 - 30 (45.45%), TG 5 - 3 (4.55%). The last 3 species belonging to the Taxon Group 5 are Aegilops, a wild relative of wheat. Goatgrass (Aegilops tauschii) is considered to be a donor of D genome of bread wheat genomic constitution = AABBDD (Petersen et al., 2006). The distribution area of this species is wide, however, D genomes of all forms of T. aestivum were found to be most closely related to accessions collected in Georgia, Armenia, Nakhitshevan and Azerbaijan (Dvorák et al., 1998). Thus, the germplasm of the populations of goatgrass in the South Caucasus needs conservation and should be preserved both in situ in protected areas and ex situ in seed collections. Barley is one of the oldest crops to be domesticated from its wild progenitor Hordeum spontaneum

(Badr et al., 2000, Kilian et al., 2006). We have found H. spontaneum in Georgia in three different places. This species was not included in the list of "Flora of Georgia' and it is a new species for Georgia.

Table 5: Gene pool and taxon group of wild relatives to Georgian ancient crops GP-Gene Pool; TG-Taxon Group

Family	*Crop*	*Taxon*	*GP*	*TG*
Apiaceae	*Daucus carota*	*Daucus carota* L.	GP1B	TG1B
Apiaceae	*Coriandrum sativum*	*Coriandrum sativum* L.	GP1B	TG1B
Asparagaceae	*Asparagus officinalis*	*Asparagus caspius* Schult. & Schult. fil.	GP1B	TG1B
Asparagaceae	*Asparagus officinalis*	*Asparagus officinalis* L.	GP1B	TG1B
Asparagaceae	*Asparagus officinalis*	*Asparagus verticillatus* L.	GP1B	TG1B
Betulaceae	*Corylus avellana*	*Corylus avellana* L.	GP1B	TG1B
Betulaceae	*Corylus avellana*	*Corylus iberica* Wittm. ex Kem.-Nath.	GP2	TG2
Betulaceae	*Corylus avellana*	*Corylus colchica* Albov	GP2	TG2
Brassicaceae	*Brassica oleracea*	*Brassica juncea* (L.) Czern.	GP2	TG2
Brassicaceae	*Brassica oleracea*	*Brassica napus* L.	GP2	TG2
Brassicaceae	*Brassica oleracea*	*Sinapis arvensis* L.	GP2	TG2
Cannabaceae	*Cannabis sativa*	*Cannabis sativa* L.	GP1A	TG1A
Cannabaceae	*Humulus lupulus*	*Humulus lupulus* L.	GP1A	TG1A
Chenopodiaceae	*Beta vulgaris*	*Beta maritima* L.	GP2	TG2
Fabaceae	*Pisum sativum*	*Pisum elatius* M. Bieb.	GP1B	TG1B
Fabaceae	*Cicer arietinum*	*Cicer caucasica* Bornm.	GP2	TG2
Fabaceae	*Lathyrus sativus*	*Lathyrus tuberosus* L.	GP2	TG2
Fabaceae	*Lens culinaris*	*Lens nigricans* (M. Bieb.) Webb & Berth.	GP2	TG2
Fabaceae	*Lens culinaris*	*Lens ervoides* (Brign.) Grande	GP2	TG2
Fabaceae	*Lens culinaris*	*Lens culinaris* Medik. subsp. *orientalis* (Boiss.) Ponert	GP1B	TG1B
Fabaceae	*Vicia faba*	*Vicia johannis* Tamamsh.	GP2	TG2
Fabaceae	*Vicia faba*	*Vicia narbonensis* L.	GP2	TG2
Fabaceae	*Vicia sativa*	*Vicia sativa* L.	GP1A	TG1A
Grossulariaceae	*Ribes rubrum*	*Ribes alpinum* L.	GP2	TG2
Grossulariaceae	*Ribes rubrum*	*Ribes caucasicum* M. Bieb.	GP2	TG2
Juglandaceae	*Juglans regia*	*Juglans regia* L.	GP1A	TG1A
Lamiaceae	*Satureja hortensis*	*Satureja laxiflora* K. Koch	GP2	TG2
Lamiaceae	*Satureja hortensis*	*Satureja spicigera* (K. Koch) Boiss.	GP2	TG2
Linaceae	*Linum usitatissimum*	*Linum bienne* Mill.	GP1B	TG1B
Linaceae	*Linum usitatissimum*	*Linum usitatissimum* L.	GP1A	TG1A
Moraceae	*Morus alba*	*Morus alba* L.	GP1A	TG1A
Moraceae	*Morus nigra*	*Morus nigra* L.	GP1A	TG1A
Moraceae	*Ficus carica*	*Ficus carica* L.	GP1A	TG1A
Poaceae	*Triticum aestivum*	*Aegilops cylindrica* Host	GP1B	TG5
Poaceae	*Triticum aestivum*	*Aegilops triuncialis* L.	GP2	TG5
Poaceae	*Triticum aestivum*	*Aegilops tauschii* Coss.	GP1B	TG5
Poaceae	*Hordeum hexastichon*	*Hordeum bulbosum* L.	GP1B	TG2
Poaceae	*Hordeum distichon*	*Hordeum spontaneum* K. Koch	GP1B	TG1B
Poaceae	*Avena sativa*	*Avena barbata* Pott ex Link	GP2	TG2
Poaceae	*Avena sativa*	*Avena sterilis* L.	GP2	TG2
Poaceae	*Secale cereale*	*Secale strictum* subsp. *anatolicum* (Boiss.) K. Hammer	GP2	TG2
Poaceae	*Secale cereale*	*Secale strictum* subsp. *kuprijanovii* (Grossh.) K. Hammer	GP2	TG2
Poaceae	*Secale cereale*	*Secacle cereale* L. subsp. *segetale* Zhuk.	GP1A	TG1A
Poaceae	*Panicum miliaceum*	*Panicum capillare* L.	GP2	TG2
Poaceae	*Panicum miliaceum*	*Panicum sumatrense* Roth	GP2	TG2

Family	*Crop*	*Taxon*	*GP*	*TG*
Poaceae	*Panicum miliaceum*	*Panicum dichotomiflorum* Michx.	GP2	TG2
Poaceae	*Setaria italica*	*Setaria viridis* (L.) P. Beauv.	GP2	TG2
Poaceae	*Setaria italica*	*Setaria verticillata* (L.) P. Beauv.	GP2	TG2
Poaceae	*Setaria italica*	*Setaria glauca* (L.) P. Beauv.	GP2	TG2
Poaceae	*Setaria italica*	*Setaria intermedia* Roem. & Schult.	GP2	TG2
Punicaceae	*Punica granatum*	*Punica granatum* L.	GP1A	TG1A
Rosaceae	*Pyrus communis*	*Pyrus caucasica* Fed.	GP1B	TG1B
Rosaceae	*Pyrus communis*	*Pyrus balansae* Decne.	GP1B	TG1B
Rosaceae	*Malus domestica*	*Malus orientalis* Uglitzk.	GP2	TG2
Rosaceae	*Cydonia oblonga*	*Cydonia oblonga* Mill.	GP1B	TG1B
Rosaceae	*Prunus domestica*	*Prunus domestica* L. subsp. *insititia* (L.) C. K. Schneid.	GP1A	TG1A
Rosaceae	*Prunus domestica*	*Prunus spinosa* L.	GP1B	TG1B
Rosaceae	*Prunus cerasifera*	*Prunus cerasifera* Ehrh. var. *divaricata* (Ledeb.)L.H.Bailey	GP1A	TG1A
Rosaceae	*Cerasus avium*	*Cerasus avium* (L.) Moench	GP1B	TG1B
Rosaceae	*Cornus mas*	*Cornus mas* L.	GP1A	TG1A
Rosaceae	*Mespilus germanica*	*Mespilus germanica* L.	GP1A	TG1A
Rosaceae	*Rubus idaeus*	*Rubus idaeus* L.	GP1A	TG1A
Rosaceae	*Amygdalus communis*	*Amygdalus georgica* Desf.	GP2	TG2
Staphyleaceae	*Staphylea pinnata*	*Staphylea pinnata* L.	GP1A	TG1A
Staphyleaceae	*Staphylea colchica*	*Staphylea colchica* Steven	GP1A	TG1A
Vitaceae	*Vitis vinifera*	*Vitis vinifera* subsp. *sylvestris* (C.C.Gmel.) Hegi	GP1B	TG1B

It is assumed that H. spontaneum might have evolved from H. bulbosum by fixation of the genes controlling self-compatibility and annual habit (Cass et al., 2005). This last species is widespread in Georgian regions. Most fruit trees in Georgia are wild in forests and have cultivars domesticated from these wild ancestors. An economically important Georgian fruit crop is grape, which has a wild relative species Vitis vinifera subsp. sylvestris. We have found 9 populations of wild grapevine and conducted studies to reveal genetic and morphological relations between wild grapevine and landraces in Georgia (Ekhvaia et al., 2010). Many fruits are domesticated in the Caucasus from wild ancestors representing Primary Gene Pool (GP-1B) to be the wild or weedy forms of the crops (Tab.5). The fruit crops (GP1A) and ancestor species (GP-1B) are the following: Pome fruits - pear (Pyrus communis, P. caucasica), apple (Malus domestica, M. orietalis), quince (Cydonia oblonga); stone fruits - plum (Prunus domestica, P. domestica var. insititia, P. spinosa), Myrobalan (Prunus vachushti), sourplum (Prunus cerasifera var. divaricata), cherries (Cerasus avium, C. vulgaris), Cornel cherry (Cornus mas), medlar (Mespilus germanica), mulberry (Morus alba, M. nigra), pomegranate (Punica granatum); berries - red raspberry (Rubus idaeus), currant (Ribes rubrum, R. nigra, R. alpinum, R. biebersteinii), fig (Ficus carica), bladdernut (Staphylea pinnata, used flowers for marinade), and nuts - such as hazelnut (Corylus avellana), almond (Amygdalus communis), and walnut (Juglans regia), etc. We evaluated CWRs endemic for the Caucasus (Tab.6). The endemic species of the same genus as crop were calculated. The number of endemic species from the total 479 CWRs of agricultural cultivars for food is 114 (23.8%).

DOMESTICATION EVENTS

Domestication of Grapevine in Georgia

The grapevine was among the first fruits to be cultivated in Georgia (Javakhishvili, 1987). There are many arguments to confirm the fact that domestication events of grape took place in Georgia. One of the indicators is archaeological evidence. The 1,5-Myr-old petrified specimen of wild grapevine leaf was found in Georgia in the Meskheti province (Fig.4A). A confirmation for long lasting cultivation of grapevine in Georgia is archaeological remains of berries and seeds of domesticated grapes dated ~6.000 BC (in the vicinity of village Shulaveri [Fig.1], southeast Georgia; Ramishvili 1988). Other archaeological evidence of prehistoric winemaking was found in proximity of the Caucasus in northern Iran at the Hajji Firuz Tepe site in the northern Zagros Mountains dated to about 5.400–5.000 BC (McGovern, 2003) and in the Levant where archaeological findings are dated from ca. 4.000-3.200 BC (Zohary & Hopf 2000). Georgian traditions based on winemaking and grape culture to a high degree might be considered to be a second indicator of Caucasian origin of the grapevine. However, the primary scientific argument should be premised on N. I. Vavilov's (1992) idea that the centres of origin of cultivars should be characterized by high genetic and morphological variability of both wild and cultivated taxa. Five hundred is a very high number of known autochthonous grapevine varieties found in such a small territory (Javakhishvili, 1987; Ketskhoveli et al., 1960). These cultivars showed high morphological variability of leaf form, colour and shape of berries and shape and structure of pips. They are adapted to wide array of climatic conditions (Ketskhoveli et al., 1960).

Each province of Georgia possesses unique grapevine cultivars adapted to local climate. The landraces of western Georgia grow in humid subtropical climate and other cultivars are adapted to moderate climates in eastern Georgia. Several local cultivars are growing in high mountain regions in Meskheti and Svaneti up to 1500 m a.s.l. Besides the cultivars, there is high morphological and genetic diversity of wild grapevine populations in the Caucasus. All five haplotypes detected by using cpDNA microsatellite markers have been found in the Caucasian ecoregion suggesting that this area is possibly the centre of origin of both wild and cultivated grapevines (Grassi et al., 2006). However, only one provenance from Georgia has been analyzed in this study despite the number of populations of wild grapevine found in Georgia today that display morphological diversity (Ramishvili, 1988). We carried out a detailed comparative morphometric study of nine populations of wild grapevine, V. vinifera subsp. sylvestris, growing in the four river basins of the Ajaristskali, Mtkvari, Alazani and Iori located in western, southern and eastern Georgia.

The results reveal high morphological diversity of wild grapevine growing in Georgia. Morphological characters such as shape of leaf blade, number of lobes, pubescence type, coloration of internodes, leaves and berry skin, leaf vein lengths and angles between them and form of petiole sinus show high variability both within and among populations. The variability was related to the skin colour of berries. Some wild grapes had white berries, most had blue-black coloration. White-fruited phenotype is considered to be determined by the variation present in the gene VvmybA1, a transcriptional regulator of anthocyanin biosynthesis (This et al., 2006).

Table 6: One hundred fourteen endemic species of the Caucasus ecoregion related to the ancient crops in Georgia (Schatz et al., 2009)

Family	*Taxon*	*Family*	*Taxon*
Apiaceae	*Carum alpinum* (M. Bieb.) Benth. & Hook. ex B. D. Jacks.	Lamiaceae	*Thymus nummularius* M. Bieb.
Apiaceae	*Carum grossheimii* Schischk.	Lamiaceae	*Thymus tiflisiensis* Klokov & Des. - Shost.
Apiaceae	*Carum porphyrocoleon* (Freyn & Sint.) Woronow	Liliaceae	*Allium albovianum* Vved.
Apiaceae	*Pastinaca armena* Fisch. & C. A. Mey.	Liliaceae	*Allium candolleanum* Albov
Apiaceae	*Pastinaca aurantiaca* (Albov) Kolak.	Liliaceae	*Allium gramineum* K. Koch
Apiaceae	*Pastinaca pimpinellifolia* M. Bieb.	Liliaceae	*Allium gunibicum* Miscz. ex Grossh.
Asparagaceae	*Asparagus caspius* Schult. & Schult. fil.	Liliaceae	*Allium kunthianum* Vved.
Asteraceae	*Scorzonera charadzeae* Papava	Liliaceae	*Allium leucanthum* K. Koch
Asteraceae	*Scorzonera czerepanovii* R. Kam.	Liliaceae	*Allium otschiauriae* Tscholokaschvili
Asteraceae	*Scorzonera dzhawakhetica* Sosn. ex Grossh.	Liliaceae	*Allium ponticum* Miscz. ex Grossh.
Asteraceae	*Scorzonera ketzkhowelii* Sosn. ex Grossh.	Liliaceae	*Allium szovitsii* Regel
Asteraceae	*Scorzonera kozlowskyi* Sosn. ex Grossh.	Linaceae	*Linum hypericifolium* Salisb.
Asteraceae	*Scorzonera seidlitzii* Boiss.	Poaceae	*Elymus buschianus* (Roshev.) Tzvelev
Betulaceae	*Corylus abchasica* (Kem.-Nath.) Kem.-Nath.	Poaceae	*Elymus troctolepis* (Nevski) Tzvelev
Betulaceae	*Corylus colchica* Albov	Poaceae	*Elytrigia gracillima* (Nevski) Nevski
Betulaceae	*Corylus egrissiensis* Kem.-Nath.	Poaceae	*Elytrigia sinuata* (Nevski) Nevski
Betulaceae	*Corylus imeretica* Kem.-Nath.		
Betulaceae	*Corylus kachetica* Kem.-Nath.	Poaceae	*Secale strictum* subsp. *anatolicum* (Boiss.) K. Hammer
Betulaceae	*Corylus x fominii* Kem.-Nath.	Poaceae	*Secale strictum* subsp. *kuprijanovii* (Grossh.) K. Hammer
Fabaceae	*Cicer caucasicum* Bornm.	Rosaceae	*Amygdalus georgica* Desf.
Fabaceae	*Lathyrus colchicus* Lipsky	Rosaceae	*Crataegus caucasica* K. Koch

Family	*Taxon*	*Family*	*Taxon*
Fabaceae	*Lathyrus cyaneus* (Steven) K. Koch	Rosaceae	*Crataegus eriantha* Pojark.
Fabaceae	*Lotus caucasicus* Kuprian. ex Juz.	Rosaceae	*Pyrus demetrii* Kuthatheladze
Fabaceae	*Medicago dzawakhetica* Bordz.	Rosaceae	*Pyrus eldarica* Grossh.
Fabaceae	*Medicago hemicycla* subsp. *medidaghestanica* Sinskaya	Rosaceae	*Pyrus fedorovii* Kuthatheladze
Fabaceae	*Onobrychis angustifolia* Chinth.	Rosaceae	*Pyrus georgica* Kuthatheladze
Fabaceae	*Onobrychis biebersteinii* Sirj.	Rosaceae	*Pyrus ketzkhovelii* Kuthatheladze
Fabaceae	*Onobrychis cyri* Grossh.	Rosaceae	*Pyrus oxyprion* Woronow
Fabaceae	*Onobrychis grossheimii* Kolak. ex Fed.	Rosaceae	*Pyrus sachokiana* Kuthatheladze
Fabaceae	*Onobrychis iberica* Grossh.	Rosaceae	*Pyrus salicifolia* Pall.
Fabaceae	*Onobrychis kachetica* Boiss. ex Huet.	Rosaceae	*Pyrus takhtadzhianii* Fed.
Fabaceae	*Onobrychis kemulariae* Chinth.	Rosaceae	*Rosa buschiana* Chrshan.
Fabaceae	*Onobrychis komarovii* Grossh.	Rosaceae	*Rosa didoensis* Boiss.
Fabaceae	*Onobrychis meschetica* Grossh.	Rosaceae	*Rosa doluchanovii* Manden.
Fabaceae	*Onobrychis oxytropoides* Bunge	Rosaceae	*Rosa ermanica* Manden.
Fabaceae	*Onobrychis petraea* (M. Bieb. ex Willd.) Fisch.	Rosaceae	*Rosa galushkoi* Demurova
Fabaceae	*Onobrychis sosnowskyi* Grossh.	Rosaceae	*Rosa hirtissima* Lonacz.
Fabaceae	*Onobrychis transcaucasica* Grossh.	Rosaceae	*Rosa irysthonica* Manden.
Fabaceae	*Trifolium fontanum* Bobrov	Rosaceae	*Rosa kozlowskii* Chrashan.
Fabaceae	*Trifolium ruprechtii* Tamamsch. & Fed.	Rosaceae	*Rosa marschalliana* Sosn.
Fabaceae	*Trifolium sintenisii* Freyn	Rosaceae	*Rosa oplisthes* Boiss.
Fabaceae	*Vicia abbreviata* Fisch. ex Spreng.	Rosaceae	*Rosa ossethica* Manden.
Fabaceae	*Vicia alpestris* Steven	Rosaceae	*Rosa oxyodon* Boiss.
Fabaceae	*Vicia antiqua* Grossh.	Rosaceae	*Rosa prilipkoana* Sosn.
Fabaceae	*Vicia caucasica* Ekutim.	Rosaceae	*Rosa pulverulenta* M. Bieb.
Fabaceae	*Vicia ciliatula* Lipsky	Rosaceae	*Rosa teberdensis* Chrshan.
Fabaceae	*Vicia grossheimii* Ekutim.	Rosaceae	*Rosa transcaucasica* Manden.
Fabaceae	*Vicia iberica* Grossh.	Rosaceae	*Rosa tuschetica* Boiss.
Fabaceae	*Vicia sosnowskyi* Ekutim.	Rosaceae	*Sorbus buschiana* Zinserl.
Grossulariace-ae	*Ribes biebersteinii* Berland. ex DC	Rosaceae	*Sorbus caucasica* Zinserl.
Lamiaceae	*Satureja bzybica* Woronow	Rosaceae	*Sorbus caucasigena* Kom.
Lamiaceae	*Thymus caucasicus* Willd. ex Ronniger	Rosaceae	*Sorbus colchica* Zinserl.
Lamiaceae	*Thymus collinus* M. Bieb.	Rosaceae	*Sorbus migarica* Zinserl.
Lamiaceae	*Thymus coriifolius* Ronniger	Rosaceae	*Sorbus subfusca* (Ledeb.) Boiss.
Lamiaceae	*Thymus grossheimii* Ronniger	Rosaceae	*Sorbus fedorovii* Zaikonn.
Lamiaceae	*Thymus karjaginii* Grossh.	Rosaceae	*Sorbus velutina* (Albov) C.K. Schneid.
Lamiaceae	*Thymus ladjanuricus* Kem.-Nath.	Staphyleaceae	*Staphylea colchica* Steven

The wild ancestor, however, should be considered to be black colour grapevine, which is most common in Georgian wild populations, eventhough the mutation leading to the white-fruited wild grapevine has been found in other Georgian wild populations (Ramishvili, 1988). On a phenotypical basis of our investigation (Ekhvaia & Akhalkatsi, 2010) it can be confirmed that the infraspecific evolution of Vitis vinifera subsp. sylvestris has produced three population groups south of the Great Caucasus Range. Overall, there are three phenetically distinct morphometric groups of western, southern and eastern Georgian wild grapevines. These three groups can be readily distinguished by the length of main leaf veins and lengths of nectaries in male flowers. This conclusion differs from the classical classification of Georgian wild and cultivated grapevine (Ramishvili, 1988) that considers two centres of origin of grapes in the South Caucasus region: (1) an Alazani origin with whole eastern and southern Georgia and adjacent territories of Azerbaijan and Armenia, and (2) a Colchic centre of origin which includes the entire western Georgian region with the Black Sea coastal zone. Our data clearly show a separated group in the southern Georgian population located in the territory of historical Tao-Klarjeti, a region of Georgia with many aboriginal grapevine cultivars. Therefore, it is of high importance to study aboriginal grape varieties in the place of its supposed domestication and to determine genetic relations among native grapevine cultivars and local wild populations.

A B

Figure. 4: A- The 1,5-Myr-old petrified specimen of wild grapevine leaf. National Museum of Akhaltsikhe, Georgia; B-Stone carving on the medieval church Ananuri, Mtskheta-Mtianeti region, Georgia. Photos by Maia Akhalkatsi.

Molecular study based on nuclear microsatellite (SSR) markers revealed close genetic relationships between wild grape and local cultivars in Georgia (Ekhvaia et al., 2010, 2011). Twenty-four Georgian autochthonous and 45 accessions of wild V. vinifera subsp. sylvestris were analyzed at 17 microsatellite loci (VrZAG21, VrZAG47, VrZAG62, VrZAG64, VrZAG79, VrZAG83, VVMD7, VVMD24, VVMD25, VVMD27, VVMD28, VVMD32, VVMD34, VVS2, VVS4, scu04vv and scu14vv). Six accessions of the American rootstocksFercal ('Cabernet -Sauvignon' x Vitis berlandieri), Telecki 5C (V. berlandieri x V. riparia), Malegue 44-53 (V. riparia X [V. cordifolia X V. rupestris]), Couderc 3309 (Riparia tomenteuse x Rupestris Martin), cultivar V. labrusca 'Odessa' and V. riparia naturalized in Georgia were used as outgroup. Thirty-seven accessions of wild grapevine were collected from different regions of Georgia and 8 wild accessions were sampled in Turkey's Artvin province adjacent to Georgia. All individuals within the studied populations of wild grapevine were identified as dioecious plants with male or female flowers. Genotype analysis at the most studied loci showed that Georgian cultivated and wild grapevine was characterized by high level of genetic variability. Genetic structure also was analyzed using F statistics. The low level of genetic differentiation (Fst=0.03) between Georgian cultivated and wild grapevines demonstrates that in situ domestication of wild germplasm took place within local populations. This means that autochthonous Georgian cultivars should be originated from local wild grapevine (Ekhvaia et al., 2010). The dendrogram generated using Dice coefficient identified eight major clusters within the 75 different genotypes defined at 0.22 similarity level (Fig.5). Clusters A consist of 13 Georgian local cultivars and 8 wild grapevine accessions from different regions of Georgia. One example confirming the genetic linkage between cultivated and local wild grapevine is placement of the famous Georgian red cultivar 'Tavkveri' in cluster A, where it is closely linked to the wild accession WT46Tbilisi5 (GS value 0.96; Fig. 5) due to identical alleles at 33 out of 34 alleles. Thus, the hypothesis that this cultivar could have been selected from local wild grapevine can be considered, especially because like wild grapevine 'Tavkveri' is characterized by presence of functionally female flowers. The fact that the 5 ancient Georgian cultivars 'Chvitiluri', 'Kachichi', 'Shonuri', 'Saperavi' and 'Uchakhardani' fall within cluster B (Fig.5), which mainly contains wild accessions allowed us to suppose that these cultivars were derived from the earliest local domestication events. Cluster G shows the genetic similarity of most ancient Georgian cultivars 'Meskhuri Shavi' and 'Krikina' adapted to high mountain climate conditions in Meskheti. In conclusion, it should be mentioned that the Georgian cultivated and wild grapevines represent a unique and interesting genetic resources that is characterized by a high similarity level between wild and cultivated grapevines. The admixture found among local Georgian cultivars and wild grapevine indicates the possibility that these cultivars are derived from ancestral domestication of local wild types. Thus, the obtained data are supporting that Georgia is one of the oldest centres of domestication of grapevine and harbour of valuable genetic resources for grape breeding. It should be mentioned that wild grapevine populations occurring nowadays on the territory of Georgia are threatened by different impacts in their natural habitats and need to be protected. The confirmation of threatened status of the Georgian wild grapevine might be detected low level of heterozygous individuals found for the most of the studied loci, which reflects the isolated status and the reduced number of individuals in the wild populations. Therefore, it is necessary to conserve wild forms and aboriginal cultivars of grape for the maintenance of genetic variability and to avoid genetic erosion of valuable genetic resources for grape breeding in Georgia.

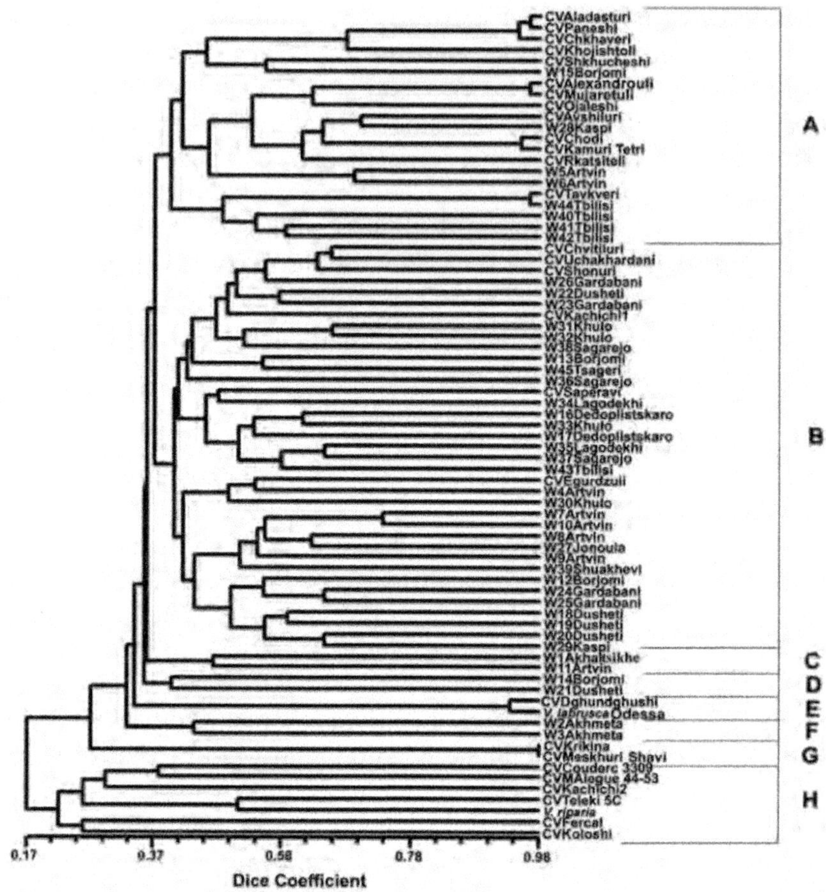

Figure. 5: Dendrogram of 75 accessions: Twenty-four Georgian local cultivars (Vitis vinifera subsp. vinifera), 45 Georgian wild grapevine (V. vinifera subsp. sylvestris), 4 rootstock cultivars (Couderc 3309, Fercal, MAlegua 44-53 and Teleki 5C), introduced cultivar V. labrusca 'Odessa' and naturalized V. riparia constructed by unweighted arithmetic average clustering (UPGMA) method based on Dice's coefficient of shared SSR polymorphisms.

Domestication of Pear in Georgia

Many pear cultivars occur in Georgia from pre-historic period indicating the early domestication event of this cultivated fruit tree (Javakhishvili, 1987). In total, 11 species of wild pear occur in Georgia (Kuthatheladze, 1980). They are distributed in different regions of Georgia, what is caused by the variable geographical relief and habitat diversity of the country. Pyrus caucasica Fed., the endemic species of the Caucasus is most widespread among the wild pears of Georgia and is considered as main progenitor species of local pear cultivars (Khomizurashvili, 1973). Moreover, P. caucasica and P. pyraster (L.) Burgsd. are regarded as the main wild progenitors, from which the cultivated European pear (P. communis L.) has probably evolved (Zohary & Hopf, 2000; Volk et al., 2006; Yamamoto & Chevreau, 2009). According to the literature data (Khomizurashvili, 1973) introduced cultivars of pear from Europe and Russia appeared in Georgia at the end of 19th century and before there were existed only the local cultivars. However, we assume that the process of cultivar introduction might have started much earlier, as Georgia had cultural contacts to Asian and European countries since antique period. It is also remarkable that Greece is considered to be a first provider of selective cultivars of pear to the ancient world (Jackson, 2003) and earliest relations between Georgia and Greece should be considered as possible way of introduction of European pear cultivars in the Caucasus. The local Georgian names of the cultivated pear Mskhali and wild Caucasian pear Panta exists in all Georgian dialects; they do not have analogues in any other languages (Javakhishvili, 1987). The Georgian names of cultivated and wild pears are linked with geographic objects such as mountains (Skhaltbis Range in Kartli, Mt. Mskhal-Gori in Kakheti's Kavkasioni), rivers (R. Skhaltba), or villages (Pantiani, Skhalta, Skhlobani, etc.; Javakhishvili, 1987). The name of wild Caucasian pear Panta is used among cultivars called 'Panta Mskhali', i.e. cultivar with name of wild pear. Moreover, the classification of Georgian pear cultivars (Khomizurashvili, 1973) contains a group of landraces with the same name. This classification system divides Georgian cultivars into four groups: 'Gulabi', 'Panta Mskhali', 'Kalos Mskhali', and 'Khechechuri'. The name of each group represents the name of a cultivar, which is considered to be a typical representative of a group. In the 'Gulabi' group, there are included both local and introduced cultivars to have most high economic values, big juicy fruits with sweet taste. The 'Panta Mskhali' group contains local varieties with small fruits becoming black after maturation. This is a character feature of wild Caucasian pear. The 'Kalos Mskhali' group includes local cultivars having bigger fruits than the second group. The 'Khechechuri' group matures in late autumn with juicy fruits containing a big amount of stone cells. According to N. Khomizurashvili (1973), the last three groups are originated by direct domestication of wild pear in Georgia. Although, some signs of selective breedings are remarkable as well. Relationships between wild P. caucasica and local cultivars are mirrored by a high morphological variability of leaf and fruit forms. This idea was for the first time confirmed by statistical methods of taxonomic identification and relationships among taxa in our study (Asanidze et al., 2011). We decided to conduct comparative morphometric study of cultivars recently occurred in Georgia and reveal their relationship to local wild pear species. The results have to determine local cultivars originated by direct domestication events in ancient time and discriminate from cultivars, which will have relationships with other wild species - P. pyraster or P. pyrifolia, considered as wild ancestors of European and Far East pear cultivars respectively. We carried out the investigation to determine morphological characteristics of leaves, young shoots and fruits differentiating local and introduced pear cultivars of Georgia to reveal the relationships between cultivars and wild ancestor species of pear by statistical methods used in plant morphology (Asanidze et al., 2011). A total of 214 wild and cultivated pear trees have been sampled in natural habitats, living collections and peasant grounds in different regions of Georgia. Wild pear species were determined according to Sh. Kuthatheladze (1980). The pear accessions evaluated in this study consisted of Caucasian endemic P. caucasica Fed. (=P. communis subsp. caucasica (Fed.) Browicz; N=100), P. balansae Decne. (= P. communis L.; N=8) and P. pyraster (L.) Burgsd. (=Pyrus communis L. subsp. pyraster (L.) Ehrh.; N=3), which has been obtained from Germany, Hessen, in surrounding of v. Erda. Eightyone individuals of 26 Georgian local and 22 individuals of 9 introduced cultivars (total 103 individuals of 35 cultivars) have been collected. Some of them are sampled in the collection of the Institute of Horticulture, Viticulture and Oenology, village Skra, Gori district, Georgia; local cultivars were sampled in the collection of aboriginal cultivars of Biological Farming Association Elkana in village Tsnisi, Akhaltsikhe district, Georgia. Many local cultivars are collected in peasant house gardens in different regions of Georgia. The individuals were evaluated by 27 morphological traits, which included one landmark analysis data, 12 leaf and shoot descriptors and 14 fruit descriptors. Morphological characters have been taken as recommended by International Union for the Protection of New Varieties of Plants (UPOV, 2000) for P. communis and J. Voltas and colleagues (Voltas et al., 2007), which delimited differences between wild and cultivated taxa of the genus Pyrus based on morphometric analysis. In total 21 morphological traits have been analysed by multivariate analysis. The Principal Components Analysis and Hierarchical Cluster Analysis (HCA) methods revealed the distance or similarity measure to be used in clustering with the Ward's method as amalgamation rules. According to HCA's results, pear cultivars, analysed in this study, are clustered into two groups (Fig.6). The first group A contains local cultivars related to P. caucasica and P. balansae by 21 morphological traits of leaves and fruits. Especially close Euclidean similarity distance is revealed between wild Caucasian pear, Panta in Georgian and a cultivar named 'Panta mskhali', which confirms etymological and taxonomic similarity within these taxa. P. balansae shows very close similarity distance with 'Tsvrili mskhali' and 'Korda'. Very closely related group of local cultivars to wild Georgian pears contains: 'Bebani', 'Samariobo', 'Tavrejuli',

'Kvichicha', 'Khinos mskhali' and 'Akiro'. The other group: 'Shavmskhala', 'Nenes mskhali', 'Borbala', 'Majara', 'Shakara' and 'Kartuli mskhali', is more distanced from wild pears, but located in the same cluster. We assume that these local cultivars must have been originated by direct domestication of wild Caucasian and Balanse's pears in Georgia. The second cluster B (Fig.6) contains introduced cultivars of pear originated in European countries and some old Georgian cultivars. The group B reveals relationship with wild pear - P. pyraster, which is distributed in Europe and does not reach Georgian territory.

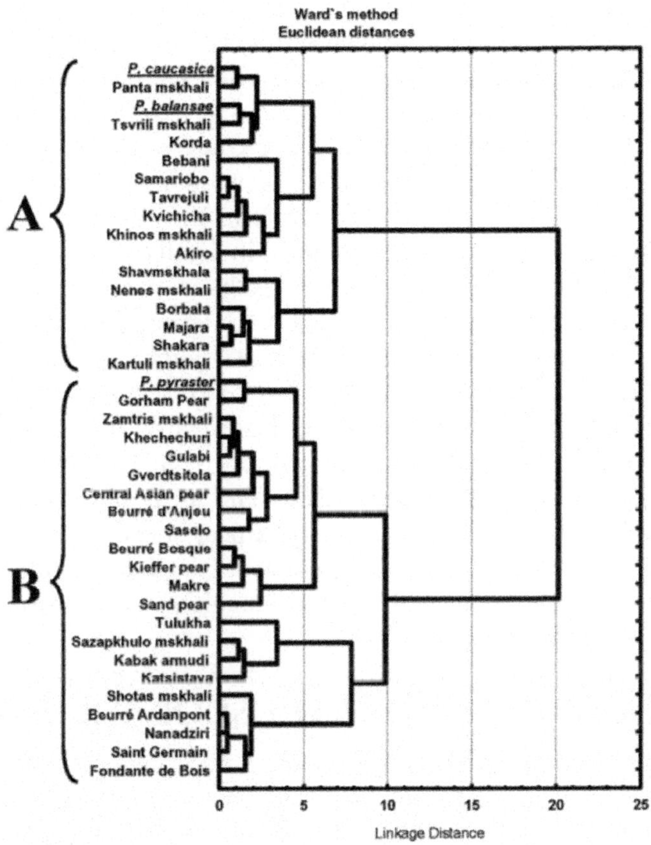

Figure. 6: HCA dendrogram of Euclidean distance with the Ward's method showing the relationships between the 35 cultivars and 3 wild species of pear based on 20 morphological traits of leaf, shoot and fruit and 20 landmark harmonics of mature leaf; The taxa in the dendrogram are clustered into two main groups - A and B. (N=214).

The area of distribution is up to the middle of Turkey. The most cultivars from intermediate group B are more widespread in Georgia than the local cultivars originated by direct domestication of wild pears from group A. Two local pears 'Gulabi' and 'Khechechuri' are the most widespread among all local pears of Georgia and there are two or more varieties of them in each localities of the country. Moreover, Georgian name of cultivar 'Gulabi', which is also used to be a name of local pear group in classification of N. Khomizurashvili (1973), means pear in Persian. We suggest that local cultivars from cluster B (Fig.6) might have appeared in Georgia very long time ago and were improved by local population using breeding procedures.

Leaf margin shape is the main morphological trait that differentiates Caucasian pear (P. caucasica) from European pear (P. communis). Leaf margins are entire in P. caucasica and serrate in P. communis. This theory was proved by the statistical analysis of the collected samples for the present study. Nowadays, P. pyraster is considered as the wild pear of Europe and cultivars are named as P. communis (Yamamoto & Chevreau, 2009). 'Communis' or the 'Common pear' group of cultivars has become the name of the cultivated pears of Europe, however, the structure and the diversity of the wild and cultivated pears of this group is not studied in details and needs further genetic and molecular investigations. Thus, the results of this study have shown that some local cultivars of Georgia are direct domesticated from the native wild pear species - P. caucasica and P. balansae. The other local cultivars might be obtained due to selective works by breeding of local landraces with introduced cultivars from different countries in historically different periods. The molecular study of these taxa will clear in more details origin of these cultivars. The results confirm the hypothesis that some local cultivars of Georgia are directly domesticated from the native wild pear species - P. caucasica and P. balansae. The other local cultivars might be obtained due to selective works by breeding of local landraces with introduced cultivars from different countries in historically different periods.

TRADITIONAL SUSTAINABLE USE OF ANCIENT CROPS

Since ancient times, agriculture in Georgia has been divided in two zones: lowlands (0-1300 m a.s.l.) and high-mountains (1300-2200 m a.s.l.). This classification was based on production of wine (Javakhishvili, 1987). Winemaking was always the major branch of agriculture in Georgia. Wine was exported from Georgia since ancient times. The vineyards were cut off to reduce income for exporting the wine in neighbour countries during the occupation of the country by the Muslim nations. The exchange of agricultural products took place between lowland and high-mountain regions and not only within the Georgians, but also with North Caucasians. This tradition remained till the end of 20th century when Dagestanian people from Didoeti region visited Kakheti lowland in eastern Georgia in late autumn. They have exchanged agricultural products from high-mountains to lowland crops. In 20th century, they brought from Dagestan cows, cheese and potato and have exchanged them on Kakhetian wine, schnapps from grapevine called chacha, bread wheat flour, fruits and vegetables.

Figure. 7: A - Satsnekheli - stone construction for the pressing of grapes in Nekresi monastery (IV century AD), Kakheti; B - Clay vessels for wine storage Kvevri, Ikalto monastery (XII century AD), Kakheti; C- Red wine Saperavi and boghlortso (wheat bread into red wine) in clay cups on Georgian table. Photos by Maia Akhalkatsi.

The information about the traditional use of wine is remained in folklore and ethnographic studies collecting this knowledge. The problem in reducing of written information was again the wars, destruction of settlements and burning of manuscripts during historical times. Therefore, the information on traditional agriculture is based on both literature data and interviews of local people obtained during our field trips in different provinces of Georgia. The ancient stone construction satsnekheli for the pressing of grape might be found near many historical monuments (Fig.7A) even in ancient caves of Chachkari near Vardzia monastry complex. Each family in lowland regions have a room for winemaking in houses called marani in which the clay vessels are buried in soil named kvevri (Fig.7B) where the wine is made and stored by traditional Georgian technology. One of kvevri in each marani was called zedashe and the wine in it might be used only in religious rituals since it belonged to the God. Zedashe was filled by schnapps called Araki derived from different fruits e.g. wild pear - P. caucasica, in high mountain regions and used for rituals (Fig. 8A-C). Wine and bread are ritual accessories of Christian religion and first cross entered in Georgia by St. Nino from Cappadocia in 4th century AD was made from vines. However, the folklore knowledge let us know that grapevine was a ritual plant in ancient religion and represented a tree of the God of sun. The remnants of grapevine images are often demonstrated as stone carvings on Christian churches (Fig.4B) and ancient golden and silver cups and Jewellery. Wine was used not only as alcohol drink but in religious rituals and as traditional food. Boghlortso (Fig.7C) was a healthy food prepared by wheat bread placed within a red wine and used by the whole family members including children and chronic invalid and consumptive people. Bread is served with all Caucasian meals. It is the same ritual food like wine in Georgia. Two landraces of bread wheat - Triticum aestivum var. erythrospermum and T. aestivum var. lutescens are used for religious rituals in Svaneti (Girgvliani, 2010). The flour of these cultivars is preserved separately from other reserves of bread wheat flour and used on religious holydays. Milled faba bean and kenaf seeds are added to the bread flour for baking ritual bread. There are barley cultivars: H. vulgare var. pallidum in Svaneti and H. vulgare var. nutans in Meskheti, used for traditional bread preparation added to the T. carthlicum 'Dika' flour.

A B C

Figure. 8: A- Murtaz Chankseliani collected wild pear (Pyrus caucasica) in forest, R. Kheledula gorge, Lower Svaneti; B - Wild Caucasian pear for preparing of alcohol schnapps Araki; C - Distillation equipment for preparing of pear schnapps. Photos by Maia Akhalkatsi.

The healthy quality of food in Georgia is connected with usage of fruit and herb sauces for roasted and fried meet. Sour plum sauce (P. cerasifera var. divaricata) is always added to spitroasted chicken and pork. Many herbs and spices are added to meet meals. This should be understand to be a modern direction in diet works when an alkaline food is recommended to neutralize acid food, which is everyday meal including meet, sugar and bread. Modern civilization eats considerably more acid-forming foods than alkalizing foods. According to well-known naturopath P. Airola (1984), acidosis, or over-acidity in the body tissues, is one of the basic causes of diseases, especially the arthritic and rheumatic diseases. There is in the internet now a lot of information on alkaline food. The lists of alkaline products show that fruits and vegetables have highest pH most of which are traditional cultivars in Georgia. There are alkaline vegetables such as - alfalfa, barley, beet greens, beets, cabbage, carrot, celery, cucumber, eggplant, garlic, green beans, green peas, herbs, lentils, lettuce, onions, radishes, spices, spinach, watercress, wild greens, etc. which are the traditional cultivars in Georgia. Alkaline traditional fruits are - apple, apricot, berries, cherries, sour cherries, figs, grapes, grapefruit, nectarine, peach, pear, strawberries, watermelon, seeds and nuts, etc. To maintain health, the diet should consist of 60% alkaline forming foods and 40% acid forming foods. To restore health, the diet should consist of 80% alkaline forming foods and 20% acid forming foods (Airola, 1984). Therefore, we should think about sustainability of traditional use of crops and wild plant species in the past time. The fact that the nature remained undisturbed in the country centuries long and population was characterized by much higher healthy features as current situation exists in Georgia should be explained as occurrence of healthy food and nature tolerant use of the plant resources.

THREATS

Threats and Conservation of Ancient Crops

There are several reasons for the genetic erosion of ancient cultivars and the wide distribution of new varieties of introduced crops. First of all, new cultivars have higher yields and are therefore preferred both as a source of food for local people and as a cash crop that determines local income. The second reason why local peasants began to prefer cultivating genetically modified (GM) plants may be explained by introduction of new diseases into Georgian agricultural fields in recent years, causing harm primarily to ancient cereals and vegetables. However, the introduction of new parasites has revealed that endemic forms of Georgian crop plants contain valuable selective disease-resistant material for genetic engineering. The tetraploid and hexaploid endemic wheat species T. timopheevii and T. zhukovskyi, for example, are characterized by a high level of resistance to a new race (TTKS, commonly known as Ug99) and many other races of Puccinia graminis f. sp. tritici due to the wheatstem rust resistance gene Sr36 (Tsilo et al., 2008). T. carthlicum is characterized by immunity to diseases, a short growing period, and resistance to cold. Intensive Genetic erosion of ancient crops started in Georgia since 1950s which was also a period of intense selection work in breeding stations in the whole of the Soviet Union, e.g. the highly productive awnless wheat cultivar Bezustaja I developed in Russia has been sown in all wheat fields in Georgia since the 1970s, and this variety eventually replaced Georgian endemic wheat species. Recently new breeder's varieties of wheat and other cereals are introduced from different countries. The process of genetic erosion of ancient crop varieties has not been a great concern for the mountain areas of Georgia, which until the 1990s acted as a depository of ancient crops. One important consideration that explains why ancient cultivars were conserved longer in mountainous regions than in the lowlands is that the local population preserved their traditional ways of life and socioeconomic structures. The traditional agricultural system is characterized by dependence on local genetic resources and locally developed technologies. Even today, peasants in mountain villages use an ox-drawn sledge made of wood for loading and transporting cereals and a threshing sledge on threshing floors to thresh wheat, oats, rye, and barley. Traditional agricultural equipment makes it possible to cultivate areas even on steep slopes and at high elevations, where modern tractors cannot be used. Moreover, some old landraces of wheat and barley are used to prepare bread and beer for religious rituals. Substitution of these landraces by others would go against centuries-old traditions (Akhalkatsi et al., 2010). Despite these conditions that support the maintenance of ancient landraces, many endemic and native representatives of crop plants are currently in danger of extinction and face severe problems of genetic erosion in all mountainous regions of Georgia. While agrobiodiversity is declining rapidly in many areas of the world due to anthropogenic pressure (Körner et al., 2007), including population growth, in Georgia the main reason for genetic erosion of ancient crop varieties is demographic decline in mountain regions due to harsh economic conditions and lack of modern infrastructure (Nakhutsrishvili et al., 2009). The shift from ancient cultivars to modern high-yielding crops such as maize and potato, which took place in the lowland areas much earlier, began in mountain villages only in the last 20 years. Greater income from marketing allows families to stay in mountain villages. Moreover, the economic security of the traditional farming systems in these mountain regions appears to be in jeopardy when traditional agriculture is replaced by cattle breeding, which causes abandonment of cultivated fields and their transformation into pastures. Several research centres maintain ex situ germplasm collections of Georgia, such as gene banks and living collections. According to the National Biodiversity Action Plan of Georgia (Jorjadze, 2005), international nature conservation institutions and Georgian scientific and nongovernmental organizations have taken care to preserve the genetic resources of local cultivars. Several gene banks and living collections occur in Georgia. There is one biggest genebank located at the Georgian Institute of Farming established in 2004 through support of International Centre for Agricultural Research in the Dry Areas (ICARDA). They owned a total 3057 accessions of local and introduced cultivars and CWRs in 2010. The other 5 gene banks are located in different research institutes unified with Agrarian University in 2011. Total number of germplasm accessions is 6286 in Georgian gene banks. However, the material kept in ex situ collections are not sufficient and need more contribution. Many seed banks worldwide contain about 7000 accessions of germplasm of Georgian cultivars and crop wild relatives. A recently initiated project, "Mountain Biodiversity in the Caucasus and its Functional Significance," supported by the Swiss National Science Foundation Program SCOPES, will build an electronic biodiversity archive for Georgia, and include data on mountain plant biodiversity in Georgia. Because it will be built in compliance with Global Biodiversity Information Facility (GBIF) standards, it will contribute to the Global Mountain Biodiversity Assessment (GMBA) mountain portal with GBIF (www.mountainbiodiversity.org). A research agenda concerned with the use of georeferenced mountain biodiversity data for science and management was developed at a GMBA workshop in Kazbegi, Central Caucasus, in July 2006 (Körner et al., 2007). Such a database of the plant species of the Caucasus will become a prominent entry in the GBIF database and highlight the current status of plant genetic resources in Georgia. It should be emphasized that establishment and maintenance of ex situ collections and databases is just a first step in the conservation process of ancient crop varieties. The next step should be return of conserved seed material to the fields of local farmers. From 2004 to 2009, the Global Environmental Facility/United Nations Development Fund (GEF/UNDP) project "Recovery, Conservation and Sustainable Use of Georgia's Agro-Biodiversity" was carried out with the aim of conservation and sustainable use of threatened local plant genetic resources in the oldest historical mountainous region of Georgia, SamtskheJavakheti. This project enabled establishment of sources of primary seed and planting material for threatened crops and fruit varieties, and assisted farmers in accessing markets for organic products from such crops as lentil, grass pea, chickpea, faba bean, common millet, Italian millet, etc. Another project was the return of the Georgian wheat variety T. aestivum var. ferrugineum 'Akhaltsikhis Tsiteli Dolis Puri' in Meskheti province, where it was sown on 10 ha and produced bread that was introduced in shops featuring organic products in Tbilisi as of 2008. Afterward, this project was supported by the Georgian church, which expressed an interest in cultivating ancient crops on monastery grounds. However, these attempts have been realized only on a small scale and not in larger areas of the country. In our opinion, the major activity of the corresponding governmental institutions should be directed on supporting local farmers in reintroducing ancient crops on the market and maintain maximum diversity of the target taxon's gene pool. The importance of agricultural achievements not should be oriented only on high yield of crops but the traditional foods to which people have adapted a long time determines their healthy lifestyle. Thus, conservation and reintroduction of ancient cultures to modern agriculture can insure longevity of people.

Threats and Conservation Needs of Crop Wild Relatives

The natural populations of many species of CWRs are increasingly at risk. The primary causes of diversity loss of wild plant species are habitat loss, degradation and fragmentation. Many cereal CWRs, including relatives of wheat and millet species, occur in arid or semi-arid lands and are severely affected by over-grazing and desertification. The forest species are affected by habitat disturbances because of illegal forest cutting occurring in 1990s in Georgia. Climate change is having significant impacts of species distributions and survival in a concrete habitat. One of the most important threats to the diversity of CWRs are genetic erosion and pollution. The threat of genetic pollution or introgression, either from genetically modified organisms (GMOs) or from conventionally bred crops, to wild species has become an increasing risk to the in situ genetic conservation of crop wild relatives. Another problem is that many species of important CWR occur in centres of plant diversity and crop diversity located mainly in developing countries, which often lack resources to invest in the necessary conservation activities. South Caucasus and Georgia, in particular, is the centre of origin and diversity of many of the world's important crop plants. There are several international projects realized by the ICARDA, the International Plant Genetic Resources Institute (IPGRI), US Department of Agriculture (USDA), United Nations Environmental Program (UNEP), etc. contributed in undertaking efforts in monitoring and conservation of plant. Although, additional resources are urgently needed in such areas of high diversity to identify priority species for conservation, determine the necessary conservation activities, monitor the status of key species, improve the use of these valuable resources. Habitat disturbances are main threats leading to the extinction of rare and endangered plant species. Deforestation took place during last decades in Georgia and caused habitat degradation. The fact detected with the population of wild grapevine has revealed the threat to the riparian forests, which is situated along rivers in very close proximity of settlements and local people uses the resources of this forest in a highest degree. We have detected that some trees were cut representing the support of clambering wild grapevine and the individuals were lying on the earth, which will cause its drying up and death. More great scale cuttings in dark coniferous forests lead to arising of forest openings with high irradiation leading to drying up the underground cover of mosses and lichens, which drastically changes habitat and determines disappearing of natural species adapted to this habitat. Overgrazing of meadows and pastures was a problem in Soviet period, when several million head of sheep were grazing summer pastures of mountainous regions of Georgia. However, now the number of cattle is reduced and does not threaten much the rare species in their natural habitats. In spite of this fact, grazing affect survival of rare species such as Hordeum spontaneum, which was found on road side and during the next visit it was grazed completely. Such disturbances as habitat degradation due to road and pipeline construction works threatens the populations but has temporary effect. These types of disturbances are especially threatening the rare and endangered species of high conservation value. The best way of in situ preservation of genetic diversity of valuable plants is creation of nature reserves on the territories, where natural populations of CWR occur. The first nature reserve of Georgia was established in Lagodekhi in 1912. In present, the protected areas occupy 7% of the country's territory, which is equal to 495.892 ha. There are 16 nature reserves, 9 national parks, 12 managed resource protected areas, 14 natural monuments and 2 protected landscapes in Georgia. The problem remains for the species, which are growing in rural habitats and on arable lands mixed with field crops have different assessment to threats. These species are depending in their existence to the monitoring of arable lands, which crop will be sown, how will be transformed filed crop to pasture or hay meadow, or what kind of herbicides and mineral fertilizers will be used in the field. The maintenance of wild populations growing as weeds in cultivated fields depends on sustainable management of agriculture in the region. The governmental institutions should control the processes which might bring to the genetic erosion of CWRs having high value of conservation. In this case the legislation bases should be effective to control local farmers not affect CWRs with ecologically unsuitable for this species actions in the field e.g. use of fertilizers or introduction of new crops leading to changing in technology of field cultivation methodology and leading to disturbances of wild weed species of high conservation value. Ex situ conservation of the germplasm of CWRs is very valuable material for improvement of crop quality and their resistance against fungal and microbial disease. It will be of interest to collect their seed material and distribute to genbanks, which will contribute to provide necessary germplasm to research centres dealing with the genetic engineering. The Tbilisi Botanical Garden and Institute of Botany has two collections of seeds. One is collection of rare endemic plant seeds, which is collected in the framework of the Millennium project managed by Kew Royal Botanical Garden, UK. The second is collection of aboriginal crop varieties collecting in different regions of Georgia. These program works together with IPK, Gatersleben Germany, where the analogy of the collected material is kept at the gene bank. The living collections of CWRs are very few. Botanical Gardens in Tbilisi and Batumi have some small collections of CWRs collected in the frameworks of international collaborative projects. However, maintenance of the collections after finishing the projects is impossible and they are cancelled in several years. The plant genetic resources documentation in Georgia is mostly computerized. There are several databases, which include all information and passport data for accessions of field crops, but so far they have no free access. Most threats to biodiversity are the results of human actions, which are expressed in the overuse of natural resources for fuel, fodder, manure, grazing and collecting of ornamental and medicinal plants. This process leads to the loss of genetic diversity including crop wild relatives.

The in situ protection measures are not easy to implement and, thus, the accent should be directed on ex situ conservation.

CONCLUSIONS

Very old archaeological findings, cultural heritages and so far existing high morphological and genetic diversity of ancient crops and their wild relatives show that Georgia has very old agricultural traditions that have preserved to our times. The fact that large-scale genetic erosion of the ancient landraces in Georgia has reached extreme levels from 1950s and almost all the local varieties of cereals (wheat, barley and millet), legumes (peas, lentils, common vetch and faba bean), and grapes are now disappear from the farms requires special analyses and development of conservation measures. Only the gene banks and living collections hold germplasm of landraces extinct in the farms. An assessment of the effectiveness of current conservation strategies to protect the diversity of ancient crops in Georgia reveals a gap in the reintroduction of conserved germplasm to the fields of local farmers. In our opinion, the corresponding governmental institutions responsible for conservation of biodiversity should refocus the strategy to require complementary in situ and ex situ conservation actions to maintain maximum diversity of the target taxon's gene pool by supporting local farmers in reintroducing ancient crops on the market and thereby filling this gap. Moreover, at present, neither field crop genebank nor live collections of the permanent crops have sufficient land and equipment in Georgia, as well as funding to carry out ex situ conservation at the modern level. Storage of the in situ collections should be improved through upgrading the present storage of the field crop genebank facilities. There is a need to improve public awareness of importance of ex situ conservation. Popularity of the data obtained by scientists should be distributed among the local population so that they themselves have contributed to the preservation of national heritages. The results of scientific investigations that some crops represent local cultivars and even domesticated landraces in this area means that this is connected with lifestyle of the local population. The fact that longevity of life in the Caucasus was very high and centenarians lived to 120 years and more should be understand that a healthy diet of mankind is not only amount of calorie but the combination of food of high quality. The modern alkaline diet almost completely coincides with the traditional Georgian cuisine. Therefore, we must appreciate the importance of conservation of local varieties to ensure the health of local people.

The data obtained in our investigations (Ekhvaia &Akhalkatsi, 2010; Ekhvaia et al., 2010, 2011; Asanidze et al., 2011) indicate importance of CWR species in Georgia as many of them represent direct ancestors of local cultivars. The fact that wild grape shows high genetic relation to local varieties of grape indicates that winemaking represents an ancient culture in Georgia, which is expressed even in religious rituals of the nation. Wild grapevine and pear representing the wild ancestors of local varieties are under threat because of wood cutting. Many other CWRs are in the same position. The legislation of species conservation is applied to rare and endemic species and in situ conservation is maintained only at protected areas. However there is no legislation that can protect CWRs growing in rural and urban areas and representing weed species. No actions of conservation are undertaken to protect these species. Many CWRs (wild wheat, rye, coriander, etc.) are grown in cultivated lands of local farmers. Many years, wild wheat species were mixed with local varieties of wheat and barley but now they are disappeared. The events which are protecting them are traditional cultivation technology of the landraces to which the local weeds are adapted by their life strategy and propagation character. The threats here will be change of traditional crops to the new varieties, which will need different cultivation events. This might lead to disappearance of the CWRs from the cultivated beds. At present, CWRs ex situ collections are almost absent in Georgia. The problem of genetic erosion of landraces and their wild relatives needs active contributions by national policies and comprehensive measures are urgently needed to avoid the complete loss of ancient crop genetic resources in Georgia. International nature conservation institutions and Georgian scientific and nongovernmental organizations should show more activity to the restoration of ancient crops, which defined the healthy life of Georgians.

ACKNOWLEDGEMENT

Research on agrobiodiversity in Georgia is supported by grant IZ73Z0_128057, in the framework of the Swiss National Science Foundation (SNSF) program SCOPES (Scientific Cooperation between Eastern Europe and Switzerland). The expeditions were funded by the Global Environmental Facility/United Nations Development Fund (GEF/UNDP) Project GEO/01/G41/A/1G/72. We are grateful to K. Pistrick, R. Fritsch, F. Blattner, and M. Gurushidze from IPK Gatersleben for supporting our investigations.

REFERENCES

1. Adler, D. S. & Bar-Oz, G. (2009). Seasonal patterns of prey acquisition during the Middle and Upper Palaeolithic of the southern Caucasus. In: The Evolution of Hominid Diets: Integrating approaches to the study of Palaeolithic subsistence, Hublin, J.-J. & Richards, M. (Eds), pp. 127-140, Springer, ISBN: 9781402096983, Leipzig.

2. Airola, P. (1984). How to Get Well: Handbook of Natural Healing. Health Plus Pub, (March 1984), ISBN:0932090036, Sherwood, OR.
3. Akhalkatsi, M., Ekhvaia, J., Mosulishvili, M., Nakhutsrishvili, G., Abdaladze, O. & Batsatsashvili, K. (2010). Reasons and processes leading to the erosion of cropgenetic diversity in mountainous regions of Georgia. Mountain Research and Development, Vol.30, No.3, (August 2010), pp.304-310. ISSN: 0276-4741.
4. Asanidze, Z., Akhalkatsi, M. & Gvritishvili, M. (2011). Comparative morphometric study and relationships between the Caucasian species of wild pear (Pyrus spp.) and local cultivars in Georgia. Flora, Vol.206, No.11, (November 2011), article in press.doi:10.1016/j.flora.2011.04.010.
5. Badr, A., Müller, K., Schäfer-Pregl, R., El Rabey, H., Effgen, S., Ibrahim, H. H., Pozzi, C., Rohde, W. & Salamini, F. (2000). On the origin and domestication history of barley (Hordeum vulgare). Molecular Biology and Evolution Vo.17, No.4, (April 2000), pp.499–510. ISSN: 0737-4038.
6. Bakhtadze, K.E. (1947). Biologia, selektsia i semenovodstvo chainogo rastenija (Biology, Selection and Seed Productivity of Tea Plants). Nauka, Moscow. (in Russian).
7. Batonishvili, V. (1991). Aghtsera samefosa sakartvelosi (Geographic Description of Georgian Kingdom). (Ed. 7), Ganatleba, Tbilisi. (In Georgian).
8. Bedoshvili, D. (2008). National Report on the State of Plant Genetic Resources for Food and Agriculture in Georgia. Available from:
9. http://www.pgrfa.org/gpa/geo/Georgian report on State of PGR Sep 29, 2008.pdf; Accessed on 21 March 2010.
10. Bockelman, H. E., Erickson, C. A. & Goates, B. J. (2002). National Small Grains Collection: wheat germplasm evaluations. Annual Wheat Newsletter, Vol.48, (10 August, 2002), pp.273-286.
11. Casas, A. M., Yahiaoui, S., Ciudad, F. & Igartua, E. (2005). Distribution of MWG699 polymorphism in Spanish European barleys. Genome, Vol.48, No.1, (February 2005), pp.41–45. ISSN: 0831-2796.
12. Davies, P. (2003). An Historical Perspective from the Green Revolution to the Gene Revolution. Nutrition Reviews, Vol.61, No.6, pp.124–134. ISSN: 0029-6643, doi:10.1301/nr.2003.jun.S124-S134.
13. Dekaprelevich, L. (1947). Saqartvelos martsvlovani kulturebis dziritadi jishebi (Main cultivars of cereals in Georgia). Proceedings of State Selective Station, Georgia, Vol.2, No.1, pp.5-47. (In Georgian).
14. Dekaprelevich, L. & Menabde, V. (1929). K izucheniu polevykh kultur

zapadnoi Gruzii. I. Racha. (Study of cereal cultivars in Georgia. I. Racha). Scientific Papers of the Applied Sections of the Tbilisi Botanical Garden. Vol.6, No.2, pp. 219-252. (In Russian).

15. Dvorák, J., Luo, M.-C. & Yang, Z.-L. (1998). Genetic Evidence on the Origin of Triticum aestivum L. In: The Origins of Agriculture and Crop Domestication. Damania, A.B., Valkoun, J., Willcox, G. & Qualset C. O. (Eds). Proceedings of the Harlan Symposium, pp. 254-267, ISBN: 92-9127-084-9. Aleppo, Syria, 10-14 May 1997.
16. Ekhvaia, J. & Akhalkatsi, M. (2010). Morphological variation and relationships of Georgian populations of Vitis vinifera L. subsp. sylvestris (C. C. Gmel.) Hegi. Flora, Vol.205, No.9, (September 2010), pp.608-617. ISSN: 0367-2530, doi:10.1016/j.flora.2009.08. 002.
17. Ekhvaia, J., Blattner, F. R. & Akhalkatsi, M. (2010). Genetic diversity and relationships between wild grapevine (Vitis vinifera subsp. sylvestris) populations and aboriginal cultivars in Georgia. Proceedings of 33rd World Congress of Vine and Wine, 8th General Assambley of the OIV, Tbilisi, Georgia, June 2010, Available from: <http://www.oiv2010.ge/index. php?page=5&lang=0>
18. Ekhvaia, J., Blattner, F.R., Gurushidze, M. & Akhalkatsi, M. (2011). Relationships between wild grapevine (Vitis vinifera subsp. sylvestris) populations and native grape cultivars in Georgia. Proceedings of BioSystematics Berlin 2011. ISBN: 978-3-921800-68-3, pp.111, Berlin, Germany, 21–27 February 2011.
19. Eliava, G. M. 1992. Dzvelkolkhuri unikaluri vazis jishta katalogi (Catalog of ancient grape varieties of Colchis). (Ed.2), Kutaisi Press N6, Georgia. (In Georgian).
20. Eritzjan, A. A. (1956). K izucheniu izmenchivosti sostava pchenits v Gruzii (Study of variability of wheat field composition). Proceedings of Tbilisi Botanical Institute, Vol.18, No.1. pp.251-277. (In Russian).
21. Finlayson, C. (2005). Biogeography and evolution of the genus Homo. Trends in Ecology and Evolution. Vol. 20, No.8, (August 2005), pp.457-463. ISSN: 0169-5347.
22. Fox, W. Sh. (2004). Living longer with phytomedicines from the Republic of Georgia. Woodland Publishing, ISBN-10: 158054388X, Health & Fitness.
23. Gabunia, L. & Vekua, A. A. (1995). Plio-Pleistocene hominid from Dmanisi, East Georgia, Caucasus. Nature, Vol.373, No.6514, (9 February 1995), pp.509-512, ISSN: 0028-0836, doi:10.1038/373509a0.
24. Gabunia, L., Vekua, A., Swisher, C. C., Ferring, R., Justus, A., Nioradze,

M., Ponce de Leon, M., Tappen, M., Tvalchrelidze, M. & Zollikofer Ch. (2000). Earliest Pleistocene hominid cranial remains from Dmanisi, Republic of Georgia: taxonomy, geological setting, and age. Science, Vol.288, No. 5578, (July 5, 2002), pp. 85-89. ISSN: 0036-8075.

25. Garson, L.K. (1991). The Centenarian Question: Old-Age Mortality in the Soviet Union, 1897 to 1970. Population Studies, Vol.45, No.2, pp. 265-278. ISSN: 0032-4728.doi:10.1080/0032472031000145436.
26. Girgvliani, T. (2010). Zemo Svanetis khorblis aborigenuli formebis istoria. (The history of aboriginal forms of wheat varieties of the Upper Svaneti). Artanuji, ISBN:978-9941-421-20-4,Tbilisi. (In Georgian).
27. Grant, K., Russel, Ch. & Everill, P. (2009). Anglo-Georgian Expedition to Nokalakevi. Interim report on excavations in July 2009. Available from: http://www.nokalakevi.org/Downloads/Reports/English/Site/2009.pdf.
28. Grassi, F., Labra, M., Imazio, S., Ocete Rubio, R., Failla, O., Scienza, A. & Sala, F. (2006).
29. Phylogeographical structure and conservation genetics of wild grapevine.
30. Conservation Genetics, Vol.7, No.6, (December 2006), pp.837–845. ISSN: 1566-0621.
31. Heun, M., Schäfer-Pregl, R., Klawan, D., Castagna, R., Accerbi, M., Borghi, B. & Salamini, F.
32. (1997). Site of einkorn wheat domestication identified by DNA fingerprinting.
33. Science, Vol.278, No.5341, (14 November 1997), pp.1312-1314. ISSN:0036-8075.
34. Jackson, J. (2003). The Biology of Apples and Pears, Series: The Biology of Horticultural Crops.
35. Cambridge University Press, ISBN: 0521380189, Cambridge.
36. Javakhishvili, I. (1987). Sakartvelos ekonomiuri istoria (Economic History of Georgia). (Ed. 2), Vol.5., Metsniereba, SB:2927, Tbilisi. (In Georgian).
37. Jorjadze, M. 2005. Agrobiodiversity, biotechnology and biosafety. In: National Biodiversity Strategy and Action Plan—Georgia. NACRES and Fauna and Flora International (FFI). Polygraph, ISBN: 99940-716-6-1. Tbilisi.
38. Ketskhoveli, N. (1928). Masalebi kulturul mtsenareta zonalobis shesastsavlad kavkasionze. (Materials on zonal distribution of cultivated plants in the Greater Caucasus). Agricultural National Committee Press, Tbilisi. (In Georgian).

39. Ketskhoveli, N. (1957). Kulturul mtsenareta zonebi sakartveloshi (Zones of cultivated plants in Georgia). Georgian Academy of Sciences Press, Tbilisi. (In Georgian).

40. Ketskhoveli , N., Ramishvili, M. & Tabidze, D. 1960. Sakartvelos ampelograpia. (Amphelography of Georgia). Georgian Academy of Sciences Press, Tbilisi. (In Georgian).

41. Khomizurashvili, N. (1973). Sakartvelos Mekhileoba (Horticulture of Georgia). Metsniereba, Tbilisi (in Georgian).

42. Kilian, B., Özkan, H., Kohl, J., von Haeseler, A., Barale, F., Deusch, O., Brandolini, A., Yucel, C., Martin, W. & Salamini, F. (2006). Haplotype structure at seven barley genes: relevance to gene pool bottlenecks, phylogeny of ear type and site of barley

43. domestication. Molecular Genetics and Genomics, Vol. 276, No.3, (September, 2006), pp. 230-241, ISSN: 1617-4615. doi 10.1007/s00438-006-0136-6.

44. Kobakhidze, A. (1965). Sakartvelos samartsvle-parkosani mtsenareebi: Phaseolus L. (cereals of Georgia: Phaseolus L.). In: Botanika (Botany), Ketskhoveli, N. (Ed), pp.87-89, Metsniereba, Tbilisi. (In Georgian).

45. Kobakhidze, A. (1974). Sakartvelos samartsvle-parkosan mtsenareta botanikur-sistematikuri shestsavlisatvis. (Botanical-systematic study of cereals in Georgia). In: Botanika (Botany), Ketskhoveli, N. (Editor), pp. 58--190, Metsniereba, Tbilisi. (In Georgian).

46. Körner, Ch., Donoghue, M., Fabbro, T., Häuse, Ch., Nogués-Bravo, D., Arroyo, M.T.K., Soberon, J., Speers, L., Spehn, E.M., Hang Sun, Tribsch, A., Tykarski, P. & Zbinden N. (2007). Creative use of Mountain Biodiversity Databases: The Kazbegi Research Agenda of GMBA-DIVERSITAS. Mountain Research and Development, Vol.27, No.3, (August 2007), pp.276–281. ISSN: 0276-4741.

47. Kuthatheladze, Sh. (1980). Pyrus L. In: Sakartvelos Flora (Flora of Georgia), Ketskhoveli, N. (Ed.), Vol. 6. Metsniereba, ISBN:5-520-00325-7, Tbilisi. (in Georgian).

48. Kvavadze, E., Bar-Yosef, O., Belfer-Cohen, A., Boaretto, E., Jakeli, N., Matskevich, Z. & Meshveliani, T. (2009). 30,000-Year-Old Wild Flax Fibers. Science, Vol.325, No.5946, (11 September 2009), pp.1359, ISSN: 0036-8075.

49. Laguna, E. (2004). The plant micro-reserve initiative in the Valencian Community (Spain) and its use to conserve populations of crop wild relatives. Crop Wild Relative, Issue 2, (July 2004), pp.10-13. ISSN 1742-3627.

50. Maghradze, D., Failla, O., Bacilieri, R., Imazio, S., Vashakidze, L., Chipashvili, R., Mdinaradze, I., Chkhartishvili, N., This, P. & Scienza, A. (2010). Georgian Vitis germplasm: usage, conservation and investigation. Bulletin de l'OIV, Vol.83, No.956-958, (October-December, 2010), pp. 485-496. ISSN: 0029-7127.

51. Maisaia, I., Shanshiashvili, T. & Rusishvili, N. (2005). Agriculture of Colchis. Metsniereba, ISBN:99940-785-3-4, Tbilisi. (In Georgian).

52. Maxted, N., Ford-Lloyd, B. V., Jury, S., Kell, Sh. & Scholten, M. (2006). Towards a definition of a crop wild relative. Biodiversity and Conservation. Vol.15, No.8, (July 2006), pp.2673-2685. ISSN: 0960-3115

53. Maxted, N., Dulloo, E., Ford-Lloyd, B. V., Iriondo, J. M & Jarvis, A. (2008). Gap analysis: a tool for complementary genetic conservation assessment. Diversity and Distributions, Vol.14, No.6, (November 2008), pp.1018–1030. ISSN:1472-4642. doi: 10.1111/j.1472-4642.2008.00512.x.

54. McGovern, P.E. (2003). Ancient wine: the search for the origins of viniculture. Princeton University Press, ISBN: 0691070806, Princeton, New Jersey.

55. Melikishvili, G. (Ed). (1970). Sakartvelos istoriis narkvevebi (Historical essays of Georgia). Sabchota Sakartvelo, Tbilisi. (In Georgian).

56. Menabde, V. (1938). Sakartvelos kerebi (Barleys of Georgia). Georgian Academy of SciencesPress, Tbilisi. (In Georgian).

57. Menabde, V. (1948). Pshenitsi Gruzii (Wheats of Georgia). Georgian Academy of Sciences Press, Tbilisi. (In Russian).

58. Menabde, V. (1961). Kartuli khorblebi da mati roli khorblis saerto evolutsiashi (Georgian wheat cultivars and their role in wheat evolution). Proceedings of Tbilisi Botanical Institute, Vol.21, No.1. pp.229-259. (In Georgian).

59. Nakhutsrishvili, G. (1999). The vegetation of Georgia (Caucasus). Braun-Blanquetia, Vol.15, pp.1-74. ISSN:0393-5434.

60. Nakhutsrishvili, G., Akhalkatsi, M. & Abdaladze, O. (2009). Main threats to the mountain biodiversity in Georgia (the Caucasus). Mountain Forum Bulletin, Vol.9, No.2, (July 2009), pp.18–19. ISSN 1815-2139.

61. Neidze, V., Bokeria, M., Kharadze, K., Kurtubadze, M. & Garsenishvili, L. (2008). Sakartvelos geographia (Geography of Georgia). LogosPress, ISBN: 978-99940-64-48-9, Tbilisi. (In Georgian).

62. Ovchinnikov, I. V., Gotherstrom, A., Romanova, G. P., Kharitonov, V. M., Liden, K. & Goodwin, W . (2000). Molecular analysis of Neanderthal

DNA from the northern Caucasus. Nature, Vol.404, No.6777, (30 March 2000), pp.490–493, ISSN: 0028-0836.doi:10.1038/35006625.

63. Petersen, G., Seberg, O., Yde, M. & Berthelsen, K. (2006). Phylogenetic relationships of Triticum and Aegilops and evidence for the origin of the A, B, and D genomes of common wheat (Triticum aestivum). Molecular Phylogenetics and Evolution, Vol.39 , No.1, (April 2006), pp.70–82. ISSN: 1055-7903.

64. Pistrick, K., Akhalkatsi M., Girgvliani, T. & Shanshiashvili, T. (2009). Collecting plant genetic resources in Upper Svaneti (Georgia, Caucasus Mountains). Journal of Agriculture and Rural Development in the Tropics and Subtropics, Supplement 92, pp.127-135.ISSN:1613-8422.

65. Ramishvili, R., (1988). Dikorastushchii vinograd Zakavkazia (Wild grape of the South Caucasus). Ganatleba, ISBN: 5-505-00690-6, Tbilisi. (In Russian).

66. Sabashvili, M. (1970). Niadagmtsodneoba (Soil Sciences). Tbilisi University Press, Tbilisi. (In Georgian).

67. Schatz, G., Shulkina, T., Nakhutsrishvili, G., Batsatsashvili, K., Tamanyan, K., Ali-zade, V., Kikodze, D., Geltman, D. & Ekim, T. (2009). Development of Plant Red List Assessments for the Caucasus Biodiversity Hotspot. In: Status and Protection of Globally Threatened Species in the Caucasus. Zazanashvili, N. & Mallon, D. (Editors).

68. Contour Ltd. pp. 188-192. Tbilisi. ISBN 978-9941-0-2203-6. Available from: <www.assets.panda.org/downloads/ cepf_caucasus_web_1.pdf >Takhtajan, A. (1986). Floristic Regions of the World. University of California Press, First English Language Edition, (September 1986), ISBN: 0520040279, Berkeley, CA.

69. This, P., Lacombe, T. & Thomas, M.R. (2006). Historical origins and genetic diversity of wine grapes. Trends in Genetics Vol.22, No.9, (September 2006), pp.511–519. ISSN: 0168-9525.

70. Tsilo, T. J., Yue Jin & Anderson, J. A. (2008). Diagnostic Microsatellite Markers for the Detection of Stem Rust Resistance Gene Sr36 in Diverse Genetic Backgrounds of Wheat. Crop Science, Vol.48, No.1, (January–February 2008) pp.253–261. ISSN: 0011-183X.

71. Tushabramishvili, N. (2011). The main problems of Palaeolithic of Georgia and the complex studies of the Western Georgian cave sites. In: Khornabuji, Pitskhelauri, K. (Ed), pp. 39-41, Ilia State University Press, ISBN: 1512-2999, Tbilisi.

72. UPOV (2000). Guidelines for the conduct of tests for distinctness, uniformity and stability. Pear (Pyrus communis L.). Geneva. Date

of access: 2000-04-05. Available from: <http://www.upov.int/en/publications/tg-rom/tg015/tg_15_3.pdf>

73. Vavilov, N. I. (1992). Origin and geography of cultivated plants. Cambridge University Press, ISBN: 0-521-40427-4, Cambridge.

74. Volk, G. M., Richards, C. M., Henk, A. D., Reilley, A. A., Bassil, N. V. & Postman, J. D. (2006). Diversity of wild Pyrus communis based on microsatellite analyses. Journal of the American Society for Horticultural Science, Vol.131, No.3, (May 2006), pp.408–417.ISSN: 0003-1062.

75. Voltas, J., Pemán, J. & Fusté, F. (2007). Phenotypic diversity and delimitation between wild and cultivated forms of the genus Pyrus in North-eastern Spain based on morphometric analyses. Genetic Resources and Crop Evolution, Vol.54, No.7,(November 2007), pp.1473–1487. ISSN:0925-9864.

76. Yamamoto, T. & Chevreau, E. (2009). Pear genomics. In: Genetics and Genomics of Rosaceae., Folta, K.M. & Gardiner, S.E. (Eds.), Springer, ISBN:0387774904, New York.

77. Zhizhizlashvili, K. & Berishvili, T. (1980). Zemo Svanetis kulturul mtsenaretashestsavlisatvis (Study of cultivated plants in Upper Svaneti). Bulletin of Georgian Academy of Sciences, Vol.100, No.2, pp.417-419. ISSN:0132-1447. (In Georgian).

78. Zohary, D., & Hopf, M. (2000). Domestication of plants in the Old World. (3rd edition.) Oxford University Press, ISBN 0-19-850356-3, New York.

CITATION

CHAPTER 1

Luciano Lourenço Nass, Mário Sérgio Sigrist, Cláudia Silva da Costa Ribeiro and Francisco José Becker Reifschneider, Genetic resources: the basis for sustainable and competitive plant breeding, http://dx.doi.org/10.1590/S1984-70332012000500009

CHAPTER 2

Rukhsar Ahmad Dar,Mushtaq Ahmad,Sanjay Kumar,Monica Reshi, Agriculture germplasm resources: A tool of conserving diversity, DOI: 10.5897/SRE2015.6206.

CHAPTER 3

Kerri L SteenwerthEmail author, Amanda K Hodson, Arnold J Bloom, Michael R Carter, Andrea Cattaneo, Colin J Chartres, Jerry L Hatfield, Kevin Henry, Jan W Hopmans, William R Horwath, Bryan M Jenkins, Ermias Kebreab, Rik Leemans, Leslie Lipper, Mark N Lubell, Siwa Msangi, Ravi Prabhu, Matthew P Reynolds, Samuel Sandoval Solis, William M Sischo, Michael Springborn, Pablo Tittonell, Stephen M Wheeler, Sonja J Vermeulen, Eva K Wollenberg, Lovell S Jarvis and Louise E Jackson, Climate-smart agriculture global research agenda: scientific basis for action, DOI: 10.1186/2048-7010-3-11.

CHAPTER 4

Mohammed Kasso and Mundanthra Balakrishnan, “Ex Situ Conservation of

Biodiversity with Particular Emphasis to Ethiopia," ISRN Biodiversity, vol. 2013, Article ID 985037, 11 pages, 2013. doi:10.1155/2013/985037.

CHAPTER 5

Pouta E, Tienhaara A and Ahtiainen H (2014) Citizens' preferences for the conservation of agricultural genetic resources. Front. Genet. 5:440. doi: 10.3389/fgene.2014.00440

CHAPTER 6

Jikun Huang and Qinfang Wang, Agricultural Biotechnology Development and Policy in China, http://www.agbioforum.org/v5n4/v5n4a01-huang.htm.

CHAPTER 7

K. Nirmal Babu, G. Yamuna, K. Praveen, D. Minoo, P.N. Ravindran and K.V. Peter (2012). Cryopreservation of Spices Genetic Resources, Current Frontiers in Cryobiology, Prof. Igor Katkov (Ed.), ISBN: 978-953-51-0191-8, InTech, DOI: 10.5772/35401.

CHAPTER 8

Maia Akhalkatsi, Jana Ekhvaia and Zezva Asanidze (2012). Diversity and Genetic Erosion of Ancient Crops and Wild Relatives of Agricultural Cultivars for Food: Implications for Nature Conservation in Georgia (Caucasus), Perspectives on Nature Conservation - Patterns, Pressures and Prospects, Prof. John Tiefenbacher (Ed.), ISBN: 978-953-51-0033-1, InTech, DOI: 10.5772/30286.

INDEX

K

L

M

N

P

S

T

V

W

Z